Biostatistical Methods

WILEY SERIES IN PROBABILITY AND STATISTICS
APPLIED PROBABILITY AND STATISTICS SECTION

Established by WALTER A. SHEWHART and SAMUEL S. WILKS

Editors: *Vic Barnett, Noel A. C. Cressie, Nicholas I. Fisher, Iain M. Johnstone, J. B. Kadane, David W. Scott, Bernard W. Silverman, Adrian F. M. Smith, Jozef L. Teugels; Ralph A. Bradley, Emeritus, J. Stuart Hunter, Emeritus, David G. Kendall, Emeritus*

A complete list of the titles in this series appears at the end of this volume.

Biostatistical Methods

The Assessment of Relative Risks

JOHN M. LACHIN
The George Washington University
Washington, D.C.

A Wiley-Interscience Publication
JOHN WILEY & SONS, INC.
New York / Chichester / Weinheim / Brisbane / Singapore / Toronto

This text is printed on acid-free paper. ♾

Published simultaneously in Canada.

For ordering and customer service, call 1-800-CALL WILEY.

Library of Congress Cataloging-in-Publication Data is available.

ISBN 0-471-36996-9

Printed in the United States of America

10 9 8 7 6 5 4 3 2 1

Contents

Preface

In 1993 to 1994 I led the effort to establish a graduate program in biostatistics at the George Washington University. The program, which I now direct, was launched in 1995 and is a joint initiative of the Department of Statistics, the Biostatistics Center (which I have directed since 1988) and the School of Public Health and Health Services. Biostatistics has long been a specialty of the statistics faculty, starting with Samuel Greenhouse, who joined the faculty in 1946. When Jerome Cornfield joined the faculty in 1972, he established a two-semester sequence in biostatistics (Statistics 225-6) as an elective for the graduate program in statistics (our 200 level being equivalent to the 600 level in other schools). Over the years these courses were taught by many faculty as a lecture course on current topics. With the establishment of the graduate program in biostatistics, however, these became pivotal courses in the graduate program and it was necessary that Statistics 225 be structured so as to provide students with a review of the foundations of biostatistics.

Thus I was faced with the question "what are the foundations of biostatistics?" In my opinion, biostatistics is set apart from other statistics specialties by its focus on the assessment of risks and relative risks through clinical research. Thus biostatistical methods are grounded in the analysis of binary and count data such as in 2×2 tables. For example, the Mantel-Haenszel procedure for stratified 2×2 tables forms the basis for many families of statistical procedures such as the G^{ρ} family of modern statistical tests in the analysis of survival data. Further, all common medical study designs, such as the randomized clinical trial and the retrospective case-control study, are rooted in the desire to assess relative risks. Thus I developed

Statistics 225, and later this text, around the principle of the assessment of relative risks in clinical investigations.

In doing so, I felt that it was important first to develop basic concepts and derive core biostatistical methods through the application of classical mathematical statistical tools, and then to show that these and comparable methods may also be developed through the application of more modern, likelihood-based theories. For example, the large sample distribution of the Mantel-Haenszel test can be derived using the large sample approximation to the hypergeometric and the Central Limit Theorem, and also as an efficient score test based on a hypergeometric likelihood.

Thus the first five chapters present methods for the analysis of single and multiple 2×2 tables for cross-sectional, prospective and retrospective (case-control) sampling, without and with matching. Both fixed and random effects (two-stage) models are employed. Then, starting in Chapter 6 and proceeding through Chapter 9, a more modern likelihood or model-based treatment is presented. These chapters broaden the scope of the book to include the unconditional and conditional logistic regression models in Chapter 7, the analysis of count data and the Poisson regression model in Chapter 8, and the analysis of event time data including the proportional hazards and multiplicative intensity models in Chapter 9. Core mathematical statistical tools employed in the text are presented in the Appendix. Following each chapter problems are presented that are intended to expose the student to the key mathematical statistical derivations of the methods presented in that chapter, and to illustrate their application and interpretation.

Although the text provides a valuable reference to the principal literature, it is not intended to be exhaustive. For this purpose, readers are referred to any of the excellent existing texts on the analysis of categorical data, generalized linear models and survival analysis. Rather, this manuscript was prepared as a textbook for advanced courses in biostatistics. Thus the course (and book) material was selected on the basis of its current importance in biostatistical practice and its relevance to current methodological research and more advanced methods. For example, Cornfield's approximate procedure for confidence limits on the odds ratio, though brilliant, is no longer employed because we now have the ability to readily perform exact computations. Also, I felt it was more important that students be exposed to over-dispersion and the use of the information sandwich in model-based inference than to residual analysis in regression models. Thus each chapter must be viewed as one professor's selection of relevant and insightful topics.

In my Statistics 225 course, I cover perhaps two-thirds of the material in this text. Chapter 9, on survival analysis, has been added for completeness, as has the section in the Appendix on quasi-likelihood and the family of generalized linear models. These topics are covered in detail in other courses. My detailed syllabus for Statistics 225, listing the specific sections covered and exercises assigned, is available at the Biostatistics Center web site (www.bsc.gwu.edu/jml/biostatmethods). Also, the data sets employed in the text and problems are available at this site or the web site of John Wiley and Sons, Inc. (www.wiley.com).

Although I was not trained as a mathematical statistician, during my career I have learned much from those with whom I have been blessed with the opportunity to collaborate (chronologically): Jerry Cornfield, Sam Greenhouse, Nathan Mantel, and Max Halperin, among the founding giants in biostatistics; and also Robert Smythe, L.J. Wei, Peter Thall, K.K. Gordon Lan and Zhaohai Li, among others, who are among the best of their generation. I have also learned much from my students, who have always sought to better understand the rationale for biostatistical methods and their application.

I especially acknowledge the collaboration of Zhaohai Li, who graciously agreed to teach Statistics 225 during the fall of 1998, while I was on sabbatical leave. His detailed reading of the draft of this text identified many areas of ambiguity and greatly improved the mathematical treatment. I also thank Costas Cristophi for typing my lecture notes, and Yvonne Sparling for a careful review of the final text and programming assistance. I also wish to thank my present and former statistical collaborators at the Biostatistics Center, who together have shared a common devotion to the pursuit of good science: Raymond Bain, Oliver Bautista, Patricia Cleary, Mary Foulkes, Sarah Fowler, Tavia Gordon, Shuping Lan, James Rochon, William Rosenberger, Larry Shaw, Elizabeth Thom, Desmond Thompson, Dante Verme, Joel Verter, Elizabeth Wright, and Naji Younes, among many.

Finally, I especially wish to thank the many scientists with whom I have had the opportunity to collaborate in the conduct of medical research over the past 30 years: Dr. Joseph Schachter, who directed the Research Center in Child Psychiatry where I worked during graduate training; Dr. Leslie Schoenfield, who directed the National Cooperative Gallstone Study; Dr. Edmund Lewis, who directed the Collaborative Study Group in the conduct of the Study of Plasmapheresis in Lupus Nephritis and the Study of Captropil in Diabetic Nephropathy; Dr. Thomas Garvey, who directed the preparation of the New Drug Application for treatment of gallstones with ursodiol; Dr. Peter Stacpoole, who directed the Study of Dichloroacetate in the Treatment of Lactic Acidosis; and especially Drs. Oscar Crofford, Saul Genuth and David Nathan, among many others, with whom I have collaborated since 1982 in the conduct of the Diabetes Control and Complications Trial, the study of the Epidemiology of Diabetes Interventions and Complications, and the Diabetes Prevention Program. The statistical responsibility for studies of such great import has provided the dominant motivation for me to continually improve my skills as a biostatistician.

JOHN M. LACHIN

Rockville, Maryland

1

Biostatistics and Biomedical Science

1.1 STATISTICS AND THE SCIENTIFIC METHOD

The aim of all biomedical research is the acquisition of new information so as to expand the body of knowledge that comprises the biomedical sciences. This body of knowledge consists of three broad components:

1. Descriptions of phenomena in terms of observable characteristics of elements or events;

2. Descriptions of associations among phenomena;

3. Descriptions of causal relationships between phenomena.

The various sciences can be distinguished by the degrees to which each contains knowledge of each of these three types. The hard sciences (e.g. physics and chemistry) contain large bodies of knowledge of the third kind — causal relationships. The soft sciences (e.g. the social sciences) principally contain large bodies of information of the first and second kind — phenomenological and associative.

None of these descriptions, however, are exact. To quote the philosopher and mathematician Jacob Bronowski (1973).

> All information is imperfect. We have to treat it with humility... Errors are inextricably bound up with the nature of human knowledge...

Thus every science consists of shared information, all of which to some extent is uncertain.

When a scientific investigator adds to the body of scientific knowledge, the degree of uncertainty about each piece of information is described through statistical assessments of the probability that statements are either true or false. Thus the language of science is statistics, for it is through the process of statistical analysis and interpretation that the investigator communicates the results to the scientific community. The syntax of this language is probability, because the laws of probability are used to assess the inherent uncertainty, errors, or precision of estimates of population parameters, and probabilistic statements are used as the basis for drawing conclusions.

The means by which the investigator attempts to control the degree of uncertainty in the research conclusions is the application of the scientific method. In a nutshell, the scientific method is a set of strategies, based on common sense and statistics, that is intended to minimize the degree of uncertainty and maximize the degree of validity of the resulting knowledge. Therefore, the scientific method is deeply rooted in statistical principles.

When considered sound and likely to be free of error, such knowledge is termed scientifically valid. The designation of scientific validity, however, is purely subjective. The soundness or validity of any scientific result depends on the manner in which the observations were collected, that is, on the design and conduct of the study, as well as the manner in which the data were analyzed.

Therefore, in the effort to acquire scientifically valid information, one must consider the statistical aspects of all elements of a study – its design, execution and analysis. To do so requires a firm understanding of the statistical basis for each type of study and for the analytic strategies commonly employed to assess a study's objectives.

1.2 BIOSTATISTICS

Biostatistics is principally characterized by the application of statistical principles to the biological/biomedical sciences; in contrast to other areas of application of statistics, such as psychometrics and econometrics. Thus biostatistics refers to the development of statistical methods for, and the application of statistical principles to, the study of biologic and medical phenomena.

Biomedical research activities range from the study of cellular biology to clinical therapeutics. At the basic physical level it includes so-called bench research or the study of genetic, biochemical, physiologic, and biologic processes, such as the study of genetic defects, metabolic pathways, kinetic models and pharmacology. Although some studies in this realm involve investigation in animals and man (*in vivo*), many of these investigations are conducted in "test tubes" (*in vitro*). The ultimate objective of these inquiries is to advance our understanding of the pathobiology or pathophysiology of diseases in man and of the potential mechanisms for their treatment.

Clinical research refers to the direct observation of the clinical features of populations. This includes *epidemiology*, which can be broadly defined as the study of the distribution and etiology of human disease. Some elements, such as infectious disease epidemiology, are strongly biologically based, whereas others are more heavily dependent on empirical observations within populations. These latter include such areas as occupational and environmental epidemiology or the study of the associations between occupational and environmental exposures with the risk of specific diseases. This type of epidemiology is often characterized as *population-based* because it relies on the observation of natural samples from populations.

Ultimately, bench research or epidemiologic observation leads to advances in medical therapeutics — the development of new pharmaceuticals (drugs), devices, surgical procedures or interventions. Such therapeutic advances are often assessed using a randomized, controlled, clinical trial. Such studies evaluate the biological effectiveness of the new agent (biological efficacy), the clinical effectiveness of the therapy in practice (the so-called intention-to-treat comparison), as well as the incidence of adverse effects.

The single feature that most sharply distinguishes clinical biomedical research from other forms of biological research is the propensity to assess the absolute and relative risks of various outcomes within populations. The *absolute risk* refers to the distribution of a disease, or risk factors for a disease, in a population. This risk may be expressed cross-sectionally as a simple probability, or it may be expressed longitudinally over time as a hazard function (or survival function) or an intensity process. The *relative risk* refers to a measure of the difference in risks among subsets of the population with specific characteristics, such as those exposed versus not to a risk factor, or those randomly assigned to a new drug treatment versus a placebo control. The relative risk of an outcome is sometimes described as a difference in the absolute risks of the outcome, the ratio of the risks, or a ratio of the odds of the outcome.

Thus a major part of biostatistics concerns the assessment of absolute and relative risks through epidemiologic studies of various types and randomized clinical trials. This, in general, is the subject of this text. This entails the study of discrete outcomes, some of which are assessed over time. This also includes many major areas of statistics that are beyond the scope of any single text. For example, the analysis of longitudinal data is another of the various types of processes studied through biostatistics. In many studies, however, interest in a longitudinal quantitative or ordinal measure arises because of its fundamental relationship to an ultimate discrete outcome of interest. For example, longitudinal analysis of serum cholesterol levels in a population is of interest because of the strong relationship between serum lipids and the risk of cardiovascular disease, not cholesterol itself. Thus this text is devoted exclusively to the assessment of the risks of discrete characteristics or events in populations.

1.3 NATURAL HISTORY OF DISEASE PROGRESSION

Underlying virtually all clinical research is some model of our understanding of the natural history of the progression of the disease under investigation. As an example,

Table 1.1 Stages of Progression of Diabetic Nephropathy

1.	Normal: Albumin excretion rate (AER) $\leq$ 40 mg/24 h
2.	Microalbuminuria: 40 $<$ AER $<$ 300 mg/24 h
3.	Proteinuria (overt albuminuria): AER $\geq$ 300 mg/24 h
4.	Renal insufficiency: Serum creatinine $>$ 2 mg/dL
5.	End-stage renal disease: Need for dialysis or renal transplant
6.	Mortality

consider the study of diabetic nephropathy (kidney disease) associated with type 1 or insulin dependent diabetes mellitus (IDDM), also known as juvenile diabetes. Diabetes is characterized by a state of metabolic dysfunction in which the subject is deficient in endogenous (self-produced) insulin. Thus the patient must administer exogenous insulin by some imperfect mechanical device, such as by multiple daily injections or a continuous subcutaneous insulin infusion (CSII) device also called a "pump". Because of technological deficiencies with the way insulin can be administered, it is difficult to maintain normal levels of blood glucose throughout the day, day after day. The resulting hyperglycemia leads to microvascular complications, the two most prevalent being diabetic retinopathy (disease of the retina in the eye) and diabetic nephropathy, and ultimately to cardiovascular disease.

Diabetic nephropathy is known to progress through a well-characterized sequence of disease states, characterized in Table 1.1. The earliest sign of emergent kidney disease is the leakage of small amounts of protein (albumin) into urine. The amount or rate of albumin excretion can be measured from a timed urine collection in which all the urine voided over a fixed period of time is collected. From the measurement of the urine volume and the concentration of albumin in the serum and urine at specific intervals of time, it is possible to compute the albumin excretion rate (AER) expressed as the mg/24 h of albumin excreted into the urine by the kidneys.

In the normal (non-diseased) subject, the AER is no greater than 40 mg/24 h, some would say no greater than 20 or 30 mg/24 h. The earliest sign of possible diabetic nephropathy is microalbuminuria, defined as an AER $>$40 mg/24 h (but $<$ 300 mg/24 h). As the disease progresses, the next landmark is the development of definite albuminuria, defined as an AER $>$300 mg/24 h. This is often termed overt proteinuria because it is at this level of albumin (protein) excretion that a simple dip-stick test for protein in urine will be positive. This is also the point at which nephropathy, and the biological processes that ultimately lead to destruction of the kidney, are considered well established.

To then chart the further loss of kidney function, a different measure is used — the glomerular filtration rate (GFR). The glomerulus is the cellular structure that serves as the body's filtration system. As diabetic nephropathy progresses, fewer and fewer intact glomeruli remain, so that the rate of filtration declines, starting with the leakage of protein and other elements into the urine. The GFR is difficult to measure accurately. In practice, a measure of creatinine clearance, also from a timed urine collection, or a simple measure of the creatinine concentration in serum are used to monitor disease progression. Renal insufficiency is often declared when the serum creatinine exceeds 2 mg/dL. This is followed by end-stage renal disease (ESRD), at which point the patient requires frequent dialysis or renal transplantation to prolong survival. Ultimately the patient dies from the renal insufficiency or related causes if a suitable donor kidney is not available for transplantation.

Thus the natural history of diabetic nephropathy is described by a collection of quantitative, ordinal and qualitative assessments. In the early stages of the disease, a study might focus entirely on quantitative measures of AER. Later, during the middle stages of the disease, this becomes problematic. For example, patients with established proteinuria may be characterized over time using a measure of GFR, but the analysis will be complicated by informatively missing observations because some patients reached ESRD or died before the scheduled completion of follow-up.

However, a study that assesses the risk of discrete outcomes, such as the incidence or prevalence of proteinuria or renal insufficiency, is less complicated by such factors and is readily interpretable by physicians. For example, if a study shows that a new drug treatment reduces the mean AER by 10 mg/24 h less than that with placebo, it is difficult to establish the clinical significance of the result. On the other hand, if the same study demonstrated a relative risk of developing proteinuria of 0.65, a 35% risk reduction with drug treatment versus placebo, the clinical significance is readily apparent to most physicians.

Therefore, we shall focus on the description of the absolute and relative risks of discrete outcomes, historically the core of biostatistics.

1.4 TYPES OF BIOMEDICAL STUDIES

Biomedical research employs various types of study designs, some of which involve formal experimentation, others not, among other characteristics. In this section the characteristics and the roles of each type of study are briefly described.

Study designs can be distinguished by three principal characteristics:

1. *Number of samples*: single versus multiple samples;

2. *Source of samples*: natural versus experimental. An experimental sample is one to which a treatment or procedure has been applied by the investigator. This may or may not involve randomization as an experimental device to assign treatments to individual patients.

3. *Time course of observation*: prospective versus retrospective versus concurrent collection of measurements and observation of responses or outcome events.

Based on these characteristics, there are basically four types of designs for biomedical studies in man: (1) the cross-sectional study, (2) the cohort study, (3) the case-control study, and (4) the randomized experiment. A more exhaustive classification was provided by Bailar, Louis, Lavori and Polansky (1984), but these four are the principal types. Examples of each type of study are described subsequently.

The **cross-sectional study** is a study of a single, natural sample with concurrent measurement of a variety of characteristics. In the review by Bailar, Louis, Lavori, and Polansky (1984), 39% of published studies were of this type. Some notable examples are the National Health and Nutritional Examination Survey (NHANES) of the relationship between health and nutrition, and the annual Health Interview Survey of the prevalence of various diseases in the general U.S. population. Such studies have provided important descriptions of the prevalence of disease in specified populations, of the co-occurrence of the disease and other factors (i.e. associations), and of the sensitivity and specificity of diagnostic procedures.

In a **cohort study** (25% of studies), one or more samples (cohorts) of individuals, either natural or experimental samples, are followed prospectively and subsequent status is evaluated.

A **case-control study** (5% of studies) employs multiple, natural samples with retrospective measurements. A sample of cases with the disease is compared to a sample of controls without the disease with respect to the previous presence of, or exposure to, some factor.

An important characteristic of cohort and case-control studies is whether or not the study employs **matching** of pairs or sets of subjects with respect to selected covariate values. Matching is a strategy to remove bias in the comparison of groups by ensuring equality of distributions of the selected matching covariates. Matching, however, changes the sample frame or the sampling unit in the analysis from the individual subject in an unmatched study to the matched set in the matched study. Thus matched studies require analytic procedures that are different from those more commonly applied to unmatched studies.

A **randomized, controlled clinical trial or parallel – comparative trial** (15% of studies) employs two or more parallel randomized cohorts, each of which receives only one treatment in the trial. Such studies provide a controlled assessment of a new drug, therapy, diagnostic procedure, or intervention procedure. Variations of this design include the multiple-period **crossover** design and the crossed **factorial** design. Since a clinical trial uses randomization to assign each subject to receive either the active treatment versus a control (e.g. drug vs. placebo), the comparison of the groups is in expectation unbiased. However, a truly unbiased study also requires other conditions such as complete and unbiased follow-up assessments.

Each of the first three types are commonly referred to as an observational or epidemiological study, in contrast to the clinical trial. It is rare, some might say impossible, that a population-based observational study will identify a single necessary and sufficient cause for a biologic effect, or a 1:1 causal relationship. Almost always, a *risk factor* is identified that has a biological effect that is associated with a change in the risk of an outcome. It is only after a preponderance of evidence is accumulated from many such studies that such a risk factor may be declared to be a *causal agent*. Such was the case with the relationship between smoking and lung cancer, and the criteria employed to declare smoking a causal agent are now widely accepted (US Surgeon General, 1964, 1982).

The principal advantage of the randomized controlled trial (RCT), on the other hand, is that it can provide conclusions with respect to causal relationships because other intervening factors are controlled through randomization. Thus the RCT provides an unbiased comparison of the effects of administering one treatment versus another on the outcome in the selected population of patients, and any differences observed can be confidently ascribed to the differences between the treatments. Therefore, the distinction between a relationship based on an observational study and one based on a randomized experiment rests in the degree to which an observed relationship might be explained by other variables or other mechanisms.

However, in no study is there an absolute guarantee that all possible influential variables are controlled, even in a randomized, controlled experiment. Also, as the extent of knowledge about the underlying natural history of a disease expands, it becomes increasingly important to account for the known or suspected risk factors in the assessment of the effects of treatments or exposures, especially in an observational cross-sectional, cohort, or case-control study. This entails the use of an appropriate statistical model for the simultaneous influence of multiple covariates on the absolute or relative risk of important outcomes or events.

Thus the principal objective of this text is to describe methods for the assessment of risk relationships derived from each type of study, and to consider methods to adjust or control for other factors in these assessments.

1.5 STUDIES OF DIABETIC NEPHROPATHY

To illustrate the different types of studies, we close this chapter with a review of selected studies on various aspects of diabetic nephropathy.

Cross-sectional surveys such as the National Health Interview Survey (NHIS) and the National Health and Nutrition Evaluation Survey (NHANES) indicate that approximately 16 million people in the United States population have some form of diabetes mellitus (Harris, Hadden, Knowler and Bennett, 1987). The majority have what is termed type 2 or non-insulin dependent diabetes mellitus (NIDDM). Approximately 10% or 1.6 million have the more severe form termed type 1 or insulin-dependent diabetes mellitus (IDDM) for which daily insulin injections or infusions are required to sustain life. Among the most important clinical features of type 1 diabetes are the development of complications related to micro- and macro-vascular abnormalities, among the most severe being diabetic nephropathy (kidney disease), which ultimately leads to end-stage renal disease (ESRD) in about a third of patients. These and other national surveys indicate that approximately 35% of all ESRD in the United States is attributed to diabetes.

As an illustration of a longitudinal observational cohort study, Deckert et al. (1978) followed a cohort of 907 Danish subjects with type 1 diabetes for many years and reported the annual incidence (proportion) of new cases of proteinuria (overt albuminuria) to appear each year. They showed that the peak incidence or greatest risk occurs approximately 15 years after the onset of diabetes. Their study also showed that over a lifetime, approximately 70% of subjects develop nephropathy whereas approximately 30% do not, suggesting that there is some mechanism that protects patients from nephropathy, possibly of a genetic nature, possibly related to the lifetime exposure to hyperglycemia, or possibly related to some environmental exposure or characteristic.

Since the discovery of insulin in the 1920s, one of the principal issues of contention in the scientific community is what was often called the Glucose Hypothesis. This hypothesis asserts that the extent of exposure to elevated levels of blood glucose or hyperglycemia is the dominant determinant of the risk of diabetic nephropathy and other microvascular abnormalities or complications of type 1 diabetes. Among the first studies to suggest an association was a large observational study conducted by Pirart (1978a, 1978b) in Belgium over the period 1947–1973. This study examined the association between the level of blood glucose and the prevalence (presence or absence) of nephropathy. The data were obtained from a retrospective examination of the clinical history of 4,400 patients treated in a community hospital over a period of up to 25 years in some patients. The rather crude analysis consisted of figures that displayed the prevalence of nephropathy by year of diabetes duration for subgroups categorized as being in good, fair or poor control of blood glucose levels. These figures suggest that as the mean level of hyperglycemia increases, the risk (prevalence) of nephropathy also increases. This type of study is clearly open to various types of sampling or selection biases. Nevertheless, this study provides evidence that hyperglycemia may be a strong risk factor, or is associated with the risk of diabetic nephropathy. Note that this study is not strictly a prospective cohort study because the cohort was identified later in time and the longitudinal observations were then obtained retrospectively.

In all of these studies, biochemical measures of renal function are used to assess the presence and extent of nephropathy. Ultimately, however, end stage renal disease is characterized by the physiologic destruction of the kidney, specifically the glomeruli, which are the cellular structures that actually perform the filtration of blood. However, the only way to determine the physical extent of glomerular damage is to conduct a morphologic evaluation of a tissue specimen obtained by a needle biopsy of the kidney. As an example of a case-control study, Chavers, Bilous, Ellis, et al. (1989) conducted a retrospective study to determine the association between established nephropathy or not (the cases vs. controls) and evidence of morphologic (structural tissue) abnormalities in the kidneys (the risk factor or exposure). They showed that approximately 69% of patients with nephropathy showed morphologic abnormalities versus 42% among those without nephropathy, for a relative risk (odds ratio) of 3.2. Other studies (*cf.* Steffes, Chavers, Bilous and Mauer (1989) show that the earliest stage of nephropathy, microalbuminuria (which they defined as an AER $\geq$ 20 mg/24 h) is highly predictive of progression to proteinuria, with a positive predictive value ranging from 83–100%. These findings established that proteinuria is indeed associated with glomerular destruction and that microalbuminuria is predictive of proteinuria. Thus a treatment that reduces the risk of microalbuminuria can be expected to reduce the risk of progression to proteinuria, and one that reduces the risk of proteinuria will also reduce the extent of physiologic damage to the kidneys.

The major question to be addressed, therefore, was whether the risk of albuminuria or nephropathy could be reduced by a treatment that consistently lowered

Fig. 1.1 Cumulative incidence of microalbuminuria (AER > 40 mg/24 h) over nine years of follow-up in the DCCT Primary Prevention Cohort.

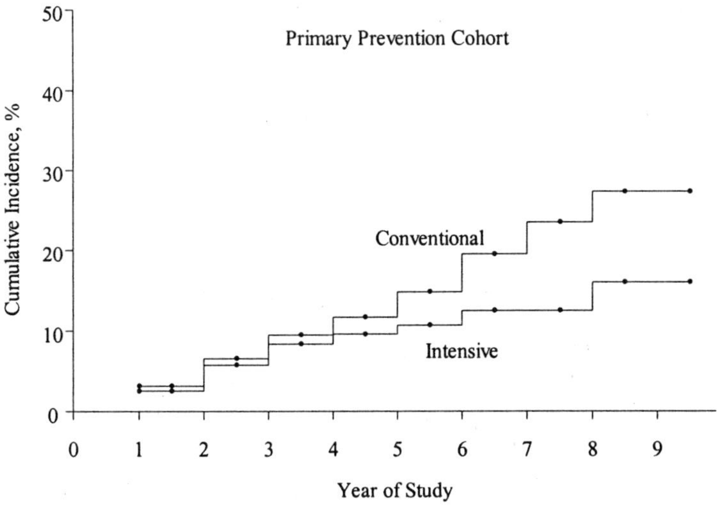

the levels of blood glucose. By the 1980s, technological developments made an experiment (clinical trial) to test this hypothesis feasible. The level of blood glucose varies continuously over the 24 hour period, with peaks following meals and troughs before meals. It was discovered that the hemoglobin (red cells) in the blood become glycosylated when exposed to blood glucose. Thus the percent of the total hemoglobin that has become glycosylated (the HbA_{1c} %) provides an indirect measure of the mean level of hyperglycemia over the preceding 4–6 weeks, the half-life of the red blood cell. This made it possible to assess the average extent of hyperglycemia in individual patients. Other developments then made it possible for patients and their health-care teams to control their blood sugar levels so as to lower the level of hyperglycemia, as reflected by the level of HbA_{1c}. Devices for self-blood glucose monitoring allowed patients to measure the current level of blood glucose (mg/dL) from a drop of blood obtained by a finger prick. Patients could then alter the amount of insulin administered to keep the level of blood glucose within a desirable range. Also, a variety of types of insulin were developed, some of which acted quickly and some over long periods of time, that could be administered using multiple daily insulin injections or a pump. The health care team could then try different algorithms to vary the amount of insulin administered in response to the current level of blood glucose.

With these advances, in 1981 the National Institute of Diabetes, Digestive and Kidney Disease launched the Diabetes Control and Complications Trial (DCCT) to

Fig. 1.2 Cumulative incidence of microalbuminuria (AER > 40 mg/24 h) over nine years of follow-up in the DCCT Secondary Intervention Cohort.

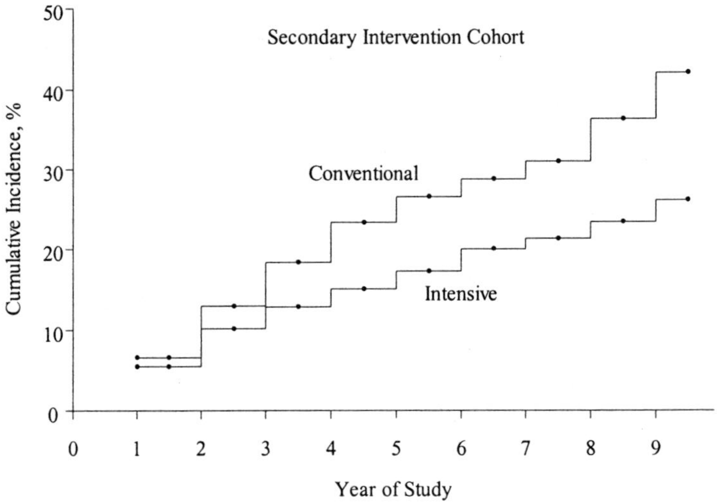

test the glucose hypothesis (DCCT 1990, 1993). This was a large scale randomized controlled clinical trial involving 1441 patients enrolled in 29 clinical centers in the United States and Canada and followed for an average of 6.5 years (4–9 years). Of these, 726 patients comprising the primary prevention cohort were free of any microvascular complications (AER $\leq$ 40 mg/dL and no retinopathy, among other features); and 715 patients comprising the Secondary Intervention Cohort may have had minimal pre-existing levels of albuminuria (AER < 200 mg/dL) and mild retinopathy. Patients were randomly assigned to receive either intensive or conventional treatment. Intensive treatment used all available means (self-monitoring four or more times a day with three or more multiple daily injections or a pump in conjunction with diet and exercise) to obtain levels of HbA_{1c} as close as possible to the normal range (< 6.05%) while attempting to avoid hypoglycemia. Hypoglycemia occurs when the blood glucose level is reduced below a physiologically safe level, resulting is dizziness and possibly coma (unconsciousness) or seizures. Conventional treatment, on the other hand, consisted of one or two daily injections of insulin with less frequent self-monitoring with the goal of maintaining the clinical well-being of the patient, but without any specific glucose targets.

Figure 1.1 presents the cumulative incidence of microalbuminuria (AER > 40 mg/24 h) among the 724 patients free of microalbuminuria at baseline in the primary cohort (adapted from DCCT, 1993); presented with permission). The average hazard ratio for intensive versus conventional treatment (I:C) over the 9 years is 0.66. This corresponds to a 34% risk reduction with intensive therapy, 95% confidence limits (2, 56%) (DCCT, 1993, 1995a). Likewise, Figure 1.2 presents the cumulative incidence of microalbuminuria among the 641 patients free of microalbuminuria at baseline in the secondary cohort (with permission). The average hazard ratio is 0.57, corresponding to a 43% (*C.I.*: 21, 58%) risk reduction with intensive therapy (DCCT, 1995a). These risk reductions are adjusted for the baseline level of log AER using the proportional hazards regression model. A model that also employed a stratified adjustment for primary and secondary cohorts yields a risk reduction of 39% (21, 52%) in the combined cohorts. Similar analyses indicate a reduction of 54% (19, 74%) in the risk of overt albuminuria or proteinuria (AER $>$ 300 mg/24 h) in the combined cohorts. Thus intensive therapy aimed at near normal blood glucose levels dramatically reduces the incidence of severe nephropathy that may ultimately lead to end-stage renal disease.

Intensive treatment, however, was associated with an increased incidence of severe episodes of hypoglycemia (DCCT, 1993, 1995b, 1997). Over the 4770 patient years of treatment and follow-up in the intensive treatment group, 271 patients experienced 770 episodes of hypoglycemia accompanied by coma and/or seizures, or 16.3 events per 100 patient years (100 PY) of follow-up. In contrast, over the 4732 patient years in the conventional treatment group, 137 patients experienced 257 episodes, or 5.4 per 100 PY. The relative risk is 3.02 with 95% confidence limits of 2.36 to 3.86 (DCCT, 1995b, 1997). Because of substantial over-dispersion of the subject-specific event rates, this confidence limit was computed using a random-effects or over-dispersed Poisson model.

Thus the DCCT demonstrated that a multifaceted intensive treatment aimed at achieving near-normal levels of blood glucose greatly reduces the risk of nephropathy. The ultimate questions, however, were whether these risk reductions are caused principally by the alterations in levels of blood glucose, as opposed to changes in diet or exercise, for example, and whether there is some threshold for hyperglycemia below which there are no further reductions in risk. Thus analyses were performed using Poisson and proportional hazards regression models, separately in the intensive and conventional treatment groups, using the current mean level of HbA_{1c} since entry into the trial as a time-dependent covariate in conjunction with numerous covariates measured at baseline. Adjusting for 25 other covariates, these models showed that the dominant determinant of the risk of proteinuria is the current level of the log mean HbA_{1c} since entry, with a 71% increase in risk per 10% increase in HbA_{1c} (such as from an HbA_{1c} of 8.1 vs. 9) in the conventional group, which explains approximately 5% of the variation in risk (DCCT, 1995c). Further analyses demonstrated that there is no statistical breakpoint or threshold in this risk relationship (DCCT, 1996).

These various studies and analyses, all of which concern the absolute and relative risks of discrete outcomes, show that microalbuminuria and proteinuria are associated with structural changes in renal tissue, that an intensive treatment regimen greatly reduces the risk of nephropathy, and that the principal risk factor is the lifetime exposure to hyperglycemia. Given that diabetes is the leading cause of end-stage renal disease, it can be anticipated that implementation of intensive therapy in the wide population of type 1 diabetes will ultimately reduce the progression of nephropathy to end-stage renal disease, with pursuant reductions in the incidence of morbidity and mortality caused by diabetic kidney disease and the costs to the public.

The following chapters describe the methods used to reach these conclusions and their statistical basis.

2

Relative Risk Estimates and Tests for Two Independent Groups

The core of biostatistics relates to the evaluation and comparison of the risks of disease and other health outcomes in specific populations. Among the many different designs, the most basic is the comparison of two independent groups of subjects drawn from two different populations. This could be a cross-sectional study comparing the current health status of those with versus those without a specific exposure of interest; or a longitudinal cohort study of the development of health outcomes among a group of subjects exposed to a purported risk factor versus a group not so exposed; or a retrospective study comparing the previous exposure risk among independent (unmatched) samples of cases of the disease versus controls; or perhaps a clinical trial where the health outcomes of subjects are compared among those randomly assigned to receive the experimental treatment versus those assigned to receive the control treatment. Each of these cases will involve the comparison of the proportions with the response or outcome between the two groups.

Many texts provide a review of the methods for comparison of the risks or probabilities of the outcome between groups. These include the classic text by Fleiss (1981) and many texts on statistical methods for epidemiology such as Breslow and Day (1980, 1987), Sahai and Khurshid (1995), Selvin (1996), and Kelsey, Whittemore, Evans and Thompson, (1996), among many. Because this book is intended principally as a graduate text, readers are referred to these texts for review of other topics not covered herein.

2.1 PROBABILITY AS A MEASURE OF RISK

2.1.1 Prevalence and Incidence

The simplest data structure in biomedical research is a sample of N independent and identically distributed ($i.i.d.$) Bernoulli observations $\{y_i\}$ from a sample of N subjects ($i = 1, \ldots, N$) drawn at random from a population with a probability π of a characteristic of interest such as death or worsening, or perhaps survival or improvement. The character of interest is often referred to as the positive response, the outcome, or the event. Thus Y is a binary random variable such that $y_i = I$(positive response for the ith observation), where $I(\cdot)$ is the indicator function, $I(\cdot) = 1$ if true, 0 if not. The total number of subjects in the sample with the positive response is $x = \sum_i y_i$ and the simple proportion with the positive response in the sample is $p = x/N$.

The *prevalence* of a characteristic is the probability π in the population, or the proportion p in a sample, with that characteristic present in a cross-section of the population at a specific point in time. For example, the prevalence of adult onset type 2 diabetes as of 1980 was estimated to be approximately 6.8% of the United States population based on the National Health and Nutrition Examination Survey (NHANES) (Harris, Hadden, Knowler and Bennett, 1987). Half of those with diabetes present on an oral glucose tolerance test (3.4%) were previously undiagnosed. In such a study, N is the total sample size of whom x have the positive characteristic (diabetes).

The *incidence* of an *event* (the positive characteristic) is the probability π in the population, or the proportion p in a sample, that acquire the positive characteristic or experience an event over an interval of time among those who were free of the characteristic at baseline. In this case, N is the sample size at risk in a prospective longitudinal follow-up study, of whom x experience the event over a period of time. For example, from the annual National Health Interview Survey (NHIS) it is estimated that the incidence of a new diagnosis of diabetes among adults in the United States population is 2.42 new cases per 1,000 in the population per year (Kenny, Aubert and Geiss, 1995).

Such estimates of the prevalence of a characteristic, or the incidence of an event, are usually simple proportions based on a sample of N $i.i.d.$ Bernoulli observations.

2.1.2 Binomial Distribution and Large Sample Approximations

Whether from a cross-sectional study of prevalence or a prospective study of incidence, the number of positive responses X is distributed as binomial with probability π, or

$$P(x) = B(x; \pi, N) = \binom{N}{x} \pi^x (1-\pi)^{N-x}, \tag{2.1}$$

where $E(X) = N\pi$ and $V(X) = N\pi(1-\pi)$. Since $E(X) = N\pi$, then a natural moment estimate of π is p, where p is the simple proportion of events $p = x/N$.

This is also the maximum likelihood estimate. From the normal approximation to the binomial, it is well known that X is normally distributed asymptotically (in large samples) as

$$X \overset{d}{\approx} \mathcal{N}\left[N\pi,\ N\pi(1-\pi)\right] \tag{2.2}$$

from which

$$p \overset{d}{\approx} \mathcal{N}\left[\pi,\ \pi(1-\pi)/N\right]. \tag{2.3}$$

These expressions follow from the Central Limit Theorem because x can be expressed as the nth partial sum of a potentially infinite series of $i.i.d.$ random variables $\{y_i\}$. (See the Section A.2 of the Appendix). Thus p is the mean of a set of $i.i.d.$ random variables, $p = \overline{y} = \sum_i y_i/N$.

As described in the Appendix, (2.3) is a casual notation for the asymptotic distribution of p or of $\overline{y}$. More precisely we would write

$$\lim_{n\to\infty} \sqrt{n}(p_n - \pi) \overset{d}{\to} \mathcal{N}\left[0,\ \pi(1-\pi)\right] \tag{2.4}$$

which indicates that as the sample size becomes infinitely large, the proportion p_n converges in distribution to the normal distribution and that p is a $\sqrt{n}$-consistent estimator for π. In this notation, the variance is a fixed quantity whereas in (2.3) the variance $\downarrow 0$ as $n \to \infty$. The expression for the variance of the statistic in (2.3), however, is the *large sample variance* that is used in practice with finite samples to compute a confidence interval and a test of significance.

Thus the large sample variance of the estimate is $V(p) = \pi(1-\pi)/N$. Since $p \overset{p}{\to} \pi$, then from Slutsky's Convergence Theorem, (A.45) in Section A.4 of the Appendix, a consistent estimate of the variance is $\widehat{V}(p) = p(1-p)/N$. This yields the usual large sample confidence interval at level $1-\alpha$ for a proportion with lower and upper confidence limits on π obtained as

$$(\widehat{\pi}_\ell, \widehat{\pi}_u) = p \pm Z_{1-\alpha/2}\sqrt{p(1-p)/N}, \tag{2.5}$$

where $Z_{1-\alpha/2}$ is the upper two-sided normal distribution percentile at level α; for example, for $\alpha = 0.05$, $Z_{0.975} = 1.96$. However, these confidence limits are not bounded by (0,1), meaning that for values of p close to 0 or 1, or for small sample sizes, the upper limit may exceed 1 or the lower limit be less than 0.

2.1.3 Asymmetric Confidence Limits

2.1.3.1 Exact Confidence Limits One approach that ensures that the confidence limits are bounded by (0,1) is an *exact computation* under the Binomial distribution, often called the Clopper-Pearson confidence limits (Clopper and Pearson, 1934). In this case the upper confidence limit π_u is the solution to the equation:

$$\sum_{a=0}^{x} B(a;\ \pi, N) = \alpha/2 \tag{2.6}$$

and the lower confidence limit π_ℓ is the solution to

$$\sum_{a=x}^{N} B(a;\, \pi, N) = \alpha/2.$$

Such confidence limits are not centered about p and thus are called *asymmetric* confidence limits.

A solution of these equations may be obtained by iterative computations. Alternately, Clopper and Pearson show that these limits may be obtained from the relationship between the cumulative F-distribution and the incomplete beta function, of which the binomial is a special case. See, for example, Wilks (1962). With a small value of Np, confidence limits may also be obtained from the Poisson approximation to the binomial distribution. Computations of the exact limits are readily obtained using commercial software such as StatXact®.

2.1.3.2 Logit Confidence Limits Another approach is to consider a function $g(\pi)$ such that the inverted confidence limits based on $g(\pi)$ are contained in the interval $(0, 1)$. One convenient function is the *logit* transformation

$$\theta = g(\pi) = \log[\pi/(1-\pi)] \tag{2.7}$$

where throughout log is the natural logarithm to the base e. The logit plays a central role in the analysis of binary (Bernoulli) data. The quantity $O = \pi/(1-\pi)$ is the *odds* of the characteristic of interest or an event in the population, such as $O = 2$ for an odds of 2:1 when $\pi = 2/3$. The inverse logit or *logistic function*

$$\pi = g^{-1}(\pi) = \frac{e^{g(\pi)}}{1+e^{g(\pi)}} = \frac{e^{\theta}}{1+e^{\theta}} \tag{2.8}$$

then transforms the odds back to the probability.

Woolf (1955) was among the first to describe the asymptotic distribution of the log odds. Using the delta (δ)-method (see Section A.3 of the Appendix), then asymptotically

$$E[g(p)] = E\left[\log\left(\frac{p}{1-p}\right)\right] \cong \log\left(\frac{\pi}{1-\pi}\right) = g(\pi) \tag{2.9}$$

and thus $\widehat{\theta} = g(p)$ provides a consistent estimate of $\theta = g(\pi)$. The large sample variance of the estimate is

$$V(\widehat{\theta}) = V\left[\log\left(\frac{p}{1-p}\right)\right] \cong \left[\frac{d}{d\pi}\log\left(\frac{\pi}{1-\pi}\right)\right]^2 V(p) \tag{2.10}$$
$$= \left(\frac{1}{\pi(1-\pi)}\right)^2 \frac{\pi(1-\pi)}{N} = \frac{1}{N\pi(1-\pi)}$$

where $\cong$ means "asymptotically equal to". Because p is a consistent estimator of π it follows from Slutsky's Theorem (A.45) that the variance can be consistently estimated by substituting p for π to yield

$$\widehat{V}(\widehat{\theta}) = \frac{1}{Np(1-p)} \, . \tag{2.11}$$

Further, from another tenet of Slutsky's Theorem (A.47) it follows that asymptotically

$$\widehat{\theta} = \log\left(\frac{p}{1-p}\right) \overset{d}{\approx} \mathcal{N}\left[\log\left(\frac{\pi}{1-\pi}\right), \; \frac{1}{N\pi(1-\pi)}\right] . \tag{2.12}$$

Further, because $\widehat{V}\left(\widehat{\theta}\right)$ is consistent for $V\left(\widehat{\theta}\right)$, it also follows from Slutsky's Theorem (A.44) that asymptotically

$$\frac{\widehat{\theta}-\theta}{\sqrt{\widehat{V}\left(\widehat{\theta}\right)}} = \frac{\log\left(\frac{p}{1-p}\right) - \log\left(\frac{\pi}{1-\pi}\right)}{\sqrt{\widehat{V}\left[\log\left(\frac{p}{1-p}\right)\right]}} \overset{d}{\approx} \mathcal{N}(0,\ 1). \tag{2.13}$$

Thus the symmetric $1-\alpha$ confidence limits on the logit θ are:

$$\left(\widehat{\theta}_\ell, \widehat{\theta}_u\right) = \widehat{\theta} \pm Z_{1-\alpha/2}\sqrt{\frac{1}{Np(1-p)}} \, . \tag{2.14}$$

Applying the inverse (logistic) function in (2.8) yields the asymmetric confidence limits on π:

$$\left(\widehat{\pi}_\ell, \widehat{\pi}_u\right) = \left[\frac{e^{\widehat{\theta}_\ell}}{1+e^{\widehat{\theta}_\ell}}, \; \frac{e^{\widehat{\theta}_u}}{1+e^{\widehat{\theta}_u}}\right] \tag{2.15}$$

that are bounded by $(0,1)$.

2.1.3.3 Complimentary log-log Confidence Limits Another convenient function is the *complimentary log-log* transformation

$$\theta = g(\pi) = \log\left[-\log(\pi)\right] \tag{2.16}$$

that is commonly used in survival analysis. It can readily be shown (see Problem 2.1) that the $1-\alpha$ confidence limits on $\theta = g(\pi)$ obtained from the asymptotic normal distribution of $\widehat{\theta} = g(p) = \log\left[-\log(p)\right]$ are:

$$\left(\widehat{\theta}_\ell, \widehat{\theta}_u\right) = \widehat{\theta} \pm Z_{1-\alpha/2}\sqrt{\frac{(1-p)}{Np(\log p)^2}} \, . \tag{2.17}$$

Applying the inverse function yields the asymmetric confidence limits on π

$$\left(\widehat{\pi}_\ell, \widehat{\pi}_u\right) = \left(\exp\left[-\exp(\widehat{\theta}_u)\right], \; \exp\left[-\exp(\widehat{\theta}_\ell)\right]\right) \tag{2.18}$$

that are also bounded by $(0,1)$. Note that because the transformation includes a reciprocal, the lower limit π_ℓ is obtained as the inverse transformation of the upper confidence limit $\widehat{\theta}_u = g(\widehat{\pi}_u)$ in (2.17).

2.1.3.4 Test Inverted Confidence Limits Another set of asymmetric confidence limits was suggested by Miettinen (1976) based on inverting the Z-test for a proportion using the usual normal approximation to the binomial. To test the null hypothesis H_0: $\pi = \pi_0$ for some specific value π_0, from (2.3) the test is of the form:

$$Z = \frac{p - \pi_0}{\sqrt{\frac{\pi_0(1-\pi_0)}{N}}} . \tag{2.19}$$

We would then reject H_0: against the two-sided alternative H_1: $\pi \neq \pi_0$ for values $|z| \geq Z_{1-\alpha/2}$. Thus setting $z^2 = (Z_{1-\alpha/2})^2$ yields a quadratic equation in π_0, the roots for which provide confidence limits for π:

$$(\widehat{\pi}_\ell, \widehat{\pi}_u) = \frac{\left[\frac{Z_{1-\alpha/2}^2}{N} + 2p\right] \pm \sqrt{\frac{Z_{1-\alpha/2}^2}{N}\left[\frac{Z_{1-\alpha/2}^2}{N} + 4p(1-p)\right]}}{2\left[\frac{Z_{1-\alpha/2}^2}{N} + 1\right]} . \tag{2.20}$$

Test inverted confidence limits have been criticized because the test statistic is based on the variance under the null hypothesis H_0 rather than under the general alternative. This is discussed further in Section 2.7.6.

Example 2.1 *Hospital Mortality*
For an example, in a particular hospital assume that $x = 2$ patients died postoperatively out of $N = 46$ patients who underwent coronary artery bypass surgery during a particular month. Then $p = 0.04348$, with $\widehat{V}(p) = 0.0009041$ and estimated standard error $S.E.(p) = 0.030068$ that yields 95% large sample confidence limits from (2.5) of $(-0.01545, 0.10241)$, the lower limit being less than 0. The exact computation of (2.6) using StatXact yields limits of (0.00531, 0.1484). The logit transformation $\widehat{\theta} = g(p) = \log[p/(1-p)]$ yields $\widehat{\theta} = \log(2/44) = -3.091$ with estimated variance $\widehat{V}(\widehat{\theta}) = 0.5227$ and estimated $S.E.(\widehat{\theta}) = 0.723$. From (2.14) this yields 95% confidence limits on θ of $(-4.508, -1.674)$. The logistic function of these limits yields 95% confidence limits for π of (0.0109, 0.1579) that differ slightly from the exact limits. Likewise, the complimentary log-log transformation $\widehat{\theta} = g(p) = \log[-\log(p)]$ yields $\widehat{\theta} = 1.1428$ with estimated $S.E.(\widehat{\theta}) = .22056$. From (2.17) this yields 95% confidence limits on θ of (0.7105, 1.5751). The inverted function of these limits yields 95% confidence limits for π of (0.00798, 0.13068) that compare favorably to the exact limits. Finally, the test inverted confidence limits from (2.20) are (0.012005, 0.14533).

With only two events in 46 subjects, clearly the exact limits are preferred. However, even in this case, the large sample approximations are satisfactory, other than the ordinary large sample limits based on the asymptotic normal approximation to the distribution of p itself.

2.1.4 Case of Zero Events

In some cases it is important to describe the confidence limits for a probability based on a sample of N observations of which none have the positive characteristic present, or experience an event, such that x and p are both zero. From the expression for the binomial probability,

$$P(X=0) = B(0;\,\pi, N) = (1-\pi)^N. \tag{2.21}$$

One then desires a one-sided confidence interval of size $1-\alpha$ of the form $(0, \widehat{\pi}_u)$ where the upper confidence limit satisfies the relation:

$$\widehat{\pi}_u = \pi : B(0;\,\pi, N) = \alpha. \tag{2.22}$$

Solving for π yields

$$\widehat{\pi}_u = 1 - \alpha^{1/N}. \tag{2.23}$$

See Louis (1981).

For example, if $N = 60$, then the 95% confidence interval for π when $x = 0$ is (0, 0.0487). Thus with 95% confidence we must admit the possibility that π could be as large as 0.049, or about 1 in 20. If, on the other hand, we desired an upper confidence limit of 1 in 100, such that $\widehat{\pi}_u = 0.01$, then the total sample size would satisfy the expression $0.01 = 1 - 0.05^{1/N}$, that yields $N = 299$ (298.07 to be exact). See Problem 2.3.

2.2 MEASURES OF RELATIVE RISK

The simplest design to compare two populations is to draw two independent samples of n_1 and n_2 subjects from each of the two populations and to then observe the numbers within each sample, x_1 and x_2, who have a positive response or characteristic of interest. The resulting data can be summarized in a simple 2×2 table to describe the association between the binary independent variable representing membership in either of two independent groups ($i = 1, 2$), and a binary dependent variable (the response), where the response of primary interest is denoted as + and its complement as −. This 2×2 table of frequencies can be expressed as:

$$\begin{array}{c} \\ \textit{Response} \\ + \\ - \\ \\ \end{array}
\begin{array}{cc} \multicolumn{2}{c}{\textit{Group}} \\ 1 & 2 \\ \hline \multicolumn{1}{|c|}{x_1} & \multicolumn{1}{c|}{x_2} \\ \hline \multicolumn{1}{|c|}{n_1 - x_1} & \multicolumn{1}{c|}{n_2 - x_2} \\ \hline n_1 & n_2 \end{array}
\quad or \quad
\begin{array}{ccc} \multicolumn{2}{c}{\textit{Group}} & \\ 1 & 2 & \\ \cline{1-2} \multicolumn{1}{|c|}{n_{11}} & \multicolumn{1}{c|}{n_{12}} & n_{1\bullet} \\ \cline{1-2} \multicolumn{1}{|c|}{n_{21}} & \multicolumn{1}{c|}{n_{22}} & n_{2\bullet} \\ \cline{1-2} n_{\bullet 1} & n_{\bullet 2} & N \end{array}
\quad or \quad
\begin{array}{ccc} \multicolumn{2}{c}{\textit{Group}} & \\ 1 & 2 & \\ \cline{1-2} \multicolumn{1}{|c|}{a} & \multicolumn{1}{c|}{b} & m_1 \\ \cline{1-2} \multicolumn{1}{|c|}{c} & \multicolumn{1}{c|}{d} & m_2 \\ \cline{1-2} n_1 & n_2 & N \end{array}
\tag{2.24}$$

where the "•" subscript represents summation over the corresponding index for rows or columns. For the most part, we shall use the notation in the last table when it is unambiguous.

Table 2.1 Measures of Relative Risk

Type θ	*Expression*	*Domain*	*Null Value*
Risk difference (RD)	$\pi_1 - \pi_2$	$[-1, 1]$	0
Relative risk (RR)	π_1/π_2	$(0, \infty)$	1
log RR	$\log(\pi_1) - \log(\pi_2)$	$(-\infty, \infty)$	0
Odds ratio (OR)	$\dfrac{\pi_1/(1-\pi_1)}{\pi_2/(1-\pi_2)}$	$(0, \infty)$	1
log OR	$\log \dfrac{\pi_1}{1-\pi_1} - \log \dfrac{\pi_2}{1-\pi_2}$	$(-\infty, \infty)$	0

Within each group ($i = 1, 2$), the number of positive responses is distributed as binomial with probability π_i, from which $p_i \overset{d}{\approx} \mathcal{N}\left[\pi_i,\ \pi_i(1-\pi_i)/n_i\right]$.

We can now define a variety of parameters to describe the differences in risk between the two populations as shown in Table 2.1. Each measure is a function of the probabilities of the positive response in the two groups. The domain refers to the parameter space for that measure while the null value refers to the value of the parameter under the null hypothesis of no difference in risk between the two populations, H_0: $\pi_1 = \pi_2$. The *risk difference (RD)* refers to the simple algebraic difference between the probabilities of the positive response in the two groups with a domain of $[-1,1]$ and with the value zero under the null hypothesis. The *relative risk (RR)* is the ratio of the probabilities in the two groups. It is also referred to as the *risk ratio*. The *odds ratio (OR)* is the ratio of the odds of the outcome of interest in the two groups. Both the relative risk and the odds ratio have a domain consisting of the positive real line and a null value of one. To provide a symmetric distribution under the null hypothesis, it is customary to use the log of each.

Each of these measures can be viewed as an index of the differential or relative risk between the two groups and will reflect a departure from the null hypothesis when an association between group membership and the probability of response exists. Thus the term relative risk is used to refer to a family of measures of the degree of association between group and response, and is also used to refer to the specific measure defined as the risk ratio.

Each of these measures is of the form $\theta = G(\pi_1, \pi_2)$ for some function $G(.,.)$. Thus each may be estimated by substituting the sample proportions p_i for the probabilities π_i to yield

$$\widehat{RD} = p_1 - p_2, \tag{2.25}$$

$$\widehat{RR} = p_1/p_2 = an_2/bn_1, \text{ and}$$

$$\widehat{OR} = \frac{p_1/(1-p_1)}{p_2/(1-p_2)} = \frac{p_1(1-p_2)}{p_2(1-p_1)} = \frac{ad}{bc}\ .$$

Fig. 2.1 Odds Ratio and relative risk over a range of values for π_2 for a fixed risk difference of -0.1.

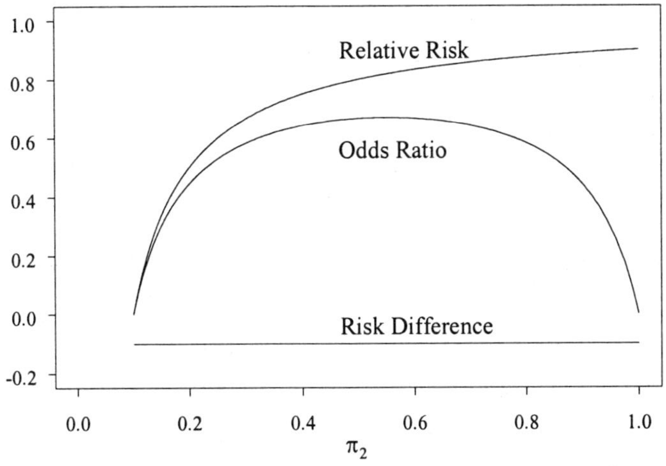

Because the p_i converge in probability to (are consistent estimates of) the probabilities π_i, then from Slutsky's Convergence Theorem (A.45) the resulting estimate $\widehat{\theta} = G(p_1, p_2)$ is a consistent estimate of the corresponding $\theta = G(\pi_1, \pi_2)$. These estimates, however, are not unbiased for finite samples.

Later in Chapter 5 we shall demonstrate that in a retrospective case-control study, the relative risk itself is not directly estimable without additional information. However, the odds ratio is estimable, and under the assumption that the disease (i.e. being a case) is rare, it provides an approximation to the relative risk. Thus in some texts, the odds ratio is called the relative risk. However, the two are distinct measures of the association between group membership and the likelihood of a positive response. At the end of this chapter we also introduce other useful measures.

The risk difference, relative risk and odds ratio clearly are non-linear functions of each other. For example, Figure 2.1 displays the values of the odds ratio and relative risk over a range of values for π_2 where the risk difference is held constant at $RD = -0.1$. As π_2 increases, the relative risk is monotonically increasing toward the null value of 1.0, indicating proportionately smaller risk reductions. For values $\pi_1 \downarrow 0$ and for values $\pi_2 \uparrow 1$ the odds ratio is $OR \cong 0$. As π_2 increases, the odds ratio increases toward the null value reaching a maximum of 0.669 at $\widetilde{\pi}_2 = 0.5(1 - RD) = 0.55$. The relative risk, however, continues to increase as $\pi_2 \uparrow 1$.

Thus if we have two separate 2×2 tables, such as from two different studies, where the risk difference is the same $(RD_{(1)} = RD_{(2)})$, but where the probabilities in the control group are different $(\pi_{2(1)} \neq \pi_{2(2)})$ then the relative risks will differ, the study with the larger value of π_2 having the smaller risk reduction (relative risk closer to 1). The odds ratios will also differ if both studies have control group probabilities that are less than $\widetilde{\pi}_2$ or both are greater than $\widetilde{\pi}_2$. It is possible, however that they may be approximately equal if $\pi_{2(1)} < \widetilde{\pi}_2 < \pi_{2(2)}$, such as $OR = 0.66033$ for $\pi_{2(1)} = 0.46$ and $\pi_{2(2)} = 0.64$.

For any single study, the choice of the measure used to describe the study results may largely be a matter of taste or clinical preference. All three measures will reflect a difference between groups when such exists. Non-statisticians such as physicians often find the relative risk most appealing. As we shall see, however, the odds ratio arises as the natural parameter under a conditional likelihood, and forms the basis for the analysis of case-control studies (see Chapter 5). It also forms the basis for logistic regression. For a single study with moderate or large sample size, the choice of the measure will not affect the conclusions reached. However, as we will show in Chapter 4, when there are multiple 2×2 tables, such as from multiple strata, the conclusions reached may indeed depend on the measure chosen to summarize the results.

Example 2.2 *Neuropathy Clinical Trial*
For an example, consider a clinical trial of a new drug (group 1) versus placebo (2) for treatment of diabetic neuropathy where the positive response of interest is improvement in peripheral sensory perception. Patients were assigned at random to receive one of the study treatments with 100 subjects in each group. Of these, 53 and 40 patients in each group, respectively, had a positive response. These data would be represented as

		Group 1	*Group* 2	
Response	+	a	b	m_1
	−	c	d	m_2
		n_1	n_2	N

=

	Group 1	*Group* 2	
+	53	40	93
−	47	60	107
	100	100	200

The risk difference $\widehat{RD} = 0.53 - 0.40 = 0.13$ shows that the excess probability of improvement is 0.13. Thus the number of patients with improved perception is increased by 13% when treated with the drug, or an additional 13 patients per 100 patients so treated. The relative risk $\widehat{RR} = 0.53/0.40 = 1.325$ indicates that the total number improved is increased by 32.5% (0.13/0.40). Finally, the odds ratio $\widehat{OR} = (0.53 \times 0.60)/(0.40 \times 0.47) = 1.691$ indicates that the odds of improvement is increased by 69.1% from an odds of $0.4/0.6 = 0.667$ to an odds of $0.53/0.47 = 1.128$.

2.3 LARGE SAMPLE DISTRIBUTION

For the $\widehat{RR}$ and $\widehat{OR}$, the domain is not symmetric about the null value. Thus the large sample distributions are better approximated using the log transformation. In this case, since the domain encompasses the entire real line, no adjustments are required to ensure that confidence intervals are bounded within the domain of the parameter.

2.3.1 Risk Difference

The asymptotic distribution of the risk difference $\widehat{RD} = p_1 - p_2$ follows directly from that of the sample proportions themselves since the $\widehat{RD}$ is a simple linear contrast of two independent proportions, each of which is asymptotically normally distributed. Thus

$$p_1 - p_2 \overset{d}{\approx} \mathcal{N}\left[\pi_1 - \pi_2,\ \frac{\pi_1(1-\pi_1)}{n_1} + \frac{\pi_2(1-\pi_2)}{n_2}\right]. \tag{2.26}$$

This is the distribution of the risk difference in general with no restrictions on the values of the probabilities π_1 and π_2. In the context of testing the null hypothesis H_0: $\pi_1 = \pi_2$ against the general alternative hypothesis H_1: $\pi_1 \neq \pi_2$, this is termed the distribution *under the alternative hypothesis*. If we let $\theta = \pi_1 - \pi_2$ and $\widehat{\theta} = p_1 - p_2$, then (2.26) is equivalent to $\widehat{\theta} \overset{d}{\approx} \mathcal{N}\left[\theta_1,\ \sigma_1^2\right]$ under the alternative H_1: $\theta = \theta_1 \neq 0$.

Under the null hypothesis, H_0: $\pi_1 = \pi_2 = \pi$, then $\theta = \theta_0 = 0$ and the variance reduces to

$$\sigma_0^2 = \pi(1-\pi)\left[\frac{1}{n_1} + \frac{1}{n_2}\right] = \pi(1-\pi)\left[\frac{N}{n_1 n_2}\right]. \tag{2.27}$$

Thus the distribution *under the null hypothesis* is $\widehat{\theta} \overset{d}{\approx} \mathcal{N}\left[\theta_0,\ \sigma_0^2\right]$.

Since the p_i are consistent estimates of the π_i , then from Slutsky's Theorem (A.45) the variance under the alternative can be consistently estimated by substituting the p_i for the π_i in the expression for σ^2. Under the null hypothesis, π is consistently estimated as

$$\widehat{\pi} = p = \frac{a+b}{n_1+n_2} = \frac{m_1}{N}, \tag{2.28}$$

which yields a consistent estimate of the null variance σ_0^2 when substituted into (2.27).

The asymptotic distribution under the alternative leads to the usual expression for the large sample $1-\alpha$ level confidence interval for the population risk difference based on the estimate of the variance under the alternative:

$$\left(\widehat{\theta}_l,\ \widehat{\theta}_u\right) = \widehat{\theta} \pm Z_{1-\alpha/2}\widehat{\sigma}_1. \tag{2.29}$$

Although common in practice, these confidence limits are not necessarily bounded by -1 and $+1$, and in rare circumstances limits are obtained that lie outside these bounds. Unlike the case of a single proportion, there is no convenient function that may be used to yield asymmetric confidence limits on the risk difference that are then bounded by (−1,1). However, by conditioning on both margins of the 2×2 table, one can obtain exact confidence limits for the risk difference that are so bounded; see Section 2.5.

The asymptotic distribution under the null hypothesis also leads to the usual expression for the large sample Z-test of the null hypothesis $H_0\colon \pi_1 = \pi_2$ described in Section 2.6 to follow.

Example 2.3 *Neuropathy Clinical Trial*

For the hypothetical clinical trial data presented in Example 2.2, $\widehat{\theta} = \widehat{RD} = 0.13$. The large sample variance from (2.27) is estimated to be $\widehat{\sigma}^2_{\widehat{\theta}} = \widehat{V}(\widehat{RD}) = (0.53 \times 0.47)/100 + (0.4 \times 0.6)/100 = 0.00489$. Thus the estimated $S.E.$ is $\widehat{\sigma}_{\widehat{\theta}} = 0.0699$, which yields a large sample 95% confidence interval of (−0.0071, 0.2671).

2.3.2 Relative Risk

We now consider the distribution of the log $\widehat{RR}$, and subsequently the log $\widehat{OR}$, under the null and the alternative hypotheses. The log $\widehat{RR}$ is the difference in the $\log(p_i)$. The variance of the $\log(p_i)$, and thus the $V[\log(\widehat{RR})]$, is obtained using the delta (δ)-method. Likewise, the asymptotic distribution of the log $\widehat{RR}$ is derived from that of the $\log(p_i)$ through the application of Slutsky's Theorem.

In summary, by application of the δ-method, it follows (Katz, Baptista, Azen and Pike, 1978) that for each group ($i = 1, 2$), asymptotically

$$E[\log(p_i)] \cong \log(\pi_i) \tag{2.30}$$

and that

$$V[\log(p_i)] \cong \left(\frac{d\log(\pi_i)}{d\pi_i}\right)^2 V(p_i) = \left(\frac{1}{\pi_i}\right)^2 \frac{\pi_i(1-\pi_i)}{n_i} = \frac{1-\pi_i}{n_i\pi_i}\ . \tag{2.31}$$

Because p_i is a consistent estimator of π_i it follows from Slutsky's Theorem (A.45) that the variance can be consistently estimated by substituting p_i for π_i to yield

$$\widehat{V}[\log(p_i)] = \frac{1-p_i}{n_i p_i}\ . \tag{2.32}$$

Again using Slutsky's Theorem (A.47) it follows that asymptotically

$$\log(p_i) \overset{d}{\approx} \mathcal{N}\left[\log(\pi_i),\ \frac{1-\pi_i}{n_i\pi_i}\right], \tag{2.33}$$

and since $\widehat{V}[\log(p_i)]$ is consistent for $V[\log(p_i)]$, that

$$\frac{\log(p_i) - \log(\pi_i)}{\sqrt{\widehat{V}[\log(p_i)]}} \overset{d}{\approx} \mathcal{N}(0,\ 1). \tag{2.34}$$

Since $\log(\widehat{RR}) = \log(p_1) - \log(p_2)$, it follows that asymptotically

$$E[\log(\widehat{RR})] \cong \log(RR) = \log(\pi_1) - \log(\pi_2). \tag{2.35}$$

Since p_1 and p_2 are independent, then under the alternative hypothesis, with no restrictions on the values of π_1 and π_2, the variance is

$$\sigma_1^2 \cong V[\log(\widehat{RR})] = V[\log(p_1)] + V[\log(p_2)] = \left(\frac{1-\pi_1}{n_1\pi_1} + \frac{1-\pi_2}{n_2\pi_2}\right) \tag{2.36}$$

that can be consistently estimated as

$$\begin{aligned} \widehat{\sigma}_1^2 = \widehat{V}[\log(\widehat{RR})] &= \left(\frac{1-p_1}{n_1 p_1} + \frac{1-p_2}{n_2 p_2}\right) \\ &= \left(\frac{1-p_1}{a} + \frac{1-p_2}{b}\right) = \left(\frac{1}{a} - \frac{1}{n_1} + \frac{1}{b} - \frac{1}{n_2}\right). \end{aligned} \tag{2.37}$$

Further, asymptotically,

$$\log(\widehat{RR}) \overset{d}{\approx} \mathcal{N}\left[\log(RR),\ V[\log(\widehat{RR})]\right] \tag{2.38}$$

and

$$\frac{\log(\widehat{RR}) - \log(RR)}{\sqrt{\widehat{V}[\log(\widehat{RR})]}} \overset{d}{\approx} \mathcal{N}(0,\ 1). \tag{2.39}$$

This distribution under the alternative hypothesis is used to derive the large sample confidence limits on $\theta = \log(RR)$ as

$$\left(\widehat{\theta}_l,\ \widehat{\theta}_u\right) = \widehat{\theta} \pm Z_{1-\alpha/2}\widehat{\sigma}_1. \tag{2.40}$$

From these, the *asymmetric confidence limits* on the relative risk are obtained as

$$\left(\widehat{RR}_l,\ \widehat{RR}_u\right) = \exp\left[\widehat{\theta} \pm Z_{1-\alpha/2}\widehat{\sigma}_1\right] = \exp\left(\widehat{\theta}_l,\ \widehat{\theta}_u\right) \tag{2.41}$$

that are contained within $[0, \infty)$.

Under the null hypothesis, H_0: $\pi_1 = \pi_2 = \pi$, then $\theta = \log(RR) = \theta_0 = 0$ and $\widehat{\theta} \overset{d}{\approx} \mathcal{N}\left[\theta_0,\ \sigma_0^2\right]$. From (2.36), the variance under the null hypothesis reduces to

$$\sigma_0^2 = \frac{1-\pi}{\pi}\left[\frac{1}{n_1} + \frac{1}{n_2}\right] = \frac{1-\pi}{\pi}\left[\frac{N}{n_1 n_2}\right] \tag{2.42}$$

that can be consistently estimated by substituting p in (2.28) for π in the above.

Example 2.4 *Neuropathy Clinical Trial*
For the data in Example 2.2, the relative risk is $\widehat{RR} = 1.325$ and the $\widehat{\theta} = \log(\widehat{RR}) = \log(0.53/0.4) = 0.2814$. The large sample variance from (2.37) is estimated to be $\widehat{\sigma}^2_{\widehat{\theta}} = \widehat{V}[\log(\widehat{RR})] = \sigma_1^2 = 0.47/(100 \times 0.53) + 0.6/(100 \times 0.4) = \$, 0.02387$. The estimated $S.E.$ is $\widehat{\sigma}_{\widehat{\theta}} = 0.1545$, that yields a large sample 95% confidence interval for the log(RR) of (−0.0214, 0.5842). Exponentiation yields the asymmetric 95% confidence limits for the RR of (0.9788, 1.7936).

Note that the relative risk herein is described as $RR_{1:2} = \pi_1/\pi_2$. If we wished to describe the relative risk as $RR_{2:1} = \pi_2/\pi_1$, then we need only invert the estimate to obtain $\widehat{RR}_{2:1} = p_2/p_1 = 1/\widehat{RR}_{1:2}$. Thus $\widehat{RR}_{2:1} = 1/1.325 = 0.755$ and $\log(\widehat{RR}_{2:1}) = \log(0.755) = -0.2814 = \log(\widehat{RR}_{1:2}^{-1}) = -\log(\widehat{RR}_{1:2})$. The estimated variance, however, is unchanged as $\widehat{V}[\log(\widehat{RR})] = 0.02387$. Thus the confidence limits for $\log(RR_{2:1})$ are the negatives of those for $\log(RR_{1:2})$, or (−0.5842, 0.0214), and the confidence limits for $RR_{2:1}$ are the reciprocals of those for $RR_{1:2}$, or (0.5575, 1.2166) = (1.7936^{-1}, 0.9788^{-1}).

2.3.3 Odds Ratio

The asymptotic distribution of the log $\widehat{OR}$ may be obtained similarly to that of the log $\widehat{RR}$ whereby the distribution of the log odds is first obtained within each group, and then the distribution of the log odds ratio is obtained as that of a linear combination of two normally distributed variates. Within each group, the log odds is simply the logit of the probability as presented in Section 2.1.3.2. In the following, however, we shall derive the distribution of the log $\widehat{OR}$ through the multivariate δ-method (see Section A.3.2 of the Appendix), starting with the asymptotic bivariate distribution of p_1 and p_2.

Because the two groups are independent, then $p = (p_1,\ p_2)^T$ is asymptotically distributed as bivariate normal with mean vector $\boldsymbol{\pi} = (\pi_1,\ \pi_2)^T$ and covariance matrix under the alternative

$$\boldsymbol{\Omega}_1 = \begin{bmatrix} \dfrac{\pi_1(1-\pi_1)}{n_1} & 0 \\ 0 & \dfrac{\pi_2(1-\pi_2)}{n_2} \end{bmatrix}. \tag{2.43}$$

See Example A.2 of the Appendix. The log odds ratio is $G(\boldsymbol{\pi}) = \log[\pi_1/(1-\pi_1)] - \log[\pi_2/(1-\pi_2)]$ with the corresponding matrix of partial derivatives (here a vector)

$$\begin{aligned} \boldsymbol{H}(\boldsymbol{\pi}) &= [\partial G(\boldsymbol{\pi})/\partial\pi_1 \;\; \partial G(\boldsymbol{\pi})/\partial\pi_2]^T \\ &= \left[\frac{1}{\pi_1(1-\pi_1)} \;\; \frac{1}{\pi_2(1-\pi_2)}\right]^T. \end{aligned} \tag{2.44}$$

By application of the multivariate δ-method, the asymptotic variance of the log $\widehat{OR}$ under the alternative hypothesis is

$$\begin{aligned}\sigma_1^2 = V[\log(\widehat{OR})] &\cong \boldsymbol{H}(\boldsymbol{\pi})'\boldsymbol{\Omega}_1\boldsymbol{H}(\boldsymbol{\pi}) \\ &= \frac{1}{n_1\pi_1(1-\pi_1)} + \frac{1}{n_2\pi_2(1-\pi_2)}\end{aligned} \tag{2.45}$$

that can be consistently estimated as

$$\begin{aligned}\widehat{\sigma}_1^2 = \widehat{V}[\log(\widehat{OR})] &= \frac{1}{n_1p_1(1-p_1)} + \frac{1}{n_2p_2(1-p_2)} \\ &= \frac{1}{a} + \frac{1}{b} + \frac{1}{c} + \frac{1}{d}\,.\end{aligned} \tag{2.46}$$

This is Woolf's (1955) estimate of the variance of the log odds ratio.

From Slutsky's Theorem (A.47) it follows that asymptotically

$$\log(\widehat{OR}) \overset{d}{\approx} \mathcal{N}\left[\log(OR),\ \sigma_1^2\right] \tag{2.47}$$

and that

$$\frac{\log(\widehat{OR}) - \log(OR)}{\sqrt{\widehat{V}[\log(\widehat{OR})]}} \overset{d}{\approx} \mathcal{N}(0,\ 1). \tag{2.48}$$

This yields large sample confidence limits on $\theta = \log(OR)$ as

$$\left(\widehat{\theta}_l,\ \widehat{\theta}_u\right) = \widehat{\theta} \pm Z_{1-\alpha/2}\widehat{\sigma}_1 \tag{2.49}$$

and asymmetric confidence limits on the odds ratio

$$\left(\widehat{OR}_l,\ \widehat{OR}_u\right) = \exp\left[\widehat{\theta} \pm Z_{1-\alpha/2}\widehat{\sigma}_1\right] = \exp\left(\widehat{\theta}_l,\ \widehat{\theta}_u\right) \tag{2.50}$$

that again are bounded by -1 and $+1$.

Under the null hypothesis, H_0: $\pi_1 = \pi_2 = \pi$, then $\theta = \log(OR) = \theta_0 = 0$ and $\widehat{\theta} \overset{d}{\approx} \mathcal{N}\left[\theta_0,\ \sigma_0^2\right]$ where the null variance reduces to

$$\sigma_0^2 = \frac{1}{\pi(1-\pi)}\left[\frac{1}{n_1} + \frac{1}{n_2}\right] = \frac{N}{\pi(1-\pi)n_1n_2} \tag{2.51}$$

that can be consistently estimated by substituting p in the above for π.

Example 2.5 *Neuropathy Clinical Trial*

Again for the data in Example 2.2, the odds ratio is $\widehat{OR} = 1.691$, and $\widehat{\theta} = \log(\widehat{OR}) = \log(1.691) = 0.526$. The estimated large sample variance from (2.46) is $\widehat{\sigma}_{\widehat{\theta}}^2 = \widehat{V}[\log(\widehat{OR})] = \widehat{\sigma}_1^2 = [1/53 + 1/47 + 1/40 + 1/60] = 0.0818$. The estimated *S.E.* is $\widehat{\sigma}_{\widehat{\theta}} = 0.286$, which yields a large sample 95% confidence interval for the $\log(OR)$ of $(-0.035,\ 1.0862)$. Exponentiation yields the asymmetric 95% confidence limits for the OR of $(0.9656,\ 2.963)$.

Also, as was the case for the relative risk, the odds ratio herein is defined as $OR_{1:2} = [\pi_1/(1-\pi_1)]/[\pi_2/(1-\pi_2)]$. If we wished to describe the odds ratio as $OR_{2:1} = [\pi_2/(1-\pi_2)]/[\pi_1/(1-\pi_1)]$, then we would follow the same steps as described for the estimation of the inverse relative risk $RR_{2:1}$ and its confidence limits.

2.4 SAMPLING MODELS: - LIKELIHOODS

In the preceding section, we derived the large sample distribution of measures of relative risk starting from basic principles. Now we consider the development of the underlying likelihood based on models of sampling from a population.

2.4.1 Unconditional Product Binomial Likelihood

We can view the table of frequencies in (2.24) as arising from two independent samples of sizes n_1 and n_2 from two separate populations, of whom x_1 and x_2 are observed to have the positive characteristic of interest with probabilities π_1 and π_2. Using the shorthand notation $a = x_1$ and $b = x_2$, the likelihood of the observations is a *product binomial likelihood*

$$\begin{aligned} L_u(\pi_1, \pi_2) &= P(a, b \mid n_1, n_2, \pi_1, \pi_2) = B(a; n_1, \pi_1) B(b; n_2, \pi_2) \qquad (2.52) \\ &= \binom{n_1}{a} \pi_1^a (1-\pi_1)^{n_1-a} \binom{n_2}{b} \pi_2^b (1-\pi_2)^{n_2-b}. \end{aligned}$$

This is also termed the unconditional likelihood.

This likelihood applies to any study in which independent samples are drawn from two populations. In some cases, however, as in a cross-sectional study, one draws a single sample of size N from the population, and the sample is then cross-classified with respect to a binary independent variable that forms the two groups, and with respect to a binary dependent variable that forms the positive or negative responses. In this case the likelihood is a multinomial (quadrinomial) with four cells. However, if we condition on the group margin totals (the n_1 and n_2), then the product binomial likelihood results.

2.4.2 Conditional Hypergeometric Likelihood

The unconditional likelihood (2.52) can also be expressed in terms of the total number of positive responses m_1 since $b = m_1 - a$. This yields

$$P(a, m_1 \mid n_1, n_2, \pi_1, \pi_2) \tag{2.53}$$
$$= \binom{n_1}{a}\binom{n_2}{m_1 - a}\left(\frac{\pi_1/(1-\pi_1)}{\pi_2/(1-\pi_2)}\right)^a (1-\pi_1)^{n_1}\pi_2^{m_1}(1-\pi_2)^{n_2-m_1}$$
$$= \binom{n_1}{a}\binom{n_2}{m_1 - a}\varphi^a (1-\pi_1)^{n_1}\pi_2^{m_1}(1-\pi_2)^{n_2-m_1}$$

where $\varphi = \frac{\pi_1/(1-\pi_1)}{\pi_2/(1-\pi_2)}$ is the odds ratio.

In general, a conditional distribution is obtained as $f(x|y) = f(x,y)/f(y) = f(x,y)/[\int f(x,y)dx]$. Thus we can obtain the conditional likelihood, conditioning on m_1 being fixed, as

$$L_c(\varphi) = \frac{P(a, m_1 \mid n_1, n_2, \pi_1, \pi_2)}{P(m_1 \mid n_1, n_2, \pi_1, \pi_2)} \tag{2.54}$$

where

$$P(m_1 \mid n_1, n_2, \pi_1, \pi_2) = \sum_{a=a_\ell}^{a_u} P(a, m_1 \mid n_1, n_2, \pi_1, \pi_2) \tag{2.55}$$

is the summation over all possible values of the index frequency $a_\ell \leq a \leq a_u$ given the row and column margins.

It is clear that the maximum possible value of a given the fixed margins is $a_u = \min(n_1, m_1)$. Likewise, the maximum value of b is $b_u = \min(n_2, m_1)$. Since $a = m_1 - b$, then the minimum possible value of a is $a_\ell = m_1 - b_u$. Thus the limits of the summation over a are:

$$a_\ell = \max(0, m_1 - n_2) \tag{2.56}$$
$$a_u = \min(m_1, n_1).$$

Then, substituting (2.53) and (2.55) into (2.54) yields the *conditional non-central hypergeometric likelihood*

$$L_c(\varphi) = P(a \mid n_1, m_1, N, \varphi) = \frac{\binom{n_1}{a}\binom{N-n_1}{m_1-a}\varphi^a}{\sum_{i=a_\ell}^{a_u}\binom{n_1}{i}\binom{N-n_1}{m_1-i}\varphi^i}, \tag{2.57}$$

which is a function only of the odds ratio φ.

2.4.3 Maximum Likelihood Estimates

Because the product binomial likelihood (2.52) is the product of two independent binomial likelihoods it then follows that the maximum likelihood estimate *(MLE)* of $\pi = (\pi_1, \pi_2)$ is $p = (p_1, p_2)$, and that the covariance matrix is the diagonal matrix in (2.43). From the invariance principle (Section A.6.8 of the Appendix), the *MLEs* of the relative risk and the odds ratio, or the logs thereof, based on the unconditional likelihood are the corresponding functions of p_1 and p_2 with large sample variances as obtained above using the δ-method.

Through the conditional hypergeometric distribution in (2.57), one can also derive the maximum likelihood estimating equation for φ, the odds ratio. However, the equation does not provide a closed-form expression for the *MLE* $\widehat{\varphi}$, and thus an iterative procedure such as Newton-Raphson is required; see Birch (1964). Because this estimate of the odds ratio is based on the conditional hypergeometric likelihood, it is sometimes called the conditional *MLE* of the odds ratio to distinguish it from the unconditional $MLE = ad/bc$. These estimators are described in Chapter 6.

2.4.4 Asymptotically Unbiased Estimates

Because of the discreteness of the binomial distribution from which the frequencies are sampled, the sample estimates of the risk difference, relative risk and odds ratio are biased. For example, consider the sample estimate of the odds ratio $\widehat{\varphi} = \widehat{OR}$. From (2.57)

$$E(\widehat{\varphi}) = \sum_{a=a_\ell}^{a_u} \frac{a(n_2 - m_1 + a)}{(n_1 - a)(m_1 - a)} P(a \mid n_1, m_1, N, \varphi), \tag{2.58}$$

which is undefined since $\widehat{\varphi}$ is undefined for those tables where $a = a_\ell$ or $a = a_u$ such that one of the cell frequencies is zero. Alternately, one could evaluate $E(a)$ from which one could obtain an unbiased moment estimate of φ. However, no closed-form solution exists. Similar results also apply to the expected value of the log odds ratio for which no simple bias corrected moment estimator exists. However, simple bias corrections have been suggested.

The bias of various estimators, say $\widehat{\theta}(x)$, of the logit $\log[\pi/(1-\pi)]$ and of $\log(\pi)$ for a single binomial have been evaluated using

$$E\left[\widehat{\theta}(x)\right] = \sum_{x=0}^{n} \widehat{\theta}(x) B(x; n, \pi) \tag{2.59}$$

for fixed value of π. Gart and Zweifel (1967) present a review of the properties of various estimators. For estimation of the logit, or the odds ratio, the most common suggestion is to add 1/2 to each cell. The resulting bias-corrected estimate of the logit $\theta = \log[\pi/(1-\pi)]$ is

$$\widehat{\theta} = \log\left[\frac{x + \frac{1}{2}}{n - x + \frac{1}{2}}\right], \tag{2.60}$$

which is unbiased except for terms of $O(n^{-2})$, or of order $1/n^2$. From the δ-method (see Section A.3 of the Appendix), the simple logit $\log[p/(1-p)]$ is unbiased except for the remainder that is $O(n^{-1})$ or of order $1/n$. Thus the estimator in (2.60) converges to the true value $\theta = \log[\pi/(1-\pi)]$ at a faster rate than does the simple logit. The estimated large sample variance of this estimate is

$$\widehat{V}(\widehat{\theta}) = \frac{1}{x+\frac{1}{2}} + \frac{1}{n-x+\frac{1}{2}}, \tag{2.61}$$

which is unbiased for the true asymptotic variance to $O(n^{-3})$. For computations over a range of values of the parameters, Gart and Zweifel (1967) also showed that the bias of $\widehat{\theta}$ is $< 1\%$ when $np > 2$.

These results suggest that the following estimate of $\theta =$ the log odds ratio, widely known as the *Haldane-Anscombe estimate* (Haldane, 1956; Anscombe, 1956), is nearly unbiased:

$$\begin{aligned}\widehat{\theta} = \log(\widehat{OR}) &= \log\left[\frac{x_1+\frac{1}{2}}{n_1-x_1+\frac{1}{2}}\right] - \log\left[\frac{x_2+\frac{1}{2}}{n_2-x_2+\frac{1}{2}}\right] \\ &= \log\left[\frac{a+\frac{1}{2}}{c+\frac{1}{2}}\right] - \log\left[\frac{b+\frac{1}{2}}{d+\frac{1}{2}}\right]\end{aligned} \tag{2.62}$$

with a nearly unbiased estimate of the large sample variance most easily expressed as a modified Woolf's estimate:

$$\widehat{V}(\widehat{\theta}) = \frac{1}{a+\frac{1}{2}} + \frac{1}{b+\frac{1}{2}} + \frac{1}{c+\frac{1}{2}} + \frac{1}{d+\frac{1}{2}}. \tag{2.63}$$

Similarly, Walter (1975) showed that a nearly unbiased estimator of $\theta = \log(\pi)$ is provided by

$$\widehat{\theta} = \log(\widehat{\pi}) = \frac{x+\frac{1}{2}}{n+\frac{1}{2}} \tag{2.64}$$

and Pettigrew, Gart and Thomas (1986) showed that the estimated large sample variance

$$\widehat{V}(\widehat{\theta}) = \frac{1}{x+\frac{1}{2}} - \frac{1}{n+\frac{1}{2}} \tag{2.65}$$

is also nearly unbiased. Thus the following estimate of $\theta =$ the log relative risk is nearly unbiased

$$\widehat{\theta} = \log(\widehat{RR}) = \log\left[\frac{x_1+\frac{1}{2}}{n_1+\frac{1}{2}}\right] - \log\left[\frac{x_2+\frac{1}{2}}{n_2+\frac{1}{2}}\right] \tag{2.66}$$

with a nearly unbiased estimate of the large sample variance:

$$\widehat{V}(\widehat{\theta}) = \frac{1}{x_1+\frac{1}{2}} - \frac{1}{n_1+\frac{1}{2}} + \frac{1}{x_2+\frac{1}{2}} - \frac{1}{n_2+\frac{1}{2}}. \tag{2.67}$$

These estimates can then be used to compute less biased confidence interval estimates for the parameters on each scale. Clearly, their principal advantage is with small samples. In such cases, however, exact confidence limits are readily calculated using statistical software such as StatXact.

Example 2.6 *Neuropathy Clinical Trial*
For the simple data table in Example 2.2, the bias-corrected or Haldane-Anscombe estimate of the odds ratio is $(53.5 \times 60.5)/[(40.5 \times 47.5)] = 1.68$, yielding a log odds ratio of 0.520 with an estimated modified Woolf's variance of 0.08096. All three quantities are slightly less than those without the added 1/2 bias correction (1.69, 0.525 and 0.0818, respectively). Likewise, the bias-corrected estimate of the log relative risk is 0.278 with an estimated variance of 0.0235, nearly equivalent to the unadjusted values of 0.28 and 0.0239, respectively. Thus these adjustments are principally advantageous with small sample sizes.

2.5 EXACT INFERENCE

2.5.1 Confidence Limits

From the non-central hypergeometric conditional likelihood (2.57), by recursive calculations one can determine the exact $1 - \alpha$ level confidence limits $(\widehat{\varphi}_\ell, \widehat{\varphi}_u)$ on the odds ratio φ. Using the limits of summation for the index cell (a_ℓ, a_u), then the lower one-sided confidence limit at level α is that value $\widehat{\varphi}_\ell$ which satisfies the equation:

$$\alpha = \sum_{x=a}^{a_u} P(x \mid n_1, m_1, N, \widehat{\varphi}_\ell). \tag{2.68}$$

Likewise, the upper one-sided confidence limit at level α satisfies the equation

$$\alpha = \sum_{x=a_\ell}^{a} P(x \mid n_1, m_1, N, \widehat{\varphi}_u) \tag{2.69}$$

(Cornfield, 1956). For two-sided limits, $\alpha/2$ is employed in the above for each limit. These limits are exact in the sense that the coverage probability is at least $1 - \alpha$. Either limit can be readily computed using the recursive secant method (*cf.* Thisted, 1988) with the usual asymmetric large sample confidence limits from (2.50) as the starting values. When $a = a_u$ there is no solution and the lower confidence limit is 0. Likewise, when $a = a_\ell$ the upper limit is ∞. An example is described below.

Before the advent of modern computing, the calculation of these exact confidence limits was tedious. In an important paper, Cornfield (1956) derived a simple, recursive approximate solution for these exact limits. Gart (1971) presented a slight modification of Cornfield's method that reduces the number of iterations required to reach the solution. Fleiss (1979), among others, has shown that Cornfield's method is often the most accurate procedure, relative to the exact limits, among the various approximations available, including the non-iterative large sample confidence limits presented earlier. However, with the widespread availability of StatXact and other programs that readily perform the exact computations, there is little use today of Cornfield's procedure.

Since the exact confidence limits for the odds ratio are obtained by conditioning on both margins as fixed, then these also yield exact confidence limits on the relative risk and the risk difference as described by Thomas and Gart (1977). For example, for a given odds ratio φ and fixed values of n_1, m_1, and N, then it can be shown that the odds ratio of the expected frequencies is a quadratic function in π_1. The solution for π_1 is the root bounded by $(0,1)$, from which the value of π_2 is obtained by subtraction (see Problem 2.7). Thus given the upper exact confidence limit on the odds ratio $\widehat{\varphi}_u$, there are corresponding probabilities $\widehat{\pi}_{u1}$ and $\widehat{\pi}_{u2}$ for which the odds ratio of the expected frequencies equals $\widehat{\varphi}_u$. Then the corresponding exact upper limit on the risk difference is $\widehat{RD}_u = \widehat{\pi}_{u1} - \widehat{\pi}_{u2}$, and that for the relative risk is $\widehat{RR}_u = \widehat{\pi}_{u1}/\widehat{\pi}_{u2}$. Likewise, exact lower confidence limits for the probabilities $\widehat{\pi}_{\ell 1}$ and $\widehat{\pi}_{\ell 2}$ are obtained from the lower limit on odds ratio $\widehat{\varphi}_\ell$ from which the lower limits $\widehat{RD}_\ell$ and $\widehat{RR}_\ell$ are obtained.

The Thomas-Gart approach, however, has been criticized and alternate approaches proposed wherein the confidence limits for the risk difference or risk ratio are based on an exact distribution for which each is the corresponding non-centrality parameter. These limits can differ substantially from the Thomas-Gart limits because the corresponding exact test on which they are based is some variation of an exact test based on the product binomial likelihood originally due to Barnard (1945) which is different from the Fisher-Irwin test described below. Since the product binomial likelihood involves a nuisance parameter (π_2) in addition to the risk difference or risk ratio, where $\pi_1 = \theta + \pi_2$ or $\pi_1 = \theta\pi_2$, respectively, then the exact test involves a maximization over the value of the nuisance parameter, π_2. Despite the fact that Barnard later retracted his test, exact confidence limits based on his procedure have been implemented in StatXact. The advantage of the Thomas-Gart confidence limits, however, is that they agree exactly with the Fisher-Irwin test, which is the most widely accepted exact test for 2×2 tables.

2.5.2 Fisher-Irwin Exact Test

Fisher (1935) and Irwin (1935) independently described an exact statistical test of the hypothesis of no association between the treatment or exposure group and the probability of the positive characteristic or response. This is expressed by the null hypothesis H_0: $\pi_1 = \pi_2$ that is equivalent to H_0: $\varphi = \varphi_0 = 1$, where φ_0 represents the odds ratio under the null hypothesis. In this case, the conditional likelihood from (2.57) reduces to the *central hypergeometric distribution*:

$$L_c(\varphi_0) = P(a \mid n_1, m_1, N, \varphi_0) = \frac{\binom{n_1}{a}\binom{N-n_1}{m_1-a}}{\sum_{i=a_\ell}^{a_u}\binom{n_1}{i}\binom{N-n_1}{m_1-i}} \,. \tag{2.70}$$

As shown in Problem 2.6, the denominator equals $\binom{N}{m_1}$ so that the conditional likelihood reduces to

$$L_c(\varphi_0) = \frac{\binom{n_1}{a}\binom{n_2}{b}}{\binom{N}{m_1}} = \frac{\binom{m_1}{a}\binom{m_2}{c}}{\binom{N}{n_1}} = \frac{n_1!\; n_2!\; m_1!\; m_2!}{N!\; a!\; b!\; c!\; d!} \,. \tag{2.71}$$

Thus the probability of the observed table can be considered to arise from a collection of N subjects of whom m_1 have a positive response, with a of these being drawn from the n_1 subjects in group 1 and b from among the n_2 subjects in group 2 ($a + b = m_1$; $n_1 + n_2 = N$). The probability can also be considered to arise by selecting n_1 of N subjects to be members of group 1, with a of these drawn from among the m_1 with a positive response and c from among the m_2 with a negative response ($a + c = n_1$; $m_1 + m_2 = N$).

For a test of H_0 against the one-sided left or lower-tailed alternative hypothesis $H_{1<}$: $\pi_1 < \pi_2$, the exact one-sided left-tailed P-value is

$$p_\ell = \sum_{x=a_\ell}^{a} P(x \mid n_1, m_1, N, \varphi_0). \tag{2.72}$$

Likewise, for a test of H_0 against the one-sided right or upper-tailed alternative $H_{1>}$: $\pi_1 > \pi_2$, the exact one-sided right-tailed P-value is

$$p_u = \sum_{x=a}^{a_u} P(x \mid n_1, m_1, N, \varphi_0). \tag{2.73}$$

The computation of the exact P-value for a test of H_0 against the two-sided alternative, $H_{1\neq}$: $\pi_1 \neq \pi_2$, requires that consideration be given to the sample space under such an alternative. Among the various approaches to this problem that have been suggested, the most widely accepted is based on consideration of the total probability of all tables with an individual probability no greater than that for the observed table. To simplify notation, let $P_c(x)$ refer to the central hypergeometric probability in (2.70) for a 2×2 table with index cell frequency x. Then, the two-sided P-value is computed as

$$p = \sum_{x=a_\ell}^{a_u} I\left[P_c(x) \le P_c(a)\right] P_c(x) , \tag{2.74}$$

where $I(\cdot)$ is the indicator function defined in Section 2.1.1. Note that this will not equal double the one-tailed upper P-value, or double the lower P-value, or the addition of the two. These and other exact calculations are readily performed by SAS PROC FREQ and by StatXact, among many available programs.

The Fisher-Irwin exact test has been criticized as being too conservative because other *unconditional* tests such as that originally proposed by Barnard (1945) have been shown to yield a smaller P-value and thus are more powerful. The two-sided exact P-value for Barnard's test is defined as

$$p = \max_{\pi} \sum_{x_1=0}^{n_1} \sum_{x_2=0}^{n_2} I\left[\left|\frac{x_1}{n_1} - \frac{x_2}{n_2}\right| \ge |p_1 - p_2|\right] P_u\left[x_1, x_2 \mid \pi\right] , \tag{2.75}$$

where $P_u\left[x_1, x_2 \mid \pi\right] = L_u(\pi_1, \pi_2 \,|\, \pi_1 = \pi_2 = \pi)$ is the unconditional probability of each 2×2 table in (2.52). This test deals with the nuisance parameter π by computing the P-value using that value π for which the P-value is maximized. The principal reason that this test is more powerful than the Fisher-Irwin exact test is because the exact P-value includes the probability of the observed table. Because the conditional hypergeometric probability of the observed table in (2.71) is generally greater than the product binomial unconditional probability in (2.75) for the observed table, then the contribution to the Fisher-Irwin exact test P-value is greater than that to the unconditional P-value.

Such tests, however, consider all possible 2×2 tables with n_1 and n_2 individuals in each group, and with the total number of successes m_1 ranging from 0 to $\min(n_1, n_2)$. Barnard (1949) later retracted his proposal because the set of all possible 2×2 tables includes specific tables that are clearly irrelevant. Nevertheless, many have continued to advocate his approach and his approach has been generalized. Many of these procedures are implemented in StatXact (see Mehta and Patel, 1999).

Example 2.7 *Exact Inference*
For example, consider the following 2×2 table with a small total sample size $N = 29$

		Group 1	2	
Response	+	7	8	15
	−	12	2	14
		19	10	29

(2.76)

where $p_1 = 0.37$ (7/19) and $p_2 = 0.8$ (8/10). The sample odds ratio is $\widehat{OR} = 0.1458$. The iterative conditional *MLE* of the odds ratio is $\widehat{\varphi} = 0.1566$, as provided by StatXact (see Chapter 6).

To compute the exact confidence limits, or conduct an exact test, we then consider the set of possible tables with values $a \in [a_\ell, a_u]$, where from (2.56) the range of possible values for a is from $a_\ell = 5$ to $a_u = 15$. A one-sided exact test of H_0 versus the one-sided alternative $H_{1<}$: $\pi_1 < \pi_2$ is conducted by evaluating the probability associated with all tables for which $p_1 \leq 0.37$ (and $p_2 > 0.8$). For this example, this corresponds to the probability of the set of tables for which the index frequency is $a \leq 7$. These are the tables

5	10	15		6	9	15		7	8	15	
14	0	14	*or*	13	1	14	*or*	12	2	14	(2.77)
19	10	29		19	10	29		19	10	29	

with corresponding probabilities 0.00015, 0.00350, and 0.02924. Note that the lower limit for the index cell is $a_\ell = 5$ that is determined by the margins of the table. Thus the lower tailed P-value is $p_\ell = \sum_{a=5}^{7} P(x \mid 19, 15, 29, \varphi_0) = 0.03289$. This would lead to rejection of H_0 in favor of $H_{1<}$ at the $\alpha = 0.05$ level, one-sided. Evaluating (2.69) recursively at the one-sided 0.05 level yields an exact 95% one-sided confidence interval $0 < \varphi < 0.862$.

Conversely, a one-sided exact test against the one-sided alternative $H_{1>}$: $\pi_1 > \pi_2$ in the opposite tail is based on the probability associated with all tables for which $p_1 \geq 0.37$. For this example, this yields the upper tailed P-value $p_u = \sum_{x=7}^{15} P(x \mid 19, 15, 29, \varphi_0) = 0.993$. This would fail to lead to rejection of H_0 in favor of $H_{1>}$ at the $\alpha = 0.05$ level, one-sided. Evaluating (2.68) recursively at the 0.05 level (using $1 - \alpha = 0.95$) yields an exact 95% one-sided confidence interval $0.0190 < \varphi < \infty$.

A two-sided test against the two-sided alternative $H_{1\neq}$: $\pi_1 \neq \pi_2$ is based on the total probability of all tables with an individual probability no greater than that for the observed table, which in this example is $P(7 \mid 19, 15, 29, \varphi_0) = 0.02924$. Examination of the table probabilities for possible values of the index cell in the opposite tail (upper in this case) yields 0.00005 for $a = 15$, 0.0015 for $a = 14$, 0.01574 for $a = 13$, and 0.078 for $a = 12$, the last of which exceeds the probability for the observed table. Thus for this example, the two-sided P-value equals

$$p = \sum_{a=5}^{7} P(a \mid 19, 15, 29, \varphi_0) + \sum_{x=13}^{15} P(a \mid 19, 15, 29, \varphi_0) = 0.0502 \qquad (2.78)$$

for which we would fail to reject H_0 at the 0.05 level.

Evaluating (2.68) and (2.69) at $\alpha = 0.025$ yields exact 95% two-sided confidence limits for the odds ratio of (0.0127, 1.0952). From these values, the Thomas and Gart (1977) exact confidence limits (−0.6885, 0.0227) are obtained for the risk difference, and (0.2890, 1.0452) for the relative risk (see Problem 2.7). These limits agree with the two-sided P-value of $p \leq 0.0502$.

As described in Section 2.5.1, the program StatXact takes a different approach to constructing confidence limits for the risk difference and the relative risk, in each case using an unconditional exact test. For the risk difference, in this example StatXact provides an exact P-value based on Barnard's test of $p \leq 0.0353$ with 95% exact confidence limits of (−0.7895, −0.0299). The P-value is smaller than that from Fisher's test and thus the confidence limits test disagree with the Thomas-Gart limits derived from the Fisher-Irwin test. For the relative risk (risk ratio), StatXact computes an exact P-value of $p \leq 0.2299$ with 95% exact confidence limits of (0.1168, 1.4316) that differ substantially from the P-value and the confidence limits from the Fisher-Irwin test. The precise methods employed by StatXact are described in Mehta and Patel (1999).

2.6 LARGE SAMPLE TESTS

2.6.1 General Considerations

With large samples, a variety of approaches can be employed to yield a test of the null hypothesis H_0: $\pi_1 = \pi_2$ based on the asymptotic distribution of a test criterion. Because H_0: $\pi_1 - \pi_2 = 0$ implies, and is implied by, the null hypothesis H_0: $\theta = [g(\pi_1) - g(\pi_2)] = 0$ for any smooth function $g(\pi)$, then a family of tests based on any family of such smooth functions will yield asymptotically equivalent results (see Problem 2.9). Note that such a family includes the log relative risk and log odds ratio, but not the relative risk and odds ratio themselves; the log relative risk and log odds ratio scales being preferred because the null distributions are then symmetric about $\theta = \theta_0 = 0$. In this section, because all such tests can be shown to be asymptotically equivalent, we will only consider the test based on the risk difference $\widehat{RD} = p_1 - p_2$, which is the basis for the usual large sample test for two proportions.

Under the Neyman-Pearson construction, one first determines *a priori* the significance level α to be employed, that is, the probability of a Type I false positive error of falsely rejecting the tested or null hypothesis H_0: $\theta = 0$ when, in fact, it is true. One then specifies the nature of the alternative hypothesis of interest. This can either be a one-sided left-tail alternative $H_{1<}$: $\theta < 0$, a one-sided right-tail alternative $H_{1>}$: $\theta > 0$, or a two-sided alternative $H_{1\neq}$: $\theta \neq 0$. Each alternative then implies a different rejection region for the statistical test. For a one-sided left-tail alternative, the rejection region consists of the lower left area of size α under the probability distribution of the test statistic under the null hypothesis. For a one-sided right-tail alternative the rejection region consists of the upper area of size α under the null hypothesis distribution. For a two-sided test the rejection region consists of the upper and lower tail areas of size $\alpha/2$ in each tail under the null hypothesis. Although one-sided tests may be justifiable in some situations, the two-sided test is more widely used. Also, some tests are inherently two-sided. If the observed value of the test statistic falls in the rejection region for the specified alternative hypothesis, then the null hypothesis is rejected with type I error probability α.

For example, consider that we wish to test H_0: $\theta = \theta_0$ for some parameter θ. Let T be a test statistic that is asymptotically normally distributed and consistent for θ, with large sample variance $\sigma_T^2(\theta)$ that may depend on θ, which is the case for proportions. For a two-sided test of H_0 versus H_1: $\theta \neq \theta_0$, the rejection region consists of all values $|T| \geq T_\alpha$ where $T_\alpha = Z_{1-\alpha/2}\widehat{\sigma}_T(\theta_0)$. Thus the test can also be based on the standardized normal deviate, or the Z-test

$$Z = \frac{T - \theta_0}{\widehat{\sigma}_T(\theta_0)}, \tag{2.79}$$

where H_0 is rejected in favor of H_1, two-sided, when $|z| \geq Z_{1-\alpha/2}$, z being the observed value of the test statistic. Alternately, H_0 is rejected when the P-value is $p \leq \alpha$, where $p = 2[1 - \Phi(|z|)]$ and $\Phi(z)$ is the standard normal cumulative distribution function. To test H_0 against the one-sided lower or left-tailed alternative hypothesis $H_{1<}$: $\pi_1 < \pi_2$ one rejects H_0 when $z < Z_\alpha$ or the one sided P-value is $p \leq \alpha$ where $p = \Phi(z)$. Likewise, to test H_0 against the one-sided upper or right-tailed alternative hypothesis $H_{1>}$: $\pi_1 > \pi_2$ one rejects H_0 when $z > Z_{1-\alpha}$ or the one sided P-value is $p \leq \alpha$ where $p = 1 - \Phi(z)$.

Tests of this form based on a consistent estimate are asymptotically most powerful, or fully efficient against H_1. It is important to note that such tests are constructed using the estimated standard error $\widehat{\sigma}_T(\theta_0)$ under the null hypothesis and not using an estimate under the alternative hypothesis such as $\widehat{\sigma}_T(\theta_1)$ or $\widehat{\sigma}_T(\widehat{\theta})$. Asymptotically, a Z-test using these variance estimates obtained under the alternative also converges to $\mathcal{N}(0,1)$ because each of these variance estimates also converges to $\sigma_T^2(\theta_0)$ under the null hypothesis H_0. However, with small sample sizes, the size of the test may be inflated (or deflated) depending on whether the null hypothesis variance is under- (or over-) estimated by these alternative variance estimates. Thus in general, for a fixed sample size, one would expect a test based on the null hypothesis variance to have a true size closer to the desired significance level α than one based on a variance estimate under the alternative hypothesis, although both are asymptotically $\mathcal{N}(0,1)$ under H_0.

Thus to test H_0: $\pi_1 = \pi_2$ in a 2×2 table, the asymptotic distribution under the null hypothesis of the risk difference presented in (2.26) and (2.27) leads to the usual expression for the large sample Z-test for two proportions based on the standardized deviate

$$Z = \frac{p_1 - p_2}{\widehat{\sigma}_0} = \frac{p_1 - p_2}{\sqrt{p(1-p)\left[N/n_1 n_2\right]}} \, . \tag{2.80}$$

Since p_1, p_2 and p are asymptotically normally distributed, and since each has expectation π under H_0, then from Slutsky's Theorem (A.45), $\widehat{\sigma}_0 \xrightarrow{p} \sigma_0$ and asymptotically, $Z \overset{d}{\approx} \mathcal{N}(0,\ 1)$ under H_0. This Z-test is asymptotically fully efficient because it is based on the large sample estimate of the variance of $\widehat{RD}$ under the null hypothesis H_0. In Problem 2.8 we show that Z^2 in (2.80) equals the usual Pearson contingency chi-square statistic for a 2×2 table presented in the next section.

Another common approach to conducting a two-sided test of significance is to evaluate the $1-\alpha$ level confidence limits, computed as $T \pm Z_{1-\alpha/2}\widehat{\sigma}_T(\widehat{\theta})$ where $T = \widehat{\theta}$ is consistent for θ. If these limits include θ_0, then the test fails to reject H_0 at level α, otherwise, one rejects H_0 at Type I error probability level α. This approach is equivalent to a two-sided Z-test using the $S.E.$ of T, $\widehat{\sigma}_T(\widehat{\theta})$, estimated under the alternative hypothesis in the denominator of (2.79) rather than the $S.E.$ estimated under the null hypothesis, $\widehat{\sigma}_T(\theta_0)$. Since under the null hypothesis, $\widehat{\sigma}_T^2(\widehat{\theta}) \xrightarrow{p} \sigma_T^2(\theta_0)$, then a test based on confidence intervals is asymptotically valid. However, the test based on the estimated $S.E.$ under the null hypothesis, $\widehat{\sigma}_T(\theta_0)$ is, in general, preferred because the Type I error probability more closely approximates the desired level α. Thus in cases where the variance of the test statistic depends on the expected value, as is the case for the test of proportions, there may be a discrepancy between the results of a significance test based on the Z-test of (2.80) and the corresponding two-sided confidence limits, in which case the test should be based on the former, not the latter. For the test of two proportions, since $\widehat{\sigma}_T(\theta_0) > \widehat{\sigma}_T(\widehat{\theta})$ (see Lachin, 1981), then it is possible that the $1-\alpha$ level confidence limits for the risk difference, or $\log(RR)$ or $\log(OR)$, would fail to include zero (implying significance) while the two-sided Z-test is not significant. In this case, one should use the Z-test and would fail to reject H_0.

2.6.2 Unconditional Test

The Z-test in (2.80) is one of many common representations for the test of association in a 2×2 table, all of which are algebraically equivalent. Perhaps the most common is the usual Pearson contingency X_P^2 test for an $R \times C$ table with R rows and C columns

$$X_P^2 = \sum_{i=1}^{R} \sum_{j=1}^{C} \frac{\left(O_{ij} - \widehat{E}_{ij}\right)^2}{\widehat{E}_{ij}}, \tag{2.81}$$

where O_{ij} is the observed frequency in the ith row and jth column and $\widehat{E}_{ij}$ is the estimated expected frequency under the null hypothesis. This test arises as a test of the null hypothesis of statistical independence between the row and column classification factors, or H_0: $\eta_{ij} = \eta_{i\bullet}\eta_{\bullet j}$ where $\eta_{ij} = P(i,j)$ is the probability of falling in the ith row and jth column, the $\eta_{i\bullet} = P(i)$ is the ith row marginal probability and $\eta_{\bullet j} = P(j)$ is the jth column marginal probability, the "•" representing summation over the respective index. Thus the alternative hypothesis H_1: $\eta_{ij} \neq \eta_{i\bullet}\eta_{\bullet j}$ specifies that there is some degree of association between the row and column factors. Under H_0, the expected frequency in any cell is $E(O_{ij}) = E_{ij} = N\eta_{i\bullet}\eta_{\bullet j}$ that is a function of the marginal probabilities associated with the ith row and jth column. Substituting the sample estimates $p_{i\bullet} = n_{i\bullet}/N$ and $p_{\bullet j} = n_{\bullet j}/N$ of the marginal probabilities yields the estimated expected frequencies $\widehat{E}_{ij} = n_{i\bullet}n_{\bullet j}/N$. Asymptotically, $X_P^2 \overset{d}{\approx} \chi^2_{(R-1)(C-1)}$ which designates the central chi-square distribution on $(R-1)(C-1)$ degrees of freedom (*df*).

The expression for the degrees of freedom arises from the fact that under H_0 there are $R-1$ row marginal parameters that must be estimated from the data, the last obtained by subtraction since the set must sum to 1. Likewise, $C-1$ column parameters must be estimated. Thus since there are RC cells, the degrees of freedom are $RC - 1 - (R-1) - (C-1) = (R-1)(C-1)$.

For the 2×2 table, since the margins are fixed, $\left|O_{ij} - \widehat{E}_{ij}\right|$ is a constant for all cells of the table. Thus the expression for X_P^2 reduces to

$$X_P^2 = \frac{(ad-bc)^2 N}{n_1 n_2 m_1 m_2}, \tag{2.82}$$

which is asymptotically distributed as the central chi-square distribution on 1 *df*, designated simply as χ^2. In the above notation, $\pi_j = \eta_{1j}/\eta_{\bullet j}$ where as shown in (2.24) the columns $(j = 1,2)$ represent the treatment or exposure groups. In this case, the null and alternative hypotheses are equivalent to H_0: $\pi_1 = \pi_2$ versus H_1: $\pi_1 \neq \pi_2$ two-sided. Thus one rejects H_0 in favor of H_1 whenever $X_P^2 \geq \chi^2_{1-\alpha}$, the upper $1-\alpha$ percentile of the central χ^2 distribution.

As stated previously, it is readily shown that $Z^2 = X_P^2$. Because $\chi^2_{1-\alpha} = (Z_{1-\alpha/2})^2$ it follows that the two-sided Z-test of H_0 versus H_1 using (2.80) is equivalent to the contingency chi-squared test that is inherently two-sided.

2.6.3 Conditional Mantel-Haenszel Test

Alternately, as was originally suggested by Mantel and Haenszel (1959) and extended by Mantel (1963), the test criterion could be based on the conditional central hypergeometric likelihood (2.70) rather than the product binomial. This likelihood involves a single random variable, the frequency a in the index cell of the 2×2 table. Thus the *Mantel-Haenszel test* for the 2×2 table is most conveniently expressed in terms of the deviation of the observed value of the index frequency a from its expectation as

$$X_c^2 = \frac{[a - E(a)]^2}{V_c(a)} . \tag{2.83}$$

Using the factorial moments, or by direct solution, the moments of the central hypergeometric distribution (not merely their estimates) are

$$E(a) = \frac{n_1 m_1}{N} \tag{2.84}$$

and

$$V_c(a) = \frac{n_1 n_2 m_1 m_2}{N^2(N-1)} \tag{2.85}$$

(*cf.* Cornfield, 1956) where $V_c(a)$ is termed the *conditional variance*. The corresponding Z-statistic is

$$Z_c = \frac{a - E(a)}{\sqrt{V_c(a)}} . \tag{2.86}$$

Since a is the sum of *i.i.d.* Bernoulli variables, then asymptotically under H_0, $Z_c \overset{d}{\approx} \mathcal{N}(0,1)$ which can be used for a one or two-sided test. Thus, $X_c^2 \overset{d}{\approx} \chi^2$ on 1 *df* under H_0.

2.6.4 Cochran's Test

The unconditional contingency X_P^2 and Z-tests may also be expressed in terms of $[a - E(a)]$. Under the null hypothesis H_0: $\pi_1 = \pi_2 = \pi$, from the unconditional product binomial likelihood (2.52), $E(a) = n_1\pi$ and $V(a) = n_1\pi(1-\pi)$, each of which may be consistently estimated from (2.28) as $\widehat{E}(a) = n_1 p = n_1 m_1/N$ and $\widehat{V}(a) = n_1 p(1-p) = n_1 m_1 m_2/N^2$. Likewise, $V(b) = n_2\pi(1-\pi)$ and $\widehat{V}(b) = n_2 p(1-p)$. Since $m_1 = (a+b)$ is not fixed, then

$$a - \widehat{E}(a) = a - \frac{m_1 n_1}{N} = \frac{n_2 a - n_1 b}{N} . \tag{2.87}$$

Thus the *unconditional variance* is

$$V_u = V[a - \widehat{E}(a)] = \frac{n_2^2 V(a) + n_1^2 V(b)}{N^2} = \frac{n_1 n_2 \pi(1-\pi)}{N} , \tag{2.88}$$

which can be consistently estimated as

$$\widehat{V}_u = \widehat{V}[a - \widehat{E}(a)] = \frac{n_1 n_2 p(1-p)}{N} = \frac{n_1 n_2 m_1 m_2}{N^3} . \tag{2.89}$$

Therefore, the unconditional chi-square test for the 2×2 table, X_u^2, can be expressed as

$$X_u^2 = \frac{\left[a - \widehat{E}(a)\right]^2}{\widehat{V}_u} , \tag{2.90}$$

where it is easily shown that $X_u^2 = X_P^2$ in (2.82). Likewise the unconditional Z-test is

$$Z_u = \frac{a - \widehat{E}(a)}{\sqrt{\widehat{V}_u}} , \tag{2.91}$$

which is easily shown to equal the usual Z-test in (2.80). Thus under H_0 Z_u is asymptotically distributed as standard normal and X_u^2 as χ^2 on 1 df.

It is also instructive to demonstrate this result as follows. Asymptotically, assume that $n_1/N \rightarrow \xi$, $n_2/N \rightarrow (1-\xi)$, ξ being the sample fraction for group 1. Also, since $m_1/N \xrightarrow{p} \pi$ and $m_2/N \xrightarrow{p} (1-\pi)$ under H_0, then from Slutsky's Convergence Theorem (A.45), $\widehat{E}(a) \xrightarrow{p} E(a)$ and $\widehat{V}(u) \xrightarrow{p} V(u)$. Thus, from Slutsky's Theorem (A.43) and (A.44), Z_u is asymptotically distributed as standard normal.

The unconditional test in this form is often called *Cochran's test*. Cochran (1954a) described generalizations of the common chi-square test for the 2×2 table to various settings, most notably to the analysis of stratified 2×2 tables that is described in Chapter 4. Although Cochran's developments were generalizations of the Z-test of the form (2.80), his results are often presented as in (2.90) to contrast his test with that of Mantel and Haenszel.

Since

$$\widehat{V}_u = \frac{N-1}{N} V_c(a) < V_c(a) \tag{2.92}$$

then the conditional Mantel-Haenszel test tends to be slightly more conservative than the unconditional test. Clearly the difference vanishes asymptotically, and thus the test using either the unconditional or conditional variance is often referred to as the *Cochran-Mantel-Haenszel*, or *CMH* test.

Example 2.8 *Exact Inference Data and Neuropathy Clinical Trial*
For the small sample size data in Example 2.7, $a = 7$, $E(a) = (19 \times 15)/29 = 9.8276$, $V_c(a) = (19 \times 10 \times 15 \times 14)/(29^2 \times 28) = 1.6944$ and $\widehat{V}_u = (28/29)V_c(a) = 1.6360$. Thus the contingency χ^2 test or Cochran's test yields $X_P^2 = X_u^2 = (2.8276)^2/1.6360 = 4.8871$, with a corresponding Z_u value of -2.211 and a one-sided left-tailed P-value of $p \leq 0.0135$. The Mantel-Haenszel test yields $X_c^2 =$

$(2.8276)^2/1.6944 = 4.7186$, with Z_c of -2.172 and $p \leq 0.0149$ (one-sided). Both tests yield P-values less than that of the exact test ($p = 0.033$), slightly less so for the Mantel test.

For such small samples, with total $N = 29$ in this example, the large sample tests are not valid. They are computed only as a frame of reference to the exact test presented previously. In general, for a 2×2 table, the minimum expected frequency for all four cells of the table, the E_{ij} in (2.81) should be at least five (some suggest four) for the asymptotic distribution to apply with reasonable accuracy (Cochran, 1954a).

For the clinical trial data in Example 2.2, the conditions are clearly satisfied. Here $a = 53$, $E(a) = 46.5$, $V_c(a) = 12.5013$ and $\widehat{V}_u = 12.4388$. Thus Cochran's test yields $X_u^2 = 3.3966$ with $p \leq 0.0653$, inherently two-sided. The Mantel-Haenszel test yields $X_c^2 = 3.3797$ and $p \leq 0.066$. The large sample Z-test in (2.80) yields a value $z = 0.13/0.0699 = 1.843$ that equals $\sqrt{3.3966}$ from Cochran's test. These tests are not significant at the $\alpha = 0.05$ significance level 2-sided.

2.6.5 Likelihood Ratio Test

Another test that is computed by many software packages, such as SAS, is the likelihood ratio or G^2-test. Like the Pearson test, this test arises from the consideration of the null hypothesis of statistical independence of the row and column factors. Using the notation of Section 2.6.2, under H_1, the likelihood is a multinomial with a unique probability for each cell η_{ij} that can be estimated as $\widehat{\eta}_{ij} = p_{ij} = n_{ij}/N$. Under H_0, the likelihood is again a multinomial with cell probabilities that satisfy H_0: $\eta_{ij0} = \eta_{i\bullet}\eta_{\bullet j}$, where the joint probabilities can be estimated from the sample marginal proportions as $\widehat{\eta}_{ij0} = p_{i\bullet}p_{\bullet j}$ to yield the same estimated expected frequencies $\widehat{E}_{ij}$ as in (2.81). Using $G^2 = -2\log[L(\widehat{\eta}_{ij0})/L(\widehat{\eta}_{ij})]$, for the 2×2 table yields

$$G^2 = 2\sum_{i=1}^{2}\sum_{j=1}^{2} O_{ij} \log \frac{O_{ij}}{\widehat{E}_{ij}}, \tag{2.93}$$

where O_{ij} and $\widehat{E}_{ij}$ are the observed and expected frequencies as in (2.81). Since this is a likelihood ratio test (see Appendix Section A.7.2), then asymptotically $G^2 \overset{d}{\approx} \chi^2$ on 1 df. For the 2 × 2 table, this expression reduces to

$$\chi^2 = 2\log\left[\frac{a^a b^b c^c d^d N^N}{n_1^{n_1} n_2^{n_2} m_1^{m_1} m_2^{m_2}}\right]. \tag{2.94}$$

The principal application of this test is to the analysis of log-linear models for multi-dimensional contingency tables. Further, the G^2 test has been shown to converge to the asymptotic chi-square distribution at a slower rate than the contingency chi-square (or Cochran's) test (*cf.* Agresti, 1990). The relative rates of convergence of Cochran's versus the Mantel-Haenszel test have not been explored. However, given that the exact test is the yardstick against which the other tests have been compared, and given that the Mantel-Haenszel test is based on the normal approximation to the central hypergeometric distribution, one might expect that the Mantel test should be preferred.

2.6.6 Test-Based Confidence Limits

In some cases, we may wish to construct a confidence interval for a parameter θ based on a consistent estimator $\widehat{\theta}$, but where no convenient estimator of the variance of the estimate $\widehat{V}(\widehat{\theta}) = \widehat{\sigma}^2_{\widehat{\theta}}$ exists. Extending the approach described in Section 2.1.3 attributed to Miettinen (1976), suppose that a simple 1 *df* chi-square test, say X^2, is available to test H_0: $\theta = 0$ versus H_1:$\theta \neq 0$. Now, if we could estimate $V(\widehat{\theta})$, we could construct an estimate-based test of the form $Q^2 = \widehat{\theta}^2/\widehat{\sigma}^2_{\widehat{\theta}}$. Thus a test-based estimate of $\sigma^2_{\widehat{\theta}}$ can be obtained by equating $Q^2 = X^2$ and solving for $\widehat{\sigma}^2_{\widehat{\theta}}$ to obtain

$$\widehat{\sigma}^2_{\widehat{\theta}} = \frac{\widehat{\theta}^2}{X^2} \tag{2.95}$$

and $S.E.(\widehat{\theta}) = \widehat{\theta}/X$. From this, the usual large sample confidence limits are obtained as $\widehat{\theta} \pm Z_{1-\alpha/2}\widehat{\theta}/X$.

Generalizations of this approach are widely used, for example, to compute the confidence limits for the Mantel-Haenszel estimate of the common log odds ratio for a set of 2×2 tables, as described in Chapter 4. However, the X^2 test usually employs an estimate of the variance of the test statistic under the null hypothesis, in which case this approach provides an estimate of the variance of $\widehat{\theta}$ under the null hypothesis as well. This leads to some inaccuracy in the coverage probabilities of the confidence limits. See Greenland (1984) for a review of the controversies regarding this approach. In general, when appropriate alternatives exist, this approach should not be the first choice. However, it can be employed to obtain approximate confidence limits for parameters in other cases where the variance of the statistic under the alternative may not be tractable.

Example 2.9 *Neuropathy Clinical Trial*
Consider θ = the log odds ratio for a 2×2 table. Suppose we wished to estimate the variance of the sample log odds ratio $\widehat{\theta}$ using this approach. The Mantel-Haenszel test X^2_c in (2.83) is an asymptotically fully efficient test for H_0: $\theta = 0$ versus H_1: $\theta \neq 0$ since it is based on the large sample approximation to the hypergeometric distribution for which the odds ratio is the non-centrality parameter. Thus the $S.E.(\widehat{\theta}) = \widehat{\theta}/X_c$.

For the data in Example 2.2, the sample log odds ratio is $\widehat{\theta} = 0.52561$, and the value of the Mantel-Haenszel test is X^2_c = 3.37966. Thus the estimated $S.E.$ = $0.52561/\sqrt{3.37966} = 0.2859$, which is nearly identical to the large sample estimated $S.E. = \widehat{\sigma}_{\widehat{\theta}} = 0.2860$ obtained from (2.46). Thus in this case the resulting confidence limits are nearly identical to those presented previously.

2.6.7 Continuity Correction

The large sample tests X_P^2, X_u^2 and X_c^2 presented in (2.82), (2.83), and (2.90), respectively, and their corresponding Z-tests, are based on the normal approximations to the binomial and the central hypergeometric distributions in which these discrete distributions are approximated by the smooth continuous normal distribution. For small sample sizes, the adequacy of the approximation is vastly improved through the use of the continuity correction. For example, the central hypergeometric distribution for a 2×2 table with the fixed margins of the example in (2.76) could be plotted as a histogram with masses of probability at discrete values $5 \leq a \leq 15$. For the observed table with a =7, the exact one-sided lower tail P-value is obtained as the sum of the histograms for $5 \leq a \leq 7$. Clearly, to approximate this tail area under the smooth normal curve, we would use $P(a \leq 7.5)$ so as to capture the total probability associated with $a = 7$.

Thus the continuity corrected conditional and unconditional tests X_c^2 in (2.83) and X_u^2 in (2.90), and their corresponding Z-tests in (2.86) and (2.91), would employ

$$|a - E(a)| - 1/2 \tag{2.96}$$

in the numerator in place of simply $[a - E(a)]$.

There has been substantial debate over the utility of the continuity correction. Those interested are referred to Conover (1974), and the discussion thereto by Starmer, Grizzle and Sen (1974), Mantel (1974) and Miettinen (1974a); and the references therein. Mantel and Greenhouse (1968) and Mantel (1974) have argued that the exact test P-value from the conditional hypergeometric distribution should be used as the basis for the comparison of the continuity corrected and uncorrected tests. In their comparisons, the continuity corrected statistic is preferred. Conover (1974) points out that this advantage principally applies when either margin is balanced ($n_1 = n_2$ or $m_1 = m_2$), but not so otherwise.

However, Grizzle (1967) and Conover (1974), among others, have shown that the uncorrected statistic more closely approximates the complete exact distribution, not only the tails. This is especially true if one considers the significance level obtained from the marginal expectation of the corrected and non-corrected test statistics on sampling from two binomial distributions without the second margin (the m_1 and m_2) fixed. Starmer, Grizzle and Sen (1974) argued that since the exact significance level is some value $\alpha^* \leq \alpha$ due to the discreteness of the binomial and hypergeometric distributions, then the basis for comparison should be the uniformly most powerful test under both product-binomial and hypergeometric sampling. This is the randomized exact test of Tocher (1950), wherein randomization is employed to yield an exact Type I error probability exactly equal to α. Compared to the randomized exact test tail areas, the continuity corrected tests are far too conservative.

Tocher's test, however, is not widely accepted. Thus the continuity correction is most useful with small sample sizes. As the sample size increases, and the central limit theorem takes hold, the uncorrected statistic is preferred, except perhaps in those cases where one of the margins of the table is perfectly balanced. Also, with the advent of modern computer systems, exact computations are readily performed with small sample sizes where the continuity correction is most valuable. In general, therefore, the continuity correction should be considered with small sample sizes for cases where exact computations are not available.

Example 2.10 *Exact Inference Data and Neuropathy Clinical Trial*
For the 2×2 table with small sample sizes in Example 2.7, the continuity corrected conditional Z_c-test is $[|-2.8276| - 0.5]/\sqrt{1.69444}$. Retaining the sign of the difference, this yields $Z_c = -2.3276/1.3017 = -1.788$ with a one-sided lower P-value of $p \leq 0.0369$, which compares more favorably to the one-sided exact P-value of $p \leq 0.0329$ (see Example 2.7). The uncorrected Z-test yields a P-value of $p \leq 0.0149$, much smaller than the exact P-value.

For the clinical trial data in Example 2.2 with much larger sample sizes, the two-sided P-values for the uncorrected and corrected conditional X_c^2 are 0.06601 and 0.0897. The corrected statistic P-value compares favorably with the exact two-sided P-value of 0.0886, principally because the group margin is balanced, $n_1 = n_2 = 100$.

Now consider the following slight modification of the data in Example 2.2 where the margins are unbalanced:

		Group 1	2	
Response	+	60	36	96
	−	53	54	107
		113	90	203

(2.97)

The pivotal quantities are virtually unchanged ($OR = 1.698$, $a - E(a) = 6.56$, and $V_c(a) = 12.5497$), so that the uncorrected and corrected X_c^2 P-values are 0.064 and 0.087, nearly the same as those for the example with the margins balanced. Nevertheless, the exact two-sided P-value changes to 0.0677, so that the corrected X_c^2 is too conservative.

In general, therefore, with small sample sizes, the exact calculation is preferred. With large sample sizes, the continuity correction should not be used except in the cases where one of the two margins of the table is perfectly balanced, or nearly so.

2.7 SAS PROC FREQ

Many of the computations described in this section are provided by the Statistical Analysis System (SAS) procedure for the analysis of cross-tabulations of frequency data, PROC FREQ. To conduct an analysis of the data from a 2×2 table such as those presented herein, the simple SAS program presented in Table 2.2 could be employed. Each 2×2 table is input in terms of the frequencies in each cell as in (2.24). However, in order to analyze the 2×2 table in SAS, it is necessary to create a data set with four observations, one per cell of the 2×2 table, and three variables: one representing group and another response, each with distinct

Table 2.2 SAS Program for Analysis of 2×2 Tables

```
title1 'SAS PROC FREQ Analysis of 2x2 tables';
data one; input a b c d; cards;
53 40 47 60
;
title2 'Neuropathy Clinical Trial, Example 2.2';
data two; set one;
  group=1; response=1; x=a; output;
  group=2; response=1; x=b; output;
  group=1; response=2; x=c; output;
  group=2; response=2; x=d; output;
proc freq; tables group*response / all nopercent nocol riskdiff;
exact or; weight x;
run;
```

numerical or categorical values, and one representing the frequency within that cell. Then PROC FREQ is called using this data set. Tables 2.3–2.5 present the results of the analysis.

The 2×2 table is presented in Table 2.3. The SAS program uses the table statement for `group*response` so that Group forms the rows and Response the columns. This is necessary because SAS computes the relative risks across rows, rather than across columns as presented throughout this text. The `nocol` and `nopercent` options suppress printing of irrelevant percentages.

The `all` option requests statistical tests, the measures of the strength of the association between group and response, the odds ratio and relative risks. These results are presented in Table 2.4. The test statistics include the contingency chi-square, with and without the correction for continuity, the likelihood ratio test, the Mantel-Haenszel test, the one-sided lower and upper-tailed exact test *P*-values, and the two-sided exact *P*-value. These are followed by various measures of the degree of association between group and response. Most have been deleted from the text in the table since they are not discussed herein. Some are like a R^2 as a measure of explained variation in a linear regression model. In particular, the uncertainty coefficient for response given group, designated as (C|R) in the table, is equivalent to the entropy R^2 from a logistic regression model that is described in Chapter 7.

The `riskdiff` option generates the "column 1 risk estimates" that are the conditional proportions within the first column, in this case the conditional proportions of a positive response within each group (row). The output also contains the risk estimates (proportions) within column 2 (a negative response) that are not of interest and thus are not shown. In addition to the large sample confidence limits for each probability, the exact limits for each are also presented. The program then computes the difference in the proportions between the two rows (groups), its large sample (asymptotic) standard error (ASE) and confidence limits.

Table 2.3 2×2 Table for Neuropathy Clinical Trial

```
SAS PROC FREQ Analysis of 2x2 tables
Neuropathy Clinical Trial, Example 2.2

TABLE OF GROUP BY RESPONSE

GROUP     RESPONSE

Frequency|
Row Pct  |       1|       2|  Total
---------+--------+--------+
       1 |     53 |     47 |    100
         |  53.00 |  47.00 |
---------+--------+--------+
       2 |     40 |     60 |    100
         |  40.00 |  60.00 |
---------+--------+--------+
Total          93      107      200
```

SAS then presents estimates of the relative risk for row1/row2 (group 1 versus group 2 in this instance) for three different study types. The first is for a `case-control` study. As will be shown in Chapter 5, the relative risk in a case-control study may be approximated by the odds ratio so that this estimated relative risk is, in fact, the sample odds ratio. However, in other types of studies, such as this clinical trial, the odds ratio is a distinct measure of the nature of the association between group and response in its own right.

This is followed by the `cohort (col1 risk)` that is the relative risk of the proportions in the first column. The other is the `cohort (col2 risk)` that is the ratio of the response proportions in the second column. Of the two, for this example the `col1` relative risk is the preferred statistic because the first column represents the positive category of interest. For each measure, the asymmetric confidence limits based on the δ-method are presented, using the log of the odds ratio and the log of the relative risks. The `exact or` option also generates the exact confidence limits for the odds ratio.

Thus it would be clearer if the SAS output were labeled as "Measures of Association," and the different measures were labeled as the odds ratio, the Col1 risk ratio and the Col2 risk ratio, respectively, rather than "Type of Study".

The `all` option also generates the Cochran-Mantel-Haenszel (CMH) statistics that are presented in Table 2.5. For a single 2×2 table such as this, the results are redundant. These computations are principally used for the stratified analysis of a set of 2×2 tables as described in Chapter 4. Three test statistics are presented, all equivalent to the Mantel-Haenszel test presented in Table 2.4. Then the measures

Table 2.4 2×2 Table Statistics

STATISTICS FOR TABLE OF GROUP BY RESPONSE

Statistic	DF	Value	Prob
Chi-Square	1	3.397	0.065
Likelihood Ratio Chi-Square	1	3.407	0.065
Continuity Adj. Chi-Square	1	2.894	0.089
Mantel-Haenszel Chi-Square	1	3.380	0.066
Fisher's Exact Test (Left)			0.977
(Right)			0.044
(2-Tail)			0.089

Statistic	Value	ASE
Uncertainty Coefficient C|R	0.012	0.013

Column 1 Risk Estimates

	Risk	ASE	95% Confidence Bounds (Asymptotic)	
Row 1	0.530	0.050	0.432	0.628
Row 2	0.400	0.049	0.304	0.496
Difference (Row 1 - Row 2)	0.130	0.070	-0.007	0.267

	95% Confidence Bounds (Exact)	
Row 1	0.428	0.631
Row 2	0.303	0.503

Estimates of the Relative Risk (Row1/Row2)

Type of Study	Value	95% Confidence Bounds	
Case-Control	1.691	0.966	2.963
Cohort (Col1 Risk)	1.325	0.979	1.794
Cohort (Col2 Risk)	0.783	0.602	1.019

Type of Study	Value	95% Confidence Bounds (Exact)	
Case-Control	1.691	0.930	3.080

Table 2.5 Cochran-Mantel-Haenszel Statistics

```
              SAS PROC FREQ Analysis of 2x2 tables
             Neuropathy Clinical Trial, Example 2.2

              SUMMARY STATISTICS FOR GROUP BY RESPONSE

 Cochran-Mantel-Haenszel Statistics (Based on Table Scores)

Statistic    Alternative Hypothesis    DF       Value       Prob
-----------------------------------------------------------------
    1        Nonzero Correlation        1       3.380      0.066
    2        Row Mean Scores Differ     1       3.380      0.066
    3        General Association        1       3.380      0.066


     Estimates of the Common Relative Risk (Row1/Row2)
                                                       95%
Type of Study     Method               Value   Confidence Bounds
-----------------------------------------------------------------
Case-Control      Mantel-Haenszel      1.691      0.966     2.962
  (Odds Ratio)    Logit                1.691      0.966     2.963

Cohort            Mantel-Haenszel      1.325      0.982     1.789
  (Col1 Risk)     Logit                1.325      0.979     1.794

Cohort            Mantel-Haenszel      0.783      0.604     1.016
  (Col2 Risk)     Logit                0.783      0.602     1.019

The confidence bounds for the M-H estimates are test-based.
```

of relative risk for a case-control study (actually the odds ratio) and the `col1` and `col2` relative risks are presented. In each case the `Mantel-Haenszel` and `logit` estimates are presented with 95% confidence limits. For a single 2×2 table, the Mantel-Haenszel and logit point estimates are the same as the simple estimates described herein. For the `Mantel-Haenszel` estimates, the confidence limits are the test-based confidence limits described in (2.95). For the `logit` estimates, the confidence limits are the asymmetric limits based on the δ-method estimated variance. Note that SAS refers to these as "logit" estimates although the logit only applies to the odds ratio.

Exact limits for the odds ratio, risk difference and the relative risk (risk ratio) are provided by StatXact.

2.8 OTHER MEASURES OF DIFFERENTIAL RISK

There are many other possible measures of the difference in risk between two groups of subjects. Three in particular are becoming more widely used in the presentation of research results: the attributable risk, the population attributable risk, and the number needed to treat. The large sample distribution of each, and a large sample estimate of the variance of each, is readily obtained from the results presented in this chapter. Details are left to Problems.

2.8.1 Attributable Risk Fraction

Consider an observational study comparing the risk of an adverse outcome, say, the incidence of a disease, in a sample of individuals exposed to a putative risk factor (E or group 1) to that in a sample of non-exposed individuals ($\overline{E}$ or group 2). In this setting the risk difference $\pi_1 - \pi_2$ is at times referred to as the attributable risk since it is a measure of the absolute excess risk of the outcome that can be attributed to the exposure in the population. However, a more useful measure is the attributable risk fraction

$$AR = \frac{\pi_1 - \pi_2}{\pi_2} = RR - 1 \, , \tag{2.98}$$

(MacMahon and Pugh, 1970), where $RR = \pi_1/\pi_2$ is the relative risk of the disease among exposed versus non-exposed individuals, and where it is assumed that exposure leads to an increase in the risk of the disease, $\pi_1 > \pi_2$. This is a measure of the fractional or proportionate increase in the risk of the disease caused by exposure to the risk factor. Since this is a simple function of the RR, asymmetric confidence limits on the AR are readily obtained from the asymmetric limits on RR computed using the log transformation as described in Section 2.3.2.

2.8.2 Population Attributable Risk

The attributable risk fraction is the proportionate excess risk associated with the exposure. However, it does not account for the prevalence of the exposure in the population, and thus is lacking from the perspective of the overall effects upon the public health. Thus Levin (1953) proposed the population attributable risk fraction (*PAR*), which is the proportion of all cases of the disease in the population that are attributable to exposure to the risk factor. Equivalently, it is the proportion of all cases of the disease in the population that would be avoided if the adverse exposure could be eliminated in the population. For example, it addresses a question such as "What fraction of lung cancer deaths could be eliminated if smoking were completely eliminated in the population?" Such questions can be answered when we have a random sample of N individuals from the general population such that the fraction exposed in the sample is expected to reflect the fraction exposed in the population. This quantity was also termed the etiologic fraction by Miettinen (1974b), and was studied by Walter (1975, 1976).

The *PAR* can be derived as follows: In the general population, let $\alpha_1 = P(E)$ be the fraction exposed to the risk factor, and the complement $\alpha_2 = 1 - \alpha_1 = P(\overline{E})$ the proportion not exposed. Among those exposed, the fraction who develop the disease, say D, is $\pi_1 = P(D \mid E)$; and among those not exposed, the proportion who do so is $\pi_2 = P(D \mid \overline{E})$. From these quantities we can then describe the probability of cases of the disease associated with exposure versus not in the following table:

Group

	E	$\overline{E}$	
D	$\alpha_1\pi_1$	$\alpha_2\pi_2$	$\pi_\bullet$
$\overline{D}$	$1-\alpha_1$	$1-\alpha_2$	$1-\pi_\bullet$
	α_1	α_2	1.0

The total probability of the disease in the population is $\pi_\bullet = \alpha_1\pi_1 + \alpha_2\pi_2$. Of these, $\alpha_1\pi_1$ is the fraction associated with exposure to the risk factor and $\alpha_2\pi_2$ that associated with non-exposure. However, if the exposure to the risk factor could be eliminated in the population, then the fraction that previously was exposed (α_1) would have the same probability of developing the disease as those not exposed (π_2), so that the fraction with the disease would be $\alpha_1\pi_2$. Thus the population attributable risk is the fraction of all cases of the disease that would be prevented if the exposure to the risk factor were eliminated in the population:

$$\begin{aligned} PAR &= \frac{\alpha_1\pi_1 - \alpha_1\pi_2}{\alpha_1\pi_1 + \alpha_2\pi_2} = \frac{\alpha_1(\pi_1 - \pi_2)}{\pi_\bullet} = \frac{\pi_\bullet - \pi_2}{\pi_\bullet} \\ &= \frac{\alpha_1(RR-1)}{\alpha_1 RR + \alpha_2} = \frac{\alpha_1(RR-1)}{1+\alpha_1(RR-1)} \,. \end{aligned} \tag{2.99}$$

Under H_0, $RR = 1$ and $PAR = 0$.

For example, consider that we have a random sample of N individuals from the general population of whom n_1 have been exposed to the risk factor. Of these $n_1\pi_1$ are expected to develop the disease. However, if these n_1 individuals were not exposed to the risk factor, then only $n_1\pi_2$ of these would be expected to develop the disease. Thus the expected number of disease cases saved or eliminated if the exposure could be eliminated is $n_1\pi_1 - n_1\pi_2$. Since $\pi_\bullet N$ is the total expected number of cases, then given n_1 of N exposed individuals, the proportion of cases eliminated or saved by complete eradication of the exposure is

$$PAR = \frac{n_1\pi_1 - n_1\pi_2}{\pi_\bullet N} \,. \tag{2.100}$$

In such a random sample, $\widehat{\alpha}_1 = n_1/N$, $\widehat{\pi}_1 = p_1$, $\widehat{\pi}_2 = p_2$, and $\widehat{\pi}_\bullet = m_1/N$. Substituting into (2.99), then the *PAR* can be consistently estimated as

$$\widehat{PAR} = \frac{\widehat{\alpha}_1(\widehat{RR}-1)}{1+\widehat{\alpha}_1(\widehat{RR}-1)} = \frac{a - n_1 p_2}{m_1} \,. \tag{2.101}$$

Implicit in this expression is the requirement that the sample fraction of exposed individuals provides an unbiased estimate of the fraction exposed in the population, that is, $\widehat{\alpha}_1 = \widehat{P}(E) = n_1/N$ is unbiased for $\alpha_1 = P(E)$.

Walter (1976) described the computation of confidence limits for the *PAR* based on a large sample estimate of the variance of the $\widehat{PAR}$. However, because the *PAR* is a probability bounded by (0,1) when $\pi_1 \geq \pi_2$, it is preferable that confidence limits be based on the asymptotic distribution of the logit of $\widehat{PAR}$. If we assume that α_1 is known (or fixed), using the δ-method (see Problem 2.10) it is readily shown that the logit of $\widehat{PAR}$ is asymptotically normally distributed with expectation logit(PAR) and with a large sample variance

$$V\left[\log\frac{\widehat{PAR}}{1-\widehat{PAR}}\right] = \frac{\pi_1}{(\pi_1-\pi_2)^2}\left(\frac{n_2\pi_2(1-\pi_1)+n_1\pi_1(1-\pi_2)}{n_1 n_2 \pi_2}\right). \tag{2.102}$$

This quantity does not involve the population exposure probability α_1. Using Slutsky's theorem the variance can be consistently estimated as

$$\widehat{V}\left[\log\frac{\widehat{PAR}}{1-\widehat{PAR}}\right] = \frac{p_1}{(p_1-p_2)^2}\left(\frac{n_2 p_2(1-p_1)+n_1 p_1(1-p_2)}{n_1 n_2 p_2}\right). \tag{2.103}$$

From this expression, the large sample $1-\alpha$ level confidence limits for the logit and the asymmetric confidence limits on *PAR* are readily obtained.

Leung and Kupper (1981) derived a similar estimate of $V[\text{logit}(\widehat{PAR})]$ based on the estimate in (2.101) which also involves the sample estimate of α_1.

While the PAR is frequently used in the presentation of epidemiologic research, it may not be applicable in many instances. As cited by Walter (1976), among others, the interpretation of the *PAR* as a fraction of disease cases attributable to exposure to a risk factor requires the assumption that the risk factor is in fact a cause of the disease, and that its removal alters neither the distribution of other risk factors in the population, nor their effects on the prevalence of the disease. Rarely can it be asserted that these conditions apply exactly.

Example 2.11 *Coronary Heart Disease in the Framingham Study*
Walter (1978) described the attributable risk using data from Cornfield (1962). In an early analysis of the Framingham study, Cornfield (1962) showed that the risk of coronary heart disease (CHD) during six years of observation among those with and without an initial serum cholesterol level $\geq$ 220 mg/dL were

		Serum Cholesterol ≥ 220	*Serum Cholesterol* < 220		
Response	*CHD*	72	20	92	(2.104)
	CHD free	684	553	1237	
		756	573	1329	

(reproduced with permission). Among those with elevated cholesterol ($\geq$ 220 mg/dL), 9.52% of subjects (p_1) developed CHD versus 3.49% (p_2) among those without an elevated cholesterol (<220 mg/dL), with a relative risk of 2.729 and 95% *C.I.* (1.683, 4.424) that is highly significant at $p < 0.001$. Thus the attributable risk in (2.98) is 1.729 with a 95% *C.I.* of (0.683, 3.424), which indicates that the incidence of CHD is 173% greater among those with elevated cholesterol values (95% *C.I.* 68, 342%).

The estimated population attributable risk is $\widehat{PAR} = 0.4958$, which indicates that nearly 50% of the CHD events in the population may be attributable to elevated serum cholesterol levels. Using the equations provided by Walter (1978) based on an estimate of the $S.E.$ of the $\widehat{PAR}$, the 95% confidence limits are (0.3047, 0.6869). The logit of the $\widehat{PAR}$ is -0.01685 with estimated variance from (2.103) of 0.15155. This yields 95% confidence limits on $\text{logit}(\widehat{PAR}) = (-0.7798, 0.7462)$. Taking the inverse logits or applying the logistic function yields asymmetric confidence limits on the *PAR* of (0.3144, 0.6783). In this example, since the $\widehat{PAR}$ is close to 0.5, the two sets of limits agree closely; however, with values approaching 0 or 1, the logit limits are expected to be more accurate and remain bounded by (0,1).

2.8.3 Number Needed to Treat

The attributable risk and the population attributable risk are principally intended to aid in the interpretation of observational epidemiologic studies. For a randomized clinical trial of a new therapy, the number needed to treat (NNT) has been suggested as a summary of the effectiveness of the therapy with the aim of preventing an adverse outcome or promoting a favorable outcome (Laupacis, Sackett and Roberts, 1988; Cook and Sackett, 1995). The question is how many patients must be treated with the new therapy in order to prevent a single adverse outcome, or to promote a single favorable outcome. First, consider the case where the probability of a positive favorable outcome with the experimental treatment π_1 is greater than that with the control treatment, π_2. Then the NNT satisfies the expression $NNT(\pi_1 - \pi_2) = 1$, so that

$$NNT = \frac{1}{\pi_1 - \pi_2} = \frac{1}{RD} \tag{2.105}$$

is the number needed to treat to yield a positive outcome in a single patient. Conversely, in the case where the probability of a negative outcome with the experimental treatment π_1 is less than that with the control treatment, π_2, the NNT satisfies $NNT(\pi_2 - \pi_1) = 1$, so that $NNT = -1/RD$. Thus confidence limits on NNT are readily obtained from confidence limits on the risk difference RD. In other settings, such as survival analysis, this concept may still be applied using the difference in cumulative incidence probabilities as of a fixed point in time.

Example 2.12 *Cholesterol and CHD*
Assume now that we could treat elevated cholesterol values effectively with a drug so that among the drug treated patients the six-year incidence of CHD is $\pi_1 = 0.040$ whereas that among the controls, whose cholesterol remains elevated, is $\pi_2 = 0.095$.

Then the number needed to treat with the new drug in order to prevent a single case of CHD is $1/0.055 = 18.18$ or 19 patients.

2.9 PROBLEMS

2.1 In Section 2.1.3 the logit was introduced as a way of constructing confidence limits for a proportion that are contained in the unit interval (0,1). Another convenient function, widely used in survival analysis, is the *complimentary log-log* transformation $\theta = g(\pi) = \log(-\log \pi)$.

2.1.1. Use the δ-method in conjunction with Slutsky's Theorem to show that the asymptotic distribution of $\widehat{\theta} = g(p) = \log(-\log p)$ is

$$\log(-\log p) \overset{d}{\approx} \mathcal{N}\left[\log(-\log \pi),\ \frac{(1-\pi)}{N\pi\,(\log \pi)^2}\right]. \tag{2.106}$$

where

$$V[\log(-\log p)] \cong \frac{(1-\pi)}{N\pi\,(\log \pi)^2}. \tag{2.107}$$

that can be consistently estimated as

$$\widehat{V}[\log(-\log p)] = \frac{(1-p)}{Np\,(\log p)^2}. \tag{2.108}$$

2.1.2. From this result, derive the expression in (2.18) for the asymmetric $1-\alpha$ level confidence limits for π based on taking the inverse function of the confidence limits for $\theta = \log(-\log \pi)$.

2.1.3. For the case where $x = 2$ and $n = 46$, compute these asymmetric confidence limits and compare them to the values presented in Example 2.1 using the exact computation and those based on the inverse logit transformation.

2.2 Derive the expression for the test-based confidence limits for a single proportion presented in (2.20). Also apply this expression to compute the confidence limits for π for the sample with 2/46 positive outcomes.

2.3 Consider the case of zero events in a sample of size N described in Section 2.4.

2.3.1. Show that the upper one-sided confidence limit equals (2.23).

2.3.2. Show that the sample size required to have an upper confidence limit of π_u at level of confidence level $1-\alpha$ is provided by

$$N = \frac{\log(\alpha)}{\log(1-\pi_u)}. \tag{2.109}$$

2.3.3. For some new drugs, the total number of patients exposed to the agent may be only $N = 500$ patients. With this sample size, if there were no drug related deaths, show that the 95% confidence interval for the probability of drug related deaths is (0, 0.00597) or as high as 6 per 1000 patients treated. Given that some drugs are administered to hundreds of thousands of patients each year, is this reassuring?

2.3.4. What is the upper 80% confidence limit?

2.3.5. What sample size would be required to provide an upper 95% confidence limit of 1 in 10000?

2.4 Show that the cell probabilities (π_1, π_2) in the 2×2 table are a function of the odds ratio and relative risk as follows:

2.4.1. For a given value of π_2 and a given value of the odds ratio OR, show that the corresponding value of π_1 is

$$\pi_1 = \frac{(OR)\pi_2}{(1-\pi_2)+(OR)\pi_2} \,. \tag{2.110}$$

2.4.2. Show that the estimated OR and RR are related as follows:

$$\widehat{RR} = \widehat{OR}\left[\frac{1+(b/d)}{1+(a/c)}\right] \,. \tag{2.111}$$

2.4.3. Show that $\widehat{OR}$ approximates $\widehat{RR}$ when either the positive or negative outcome is rare in the population sampled, assuming $E(n_1/N) = \xi$ for some constant ξ bounded away from 0 and 1.

2.5 Starting from the asymptotic bivariate normal distribution of (p_1, p_2) with covariance matrix as in (2.43),

2.5.1. Use the δ-method to derive the expression for the large sample variance of $\widehat{\theta}$, say $\sigma^2_{\widehat{\theta}}$, and a consistent estimator of this variance, $\widehat{\sigma}^2_{\widehat{\theta}}$, for:

1 The risk difference presented in (2.26),

2. Log relative risk in (2.36), and

3. The log odds ratio in (2.45).

2.5.2. For the log relative risk scale show that the estimated variance $\widehat{\sigma}^2_{\widehat{\theta}}$ can be expressed as shown in (2.37).

2.5.3. For the log odds scale, show that the estimated variance $\widehat{\sigma}^2_{\widehat{\theta}}$ is Woolf's estimated variance in (2.46).

2.5.4. In practice, the estimated variance of the relative risk and of the odds ratio (without the log transformation) should not be used for computations of confidence limits. In the following, they are derived principally as an exercise, although similar results are employed later in Chapter 9. Apply the δ-method, starting from the results in Problem 2.5.1, to show that the variance of the relative risk (without the log transformation) is asymptotically:

$$V(\widehat{RR}) \cong (RR)^2 V[\log(\widehat{RR})] \tag{2.112}$$

and of the odds ratio is

$$V(\widehat{OR}) \cong (OR)^2 V[\log(\widehat{OR})] \,. \tag{2.113}$$

2.5.5. Under the null hypothesis it is assumed that H_0: $\pi_1 = \pi_2 = \pi$, which is equivalent to H_0: $\theta = \theta_0$, where θ_0 is the null value, $\theta_0 = 0$ for the risk difference, the log relative risk and log odds ratio. Under the null hypothesis derive the corresponding expression for the variance of $\widehat{\theta}$ and a consistent estimator thereof for each of these measures.

2.5.6. Redefining the relative risk herein as $RR_{2:1} = \pi_2/\pi_1$, show that

$$V[\log(\widehat{RR}_{2:1})] = V[\log(\widehat{RR}_{1:2})] \,. \tag{2.114}$$

2.5.7. Also show that the confidence limits for $\log(RR_{2:1})$ are the negatives of those for $\log(RR_{1:2})$, and that the confidence limits for $RR_{2:1}$ are the reciprocals of those for $RR_{1:2}$.

2.6 Starting from the product-binomial likelihood in (2.52), under the null hypothesis H_0: $\pi_1 = \pi_2 = \pi$, show the following:

2.6.1. The probability of the 2×2 table is

$$P(a, b|n_1, n_2, \pi) = \binom{n_1}{a}\binom{n_2}{b}\pi^{m_1}(1-\pi)^{m_2} \,. \tag{2.115}$$

2.6.2. In group 1, show that

$$E(a) = \pi n_1 \mathrel{\widehat{=}} \frac{m_1 n_1}{N} \tag{2.116}$$

and that

$$V(a) = \pi(1-\pi)n_1 \mathrel{\widehat{=}} \frac{m_1 m_2 n_1}{N^2} \,, \tag{2.117}$$

where $\widehat{=}$ means "estimated as".

2.6.3. For Cochran's unconditional test, then derive the variance $V[a - \widehat{E}(a)]$ in (2.88) and its estimate in (2.89).

2.6.4. Use the conditioning arguments in Section 2.4.2 to derive the conditional hypergeometric likelihood in (2.57).

2.6.5. Under the null hypothesis with odds ratio $\varphi = 1$, show that this likelihood reduces to the expression in (2.70).

2.6.6. Using basic probability theory, show that the probability

$$P(a|n_1, m_1, N) = \frac{\binom{n_1}{a}\binom{N-n_1}{m_1-a}}{\binom{N}{m_1}} \tag{2.118}$$

can be derived as the probability of choosing m_1 of N elements that includes a of n_1 positive elements, for fixed N and n_1.

2.6.7. Since $\sum_a P(a|n_1, m_1, N) = 1$, then show that

$$\sum_{i=a_\ell}^{a_u}\binom{n_1}{i}\binom{N-n_1}{m_1-i} = \binom{N}{m_1} \,. \tag{2.119}$$

2.6.8. Then derive the simplifications in (2.71).

2.6.9. Using these results, show that $E(a) = m_1 n_1/N$ as presented in (2.84). *Hint*: Note that for total sample size $N-1$, the sum of the possible hypergeometric probabilities equals 1.

2.6.10. Show that

$$E\left[a(a-1)\right] = \frac{m_1(m_1-1)n_1(n_1-1)}{N(N-1)} \,. \tag{2.120}$$

Given that $E(a^2) = E\left[a(a-1)\right] + E(a)$ then show that the conditional hypergeometric variance is $V_c(a)$ in (2.85).

2.7 Given a sample of N observations with fixed margins (n_1, n_2, m_1, m_2):

2.7.1. Show that the expected value of the odds ratio, say $\widetilde{\varphi}$, can be defined in terms of the probability of the outcome in group 1, π_1, as

$$\widetilde{\varphi} = \frac{\pi_1(n_2 - m_1 + \pi_1 n_1)}{(1-\pi_1)(m_1 - \pi_1 n_1)} \,, \tag{2.121}$$

where $E(a) = n_1\pi_1$ and $E(b) = m_1 - n_1\pi_1$, etc.

2.7.2. Show that the quadratic equation in π_1 as a function of $\widetilde{\varphi}$ is

$$\left[n_1(\widetilde{\varphi}-1)\right]\pi_1^2 - \left[\widetilde{\varphi}(n_1+m_1) + (n_2 - m_1)\right]\pi_1 + \widetilde{\varphi} m_1 = 0 \,. \tag{2.122}$$

2.7.3. Then show that the value of π_1 which satisfies this equation equals

$$\pi_1 = \frac{C}{2\left[n_1(\widetilde{\varphi}-1)\right]} \tag{2.123}$$

with

$$\begin{aligned} C &= \varphi(n_1+m_1) + (n_2 - m_1) \\ &\quad - \sqrt{\left[\widetilde{\varphi}(n_1+m_1) + (n_2-m_1)\right]^2 - 4\widetilde{\varphi} m_1 \left[n_1(\widetilde{\varphi}-1)\right]} \,, \end{aligned} \tag{2.124}$$

Thomas and Gart (1977).

2.7.4. Also show that given the fixed margins, the corresponding value of π_2 that satisfies this relationship is

$$\pi_2 = \frac{m_1 - n_1\pi_1}{n_2} \,. \tag{2.125}$$

Thus one can determine the probabilities $\widehat{\pi}_{1u}$ and $\widehat{\pi}_{1\ell}$ corresponding to the exact confidence limits on the odds ratio, from which the values of $\widehat{\pi}_{2u}$ and $\widehat{\pi}_{2\ell}$ can be obtained. These then provide exact limits on the risk difference and relative risk, such as $\widehat{RR}_\ell = \widehat{\pi}_{1\ell}/\widehat{\pi}_{2\ell}$.

2.7.5. For the data in Example 2.7, given the exact 95% confidence limits for the odds ratio of (0.0127, 1.0952), use the above expressions to determine the corresponding probabilities. Then show that the exact confidence limits for the risk difference are $(-0.6885, 0.0227)$ and for the relative risk are $(0.2890, 1.0452)$.

2.8 Consider the large sample tests for the 2×2 table.

2.8.1. Show that the Pearson X_P^2 in (2.81) equals Cochran's test X_u^2 in (2.90).

2.8.2. Also show that X_u^2 equals Z^2 where Z is the usual Z-test for two proportions in (2.80).

2.9 Now consider a test based on some smooth function $g(p)$, such as a test for the log relative risk or log odds ratio with H_0: $g(\pi_1) = g(\pi_2)$, or equivalently H_0: $\pi_1 = \pi_2$. The corresponding Z-test then is of the form

$$Z_g = \frac{g(p_1) - g(p_2)}{\sqrt{\widehat{V}\left[g(p_1) - g(p_2) \mid H_0\right]}} \,. \tag{2.126}$$

Using the δ-method, show that Z_g is asymptotically equal to the usual Z-test in (2.80).

2.10 Consider the Population Attributable Risk (PAR) in (2.100) and the consistent estimate in (2.101). Note that the PAR is a proportion.

2.10.1. For α_1 specified (fixed) show that the PAR estimated odds are

$$\frac{\widehat{PAR}}{1 - \widehat{PAR}} = \frac{\alpha_1(p_1 - p_2)}{p_2} \,. \tag{2.127}$$

2.10.2. Using the δ-method, show that $V\left[\log \frac{\widehat{PAR}}{1-\widehat{PAR}}\right]$ is as expressed in (2.102) and that this variance can be estimated consistently as shown in (2.103).

2.11 Hypothetically, in a clinical trial of ursodiol (a bile acid drug) versus a placebo for the treatment of patients with gallstones, the following 2×2 table was obtained giving the frequencies of a positive response (disappearance of gallstones) versus no response after 12 months of treatment:

		Group			
		Urso	*Placebo*		
Response	+	23	8	31	(2.128)
	−	87	92	179	
		110	100	210	

2.11.1. For this table, compute the estimates of the risk difference, relative risk and its logarithm, and the odds ratio and its logarithm. Also compute the estimated variance and standard error, and the 95% confidence intervals for the risk difference, log relative risk and the log odds ratio.

2.11.2. From the log odds ratio and log relative risk, also compute the asymmetric 95% confidence limits on the odds ratio and relative risk, respectively.

2.11.3. Also compute the asymmetric confidence limits for the relative risk and odds ratio for the placebo group to the Ursodiol group, e.g. $RR_{2:1}$.

2.11.4. Then compute the Pearson contingency Chi-square test (Cochran's test) and the Mantel-Haenszel test, with and without the continuity correction.

2.11.5. Now use the results from the Mantel-Haenszel conditional test without the continuity correction to compute test-based confidence limits on the log odds ratio and the log relative risk. From these compute the asymmetric confidence limits on the odds ratio and the relative risk. Compare these to the asymmetric confidence limits obtained above.

2.11.6. Now use PROC FREQ in SAS with the CMH CHISQ options to perform an analysis of the above 2×2 table. This will require that you input the data as shown in Table 2.2. Alternately, use a program such as StatXact to compute the exact P-values. Compare the respective large sample tests to the exact two-sided P-value.

2.11.7. Compute the number needed to treat with ursodiol (NNT) so as to yield one treatment success. Use the 95% confidence limits on the risk difference to compute the confidence limits on the NNT.

2.12 Mogensen (1984) examined the influence of early diabetic nephropathy on the premature mortality associated with type 2 or adult onset diabetes mellitus in a sample of 204 men 50–75 years of age with onset of diabetes after 45 years of age. At the time of entry into the cohort, each subject was characterized as having normal levels of albumin excretion defined as AER <30 μg/min (micrograms per minute) versus those with microalbuminuria ($30 \leq$,AER <140 μg/min). See Section 1.5 for a description of these terms. The cohort was then followed for 10 years. The following results were obtained (reproduced with permission):

	Albuminuria Group			
	Micro	*Normal*		
Died	55	59	114	(2.129)
Survived	73	17	90	
	128	76	204	

2.12.1. Use the relative risk (risk ratio) for mortality, with an appropriate confidence interval, and the Mantel-Haenszel test to describe the relationship between the presence of microalbuminuria versus not and the 10 year mortality.

2.12.2. Compute the population attributable risk and its 95% confidence limits. Describe these results in terms of the impact of diabetic nephropathy on 10 year mortality among those with type 2 diabetes mellitus.

2.13 Starting from the unconditional product-binomial likelihood (2.52) we now use a *logit (logistic) model* to derive the conditional hypergeometric distribution for a 2×2 table. Conditional on n_1 and n_2 (or n_1, N), under the logistic model we can express the logit of π_1 and π_2 as

$$\log\left(\frac{\pi_1}{1-\pi_1}\right) = \alpha + \beta \qquad \log\left(\frac{\pi_2}{1-\pi_2}\right) = \alpha\,. \tag{2.130}$$

2.13.1. Show that the *inverse logits* are provided by logistic functions of the form

$$\pi_1 = \frac{e^{\alpha+\beta}}{1+e^{\alpha+\beta}}\;; \qquad \pi_2 = \frac{e^{\alpha}}{1+e^{\alpha}} \tag{2.131}$$

and

$$1-\pi_1 = \frac{1}{1+e^{\alpha+\beta}}\;; \qquad 1-\pi_2 = \frac{1}{1+e^{\alpha}}\,. \tag{2.132}$$

2.13.2. Then show that

$$\varphi = \frac{\pi_1/\left(1-\pi_1\right)}{\pi_2/\left(1-\pi_2\right)} = e^{\beta}. \tag{2.133}$$

2.13.3. Substituting this logistic model representation into (2.52) show that the probability of the 2×2 table can be expressed as

$$P(a, m_1 \mid n_1, n_2) = \binom{n_1}{a}\binom{n_2}{m_1-a}\frac{e^{a\beta}e^{m_1\alpha}}{\left(1+e^{\alpha+\beta}\right)^{n_1}\left(1+e^{\alpha}\right)^{n_2}}\,, \tag{2.134}$$

where a and m_1 are random.

2.13.4. Conditioning on m_1, show that

$$P(a \mid m_1) = \frac{P(a, m_1)}{\sum_{i=a_\ell}^{a_u} P(i, m_1)} = \frac{P(a, m_1)}{P(m_1)} \tag{2.135}$$
$$= \frac{\binom{n_1}{a}\binom{n_2}{m_1-a}e^{a\beta}}{\sum_{i=a_\ell}^{a_u}\binom{n_1}{i}\binom{n_2}{m_1-i}e^{i\beta}}\,,$$

where (a_l, a_u) are the limits for the possible values for the index frequency a.

2.13.5. Show that this expression is equivalent to (2.57).

2.14 Consider a set of Bernoulli variables (x_{i1}, x_{i2}, x_{i3}) that denotes whether the ith element in a sample of N observations falls in category 1, 2 or 3 of some characteristic. Using basic principles, derive the covariance matrix of the vector $(x_{i1}\; x_{i2}\; x_{i3})$ as shown in Example A.2 of the Appendix.

3

Sample Size, Power, and Efficiency

In the previous chapter we described the large sample distribution of estimates of the differential in risk between two groups, and tests of the significance of these differences. In this chapter we consider the evaluation of the sample size required for a study based on either the desired precision of an estimate of the risk difference, or based on the power function of the Z-test for two proportions. Since this test is a special case of a normal deviate Z-test, we begin with the derivation of the power function for the general family of Z-tests, which will also be employed in subsequent chapters to evaluate the power of other normal deviate Z-tests. This is followed by an introduction of the concepts of asymptotic efficiency under local alternatives that is used in subsequent chapters to derive asymptotically efficient, or most powerful, tests.

There is a large literature on methods for evaluation of the precision of estimates of parameters and the power of specific statistical tests. Since estimation precision and the power of a test largely depend on sample size, consideration of these properties allows the determination of an adequate, if not optimal, sample size when a study is designed. Interested readers are referred to the following references for additional materials. McHugh and Le (1984) present a review of procedures for determination of sample size from the perspective of the precision of an estimator. Lachin (1981), Donner (1984), and Lachin (1998), among others, likewise review methods for sample size determination from the perspective of the power of commonly used statistical tests. General reference texts on the topic include Machin and Campbell (1987), Cohen (1988), Desu and Raghavarao (1990), Schuster (1990), and Odeh and Fox (1991), among others.

3.1 ESTIMATION PRECISION

Consider that we wish to estimate a parameter θ in a population, such as the log odds ratio, based on a simple random sample that yields an estimator $\widehat{\theta}$ that is normally distributed, at least asymptotically, as $\widehat{\theta} \sim \mathcal{N}(\theta, \sigma^2)$. In most instances, the large sample variance σ^2 can be expressed as a function of the sample size N, such as $\sigma^2 = \phi^2/N$ where ϕ^2 is a *variance component*. Then the $1-\alpha$ confidence interval (*C.I.*) for θ is of the form $\widehat{\theta} \pm e_\alpha$, where the precision of the estimate at level $1-\alpha$ is

$$e_\alpha = Z_{1-\alpha/2}\sigma \tag{3.1}$$

and $Z_{1-\alpha/2}$ is the upper two-sided standard normal deviate at level $1-\alpha/2$. Since σ is a function of the sample size N, then so also is the precision of the estimate e_α at confidence level $1-\alpha$. The smaller the value of e_α, the greater the precision of the estimate.

This relationship also can be inverted to describe the level of confidence $1-\alpha$ corresponding to any stated degree of precision e of the estimate. The confidence level $1-\alpha$ is provided by the standardized deviate $Z_{1-\alpha/2} = e/\sigma = \sqrt{N}e/\phi$. This allows one to evaluate the relationship between the level of confidence and the degree of precision of the estimate for different sample sizes.

Often the sample size is under the control of the investigator when a study is designed or planned. In this case, the above relationship can be inverted to solve for the sample size N required to provide a confidence interval estimate with precision $\pm e$ at confidence level $1-\alpha$ for a given variance component ϕ. The required sample size is given by

$$N = \left(\frac{Z_{1-\alpha/2}\phi}{e}\right)^2 . \tag{3.2}$$

One can likewise consider the estimation precision and sample size for a study comparing two populations (groups) with respect to the risk difference, the log relative risk or the log odds ratio, using the large sample variance and the asymptotic normal distribution for each derived in the preceding chapter. These specific cases, however, will be left to a Problem. Those interested in other common instances are referred to the article by McHugh and Le (1984), or to texts on sampling.

Such computations, however, assume that the variance is known a priori. In the case of functions of probabilities, such as the log odds ratio, this requires that the probabilities in each population are known or are specified because the variance is a function of the probabilities. This approach can be further generalized to allow for sampling variation in the estimated variance, and thus in the precision of the estimate, by determining the sample size needed to provide probability $1-\beta$ that the realized $1-\alpha$ confidence interval will have precision no greater than e (*cf.* Kupper and Hafner, 1989).

Example 3.1 *Simple proportion*
For example, the simple proportion with a characteristic of interest, say p, from a sample of N observations is asymptotically distributed as $p \overset{d}{\approx} \mathcal{N}[\pi, \pi(1-\pi)/N]$, where π is the probability of the characteristic in the population and the large sample variance of p is $\sigma^2 = \pi(1-\pi)/N$ with $\phi^2 = \pi(1-\pi)$. To estimate a probability π that is assumed to be ≤ 0.3 (or $\pi \geq 0.7$) with precision $e = 0.02$ at 90% confidence, then the required sample size is $N = 1.645\sqrt{(0.3 \times 0.7)}/0.02 = 1421$ (rounded up to the next whole integer). If this sample size is too large but a sample size of 1000 is feasible, then one can determine that a 90% confidence interval with $N = 1000$ provides a degree of precision of $e = 1.645\sqrt{(0.3 \times 0.7)/1000} = 0.024$. Alternately, a sample size of $N = 1000$ provides 83% confidence of estimating π with a precision of $e = 0.02$, where the corresponding normal deviate is $Z_{1-\alpha/2} = \sqrt{1000}(0.02)/\sqrt{(0.3 \times 0.7)} = 1.38$.

3.2 POWER OF Z-TESTS

When the object of an investigation is to compare two populations then the required sample size, or the adequacy of the available sample size, is usually determined on the basis of a test of significance. In reaching an inference as to the difference between two populations, we usually first conduct a statistical test of significance, and then present the large sample confidence interval. The test statistic assesses the probability that the observed results could have occurred by chance, whereas the confidence limits provide a description of the precision of the sample estimate. In this setting, it is usually important to determine the sample size so as to provide a suitably high probability of obtaining a statistically significant result when a true difference of some magnitude exists between the populations being compared. This probability is termed the *power* of the test.

In this section we first define the relationship between the Type I and Type II error probabilities and power. We then derive the general expressions for the power of any normal deviate Z-test that can be applied to specific tests, such as the large sample Z-test for two proportions. From this we derive equations that provide the sample size necessary to achieve a desired level of power.

3.2.1 Type I and II Errors and Power

Consider a Z-test based on a statistic T that is a consistent estimator of the mean μ of a normal distribution. Using T we wish to test the null hypothesis

$$H_0\colon\ \mu = \mu_0\ , \tag{3.3}$$

where usually $\mu_0 = 0$. For now consider an upper-tail one-sided test of H_0 versus the one-sided alternative hypothesis

$$H_1\colon\ \mu = \mu_1 > \mu_0. \tag{3.4}$$

Fig. 3.1 Distribution of a test statistic under the null and alternative hypotheses, with the rejection region of size α and a type II error probability of size β.

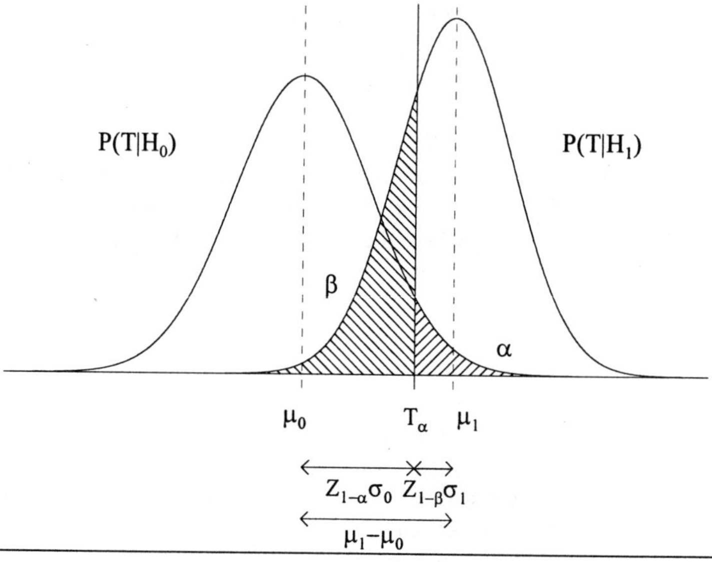

This implies H_0: $\Delta = \mu_1 - \mu_0 = 0$ versus and H_1: $\Delta > 0$. To allow for cases where the variance of T is a function of μ, $V(T)$ is expressed as

$$V(T) = \sigma^2(\mu). \tag{3.5}$$

Therefore, we have two possible distributions of T, that under the null hypothesis H_0 and that under the alternative hypothesis H_1, designated as:

$$Null: \quad T_{(H_0)} \sim \mathcal{N}(\mu_0, \sigma_0^2), \quad \sigma_0^2 = \sigma^2(\mu_0) \tag{3.6}$$

$$Alternative: \quad T_{(H_1)} \sim \mathcal{N}(\mu_1, \sigma_1^2), \quad \sigma_1^2 = \sigma^2(\mu_1)\ , \tag{3.7}$$

where $\mu_1 > \mu_0$. An example is depicted in Figure 3.1.

To conduct the test of significance, we determine whether the observed value of the statistic t falls within the upper-tail rejection region of size α. Thus H_0 is rejected when

$$t \geq T_\alpha = \mu_0 + Z_{1-\alpha}\sigma_0. \tag{3.8}$$

When the null hypothesis is true, the probability of rejection is only α and the probability of failing to reject it is $1 - \alpha$. Both quantities are determined by the investigator through the specification of the size α of the rejection region. However, when the alternative hypothesis H_1 is true, the probabilities of rejecting or failing to reject H_0 are not fixed by the investigator uniquely, but rather are determined by other factors that include the magnitude of the true difference Δ.

Thus for any fixed $\Delta = \mu_1 - \mu_0$, two types of errors can occur, a false positive Type I error with probability α and a false negative Type II error with probability β as presented by the following:

	H_0	H_1: $\mu_1 - \mu_0 = \Delta$
Reject: +	α	$1 - \beta(\Delta, N, \alpha)$
Fail to Reject: +	$1 - \alpha$	$\beta(\Delta, N, \alpha)$
	1.0	1.0

(3.9)

Because α is fixed arbitrarily, it does not depend on either the value of Δ or on the value of N. Conversely, the probability of a false negative error, designated as $\beta(\Delta, N, \alpha)$ depends explicitly upon Δ and on the total sample size N, as well as the size of the test α. The complement of the Type II error, $1 - \beta(\Delta, N, \alpha)$, is the *power* of the test to detect a difference Δ with total sample size N.

The relationship between α and β is illustrated in Figure 3.1. It is clear that one cannot fix the levels of both the type I and type II error probabilities α and β. Rather, α and β are inversely related. As the critical value T_α is shifted toward μ_0, then α increases and β decreases, and vice versa. Also, as the magnitude of the difference under the alternative Δ increases, then the distribution under the alternative is shifted to the right and β decreases, even though α remains a constant. Finally, for fixed Δ, as the variance of the statistic σ^2 decreases, the curves shrink. This is readily accomplished by increasing the sample size. This has two effects. First, in order to retain the upper tail area of size α under the null hypothesis distribution, the value T_α shrinks toward the mean μ_0 as the variance $\sigma^2(\mu_0)$ becomes smaller. Second, as the variance $\sigma^2(\mu_1)$ decreases under the alternative hypothesis, the curve under H_1 also shrinks. Each factor contributes to a decrease in the value of β. Thus while the value of the Type I error probability is specified *a priori,* the value for the Type II error probability $\beta(\Delta, N, \alpha)$ is determined by other factors.

We now describe these relationships algebraically. Under the null hypothesis, the significance test leads to rejection of H_0 if the standardized Z-test satisfies:

$$z = \frac{t - \mu_0}{\sigma_0} \geq Z_{1-\alpha}. \tag{3.10}$$

Therefore, the Type II error probability is

$$\beta = P\left[z < Z_{1-\alpha} \mid H_1\right] \tag{3.11}$$

and its complement, power, is

$$1 - \beta = P\left[z \geq Z_{1-\alpha} \mid H_1\right] , \tag{3.12}$$

where each is evaluated under the alternative hypothesis distribution. These quantities are identical to the areas depicted in Figure 3.1 noting the change of scale from T to Z.

Under $T_{(H_1)}$ in (3.7), where $\Delta = \mu_1 - \mu_0 \neq 0$, then

$$T - \mu_0 \sim \mathcal{N}\left[\mu_1 - \mu_0,\ \sigma_1^2\right] \tag{3.13}$$

and

$$Z = \frac{T - \mu_0}{\sigma_0} \sim \mathcal{N}\left[\frac{\mu_1 - \mu_0}{\sigma_0},\ \frac{\sigma_1^2}{\sigma_0^2}\right]. \tag{3.14}$$

Therefore,

$$\beta = P\left[z < Z_{1-\alpha} \mid H_1\right] = \Phi\left[\frac{Z_{1-\alpha} - \left(\frac{\mu_1 - \mu_0}{\sigma_0}\right)}{\sigma_1/\sigma_0}\right] = \Phi\left[Z_\beta\right]. \tag{3.15}$$

Thus

$$Z_\beta = \frac{Z_{1-\alpha} - \left(\frac{\mu_1 - \mu_0}{\sigma_0}\right)}{\sigma_1/\sigma_0}. \tag{3.16}$$

However, β is the area to the left of T_α in Figure 3.1, and we desire the expression for $1 - \beta$ that is the area to the right. Since $Z_{1-\beta} = -Z_\beta$, then

$$\begin{aligned} Z_{1-\beta} &= \left(\frac{\sigma_0}{\sigma_1}\right)\left[\frac{(\mu_1 - \mu_0)}{\sigma_0} - Z_{1-\alpha}\right] \\ &= \frac{(\mu_1 - \mu_0) - Z_{1-\alpha}\sigma_0}{\sigma_1} \\ &= \frac{\Delta - Z_{1-\alpha}\sigma_0}{\sigma_1}. \end{aligned} \tag{3.17}$$

Thus $(Z_{1-\beta} > 0) \Rightarrow (1 - \beta > 0.5)$.

For a one-sided left tail alternative hypothesis, H_1: $\mu_1 < \mu_0$ with $\mu_1 - \mu_0 = \Delta < 0$, a similar derivation again leads to this result but with the terms $-(\mu_1 - \mu_0)$ and $-\Delta$ (see Problem 3.1). In general, therefore, for a one-sided test in either tail, the basic equation relating Δ, α, and β is of the form

$$|\Delta| = Z_{1-\alpha}\sigma_0 + Z_{1-\beta}\sigma_1 \tag{3.18}$$

for values of Δ that are in the specified direction under the one-sided alternative hypothesis.

This expression can also be derived heuristically as shown in Figure 3.1. The line segment distance $\Delta = \mu_1 - \mu_0$ is partitioned into the sum of two line segment distances. The first is the distance $|T_\alpha - \mu_0| = Z_{1-\alpha}\sigma_0$ under the null distribution given H_0. The second is the distance $|T_\alpha - \mu_1| = Z_{1-\beta}\sigma_1$ under the alternative distribution given H_1.

For a two-sided test, Figure 3.1 would be modified as follows. The distribution under the null hypothesis would have a two-sided rejection region with an area of size $\alpha/2$ in each tail. Then, for a fixed alternative, in this case a positive value for Δ, there would be a contribution to the type II probability β from the corresponding rejection area in the far left tail under the alternative hypothesis distribution H_1. For an alternative such as that shown in the figure, where μ_1 is a modest distance from μ_0, then this additional contribution to the probability β is negligible and can be ignored. In this case, the general expression (3.18) is obtained with the value $Z_{1-\alpha/2}$.

3.2.2 Power and Sample Size

The general expression (3.18) provides an explicit equation relating Δ, N and $1-\beta$ that forms the basis for the evaluation of power and the determination of sample size for any test statistic that is asymptotically normally distributed. If we can decompose the variance expressions such that $\sigma_0^2 = \phi_0^2/N$ and $\sigma_1^2 = \phi_1^2/N$, then the basic equation becomes

$$|\Delta| = \frac{Z_{1-\alpha}\phi_0 + Z_{1-\beta}\phi_1}{\sqrt{N}}\,. \tag{3.19}$$

Therefore,

$$Z_{1-\beta} = \frac{\sqrt{N}|\Delta| - Z_{1-\alpha}\phi_0}{\phi_1} \tag{3.20}$$

and

$$N = \left[\frac{Z_{1-\alpha}\phi_0 + Z_{1-\beta}\phi_1}{\Delta}\right]^2. \tag{3.21}$$

Equation (3.20) is used to compute the power function for given N and Δ, while (3.21) allows the *a priori* determination of the N needed to provide power $1-\beta$ to detect a difference Δ with an α level test. For a two-sided test, $Z_{1-\alpha/2}$ is employed in these expressions.

Lachin (1981) presented a simplification of these general expressions for cases where the variances are approximately equal, that is, $\sigma_0^2 \doteq \sigma_1^2 \doteq \sigma^2 = \phi^2/N$. Substituting into (3.18) yields the simplified general expression

$$|\Delta| = (Z_{1-\alpha} + Z_{1-\beta})\frac{\phi}{\sqrt{N}} \tag{3.22}$$

or

$$\sqrt{N}K = Z_{1-\alpha} + Z_{1-\beta} \tag{3.23}$$

as a function of the *non-centrality factor*

$$K = \frac{|\Delta|}{\phi} . \tag{3.24}$$

This yields a somewhat simpler expression for power

$$Z_{1-\beta} = K\sqrt{N} - Z_{1-\alpha} \tag{3.25}$$

and sample size

$$N = \left(\frac{Z_{1-\alpha} + Z_{1-\beta}}{K} \right)^2 . \tag{3.26}$$

Lachin (1981) used these expressions to derive equations for the power functions of many commonly used tests, such as the test for mean values, proportions, paired mean differences, paired proportions, exponential hazard ratios and correlations, among others.

As will be shown in Section 3.4, these latter expressions also arise from a consideration of the *non-centrality parameter* that is the expected value of the Z statistic under the alternative hypothesis, or $\psi = E(Z \mid H_1)$. When the variances are approximately equal, from (3.14),

$$|\psi| = \frac{|\mu_1 - \mu_0|}{\sigma} \doteq \frac{|\mu_1 - \mu_0|}{\phi/\sqrt{N}} = \sqrt{N}K , \tag{3.27}$$

where K is the non-central factor. Then the basic equation (3.18) is based on the relationship

$$|\psi| = Z_{1-\alpha} + Z_{1-\beta} \tag{3.28}$$

which relates the non-centrality parameter of the test to the levels of α and β.

In general, these relationships can be expressed using a set of power curves as in Figure 3.2. Such curves depict the increase in power as the value of Δ (or K) increases, and also as N increases. For a 2-sided test, the power curves are symmetric about $\Delta = 0$. Such power functions are computed using (3.20) over a range of values for Δ and N, given the type I error probability α and values for the other parameters involved, such as the variance components ϕ_0 and ϕ_1. This figure presents the power function for a one-sided test at level $\alpha = 0.05$ of the differences between two means with unit variance ($\phi = 1$). Equation (3.21), therefore, simply determines the sample size N that provides a power function (curve) that passes through the point (Δ, $1 - \beta$). The essential consideration is that there is a power function associated with any given sample size N that describes the level of power $1 - \beta$ with which any difference Δ can be detected.

3.3 TEST FOR TWO PROPORTIONS

The Z-test for two proportions described in Section 2.6.1 falls into the class of tests employed in the previous section. Thus the above general expressions can

Fig. 3.2 Power as a function of the non-central factor (*K*) and the total sample size (*N*) for $\alpha = 0.05$, two-sided.

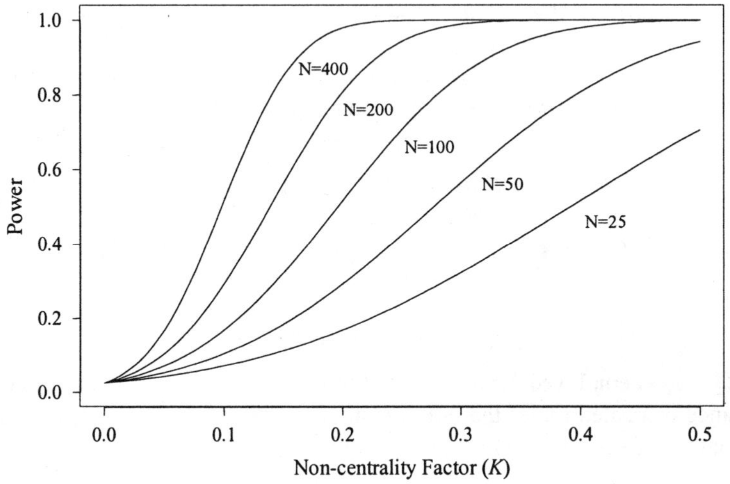

be used to derive the specific expressions required to evaluate the power of this test and to determine the required sample size. Because the square of the Z-test is algebraically equal to the contingency chi-square test (see Problem 2.8), the power of the latter is also provided by the power of the two-sided Z-test.

3.3.1 Power of the Z-Test

The Z-test for two proportions in (2.80) is based on the test statistic $T = p_1 - p_2$, where $E(T) = \pi_1 - \pi_2$. Under H_1: $\pi_1 \neq \pi_2$, from (3.7), $T \overset{d}{\approx} \mathcal{N}\left[\mu_1, \sigma_1^2\right]$ with $\mu_1 = \pi_1 - \pi_2$ and

$$\sigma_1^2 = \frac{\pi_1 (1 - \pi_1)}{n_1} + \frac{\pi_2 (1 - \pi_2)}{n_2} . \tag{3.29}$$

Under H_0: $\pi_1 = \pi_2 = \pi$, then from (3.6), $T \overset{d}{\approx} \mathcal{N}\left[\mu_0,\ \sigma_0^2\right]$ with $\mu_0 = 0$ and

$$\sigma_0^2 = \pi (1 - \pi) \left(\frac{1}{n_1} + \frac{1}{n_2} \right) . \tag{3.30}$$

To allow for unequal sample sizes, denote the expected sample fraction in the ith group as ξ_i where $\xi_i = E(n_i/N)$ $(i = 1, 2)$ and $\xi_1 + \xi_2 = 1$. Then, the variances can be factored as

$$\sigma_0^2 = \frac{\phi_0^2}{N}\,; \qquad \phi_0^2 = \pi\left(1-\pi\right)\left(\frac{1}{\xi_1} + \frac{1}{\xi_2}\right) \tag{3.31}$$

and

$$\sigma_1^2 = \frac{\phi_1^2}{N}\,; \qquad \phi_1^2 = \frac{\pi_1\left(1-\pi_1\right)}{\xi_1} + \frac{\pi_2\left(1-\pi_2\right)}{\xi_2}\,. \tag{3.32}$$

Therefore, $\Delta = \mu_1 - \mu_0 = \pi_1 - \pi_2$ and from (3.18) the basic equation relating the size of the test α, the power $1-\beta$, and sample size N is

$$\begin{aligned}\sqrt{N}\left|\mu_1 - \mu_0\right| &= Z_{1-\alpha}\sqrt{\pi\left(1-\pi\right)\left(\frac{1}{\xi_1} + \frac{1}{\xi_2}\right)} \\ &+ \quad Z_{1-\beta}\sqrt{\frac{\pi_1\left(1-\pi_1\right)}{\xi_1} + \frac{\pi_2\left(1-\pi_2\right)}{\xi_2}} \\ &= Z_{1-\alpha}\phi_0 + Z_{1-\beta}\phi_1\,,\end{aligned} \tag{3.33}$$

where $Z_{1-\alpha/2}$ is employed for a two-sided test. Since the Z-test in (2.80) employs the variance estimated under the null hypothesis, where $p = \xi_1 p_1 + \xi_2 p_2$, then for the evaluation of the above equations, σ_0^2 and ϕ_0^2 are computed using

$$\pi = \xi_1\pi_1 + \xi_2\pi_2. \tag{3.34}$$

Thus only the values π_1 and π_2 need be specified.

Solving for $Z_{1-\beta}$, the level of power provided by a given sample size N to detect the difference between proportions with specified probabilities π_1 and π_2 is obtained from

$$Z_{1-\beta} = \frac{\sqrt{N}\left|\pi_1 - \pi_2\right| - Z_{1-\alpha}\phi_0}{\phi_1}\,. \tag{3.35}$$

Likewise, the sample size required to provide power $1-\beta$ to detect a difference $\Delta = \pi_1 - \pi_2$ is provided by

$$N = \left[\frac{Z_{1-\alpha}\phi_0 + Z_{1-\beta}\phi_1}{\pi_1 - \pi_2}\right]^2. \tag{3.36}$$

With equal sample sizes $\xi_1 = \xi_2 = 1/2$, these expressions simplify slightly using

$$\begin{aligned}\phi_0^2 &= 4\pi\left(1-\pi\right) \\ \phi_1^2 &= 2\pi_1\left(1-\pi_1\right) + 2\pi_2\left(1-\pi_2\right).\end{aligned} \tag{3.37}$$

Finally, for this test, Lachin (1981) shows that $\sigma_0^2 > \sigma_1^2$, in which case we can use the conservative simplifications presented in (3.22 – 3.24) that yield the non-centrality factor

$$K = \frac{\left|\pi_1 - \pi_2\right|}{\sqrt{\pi\left(1-\pi\right)\left(\frac{1}{\xi_1} + \frac{1}{\xi_2}\right)}} = \frac{\left|\pi_1 - \pi_2\right|}{\phi_0}\,. \tag{3.38}$$

While (3.35) and (3.36) are preferred for the assessment of power or sample size, the non-central factor (3.38) will be used below to explore other relationships.

Example 3.2 *Planning a Study*
For example, suppose that we wish to plan a study with two equal sized groups ($n_1 = n_2$) to detect a 30% reduction in mortality associated with congestive heart failure, where the 1-year mortality in the control group is assumed to be no greater than 0.40. Thus $\pi_2 = 0.40$ and $\pi_1 = 0.28$ ($= 0.70 \times 0.40$). Under the null hypothesis we assume $\pi_1 = \pi_2 = \pi = 0.34$. We desire 90% power for a two-sided test for two proportions at $\alpha = 0.05$. Using (3.36) the required total N is obtained as

$$N = \left[\frac{1.96[4(0.34 \times 0.66)]^{\frac{1}{2}} + 1.282[2(0.28 \times 0.72) + 2(0.4 \times 0.6)]^{\frac{1}{2}}}{0.40 - 0.28} \right]^2 = 652$$

rounded up to the nearest even integer.

Alternately one could solve for $Z_{1-\beta}$ to determine the power to detect a difference with a specified sample size, or the magnitude of the difference that could be detected with a given power for a specific sample size. For example, the power to detect this same difference with a smaller sample size of $N = 500$ using (3.35) is provided by

$$Z_{1-\beta} = \frac{\sqrt{500}(0.40 - 0.28) - 1.96[4(0.34 \times 0.66)]^{\frac{1}{2}}}{[2(0.28 \times 0.72) + 2(0.40 \times 0.60)]^{\frac{1}{2}}} = 0.879$$

yielding 81% power.

Note that for a fixed N and a fixed value of π_2, such computations could be used to generate the power curve as a function of increasing values of π_1 (or Δ). However, as π_1 changes, so do the values of π from (3.34).

3.3.2 Relative Risk and Odds Ratio

Since the test statistic is $T = p_1 - p_2$, then the power function is naturally expressed in terms of the risk difference. However, in many cases, it is more desirable to express power as a function of the odds ratio or relative risk. To do so, one specifies the probability in the control group π_2, and then determines the value of π_1 in the exposed or treated group which corresponds to the specified relative risk or odds ratio.

Since the relative risk is $RR = \pi_1/\pi_2$, then for any value of π_2

$$\pi_1 = \pi_2(RR). \tag{3.39}$$

Likewise, since the odds ratio is $OR = \frac{\pi_1(1-\pi_2)}{\pi_2(1-\pi_1)}$, then solving for π_1 yields

$$\pi_1 = \frac{(OR)\pi_2}{(1 - \pi_2) + (OR)\pi_2} \tag{3.40}$$

Table 3.1 Non-Centrality Factors for a Test of the Difference Between Two Proportions with Probabilities π_1 and π_2

$\pi_2 \backslash \pi_1$	0.2	0.3	0.4	0.5	0.6	0.7	0.8	0.9
0.1	0.140	0.250	0.346	0.436	0.524	0.612	0.704	0.800
0.2	–	0.115	0.218	0.314	0.408	0.503	0.600	0.704
0.3		–	0.105	0.204	0.302	0.40	0.503	0.612
0.4			–	0.101	0.20	0.302	0.408	0.524
0.5				–	0.101	0.204	0.314	0.436
0.6					–	0.105	0.218	0.346
0.7						–	0.115	0.250
0.8							–	0.140

(see Problem 2.4.1). Therefore, the power function for the test of two proportions, or the 2×2 table for two independent groups, is readily expressed in terms of relative risks or odds ratios.

It is then of interest to describe the factors that affect the power of a test to detect a given risk difference, relative risk or odds ratio. This is easily done by considering the non-centrality factor $|K|$ in (3.38) as a function of the probabilities π_1 and π_2 in the two groups, as presented in Table 3.1 for the case of equal sample sizes.

Examination of this table shows that for a fixed value of the risk difference $\Delta = \pi_1 - \pi_2$, as $\pi = (\pi_1 + \pi_2)/2$ approaches 0.5, then K decreases, power $1 - \beta$ decreases and thus the N required to achieve a given level of power increases. For example, consider any diagonal of the table corresponding to a fixed risk difference, such as the entries one step off the main diagonal that correspond to differences of $\Delta = 0.1$ with π_2 ranging from 0.1 to 0.8. Therefore, the power to detect a fixed difference is greater when one of the outcomes (positive or negative) is rare than when the two outcomes occur with equal probability (0.5). The reason is that the fixed difference becomes proportionately smaller, and the variance larger, as the probability of either outcome approaches 0.5.

Even though the expressions for the non-centrality factor, sample size and power are stated in terms of the risk difference, it is common to describe the change in power as a function of the relative risk, or odds ratio. For a fixed value of the relative risk π_1/π_2, the table shows that as π_2 increases the total number of positive outcomes also increases, K increases, power $1 - \beta$ increases and the N required to provide a fixed level of power decreases. For example, compare the value of K for $\pi_1/\pi_2 = 2$ and values of π_2 ranging from 0.1 to 0.4. Thus the power to detect a given relative risk appears to depend directly on the expected number of outcomes. However, the reason is that the magnitude of the difference is growing as π_2 increases, and thus π_1 approaches 1 faster than π_2 approaches 0.5.

When the non-centrality parameters are examined in terms of odds ratios, a pattern similar to that for the risk difference is observed. For example, $OR = 6$ for values of $(\pi_1, \pi_2) = (0.4, 0.1)$, (0.6,0.2), (0.8,0.4) and (0.9,0.6). The respective values of the non-centrality parameter are 0.346, 0.408, 0.408 and 0.346.

Thus when using power functions such as those in Figure 3.2 to evaluate the power over a range of parameters, the power curves using the risk difference, the relative risk or the odds ratio all depend on the value of π_2. In each case, the actual values of the probabilities π_1 and π_2 determine power.

3.4 POWER OF CHI-SQUARE TESTS

For test statistics that are distributed as chi-square on $p \geq 1$ *df*, and also as t or F, among others, the power of the test is likewise a function of the non-centrality parameter, ψ^2, of the non-central distribution of the test statistic. The non-centrality parameter is the expected value of the chi-square statistic X_p^2 on p *df* under the alternative hypothesis, or $\psi^2 = E(X_p^2 \,|\, H_1)$. For a test at level α on p *df*, the power of the test $1 - \beta$ is a monotonically increasing function of ψ^2. As for a Z-test, the non-centrality parameter can be factored as $\psi^2 = NK^2$, where K^2 is the non-centrality factor. For a χ^2 test on p *df*, the value of the non-centrality parameter that provides power $1 - \beta$ for a test at level α is denoted as $\psi^2(p,\, \alpha,\, \beta) = NK(p,\, \alpha,\, \beta)^2$.

For example, for a 1 *df* χ^2 test statistic, from (3.25) the value of the non-centrality parameter that provides power $1-\beta$ for a 1 *df* two-sided test at level α, designated as $\psi^2(1,\, \alpha,\, \beta) = NK(1,\, \alpha,\, \beta)^2 = (Z_{1-\alpha/2}+Z_{1-\beta})^2$. For example, the non-centrality parameter that provides Type II error probability $\beta = 0.1$ and power $1 - \beta = 0.9$ for a 1 *df* χ^2 test at $\alpha = 0.05$ is $\psi^2(1,\, 0.05,\, 0.10) = (1.96 + 1.645)^2 = 10.507$.

Values of the non-centrality parameter $\psi^2(p,\, \alpha,\, \beta)$ providing various levels of power for the non-central chi-square distribution on p *df* are widely tabulated. Programs are also available, such as the SAS functions PROBCHI for the cumulative probabilities and CINV for quantiles of the chi-square distribution, both of which provide computations under the non-central distribution. The SAS function CNONCT then provides the value of the non-centrality parameter ψ^2 for specific levels of α and β (Hardison, Quade and Langston, 1986).

To determine sample size using this approach, one first obtains the value $\psi^2(p,\, \alpha,\, \beta)$ of the non-centrality parameter that will provide the desired level of power for the non-central chi-square distribution. The value of the non-centrality factor K under the alternative hypothesis is then specified or evaluated under an appropriate model. The non-centrality factor is usually defined using the variance under the null hypothesis, often because the expected value of the statistic (the non-centrality parameter) is derived under a sequence of local alternatives (defined subsequently). Given the value of K, the N required to provide power $1 - \beta$ is that value for which $\psi^2(p,\alpha,\beta) = NK^2$, yielding

$$N = \frac{\psi^2(p,\alpha,\beta)}{K^2} \,. \tag{3.41}$$

For a 1 *df* χ^2 or a two-sided Z-test, this is equivalent to the simplifications presented in (3.22)–(3.27).

For any specific test statistic, the expression for the non-centrality parameter is obtained as the expected value of the test under the alternative hypothesis. In some cases, these expressions are obtained under a local alternative in which case the calculations describe the limiting power function of the test. These concepts are described in Section 3.5.

Example 3.3 *Test for Proportions*
The non-centrality parameter of the 1 *df* Pearson contingency chi-square test for the 2×2 table may also be used to evaluate the power function for the two-sided test for two proportions since $X^2 = Z^2$ as shown in Problem 2.8. Meng and Chapman (1966) described the limiting power function of chi-square tests for $R \times C$ contingency tables. These results were employed by Lachin (1977) and Guenther (1977) to describe the power function and to determine the sample size needed to provide a desired level of power for the contingency chi-square test for an $R \times C$ table. The simplest case is the non-centrality parameter for the Pearson chi-square test for the 2×2 table.

Using (2.81) for the 2×2 table, then the expected value of the observed frequency in cell ij is $E(O_{ij}) = N\eta_{ij}$ where η_{ij} refers to the probability of an observation falling in that cell under the alternative hypothesis, $\sum_{i=1}^{2}\sum_{j=1}^{2}\eta_{ij} = 1$. As described in Section 2.6.2, the expected frequency under the null hypothesis is $E(O_{ij}|H_0) = N\eta_{0ij}$ where η_{0ij} is the probability for the $ijth$ cell under the null hypothesis that is determined from the marginal probabilities as $\eta_{0ij} = \eta_{i\bullet}\eta_{\bullet j}$. Then the expected value of the chi-square test statistic under the alternative hypothesis is

$$E(X^2) = N\sum_{i=1}^{2}\sum_{j=1}^{2}\frac{[\eta_{ij} - \eta_{0ij}]^2}{\eta_{0ij}} \,. \tag{3.42}$$

Since the sum of the two sets of probabilities within rows or columns is one, then for all four cells of the table, $[\eta_{ij} - \eta_{0ij}]^2 = \delta^2$ for some value $\delta \neq 0$. Thus

$$E(X^2) = N\delta^2\sum_{i=1}^{2}\sum_{j=1}^{2}\frac{1}{\eta_{0ij}} \,. \tag{3.43}$$

Now adopt the notation for the test for two proportions as in (2.24), where columns (j) represent treatment group and rows the response (i) such that $\xi_1 = \eta_{\bullet 1}$, $\xi_2 = \eta_{\bullet 2}$, $\pi = \eta_{1\bullet}$ and $1 - \pi = \eta_{2\bullet}$. Then is readily shown that $E(X^2) = NK^2$ for K as shown in (3.38).

Example 3.4 *Multiple Regression Model Test*
In a homoscedastic normal errors regression model with p covariates, a variety of different tests may be conducted, such as an overall model test on p *df* and tests of

each of the individual regression coefficients, each of which will have a different power function. The power for these tests depends on the total N, the error variance σ_ϵ^2 and on the joint distribution of the p covariates.

To illustrate this, consider the Wald test of H_0: $\boldsymbol{\beta} = 0$ (including the intercept). As shown in Section A.5.1 of the Appendix, the coefficient estimates (including the intercept) are obtained as $\widehat{\boldsymbol{\beta}} = (\mathbf{X}'\mathbf{X})^{-1}\mathbf{X}'Y$ and $V(\widehat{\boldsymbol{\beta}}) = \Sigma_{\widehat{\boldsymbol{\beta}}} = (\mathbf{X}'\mathbf{X})^{-1}\sigma_\epsilon^2$. Then, assuming σ_ϵ^2 is known, or relying on a large sample consistent estimate, the Wald test is the T^2-like quadratic form $X^2 = \widehat{\boldsymbol{\beta}}'\Sigma_{\widehat{\boldsymbol{\beta}}}^{-1}\widehat{\boldsymbol{\beta}}$ described in Section A.7.1 of the Appendix. Under H_1: $\boldsymbol{\beta} \neq 0$, X^2 is distributed as non-central χ^2 on $p+1$ *df* with non-centrality parameter

$$\psi^2 = \boldsymbol{\beta}'\Sigma_{\widehat{\boldsymbol{\beta}}}^{-1}\boldsymbol{\beta} = \frac{E[\boldsymbol{\beta}'(\mathbf{X}'\mathbf{X})\boldsymbol{\beta}]}{\sigma_\epsilon^2} \tag{3.44}$$

$$= \frac{E\left(\boldsymbol{\beta}'(\mathbf{X}'\mathbf{X})E[(\mathbf{X}'\mathbf{X})^{-1}\mathbf{X}'\mathbf{Y}]\right)}{\sigma_\epsilon^2} = \frac{\boldsymbol{\beta}'E(\mathbf{X}'\mathbf{Y})}{\sigma_\epsilon^2}$$

where the expectation is with respect to the joint distribution of $\mathbf{X}$ and $\mathbf{Y}$. Thus to evaluate power or sample size *a priori* for a test given a true parameter vector $\boldsymbol{\beta}$, it is necessary to specify the covariance matrix of $\mathbf{X}$ or $E(\mathbf{X}'\mathbf{Y})$ in order to determine the non-centrality parameter for this test.

The same developments would also apply to the model test that the p coefficients $(\beta_1, \ldots, \beta_p) = \mathbf{0}$, or to the test of an individual coefficient or a subset of the coefficient vector since all are special cases of a linear contrast of the elements of the vector $\boldsymbol{\beta}$.

3.5 EFFICIENCY

3.5.1 Pitman Efficiency

Generalizations of the expression for the power function of a test may also be used to describe the asymptotic efficiency of a test statistic under a family of local alternatives, an approach originally due to Pitman (1948) and elaborated by Noether (1955). This approach is important when a variety of competing statistics could be used to test a particular null versus alternative hypothesis, in which case it is desirable to employ whichever test is most powerful. However, many such families of tests are based on asymptotic distribution theory, and asymptotically any test in the family will provide a level of power approaching 1.0 against any alternative hypothesis away from the null hypothesis. Thus Pitman introduced the concept of the asymptotic power (efficiency) against a family of *local alternatives* that remain "close" to the null hypothesis value for large values of N so that the power remains below 1.0 except in the limiting case $(n \to \infty)$.

Let T be a test statistic for H_0: $\theta = \theta_0$ versus H_1: $\theta \neq \theta_0$, where the power function of T increases in θ, so that T is an unbiased test, and where the power also increases in the total sample size n so that T is also consistent (*cf.* Lehman,

1986). Let $\alpha_n(\theta_0)$ designate the Type I error probability of the test for a given sample size n, and let $\gamma_n(\theta) = 1 - \beta_n(\theta)$ designate the power of the test to detect a specific value $\theta \neq \theta_0$ for a given n as n increases indefinitely. Asymptotically, for any such test, $\lim_{n\to\infty} \gamma_n(\theta) \to 1$ for any $\theta \neq \theta_0$ and thus the limiting power in a broad sense is not useful for evaluating any particular test, nor for the comparison of different tests at the same value of θ under H_1. Therefore, we use the concept of *local power* defined under a sequence of local alternatives as n increases of the form

$$\theta_n = \theta_0 + \frac{\delta}{\sqrt{n}} \tag{3.45}$$

such that $\lim_{n\to\infty} \theta_n = \theta_0$ for any finite δ. Since $Z_{1-\beta} = Z_{\gamma_n}$ increases in $\sqrt{n}$ then Z_{γ_n} approaches a constant at the same rate that $\theta_n \to \theta_0$ so that $\alpha < \gamma_n(\theta_n) < 1$.

For such a sequence of local alternatives, the efficiency of the test is a function of the limiting power

$$\lim_{n\to\infty} \gamma_n(\theta_n) = \lim_{n\to\infty} 1 - \beta_n(\theta_n). \tag{3.46}$$

or

$$\lim_{n\to\infty} P\left[T_n > T_{n,\alpha} \mid \theta_n\right], \tag{3.47}$$

where $T_{n,\alpha}$ is the α level critical value. If we restrict consideration to test statistics that are asymptotically normally distributed, under the alternative hypothesis the asymptotic distribution can be described as:

$$T_n|\theta_n \overset{d}{\approx} N\left[\mu(\theta_n), \phi^2(\theta_n)/n\right] \tag{3.48}$$

with mean $E(T_n|\theta_n) = \mu(\theta_n)$ and variance $V(T_n|\theta_n) = \sigma^2(\theta_n)$ that may depend on the value of θ_n with variance component $\phi^2(\theta_n)$. Then under H_0: $\theta = \theta_0$, asymptotically

$$Z_n = \frac{T_n - \mu(\theta_0)}{\phi(\theta_0)/\sqrt{n}} \overset{d}{\to} N(0,1). \tag{3.49}$$

Therefore, for an upper tailed $1-\alpha$ level test,

$$\gamma_n(\theta_n) = P\left[Z_n \geq Z_{1-\alpha} \mid \theta_n\right]. \tag{3.50}$$

From the description of the power function of such a test (3.20), where $1-\beta_n(\theta_n) = \gamma_n(\theta_n) = \Phi\left[Z_{\gamma_n(\theta_n)}\right]$, yields

$$Z_{\gamma_n(\theta_n)} = \frac{\mu(\theta_n) - \mu(\theta_0)}{\phi(\theta_n)/\sqrt{n}} - Z_{1-\alpha}\frac{\phi(\theta_0)}{\phi(\theta_n)}. \tag{3.51}$$

Now assume that $\mu(\theta_n)$ and $\phi(\theta_n)$ are continuous and differentiable at $\theta = \theta_0$. Evaluating γ_n under a sequence of local alternatives, the limiting power is provided by $\lim_{n\to\infty} Z_{\gamma_n(\theta_n)}$. For the second term on the r.h.s., evaluating $\phi(\theta_n)$ under the local alternative yields

$$\lim_{n\to\infty} \frac{Z_{1-\alpha}\phi(\theta_0)}{\phi\left(\theta_0 + \frac{\delta}{\sqrt{n}}\right)} \to Z_{1-\alpha} \,. \tag{3.52}$$

Therefore, the limiting value of the non-centrality parameter for the test is

$$\lim_{n\to\infty} \left[Z_{\gamma_n(\theta_n)}\right] + Z_{1-\alpha} = \lim_{n\to\infty} \left[\frac{\mu(\theta_n) - \mu(\theta_0)}{\phi(\theta_0)/\sqrt{n}}\right] . \tag{3.53}$$

Multiplying the numerator and denominator by δ then,

$$\lim_{n\to\infty} \left[Z_{\gamma_n(\theta_n)}\right] + Z_{1-\alpha} = \left(\frac{\delta}{\phi(\theta_0)}\right) \lim_{n\to\infty} \left[\frac{\mu\left(\theta_0 + \frac{\delta}{\sqrt{n}}\right) - \mu(\theta_0)}{\delta/\sqrt{n}}\right] . \tag{3.54}$$

Now let $\varepsilon = \delta/\sqrt{n}$, so that $\lim_{n\to\infty} H(\delta/\sqrt{n}) \equiv \lim_{\varepsilon\to 0} H(\varepsilon)$ for any function $H(\cdot)$. Then

$$\begin{aligned} \lim_{n\to\infty} \left[Z_{\gamma_n(\theta_n)}\right] + Z_{1-\alpha} &= \left(\frac{\delta}{\phi(\theta_0)}\right) \lim_{\varepsilon\to 0} \left[\frac{\mu(\theta_0 + \varepsilon) - \mu(\theta_0)}{\varepsilon}\right] \\ &= \left(\frac{\delta}{\phi(\theta_0)}\right) \left(\frac{d\mu(\theta_0)}{d\theta_0}\right) = \left[\frac{\delta\mu'(\theta_0)}{\phi(\theta_0)}\right] . \end{aligned} \tag{3.55}$$

Since δ is a constant by construction, then the efficiency of any test at level α will be proportional to

$$\lim_{n\to\infty} \left[Z_{\gamma_n(\theta_n)}\right] \propto \frac{\mu'(\theta_0)}{\phi(\theta_0)} \propto \frac{[\mu'(\theta_0)]^2}{\phi^2(\theta_0)} \,. \tag{3.56}$$

Therefore, the efficiency of any given statistic T as a test of H_0 versus H_1 is usually defined as

$$Eff(T) = \left[\left(\frac{dE(T)}{d\theta}\right)^2 / \phi^2(\theta)\right]_{|\theta=\theta_0} . \tag{3.57}$$

The numerator is the derivative of the expectation of T evaluated at the null hypothesis value. When $\mu'(\theta) = 1$, such as when T is unbiased or consistent for θ itself such that $E(T) = \mu(\theta) = \theta$, then

$$Eff(T) = \frac{1}{\phi^2(\theta_0)} \propto \frac{1}{\phi^2(\theta_0)/n} = \frac{1}{V(T|\theta_0)} \,, \tag{3.58}$$

which is the reciprocal of the large sample variance of T under the null hypothesis (the null variance).

3.5.2 Asymptotic Relative Efficiency

If there are two alternative tests for H_0: $\theta = \theta_0$, say T_1 and T_2, then the locally more powerful test of the two will be that for which the efficiency is greatest. The difference in efficiency is reflected by the *asymptotic relative efficiency* (ARE) of T_1 to T_2

$$ARE(T_1, T_2) = \frac{Eff(T_1)}{Eff(T_2)}. \tag{3.59}$$

A value of $ARE(T_1, T_2) > 1$ indicates that T_1 is the more powerful test asymptotically, whereas $ARE(T_1, T_2) < 1$ indicates that T_2 is more powerful.

If T_1 and T_2 are both unbiased such that $E(T_i) = \theta$ for $i = 1, 2$, then

$$ARE(T_1, T_2) = \frac{\left[\phi^2_{T_1}(\theta_0)\right]^{-1}}{\left[\phi^2_{T_2}(\theta_0)\right]^{-1}} = \frac{\sigma^2_{T_2}(\theta_0)}{\sigma^2_{T_1}(\theta_0)} = \frac{V(T_2|\theta_0)}{V(T_1|\theta_0)},$$

where T_1 is the more powerful test when $\sigma^2_{T_1} < \sigma^2_{T_2}$, and vice versa. From (3.21), assuming $\sigma_0^2 \doteq \sigma_1^2 \doteq \sigma^2$, the sample size required to provide power $1-\beta$ to detect a difference Δ using a test at level α is of the form

$$N = \left[\frac{(Z_{1-\alpha} + Z_{1-\beta})^2 \phi^2}{\Delta^2}\right] \propto \phi^2, \tag{3.60}$$

which is directly proportional to the variance of the statistic. Thus for such tests, the $ARE(T_1, T_2)$ can be interpreted in terms of the relative sample sizes required to provide a given level of power against a sequence of local alternatives:

$$ARE(T_1, T_2) \doteq \frac{N(T_2)}{N(T_1)}, \tag{3.61}$$

where $N(T_i)$ is the sample size required to achieve power $1-\beta$ for a fixed Δ with test T_i.

Example 3.5 *ARE of Normal Median: Mean*
Consider sampling from a normal distribution, $X \sim \mathcal{N}(\mu, \phi^2)$, where we wish to test H_0: $\mu = 0$ versus H_1: $\mu \neq 0$. Two consistent estimators of μ are

$$T_1 = \overline{x} \sim \mathcal{N}(\mu, \phi^2/n) \tag{3.62}$$

$$T_2 = x_{(0.5n)} = \text{median} \sim \mathcal{N}\left[\mu, (\pi\phi^2)/(2n)\right] \tag{3.63}$$

where $\pi/2 = 1.57$. Thus

$$ARE(T_1, T_2) = \frac{\left[\phi^2/n\right]^{-1}}{\left[(1.57)\,\phi^2/n\right]^{-1}} = 1.57. \tag{3.64}$$

Therefore, in order to provide the same level of power against a sequence of local alternatives, the test using the median would require a 57% greater sample size than the test using the mean. Thus the test based on the mean is 57% more efficient than the test using the median.

3.5.3 Estimation Efficiency

An equivalent expression for the efficiency of a statistic may also be obtained from the perspective of the estimation efficiency of that statistic as an estimate of the parameter of interest. Let $\widehat{\theta}$ be a consistent estimator of θ, where $\widehat{\theta}$ is a function of a sample statistic T of the form $\widehat{\theta} = f(T)$. Then, from the Law of Large Numbers and Slutsky's Theorem (A.47), as $n \to \infty$, then $T \xrightarrow{p} \tau = E(T)$ and $\widehat{\theta} \xrightarrow{p} \theta = f(\tau)$. Therefore, by a Taylor's expansion of $f(T)$ about τ, asymptotically

$$\widehat{\theta} \cong \theta + (T - \tau)\left[\frac{df(T)}{dT}\right]_{|T=\tau} \tag{3.65}$$

and

$$\left(\widehat{\theta} - \theta\right) \cong \frac{T - \tau}{\left(\frac{dE(T)}{d\theta}\right)} . \tag{3.66}$$

Therefore, asymptotically

$$V(\widehat{\theta}) = E\left(\widehat{\theta} - \theta\right)^2 \cong \frac{E(T-\tau)^2}{\left(\frac{dE(T)}{d\theta}\right)^2} \cong \frac{V(T)}{\left(\frac{dE(T)}{d\theta}\right)^2} = \frac{1}{\mathit{Eff}(T)} . \tag{3.67}$$

Thus $V(\widehat{\theta})^{-1}$ is equivalent to the Pitman efficiency of the test statistic T in (3.56) evaluated at the true value of θ.

In some cases these concepts are used to evaluate the properties of a set of tests within a specific family that may not be all inclusive. For example, if the estimator T is defined as a linear combination of other random variables, then we refer to the efficiency of the family of linear estimators. In this case, another test that is not a member of this family, such as a non-linear combination of the random variables, may have greater efficiency. The minimum variance, and thus maximum efficiency, for any estimation-based test then is provided by the Cramer-Rao lower bound for the variance of the estimate as given in (A.109)

$$V(T \mid \theta) \geq \frac{\left[\mu_T'(\theta)\right]^2}{I(\theta)} , \tag{3.68}$$

where $E(T|\theta) = \mu_T(\theta)$ is some function of θ and $I(\theta)$ is the Information function derived from the likelihood function for θ (see Section A.6.4 of the Appendix). For an unbiased estimator of θ, then the maximum possible efficiency of the estimator or test is $I(\theta)$.

These concepts of efficiency and asymptotic relative efficiency, in general, are widely used in the study of distribution-free statistics. Hájek and Śidak (1967) have developed general results that provide the asymptotically most powerful test for location or scale alternatives from any specified distribution. Another important application is in the development of asymptotically fully efficient tests for multiple or stratified 2×2 tables under various "scales" or measures of association. This application is addressed in Chapter 4. The following section presents an illustration of these concepts. Additional problems involving evaluation of the power and sample size for specific tests, and the evaluation of the Pitman efficiency, are presented in future chapters.

3.5.4 Stratified Versus Unstratified Analysis of Risk Differences

To illustrate these concepts, consider a study where we wish to assess the differences in proportions between two treatment or exposure groups where we may also be interested in adjusting for the differences in risk between subgroups or strata. Such analyses are the subject of the next chapter. Here, we consider the simplest case of two strata, such as men and women, within which the underlying probabilities or risks of the outcome may differ. We then wish to test the null hypothesis of no difference within both strata H_0: $\pi_{1j} = \pi_{2j} = \pi_j$ for $j = 1, 2$, versus the alternative H_1: $\theta = \pi_{1j} - \pi_{2j} \neq 0$ of a constant difference across strata, where the probability of the positive outcome in the control group differs between strata, $\pi_{21} \neq \pi_{22}$. Under this model we then wish to evaluate the ARE of an unstratified Z-test based on the pooled 2×2 table versus a stratified-adjusted test.

Under this model, the exposed or treated group probability in the jth stratum is $\pi_{1j} = \theta + \pi_{2j}$ for $j = 1, 2$. Let ζ_j denote the expected sample fraction in the jth stratum, $E(N_j/N) = \zeta_j$, where N_j is the total sample size in the jth stratum ($j = 1, 2$) and N is the overall total sample size. Within the jth stratum we assume that there are equal sample sizes within each group by design such that $n_{ij} = N_j/2$ and $E(n_{ij}) = \zeta_j N/2$ ($i = 1, 2$; $j = 1, 2$). The underlying model can then be summarized as follows:

Stratum	*Stratum Fraction*	*Probability + Exposed*	*Probability + Control*
1	ζ_1	$\theta + \pi_{21}$	π_{21}
2	ζ_2	$\theta + \pi_{22}$	π_{22}

where $\zeta_1 + \zeta_2 = 1$.

Within each stratum let p_{ij} refer to the proportions positive in the ith group in the jth stratum, where $E(p_{ij}) = \pi_{ij}$ ($i = 1, 2$; $j = 1, 2$). Then the unstratified Z-test simply employs (2.80) using the $\{p_{i\bullet}\}$ and $\{n_{i\bullet}\}$ from the pooled 2×2 table where $n_{i\bullet} = N/2$ by design and

$$p_{i\bullet} = \frac{n_{i1}p_{i1} + n_{i2}p_{i2}}{n_{i1} + n_{i2}} = \frac{N_1 p_{i1} + N_2 p_{i2}}{N}, \tag{3.69}$$

since $n_{ij} = N_j/2$ and $N_1 + N_2 = N$. Thus the test statistic in the marginal unadjusted analysis is

$$T_\bullet = p_{1\bullet} - p_{2\bullet} = \sum\nolimits_{j=1}^{2} \frac{N_j}{N}(p_{1j} - p_{2j}). \tag{3.70}$$

Under this model, the underlying parameters are the probabilities within each stratum, the $\{\pi_{ij}\}$. Thus $E(p_{i\bullet})$ is a function of the $\{\pi_{ij}\}$ and the sampling fractions of the two strata:

$$E(p_{2\bullet}) = \sum_{j=1}^{2} \zeta_j \pi_{2j} = \pi_{2\bullet}, \text{ and} \tag{3.71}$$
$$E(p_{1\bullet}) = \sum_{j=1}^{2} \zeta_j (\theta + \pi_{2j}) = \pi_{1\bullet} = \theta + \pi_{2\bullet} \,.$$

Thus

$$E(T_{\bullet}) = E(p_{1\bullet}) - E(p_{2\bullet}) = \theta \tag{3.72}$$

and the statistic is asymptotically unbiased.

The variance of this pooled statistic is $V(T_{\bullet}) = V(p_{1\bullet}) + V(p_{2\bullet})$. Under H_0: $\pi_{1j} = \pi_{2j} = \pi_j$, then

$$V(p_{ij}|H_0) = \frac{\pi_j(1-\pi_j)}{N\zeta_j/2} \,. \tag{3.73}$$

Thus

$$V(p_{i\bullet}|H_0) = \sum_{j=1}^{2} \frac{\zeta_j^2 \pi_j(1-\pi_j)}{N\zeta_j/2} \tag{3.74}$$

and

$$V(T_{\bullet}|H_0) = \sum_{j=1}^{2} \zeta_j^2 \pi_j(1-\pi_j)\left(\frac{4}{N\zeta_j}\right) = \sum_{j=1}^{2} \zeta_j^2 \sigma_{0j}^2 \,, \tag{3.75}$$

where

$$\sigma_{0j}^2 = \pi_j(1-\pi_j)\left(\frac{4}{N\zeta_j}\right) \tag{3.76}$$

is the variance of the difference within the jth stratum under the null hypothesis with variance component $\phi_{0j}^2 = 4\pi_j(1-\pi_j)/\zeta_j$.

Now consider the stratified adjusted test where the test statistic is an optimally weighted average of the differences within each of the two strata of the form

$$T = \sum_{j=1}^{2} w_j (p_{1j} - p_{2j}) \,, \tag{3.77}$$

where $w_1 + w_2 = 1$. First, the weights must be determined that provide maximum efficiency (power) under the specified model that there is a constant difference within the two strata. Then, given the optimal weights, we obtain the asymptotic efficiency of the test and the asymptotic relative efficiency versus that of another test, such as the pooled unstratified test above.

Since $E(p_{1j} - p_{2j}) = \theta$ within each stratum ($j = 1, 2$), then asymptotically $E(T) = \theta$ for any set of weights that sum to 1. The variance of the statistic, however, depends explicitly on the chosen weights through

$$V(T|H_0) = \sum\nolimits_{j=1}^{2} w_j^2 \pi_j (1 - \pi_j) \left(\frac{4}{N\zeta_j} \right) = \sum\nolimits_{j=1}^{2} w_j^2 \sigma_{0j}^2. \tag{3.78}$$

Since the statistic provides a consistent estimate of θ, then the asymptotic efficiency of the test is

$$Eff(T) = \frac{1}{V(T|H_0)} = \frac{1}{w_1^2 \sigma_{01}^2 + w_2^2 \sigma_{02}^2}. \tag{3.79}$$

To obtain the test with greatest power asymptotically, we desire that value of w_1 (and w_2) for which the asymptotic efficiency is maximized. Using the calculus of maxima/minima, it is readily shown (see Problem 3.6) that the optimal weights are defined as

$$w_j = \frac{\sigma_{0j}^{-2}}{\sigma_{01}^{-2} + \sigma_{02}^{-2}}, \tag{3.80}$$

which are inversely proportional to the variance of the difference within the jth stratum. With these weights, it follows that

$$V(T|H_0) = \frac{1}{\sigma_{01}^{-2} + \sigma_{02}^{-2}}, \tag{3.81}$$

which is the minimum variance for any linear combination over the two strata. These results generalize to the case of more than two strata. Such weights that are *inversely proportional to the variance* play a central role in optimally weighted estimates and tests.

We can now assess the asymptotic relative efficiency of the stratified-adjusted test with optimal weights versus the pooled test. From (3.71)–(3.72), the weights in the pooled test asymptotically are simply the sample fractions $\{\zeta_j\}$. Since both test statistics are based on a consistent estimator of θ, then asymptotically $dE(T_\bullet)/d\theta = dE(T)/d\theta = 1$ when evaluated at any value of θ. Then

$$ARE(T, T_\bullet) = \left[\frac{V(T_\bullet|\theta)}{V(T|\theta)} \right]_{\theta=0} = \frac{w_1^2 \sigma_{01}^2 + w_2^2 \sigma_{02}^2}{\zeta_1^2 \sigma_{01}^2 + \zeta_2^2 \sigma_{02}^2} \tag{3.82}$$

$$= \frac{\left[\sigma_{01}^{-2} + \sigma_{02}^{-2}\right]^{-1}}{\zeta_1^2 \sigma_{01}^2 + \zeta_2^2 \sigma_{02}^2} = \frac{\left[\phi_{01}^{-2} + \phi_{02}^{-2}\right]^{-1}}{\zeta_1^2 \phi_{01}^2 + \zeta_2^2 \phi_{02}^2}.$$

Because the total sample size N cancels from each variance expression, the ARE is not a function of the sample size.

By construction, the optimal weights minimize the variance of a linear combination over strata, so that $ARE(T, T_\bullet) \leq 1$ and the stratified adjusted test will always provide greater power when there is a constant difference over strata. However, when the difference between groups is not constant over strata and the model assumptions do not apply, then the stratified-adjusted test may be less powerful than the pooled test. In the next chapter we consider these issues in greater detail. There we generalize the analysis to allow for any number of strata and we consider hypotheses that there is a constant difference on other scales, such as a constant relative risk or constant odds ratio over multiple 2×2 tables.

Example 3.6 *Two Strata*
Consider the case of two strata with the following parameters:

Stratum (j)	ζ_j	π_{1j}	π_{2j}	ϕ_{0j}^2	w_j
1	0.3	0.35	0.2	2.6583	0.65102
2	0.7	0.55	0.4	1.4250	0.34898

where ϕ_{0j}^2 is the variance component from (3.76) that does not depend on the total sample size. In this case, $ARE(T, T_\bullet) = 0.98955$, indicating a slight loss of efficiency when using the pooled test versus the stratified adjusted test. Note that the optimal weights are close to the fractions within each stratum, $w_j \doteq \zeta_j$ so that there is similar efficiency of the two tests.

In Section 4.4.3 of Chapter 4, however, we show that the marginal unadjusted statistic $T_\bullet = p_{1\bullet} - p_{2\bullet}$ is biased when there are differences in the sample fractions within each treatment group among strata, that is, $n_{i1}/N_1 \neq n_{i2}/N_2$ for the ith treatment group with the two strata. In this case, the marginal unadjusted analysis would use a bias-corrected statistic, and the relative efficiency of the stratified adjusted test would be much greater.

3.6 PROBLEMS

3.1 Consider a one-sided Z-test against a left tail alternative hypothesis. Derive the basic equation (3.18) that describes the power for this test.

3.2 Show that non-central factor for the X^2 test in a 2×2 table presented in (3.43) is equivalent to $\varphi^2 = NK^2$, where K is provided by the simplification in (3.38).

3.3 Consider the case of two simple proportions with expectations π_1 and π_2. We wish to plan a study to assess the incidence of improvement (healing) among those treated with a drug versus placebo, where prior studies suggest that the placebo control group probability of improvement is on the order of $\pi_2 = 0.20$. We then wish to detect an increase in the probability of healing with drug treatment where the investigators feel that it is important to detect a minimal risk difference on the order of 0.10; that is, a probability of $\pi_1 = 0.30$. Perform the following calculations needed to design the trial.

3.3.1. For a two-sided test at the $\alpha = 0.05$ level with $Z_{0.975} = 1.96$, what total sample size N would be needed (with equal sized groups) to provide power $1 - \beta = 0.90$ ($Z_{0.90} = 1.282$) to detect this difference?

3.3.2. Now suppose that a total sample size of only $N = 400$ is feasible. With $\pi_2 = 0.2$, what level of power is there to detect

1. A difference of 0.10?
2. A relative risk of 1.5?
3. An odds ratio of 2.0?

When doing this, note that as π_1 changes, so also does π.

3.3.3. Now suppose that the control healing rate is actually higher than originally expected, say $\pi_2 = 0.30$ rather than the initial projection of 0.20. For a total sample size of $N = 400$ with equal sized groups, recompute the power to detect a difference of 0.10, a relative risk of 2.0 and an odds ratio of 2.5.

3.3.4. Also, suppose that the new treatment is very expensive to administer. To reduce costs the sponsor requires that only 1/3 of the total N be assigned to the experimental treatment ($Q_1 = 1/3$). Recompute the power for the conditions in Problems 3.1.2 and 3.1.3. What effect does the unbalanced design have on the power of the test?

3.4 Consider the case of the large sample Z-test for the difference between the means of two populations based on the difference between two sample means $\overline{x}_1$ and $\overline{x}_2$ that are based on samples drawn from some distribution with equal variances φ^2 in each population. Then asymptotically

$$Z = \frac{\overline{x}_1 - \overline{x}_2}{\varphi\sqrt{\frac{1}{n_1} + \frac{1}{n_2}}} \overset{d}{\approx} \mathcal{N}(0,1) \tag{3.83}$$

under the null hypothesis H_0: $E(\overline{x}_1) = E(\overline{x}_2)$. Let $E(\overline{x}_i) = \upsilon_i$ for $i = 1, 2$.

3.4.1. Show the asymptotic distribution of Z under H_1: $\upsilon_1 \neq \upsilon_2$

3.4.2. Derive the equation to compute the sample size N required to detect a difference $\mu_1 = \upsilon_1 - \upsilon_2 \neq 0$ expressed as

$$N = \frac{\left(Z_{1-\alpha} + Z_{1-\beta}\right)^2 \varphi^2 \left(\frac{1}{\xi_1} + \frac{1}{\xi_2}\right)}{\left(\upsilon_1 - \upsilon_2\right)^2}. \tag{3.84}$$

with sample fractions $\{\xi_j\}$.

3.4.3. Derive the equation to compute the power of the test expressed as

$$Z_{1-\beta} = \frac{\sqrt{N}\left(\upsilon_1 - \upsilon_2\right)}{\varphi\sqrt{\frac{1}{\xi_1} + \frac{1}{\xi_2}}} - Z_{1-\alpha} = \frac{\upsilon_1 - \upsilon_2}{\varphi\sqrt{\frac{1}{n_1} + \frac{1}{n_2}}} - Z_{1-\alpha}. \tag{3.85}$$

3.4.4. What sample size is required to detect a difference $\mu_1 = 0.20$ with power $= 0.9$ where $\varphi^2 = 1$ with two equal sized groups using a two-sided test at level $\alpha = 0.05$ two-sided?

3.4.5. What power would be provided to detect this difference with $N = 120$?

3.5 Consider the case of the large sample Z-test of the difference between the rate or intensity parameters for two populations based on samples drawn from populations with a homogeneous Poisson distribution with rate parameters λ_1 and λ_2, where there is an equal period of exposure per observation (see Section 8.1.2). For a sample of n_i observations in the ith group ($i = 1, 2$), the sample estimate of the rate is $\widehat{\lambda}_i = d_i/n_i$ where d_i is the number of events observed among the n_i observations in the ith group. Under H_0: $\lambda_1 = \lambda_2 = \lambda$, the sample estimate of the assumed common rate is $\widehat{\lambda} = (d_1 + d_2)/(n_1 + n_2)$.

3.5.1. Within the ith group, from the normal approximation to the Poisson, $d_i \overset{d}{\approx} \mathcal{N}(n_i\lambda_i,\, n_i\lambda_i)$, show that under H_0 the large sample Z-test is

$$Z = \frac{\widehat{\lambda}_1 - \widehat{\lambda}_2}{\sqrt{\widehat{\lambda}\left(\frac{1}{n_1} + \frac{1}{n_2}\right)}} \sim \mathcal{N}(0, 1). \tag{3.86}$$

3.5.2. Show the asymptotic distribution of Z under H_1: $\lambda_1 \neq \lambda_2$.

3.5.3. Derive the equation to compute the sample size N required to detect a difference $\mu_1 = \lambda_1 - \lambda_2 \neq 0$ expressed as

$$N = \frac{\left(Z_{1-\alpha}\sqrt{\lambda\left(\frac{1}{\xi_1} + \frac{1}{\xi_2}\right)} + Z_{1-\beta}\sqrt{\frac{\lambda_1}{\xi_1} + \frac{\lambda_2}{\xi_2}}\right)^2}{\left(\lambda_1 - \lambda_2\right)^2}. \tag{3.87}$$

3.5.4. Derive the equation to compute the power of the test expressed as

$$Z_{1-\beta} = \frac{\sqrt{N}\,|\lambda_1 - \lambda_2| - Z_{1-\alpha}\sqrt{\lambda\left(\frac{1}{\xi_1} + \frac{1}{\xi_2}\right)}}{\sqrt{\frac{\lambda_1}{\xi_1} + \frac{\lambda_2}{\xi_2}}}. \tag{3.88}$$

3.5.5. What sample size is required to detect a difference $\lambda_1 = 0.20$ versus $\lambda_2 = 0.35$ with 90% power in a study with two equal sized groups using a two-sided test at level $\alpha = 0.05$?

3.5.6. What power would be provided to detect this difference with $N = 200$?

3.6 Consider the ARE of the stratified versus the marginal analysis described in Section 3.5.4.

3.6.1. Show that for the stratified adjusted test, the efficiency of the test is maximized and the variance in (3.79) is minimized using the optimal weights in (3.80).

3.6.2. Then show that $V(T|H_0)$ equals the expression in (3.81).

3.6.3. Now consider three strata where the variance of the stratified adjusted statistic is as in (3.78) with summation over the three strata. Then we desire the weights w_1 and w_2 that minimize the function

$$V(T|H_0) = w_1\sigma_{01}^2 + w_2\sigma_{02}^2 + (1 - w_1 - w_2)\sigma_{03}^2. \tag{3.89}$$

Differentiating with respect to w_1 and w_2 yields two simultaneous equations. Solve these and then simplify to show that the optimal weights are

$$w_j = \frac{\sigma_{0j}^{-2}}{\sigma_{01}^{-2} + \sigma_{02}^{-2} + \sigma_{03}^{-2}} \tag{3.90}$$

for $j = 1, 2, 3$.

4

Stratified-Adjusted Analysis for Two Independent Groups

4.1 INTRODUCTION

In many studies, it is important to account for or adjust for the potential influence of other covariates on the observed association between the treatment or exposure groups and the response. This observed association may be biased when there is an imbalance in the distributions of an important covariate between the two groups. One approach to adjust for such imbalances is to conduct a stratified analysis, stratifying on the other covariates of importance. Another, introduced in later chapters, is to employ an appropriate regression model. When the regression model is equivalent to the model adopted in the stratified analysis, then the two approaches are equivalent, at least asymptotically.

In a stratified analysis the original samples of n_1 and n_2 observations are divided into strata, each an independent subdivision of the study, for example, males and females, and for each a separate 2×2 table is constructed. The stratified-adjusted analysis then aggregates the measures of association across all strata to provide a stratified-adjusted measure of association. The analysis also provides an aggregate overall stratified-adjusted test of significance. The stratified-adjusted analysis is also called an analysis of *partial association* because it assesses the influence of the treatment or exposure group on the response, after allowing for or adjusting for the association between the stratifying covariate with both group membership and with the response.

In the previous chapter we explored a stratified test of significance of risk differences over two strata to illustrate the concept of the asymptotic relative efficiency of two possible tests for the same hypothesis. We now extend this concept to include estimators of an assumed common or average parameter over strata and asymptotically efficient tests of significance. In addition to the risk difference, we also consider the log relative risk and the log odds ratio. We do so first from the perspective of a fixed effects model and then generalize these developments to a random effects model.

A stratified analysis may be performed for various reasons. The most common is to adjust for the potential influence of an important covariate that may be associated with group membership. In observational studies, it is possible that such a covariate may explain some of the observed association between group and response, or the lack thereof. This can occur when there is an imbalance between groups with respect to the distribution of a covariate, such as when there is a larger fraction of males in one group than in the other. However, in a stratified analysis this imbalance can be accounted for by comparing the groups separately within each stratum, such as separately among males and separately among females. In this manner the influence of the covariate on the observed association, if any, between group and response is accounted for or adjusted for.

Another common application of a stratified analysis is a *meta-analysis*. This refers to the combination of the results of separate studies or sub-studies, each constituting a separate stratum in the analysis. The analysis then provides a stratified-adjusted combined estimate of an overall group effect, and assesses its statistical significance. For example, in a multicenter clinical trial, it may be desirable to conduct a combined assessment of the overall treatment effect adjusting for center-to-center differences in the nature of the patient populations or other center-specific differences. In this case, the results within each clinical center form an independent stratum and the adjusted analysis combines the results over strata. Similarly, a meta-analysis of a specific treatment effect may be obtained by conducting a stratified-adjusted analysis in which the results of several published studies of the same treatment versus control are combined. For example, the Early Breast Cancer Trialist's Collaborative Group (1998) presented a meta-analysis of 37,000 women from 55 studies of the effects of adjuvant tamoxifen on the risk of recurrence of breast cancer following surgical treatment. The majority of these studies had produced inconclusive (not statistically significant) results. However, combining all of these studies into a single analysis greatly increased the power to detect an important therapeutic effect. Among all women, the risk of cancer recurrence was reduced by 26%, and among those treated for three or more years it was reduced by 42%, each $p \leq 0.0001$.

The most commonly used and well-known method for conducting a stratified analysis of multiple independent 2×2 tables is the procedure of Mantel and Haenszel (1959). The Mantel-Haenszel procedure yields an aggregate or combined test of partial association that is optimal under the alternative hypothesis of a common odds ratio. An asymptotically equivalent test for a common odds ratio is the test of Cochran (1954a). Cochran's test is also a member of a family of tests described by Radhakrishna (1965), which also includes tests of a common relative risk or a common risk difference, among others. Mantel and Haenszel (1959) also describe an estimate of the assumed common odds ratio. Other estimators are the maximum likelihood estimates, and a family of efficient estimates described by Gart (1971) that can be derived using weighted least squares. These and other related procedures are the subject of this chapter.

First we consider the analyses within strata and the unadjusted marginal analysis. We then consider the asymptotically equivalent Mantel-Haenszel and Cochran stratified-adjusted tests of significance.

4.2 MANTEL-HAENSZEL TEST AND COCHRAN'S TEST

4.2.1 Conditional Within-Strata Analysis

The analysis begins with a conditional within-strata analysis in which a separate 2×2 table is constructed from the observations within each strata. Let K refer to the total number of strata, indexed by $j = 1, \ldots, K$. The strata may be defined by the categories of a single covariate, for example, gender, or by the intersection of the categories of two or more covariates considered jointly, such as by four categories defined by gender and the presence versus absence of a family history of diabetes. Conditionally within the jth strata, the observed frequencies in the 2×2 table and the corresponding probabilities of a positive response with each group are denoted as

jth stratum (4.1)

Frequencies

Response	Group 1	Group 2	
+	a_j	b_j	m_{1j}
−	c_j	d_j	m_{2j}
	n_{1j}	n_{2j}	N_j

Probabilities

Response	Group 1	Group 2
+	π_{1j}	π_{2j}
−	$1-\pi_{1j}$	$1-\pi_{2j}$
	1	1

Within the jth stratum any of the measures of group-response association described previously may be computed: the risk difference $\widehat{RD}_j = p_{1j} - p_{2j}$, relative risk $\widehat{RR}_j = p_{1j}/p_{2j} = (a_j/n_{1j})/(b_j/n_{2j}) = a_j n_{2j}/b_j n_{1j}$, and the odds ratio $\widehat{OR}_j = p_{1j}/(1-p_{1j})/[p_{2j}/(1-p_{2j})] = a_j d_j / b_j c_j$. One could also conduct a test of association using the contingency chi-square test (Cochran's test), Mantel-Haenszel test, or an exact test, separately for each of the K tables. However, this leads to the problem of multiple tests of significance, with pursuant inflation of the Type I error probability of the set of tests. Thus it is preferable to obtain a single overall test of significance for the adjusted association between group and response, and a single overall estimate of the degree of association.

4.2.2 Marginal Unadjusted Analysis

In Chapter 2 we considered the case of a single 2×2 table. This is also called a marginal or unadjusted analysis because it is based on the pooled data for all strata combined. Using the "•" notation to refer to summation over the K strata, the observed frequencies and the underlying probabilities for the marginal 2×2 table are

$$
\begin{array}{c|c|c|c}
\multicolumn{4}{c}{\textit{Frequencies}} \\
\multicolumn{4}{c}{\textit{Group}} \\
 & 1 & 2 & \\
\hline
+ & a_{\bullet} & b_{\bullet} & m_{1\bullet} \\
\hline
- & c_{\bullet} & d_{\bullet} & m_{2\bullet} \\
\hline
 & n_{1\bullet} & n_{2\bullet} & N_{\bullet}
\end{array}
\qquad
\begin{array}{c|c|c|}
\multicolumn{3}{c}{\textit{Probabilities}} \\
\multicolumn{3}{c}{\textit{Group}} \\
 & 1 & 2 \\
\hline
+ & \pi_{1\bullet} & \pi_{2\bullet} \\
\hline
- & 1-\pi_{1\bullet} & 1-\pi_{2\bullet} \\
\hline
 & 1 & 1
\end{array}
\, . \tag{4.2}
$$

This provides the basis for the unadjusted conditional Mantel-Haenszel test statistic $X^2_{c\bullet}$ in (2.83) and the unconditional Cochran test statistic $X^2_{u\bullet}$ in (2.90), the "•" designating the marginal test. The measures of association computed for this table are likewise designated as $\widehat{RD}_{\bullet}$, $\widehat{RR}_{\bullet}$, and $\widehat{OR}_{\bullet}$. These tests and estimates, however, ignore the possible influence of the stratifying covariate.

Example 4.1 *Clinical Trial in Duodenal Ulcers*
Blum (1982) describes a hypothetical clinical trial of the effectiveness of a new drug versus placebo. Through randomization, any randomized study is expected to provide an unbiased assessment of the difference between treatments, so that the marginal unadjusted analysis is expected to be unbiased. Even in this case, however, it is sometimes instructive to assess the treatment effect adjusting for an important covariate that may have a strong influence on the risk of the outcome, or the effectiveness of the treatment. In this example we assess the effectiveness of a new drug for the treatment of duodenal ulcers where the drug is expected to promote healing of the ulcers by retarding the excretion of gastric juices that leads to ulceration of the duodenum.

Ulcers typically have three classes of etiology. *Acid-dependent* ulcers are principally caused by gastric secretion and it is expected that these ulcers will be highly responsive to treatment if the drug is effective. *Drug-dependent* ulcers are usually formed by excessive use of drugs, such as aspirin, that may irritate the lining of the duodenum. Since gastric secretion plays a minor role in drug-induced ulcer formation, it is expected that these ulcers will be resistant to treatment. The third category consists of ulcers of *intermediate origin* where it is difficult to determine whether the ulcer is principally caused by a defect in acid secretion or excessive use of a drug irritant.

Initially 100 patients were assigned to each treatment (drug versus placebo). When stratified by ulcer type, the following 2×2 tables for drug *(D)* versus placebo *(P)* and healing (+) versus not (−) are formed.

1. *Acid-Dependent*

	D	P	
+	16	20	36
−	26	27	53
	42	47	89

2. *Drug-Dependent*

	D	P	
+	9	4	13
−	3	5	8
	12	9	21

3. *Intermediate*

	D	P	
+	28	16	44
−	18	28	46
	46	44	90

(4.3)

(reproduced with permission). The unadjusted analysis is based on the marginal 2×2 table obtained as the sum of the stratum-specific tables to yield

Marginal

	D	P	
+	53	40	93
−	47	60	107
	100	100	200

Within each stratum, and in the marginal unadjusted analysis among all strata combined, the proportions that healed in each group are

Group Proportion	*Stratum* 1	2	3	*Marginal Unadjusted*
Drug (p_1)	0.381	0.750	0.609	0.530
Placebo (p_2)	0.426	0.619	0.364	0.400

The estimates of the three principal summary measures and the 95% confidence limits are presented in Table 4.1. For the relative risk and odds ratios, the asymmetric confidence limits are presented. The Cochran and Mantel-Haenszel test statistics within each stratum and marginally (unadjusted) are

Test	*Stratum* *1*	2	3	*Marginal Unadjusted*
Mantel-Haenszel X_c^2	0.181	1.939	5.345	3.380
$p \leq$	0.671	0.164	0.021	0.067
Cochran X_u^2	0.183	2.036	5.405	3.397
$p \leq$	0.669	0.154	0.021	0.066

Among those treated with the drug, the highest healing rates (proportions) were observed in the drug dependent (2) and intermediate (3) ulcer strata. The beneficial effect of drug treatment was greatest among those with drug dependent ulcers, whether measured as the risk difference, relative risk or odds ratio. Although the tests of significance within strata are presented, these are not usually employed because of the problem of multiple tests of significance and the pursuant increase in the Type I error probability.

Table 4.1 Measures of Association Within Each Stratum and in the Marginal Unadjusted Analysis

Measure	*Stratum* 1	2	3	*Marginal*
Risk difference $(\widehat{RD})$	−0.045	0.306	0.245	0.130
$\widehat{V}(\widehat{RD})$	0.011	0.043	0.010	0.0049
95% *C.I.* for *RD*	−0.25, 0.16	−0.10, 0.71	0.04, 0.45	−0.007, 0.27
Relative risk $(\widehat{RR})$	0.895	1.688	1.674	1.325
log relative risk	−0.111	0.523	0.515	0.281
$\widehat{V}[\log \widehat{RR}]$	0.067	0.167	0.054	0.0239
95% *C.I.* for *RR*	0.54, 1.49	0.76, 3.76	1.06, 2.64	0.98, 1.79
Odds ratio $(\widehat{OR})$	0.831	3.750	2.722	1.691
log odds ratio	−0.185	1.322	1.001	0.526
$\widehat{V}[\log \widehat{OR}]$	0.188	0.894	0.189	0.0818
95% *C.I.* for *OR*	0.36, 1.94	0.59, 23.9	1.16, 6.39	0.97, 2.96

Marginally, the differences between groups is not statistically significant. This test, however, ignores the imbalances between groups in the numbers of subjects within each of the three ulcer-type strata. One might ask whether the nature or significance of the treatment group effect is altered in any way after adjusting for these imbalances within strata.

4.2.3 Mantel-Haenszel Test

Mantel and Haenszel (1959) advanced the following test for a common odds ratio among the set of K tables that adjusts for the influence of the stratifying covariate(s). This provides a test of the global null hypothesis H_0: $\pi_{1j} = \pi_{2j}$ $(OR_j = 1)$ for all j versus the alternative that the probabilities within strata differ such that there is a common odds ratio. That is, H_1: $OR_j = OR \neq 1$ for all $j = 1, ..., K$. As shown in Section 2.6.3, from the conditional hypergeometric likelihood for a single 2×2 table under the null hypothesis within the jth stratum, the expected frequency for the index cell, $E(a_j)$, is simply

$$E_j = E(a_j) = n_{1j} m_{1j} / N_j \tag{4.4}$$

and central hypergeometric variance of a_j under H_0 is

$$V_{cj} = \frac{m_{1j} m_{2j} n_{1j} n_{2j}}{N_j^2 (N_j - 1)} \,. \tag{4.5}$$

Within the jth stratum, these provide a within-stratum test. However, we are principally interested in the aggregate stratified-adjusted test. The Mantel-Haenszel test and its asymptotic distribution are obtained as follows.

Within the jth stratum, under H_0 asymptotically for large N_j and for fixed K,

$$a_j - E_j \overset{d}{\approx} \mathcal{N}\left[0,\ V_{cj}\right]. \tag{4.6}$$

Since the strata are independent

$$\sum\nolimits_j (a_j - E_j) \overset{d}{\approx} \mathcal{N}\left[0,\ \sum\nolimits_j V_{cj}\right]. \tag{4.7}$$

Therefore, the stratified-adjusted Mantel-Haenszel test, conditional on all margins fixed, is

$$X^2_{C(MH)} = \frac{\left[\sum_j (a_j - E_j)\right]^2}{\sum_j V_{cj}} = \frac{\left[\sum_j (a_j - n_{1j}m_{1j}/N_j)\right]^2}{\sum_j \left(\dfrac{m_{1j}m_{2j}n_{1j}n_{2j}}{N_j^2(N_j-1)}\right)} = \frac{\left[a_+ - E_+\right]^2}{V_{c+}}, \tag{4.8}$$

where $a_+ = \sum_j a_j$, $E_+ = \sum_j E_j$ and $V_{c+} = \sum_{j=i}^{s} V_{cj}$. Since $(a_+ - E_+)$ is the sum of asymptotically normally distributed stratum-specific variates, then asymptotically $(a_+ - E_+) \overset{d}{\approx} \mathcal{N}[0, V_{c+}]$ and asymptotically $X^2_{C(MH)} \overset{d}{\approx} \chi^2$ on 1 df. Note that while $a_+ = a_\bullet$ is also the index frequency in the marginal 2×2 table in (4.2), the other components E_+ and V_{c+} are not based on the marginal table at all.

The asymptotic distribution can also be demonstrated for the case where the sample size within each stratum is small but the number of strata increases indefinitely (Breslow, 1981).

4.2.4 Cochran's Test

An asymptotically equivalent test was also suggested by Cochran (1954a) using a linear combination of the differences between the proportions $p_{1j} - p_{2j}$ over the K strata. The representation of this and other tests in this form is described subsequently. Algebraically, Cochran's stratified adjusted test can also be expressed in terms of $(a_+ - E_+)$.

Starting from the unconditional product binomial likelihood, as shown in Section 2.6.4 for a single 2×2 table, then under the null hypothesis, $E(a_j) = n_{1j}\pi_j$ within the jth stratum that can be estimated consistently as

$$\widehat{E}_j = n_{1j}m_{1j}/N_j. \tag{4.9}$$

As shown in (2.88), the unconditional variance of a_j is

$$V_{uj} = \frac{n_{1j}n_{2j}\pi_j(1-\pi_j)}{N_j}, \tag{4.10}$$

which can be consistently estimated as

$$\widehat{V}_{uj} = \frac{m_{1j} m_{2j} n_{1j} n_{2j}}{N_j^3} . \tag{4.11}$$

Therefore, when aggregated over strata, the stratified-adjusted unconditional test is provided by

$$X^2_{U(C)} = \frac{\left[\sum_j \left(a_j - \widehat{E}_j\right)\right]^2}{\sum_j \widehat{V}_{uj}} = \frac{\left[\sum_j \left(a_j - n_{1j} m_{1j}/N_j\right)\right]^2}{\sum_j \left(m_{1j} m_{2j} n_{1j} n_{2j}/N_j^3\right)} = \frac{\left[a_+ - \widehat{E}_+\right]^2}{\widehat{V}_{u+}} . \tag{4.12}$$

Analogously to (4.6) and (4.7), asymptotically $(a_j - E_j) \sim N\left[0, V_{uj}^2\right]$ within the jth stratum and $(a_+ - E_+) \overset{d}{\approx} \mathcal{N}\left[0, V_{u+}^2\right]$. Extending the approach used in Section (2.6.4), asymptotically $\widehat{E}_+ \overset{p}{\to} E_+$ and $\widehat{V}_{u+} \overset{p}{\to} V_{u+}$, and from Slutsky's Theorem (Appendix, Section A.4.3) $X^2_{U(C)} \overset{d}{\approx} \chi^2$ on 1 *df*.

Since the only difference between the conditional Mantel-Haenszel test $X^2_{C(MH)}$ and the unconditional Cochran test $X^2_{U(C)}$ is in the denominators, and since $\widehat{V}_{uj} = V_{cj}(N_j - 1)/N_j$, then the two tests are asymptotically equivalent. Thus the two are often referred to interchangeably as the *Cochran-Mantel-Haenszel Test.*

Formally, as will be shown in Section 4.7 to follow, these tests address the following null hypothesis:

$$H_0\colon OR = 1 \text{ assuming } E(\widehat{OR}_j) = OR_j = OR, \forall j \tag{4.13}$$

versus the alternative

$$H_1\colon OR \neq 1 \text{ assuming } E(\widehat{OR}_j) = OR_j = OR, \forall j.$$

Thus these tests are described explicitly in terms of odds ratios. This also implies that different test statistics could be used to detect an adjusted difference in relative risks or risk differences. Such tests are also described in Section 4.7.

Example 4.2 *Clinical Trial in Duodenal Ulcers (continued)*
For the above example, the Mantel-Haenszel test statistic is $X^2_{C(MH)} = 3.0045$ with $p \leq 0.083$. Cochran's test is $X^2_{U(C)} = 3.0501$ with $p \leq 0.081$. Compared to the marginal unadjusted analysis, the stratified adjusted analysis yields a slightly less significant test statistic value, $p \leq 0.083$ versus $p \leq 0.067$ (using the Mantel-Haenszel tests). These results suggest that some of the association between treatment and response is now accounted for through the stratification adjustment. Stratified-adjusted estimates of the degree of association are then required to further describe the nature of the stratification adjustment.

4.3 STRATIFIED-ADJUSTED ESTIMATORS

4.3.1 Mantel-Haenszel Estimates

Mantel and Haenszel (1959) presented a "heuristic" estimate of the assumed common odds ratio, and also of the assumed relative risks when a common relative risk is assumed to exist rather than a common odds ratio. There was no formal proof of the asymptotic properties of these estimators, nor were they derived so as to provide any particular statistical properties. Rather, Mantel and Haenszel simply stated that these estimators seemed to work well compared to other possible estimators. In fact, it was not until recently that the asymptotic variances of the estimates were actually derived.

The Mantel-Haenszel estimate of the common odds ratio is

$$\widehat{OR}_{MH} = \frac{\sum_j a_j d_j / N_j}{\sum_j b_j c_j / N_j}\,, \tag{4.14}$$

which can be expressed as a weighted combination of the stratum-specific estimates of the form

$$\widehat{OR}_{MH} = \sum\nolimits_j \widehat{v}_j (\widehat{OR}_j) \tag{4.15}$$

with weights

$$\widehat{v}_j = \frac{b_j c_j / N_j}{\sum_\ell b_\ell c_\ell / N_\ell} \tag{4.16}$$

that sum to unity.

Likewise, when a constant relative risk is assumed, the Mantel-Haenszel estimate of the common relative risk is

$$\widehat{RR}_{MH} = \frac{\sum_j a_j n_{2j} / N_j}{\sum_j b_j n_{1j} / N_j}\,. \tag{4.17}$$

This estimate can also be expressed as a weighted average of the stratum-specific relative risks:

$$\widehat{RR}_{MH} = \sum\nolimits_j \widehat{v}_j (\widehat{RR}_j) \tag{4.18}$$

with weights

$$\widehat{v}_j = \frac{b_j n_{1j} / N_j}{\sum_\ell b_\ell n_{1\ell} / N_\ell}\,. \tag{4.19}$$

Because these estimators were derived heuristically, their large sample properties were not derived; nor did Mantel and Haenszel describe an estimate of the large sample variance, or the computation of confidence limits.

4.3.2 Test-Based Confidence Limits

One of the first approaches to constructing confidence limits for the assumed common odds ratio, which may also be applied to the Mantel-Haenszel estimate of the common relative risk, is the suggestion by Miettinen (1976) that the aggregate Mantel-Haenszel test be inverted to yield so-called test-based confidence limits. These are a simple generalization of the test-based limits described in Section 2.6.6.

Let $\widehat{\theta}$ denote the log of the Mantel-Haenszel estimate, either $\widehat{\theta} = \log\left(\widehat{OR}_{MH}\right)$ or $\widehat{\theta} = \log\left(\widehat{RR}_{MH}\right)$. Then assume that the Mantel-Haenszel test could also be expressed as an estimator-based test. If the variance of the estimate $\sigma^2_{\widehat{\theta}}$ were known, then the test statistic would be constructed using the variance estimated under the null hypothesis of no partial association, that is, using the variance estimate $\widehat{\sigma}^2_{\widehat{\theta}|H_0}$. This would yield a test of the form

$$[Z_{C(MH)}]^2 = X^2_{C(MH)} = \widehat{\theta}^2 / \widehat{\sigma}^2_{\widehat{\theta}|H_0}. \tag{4.20}$$

Given the observed values $z_{C(MH)}$ and $\widehat{\theta}$, the test statistic can be inverted to yield a test-based estimate of the variance of $\widehat{\theta}$ as

$$\widehat{\sigma}^2_{\widehat{\theta}|H_0} = [\widehat{\theta}/z_{C(MH)}]^2. \tag{4.21}$$

This test-based variance could then be used to construct confidence limits for the parameter of interest. For example, for $\widehat{\theta} = \log\left(\widehat{OR}_{MH}\right)$, the resulting test-based $1-\alpha$ level confidence limits on $\log(OR)$ are provided by

$$\widehat{\theta} \pm Z_{1-\alpha/2}\widehat{\theta}/z_{C(MH)} \tag{4.22}$$

and those on OR as

$$\begin{aligned}(\widehat{OR}_\ell, \widehat{OR}_u) &= \exp\left[\widehat{\theta} \pm Z_{1-\alpha/2}\widehat{\theta}/z_{C(MH)}\right] \qquad (4.23)\\ &= \exp\left[\log\left(\widehat{OR}_{MH}\right) \pm \frac{Z_{1-\alpha/2}}{z_{C(MH)}} \log\left(\widehat{OR}_{MH}\right)\right]\\ &= \widehat{OR}_{MH}^{\;1 \pm \frac{Z_{1-\alpha/2}}{z_{C(MH)}}}.\end{aligned}$$

These test-inverted confidence limits are inherently incorrect because they are based on an estimate of the variance derived under the null hypothesis, whereas the proper limits are defined under the alternative hypothesis. However, they often work well in practice. See Halperin (1977) and Greenland (1984), among others, for a further discussion of the properties of these confidence limits.

4.3.3 Large Sample Variance of Log Odds Ratio

In general, it is preferable to obtain asymmetric confidence limits on an odds ratio or relative risk as the exponentiation of the symmetric confidence limits from the log odds ratio or log relative risk. Since the Mantel-Haenszel estimates are weighted averages of the odds ratios and relative risks, not the logs thereof, a large sample variance of the log odds ratio, and also the log relative risk, can be obtained as follows using the δ-method (Hauck, 1979).

Consider the estimate of the common odds ratio OR_{MH} in (4.14), where $\theta = \log(OR_{MH})$. As shown in Problem 2.5.4, given an estimate of the variance of the log odds ratio, $V(\widehat{\theta})$, then asymptotically

$$V\left(\widehat{OR}_{MH}\right) \cong (OR_{MH})^2 V(\widehat{\theta}) \triangleq \left(\widehat{OR}_{MH}\right)^2 \widehat{V}(\widehat{\theta}) \,, \tag{4.24}$$

where "$\triangleq$" means "estimated as." However, the Mantel-Haenszel estimate is a weighted average of the stratum-specific odds ratios. If we treat the weights $\{v_j\}$ as known (fixed), then the asymptotic variance $V\left(\widehat{OR}_{MH}\right)$ can be obtained directly from (4.15) as

$$\begin{aligned} V\left(\widehat{OR}_{MH}\right) &= \sum\nolimits_j v_j^2 V(\widehat{OR}_j) \cong \sum\nolimits_j v_j^2 (OR_j)^2 V\left[\log(\widehat{OR}_j)\right] \\ &\triangleq \sum\nolimits_j \widehat{v}_j^2 (\widehat{OR}_j)^2 \widehat{V}\left[\log(\widehat{OR}_j)\right] . \end{aligned} \tag{4.25}$$

Substituting into (4.24) yields

$$V(\widehat{\theta}) \cong \frac{V\left(\widehat{OR}_{MH}\right)}{(OR_{MH})^2} \triangleq \frac{\widehat{V}\left(\widehat{OR}_{MH}\right)}{\left(\widehat{OR}_{MH}\right)^2} \,, \tag{4.26}$$

from which the estimate $\widehat{V}(\widehat{\theta})$ can then be computed. This then yields confidence limits on the log odds ratio, and asymmetric confidence limits on the odds ratio. The expressions for the relative risk and log thereof are similar.

Guilbaud (1983) presents a precise derivation of this result also taking into account the fact that the estimated odds ratio is computed using estimated weights $\{\widehat{v}_j\}$ rather than the true weights. However, from Slutsky's Convergence Theorem (A.45), since the estimated weights in (4.16) and (4.19) can be shown to converge to constants, the above simple derivation applies.

Various authors have derived other approximations to the variance of the Mantel-Haenszel estimate of the common log odds ratio $\widehat{\theta} = \log\left(\widehat{OR}_{MH}\right)$. The most accurate of these is the expression given by Robins, Breslow and Greenland (1986) and Robins, Greenland and Breslow (1986). The derivation is omitted. The estimate is based on the following five sums:

$$S_1 = \sum_{j=1}^{S} a_j d_j / n_j; \quad S_2 = \sum_{j=1}^{S} b_j c_j / n_j; \quad S_3 = \sum_{j=1}^{S} (a_j + d_j)\, a_j d_j / n_j^2; \tag{4.27}$$

$$S_4 = \sum_{j=1}^{S} (b_j + c_j)\, b_j c_j / n_j^2; \quad S_5 = \sum_{j=1}^{S} \left[(a_j + d_j)\, b_j c_j + (b_j + c_j)\, a_j d_j\right] / n_j^2$$

Table 4.2 Relative Risks and Odds Ratios Within Strata and the Mantel-Haenszel Adjusted Estimates.

Association Measure	*Stratum* 1	2	3	*Mantel-Haenszel Estimate*	95% *C.I.*
Odds Ratio	0.831	3.750	2.722	1.634	0.90, 2.97
$\widehat{v}_j$	0.608	0.059	0.333		
$\widehat{V}(\widehat{OR}_{MH})$				0.248	
Relative Risk	0.895	1.688	1.674	1.306	0.68, 2.50
$\widehat{v}_j$	0.474	0.115	0.411		
$\widehat{V}(\widehat{RR}_{MH})$				0.188	

The estimate of the large sample variance is then calculated in terms of these five sums as follows:

$$\widehat{\sigma}^2_{\widehat{\theta}} = \widehat{V}\left[\log\left(\widehat{OR}_{MH}\right)\right] = \frac{S_3}{2S_1^2} + \frac{S_5}{2S_1S_2} + \frac{S_4}{2S_2^2} \,. \tag{4.28}$$

Example 4.3 *Clinical Trial in Duodenal Ulcers (continued)*
For this example, Table 4.2 presents a summary of the calculation of the Mantel-Haenszel estimate of the stratified-adjusted odds ratio, and the adjusted relative risk, along with the Hauck estimate of the variance and the corresponding 95% confidence limits. The Mantel-Haenszel estimate of the common odds ratio is $\widehat{OR}_{MH} = 1.634$, and its log is 0.491. The Robins, Breslow and Greenland estimate of the variance of $\log\left(\widehat{OR}_{MH}\right)$ is 0.0813, which yields asymmetric 95% confidence limits for the common odds ratio of (0.934, 2.857). The terms entering into this computation are $S_1 = 15.708$, $S_2 = 9.614$, $S_3 = 9.194$, $S_4 = 4.419$ and $S_5 = 11.709$. For comparison, the test-based confidence limits for the common odds ratio are (0.938, 2.846). Hauck's method yields $\widehat{V}\left(\widehat{OR}_{MH}\right) = 0.24792$ with $\widehat{V}\left(\log\widehat{OR}_{MH}\right) = 0.09287$, somewhat larger than the Robins, Breslow and Greenland estimate. The resulting confidence limits for OR are (0.899, 2.97) that are somewhat wider.

Compared to the marginal unadjusted analysis, the stratified adjusted analysis yields a slightly smaller odds ratio in favor of the new drug treatment (1.63 versus 1.69). Thus a small part of the original unadjusted estimate of the difference in effectiveness of drug versus placebo was caused by the imbalances in the numbers of subjects from each stratum within each group.

The weights $\{\widehat{v}_j\}$ show that Mantel-Haenszel estimates give greatest weight to stratum 1, the least to stratum 2, more so for the odds ratios than the relative risks.

4.3.4 Maximum Likelihood Estimates of the Common Odds Ratio

An adjusted estimate of the assumed common odds ratio can also be obtained through maximum likelihood estimation. Under the hypothesis that $E(\widehat{OR}_j) = OR = \varphi$ for all K strata, then the total likelihood is the product of the K stratum-specific conditional hypergeometric likelihoods presented in (2.57)

$$L_c(\varphi) = \prod_{j=1}^{K} \frac{\binom{n_{1j}}{a_j}\binom{n_{2j}}{m_{1j}-a_j}\varphi^{a_j}}{\sum_{i=a_{\ell j}}^{a_{uj}} \binom{n_{1j}}{i}\binom{n_{2j}}{m_{1j}-i}\varphi^{a_j}} , \tag{4.29}$$

where $a_{\ell j}$ and a_{uj} are the limits on the sample space for a_j given the margins in the jth stratum as described in Section 2.4.2. Taking the derivative with respect to φ, the estimating equation for the *MLE* for φ cannot be expressed in a closed-form expression. This approach is described by Birch (1964) and is often called the *conditional maximum likelihood estimate* of the common odds ratio. Example 6.7 of Chapter 6 describes Newton-Raphson iteration to obtain the *MLE* for φ.

Then in Chapter 7 we also show that the *MLE* of the common log odds ratio can be obtained through a logistic regression model. This latter estimate is often termed the *unconditional MLE* of the common odds ratio.

4.3.5 Minimum Variance Linear Estimators (MVLE)

A third family of estimators is described by Gart (1971) using the principle of weighting inversely proportional to the variances (Meier, 1953), which is derived from weighted least squares estimation. This approach provides an adjusted estimator that is a minimum variance linear estimator (*MVLE*) of θ for measures of association on any "scale" $\theta = G(\pi_1, \pi_2)$ for some smooth function $G(\cdot,\cdot)$. Therefore, these are asymptotically efficient within the class of linear estimators. However, since the estimates within each table are consistent, then the *MVLE* is also a consistent estimator and is asymptotically fully efficient, and its asymptotic variance is easy to derive.

Using the framework of weighted least squares (Section A.5.2 of the Appendix) we have a vector of random variables $\widehat{\boldsymbol{\theta}} = \left(\widehat{\theta}_1 \dots \widehat{\theta}_K\right)^T$, where the assumed model specifies that a common θ applies to all strata such that $E(\widehat{\theta}_j) = \theta$ for $j = 1, \dots, K$. Further, the variance of the estimate within the jth stratum is $V(\widehat{\theta}_j) = E(\widehat{\theta}_j - \theta)^2 = \sigma^2_{\widehat{\theta}_j}$, which will be designated simply as σ_j^2. For each common measure of association, these variances are presented in Section 2.3. For now, assume that the $\{\sigma_j^2\}$ are known (fixed). Therefore, under this model, asymptotically

$$\widehat{\boldsymbol{\theta}} \cong \begin{bmatrix} 1 \\ \vdots \\ 1 \end{bmatrix} \theta + \boldsymbol{\varepsilon} = \boldsymbol{J}\theta + \boldsymbol{\varepsilon} , \tag{4.30}$$

where $\boldsymbol{J}$ is a $K \times 1$ unit vector of ones, $\boldsymbol{\varepsilon}$ is a $K \times 1$ vector, and where

$$E(\boldsymbol{\varepsilon}) = \mathbf{0}, \quad \text{and} \quad V(\boldsymbol{\varepsilon}) = diag\left(\sigma_1^2 \;\ldots\; \sigma_K^2\right) = \boldsymbol{\Sigma}_{\boldsymbol{\varepsilon}}. \tag{4.31}$$

We express the relationship only asymptotically because $\widehat{\theta}_j \xrightarrow{p} \theta_j = \theta$ for all j. Also $V(\varepsilon_j) = \sigma_j^2$ represents the random sampling variation of the estimate $\widehat{\theta}_j$ in the jth stratum about the assumed common parameter θ.

Then from (A.70), the WLS estimate of the common parameter is

$$\widehat{\theta} = \left(\boldsymbol{J}'\boldsymbol{\Sigma}_{\boldsymbol{\varepsilon}}^{-1}\boldsymbol{J}\right)^{-1}\left(\boldsymbol{J}'\boldsymbol{\Sigma}_{\boldsymbol{\varepsilon}}^{-1}\widehat{\boldsymbol{\theta}}\right). \tag{4.32}$$

Since $\boldsymbol{\Sigma}_{\boldsymbol{\varepsilon}}^{-1} = diag\left(\sigma_1^{-2} \ldots \sigma_K^{-2}\right)$, this estimator can be expressed as a weighted average of the stratum-specific estimates

$$\widehat{\theta} = \frac{\sum_j \sigma_j^{-2}\widehat{\theta}_j}{\sum_j \sigma_j^{-2}} = \sum\nolimits_j \omega_j \widehat{\theta}_j \,, \tag{4.33}$$

where

$$\omega_j = \frac{\sigma_j^{-2}}{\sum_\ell \sigma_\ell^{-2}} = \frac{\tau_j}{\sum_\ell \tau_\ell} \,, \tag{4.34}$$

$\tau_j = \sigma_j^{-2}$ and $\sum_j \omega_j = 1$. Also, from (A.72), the variance of the estimate is

$$V(\widehat{\theta}) = \sigma_{\widehat{\theta}}^2 = \left(\boldsymbol{J}'\boldsymbol{\Sigma}_{\boldsymbol{\varepsilon}}^{-1}\boldsymbol{J}\right)^{-1} = \frac{1}{\sum_j \sigma_j^{-2}} \,. \tag{4.35}$$

In practice, the estimate is computed using estimated weights $\{\widehat{\omega}_j\}$ obtained by substituting the large sample estimate of the stratum-specific variances $\{\widehat{\sigma}_j^2\}$ in (4.33) so that

$$\widehat{\theta} = \sum\nolimits_j \widehat{\omega}_j \widehat{\theta}_j. \tag{4.36}$$

Since $\widehat{\sigma}_j^2 \xrightarrow{p} \sigma_j^2$, $\widehat{\tau}_j \xrightarrow{p} \tau_j$, and $\widehat{\omega}_j \xrightarrow{p} \omega_j$, then from Slutsky's Convergence Theorem (A.45) the resulting estimate $\widehat{\theta}$ is consistent for θ, or $\widehat{\theta} \xrightarrow{p} \theta$, and asymptotically $\widehat{\theta}$ is distributed as

$$\left(\widehat{\theta} - \theta\right) \overset{d}{\approx} N\left(0, \sigma_{\widehat{\theta}}^2\right). \tag{4.37}$$

Likewise, it follows from Slutsky's theorem that a consistent estimate of the variance is obtained by substituting the stratum-specific variance estimates $\{\widehat{\sigma}_j^2\}$ into (4.35) so that

$$\widehat{V}(\widehat{\theta}) = \frac{1}{\sum_j \widehat{\sigma}_j^{-2}} \,. \tag{4.38}$$

This is another derivation of the principle that weighting inversely proportional to the variances provides a minimum variance estimate under an appropriate model. Also, this is a *fixed effects model* because we assume a common value of the parameter for all strata, or that $\theta_1 = \theta_2 = \ldots = \theta_S = \theta$ for all K strata, that is, $E(\widehat{\theta}_j) = \theta$ for all j. Therefore, the model assumes that all of the variation *between* the values of the observed $\{\widehat{\theta}_j\}$ is caused by random sampling variation about a common value θ.

The explicit expressions for $\widehat{\theta}$ and $\widehat{V}(\widehat{\theta})$ for the risk difference, log odds ratio and log relative risk are derived in Problem 4.2.3. The following is an example of the required computations.

Example 4.4 *Clinical Trial in Duodenal Ulcers (continued)*
For the data in Example 4.1, the following is a summary of the computation of the *MVLE* for the stratified-adjusted log odds ratio, where $\widehat{\theta}_j = \log\left(\widehat{OR}_j\right)$. The estimates, variances and estimated weights within each stratum are

Stratum	$\widehat{\theta}_j$	$\widehat{\sigma}_j^2$	$\widehat{\sigma}_j^{-2}$	$\widehat{\omega}_j$
1	-0.185	0.188	5.319	0.454
2	1.322	0.894	1.118	0.095
3	1.001	0.189	5.277	0.451
Total			11.715	1.0

so that the *MVLE* is $\widehat{\theta} = \log(\widehat{OR}) = (-0.185 \times 0.454) + (1.322 \times 0.095) + (1.001 * 0.451) = 0.493$ and $\widehat{OR} = 1.637$. The estimated variance of the log odds ratio is $\widehat{V}(\widehat{\theta}) = 1/11.715 = 0.0854$, which yields an asymmetric 95% *C.I.* on OR of $\exp\left(\widehat{\theta} \pm 1.96\widehat{\sigma}_{\widehat{\theta}}\right) = \exp\left(0.493 \pm (1.96)\sqrt{0.0854}\right) = (0.924, 2.903)$. Note that the *MVLE* estimate and the asymmetric confidence limits are very close to the Mantel-Haenszel estimate and its confidence limits.

Table 4.3 presents a summary of the computations of the *MVLE* of the common parameter θ and its estimated large sample variance for the risk difference, relative risk and odds ratio. Note that each of the adjusted estimates is based on a slightly different set of weights. All three estimates, however, give less weight to the second stratum because of its smaller sample size.

4.3.6 MVLE versus Mantel Haenszel Estimates

The Mantel-Haenszel estimate of the common odds ratio $\widehat{OR}_{MH}$ can be expressed as a linear combination of the stratum-specific odds ratios using weights $\widehat{v}_j$ as shown in (4.15). However, the weights are not proportional to the inverse of the variance of the estimate within each stratum, that is,

$$\widehat{v}_j \neq \frac{\widehat{\sigma}_j^{-2}}{\sum_\ell \widehat{\sigma}_\ell^{-2}}, \tag{4.39}$$

Table 4.3 *MVLE* of the Common Parameter θ and Its Estimated Large Sample Variance for Each Measure

	Stratum			Adjusted	
Measure	1	2	3	$\widehat{\theta}$	95% C.I.
$\widehat{\theta}_j =$ *Risk Difference*	−0.045	0.306	0.245	0.125	−0.01, 0.26
$\widehat{V}(\widehat{\theta}_j)$	0.011	0.043	0.010	0.0047	
$\widehat{\omega}_j$	0.437	0.110	0.453		
$\widehat{\theta}_j =$ log *Relative Risk*	−0.111	0.523	0.515	0.281	
$\widehat{V}(\widehat{\theta}_j)$	0.067	0.167	0.054	0.0254	
$\widehat{\omega}_j$	0.376	0.152	0.472		
Relative Risk				1.324	0.97, 1.81
$\widehat{\theta}_j =$ log *Odds Ratio*	−0.185	1.322	1.000	0.493	
$\widehat{V}(\widehat{\theta}_j)$	0.188	0.894	0.189	0.0854	
$\widehat{\omega}_j$	0.454	0.095	0.451		
Odds Ratio				1.637	0.92, 2.90

where in this context $\sigma_j^2 = V(\widehat{OR}_j)$. Thus the Mantel-Haenszel estimate is not a minimum variance linear estimator of the common odds ratio, and will have a larger variance of the estimate than the *MVLE* of the common odds ratio. Note that in the preceding section we described the *MVLE* of the common log odds ratio, which is preferable for the purpose of constructing confidence intervals. However, the *MVLE* of the common odds ratio, without the log transformation, may also be readily obtained.

Nevertheless, the Mantel-Haenszel estimate still has a favorable total mean square error (MSE) when compared to the *MVLE* and the *MLE* because the individual odds ratios are not unbiased with finite samples, and thus the adjusted estimators are also not unbiased. From the principle of partitioning of variation in Section A.1.3 of the Appendix, (A.5), the total MSE of an estimator can be partitioned as $MSE(\widehat{\theta}) = V\left(\widehat{\theta}\right) + Bias^2$. Gart (1971), McKinlay (1978), Breslow (1981) and Hauck (1989), among others, have shown that the MSE of the Mantel-Haenszel estimate is close to that of the *MLE* in various settings. Thus there is a trade-off between the slightly larger variance of the Mantel-Haenszel estimate and its slightly smaller bias with finite samples than is the case with the *MVLE* estimates.

Table 4.4 SAS PROC FREQ Analysis of Example 4.1

```
data one;
input k a b c d;
cards;
1  16  20  26  27
2   9   4   3   5
3  28  16  18  28
;
Title1 'Example 4.1: Ulcer Clinical Trial';
data two; set one;
keep i j k f;
K=Stratum, I=Group, J=Response, F=Frequency;
  i = 1; j = 1; f =a; output;
  i = 2; j = 1; f =b; output;
  i = 1; j = 2; f =c; output;
  i = 2; j = 2; f =d; output;
proc freq; table k*(i j) / chisq nocol nopercent; weight f;
Title2
'Association Of Stratum By Group (k*i) And By Response (k*j)';
proc freq; table k*i*j / cmh; weight f;
Title2 'SAS Mantel-Haenszel Analysis';
run;
```

4.3.7 SAS PROC FREQ

Some, but not all, of the preceding methods are provided by the SAS procedure PROC FREQ for the analysis of cross-classified frequency data. Section 2.7 describes the SAS output and computations for a single 2×2 table. This procedure also conducts a Cochran-Mantel-Haenszel analysis of multiple 2×2 tables. For illustration, Table 4.4 presents SAS statements to conduct an analysis of the data from Example 4.1. The results are presented in Table 4.5. Note that a data set is required that includes the level of stratum (k), group (i) and response (j) along with the frequency (f) within each cell of each 2×2 table.

The first call of PROC FREQ generates the tables that assess the association of the strata with group membership and with the response. These analyses are discussed in the following section. Thus these pages of SAS output are not displayed. The second call of PROC FREQ conducts the analysis of the association between group and response within each stratum, and the Mantel-Haenszel analysis over strata. The 2×2 tables of group by response within each stratum also are not displayed because this information is presented in Section 4.2 above. The Mantel-Haenszel stratified adjusted analysis is shown in Table 4.5.

The Mantel-Haenszel analysis first presents three stratified-adjusted tests; the test of non-zero correlation, the test that the row mean scores differ and the test of

Table 4.5 SAS PROC FREQ Mantel-Haenszel Analysis of Example 4.1

SUMMARY STATISTICS FOR I BY J
CONTROLLING FOR K

Cochran-Mantel-Haenszel Statistics (Based on Table Scores)

Statistic	Alternative Hypothesis	DF	Value	Prob
1	Nonzero Correlation	1	3.005	0.083
2	Row Mean Scores Differ	1	3.005	0.083
3	General Association	1	3.005	0.083

Estimates of the Common Relative Risk (Row1/Row2)

Type of Study	Method	Value	95% Confidence Bounds	
Case-Control	Mantel-Haenszel	1.634	0.938	2.846
(Odds Ratio)	Logit	1.637	0.924	2.903
Cohort	Mantel-Haenszel	1.306	0.966	1.767
(Col1 Risk)	Logit	1.324	0.969	1.810
Cohort	Mantel-Haenszel	0.796	0.615	1.030
(Col2 Risk)	Logit	0.835	0.645	1.083

The confidence bounds for the M-H estimates are test-based.

Breslow-Day Test for Homogeneity of the Odds Ratios

Chi-Square = 4.626 DF = 2 Prob = 0.099

general association. For multiple 2×2 tables, all three tests are equivalent to the Mantel-Haenszel test described herein. For $R \times C$ tables with $R > 2$ or $C > 2$, the three tests differ (see Stokes, Davis and Koch, 1995).

This analysis is followed by measures of association. SAS labels these as measures of relative risk. Three measures are computed: "Case-control", "Cohort (Column 1 risk)" and "Cohort (Column 2 risk)". As described in Section 2.7, these refer to the odds ratio, column 1 relative risk and the column 2 relative risk, respectively. For each, the Mantel-Haenszel estimate and a "logit" estimate are computed. For the odds ratio (Case-control relative risk) the logit estimate refers to the *MVLE*, actually the exponentiation of the *MVLE* of the common log odds ratio. The logit confidence limits are the asymmetric confidence limits for the odds ratio obtained from the *MVLE* confidence limits for the common log odds ratio. For the "Column 1" and the "Column 2" relative risks, the Mantel-Haenszel estimate of the stratified-adjusted relative risk and the test-inverted confidence limits are presented. The *MVLE* of the common relative risk and the corresponding asymmetric confidence limits are also presented. These are also labeled as the "logit" estimates, although they are based on the log transformation, not the logit.

4.4 NATURE OF COVARIATE ADJUSTMENT

For whichever measure is chosen to describe the association between the treatment or exposure group and the response, the marginal unadjusted estimate of the parameter $\widehat{\theta}_{\bullet}$ is often different from the stratified-adjusted estimate $\widehat{\theta}$. Also the value of the marginal unadjusted chi-square test, and its level of significance, are often different from those in the stratified-adjusted analysis. Part of this phenomenon may be explained by simple sampling variation, but a more important consideration is that the parameters in the conditional within-stratum analysis (the $\{\theta_j\}$), the marginal unadjusted analysis ($\theta_{\bullet}$), and the stratified-adjusted analysis (θ) are all distinct. Thus, in general, one expects that the marginal parameter in the population will differ conceptually, and thus in value, from the assumed common value among a defined set of strata, that is, $\theta_{\bullet} \neq \theta$.

The marginal parameter $\theta_{\bullet}$, such as the odds ratio, is the expected value of the sample estimate on sampling n_1 and n_2 observations from a large (infinite) population. Within the jth of K strata, for K fixed, the stratum-specific value θ_j is the expectation $E\left(\widehat{\theta}_j\right)$ on sampling n_{1j} and n_{2j} observations from the large population defined by stratum j. In the stratified-adjusted model, it is assumed that the stratum-specific parameters share a common value θ that is distinct from $\theta_{\bullet}$. Thus if one changes the nature of the strata then the meaning of θ changes. In general, one should expect some difference between the marginal unadjusted analysis and the stratified-adjusted analysis.

Thus there is no unique adjusted estimate of the parameter of interest nor a unique *P*-value for a statistical test that may be considered "correct", or "the truth". The results obtained depend on the model employed, which, in this case, refers to the scale used to measure the effect (RD, RR, OR, etc.), and on the set of covariates used to define the strata. Therefore, the conclusion of the analysis may depend on the measure chosen to reflect the difference between groups and the covariate(s) chosen to adjust for. In fact, this is true of all models.

4.4.1 Confounding and Effect Modification

When epidemiologists contrast the marginal unadjusted estimate versus the stratum specific estimates versus the stratified-adjusted estimate, usually in terms of the odds ratio or relative risk, they often draw a distinction between the influence of a confounding variable versus effect modification. See Schlesselman (1982), Kleinbaum, Kupper and Morgenstern (1982), Rothman (1986) and Kelsey, Whittemore, Evans and Thompson (1996), among many, for further discussion.

Confounding usually refers to the situation where the stratification variable is the true causal agent and the treatment or exposure group (the independent variable) is indirectly related to the outcome through its association with the causal stratification variable. For example, individuals who drink large amounts of coffee have a much higher prevalence of smoking than is found in the general population. Thus any association between the amount of coffee drunk per day and the risk of heart disease or cancer will likely be confounded with the association between smoking and coffee drinking. In this case, one expects that the unadjusted association between coffee drinking and heart disease is different from the adjusted association after stratification by smoking versus not. In fact, one expects the adjusted odds ratio to be substantially smaller. Therefore, it is now common to refer to "confounding" whenever an adjustment for another variable (the covariate) leads to a change in the nature of the association between the independent variable and the response. Such informal usage, however, is not precise unless the covariate can in fact be viewed as a true causal agent.

In some cases the stratification adjustment may result in an increase in the strength of association between the independent and dependent variables. In extreme cases this is referred to as *Simpson's paradox*. However, this is a misnomer. In multiple regression, for example, it is well known that adjustment for other covariates may result in a substantial increase in the strength of the association between the independent variable and the response.

Confounding is an example of an *antagonistic* effect of adjustment where some of the association of the independent variable with the response is explained by the association between the independent variable and the covariate and between the covariate and the response. However, it is also possible that the adjustment may introduce a *synergistic* effect between the independent variable and the covariate such that the covariate-adjusted association between the independent variable and the response is greater than the marginal unadjusted association.

Basically this can be viewed as follows: Consider the multiple regression case where all variates are quantitative. Let X be the independent variable, Y the dependent variable (response), and Z the covariate. If Z has a strong association with X but a weak association with Y, then including Z in the model will help explain some of the residual error or noise in the measurements of X, thus allowing the signal in X to better "predict" or be more strongly correlated with the values of Y. In the social sciences, the covariate Z in this case would be called a *suppressor variable*.

Effect modification, on the other hand, refers the observation that the stratum-specific estimates, the $\{\theta_j\}$, differ among the specific strata. Subsequently this is referred to as *heterogeneity* among strata, or an *interaction* between the group and strata effects. For example, assume that the association between body weight and the risk of developing diabetes were found to differ significantly between men and women, such as odds ratios of 1.7 per 5 kg greater weight for women and 2.8 for men, where the difference between men versus women (between strata) is

Fig. 4.1 Covariance adjustment of the mean value in each group ($\overline{y}_i$) by removing the bias $\widehat{\beta}(\overline{z}_i - \overline{z})$ caused by the imbalance in the covariate means ($\overline{z}_i$), $i = 1, 2$.

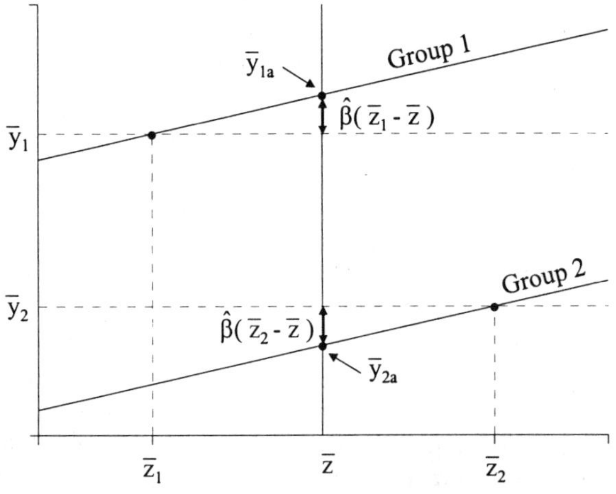

statistically significant. In this case we would say that gender modifies or interacts with the effect of body weight on the risk of developing diabetes.

4.4.2 Stratification Adjustment and Regression Adjustment

Insight into the nature of covariate adjustment is also provided by the Analysis of Covariance (ANCOVA). Consider that we have two groups represented by the independent variable $X = 1$ or 2, and we wish to estimate the difference between groups in the means of a quantitative dependent variable Y. However, we also wish to adjust this estimate for any differences between groups with respect to a quantitative covariate Z based on the regression of Y on Z. The elements of an analysis of covariance are depicted in Figure 4.1. Assume that the slope of the relationship of the response (Y) with the covariate (Z) is the same in both groups so that the regression lines are parallel. By chance, or because of confounding, the mean value of the covariate may differ between the two groups, that is, $\overline{z}_1 < \overline{z}_2$. Thus some of the difference between the unadjusted Y means in the two groups ($\widehat{\theta}_u = \overline{y}_1 - \overline{y}_2$) is attributable to a bias introduced by the difference in the covariate means ($\overline{z}_1 - \overline{z}_2$) between the groups.

Based on the estimated slope of the regression of Y on Z, that is assumed to be the same in the two groups, the bias introduced into the estimate of each Y mean can be estimated and then removed. Usually this is expressed by assessing the expected difference between groups had both groups had the same mean for Z, that is, when $\overline{z}_1 = \overline{z}_2 = \overline{z}$. This adjusted difference is provided by $\widehat{\theta}_a = \overline{y}_{1a} - \overline{y}_{2a}$ where the adjusted Y means $\overline{y}_{1a}$ and $\overline{y}_{2a}$ are the values along the regression lines for which $z = \overline{z}$. Thus in the ith group

$$\overline{y}_{ia} = \overline{y}_i - (\overline{z}_i - \overline{z})\widehat{\beta} \tag{4.40}$$

and the difference between each unadjusted mean ($\overline{y}_i$) and the adjusted mean ($\overline{y}_{ia}$) is the magnitude of the estimated bias introduced by the difference in the covariate mean from its overall mean ($\overline{z}_i - \overline{z}$) based on the estimated slope $\widehat{\beta}$ that is assumed to be the same in both groups.

The assumption of a common slope between groups implies that the difference between the conditional expectations $\theta_z = E(y_1|z) - E(y_2|z)$ is constant for all values of the covariate Z. This is also called the assumption of parallelism, or of homogeneity of the group difference for all values of the covariate. This is directly equivalent to the assumption of a common difference between groups on some scale when z takes on discrete values rather than being continuous. Thus for a discrete covariate, an adjustment using a regression model is conceptually equivalent to an adjustment using direct stratification, although the methods of estimation may differ, for example, iterative maximum likelihood for a logistic regression model, and one-step weighted least squares for the *MVLE* estimates of a common odds ratio. Later in Chapter 7, we show more directly the asymptotic equivalence between a stratification-adjusted Mantel-Haenszel analysis and an adjustment for the same covariate using a logistic regression model.

It should be noted that the above covariate adjustment may also be obtained from an analysis of the residuals of the relationship of Y on Z. Assuming a quantitative covariate Z with estimated common slope $\widehat{\beta}$ in each group, then the residual for the jth observation in the ith group is

$$e_{ij} = y_{ij} - z_{ij}\widehat{\beta}. \tag{4.41}$$

Thus the adjusted difference between groups is also provided by the difference in the group means of the residuals $\widehat{\theta}_a = \overline{y}_{1a} - \overline{y}_{2a} = \overline{e}_1 - \overline{e}_2$, the constant $\overline{z}\widehat{\beta}$ cancelling.

4.4.3 When Does Adjustment Matter?

Further insight into the nature of covariate adjustment is provided by contrasting a simple versus a partial correlation coefficient for quantitative observations, as in multiple regression. Again consider three quantitative variates X, Y and Z, analogous to group, response and the stratifying covariate, respectively. The unadjusted correlation p_{xy} is a measure of the marginal association between the independent (X) and dependent (Y) variables, without consideration of the association between either variable with the possible mediating covariate (Z). The partial or adjusted correlation, however, examines the association between X and Y after removing the association of each with Z. The partial correlation can be expressed as

$$\rho_{xy,z} = corr\left[e\left(x|z\right),\, e(y|z)\right] \;, \tag{4.42}$$

where $e(x|z) = x - E(x|z)$ is the residual of x from its conditional expectation given the value z for each observation, and $e(y|z) = y - E(y|z)$ is the corresponding residual of y given the value z. Algebraically the partial correlation reduces to

$$\rho_{xy,z} = \frac{\rho_{xy} - \rho_{xz}\rho_{yz}}{\sqrt{\left(1-\rho_{xz}^2\right)\left(1-\rho_{yz}^2\right)}}, \tag{4.43}$$

where ρ_{yz} is the correlation between the mediating or stratification variable Z and the response Y, and where ρ_{xz} the correlation between the independent variable X and Z. Thus $\rho_{xy,z} = \rho_{xy}$ if and only if both $\rho_{xz} = 0$ and $\rho_{yz} = 0$. If $\rho_{xz} \neq 0$ then there is some association between the covariate and the independent variable, and if $\rho_{yz} \neq 0$ then there is some association between the covariate and the response.

Cochran (1983) illustrated an equivalent relationship in a stratified analysis of 2×2 tables with only two strata. As in Section 3.5.4, assume that there is a constant risk difference θ within each stratum, but the probability of the positive outcome in the control group differs between strata, $\pi_{21} \neq \pi_{22}$. Thus the exposed or treated group probability in the jth stratum is $\pi_{1j} = \theta + \pi_{2j}$ for $j = 1, 2$. Also assume that the expected sample fraction in the jth stratum is $\zeta_j = E(N_j/N)$ for $j = 1, 2$. We also generalize the model beyond that in Section 3.5.4 to allow unequal treatment group sample fractions among strata such that within the jth stratum the expected sample fractions within each group are $\xi_{ij} = E(n_{ij}/N_j)$ $(i = 1, 2;\ j = 1, 2)$. It is instructive to summarize the model as follows:

	Stratum	Group Sample Fractions		Probability +	
Stratum	*Fraction*	*Exposed*	*Control*	*Exposed*	*Control*
1	ζ_1	ξ_{11}	ξ_{21}	$\theta + \pi_{21}$	π_{21}
2	ζ_2	ξ_{12}	ξ_{22}	$\theta + \pi_{22}$	π_{22}

Now let κ_i refer to the fraction of all subjects in the ith group who are from the first stratum ($j = 1$), where

$$\kappa_i = \frac{\xi_{i1}\zeta_1}{\xi_{i1}\zeta_1 + \xi_{i2}\zeta_2} \tag{4.44}$$

for $i = 1, 2$. Then the expected difference in the marginal unadjusted analysis is

$$E\left(p_{1\bullet} - p_{2\bullet}\right) = \frac{\left(\theta + \pi_{21}\right)\xi_{11}\zeta_1 + \left(\theta + \pi_{22}\right)\xi_{12}\zeta_2}{\xi_{11}\zeta_1 + \xi_{12}\zeta_2} - \frac{\pi_{21}\xi_{21}\zeta_1 + \pi_{22}\xi_{22}\zeta_2}{\xi_{21}\zeta_1 + \xi_{22}\zeta_2} \tag{4.45}$$

$$= \left(\theta + \pi_{21}\right)\kappa_1 + \left(\theta + \pi_{22}\right)\left(1 - \kappa_1\right) - \pi_{21}\kappa_2 - \pi_{22}\left(1 - \kappa_2\right)$$
$$= \theta + \left(\kappa_1 - \kappa_2\right)\left(\pi_{21} - \pi_{22}\right).$$

Therefore, the marginal unadjusted estimate of the risk difference is biased by the quantity $(\kappa_1 - \kappa_2)(\pi_{21} - \pi_{22})$, where $(\kappa_1 - \kappa_2)$ reflects the degree of exposure or treatment group imbalance between the strata, and where $(\pi_{21} - \pi_{22})$ reflects the differential in risk between strata. Conversely, a stratified analysis using a weighted average of the differences between groups is unbiased since within each stratum $E(p_{1j} - p_{2j}) = \theta$ for $j = 1, 2$.

For example, consider a study with the following design elements:

	Stratum	*Group Sample Fractions*		*Probability +*	
Stratum	*Fraction*	*Exposed*	*Control*	*Exposed*	*Control*
1	0.30	0.75	0.25	0.35	0.20
2	0.70	0.40	0.60	0.55	0.40

where $\kappa_1 = (0.3 \times 0.75)/[(0.3 \times 0.75) + (0.7 \times 0.4)] = 0.4554$ and $\kappa_2 = 0.1515$. Because of the imbalance in the treatment group sample fractions and the difference in risk between strata, the difference in the marginal proportions from the unadjusted analysis will be biased. In this case, the bias of the marginal test statistic is $(0.4554 - 0.1515) \times (0.2 - 0.4) = -0.061$, which is substantial relative to the true difference of 0.15 within each stratum.

Beach and Meier (1989), Canner (1991) and Lachin and Bautista (1995), among others, have demonstrated that the same relationships apply to other marginal unadjusted measures of association, such as the log odds ratio. Therefore the difference between the unadjusted measure of association between group and response, equivalent to ρ_{xy}, and the stratified-adjusted measure of association, equivalent to $\rho_{xy,z}$, is a function of the degree of association between the covariate with group membership and with the outcome. In a stratified analysis, these associations are reflected by the $K \times 2$ contingency table for strata by group and the $K \times 2$ table for stratum by response.

Example 4.5 *Clinical Trial in Duodenal Ulcers (continued)*

The following tables describe the association between the stratification covariate ulcer type and treatment group (Drug vs. Placebo) and the association between the covariate and the likelihood of healing (+) versus not (−):

	Stratum by Group				*Stratum by Response*			
	#		%		#		%	
Stratum	*D*	*P*	*D*	*P*	+	−	+	−
1	42	47	47.2	52.8	36	53	40.5	59.5
2	12	9	57.1	42.9	13	8	61.9	38.1
3	46	44	51.1	48.9	44	46	48.9	51.1

These tables would be provided by the SAS statements in Table 4.4.

In the table of stratum by group, if there are no group imbalances, then the same proportion of subjects should be from each group within each stratum. However, proportionately fewer patients from the drug treated group fall in the first stratum (47.2%) and more in the second stratum (57%). The contingency test of association for this table is $X^2 = 0.754$ on 2 df, $p \leq 0.686$. Similarly, in the table of stratum by response, if the covariate was not associated with the likelihood of healing, that is, were not a risk factor for healing, the same proportion of patients would have a

Table 4.6 Stratified-Adjusted Analysis of the Data from Example 4.6

Measure $\widehat{\theta}_j$	*Stratum* 1	2	3	4	*Marginal* $\widehat{\theta}_{\bullet}$	*Adjusted* $\widehat{\theta}$
RD	−0.0048	−0.2288	−0.0098	−0.0904	−0.0785	−0.0265
$\widehat{V}(\widehat{\theta}_j)$	0.0054	0.0152	0.0008	0.0049	0.0011	0.0006
$\log RR$	−0.0408	−0.7892	−0.3615	−0.6016	−0.6685	−0.5303
$\widehat{V}(\widehat{\theta}_j)$	0.3976	0.2471	0.9726	0.2002	0.0799	0.0795
RR	0.9600	0.4542	0.6966	0.5480	0.5125	0.5885
$\log OR$	−0.0462	−1.1215	−0.3716	−0.7087	−0.7580	−0.6297
$\widehat{V}(\widehat{\theta}_j)$	0.5093	0.4413	1.0282	0.2793	0.1017	0.1139
OR	0.9548	0.3258	0.6897	0.4923	0.4686	0.5327

"+" response within each stratum. However, there is proportionately less healing in the first stratum (40.5%) and more in the second stratum (61.9%). For this table, $X^2 = 3.519$ on 2 df, $p \leq 0.172$.

Neither table shows a significant association with the stratification covariate, although in this regard statistical significance is less important than the degree of proportionate differences because significance will largely be a function of the sample size. These proportionate differences are relatively minor, so the covariate adjusted analysis differs only slightly from the unadjusted analysis. This should be expected in a randomized study because, through randomization, it is unlikely that a substantial imbalance in the treatment group proportions among strata will occur. The following are additional examples of non-randomized studies in which the adjustment does make a difference.

Example 4.6 *Religion and Mortality*

Zuckerman, Kasl and Ostfeld (1984) present the following stratified analysis of a prospective cohort study of the association between having religious beliefs versus not on mortality over a period of two years among an elderly population that was forced to relocate to nursing homes. The four strata comprised 1) healthy males, 2) ill males, 3) healthy females and 4) ill females. Note that the strata are simultaneously adjusting for gender and healthy versus ill. The 2×2 tables for the four

Table 4.7 Mantel-Haenszel Analysis of Odds Ratios for Data in Example 4.6

Mantel-Haenszel Analysis	*Marginal Unadjusted*	*Stratified-Adjusted*
$\widehat{OR}$	0.4686	0.5287
$\log \widehat{OR}$	-0.758	-0.637
$\widehat{V}\left[\log \widehat{OR}\right]$	0.1017	0.1116
95% *C.I.* for OR	0.251, 0.875	0.275, 1.018
X^2_{MH}	5.825	3.689
$p \leq$	0.0158	0.0548

strata, in abbreviated notation, are

	Religious			*Non-Religious*		
Stratum	a_j	n_{1j}	p_{1j}	b_j	n_{2j}	p_{2j}
1: Healthy females	4	35	0.114	5	42	0.119
2: Ill females	4	21	0.190	13	31	0.419
3: Healthy males	2	89	0.022	2	62	0.032
4: Ill males	8	73	0.110	9	45	0.200
Marginal	18	218	0.083	29	180	0.161

where a_j and b_j, respectively, are the numbers of patients who died in the religious and non-religious groups (reproduced with permission). The measures of association within each stratum, for the pooled data marginally with no stratification adjustment, and the *MVLE* stratified-adjusted estimates and their variances are presented in Table 4.6. For all measures of association, the magnitude of the group effect in the marginal unadjusted analysis is greater than that in the stratified-adjusted analysis; that is, the $\widehat{\theta}$ is closer to zero than is the $\widehat{\theta}_{\bullet}$.

Likewise, the Mantel-Haenszel analysis of the odds ratios, compared to the marginal unadjusted analysis, is presented in Table 4.7. The stratified-adjusted Mantel-Haenszel estimate is close to that provided by the *MVLE* and also shows less degree of association between religious versus not with mortality. The net result is that whereas the unadjusted Mantel-Haenszel test is significant at $p \leq 0.02$, the adjusted statistic is not significant at the usual 0.05 significance level.

To determine why the stratified-adjusted analysis provides a somewhat different conclusion from the unadjusted analysis, it is necessary to examine the degree of association of the stratification variables with group membership, and that with mortality. The corresponding 4×2 tables are

	By Group			By Response		
	#			#		
Stratum	R	$\overline{R}$	$\%R$	*Died*	*Alive*	*%Died*
1: Healthy females	35	42	45.5	9	68	11.7
2: Ill females	21	31	40.4	17	35	32.7
3: Healthy males	89	62	58.9	4	147	2.7
4: Ill males	73	45	61.9	17	101	14.4

The chi-square test of association between stratum and group membership is $X^2 = 10.499$ on 3 *df* with $p \leq 0.015$. The proportion of religious subjects within strata 3 and 4 (all males) is higher than that in strata 1 and 2 (all females), so that the effect of religion versus not is somewhat associated with the effect of gender. The test of association between stratum and mortality is also highly significant ($X^2 = 334.706$, $p \leq 0.001$). The mortality among ill patients of both genders (strata 2 and 4) is much higher than that among healthy patients (strata 1 and 3), and the mortality among females is less than that among males.

Thus some of the association between religious versus not in the marginal analysis is explained by the imbalances in the proportions of religious subjects among the four strata, and the corresponding differences in mortality between strata. The stratified-adjusted analysis eliminates this "confounding" by comparing the religious versus non-religious subjects within strata and then averaging these differences over strata. Mantel and Haenszel (1959) refer to this as the principle of comparing like-to-like in the stratified analysis.

Example 4.7 *Simpson's Paradox*
In some cases the stratification adjustment yields not only an adjustment in the magnitude of the association between the independent variable and the response (the quantity of the effect), but also in the quality or direction of the association (the quality of the effect). Simpson's Paradox refers to cases where marginally there is no group effect but after adjustment there is a big group effect. This is illustrated by the following hypothetical data from Stokes, Davis and Koch (1995) of a health policy opinion survey of the association between stress (No/Yes) and the risk (probability) of favoring a proposed new health policy among residents of urban and rural communities.

The 2×2 tables for the two strata and marginally are

	Not-Stressed			Stressed		
Stratum	a_j	n_{1j}	p_{1j}	b_j	n_{2j}	p_{2j}
1: Urban	48	60	0.800	96	190	0.505
2: Rural	55	190	0.289	7	60	0.117
Marginal	103	250	0.412	103	250	0.412

where a_j and b_j are the numbers favoring the new health policy in each stress group. The marginal unadjusted analysis shows equal proportions favoring the new policy so that the unadjusted odds ratio is 1.0. However, when stratified by urban versus rural residence the following odds ratios and *MVLE* of the common odds ratio $\widehat{\theta}$ are obtained.

	Stratum		
Measure	*1*	*2*	$\widehat{\theta}$
$\widehat{\theta}_j = \log$ *Odds ratio*	1.365	1.126	1.270
$\widehat{V}(\widehat{\theta}_j)$	0.125	0.187	0.075
Odds ratio	3.917	3.085	3.559

Note that the n_{1j} and n_{2j} are reversed in the two strata so that there is a large group imbalance between strata. Among the urban subjects, 76% are stressed versus only 24% among the rural subjects, $X^2 = 135.2$ on 1 *df*, $p \leq 0.001$. Also, 57.6% of the urban subjects favored the new health policy versus 24.8% among rural subjects, $X^2 = 55.5$ on 1 *df*, $p \leq 0.001$. Comparing like-to-like within strata, the resulting odds ratios within the two strata are 3.917 and 3.085 so that the stratified-adjusted *MVLE* of the odds ratio is 3.559. The Mantel-Haenszel test statistic is also highly significant, $X^2_{MH} = 23.05$, $p \leq 0.0001$.

Thus we observe a strong positive association within strata, which yields a significant association when adjusted over strata, whereas marginally absolutely no association is seen. This is an illustration of a synergistic effect where the covariate adjustment enhances the estimate of the association between the independent and dependent variables.

4.5 MULTIVARIATE TESTS OF HYPOTHESES

4.5.1 Multivariate Null Hypothesis

The Mantel-Haenszel test is in fact a multivariate test of the joint multivariate null hypothesis of no association in any of the K strata versus a *restricted* alternative hypothesis that a non-zero common log odds ratio applies to all strata. However, there are other multivariate tests that could be constructed that would provide greater power against other alternative hypotheses.

In a stratified analysis with K strata, we wish to conduct a test for a vector of K association parameters $\boldsymbol{\theta} = (\theta_1 \ldots \theta_K)^T$. The $\{\theta_j\}$ can be measured on any scale of our choosing such as $\theta_j = G(\pi_{1j}, \pi_{2j})$ for some differentiable function $G(\cdot, \cdot)$. The vector of sample estimates is assumed to be asymptotically distributed as

$$\widehat{\boldsymbol{\theta}} \overset{d}{\approx} \mathcal{N}_K\left(\boldsymbol{\theta},\, \boldsymbol{\Sigma}_{\widehat{\boldsymbol{\theta}}}\right) \tag{4.46}$$

or

$$\begin{bmatrix} \widehat{\theta}_1 - \theta_1 \\ \widehat{\theta}_2 - \theta_2 \\ \vdots \\ \widehat{\theta}_K - \theta_K \end{bmatrix} \overset{d}{\approx} \mathcal{N}_K\left(\mathbf{0}, \boldsymbol{\Sigma}_{\widehat{\boldsymbol{\theta}}}\right) , \tag{4.47}$$

with a known (or consistently estimable) covariance matrix $\boldsymbol{\Sigma}_{\widehat{\boldsymbol{\theta}}} = V(\widehat{\boldsymbol{\theta}})$. For the case of K 2×2 tables, $\boldsymbol{\Sigma}_{\widehat{\boldsymbol{\theta}}} = diag(\sigma^2_{\widehat{\theta}_1} \ldots \sigma^2_{\widehat{\theta}_K})$, which is by definition positive definite. Under the null hypothesis H_0: $\pi_{1j} = \pi_{2j} = \pi_j$, let θ_0 designate the null value $\theta_j = \theta_0 = G(\pi_j, \pi_j)$ for all j. The equivalent null hypothesis in terms of the values of $\{\theta_j\}$ is

$$H_0\text{: } \theta_1 = \theta_2 = \ldots = \theta_K = \theta_0, \quad \text{or} \quad H_0\text{: } \boldsymbol{\theta} = \boldsymbol{J}\theta_0 , \tag{4.48}$$

which is a joint multivariate null hypothesis for the vector $\boldsymbol{\theta}$, where $\boldsymbol{J}$ is the unit vector as in (4.30). A variety of statistical tests can be employed to test H_0, each of which will be optimal against a specific type of alternative hypothesis.

4.5.2 Omnibus Test

The omnibus alternative hypothesis specifies that at least one of the elements of $\boldsymbol{\theta}$ differs from the null value, or

$$H_{1O}\text{: } \theta_j \neq \theta_0 \text{ for some } j, 1 \leq j \leq K, \quad \text{or} \quad H_{1O}\text{: } \boldsymbol{\theta} \neq \boldsymbol{J}\theta_0. \tag{4.49}$$

Thus H_0 specifies that there is no association between group and response in any table versus H_1 that there is some association in either direction in at least one table, that is, perhaps favoring either group 1 or group 2 for any table.

The asymptotically most powerful test of H_0 versus H_{1O} is the T^2-like test statistic attributed to Wald (1943)

$$X_O^2 = \left(\widehat{\boldsymbol{\theta}} - \boldsymbol{J}\theta_0\right)' \boldsymbol{\Sigma}_0^{-1} \left(\widehat{\boldsymbol{\theta}} - \boldsymbol{J}\theta_0\right) , \tag{4.50}$$

which uses the covariance matrix defined under the null hypothesis, $\boldsymbol{\Sigma}(\theta_0) = \boldsymbol{\Sigma}_0$, and is asymptotically distributed as χ^2_K under H_0. In practice, a consistent estimate of the covariance matrix $\widehat{\boldsymbol{\Sigma}}_0$ is employed.

To illustrate the nature of this test, consider the case of a bivariate statistic $(K = 2)$, or the case of two strata, where we wish to test H_0: $\theta_1 = \theta_2 = \theta_0 = 0$. Figure 4.2.A describes the null and alternative hypotheses (H_0 and H_{1O}) for this test. The alternative parameter space is *omnibus* or all-inclusive and consists of all points in the two-dimensional space other than the origin.

To describe the rejection region for the test, consider the case of a bivariate statistic vector with correlated elements where the elements of Σ_0 are

$$\boldsymbol{\Sigma}_0 = \frac{1}{N}\begin{bmatrix} 1 & 0.9428 \\ 0.9428 & 2 \end{bmatrix}$$

Fig. 4.2 The null (H_0) and alternative (H_1) hypotheses for the omnibus test with two parameters $\theta = (\theta_1, \theta_2)$ (A) and the corresponding rejection region (B).

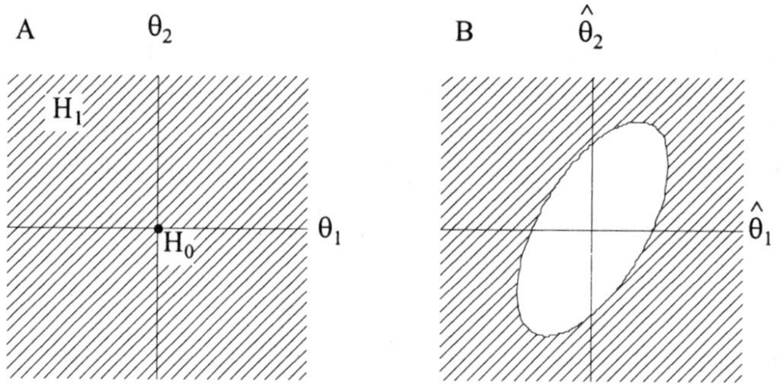

such that the correlation between $\widehat{\theta}_1$ and $\widehat{\theta}_2$ is 2/3, and the variance of $\widehat{\theta}_1$ is half that of $\widehat{\theta}_2$. In a stratified analysis, the estimates are independent, and thus uncorrelated. However, it is instructive to consider the more general case of correlated estimates as may apply to the analysis of repeated measures, as one example.

The rejection region for the omnibus test in this instance is shown in Figure 4.2.B for a sample of $N = 100$ and a Type I error probability of $\alpha = 0.05$. The rejection region for this test is defined by an ellipse that specifies all values of $\widehat{\theta}_1$ and $\widehat{\theta}_2$ such that the test statistic $X_O^2 = \widehat{\boldsymbol{\theta}}' \widehat{\Sigma}_0^{-1} \widehat{\boldsymbol{\theta}} = \chi^2_{2(0.95)} = 5.991$. The rejection ellipse is defined along the 45° axis of values for $\widehat{\theta}_1$ and $\widehat{\theta}_2$ such that the longevity of the ellipse is determined by the relative variances of $\widehat{\theta}_1$ and $\widehat{\theta}_2$ and the direction of their correlation (positive in this case). All points $\widehat{\theta}_1$ and $\widehat{\theta}_2$ interior to the ellipse lead to failure to reject the null hypothesis; those on or exterior to the ellipse lead to rejection of H_0 in favor of H_{1O} that a difference from zero of some magnitude in either direction exists for at least one of the values θ_1 or θ_2. Thus the statistic can lead to rejection of H_0 when $\widehat{\theta}_1$ is positive, say favoring group 2, and $\widehat{\theta}_2$ is negative, favoring group 1.

In the case of K stratified 2×2 tables, since Σ_0 is a diagonal matrix then the computation of the omnibus test X_O^2 in (4.50) simplifies to

$$X_O^2 = \sum\nolimits_j \frac{\left(\widehat{\theta}_j - \theta_0\right)^2}{\sigma_{0j}^2}, \tag{4.51}$$

where $\sigma_{0j}^2 = V(\widehat{\theta}_j|H_0)$. In this setting, $\theta_0 = 0$ for θ defined as the risk difference, the log relative risk or the log odds ratio.

When expressed in terms of the risk difference $\theta_j = \pi_{1j} - \pi_{2j}$, then under H_0: $\theta_0 = 0$,

$$V_{0j} = \sigma_{0j}^2 = \pi_j\left(1 - \pi_j\right)\left(\frac{1}{n_{1j}} + \frac{1}{n_{2j}}\right). \tag{4.52}$$

Based on the consistent estimate $\widehat{\theta}_j = p_{1j} - p_{2j}$, then the variance is consistently estimated as

$$\widehat{V}_{0j} = \widehat{\sigma}_{0j}^2 = p_j\left(1 - p_j\right)\left(\frac{1}{n_{1j}} + \frac{1}{n_{2j}}\right) \tag{4.53}$$

with $p_j = m_{1j}/N_j$ (see Section 2.3.1). Using the simpler notation $\widehat{V}_{0j}$ to refer to $\widehat{\sigma}_{0j}^2$,

$$X_O^2 = \sum_j \frac{\left(p_{1j} - p_{2j}\right)^2}{\widehat{V}_{0j}} = \sum_j X_j^2\,, \tag{4.54}$$

where X_j^2 is the Pearson contingency (Cochran) X^2 value for the jth stratum, which is asymptotically distributed as χ_K^2 on K *df*.

In Problem 2.9 we showed that the Z-test (and thus the chi-square test) for a single 2×2 table is asymptotically invariant to the choice of scale. Thus the omnibus test using risk differences, log relative risks or log odds ratios are all asymptotically equivalent. In practice, therefore, the test is usually computed only in terms of the risk differences.

4.5.3 Bonferroni Inequality

Another approach to a simultaneous test of the K-parameter joint null hypothesis H_0 is to use the Bonferroni inequality, which is based on the first term in the Bonferroni expansion for the probability of a joint "event" such that

$$P(\text{at keast one of } K \text{ tests significant at level } \alpha'\ |H_0) \leq \alpha' K.$$

Therefore, to achieve a total Type I error probability no greater than the desired level α, we would use the significance level $\alpha' = \alpha/K$ for each of the K separate tests of significance. Likewise, to construct K confidence intervals with the desired coverage probabilities no less than the desired confidence level $1 - \alpha$, the K confidence limits would be constructed using the confidence level $(1 - \alpha')$. For example, for $K = 4$ we would use $\alpha' = 0.05/4 = 0.0125$ to conduct each of the tests of significance and to compute the confidence limits.

A variety of less conservative adjustments for multiple tests of significance have been proposed, such as the procedures of Holm (1979) and Hochberg (1988) and the general family of closed multiple comparison procedures of Marcus, Peritz and Gabriel (1976). All of these conduct the multiple tests after ordering the test statistics from largest to smallest and then applying less restrictive significance levels to the second, third, and so on, test conducted. For example, the Holm procedure requires a significance level of α/K for the first test (maximum statistic, minimum P-value), $\alpha/(K-1)$ for the second largest test statistic, $\alpha/(K-2)$ for the third largest test statistic, and so on. When any one test is not significant, the procedure stops and all further tests are also declared non-significant.

All of these procedures are still conservative compared to the T^2-type multivariate test. However, these adjustments can be applied in instances where the T^2-test does not apply, such as for a K-statistic vector that is not normally distributed.

Example 4.8 *Religion and Mortality (continued)*
For the four strata in Example 4.6, the multivariate omnibus test in terms of the risk differences $\theta_j = \pi_{1j} - \pi_{2j}$ provides a test of

$$H_0\colon \begin{pmatrix} \theta_1 \\ \theta_2 \\ \theta_3 \\ \theta_4 \end{pmatrix} = \begin{pmatrix} 0 \\ 0 \\ 0 \\ 0 \end{pmatrix}$$

against the alternative

$$H_{1O}\colon \theta_j \neq 0 \text{ for some } 1 \leq j \leq 4.$$

The within-stratum values of the contingency (Cochran) 1 *df* X^2 tests are: 0.00419, 2.98042, 0.13571 and 1.84539. Thus the omnibus test is $X_O^2 = \sum_j X_j^2 = 4.8436$ on 4 *df*, with $p \leq 0.304$ that is not statistically significant. This is in distinct contrast to the significant unadjusted analysis and the nearly significant stratified-adjusted Mantel-Haenszel analysis.

Example 4.9 *Clinical Trial in Duodenal Ulcers (continued)*
Similarly, for Example 4.1, the omnibus test is $X_O^2 = 0.18299 + 2.03606 + 5.40486 = 7.6239$ on 3 *df* with $p \leq 0.0545$ that approaches significance at the 0.05 level. The P-value for this omnibus test is smaller than that for the unadjusted and Mantel-Haenszel stratified-adjusted test of a common odds ratio.

4.5.4 Partitioning of the Omnibus Alternative Hypothesis

In general, the omnibus test and the Mantel-Haenszel test differ because of the difference between the alternative hypotheses of each test and the corresponding rejection regions. The omnibus test is directed toward any point θ in the parameter space $\boldsymbol{R}^K$ away from the origin. In many analyses, however, it may be of scientific interest to use a test directed toward a restricted alternative hypothesis represented by a sub-region of $\boldsymbol{R}^K$. In this case, the omnibus test will not be as powerful as a test of such a restricted alternative, when that alternative is true.

The Mantel-Haenszel test is often called a test of no partial association, or just association, because it is directed to such a restricted alternative hypothesis. To show how this test relates to the omnibus test, and as a prelude to developing other tests of the joint null hypothesis H_0 in (4.48), it is instructive to show that the null hypothesis can be partitioned into the intersection of two sub-hypotheses as follows:

Omnibus		*Homogeneity*		*No Partial Association*
$H_0\colon \boldsymbol{\theta} = \boldsymbol{J}\theta_0$ or $\theta_j = \theta_0\ \forall j$	$\equiv$	$H_{0H}\colon \theta_j = \theta \quad \forall j$	*and*	$H_{0A}\colon \theta = \theta_0$
K *df*	$=$	$K-1$ *df*	$+$	1 *df*
$H_1\colon \boldsymbol{\theta} \neq \boldsymbol{J}\theta_0$	$\equiv$	$H_{1H}\colon \theta_j \neq \theta_\ell,\ j \neq \ell$	*or*	$H_{1A}\colon \theta_j = \theta \neq \theta_0\ \forall j$

(4.55)

The omnibus joint null hypothesis H_0 in (4.48) states that the K components of $\boldsymbol{\theta}$ are all equal to θ_0 and thus the test has K degrees of freedom. The null hypothesis of *homogeneity* H_{0H} specifies that the components of $\boldsymbol{\theta}$ share a common value $\theta \neq \theta_0$. This, in turn, implies that we can define $K-1$ independent contrasts among the values of $\boldsymbol{\theta}$, each of which has expectation zero under H_{0H}, and thus this test has $K-1$ *df.* The null hypothesis of no *partial association* H_{0A} specifies that the assumed common value for all the components of $\boldsymbol{\theta}$ equals the null hypothesis value. Clearly $H_0 \equiv H_{0H} \cap H_{0A}$. In terms of a linear model, such as a simple ANOVA, the hypothesis of homogeneity H_{0H} corresponds to the hypothesis of no group by stratum interaction effect, and the hypothesis of no association H_{0A} corresponds to the hypothesis of no overall group effect.

For example, for $\theta = \log(\text{odds ratio})$, the null hypothesis for the Mantel-Haenszel test in (4.13) can be formally expressed as

$$\begin{aligned} H_{0(MH)}&\colon (\theta_1 = \ldots = \theta_K = \theta = \theta_0 = 0) \\ &\equiv (\theta_1 = \ldots = \theta_K = \theta) \cap (\theta = \theta_0 = 0) \\ &\equiv H_{0H} \cap H_{0A}. \end{aligned} \tag{4.56}$$

Therefore, the Mantel-Haenszel test assumes that the hypothesis H_{0H} of homogeneity is true, in which case the test is maximally efficient (as we shall see shortly).

The general alternative hypothesis, that the components do not equal the null hypothesis value for all strata, then is the union of the alternative hypothesis of some heterogeneity among the components, or the assumed common component being different from the null value. If there is heterogeneity of the odds ratios in the population, meaning that $\theta_j \neq \theta_\ell$ for some $j \neq \ell$ so that H_{0H} is false, then the Mantel-Haenszel test might be viewed as testing the hypothesis that the *average* of the stratum-specific log odds ratios is zero, or solely testing the hypothesis H_{0A}: $\theta = \theta_0 = 0$. In this case, however, the lack of homogeneity in the stratum-specific parameters will reduce the power of the stratified-adjusted Mantel-Haenszel test as a test of H_{0A} alone. In fact, when H_{0H} is false and there is heterogeneity of the odds ratios over strata in the population, then the Mantel-Haenszel test is no longer appropriate. Alternative models in this case are described later in Section 4.10.

Fig. 4.3 The null (H_0) and alternative (H_1) hypotheses for the test of homogeneity with two parameters $\theta = (\theta_1,\ \theta_2)$ (A) and the corresponding rejection region (B).

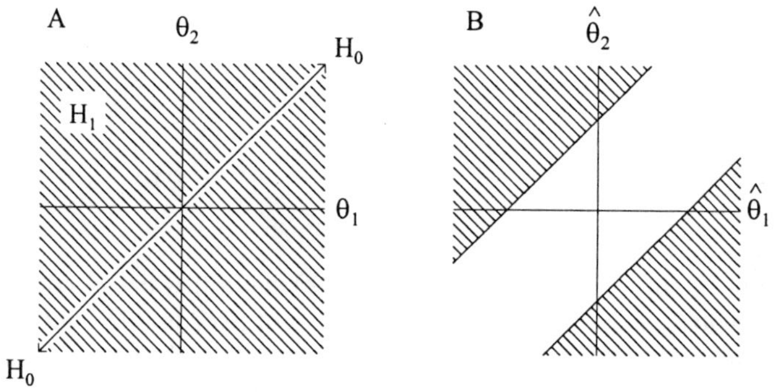

4.6 TESTS OF HOMOGENEITY

A variety of methods are available to test the hypothesis of homogeneity among the measures of association $\{\theta_j\}$ on some scale $\theta_j = G(\pi_{1j}, \pi_{2j})$

$$H_{0H}\colon\ \theta_1 = \theta_2 = \ldots = \theta_K \tag{4.57}$$

against the alternative that there is a difference between at least two strata

$$H_{1H}\colon\ (\theta_j \neq \theta_\ell)\ \text{ some } 1 \leq j < \ell \leq K. \tag{4.58}$$

Note that in this case, it is irrelevant whether $\theta_j = \theta_0$ for any or all strata.

For the case of only two strata, these hypotheses are depicted in Figure 4.3.A. The null hypothesis is that the parameter values correspond to one of the possible points that lie on the line of equality. The alternative hypothesis is that the values lie somewhere in the two-dimensional space away from this line of equality.

4.6.1 Contrast Test of Homogeneity

The null hypothesis of homogeneity H_{0H} implies that the difference between any two strata is zero, $(\theta_j - \theta_\ell) = 0$ for all $j \neq \ell$. Thus it is possible to define $K - 1$ orthogonal contrasts among the sample estimates using a $K \times (K - 1)$ contrast matrix $\boldsymbol{C}$ of rank $K-1$ to test the hypothesis H_{0C}: $\boldsymbol{C}'\boldsymbol{\theta}=\mathbf{0}$ where $\boldsymbol{C}'\boldsymbol{\theta}$ is a $1\times(K-1)$ vector. A variety of such contrast matrices could be defined such as the contrast matrix of successive differences $\theta_j-\theta_{j+1}$ for $j=1,\ldots,(K-1)$

$$\boldsymbol{C}'=\begin{bmatrix} 1 & -1 & 0 & \cdots & 0 & 0 \\ 0 & 1 & -1 & \cdots & 0 & 0 \\ \vdots & \vdots & \vdots & \vdots & \vdots & \vdots \\ 0 & 0 & 0 & \cdots & 1 & -1 \end{bmatrix} \tag{4.59}$$

or such as each stratum versus the last $\theta_j-\theta_K$

$$\boldsymbol{C}'=\begin{bmatrix} 1 & 0 & 0 & \cdots & 0 & -1 \\ 0 & 1 & 0 & \cdots & 0 & -1 \\ \vdots & \vdots & \vdots & \vdots & \vdots & \vdots \\ 0 & 0 & 0 & \cdots & 1 & -1 \end{bmatrix} \tag{4.60}$$

or such as each stratum versus the simple average value over all strata $\theta_j-\theta$,

$$\boldsymbol{C}'=\frac{1}{K}\begin{bmatrix} K-1 & -1 & \cdots & -1 & -1 \\ -1 & K-1 & \cdots & -1 & -1 \\ \vdots & \vdots & \vdots & \vdots & \vdots \\ -1 & -1 & \cdots & K-1 & -1 \end{bmatrix} \tag{4.61}$$

where

$$\theta_j-\theta=\theta_j-\frac{\sum_\ell \theta_\ell}{K}=\frac{K\theta_j-\sum_\ell\theta_\ell}{K}\propto(K-1)\,\theta_j-\sum_{\ell\neq j}\theta_\ell . \tag{4.62}$$

Then, for any such contrast matrix

$$H_{0H}\Leftrightarrow H_{0C}\colon\ \boldsymbol{C}'\boldsymbol{\theta}=\mathbf{0} \tag{4.63}$$

$$H_{1H}\Leftrightarrow H_{1C}\colon\ \boldsymbol{C}'\boldsymbol{\theta}\neq\mathbf{0}\,.$$

Since $\widehat{\boldsymbol{\theta}}$ is asymptotically normally distributed as in (4.46), then the test of homogeneity is provided by the T^2-like Wald statistic defined as the quadratic form

$$X_H^2=\left(\boldsymbol{C}'\widehat{\boldsymbol{\theta}}\right)'\left(\boldsymbol{C}'\widehat{\boldsymbol{\Sigma}}_{\widehat{\boldsymbol{\theta}}}\boldsymbol{C}\right)^{-1}\boldsymbol{C}'\widehat{\boldsymbol{\theta}}\,, \tag{4.64}$$

which is asymptotically distributed as χ^2_{K-1} on $K-1$ *df.* Note that the test is computed using the estimate of the covariance matrix defined under the general hypothesis of homogeneity, with no restrictions that the parameter values equal θ_0, unlike the omnibus test statistic presented in (4.50).

The value of the test statistic is the same for all such contrast matrices that are of rank $K-1$ and that satisfy $\boldsymbol{C}'\boldsymbol{J}=\mathbf{0}$ under H_{0H}, where $\boldsymbol{J}$ is the unit vector (*cf.* Anderson, 1984, p. 170). This test can be computed directly using a custom routine programmed in a matrix language such as SAS PROC IML, or by using the SAS procedure PROC CATMOD with an appropriate response function. PROC CATMOD uses the method of weighted least squares described by Grizzle, Starmer and Koch (1969) to fit models describing the response probabilities associated with covariate values for each sub-population (stratum). The test obtained from the non-iterative weighted least squares fit is algebraically equivalent to that above for the same response function. For example, if θ is defined as the log relative risk, then the equivalent response function in the GSK procedure (PROC CATMOD) is the log of the probability. The test results then are identical.

For the case of two correlated measures such as those described in Figure 4.2.B, the test statistic X_H^2 in (4.64) using the successive difference contrast matrix (4.59) reduces to $X_H^2 = \left(\widehat{\theta}_1 - \widehat{\theta}_2\right)^2 / \widehat{V}\left(\widehat{\theta}_1 - \widehat{\theta}_2\right)$, where $\widehat{V}\left(\widehat{\theta}_1 - \widehat{\theta}_2\right) = \widehat{\sigma}^2_{\widehat{\theta}_1} + \widehat{\sigma}^2_{\widehat{\theta}_2} - 2\widehat{Cov}\left(\widehat{\theta}_1, \widehat{\theta}_2\right)$. Thus the null hypothesis reduces to H_{0H}: $\theta_1 = \theta_2$, in which case the rejection region is defined as all points $\left(\widehat{\theta}_1, \widehat{\theta}_2\right)$ for which $X_H^2 = X^2_{1(1-\alpha)}$ or for which $\left(\widehat{\theta}_1 - \widehat{\theta}_2\right) = \sqrt{3.841 \times \widehat{V}\left(\widehat{\theta}_1 - \widehat{\theta}_2\right)}$, 3.841 being the $95th$ percentile of the 1 *df* chi-square distribution. For the bivariate example, the rejection region is defined by the parallel lines shown in Figure 4.3.B for which the observations leading to rejection of H_{0H} principally fall in Quadrants II and IV, wherein the values of $\widehat{\theta}_1$ and $\widehat{\theta}_2$ differ in sign.

4.6.2 Cochran's Test of Homogeneity

Cochran (1954b) proposed a test of homogeneity for odds ratios based on the sums of squares of deviations of the stratum specific odds ratios about the mean odds ratio. This idea can be generalized to a test for homogeneity on any scale. To obtain the *MVLE* of the assumed common value for measures of association $\{\theta_j\}$ on some scale $\theta_j = G(\pi_{1j}, \pi_{2j})$ we used weights inversely proportional to the variances as described in (4.33) through (4.36). Under the hypothesis of homogeneity H_{0H} in (4.57), the stratum-specific estimate for the jth stratum is asymptotically distributed as

$$\left(\widehat{\theta}_j - \theta\right) \overset{d}{\approx} N\left(0, \sigma^2_{\widehat{\theta}_j}\right) \tag{4.65}$$

for $1 \leq j \leq K$. Since the hypothesis of homogeneity does not require that $\pi_{1j} = \pi_{2j}$, the variance of $\widehat{\theta}_j$ is evaluated assuming $\pi_{1j} \neq \pi_{2j}$ as described in Section 2.3.

For now assume that the variances, and thus their inverse ($\tau_j = \sigma^{-2}_{\widehat{\theta}_j}$) are known. From (4.37), since $\widehat{\theta} \xrightarrow{p} \theta$ then asymptotically, from Slutsky's Theorem (A.44),

$$\sqrt{\tau_j}\left(\widehat{\theta}_j - \widehat{\theta}\right) \xrightarrow{p} \sqrt{\tau_j}\left(\widehat{\theta}_j - \theta\right) \overset{d}{\approx} \mathcal{N}(0, 1). \tag{4.66}$$

Further, since $\widehat{\sigma}^2_{\widehat{\theta}_j} \xrightarrow{p} \sigma^2_{\widehat{\theta}_j}$ and $\widehat{\tau}_j \xrightarrow{p} \tau_j$, then

$$\widehat{\tau}_j \left(\widehat{\theta}_j - \widehat{\theta}\right)^2 \overset{d}{\approx} \chi^2_1. \tag{4.67}$$

Therefore,

$$X^2_{H,C} = \sum\nolimits_j \widehat{\tau}_j \left(\widehat{\theta}_j - \widehat{\theta}\right)^2 \tag{4.68}$$

is distributed asymptotically as chi-square on $K - 1$ *df.* The degrees of freedom are $K - 1$ because we estimate θ as a linear combination of the K stratum-specific estimates. Algebraically, Cochran's test is equivalent to the contrast test X^2_H in (4.64), which, in turn, is equivalent to the GSK test obtained from the SAS PROC CATMOD.

Example 4.10 *Clinical Trial in Duodenal Ulcers (continued)*
For the ulcer drug clinical trial example, Cochran's test of homogeneity for the odds ratios is based on the elements

Stratum	$\widehat{\theta}_j = \log(\widehat{OR}_j)$	$\widehat{\tau}_j = \widehat{\sigma}^{-2}_{\widehat{\theta}_j}$
1	−0.1854	5.31919
2	1.3218	1.11801
3	1.0015	5.27749

and $X^2_{H,W} = 4.5803$ on 2 *df* with $p \leq 0.102$. The tests of homogeneity for the three principal measures of association, each on 2 *df*, are

Measure	$X^2_{H,W}$	$p \leq$
log Risk differences	4.797	0.091
log Relative risks	3.648	0.162
log Odds ratios	4.580	0.102

For this example, based on the relative values of the test statistics, there is the least heterogeneity among the relative risks, and the most heterogeneity among the risk differences. However, the difference in the extent of heterogeneity among the three scales is slight.

Example 4.11 *Religion and Mortality (continued)*
For the study of religion versus mortality, the tests of homogeneity on 3 *df* for each of the three scales likewise are

Measure	$X^2_{H,W}$	$p \leq$
Risk differences	3.990	0.263
Relative risks	0.929	0.819
Odds ratios	1.304	0.728

In this example, there is a some heterogeneity among the risk differences, but far from significant, and almost none among the relative risks or odds ratios.

4.6.3 Zelen's Test

A computationally simple test of homogeneity was proposed by Zelen (1971). Let X_A^2 designate a test of the hypothesis H_{0A} of no partial association (or just association) for the average measure of association among the K strata, such as the Cochran-Mantel-Haenszel test on 1 *df* of the common odds ratio. Since the omnibus null hypothesis can be partitioned as shown in (4.55), Zelen (1971) proposed that the test of homogeneity

$$X_{H,Z}^2 = X_O^2 - X_A^2 \tag{4.69}$$

be obtained as the difference between the omnibus chi-square test X_O^2 on K *df*, and the test of association X_A^2 on 1 *df*. For the ulcer clinical trial example using the conditional Mantel-Haenszel test yields $X_{H,Z}^2 = 7.4648 - 3.00452 = 4.46$, which is slightly less than the Cochran test value $X_{H,C}^2 = 4.58$.

Mantel, Brown, and Byar (1977) and Halperin, Ware, Byar, et al. (1977) criticize this test and present examples that show that this simple test may perform poorly in some situations. The problem is that an optimal test of the null hypothesis of homogeneity H_{0H} should use the variances estimated under that hypothesis. Thus both the contrast test X_H^2 in (4.64) and Cochran's test $X_{H,C}^2$ in (4.68) use the variances $\widehat{\Sigma}_{\widehat{\theta}}$ estimated under the general alternative hypothesis H_{1O} in (4.49) that some of the $\{\theta_j\}$, if not all, differ from the null value θ_0. However, both X_A^2 and X_O^2 use the variances defined under the general null hypothesis H_0, Σ_0, so that equality does not hold, that is, $X_O^2 \neq X_H^2 + X_A^2$. In general, therefore, this test should be avoided.

4.6.4 Breslow-Day Test for Odds Ratios

Breslow and Day (1980) also suggested a test of homogeneity for odds ratios for use with a Mantel-Haenszel test that is based on the Mantel-Haenszel estimate of the common odds ratio $\widehat{OR}_{MH}$ in (4.14). In the jth stratum, given the margins for that 2×2 table $(m_{1j}, m_{2j}, n_{1j}, n_{2j})$ then the expectation of the index frequency a_j under the hypothesis of homogeneity $OR_j = OR$ can be estimated as

$$\widehat{E}(a_j|\widehat{OR}_{MH}) = \widetilde{a}_j \text{ such that } OR_j = \widehat{OR}_{MH}. \tag{4.70}$$

This expected frequency is the solution to

$$\frac{(\widetilde{a}_j)(n_{2j} - m_{1j} + \widetilde{a}_j)}{(m_{1j} - \widetilde{a}_j)(n_{1j} - \widetilde{a}_j)} = \widehat{OR}_{MH} \tag{4.71}$$

(see also Problem 2.7.2). Solving for $\widetilde{a}_j$ yields

$$m_{1j} n_{1j} \widehat{OR}_{MH} = \tag{4.72}$$

$$\widetilde{a}_j \left[n_{2j} - n_{1j} + \widehat{OR}_{MH}(n_{1j} + m_{1j}) \right] + \widetilde{a}_j^2 \left(1 - \widehat{OR}_{MH} \right),$$

which is a quadratic function in $\widetilde{a}_j$. The root such that $0 < \widetilde{a}_j \leq \min(n_{1j}, m_{1j})$ yields the desired estimate. Given the margins of the table $(n_{1j}, n_{2j}, m_{1j}, m_{2j})$ the expected values of the other cells of the table are obtained by subtraction, such as $\widetilde{b}_j = m_{1j} - \widetilde{a}_j$.

Then the Breslow-Day test of homogeneity of odds ratios is a Pearson contingency test of the form

$$X^2_{H,BD} = \sum_{j=1}^{K} \left[\frac{(a_j - \widetilde{a}_j)^2}{\widetilde{a}_j} + \frac{(b_j - \widetilde{b}_j)^2}{\widetilde{b}_j} + \frac{(c_j - \widetilde{c}_j)^2}{\widetilde{c}_j} + \frac{(d_j - \widetilde{d}_j)^2}{\widetilde{d}_j} \right]. \quad (4.73)$$

Since the term in the numerator is the same for each cell of the jth stratum, for example, $(b_j - \widetilde{b}_j)^2 = (a_j - \widetilde{a}_j)^2$, this statistic can be expressed as

$$X^2_{H,BD} = \sum_{j=1}^{K} \left[\frac{(a_j - \widetilde{a}_j)^2}{\widehat{V}(a_j|\widehat{OR}_{MH})} \right] \quad (4.74)$$

where

$$\widehat{V}(a_j|\widehat{OR}_{MH}) = \left[\frac{1}{\widetilde{a}_j} + \frac{1}{\widetilde{b}_j} + \frac{1}{\widetilde{c}_j} + \frac{1}{\widetilde{d}_j} \right]^{-1}. \quad (4.75)$$

This test for homogeneity of odds ratios is used in SAS PROC FREQ as part of the Cochran-Mantel-Haenszel analysis with the CMH option. For the data from Example 4.1, this test yields the value $X^2_{H,BD} = 4.626$ on 2 *df* with $p \leq 0.099$ and for Example 4.6, this test value is $X^2_{H,BD} = 1.324$ on 3 *df* with $p \leq 0.7234$. In both cases, the test value is slightly larger than the Cochran test value, the *P*-values smaller.

Breslow and Day (1980) suggested that $X^2_{H,BD}$ is distributed asymptotically as χ^2_{K-1} on $K-1$ *df.* Tarone (1985) showed that this would be the case if a fully efficient estimate of the common odds ratio, such as the *MLE*, were used as the basis for the test. Since the Mantel-Haenszel estimate is not fully efficient, then $X^2_{H,BD}$ is stochastically larger than a variate distributed as χ^2_{K-1}. Tarone also showed that a corrected test can be obtained as

$$X^2_{H,BD,T} = X^2_{H,BD} - \frac{\left(\sum_{j=1}^{K} a_j - \sum_{j=1}^{K} \widetilde{a}_j\right)^2}{\sum_{j=1}^{K} \widehat{V}(a_j|\widehat{OR}_{MH})} \quad (4.76)$$

which is asymptotically distributed as χ^2_{K-1} on $K-1$ *df.* Breslow (1996) recommends that in general the corrected test should be preferred to the original Breslow-Day test, but also points out that the correction term is often negligible. For the data in Example 4.1, the corrected test value $X^2_{H,BD,T} = 4.625$ on 2 *df* with $p \leq 0.100$; and for Example 4.6 $X^2_{H,BD,T} = 1.3236$ on 3 *df* with $p \leq 0.7235$; in both cases, nearly identical to the original Breslow-Day test.

4.7 EFFICIENT TESTS OF NO PARTIAL ASSOCIATION

From the partitioning of hypotheses presented in (4.55), the global null hypothesis in (4.48) has two components, that the measures $\{\theta_j\}$ among strata are all equal$\{\theta_j = \theta\}$ (homogeneity) and that they are all equal to the null value $\{\theta = \theta_0\}$ (association). Clearly, if the hypothesis of homogeneity is rejected, then so also is the global null hypothesis because if $\theta_j \neq \theta_\ell$ for any two strata j and ℓ then both can not also equal θ_0. However, the hypothesis of homogeneity could be satisfied when all strata have a common measure that differs from the null value. Thus we desire an efficient test that is directed specifically toward the association hypothesis under the assumption of homogeneity.

4.7.1 Restricted Alternative Hypothesis of Association

The principal disadvantage of omnibus T^2-like test of the joint null hypothesis for the K strata is that it is directed to the global alternative hypothesis that a difference of some magnitude in either direction exists for at least one strata. Usually, however, one is interested in declaring that an overall difference of some magnitude exists over all strata, or on average in some sense in the overall population. Thus a statistically significant omnibus test may not be scientifically relevant or clinically compelling.

Alternately a test directed against the restricted alternative hypothesis of a common difference on some scale for all strata has more scientific appeal, and will be more powerful than the omnibus test to detect a common difference when such is the case. The null and alternative hypotheses in this case are

$$\begin{aligned} H_{0A}: \ \theta_1 = \theta_2 = \ldots = \theta_K = \theta = \theta_0 \\ H_{1A}: \ \theta_1 = \theta_2 = \ldots = \theta_K = \theta \neq \theta_0. \end{aligned} \tag{4.77}$$

The null hypothesis is the same as that for the K *df* omnibus test, however this test has a restricted alternative hypothesis that a constant difference $\theta \neq \theta_0$ exists on some scale. Thus $H_{1A} \subset H_{1O}$ for the omnibus test.

For the case of a bivariate analysis as described in Figure 4.2 for the omnibus test, Figure 4.4.A depicts the null and alternative hypothesis for the test of association. As for the omnibus test, the null hypothesis corresponds to the origin in the two-dimensional parameter space. The alternative hypothesis, however, consists of all points falling on the positive and negative projections corresponding to the line of equality $\theta_1 = \theta_2$. Points in the parameter space for which $\theta_1 \neq \theta_2$ are not of interest. Thus when H_{1A} is true for θ defined on some scale $G(\pi_1, \pi_2)$, this test will have greater power than the K *df* omnibus test.

For illustration, as shown in (4.20), the Mantel-Haenszel test can be viewed as an estimation-based test using an estimate of the common value of the parameter $\widehat{\theta}$ and an estimate of its variance under H_0, $\widehat{\sigma}^2_{\widehat{\theta}|H_0}$. In such a construction, $\widehat{\theta}$ would

Fig. 4.4 The null (H_0) and alternative (H_1) hypotheses for the test of association with two parameters $\theta = (\theta_1,\ \theta_2)$ (A) and the corresponding rejection region (B).

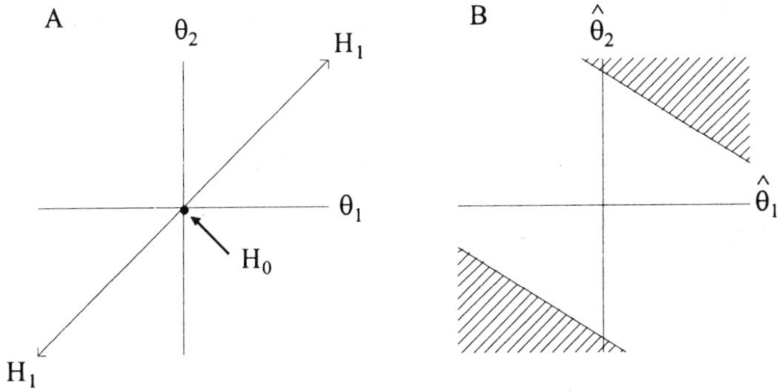

be a linear function of the $\{\widehat{\theta}_j\}$ and the test could be expressed as

$$Z_A = \frac{\widehat{\theta}}{\widehat{\sigma}_{\widehat{\theta}|H_0}} = \sum\nolimits_j a_j \widehat{\theta}_j \ , \tag{4.78}$$

where asymptotically Z_A is distributed as $N(0, 1)$. As with the *MVLE*, the estimate $\widehat{\theta}$ would be obtained as the weighted average of the $\{\theta_j\}$ using weights $\{a_j\}$ inversely proportional to the variances, in which case the variances would be estimated under the null hypothesis $\{\widehat{\sigma}^2_{0j}\}$. Thus the elements of $\{\widehat{\theta}_j\}$ with smaller variance would receive greater weight.

For the bivariate example in Figure 4.2.B, where the two sample statistics are correlated, a similar statistic $Z_A = a_1\widehat{\theta}_1 + a_2\widehat{\theta}_2$ is employed. (Because of the correlation, the actual weights are slightly different from those with independent strata.) Setting $Z_A = \pm Z_{1-\alpha/2} = \pm 1.96$ for $\alpha = 0.05$, we can solve for the values of $\widehat{\theta}_1$ and $\widehat{\theta}_2$ that satisfy this equality to determine the lines defining the upper and lower rejection regions. Figure 4.4.B shows the corresponding rejection regions for this bivariate example. These rejection regions contain observations principally in Quadrants I and III, and even admit cases in Quadrants II and IV where either statistic $\widehat{\theta}_1$ or $\widehat{\theta}_2$ is quite distant from zero. When the bivariate statistics $\widehat{\theta}_1$ and $\widehat{\theta}_2$ have equal variances, or are assigned equal weight, then the lines defining the rejection region lie perpendicular to the projection of the alternative hypothesis. However, when the weights differ, as in this example, then the line of rejection is tilted toward the origin for the sample statistics that receive the greater weight or which have the smaller variance.

4.7.2 Radhakrishna Family of Efficient Tests of Association

We now show that the Cochran and Mantel-Haenszel tests are such tests of association under the hypothesis that a common odds ratio applies to all K strata. These tests were derived rather heuristically, although Cochran does provide a justification for his test.

Radhakrishna (1965) considered the problem more generally from the perspective of deriving a family of asymptotically fully efficient tests of H_{0A} versus H_{1A} for multiple 2×2 tables with measures of association $\{\theta_j\}$ on some scale $\theta = g(\pi_1) - g(\pi_2)$ such that H_0: $\pi_{1j} = \pi_{2j} = \pi_j$ implies, and is implied by

$$H_{0A(g)}\colon\ g(\pi_{1j}) - g(\pi_{2j}) = \theta_0 = 0. \tag{4.79}$$

Specifically we desire an asymptotically efficient test of H_0 against a restricted alternative hypothesis of the form

$$H_{1A(g)}\colon\ g(\pi_{1j}) = g(\pi_{2j}) = \theta \text{ for } \forall j \tag{4.80}$$

or that there is a constant difference θ on the specified scale $g(\pi)$. This family includes a test of a common log odds ratio for which $g(\pi) = \log[\pi/(1-\pi)]$ (the logit), a test of a common log relative risk for which $g(\pi) = \log(\pi)$, and a test of a common risk difference for which $g(\pi) = \pi$ (the identity function), among others. Radhakrishna's initial intent was to explore the asymptotic efficiency of Cochran's stratified-adjusted test, X_U^2, but in doing so he described a much broader family of tests.

In Problem 4.1 it is readily shown that the Cochran test of association in (4.12), using the unconditional variance, can be expressed as

$$X_{U,C}^2 = \frac{\left[\sum_j \widehat{w}_j (p_{1j} - p_{2j})\right]^2}{\sum_j \widehat{w}_j^2 \widehat{V}_{0j}} \tag{4.81}$$

with $\widehat{V}_{0j} = \widehat{\sigma}_{0j}^2 = \widehat{V}(p_{1j} - p_{2j}|H_0)$ as presented in (4.53) and where

$$\widehat{w}_j = \left(\frac{1}{n_{1j}} + \frac{1}{n_{2j}}\right)^{-1} = \frac{n_{1j}n_{2j}}{N_j}. \tag{4.82}$$

It is not obvious, however, that these weights are optimal in any statistical sense. Note that these weights $\{w_j\}$ differ from the weights $\{\omega_j\}$ used earlier to obtain the *MVLE* of the common log odds ratio.

Radhakrishna (1965) proposed that the weights be derived so as to maximize the asymptotic efficiency of the test. To a first-order Taylor's approximation about an intermediate value $\pi_j \in (\pi_{1j}, \pi_{2j})$ in the jth stratum under a local alternative it can be shown (see Problem 4.4) that asymptotically

$$\theta_j \cong g'(\pi_j)(\pi_{1j} - \pi_{2j}) \tag{4.83}$$

so that

$$(\pi_{1j} - \pi_{2j}) \cong \theta_j / g'(\pi_j). \tag{4.84}$$

Under H_{1A}: $\theta_j = \theta$ for all strata this implies that a test based on a weighted average of the $\{p_{1j} - p_{2j}\}$ should be asymptotically fully efficient to test H_{0A} versus H_{1A} above.

For a measure of association θ corresponding to a given scale with function $g(\pi)$, we then desire an asymptotically most powerful or fully efficient test statistic of the form

$$T = \sum\nolimits_j w_j (p_{1j} - p_{2j}). \tag{4.85}$$

The asymptotic variance of the statistic evaluated under the null hypothesis is $V(T|\theta_0) = \sum_j w_j^2 \sigma_{0j}^2$ where $\sigma_{0j}^2 = V[(p_{1j} - p_{2j})|H_0]$ as defined in (4.52).

From (3.57) the Pitman efficiency of such a test statistic is

$$Eff(T) = \left[\left(\frac{dE(T)}{d\theta} \right)^2 / V(T|\theta) \right]_{|\theta=\theta_0}. \tag{4.86}$$

Thus we seek the expression for the weights $\{w_j\}$ that will maximize the efficiency of the test for a given scale $g(\pi)$. From the first-order approximation in (4.84) under the restricted alternative hypothesis that $\theta_j = \theta \; \forall j$, it follows that asymptotically the expected value of the statistic is

$$E(T|\theta) = \sum\nolimits_j w_j (\pi_{1j} - \pi_{2j}) \cong \sum\nolimits_j \frac{w_j \theta_j}{g'(\pi_j)} = \theta \sum\nolimits_j \frac{w_j}{g'(\pi_j)}. \tag{4.87}$$

Therefore,

$$\frac{dE(T)}{d\theta} = \sum\nolimits_j w_j g'(\pi_j)^{-1} \tag{4.88}$$

and

$$Eff(T) = \frac{\left[\sum_j w_j g'(\pi_j)^{-1}\right]^2}{\sum_j w_j^2 \sigma_{0j}^2}. \tag{4.89}$$

To obtain the set of weights for which the $Eff(T)$ is maximized it is convenient to express this result in matrix terms. Let $\mathbf{W} = (w_1 \ldots w_K)^T$, $\mathbf{G} = \left[g'(\pi_1)^{-1} \ldots g'(\pi_K)^{-1}\right]^T$, and $\mathbf{\Sigma}_0 = diag\left(\sigma_{01}^2 \ldots \sigma_{0K}^2\right)$. Then

$$Eff(T) = \frac{\left(\mathbf{W}^T \mathbf{G}\right)^2}{\mathbf{W}^T \mathbf{\Sigma}_0 \mathbf{W}}. \tag{4.90}$$

The value of w for which the *Eff*(T) is maximized can be obtained from a basic theorem on the extrema of quadratic forms that follows from the *Cauchy-Schwartz Inequality:* $(x^T y)^2 \leq (x^T x)(y^T y)$. Let $\boldsymbol{\Sigma}_0^{1/2}$ be the root matrix to $\boldsymbol{\Sigma}_0$ defined such that $\boldsymbol{\Sigma}_0^{1/2}\boldsymbol{\Sigma}_0^{1/2} = \boldsymbol{\Sigma}_0$ and $\boldsymbol{\Sigma}_0^{-1/2}\boldsymbol{\Sigma}_0\boldsymbol{\Sigma}_0^{-1/2} = \boldsymbol{I}_K$, the identity matrix of order K. Then from the Cauchy-Schwartz inequality

$$\left(\boldsymbol{W}^T\boldsymbol{G}\right)^2 = \left(\boldsymbol{W}^T\boldsymbol{\Sigma}_0^{1/2}\boldsymbol{\Sigma}_0^{-1/2}\boldsymbol{G}\right)^2 \leq \left(\boldsymbol{W}^T\boldsymbol{\Sigma}_0\boldsymbol{W}\right)\left(\boldsymbol{G}^T\boldsymbol{\Sigma}_0^{-1}\boldsymbol{G}\right) \tag{4.91}$$

and

$$\frac{\left(\boldsymbol{W}^T\boldsymbol{G}\right)^2}{\boldsymbol{W}^T\boldsymbol{\Sigma}_0\boldsymbol{W}} \leq \left(\boldsymbol{G}^T\boldsymbol{\Sigma}_0^{-1}\boldsymbol{G}\right). \tag{4.92}$$

The equality, and thus the maxima, is obtained using a vector proportional to $\boldsymbol{W} = \boldsymbol{\Sigma}_0^{-1}\boldsymbol{G}$ that yields

$$\frac{\left(\boldsymbol{W}^T\boldsymbol{G}\right)^2}{\boldsymbol{W}^T\boldsymbol{\Sigma}_0\boldsymbol{W}} = \frac{\left(\boldsymbol{G}^T\boldsymbol{\Sigma}_0^{-1}\boldsymbol{G}\right)^2}{\left(\boldsymbol{G}^T\boldsymbol{\Sigma}_0^{-1}\boldsymbol{\Sigma}_0\boldsymbol{\Sigma}_0^{-1}\boldsymbol{G}\right)} = \frac{\left(\boldsymbol{G}^T\boldsymbol{\Sigma}_0^{-1}\boldsymbol{G}\right)^2}{\left(\boldsymbol{G}^T\boldsymbol{\Sigma}_0^{-1}\boldsymbol{G}\right)} = \boldsymbol{G}^T\boldsymbol{\Sigma}_0^{-1}\boldsymbol{G}. \tag{4.93}$$

Thus we have the well-known result (*cf.* Rao, 1973, p.60)

$$\max_{\boldsymbol{W}}\frac{\left(\boldsymbol{W}^T\boldsymbol{G}\right)^2}{\boldsymbol{W}^T\boldsymbol{\Sigma}_0\boldsymbol{W}} = \boldsymbol{G}^T\boldsymbol{\Sigma}_0\boldsymbol{G} \tag{4.94}$$

with the maximum obtained using a vector of weights $\boldsymbol{W} \propto \boldsymbol{\Sigma}_0^{-1}\boldsymbol{G}$. Note that the weights need not sum to unity because multiplying by any norming constant will cancel from the numerator and denominator.

For the case of K independent strata, $\boldsymbol{\Sigma}_0$ is a diagonal matrix so that the vector of weights which maximizes *Eff*(T) is

$$\boldsymbol{W} = \boldsymbol{\Sigma}_0^{-1}\boldsymbol{G} = \left[\frac{g'(\pi_1)^{-1}}{\sigma_{01}^2} \quad \cdots \quad \frac{g'(\pi_K)^{-1}}{\sigma_{0K}^2}\right]^T. \tag{4.95}$$

Thus the optimal test weight for the jth stratum is

$$w_j = \frac{1}{g'(\pi_j)\,\sigma_{0j}^2}. \tag{4.96}$$

A consistent estimate of $\boldsymbol{W}$ is provided by $\widehat{\boldsymbol{W}}$ with elements

$$\widehat{w}_j = \left[g'(p_j)\,\widehat{\sigma}_{0j}^2\right]^{-1} = \left[\left(\frac{dg(p_j)}{dp_j}\right)\widehat{V}_{0j}\right]^{-1}, \tag{4.97}$$

where $\widehat{V}_{0j} = \widehat{\sigma}_{0j}^2$ is presented in (4.53). Then the test that maximizes the asymptotic efficiency for the scale $g(\pi)$ is

$$X_{A(g)}^2 = \frac{\left[\sum_j \widehat{w}_j\,(p_{1j} - p_{2j})\right]^2}{\sum_j \widehat{w}_j^2\widehat{V}_{0j}} = \frac{T_g^2}{\widehat{V}(T_g|H_0)}, \tag{4.98}$$

Table 4.8 Radhakrishna Test Weights for Each Scale of Association

Scale $\widehat{\theta}_j = g(p_{1j}) - g(p_{2j})$	*Derivative* $g'(p_j)$	*Optimal Weight* $\widehat{w}_j = \left[\widehat{V}_{0j} g'(p_j)\right]^{-1}$
Risk Difference		
$p_{1j} - p_{2j}$	1.0	$\widehat{V}_{0j}^{-1} = \dfrac{n_{1j} n_{2j}}{N_j p_j (1 - p_j)}$
log *Relative Risk*		
$\log(p_{1j}) - \log(p_{2j})$	$\dfrac{1}{p_j}$	$\dfrac{p_j}{\widehat{V}_{0j}} = \dfrac{n_{1j} n_{2j}}{N_j (1 - p_j)}$
log *Odds Ratio*		
$\log \dfrac{p_{1j}}{1 - p_{1j}} - \log \dfrac{p_{2j}}{1 - p_{2j}}$	$\dfrac{1}{p_j (1 - p_j)}$	$\dfrac{p_j (1 - p_j)}{\widehat{V}_{0j}} = \dfrac{n_{1j} n_{2j}}{N_j}$

where

$$\widehat{V}(T_g | H_0) = \sum\nolimits_j \frac{\widehat{V}_{0j}}{\left[g'(p_j)\, \widehat{V}_{0j}\right]^2} = \sum\nolimits_j \frac{1}{g'(p_j)^2\, \widehat{V}_{0j}}, \tag{4.99}$$

and where $X^2_{A(g)}$ is asymptotically distributed as χ^2 on 1 *df* under H_0. This provides an asymptotically efficient test of H_{0A} versus H_{1A} for any parameter that can be expressed as a scale transformation of the form $\theta_j = g(\pi_{1j}) - g(\pi_{2j})$.

Note that because the test employs estimated weights $\{\widehat{w}_j\}$, there is a small loss of efficiency in the finite sample case; see Bloch and Moses (1988). However, since the estimated weights $\{\widehat{w}_j\}$ are consistent estimates of the true optimal weights, then from Slutsky's Theorem (A.45) the test is asymptotically fully efficient.

For the principal measures of association employed in Chapter 2, Table 4.8 summarizes the elements of the optimal weights $\widehat{w}_j$. The test for a specific scale can then be obtained by substituting the corresponding weights into the weighted Cochran-Mantel-Haenszel statistic in (4.98). This leads to a specific test that is directed to the specific alternative hypothesis of association on a specific scale $g(\pi)$. The precise expressions for the tests for a common risk difference, odds ratio and relative risk are derived in Problem 4.4.3.

Each of these tests is a weighted linear combination of the risk differences within the K strata. Therefore, any one test is distinguished from the others by the *relative* magnitudes of the weights for each strata, or $w_j / \sum_\ell w_\ell$. Also, under H_0: $\pi_{1j} = \pi_{2j}$ for all strata (j), any non-zero set of weights $\{\widehat{w}_j\}$ yields a weighted test that is asymptotically distributed as χ_1^2 since asymptotically

$$E\left[\sum_j \widehat{w}_j (p_{1j} - p_{2j})\right] \cong \sum_j w_j (\pi_{1j} - \pi_{2j}) = 0, \qquad (4.100)$$
$$V\left[\sum_j \widehat{w}_j (p_{1j} - p_{2j})\right] \cong \sum_j w_j^2 \sigma_{0j}^2$$

and asymptotically $X_A^2 \overset{d}{\approx} \chi^2$ on 1 df. Therefore, the test for each scale $g(\pi)$ is of size α under H_0. However, the tests differ in their power under specific alternatives.

Example 4.12 *Clinical Trial in Duodenal Ulcers (continued)*
For the ulcer drug clinical trial example, the differences within each stratum, the optimal weights $\{\widehat{w}_j\}$ for the test of association on each scale (RD, log RR, and log OR), and the relative magnitudes of the weights for each scale expressed as a percentage of the total, are

		$\widehat{w}_j$			*Proportional Weights*		
j	$p_{1j} - p_{2j}$	RD	$\log RR$	$\log OR$	RD	$\log RR$	$\log OR$
1	−0.04458	92.08	37.25	22.18	0.45	0.39	0.46
2	0.30556	21.81	13.50	5.14	0.11	0.14	0.10
3	0.24506	90.00	44.00	22.49	0.44	0.46	0.45
Total		203.89	94.75	49.81	1.00	1.00	1.00

For example, for the first stratum ($j = 1$)

$$\widehat{w}_{(RD)1} = [p_1(1-p_1)]^{-1}\left(\frac{n_{11}n_{21}}{N_1}\right)$$
$$= [0.40440(1 - 0.40449)]^{-1}(22.18) = 92.08,$$
$$\widehat{w}_{(RR)1} = (1-p_1)^{-1}\left(\frac{n_{11}n_{21}}{N_1}\right)$$
$$= (1 - 0.40449)^{-1}(22.18) = 37.25,$$
$$\widehat{w}_{(OR)1} = \left(\frac{n_{11}n_{21}}{N_1}\right) = \left(\frac{42 \times 47}{89}\right) = 22.18.$$

Because the relative values of the weights distinguish each test from the others, in this example the tests differ principally with respect to the weights for the first two strata.

The terms that enter into the numerator and denominator of the test statistic for each scale are

	$\widehat{w}_j(p_{1j} - p_{2j})$			$\widehat{w}_j^2 \widehat{V}_{0j}$		
Stratum	RD	$\log RR$	$\log OR$	RD	$\log RR$	$\log OR$
1	−4.1048	−1.6604	−0.9888	92.0786	15.0655	5.3426
2	6.6635	4.1250	1.5714	21.8077	8.3571	1.2128
3	22.0553	10.7826	5.5111	90.0000	21.5111	5.6195
Total	24.6140	13.2472	6.0938	203.8863	44.9338	12.1749

and the resulting test statistics are

Scale	T	$\widehat{V}(T)$	X_A^2	$p \leq$
Risk difference	24.6140	203.8863	2.9715	0.085
log Relative risk	13.2472	44.9338	3.9055	0.049
log Odds ratio	6.0938	12.1749	3.0501	0.081

The test designed to detect a common log relative risk different from zero is nominally statistically significant at $p \leq 0.05$, whereas the tests designed to detect a common risk difference or a common log odds ratio are not significant. Since the tests of homogeneity for the three strata indicated the least heterogeneity among the relative risks, then we expect the test statistic for relative risks to be greater than that for the other scales.

Example 4.13 *Religion and Mortality (continued)*
For the assessment of the association between religious versus not and mortality, the tests of association for each of the three scales are

Scale	X_A^2	$p \leq$
Risk difference	1.2318	0.267
log Relative risk	4.1154	0.043
log Odds ratio	3.7359	0.053

In this case, the results are strongly dependent on the extent of heterogeneity among the four strata on each scale. The risk differences show the most heterogeneity, and even though the test of homogeneity is not significant, the test of association based on the risk differences is substantially less than that for the relative risks or odds ratios that showed less heterogeneity.

4.8 ASYMPTOTIC RELATIVE EFFICIENCY OF COMPETING TESTS

4.8.1 Family of Tests

The preceding examples show that the efficiency of the test of association differs from one scale to the next and that the results of the test on each scale depend on the extent of heterogeneity over the various strata. The fundamental problem is that we now have a family $\mathcal{G}$ of tests, each directed to a different alternative hypothesis of a constant difference on a specified scale $g(\pi)$. However, the one test that is in fact most powerful unfortunately is not known *a priori*. This will be that test directed to the one scale $g(\pi)$ for which the corresponding measures of association $\{\theta_{gj}\}$ are indeed constant over strata or nearly so. Though tempting, it is cheating to conduct the test for all members of the family (three tests herein) and then select that one for which the *P*-value is smallest. Clearly, if one adopts this strategy, then the Type I error probability of the "test" will be increased beyond the desired size α. Thus the test of association for a given scale will only have a test size equal to the desired level α when that scale is specified *a priori*, meaning that the specific restricted alternative to which the test is directed is prespecified *a priori*.

The same problem exists if one first conducts the test of homogeneity for all three scales and then selects the test of association for whichever scale is least significant. In fact, the two strategies are equivalent because the scale for which the test of homogeneity has the maximum P-value will also be the scale for which the test of association has the minimum P-value. This is an instance of the problem of two-stage inference (*cf.* Bancroft, 1972). Under the joint null hypothesis in (4.49), Lachin and Wei (1988) show that the tests of homogeneity and of association are uncorrelated so that one could partition the total Type I error probability for the two successive tests. This approach, however, is not very appealing because the size of the test of association must be some quantity less than α, thus diminishing power. Also, the specific scale to be tested must be prespecified.

Another approach to this problem is to first consider the possible loss in efficiency associated with choosing *a priori* a test on the "wrong" scale, or one for which there is more heterogeneity relative to another scale for which the corresponding test will have greater efficiency. This can be described through the assessment of the asymptotic relative efficiency (*ARE*) of two tests designed to detect a common difference among strata on different scales of association.

The Radhakrishna family can be described as a family of tests of H_0 directed toward a family of alternative hypotheses of the form $H_{1A(g)}$: $g(\pi_{1j}) - g(\pi_{2j}) = \theta_g$ as presented in (4.80), where $g(\pi) \in \mathcal{G}$ is a family of scales for which is $\theta_0 = 0$ under the null hypothesis H_0. One property of the family $\mathcal{G}$ is that if the $\{\pi_{2j}\}$ vary over strata, then strict homogeneity on one scale implies heterogeneity on another scale. For example, when $\theta_{sj} = \theta_s \; \forall j$ for scale g_s, if $\pi_{2j} \neq \pi_{2\ell}$ for two of the strata, say $j \neq \ell$, then for any other scale g_r, $\theta_{rj} \neq \theta_{r\ell}$ for the same pair of strata. In such cases, whichever scale provides strict homogeneity over strata, or the scale with the greatest homogeneity over strata, will provide the more powerful test.

Example 4.14 *Two Homogeneous Strata*

For example, consider the following case of only two strata with probabilities such that there is a common odds ratio ($g_s = logit$), and as a result, heterogeneity in the risk differences ($g_r = identity$)

			$g_s = logit$		$g_r = identity$
Stratum	π_{1j}	π_{2j}	$\theta_{sj} = \log(OR)$	OR	$\theta_{rj} = RD$
1	0.20	0.091	0.9163	2.50	0.109
2	0.30	0.146	0.9163	2.50	0.154

In this case, the test based on the logit scale, or assuming a common odds ratio, provides the most powerful test of H_0. Conversely, the test assuming a constant risk difference will provide less power as a test of H_0.

4.8.2 Asymptotic Relative Efficiency

The loss in efficiency incurred when we use a test for a scale other than the optimal scale is provided by the asymptotic relative efficiency of the two tests. Let $\mathcal{G}$ designate a family of scales, for each of which there is a different optimal test. Assume that the alternative hypothesis of a common association parameter $\theta_{sj} = \theta_s$ over all strata is true for scale $g_s(\pi) \in \mathcal{G}$ so that $X^2_{A(g_s)}$ is the asymptotically locally most powerful test. To simplify notation, designate $X^2_{A(g_s)}$ as $X^2_{A(s)}$. Then, let $g_r(\pi)$ be any other scale, $g_r(\pi) \in \mathcal{G}$, so that $\theta_{rj} \neq \theta_{rk}$ for some $1 \leq j < k \leq K$. The two competing statistics for a test of association are

$$X^2_{A(\ell)} = \frac{T_\ell^2}{\widehat{V}(T_\ell)}, \qquad \ell = r, s \tag{4.101}$$

where

$$T_\ell = \sum\nolimits_j \widehat{w}_{\ell j}(p_{1j} - p_{2j}), \qquad \ell = r, s \tag{4.102}$$

$$\widehat{V}(T_\ell) = \widehat{V}(T_\ell \mid H_0) = \sum\nolimits_j \widehat{w}^2_{\ell j} \widehat{V}_{0j}, \qquad \ell = r, s \tag{4.103}$$

and

$$\widehat{w}_{\ell j} = \frac{1}{g'_\ell(p_j)\widehat{V}_{0j}}, \qquad \ell = r, s. \tag{4.104}$$

Let $ARE\left(X^2_{A(r)}, X^2_{A(s)}\right)$ designate the ARE of $X^2_{A(r)}$ to $X^2_{A(s)}$ when the alternative hypothesis H_{1A} is true for scale $g_s(\pi)$ so that there is a common parameter $\theta_s \neq \theta_0$ for all strata. Since the tests are each based on weighted sums of the differences in proportions, T_r and T_s are each asymptotically normally distributed so that $ARE\left(X^2_{A(r)}, X^2_{A(s)}\right) = ARE(T_r, T_s)$. From (3.59) in Chapter 3, then

$$ARE(T_r, T_s) = \frac{Eff(T_r \mid \theta_s)}{Eff(T_s \mid \theta_s)} = \frac{\left[\left(\frac{dE(T_r|\theta)}{d\theta}\right)^2 / V(T_r)\right]}{\left[\left(\frac{dE(T_s|\theta)}{d\theta}\right)^2 / V(T_s)\right]}\Bigg|_{\theta=\theta_0}. \tag{4.105}$$

First consider the efficiency of the optimal test T_s. Substituting $\widehat{w}_{sj}$ into the expression for T_s then

$$E(T_s \mid \theta_s) = E\left[\sum\nolimits_j \frac{(p_{1j} - p_{2j})}{g'_s(p_j)\widehat{V}_{0j}}\right] = E\left[\sum\nolimits_j \frac{g'_s(p_j)(p_{1j} - p_{2j})}{g'_s(p_j)^2 \widehat{V}_{0j}}\right]. \tag{4.106}$$

From Slutsky's Convergence Theorem (A.45), since $g'_s(p_j) \xrightarrow{p} g'_s(\pi_j)$, then asymptotically

$$E\left[g_s'\left(p_j\right)\left(p_{1j}-p_{2j}\right)\right] \cong g_s'\left(\pi_j\right)\left(\pi_{1j}-\pi_{2j}\right)=\theta_s \quad \forall j. \tag{4.107}$$

Likewise, since $\widehat{V}_j \xrightarrow{p} \sigma_{0j}^2$ then $g_s'\left(p_j\right)^2 \widehat{V}_j \xrightarrow{p} g_s'\left(\pi_j\right)^2 \sigma_{0j}^2$. Substituting each into the above yields

$$E\left(T_s \mid \theta_s\right)=\theta_s \sum\nolimits_j\left(\frac{1}{g_s'\left(\pi_j\right)^2 \sigma_{0j}^2}\right). \tag{4.108}$$

Thus

$$\frac{dE\left(T_s \mid \theta_s\right)}{d\theta_s}=\sum\nolimits_j\left(\frac{1}{g_s'\left(\pi_j\right)^2 \sigma_{0j}^2}\right)=\sum\nolimits_j \frac{\sigma_{0j}^2}{\left[g_s'\left(\pi_j\right) \sigma_{0j}^2\right]^2}=\sum\nolimits_j w_{sj}^2 \sigma_{0j}^2. \tag{4.109}$$

Again using Slutsky's Convergence Theorem, since $\widehat{w}_{sj} \xrightarrow{p} w_{sj}$ and $\widehat{V}_j \xrightarrow{p} \sigma_{0j}^2$, then

$$\widehat{V}\left(T_s \mid \theta_s\right) \xrightarrow{p} V\left(T_s \mid \theta_s\right)=\sum\nolimits_j w_{sj}^2 \sigma_{0j}^2. \tag{4.110}$$

Therefore,

$$\mathit{Eff}\left(T_s \mid \theta_s\right)=\frac{\left[\sum_j w_{sj} \sigma_{0j}^2\right]^2}{\sum_j w_{sj} \sigma_{0j}^2}=\sum\nolimits_j w_{sj} \sigma_{0j}^2 . \tag{4.111}$$

Now consider the efficiency of the alternative test T_r. Since we assume that $\theta_{sj}=\theta_s \ \forall j$, this implies that θ_{rj} is not constant for all strata so that T_r will be suboptimal. Evaluating the efficiency of T_r under this assumption yields

$$E\left(T_r \mid \theta_s\right)=E\left[\sum\nolimits_j \frac{\left(p_{1j}-p_{2j}\right)}{g_r'\left(\pi_j\right) \sigma_{0j}^2}\right]. \tag{4.112}$$

However, from (4.107) the alternative of a constant difference on scale $g_s(\pi)$ for all strata, in turn, implies that asymptotically

$$E\left(p_{1j}-p_{2j} \mid \theta_s\right) \cong \frac{\theta_s}{g_s'\left(\pi_j\right)}. \tag{4.113}$$

Therefore,

$$E\left(T_r \mid \theta_s\right)=\theta_s\left[\sum\nolimits_j \frac{1}{g_s'\left(\pi_j\right) g_r'\left(\pi_j\right) \sigma_{0j}^2}\right] \tag{4.114}$$

and

$$\frac{dE\left(T_r\,|\,\theta_s\right)}{d\theta_s}=\left[\sum\nolimits_j \frac{1}{g_s'\left(\pi_j\right)g_r'\left(\pi_j\right)\sigma_{0j}^2}\right]=\sum\nolimits_j w_{rj}w_{sj}\sigma_{0j}^2. \tag{4.115}$$

Since

$$V\left(T_r\,|\,\theta_s\right)=\sum\nolimits_j w_{rj}^2\sigma_{0j}^2 \tag{4.116}$$

then

$$ARE\left(T_r,T_s\,|\,\theta_s\right)=\frac{\left[\sum_j w_{rj}w_{sj}\sigma_{0j}^2\right]^2}{\left[\sum_j w_{rj}^2\sigma_{0j}^2\right]\left[\sum_j w_{sj}^2\sigma_{0j}^2\right]}=\rho_{rs}^2 \tag{4.117}$$

where

$$\rho_{rs}=corr\left(T_r,T_s\right). \tag{4.118}$$

These expressions involve the asymptotic null variance σ_{0j}^2 that in turn involves the sample sizes within each stratum, the $\{n_{ij}\}$. However, we can factor N from the expression for σ_{0j}^2 by substituting $E(n_{ij})=N\zeta_j\xi_{ij}$, where $\zeta_j=E(N_j/N)$ are the expected stratum sample fractions, and $\xi_{ij}=E(n_{ij}/N_j)$ are the expected group sample fractions within the jth stratum. Then

$$\sigma_{0j}^2=\frac{\pi_j\left(1-\pi_j\right)}{N\zeta_j}\left(\frac{1}{\xi_{1j}}+\frac{1}{\xi_{2j}}\right)=\frac{\phi_{0j}^2}{N}, \tag{4.119}$$

where ϕ_{0j}^2 does not involve the value of N. Thus for each scale $\ell=r,s$ the weights $\{w_{\ell j}\}$ are proportional to

$$\widetilde{w}_{\ell j}=\frac{1}{g_\ell'\left(\pi_j\right)\phi_{0j}^2}. \tag{4.120}$$

Substituting into (4.117), the resulting expression for $ARE\left(T_r,T_s\right)$ is a function only of the $\{\pi_j\}$, $\{\zeta_j\}$, and $\{\xi_{ij}\}$.

For a given set of data, the *ARE* can be consistently estimated as the sample correlation of the two test statistics based on the estimated weights and variances, expressed as

$$\widehat{ARE}\left(T_r,T_s|\theta_s\right)=\frac{\left[\sum_j \widehat{w}_{rj}\widehat{w}_{sj}\widehat{V}_{0j}\right]^2}{\left[\sum_j \widehat{w}_{rj}^2\widehat{V}_{0j}\right]\left[\sum_j \widehat{w}_{sj}^2\widehat{V}_{0j}\right]}=\widehat{\rho}_{rs}^2. \tag{4.121}$$

The denominator is the product of the variances of the two test statistics, $\widehat{V}\left(T_r|H_0\right)$ and $\widehat{V}\left(T_s|H_0\right)$. The numerator can either be computed as the sum of cross products directly, or it can be simplified by noting that the product $\left(\widehat{w}_{rj}\widehat{w}_{sj}\widehat{V}_{0j}\right)$ includes terms that cancel (see Problem 4.6.4).

Example 4.15 *Two Homogeneous Strata*
Again consider the data from Example 4.14 with two strata for which there is homogeneity of odds ratios and heterogeneity of the risk differences. If we assume that the limiting sample fractions are $\xi_{1j} = \xi_{2j} = 1/2$ for $j = 1, 2$, then

				Odds Ratio		Risk Difference	
Stratum	π	ζ_j	ϕ^2_{0j}	$g'_s(\pi)$	$\widetilde{w}_{sj}$	$g'_r(\pi)$	$\widetilde{w}_{rj}$
1	0.146	0.6	0.829	8.04	0.15	1	1.21
2	0.223	0.4	1.733	5.77	0.10	1	0.58

where $g'_s(\pi) = [\pi(1-\pi)]^{-1}$ and $g'_r(\pi) = 1$. Therefore, asymptotically, regardless of the sample size,

$$ARE(T_r, T_s|\theta_s)$$
$$= \frac{[(0.15 \times 1.21 \times 0.8289) + (0.1 \times 0.53 \times 1.733)]^2}{\left[(0.15)^2(0.8289) + (0.1)^2(1.733)\right]\left[(1.21)^2(0.8289) + (0.58)^2(1.733)\right]}$$
$$= 0.974\ .$$

Therefore, with a large sample there is a loss of efficiency of 2.6% if we use the test designed to detect a common risk difference rather than the test designed to detect a common odds ratio that is optimal for this example.

Example 4.16 *Clinical Trial in Duodenal Ulcers (continued)*
For the ulcer clinical trial example, the terms entering into the computation of the covariance for each pair of tests $\{\widehat{w}_{rj}\widehat{w}_{sj}\widehat{V}_{0j}\}$ are as follows:

	$\widehat{w}_{rj}\widehat{w}_{sj}\widehat{V}_{0j}$		
Stratum	$RD, \log RR$	$RD, \log OR$	$\log OR, \log RR$
1	37.2453	22.1798	8.9716
2	13.5000	5.1429	3.1837
3	44.0000	22.4889	10.9946
Total	94.7453	49.8115	23.1498

Thus the estimated *AREs* $\{\widehat{\rho}^2_{rs}\}$ for each pair of tests are

Scales (r, s)	$\widehat{V}(T_r)$	$\widehat{V}(T_s)$	$\widehat{Cov}(T_r, T_s)$	$\widehat{ARE}(T_r, T_s)$
RD, $\log RR$	203.8863	44.9338	94.7453	0.9798
RD, $\log OR$	203.8863	12.1749	49.8115	0.9996
$\log OR$, $\log RR$	44.9338	12.1749	23.1498	0.9796

The tests for a common risk difference and for a common log odds ratio each have an *ARE* of about 0.98, indicating about a 2% loss of efficiency relative to the test for a common log relative risk. The *ARE* of the test for risk differences versus that for the log odds ratios is close to 1.0, neither test being clearly preferable to the other.

Example 4.17 *Religion and Mortality (continued)*
Likewise for the association between religion and mortality, the estimated *AREs* for each pair of tests are

Scales (r, s)	$\widehat{ARE}(T_r, T_s)$
Risk difference, log Relative risk	0.4517
Risk difference, log Odds ratio	0.5372
log Odds ratio, log Relative risk	0.9825

Thus there is a substantial loss in power using the risk differences relative to a test using either the relative risks or the odds ratios.

4.9 MAXIMIN EFFICIENT ROBUST TESTS

Since the efficiency of the a test of partial association depends on the extent of homogeneity, the most efficient or powerful test is not known *a priori*. Therefore, in practice it would be preferable to use a test that is robust to the choice of scale, meaning one that has good power irrespective of whichever scale is in fact optimal. One approach to developing such a robust test is to choose a maximin-efficient test with respect to the family of tests. Two such tests are the Gastwirth scale robust test of association and the Wei-Lachin test of stochastic ordering.

4.9.1 Maximin Efficiency

Again consider a family of tests $\mathcal{G}$ each of which is optimal under a specific alternative. For the Radhakrishna family we defined the family of tests $\mathcal{G}$ based on the difference on some scale $g(\pi)$ for $g \in \mathcal{G}$. Herein we have restricted consideration to the family $\mathcal{G}$ consisting of the logit, log and identity functions corresponding to the log odds ratio, log relative risk and risk difference scales, respectively. This family, however, could be extended to include other functions such as the arcsine, probit and square root scales that are explored as Problems.

Let the test T_s corresponding to the scale $g_s \in \mathcal{G}$ be the optimal test within the family for a given set of data. Then let T_r be any other test corresponding to a different scale $g_r \in \mathcal{G}$ within the family. The test T_r may in fact be optimal for another set of data, but it is suboptimal for the data at hand. Thus *ARE* $(T_r, T_s) < 1$ for $r \neq s$. However, the optimal test T_s is unknown *a priori* and whenever one prespecifies a particular test one risks choosing a suboptimal test T_r with a pursuant loss of efficiency (power). Instead of prespecifying a particular test, a *maximin efficient robust test (MERT)* can be defined as one that suffers the least loss in efficiency (power) irrespective of whichever member of the family is optimal for any given set of data.

Let Z_m designate such a *MERT* for the family $\mathcal{G}$. Formally, Z_m is chosen from a family of possible test statistics $Z \in \mathcal{M}$ such that Z_m maximizes the minimum *ARE* with respect to whichever member of the family $\mathcal{G}$ is optimal. Then Z_m satisfies the relationship

$$\sup_{Z \in \mathcal{M}} \inf_{g \in \mathcal{G}} ARE\left(Z, T_g\right) = \inf_{g \in \mathcal{G}} ARE\left(Z_m, T_g\right). \tag{4.122}$$

The expression on the r.h.s. is the minimum *ARE* of the *MERT* with respect to any member of the family $\mathcal{G}$. Thus Z_m maximizes the minimum relative efficiency, regardless of which scale $g \in \mathcal{G}$ provides the test T_g that is optimal. Since T_s is the optimal test for the data at hand, then the *ARE* of the *MERT* is

$$\inf_{g \in \mathcal{G}} ARE\left(Z_m, T_g\right) = ARE\left(Z_m, T_s\right) . \tag{4.123}$$

A comparable interpretation of maximin efficiency is in terms of power. The *MERT* Z_m suffers the least possible loss of power relative to the optimal test within the family $\mathcal{G}$. Thus the test with maximin efficiency with respect to the family of alternatives is the test with minimax loss in power, that is, the test that minimizes the maximum loss in power compared to whichever test is optimal.

4.9.2 Gastwirth Scale Robust Test

Now let $\mathcal{G}$ refer to the Radhakrishna family of tests, each of which is asymptotically most powerful for a restricted alternative hypothesis of the form $H_{1A(g)}$ in (4.80) for a specific scale $g(\pi) \in \mathcal{G}$. So far we have focused on the family containing the logit, log and identity scales corresponding to a test designed to detect a common odds ratio, relative risk and risk difference, respectively. The family could be extended to include other scales as well, some of which are described by Radhakrishna (1965).

For a given set of data, let Z_g refer to the normal deviate test corresponding to the root of $X^2_{A(g)}$ in (4.98) for scale $g(\pi) \in \mathcal{G}$. Let (Z_r, Z_s) be the *extreme pair* of tests within the family defined as

$$\rho_{rs} = \min_{r,s}(\rho_{i,j}) \qquad (g_i, g_j) \in \mathcal{G} \tag{4.124}$$

where $\rho^2_{r,s} = ARE\left(Z_r, Z_s\right) = ARE\left(T_r, T_s\right)$ from (4.117) is the *ARE* of Z_r to Z_s. Usually $Z_r = \min_{g \in \mathcal{G}}\left(Z_g\right)$ and $Z_s = \sup_{g \in \mathcal{G}}(Z_g)$ so that usually Z_r is the test for the scale with the greatest heterogeneity among the strata while Z_s is that for the scale with the greatest homogeneity among the strata with respect to the corresponding measures of association. Gastwirth (1985) then showed that if

$$\rho_{rg} + \rho_{sg} \geq 1 + \rho_{rs} \quad \forall(g \in \mathcal{G}), \quad g \neq r, \quad g \neq s\,, \tag{4.125}$$

then the maximin efficient scale robust test (*MERT*) is obtained as a convex combination of the extreme pair of tests

$$Z_m = \frac{Z_r + Z_s}{\left[2\left(1 + \rho_{rs}\right)\right]^{1/2}} \tag{4.126}$$

that is asymptotically distributed as standard normal under H_0. The maximin efficiency of the *MERT* then is

$$\inf_{g \in \mathcal{G}} ARE\,(Z_m, Z_g) = \frac{1+\rho_{rs}}{2} \tag{4.127}$$

meaning that the *ARE* of the *MERT* Z_m relative to the unknown optimal test within the family is at least this quantity.

Gastwirth (1966) also shows that if the condition (4.125) does not hold, then the *MERT* still exists but it must be obtained as a linear combination of the Z_g

$$Z_m = \sum_{g \in \mathcal{G}} a_g Z_g \tag{4.128}$$

with coefficients $\{a_g\}$ that satisfy a set of constraints. For a family of three tests as herein, these constraints are

$$\begin{aligned} a_1(1-\rho_{12}) - a_2(1-\rho_{12}) + a_3(1-\rho_{23}) &= 0 \\ a_1(1-\rho_{13}) + a_2(\rho_{12}-\rho_{23}) - a_3(1-\rho_{23}) &= 0 \end{aligned} \tag{4.129}$$

and

$$\sum_{i=1}^{3}\sum_{j=1}^{3} a_i a_j \rho_{ij} = 1\,. \tag{4.130}$$

In this case, the coefficients must be solved by an iterative procedure.

Example 4.18 *Clinical Trial in Duodenal Ulcers (continued)*
From Example 4.16, the extreme pair of scales are the log relative risk (R) and log odds ratio (O) with correlation $\widehat{\rho}_{R,O} = 0.98976$. The other pairs of correlations are $\widehat{\rho}_{R,D} = 0.98987$ and $\widehat{\rho}_{O,D} = 0.99978$. However,

$$(\widehat{\rho}_{R,D} + \widehat{\rho}_{O,D}) = 1.98965 < (1 + \widehat{\rho}_{R,O}) = 1.98976$$

and the condition in (4.125) is not satisfied. Therefore, the *MERT* cannot be readily computed using the convex combination in (4.126). Rather the iterative computation in (4.129) would be required.

Example 4.19 *Religion and Mortality (continued)*
From Example 4.17, the extreme pair of scales are the relative risk and risk difference with $\widehat{\rho}_{R,D} = 0.67205$. The other pairs of correlations are $\widehat{\rho}_{R,O} = 0.99123$ and $\widehat{\rho}_{O,D} = 0.73297$ and the condition in (4.125) is satisfied:

$$(\widehat{\rho}_{R,O} + \widehat{\rho}_{O,D}) = 1.7242 > (1 + \widehat{\rho}_{R,D}) = 1.67205.$$

Thus the MERT can be readily computed as

$$Z_m = \frac{\sqrt{4.11535} + \sqrt{1.23176}}{\sqrt{2(1.67205)}} = 1.7162$$

with two-sided $p \leq 0.087$.

4.9.3 Wei-Lachin Test of Stochastic Ordering

The efficiency of the test of association is highly dependent on the chosen scale because the alternative hypothesis $H_{1A(g)}$ in (4.80) specifies that there is a common difference within each stratum on the chosen scale, as depicted in Figure 4.4.A. Thus the test is directed to a highly restricted subset of the general K-dimensional omnibus parameter space and many meaningful values are excluded, such as where there is a risk difference of 0.2 in one stratum and 0.4 in another. In such cases there may be some quantitative differences in the measures of association among strata but no qualitative differences, meaning that the *direction* of the association is consistent among strata.

Therefore, another way to derive a more robust test is to specify a less restrictive alternative hypothesis. In an entirely different setting, that of a multivariate rank test for repeated measures, Wei and Lachin (1984) and Lachin (1992a) suggested a test of stochastic ordering that is directed toward the alternative hypothesis of a common *qualitative* degree of association on some scale rather than a strictly common quantitative value on that scale. First consider a generalized one-sided upper-tail test. In this case, the alternative hypothesis of *stochastic ordering* is

$$H_{1S}\colon \ \pi_{1j} \geq \pi_{2j} \quad (j = 1, \dots, K) \quad \text{or} \quad \pi_{1j} > \pi_{2j} \text{ for some } 1 \leq j \leq K \,. \tag{4.131}$$

This alternative specifies that the probabilities in group 1 are at least as great as those in group 2 for all strata, and are strictly greater for some. For a measure of association for some scale $g(\pi) \in \mathcal{G}$, as employed above for the Radhakrishna family, then using a simplified notation, this specifies that

$$H_{1S} \Leftrightarrow \ \theta_j \geq 0 \text{ for } \forall j \tag{4.132}$$

with a strict inequality for at least one j. Thus it is sufficient to employ a test based on the risk differences.

This test can also be used to conduct a two-sided test of stochastic ordering for which the alternative hypothesis is stated as

$$H_{1S}\colon \ \pi_{ij} \geq \pi_{(3-i)j} \qquad i = 1 \text{ or } 2;\ j = 1, \dots, K, \tag{4.133}$$

with a strict inequality for at least one j. This is a two-sided specification that the probabilities in one group (either $i = 1$ or 2) are larger than those in the other group. Equivalently, the alternative specifies that

$$H_{1S} \Leftrightarrow \ (\theta_j \geq 0 \text{ for } \forall j) \text{ or } (\theta_j \leq 0 \text{ for } \forall j) \tag{4.134}$$

with a strict inequality for at least one j.

For the case of two strata or measures, Figure 4.5.A shows the null and alternative hypothesis spaces in terms of the risk differences $\{RD_j\}$ in the two strata. The test of stochastic ordering is directed toward all points in Quadrants I and III for which

Fig. 4.5 The null (H_0) and alternative (H_1) hypotheses for the test of stochastic ordering with two parameters $\theta = (\theta_1,\ \theta_2)$ (A) and the corresponding rejection region (B).

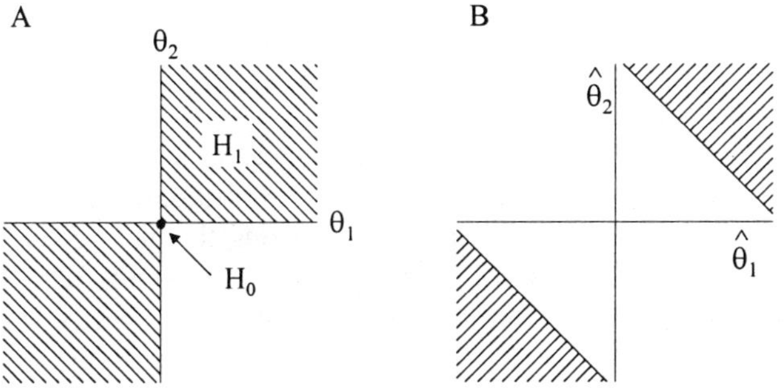

the two differences are either both positive or both negative. If either difference is zero, but not both, the convention used is to include that point in the alternative hypothesis parameter space.

Various authors have considered tests for this problem. For the case of multivariate normal variates with estimated variances and covariances, Perlman (1969) derives the computationally tedious likelihood ratio test of stochastic ordering; see also Kudo (1963). Tang, Gnecco and Geller (1989) describe an approximation to Perlman's test that is somewhat less tedious, and that follows a simplified chi-bar-squared distribution.

In the setting of a stratified analysis of 2×2 tables, the Wei and Lachin (1984) test is based on the simple unweighted sum (or the unweighted mean) of the risk differences among strata

$$X_S^2 = \frac{\left[\sum_j (p_{1j} - p_{2j})\right]^2}{\sum_j \widehat{V}_{0j}}, \tag{4.135}$$

which is asymptotically distributed as χ_1^2 under H_0. For example, consider the one-sided alternative hypothesis that the risk differences for two strata fall in the positive orthant, or the subhypothesis in H_{1S} that $RD_j \geq 0$ for $j = 1, 2$. This alternative hypothesis can be viewed as consisting of the family $\mathcal{H}$ of all possible projections in the positive orthant. For each projection $h \in \mathcal{H}$ there is a corresponding optional test statistic of the form

$$T_h = w_{h_1}\widehat{RD}_1 + w_{h_2}\widehat{RD}_2. \tag{4.136}$$

The Wei-Lachin test, however, is based on

$$T_S = \widehat{RD}_1 + \widehat{RD}_2 \tag{4.137}$$

using unit (equal) weights.

Frick (1995) showed that the Wei-Lachin test is a maximin efficient test in the following sense. For a vector of statistics $\widehat{\boldsymbol{\theta}}$ with covariance matrix $\boldsymbol{\Sigma}$ as in Section 4.7, if the vector $\boldsymbol{J}'\boldsymbol{\Sigma}$ is positive, $\boldsymbol{J}$ being the unit vector, then the Wei-Lachin test is maximin efficient with respect to the family $\mathcal{H}$ of projections in the positive or negative orthant. For independent 2×2 tables, $\widehat{\boldsymbol{\theta}} = (\widehat{RD}_1\ \widehat{RD}_2)^T$ with $\boldsymbol{\Sigma} = diag(\sigma^2_{01}, \sigma^2_{02})$, this condition obviously applies. Therefore, the Wei-Lachin test on average minimizes the loss in power relative to the optimal test corresponding to the true alternative hypothesis point (RD_1, RD_2) that lies on a particular projection in the positive or negative orthant.

For the bivariate example employed in Figure 4.2.B for the omnibus test, and in Figure 4.4.B for the test of association, Figure 4.5.B presents the rejection region for the test of stochastic ordering. Because the test is based on a linear combination of the two statistics, the rejection region is defined by the simple sum of sample statistics satisfying $\left(\widehat{\theta}_1 + \widehat{\theta}_2\right)^2 / V\left(\widehat{\theta}_1 + \widehat{\theta}_2\right) = 3.841 = \chi^2_{1(0.95)}$ for $\alpha = 0.05$. This yields a line of rejection at a 135° angle to the X axis. This differs from the line of rejection for the test of association that is tilted toward the origin for whichever $\widehat{\theta}_j$ has smaller variance, that is, whichever is more precise.

Example 4.20 *Clinical Trial in Duodenal Ulcers (continued)*
For the ulcer clinical trial the sum of the risk differences is $\sum_j \widehat{RD}_j = (-0.04458 + 0.30556 + 0.24506) = 0.50604$ and the sum of the variances estimated under the null hypothesis is $\sum_j \widehat{V}_{0j} = (0.01086 + 0.04586 + 0.01111) = 0.06783$. Thus the Wei-Lachin test is

$$X^2_S = \frac{(0.50604)^2}{0.0678} = 3.78$$

with $p \leq 0.052$. In comparison, the test of association directed toward a common risk difference yields $X^2_A = 2.97$ with $p \leq 0.085$.

Example 4.21 *Religion and Mortality (continued)*
For the observational study of religion and mortality, $\sum_j \widehat{RD}_j = -0.33384$, $\sum_j \widehat{V}_{0j} = 0.02812$ and $X^2_S = 3.96$ with $p \leq 0.0465$. In comparison, the test of association for a common risk difference yields $p \leq 0.27$ and the Gastwirth scale robust *MERT* yields $p \leq 0.086$.

4.9.4 Comparison of Weighted Tests

The omnibus test, all the members of the Radhakrishna family of weighted Cochran-Mantel-Heanszel tests of no partial association, the Gastwirth *MERT* and the Wei-Lachin test of stochastic ordering are all tests of the global null hypothesis for the K strata in (4.48). The tests differ with respect to the alternative hypotheses H_{1O}, H_{1A} and H_{1S} as depicted in Figures 4.2.A, 4.4.A and 4.5.A, respectively. Each test will have greater power than the others in specific instances. When the $\{\theta_j\}$ are homogeneous on the prespecified scale, or nearly so, the test of association will be more powerful than either the omnibus test or the test of stochastic ordering, but not necessarily otherwise. Similarly, when the $\{\theta_j\}$ are homogeneous for one scale within a family of tests $g \in \mathcal{G}$, or nearly so, then the Gastwirth scale robust *MERT* will be more powerful. Conversely, when the $\{\theta_j\}$ fall along a projection in the positive or negative orthants (Quadrants I and III), the test of stochastic ordering will tend to be more powerful, especially if the corresponding projection under H_{1S} is not close to the projection of equality under H_{1A} for which the test of association is optimal. Finally, when the $\{\theta_j\}$ fall in some other orthant where some of the $\{\theta_j\}$ differ in sign (Quadrants II and IV), then the omnibus test will tend to be more powerful than the others. See Lachin (1992a, 1996) for a further comparison of these tests in the analysis of repeated measures.

In practice, there is a trade-off between the power robustness of the omnibus test to detect group differences under the broadest possible range of alternatives, versus the increased efficiency of the other tests to detect systematic differences between the groups under specific alternatives. As one compares the omnibus test, the test of stochastic ordering, and the test of association, in turn, there is decreasing robustness to a range of alternative hypotheses, but increasing power to detect specific restricted alternatives. In general, the Wei-Lachin test will have good power for alternatives approaching a constant difference on some scale. It will also have good power for alternatives where there is some heterogeneity but the risk differences are all in the same direction. In fact, X_S^2 is not nearly as sensitive to heterogeneity as is the test of association.

4.10 RANDOM EFFECTS MODEL

All of the methods described previously in this chapter are based on a fixed effects model. This model explicitly specifies that $E(\widehat{\theta}_j) = \theta$ for $\forall j$ or that there is a common measure of association on some scale under the alternative hypothesis. This model specification is equivalent to the null hypothesis of homogeneity H_{0H} specified in (4.57) that is tested by the test of homogeneity on $K - 1$ *df*. If heterogeneity is observed, it could arise for either of two reasons.

The first is that the fixed effects model has been misspecified in some respect. Perhaps there is homogeneity on some scale other than that specified for the analysis. Alternately, perhaps an additional covariate must be adjusted for to yield homogeneity. For example, if we first adjust for ethnicity and find heterogeneity of odds ratios, then perhaps an adjustment for gender and ethnicity simultaneously would yield homogeneity over strata.

The second possibility is that the fixed effects model simply does not hold, meaning that there is some *extra-variation* or *over-dispersion* due to random differences among strata. This extra-variation leads to the formulation of a random effects model.

4.10.1 Measurement Error Model

The simplest random effects model is a simple measurement error model, where a quantitative variable such as the level of serum cholesterol is measured with random error and where the true cholesterol value varies at random from one subject to the next. These assumptions can be expressed in a *two-stage model* as follows.

Consider that we have a sample of *i.i.d.* observations $\{y_i\}$, $i = 1, \ldots, N$. At the first stage of the model we assume that $y_i = \upsilon_i + \varepsilon_i$, where $E(y_i) = \upsilon_i$ is the true value that varies at random from one subject to the next. The conditional distribution $f(y_i|\upsilon_i)$ is determined by the value of υ_i and the form of the distribution of the errors. For example, if $\varepsilon_i \sim \mathcal{N}(0, \sigma_\varepsilon^2)$, then $f(y_i|\upsilon_i)$ is $N(\upsilon_i, \sigma_\varepsilon^2)$. The second or random stage of the model then specifies that the υ_i are randomly distributed in the population with some *mixing* distribution $\upsilon_i \sim f(\upsilon)$, where $E(\upsilon_i) = \mu$ and $V(\upsilon_i) = \sigma_\upsilon^2$. Thus unconditionally the $\{y_i\}$ are distributed as $f(y) = \int_\upsilon f(y|\upsilon) f(\upsilon) d\upsilon$. Here we use f to denote the distribution of any random variable so that $f(y)$, $f(y|\upsilon)$, and $f(\upsilon)$ need not be the same distribution.

As in the usual fixed effects model, we also assume that the random errors are distributed as $\varepsilon_i \sim f(\varepsilon)$ with some distribution f, where $E(\varepsilon_i) = 0$ and $V(\varepsilon_i) = \sigma_\varepsilon^2$ for $\forall i$. In addition, we assume that $\varepsilon_i \perp \upsilon_i$ (independent of) $\forall i$ meaning that the random errors are statistically independent of the random conditional expectations. Note that the corresponding simple fixed effects model with only an overall mean (no strata or covariate effects) simply specifies that all observations share the same expectation such that $y_i = \mu + \varepsilon_i$ with $\varepsilon_i \sim h(\varepsilon)$, $E(\varepsilon_i) = 0$ and $V(\varepsilon_i) = \sigma_\varepsilon^2$ for all observations.

Using this random effects model specification, we then wish to estimate the moments of the mixing distribution μ and σ_υ^2 and the variance of the errors σ_ε^2. Intuitively, since the conditional expectations $\{\upsilon_i\}$ are assumed to be independent of the random errors $\{\varepsilon_i\}$ then $\sigma_y^2 = \sigma_\upsilon^2 + \sigma_\varepsilon^2$, or the total variation among the observed $\{y_i\}$ can be partitioned into the variation among the true values plus the variation among the random errors. This can be formally shown from the well-known result in (A.6) of the Appendix which in this measurement error model yields

$$\begin{aligned} V(Y) = \sigma_y^2 = E(y-\mu)^2 &= E[V(y|\upsilon)] + V[E(y|\upsilon)] \\ &= E\left[E(y-\upsilon)^2 \,|\upsilon\right] + V[\upsilon] \\ &= E\left[E(\varepsilon^2 \,|\upsilon)\right] + V[\upsilon] . \end{aligned} \tag{4.138}$$

If the errors have constant variance for all observations such that $E\left[E\left(\varepsilon^2 \,|\upsilon\right)\right] = \sigma_\varepsilon^2$ independently of υ, then

$$V(Y) = \sigma_\varepsilon^2 + \sigma_\upsilon^2. \tag{4.139}$$

This also demonstrates the principle of partitioning of variation as in (A.5) since $\sigma_y^2 = \sigma_\varepsilon^2 + \sigma_\upsilon^2$ specifies that

$$E\left(y-\mu\right)^2 = E\left(y-\upsilon\right)^2 + E\left(\upsilon-\mu\right)^2. \tag{4.140}$$

The mean μ is readily estimated from $\overline{y}$, and the variance of Y from the usual sample estimate s_y^2 on $N-1$ df. If we can obtain an estimate of one of the variance components, usually $\widehat{\sigma}_\varepsilon^2$, then we can obtain the other, usually $\widehat{\sigma}_\upsilon^2$, by substraction. For quantitative measurements, such as a laboratory assay like serum cholesterol, these variance components are readily estimated from a set of independent duplicate measurements using moment estimators obtained from the expected mean squares of an analysis of variance (*cf.* Fleiss, 1986), or using restricted maximum likelihood or other methods (see Harville, 1977).

Such two-stage random effects models can be viewed as an application of what is often called *the NIH model*, a device pioneered by the early group of biostatisticians at the National Institutes of Health (Cornfield, Mantel, Haenszel, Greenhouse, among others) that was never explicitly published. The NIH model was originally employed for an inference about the mean slope over time in a sample of subjects, each of whom has a unique slope (υ) that is estimated from a set of repeated measurements over time. Within the population of subjects, these slopes follow a distribution with overall mean μ and variance σ_υ^2. The key is to then employ a moment estimator for the mixing distribution variance component, which is then used to obtain the total variance of the estimated mean slope.

4.10.2 Stratified-Adjusted Estimates from Multiple 2x2 Tables

DerSimonian and Laird (1986) applied these ideas to develop a random effects model for the analysis of multiple 2×2 tables. Their objective was to obtain an overall stratified-adjusted assessment of the treatment effect from a *meta-analysis* of many studies where there is some heterogeneity, or extra-variation or overdispersion, among studies.

Under a random effects model we now assume that the true measure of association for some scale $\theta_j = g(\pi_{1j}) - g(\pi_{2j})$ varies from stratum to stratum. Thus at the first stage of the model we assume that

$$\widehat{\theta}_j = \theta_j + \varepsilon_j\ , \tag{4.141}$$

where we again assume that

$$E(\varepsilon_j) = 0, \qquad V(\varepsilon_j) = \sigma_{\widehat{\theta}_j}^2 \qquad \text{for } j = 1, \ldots, K. \tag{4.142}$$

Note that $V(\varepsilon_j) = E\left(\widehat{\theta}_j - \theta_j\right)^2 = \sigma^2_{\widehat{\theta}_j}$ and thus the variance of the estimate $\sigma^2_{\widehat{\theta}_j}$ is equivalent to the variance of the random errors σ^2_ε. Unlike the simple measurement error model, here the variance of the estimate (of the errors) is assumed to vary from unit to unit, or from stratum to stratum.

At the second stage we assume some mixing distribution

$$\theta_j \sim f(\theta \mid \mu_\theta, \sigma^2_\theta) \tag{4.143}$$

with

$$E(\theta_j) = \mu_\theta, \qquad V(\theta_j) = \sigma^2_\theta \tag{4.144}$$

and where $\varepsilon_j \perp \theta_j$. These variance components can be expressed as

$$\begin{aligned} \sigma^2_\theta &= E\left(\theta_j - \mu_\theta\right)^2 = V\left[E\left(\widehat{\theta}_j \mid \theta_j\right)\right] \\ \sigma^2_{\widehat{\theta}_j} &= E\left(\widehat{\theta}_j - \theta_j\right)^2 = E\left[V\left(\widehat{\theta}_j \mid \theta_j\right)\right], \end{aligned} \tag{4.145}$$

where $\sigma^2_\theta \equiv \sigma^2_v$ and $\sigma^2_{\widehat{\theta}_j} \equiv \sigma^2_\varepsilon$ in the measurement error example of Section 4.10.1. Therefore, *unconditionally*

$$V\left(\widehat{\theta}_j\right) = \sigma^2_\theta + \sigma^2_{\widehat{\theta}_j}. \tag{4.146}$$

If the variance component $\sigma^2_\theta = 0$, then this implies that the fixed-effects model is appropriate. On the other hand, if $\sigma^2_\theta > 0$ then there is some over-dispersion relative to the fixed-effects model. In this case, a fixed effects analysis yields stratified-adjusted tests and estimates for which the variance is under-estimated because the model assumes that $\sigma^2_\theta = 0$.

A test of homogeneity in effect provides a test of the null hypothesis H_{0H}: $\sigma^2_\theta = 0$ versus the alternative H_{1H}: $\sigma^2_\theta \neq 0$. If this test is significant, then a proper analysis using the two-stage random effects model requires that we estimate the between stratum variance component σ^2_θ. This is readily done using a simple moment estimator derived from the test of homogeneity.

Cochran's test of homogeneity $X^2_{H,C}$ in (4.68) can be expressed as a weighted sum of squares $\sum_j \widehat{\tau}_j \left(\widehat{\theta}_j - \widehat{\mu}_\theta\right)^2$, where $\widehat{\mu}_\theta$ is the *MVLE* of the mean measure of association obtained under the fixed effects model and $\widehat{\tau}_j$ is the inverse of the estimated variance of the estimate. Clearly, the expected value $E\left(X^2_{H,C}\right)$under the alternative hypothesis H_{1H} will be some function of the variance between strata, σ^2_θ. To obtain this expectation we first apply the principle of partitioning of sums of squares described in Section A.1.3 of the Appendix.

Treating the $\{\tau_j\}$ as fixed (known), then from (A.4), the sum of squares of each estimate about the overall mean can be partitioned about the estimated mean as

$$\sum_j \tau_j \left(\widehat{\theta}_j - \mu_\theta\right)^2 = \sum_j \tau_j \left(\widehat{\theta}_j - \widehat{\mu}_\theta\right)^2 + \sum_j \tau_j \left(\widehat{\mu}_\theta - \mu_\theta\right)^2 \tag{4.147}$$

so that

$$X^2_{H,C} = \sum\nolimits_j \tau_j \left(\widehat{\theta}_j - \widehat{\mu}_\theta\right)^2 = \sum\nolimits_j \tau_j \left(\widehat{\theta}_j - \mu_\theta\right)^2 - \sum\nolimits_j \tau_j \left(\widehat{\mu}_\theta - \mu_\theta\right)^2. \tag{4.148}$$

Since $E\left(\widehat{\theta}_j - \mu_\theta\right)^2 = V\left(\widehat{\theta}_j\right)$ in (4.146), then the expected value of the test statistic is

$$E\left(X^2_{H,C}\right) = \sum\nolimits_j \tau_j V\left(\widehat{\theta}_j\right) - V\left(\widehat{\mu}_\theta\right)\left(\sum\nolimits_j \tau_j\right). \tag{4.149}$$

In practice, we would use the estimated weights $\{\widehat{\tau}_j\}$, as in Section 4.6.2. However, since $\widehat{\tau}_j \xrightarrow{p} \tau_j$, then from Slutsky's Theorem the above still applies.

Using the unconditional variance of each $\widehat{\theta}_j$ in (4.146), the first term on the r.h.s. of (4.149) is

$$\sum\nolimits_j \tau_j V\left(\widehat{\theta}_j\right) = \sum\nolimits_j \tau_j \left(\sigma^2_\theta + \sigma^2_{\widehat{\theta}_j}\right). \tag{4.150}$$

For the second term on the r.h.s. of (4.149), note that the *MVLE* is obtained as $\widehat{\mu}_\theta = \sum_j \omega_j \widehat{\theta}_j$ using the *MVLE* weights $\omega_j = \tau_j / \sum_\ell \tau_\ell$, where $\tau_j = \sigma^{-2}_{\widehat{\theta}_j}$ in (4.34) is assumed known (fixed). Again using the unconditional variance of each $\widehat{\theta}_j$, given the *MVLE* weights, then

$$V\left(\widehat{\mu}_\theta\right) = \frac{\sum_j \tau_j^2 \left(\sigma^2_{\widehat{\theta}_j} + \sigma^2_\theta\right)}{\left(\sum_j \tau_j\right)^2}. \tag{4.151}$$

Thus

$$V\left(\widehat{\mu}_\theta\right) \sum\nolimits_j \tau_j = \frac{\sum_j \tau_j^2 \left(\sigma^2_{\widehat{\theta}_j} + \sigma^2_\theta\right)}{\sum_j \tau_j} \tag{4.152}$$

so that

$$E\left(X^2_{H,C}\right) = \sum\nolimits_j \tau_j \left(\sigma^2_{\widehat{\theta}_j} + \sigma^2_\theta\right) - \frac{\sum_j \tau_j^2 \left(\sigma^2_{\widehat{\theta}_j} + \sigma^2_\theta\right)}{\sum_j \tau_j}. \tag{4.153}$$

Noting that $\tau_j = \sigma^{-2}_{\widehat{\theta}_j}$, it is readily shown that

$$E\left(X^2_{H,C}\right) = (K-1) + \sigma^2_\theta \left[\sum\nolimits_j \tau_j - \frac{\sum_j \tau_j^2}{\sum_j \tau_j}\right]. \tag{4.154}$$

From Slutsky's theorem, a consistent estimate of this expectation is provided upon substituting the estimated *MVLE* weights or the $\{\widehat{\tau}_j\}$. This yields a consistent moment estimate for σ_θ^2 of the form

$$\widehat{\sigma}_\theta^2 = \max\left[0, \frac{X_{H,C}^2 - (K-1)}{\sum_j \widehat{\tau}_j - \dfrac{\sum_j \widehat{\tau}_j^2}{\sum_j \widehat{\tau}_j}}\right], \tag{4.155}$$

where the estimate is set to zero when the solution is a negative value.

Given the estimate of the between strata variance component $\widehat{\sigma}_\theta^2$ we can then update the estimate the mean $\widehat{\mu}_\theta$ by a reweighted estimate using the unconditional variance of the estimate within each stratum:

$$\widehat{V}\left(\widehat{\theta}_j\right) = \widehat{\sigma}_{\widehat{\theta}_j}^2 + \widehat{\sigma}_\theta^2. \tag{4.156}$$

The initial *MVLE* estimate is called the *initial estimate* and the reweighted estimate is called the *first-step iterative estimate.* The first-step revised weights are

$$\widehat{\omega}_j^{(1)} = \frac{\widehat{\tau}_j^{(1)}}{\sum_\ell \widehat{\tau}_\ell^{(1)}} = \frac{\widehat{V}\left(\widehat{\theta}_j\right)^{-1}}{\sum_\ell \widehat{V}\left(\widehat{\theta}_\ell\right)^{-1}} = \frac{\left(\widehat{\sigma}_{\widehat{\theta}_j}^2 + \widehat{\sigma}_\theta^2\right)^{-1}}{\sum_\ell \left(\widehat{\sigma}_{\widehat{\theta}_\ell}^2 + \widehat{\sigma}_\theta^2\right)^{-1}}. \tag{4.157}$$

The reweighted estimate of the mean over all strata then is

$$\widehat{\mu_\theta}^{(1)} = \sum_j \widehat{\omega}_j^{(1)} \widehat{\theta}_j \tag{4.158}$$

with estimated variance

$$\widehat{V}\left(\widehat{\mu_\theta}^{(1)}\right) = \sum_j \left(\widehat{\omega}_j^{(1)}\right)^2 \left(\widehat{\sigma}_{\widehat{\theta}_j}^2 + \widehat{\sigma}_\theta^2\right). \tag{4.159}$$

From these quantities one can obtain a random effects confidence interval for the mean measure of association in the population.

DerSimonian and Laird also describe an iterative convergent solution using the EM-algorithm. Alternately, the above process could be continued to obtain fully iterative estimates of the mean and its variance. The reweighted estimate of the mean $\widehat{\mu}_\theta^{(1)}$ would be used to recalculate the test of homogeneity, which, in turn, is used to update the estimate of the variance between strata $\left(\widehat{\sigma}_\theta^2\right)^{(2)}$. This updated estimate of the variance is used to obtain revised weights $\{\widehat{\omega}_j^{(2)}\}$ and then to obtain an updated estimate of the mean $\widehat{\mu}_\theta^{(2)}$, and so on. The iterative procedure continues until both the mean $\widehat{\mu}_\theta^{(m)}$ and its variance $\widehat{V}\left(\widehat{\mu}_\theta^{(m)}\right)$ converge to constants, say at the mth step. This approach is called the *fixed-point method* of solving a system of simultaneous equations. Alternately, the two iterative estimates could be obtained simultaneously using other numerical procedures such as Newton-Raphson (*cf.* Thisted, 1988).

For such calculations Lipsitz, Fitzmaurice, Orav and Laird (1994) have shown that the one-step estimates are often very close to the final iterative estimates, and that the mean square error of the one-step estimate also is close to that of the final iterative estimate. Thus the first step estimates are often used in practice.

This iteratively reweighted random effects estimate can also be described as an *empirical Bayes estimate* (see Robbins, 1963; Morris, 1983). The addition of the non-zero variance component between strata, $\widehat{\sigma}_\theta^2$, to the variance of the estimate to obtain the unconditional variance has the effect of adding a constant to all of the weights. Thus the random effects (empirical Bayes) analysis "shrinks" the weights toward the average $1/K$, so that the resulting estimate is closer to the unweighted mean of the $\{\widehat{\theta}_j\}$ than is the *MVLE*. If the estimate of this variance component is zero, or nearly so, the random effects analysis differs trivially from the fixed effects analysis.

Thus one strategy could be to always adopt a random effects model because there is usually some extra-variation or over-dispersion, and if not then $\widehat{\sigma}_\theta^2 \doteq 0$ and the fixed analysis will result. However, this could sacrifice power in those cases where a fixed effects model actually applies because even when $\sigma_\theta^2 = 0$, the estimate $\widehat{\sigma}_\theta^2$ will vary and a small value will still inflate the variance of the estimate of $\widehat{\mu}_\theta^{(1)}$. Thus it is customary to first conduct a test of homogeneity and to only conduct a random effects analysis if significant heterogeneity is detected.

In theory, an asymptotically efficient test of the hypothesis H_0: $\mu_\theta = 0$ on a given scale under a random effects model could be obtained analogously to the Radhakrishna test of Section 4.7.2 that was derived under a fixed effects model. In practice, however, inference from a random effects analysis, such as in a meta-analysis, is usually based on the 95% confidence limits for the mean μ_θ.

Example 4.22 *Clinical Trial in Duodenal Ulcers (continued)*
For the ulcer drug trial example, the following is a summary of the computation of the one-step random effects estimate of the mean stratified-adjusted log odds ratio, where $\widehat{\theta}_j = \log \widehat{OR}_j$. As shown in Table 4.3, the *MVLE* under the fixed effects model is $\widehat{\theta} = 0.493$ with estimated variance $\widehat{V}(\widehat{\theta}) = 0.085$. The corresponding estimate of the assumed common odds ratio is $\widehat{OR} = 1.637$ with asymmetric 95% *C.I.* of $(0.924, 2.903)$. The resulting test of homogeneity of the log odds ratios is $X_{H,C}^2 = 4.58028$ with $p \leq 0.102$. Although not significant, the random effects analysis is presented for purpose of illustration.

The moment estimate of the variance between strata is $\widehat{\sigma}_\theta^2 = 0.37861$. This is then used to obtain an updated (one-step) estimate of the mean μ_θ and its variance. The random effects analysis, contrasted to the original fixed effects (*MVLE*) analysis

Table 4.9 Random-Effects Model Stratified-Adjusted Analysis of the Ulcer Clinical Trial Data From Example 4.1

	Stratum			Mean	
Measure, $\widehat{\theta}_j$	1	2	3	$\widehat{\mu}_\theta$	95% *C.I.*
Risk difference	−0.045	0.306	0.245	0.142	−0.08, 0.37
$\widehat{V}(\widehat{\theta}_j)$	0.033	0.065	0.032	0.013	
$\widehat{\omega}_j^{(1)}$	0.397	0.201	0.402		
log *Relative risk*	−0.111	0.523	0.515	0.284	
$\widehat{V}(\widehat{\theta}_j)$	0.136	0.235	0.122	0.050	
$\widehat{\omega}_j^{(1)}$	0.372	0.215	0.413		
Relative risk				1.329	0.86, 2.06
log *Odds ratio*	−0.185	1.322	1.000	0.574	
$\widehat{V}(\widehat{\theta}_j)$	0.567	1.273	0.568	0.232	
$\widehat{\omega}_j^{(1)}$	0.489	0.102	0.408		
Odds ratio				1.775	0.69, 4.56

is as follows:

			MVLE		Random Effects		
Stratum	$\widehat{\theta}_j$	$\widehat{\sigma}_j^2$	$\widehat{\tau}_j$	$\widehat{\omega}_j$	$\widehat{V}(\widehat{\theta}_j)$	$\widehat{\tau}_j^{(1)}$	$\widehat{\omega}_j^{(1)}$
1	−0.185	0.188	5.319	0.454	0.567	1.765	0.409
2	1.322	0.894	1.118	0.095	1.273	0.786	0.182
3	1.001	0.189	5.277	0.451	0.568	1.760	0.408
Total			11.715	1.0		4.311	1.000

Through the addition of the variance component estimate $\widehat{\sigma}_\theta^2 = 0.37861$ to the unconditional variance of each estimate, the random effects weights for each stratum are now shrunk toward 1/3, such as from 0.454 to 0.409 for the first stratum. The resulting one-step reweighted estimate of the mean log odds ratio is $\widehat{\mu}_\theta^{(1)} = (-0.185 \times 0.409) + (1.322 \times 0.182) + (1.001 \times 0.408) = 0.574$ and $\widehat{\mu}_{OR}^{(1)} = 1.775$. The estimated variance of the log odds ratio is $\widehat{V}(\widehat{\mu}_\theta^{(1)}) = 1/4.311 = 0.2320$, which yields asymmetric 95% *C.I.* on μ_{OR} of $(0.691, 4.563)$. The point estimate of the mean log odds ratio in the random effects model is slightly greater than the *MVLE* because the negative estimate in the first stratum is given less weight. However, the variance of the random effects estimate is slightly greater because of the allowance for the extra-variation between strata, so that the confidence limits are wider.

Table 4.10 Meta-analysis of Prevention of Pre-Eclampsia With Diuretics During Pregnancy

	Diuretics Group			*Placebo Group*			
Study	a_j	n_{1j}	p_{1j}	b_j	n_{2j}	p_{2j}	OR_j
1	14	131	0.107	14	136	0.103	1.043
2	21	385	0.055	17	134	0.127	0.397
3	14	57	0.246	24	48	0.500	0.326
4	6	38	0.158	18	40	0.450	0.229
5	12	1011	0.012	35	760	0.046	0.249
6	138	1370	0.101	175	1336	0.131	0.743
7	15	506	0.030	20	524	0.038	0.770
8	6	108	0.056	2	103	0.019	2.971
9	65	153	0.425	40	102	0.392	1.145

Table 4.9 presents a summary of the computations of the random effects estimates of the mean parameter μ_θ and its estimated large sample variance for the risk difference, relative risk and odds ratio. Of the three scales, the relative risk estimates are the least affected by the random effects analysis. That is because the estimates of the relative risks among the strata showed the least heterogeneity. The corresponding estimate of the variance between strata in the log relative risks is $\widehat{\sigma}_\theta^2 = 0.06818$, which is smaller, relative to the variance of the estimates within strata, than the between stratum variances for the other scales. Thus the $\{\widehat{\omega}_j^{(1)}\}$ are minimally different in the fixed effects and random effects analyses of the relative risks.

The estimate of the variance between strata of the risk differences is $\widehat{\sigma}_\theta^2 = 0.02235$, which yields similar effects on the estimates as observed for the log odds ratios.

Example 4.23 *Religion and Mortality (continued)*
For the observational study of religion and mortality, the estimates of the variances between strata are zero for the log odds ratios and the log relative risks, as reflected by the non-significant tests of homogeneity in Example 4.11. The estimate for the variance among strata for the risk differences is $\widehat{\sigma}_\theta^2 = 0.00132$, which has a slight effect on the resulting estimates: the estimated mean being $\widehat{\mu}_\theta^{(1)} = -0.040$ with $\widehat{V}(\widehat{\mu}_\theta^{(1)}) = 0.00117$ and 95% confidence limits of (−0.107, 0.027), slightly wider than the fixed effects limits.

Example 4.24 *Meta-Analysis of Effects of Diuretics on Pre-Eclampsia*
Collins, Yusuf and Peto (1985) present a meta-analysis of nine studies of the use of diuretics during pregnancy to prevent the development of pre-eclampsia. The data are presented in Table 4.10 (reproduced with permission). Of the nine studies,

Fig. 4.6 Meta-analysis display of the odds ratio and 95% confidence limits on the log scale for the studies of pre-eclampsia, and the random-effects model combined analysis.

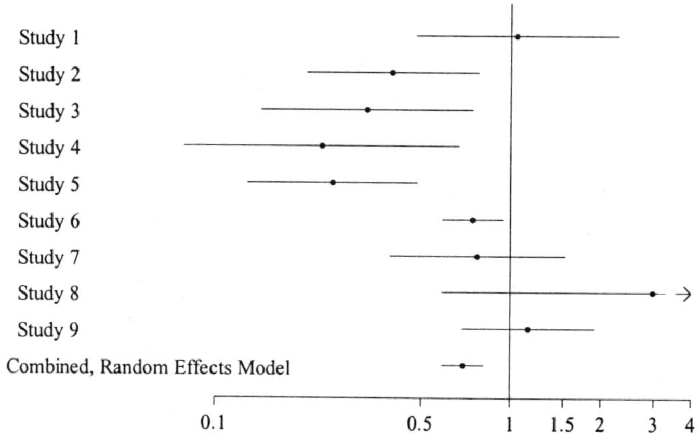

three show an increase in the odds ratio of pre-eclampsia among those treated with diuretics whereas the others show a decreased risk. The Cochran test of homogeneity of the log odds ratios yields $X^2 = 27.3$ on 8 df with $p \leq 0.0007$. The initial estimate of the variance in the log odds ratios between studies is $\widehat{\sigma}_\theta^2 = 0.2297$. The one-step random effects estimate of the average log odds ratio is $\widehat{\mu}_\theta^{(1)} = -0.517$ with estimated variance $\widehat{V}(\widehat{\mu}_\theta^{(1)}) = 0.0415$, which yields a point estimate of the average odds ratio of $\widehat{\mu}_{OR}^{(1)} = 0.596$ with asymmetric 95% confidence limits (0.400, 0.889) that is significant at the 0.05 level since the limits do not bracket 1.0.

Applying the fixed-point iterative algorithm, wherein the estimate $\widehat{\sigma}_\theta^2$ is used to obtain an updated estimate $\widehat{\mu}_\theta^{(2)}$, and so on, requires 27 iterations to reach convergence. The final estimate of the mixing distribution variance is $\widehat{\sigma}_\theta^2 = 0.1733$, somewhat less than the initial estimate. The resulting estimate of the mean log odds ratio is $\widehat{\mu}_\theta = -0.513$ with estimated variance $\widehat{V}(\widehat{\mu}_\theta) = 0.0346$. The estimate of the mean is virtually unchanged but that of the variance is slightly less. The final estimate of the mean odds ratio, therefore, is $\widehat{\mu}_{OR} = 0.599$ with asymmetric 95% confidence limits (0.416, 0.862).

In a meta-analysis it is traditional to display the results as in Figure 4.6, which shows the odds ratio and 95% confidence limits within each study and in aggregate. For illustration, the figure displays the random-effects model estimate of the aggregate combined odds ratio and its 95% confidence limits. Because the confidence interval does not bracket 1.0, the conclusion would be that the aggregate estimate is statistically significant at the 0.05 level.

4.11 POWER AND SAMPLE SIZE FOR TESTS OF ASSOCIATION

Wittes and Wallenstein (1987) describe the power function of the Mantel-Haenszel test for multiple 2×2 tables. Earlier, Gail (1973) presented the power function for a similar but different test, a Wald test based on the *MVLE*; and Birch (1964) described the conditional power function using the non-centrality parameter of the non-central chi-square distribution. Because the Mantel-Haenszel test is asymptotically equivalent to Cochran's test, and since Cochran's test is a member of the Radhakrishna family, the Wittes-Wallenstein result can be derived more generally for the Radhakrishna family.

4.11.1 Power Function of the Radhakrishna Family

The Radhakrishna family provides an asymptotically efficient test for the assumed common measure of association θ, where $\theta_j = g(\pi_{1j}) - g(\pi_{2j}) = \theta \ \forall j$ under a fixed effects model. The test statistic T in (4.85) is obtained as a weighted linear combination of the risk differences within each stratum, the weights providing an asymptotically optimal test for a measure on the specified scale $g(\pi)$. The variance of the test statistic involves the variance of the risk differences σ_{0j}^2 that involves the sample sizes within each stratum. As shown in (4.119), the variance can be factored as ϕ_{0j}^2/N using $E(n_{ij}) = N\zeta_j\xi_{ij}$, where $\zeta_j = E(N_j/N)$ are the expected stratum sample fractions, and $\xi_{ij} = E(n_{ij}/N_j)$ the group sample fractions within the jth stratum. Thus

$$\phi_{0j}^2 = \frac{\pi_j(1-\pi_j)}{\zeta_j\xi_{1j}\xi_{2j}} \,. \tag{4.160}$$

Therefore, the test statistic can be expressed as

$$X_{A(g)}^2 = \frac{T^2}{V(T|H_0)} = \frac{\sum_j \left[\widetilde{w}_j(p_{1j}-p_{2j})\right]^2}{\sum_j \widetilde{w}_j^2\sigma_{0j}^2} = \frac{\widetilde{T}^2}{V\left(\widetilde{T}|H_0\right)} \,. \tag{4.161}$$

using the weights

$$\widetilde{w}_j = \frac{1}{\phi_{0j}^2 g'(\pi_j)} \,. \tag{4.162}$$

Under H_0: $\pi_{1j} = \pi_{2j} = \pi_j$, then $\widetilde{T} \overset{d}{\approx} \mathcal{N}\left(0, \sigma^2_{\widetilde{T}_0}\right)$, where $E\left(\widetilde{T}\right) = 0$ and

$$V\left(\widetilde{T}|H_0\right) = \sigma^2_{\widetilde{T}_0} = \sum\nolimits_j^2 \widetilde{w}^2 \sigma^2_{0j} = \sum\nolimits_j \widetilde{w}_j^2 \frac{\phi^2_{0j}}{N} \tag{4.163}$$
$$= \frac{1}{N}\sum\nolimits_j \left(\frac{1}{\phi^2_{0j} g'(\pi_j)^2}\right) = \frac{\phi^2_{\widetilde{T}_0}}{N},$$

where $\phi^2_{\widetilde{T}_0}$ is the variance component remaining after factoring N from this expression for the variance under H_0.

Under the omnibus alternative H_{1O} in (4.49) that some of the stratum specific parameters differ from the null value, $\theta_j \neq \theta_0 = 0$ for some $1 \leq j \leq K$, then asymptotically $\widetilde{T} \overset{d}{\approx} \mathcal{N}\left(\mu_{\widetilde{T}}, \sigma^2_{\widetilde{T}_1}\right)$ with mean

$$\mu_{\widetilde{T}} = \sum\nolimits_j \widetilde{w}_j (\pi_{1j} - \pi_{2j}) \cong \sum\nolimits_j \left(\frac{\theta_j}{\phi^2_{0j} g'(\pi_j)^2}\right) \tag{4.164}$$

and variance

$$\sigma^2_{\widetilde{T}_1} = \sum\nolimits_j \widetilde{w}^2_{gj} \sigma^2_{1j}. \tag{4.165}$$

where $\sigma^2_{1j} = V(p_{1j} - p_{2j}|H_1)$ is the variance of the risk difference under the alternative hypothesis within the jth stratum as in (2.26). This variance likewise can be factored as $\sigma^2_{1j} = \phi^2_{1j}/N$ with

$$\phi^2_{1j} = \frac{1}{\zeta_j}\left(\frac{\pi_{1j}(1-\pi_{1j})}{\xi_{1j}} + \frac{\pi_{2j}(1-\pi_{2j})}{\xi_{2j}}\right). \tag{4.166}$$

Therefore, the asymptotic variance under the alternative is

$$\sigma^2_{\widetilde{T}_1} = \frac{1}{N}\sum\nolimits_j \frac{\phi^2_{1j}}{\left[\phi^2_{0j} g'(\pi_j)^2\right]^2} = \frac{\phi^2_{\widetilde{T}_1}}{N}. \tag{4.167}$$

From the general equation for sample size and power (3.18) we then obtain

$$\left|\mu_{\widetilde{T}}\right| = Z_{1-\alpha}\frac{\phi_{\widetilde{T}_0}}{\sqrt{N}} + Z_{1-\beta}\frac{\phi_{\widetilde{T}_1}}{\sqrt{N}} \tag{4.168}$$

and

$$\sqrt{N}\left|\mu_{\widetilde{T}}\right| = Z_{1-\alpha}\left[\sum\nolimits_j \left(\frac{1}{\phi^2_{0j} g'(\pi_j)^2}\right)\right]^{1/2} + Z_{1-\beta}\left[\sum\nolimits_j \frac{\phi^2_{1j}}{\left[\phi^2_{0j} g'(\pi_j)^2\right]^2}\right]^{1/2}. \tag{4.169}$$

This expression can be solved for N to determine the sample size required to provide any specific level of power, and can be solved for $Z_{1-\beta}$ to determine the power of a test for any given sample size.

4.11.2 Power and Sample Size for Cochran's Test

Cochran's test is directed toward alternatives specified in terms of odds ratios using the logit link, $g(\pi) = \log[\pi(1-\pi)]$, for which the optimal weights are given in (4.82). Factoring the total sample size N from this expression yields

$$w_j = \frac{N}{\phi_{0j}^2 g'(\pi_j)} = \frac{N}{\left(\frac{1}{\zeta_j \xi_{1j}} + \frac{1}{\zeta_j \xi_{2j}}\right)} \tag{4.170}$$

so that

$$\widetilde{w}_j = \left(\frac{1}{\zeta_j \xi_{1j} \xi_{2j}}\right)^{-1} = \zeta_j \xi_{1j} \xi_{2j}. \tag{4.171}$$

Therefore, the variance component under the null hypothesis is

$$\phi_{\widetilde{T}_0}^2 = \sum\nolimits_j \widetilde{w}_j^2 \phi_{0j}^2 = \sum\nolimits_j \zeta_j \xi_{1j} \xi_{2j} \pi_j (1-\pi_j). \tag{4.172}$$

Under the alternative, the expectation is

$$\mu_{\widetilde{T}} = \sum\nolimits_j \zeta_j \xi_{1j} \xi_{2j} (\pi_{1j} - \pi_{2j}) \tag{4.173}$$

with variance component

$$\phi_{\widetilde{T}_1}^2 = \sum\nolimits_j \widetilde{w}_j^2 \phi_{1j}^2 = \sum\nolimits_j (\zeta_j \xi_{1j} \xi_{2j}) \left[\xi_{2j} \pi_{1j} (1-\pi_{1j}) + \xi_{1j} \pi_{2j} (1-\pi_{2j})\right]. \tag{4.174}$$

Therefore, the expression for the power of the test is

$$Z_{1-\beta} = \frac{\left|\mu_{\widetilde{T}}\right| \sqrt{N} - Z_{1-\alpha} \phi_{\widetilde{T}_0}}{\phi_{\widetilde{T}_1}} \tag{4.175}$$

and the expression for the sample size required to provide power $1-\beta$ is

$$N = \left[\frac{Z_{1-\alpha} \phi_{\widetilde{T}_0} + Z_{1-\beta} \phi_{\widetilde{T}_1}}{\mu_{\widetilde{T}}}\right]^2. \tag{4.176}$$

To perform computations of the power or required sample size under a specific alternative hypothesis, one first specifies the expected sample fractions ζ_j, ξ_{1j}, and ξ_{2j} for each stratum $(1 \leq j \leq K)$. For each stratum the expected control group probability π_{2j} is specified and the expected log odds ratio θ_j in that stratum. By inverting the expression for the odds ratio, the expected probability in the exposed or treated group is then obtained as

$$\pi_{1j} = \frac{\pi_{2j} e^{\theta_j}}{(1-\pi_{2j}) + \pi_{2j} e^{\theta_j}}. \tag{4.177}$$

Although the test is asymptotically most powerful when there is absolute homogeneity of the odds ratios, $\theta_j = \theta$ for all strata, these expression allow for the θ_j to differ systematically among strata under a fixed effects model without a common value for θ. Also, the variance components in (4.172) and (4.174) are obtained under this fixed effects model.

Example 4.25 *Three Strata with Heterogeneity*
Consider a non-randomized study with three strata allowing for some imbalances between groups among strata ($\xi_{1j} \neq \xi_{2j}$) and some heterogeneity of odds ratios as reflected by the following parameters under the alternative hypothesis:

Age Stratum	*Stratum Fraction* ζ_j	*Group Fractions* ξ_{1j}	ξ_{2j}	π_{2j}	OR_j (e^{θ_j})	π_{1j}
20–49	0.15	0.30	0.70	0.75	4.0	0.923
50–69	0.50	0.60	0.40	0.70	3.2	0.882
70–80	0.35	0.45	0.55	0.65	1.8	0.770

As described above, the values $\{\pi_{1j}\}$ are obtained from the specification of the control group probabilities $\{\pi_{2j}\}$ and the odds ratio $\{OR_j\}$. For this example, $\mu_{\widetilde{T}} = 0.0377$, $\phi_{\widetilde{T}_0} = 0.2055$, and $\phi_{\widetilde{T}_1} = 0.2033$. For a test at $\alpha = 0.05$ (2-sided), in order to provide 90% power ($\beta = 0.10$), a total sample size $N = 306.8$ is required.

This calculation could be compared to the sample size required to provide 90% power in a marginal analysis. Weighting by the stratum specific sample fractions, $\zeta_j \xi_{ij}$ yields overall group fractions $\xi_{1\bullet} = \sum_j \zeta_j \xi_{1j} = 0.5025$, $\xi_{2\bullet} = 0.4975$, and probabilities $\pi_{2\bullet} = \sum_j \zeta_j \xi_{2j} \pi_{2j} / \xi_{2\bullet} = 0.689$, and $\pi_{1\bullet} = \sum_j \zeta_j \xi_{1j} \pi_{1j} / \xi_{1\bullet} = 0.847$. The resulting marginal odds ratio is $OR_{\bullet} = 2.505$. Then using the equation for sample size for the test of two proportions in (3.36) yields $N = 294.4$.

Therefore, the marginal analysis appears to be more powerful and to require a smaller sample size than does the stratified analysis. This is caused by the heterogeneity of the odds ratios over strata and the treatment group imbalances over strata. As shown in Section 4.4.3, the marginal analysis is biased in this case. The stratified analysis, however, is unbiased, and the larger sample size is indeed required to provide an unbiased test with the desired level of power.

The impact of these factors is described in Table 4.11 which compares the sample sizes required for a stratified versus a marginal analysis where the parameters in this example are modified so as to have either homogeneity or heterogeneity coupled with either group by covariate imbalance or balance. Under homogeneity of odds ratios, the odds ratio within each stratum is assumed to be $OR_j = 2.52 \quad \forall j$, which is the weighted average odds ratio among the three strata. Also in the case of a balanced design, the sample fractions are assumed equal in each group ($\xi_{1j} = \xi_{2j}$). The computations in the first row of the table are those described above. Such computations demonstrate that a moderate degree of heterogeneity alone has little effect on the power of the stratified versus that of the marginal analysis. The important determinant of power is the average odds ratio. The introduction of a group by covariate imbalance appears to reduce the power of the stratified test due to the pursuant increase in the variance of the test because the term

$$\left(\frac{1}{\xi_{1j}} + \frac{1}{\xi_{2j}}\right) = \frac{1}{\xi_{1j}\xi_{2j}} \tag{4.178}$$

Table 4.11 Required Sample Size for Stratified and Unstratified Analysis With Homogeneity or Heterogeneiy of Odds Ratios, and With Balanced or Unbalanced Group Fractions

Odds Ratios Among Strata	*Group Sample Fractions*	*Analysis* Stratified	Marginal
Heterogeneous:			
	Unbalanced	306.8	294.4
	Balanced	287.8	291.5
Homogeneous:			
	Unbalanced	305.6	287.3
	Balanced	292.4	293.6

is a minimum for $\xi_{1j} = 0.5$.

Lachin and Bautista (1995) provide a detailed assessment of factors influencing the power of a stratified versus unstratified analysis. They considered three factors a) the degree of heterogeneity among the odds ratios $\{\theta_j\}$; b) the strength of the association between the covariate and the response (the risk factor potential); and c) the association between the covariate and group (the extent of imbalance). In accordance with the developments in Section 4.4.3, they show that when (b) and (c) both exist, the unadjusted marginal odds ratio is positively biased whereas the stratified-adjusted estimate is not. However, a strong association in both (b) and (c) must be present for a substantial bias to exist. They also show that heterogeneity of odds ratios alone has little effect on the difference in power between the unadjusted versus adjusted analysis.

Since the marginal analysis is positively biased when (b) and (c) exist, then the marginal test appears to have greater power due to this bias, and will misleadingly suggest that a smaller N is required. However, in this case the preferred analysis is the stratified analysis that is unbiased. However, a larger N is required to provide a given level of power.

4.12 PROBLEMS

4.1 Show that Cochran's test for multiple 2×2 tables is equal to

$$X^2_{U,C} = \frac{\left(\sum_j \left[a_j - E\left(a_j\right)\right]\right)^2}{\sum_j \widehat{V}_u\left(a_j\right)} = \frac{\left[\sum_j \widehat{w}_j \left(p_{1j} - p_{2j}\right)\right]^2}{\sum_j^2 \widehat{w}_j^2 \widehat{V}_{0j}} \tag{4.179}$$

with

$$\widehat{w}_j = \frac{1}{\left(\frac{1}{n_{1j}} + \frac{1}{n_{2j}}\right)} \qquad \widehat{V}_{0j} = p_j\left(1 - p_j\right)\left(\frac{1}{n_{1j}} + \frac{1}{n_{2j}}\right). \tag{4.180}$$

4.2 Assume that there is a common value of θ for all strata, or H_0: $\theta_j = \theta$ such that $E\left(\widehat{\theta}_j\right) = \theta$ for all j. Thus we wish to combine the estimates over strata to estimate θ.

4.2.1. Under a fixed effects model (4.30) and (4.31), use weighted least squares (see Section A.5 of the Appendix) to show that

$$\widehat{\theta} = \sum\nolimits_j \widehat{\tau}_j \widehat{\theta}_j / \left(\sum\nolimits_j \widehat{\tau}_j\right) \tag{4.181}$$

and

$$\widehat{V}\left(\widehat{\theta}\right) = \left(\sum\nolimits_j \widehat{\tau}_j\right)^{-1}, \tag{4.182}$$

where $\widehat{\tau}_j = \widehat{V}(\widehat{\theta}_j)^{-1} = \widehat{\sigma}_j^{-2}$.

4.2.2. Also show that the variance of the estimate can be obtained as

$$V(\widehat{\theta}) = \sum\nolimits_j \omega_j^2 \sigma_j^2 = \frac{1}{\sum_j \sigma_j^{-2}}. \tag{4.183}$$

4.2.3. For the risk difference, log odds ratio and log relative risk, use the above to obtain the expressions for $\widehat{\tau}_j$, the weights $\widehat{\omega}_j$, the pooled *MVLE* $\widehat{\theta}$, $V\left(\widehat{\theta}\right)$ and a consistent estimate $\widehat{V}\left(\widehat{\theta}\right)$. In these derivations use the expressions for the variance of $\widehat{\theta}_j$ under the assumption that a difference exists $(\pi_{1j} \neq \pi_{2j})$.

4.2.4. Do likewise for θ_j defined as $\theta_j = g(\pi_1) - g(\pi_2)$ for the following scales:

1. The arcsin transformation $g(\pi) = \sin^{-1}(\sqrt{\pi})$.
2. The probit transformation $g(\pi) = \Phi^{-1}(\pi)$, where $\Phi(\cdot)$ is the standard normal cdf.
3. The square root transformation $g(\pi) = \sqrt{\pi}$.

4.3 The Mantel-Haenszel estimate of the common odds ratio is also a linear estimator as shown in (4.15) and (4.16).

4.3.1. Show that these $\widehat{v}_j$ are not proportional to $V(\widehat{OR}_j)^{-1}$, where $V(OR_j)$ was derived in Problem 2.5.4. Thus the variance of the Mantel-Haenszel estimate $\widehat{OR}_{MH}$ is greater than that of an *MVLE* efficient estimate of the common odds ratio.

4.3.2. Also, show that for $OR_j \doteq 1$, then the $\widehat{OR}_{MH}$ is approximately the *MVLE* of the common odds ratio (not the common log odds ratio).

4.4 The Radhakrishna family of tests $X^2_{A(g)}$ is presented in (4.98) using estimated optimal weights $\{\widehat{w}_j\}$ in (4.97) specific to the desired scale $\theta = g(\pi_1) - g(\pi_2)$, and where $X^2_{A(g)}$ is asymptotically distributed as chi-square on 1 *df* under H_0: $\pi_{1j} = \pi_{2j} = \pi_j \; \forall j$.

4.4.1. Adopt a local alternative of the form $\pi_{1j} = \pi_j + \delta n^{-\frac{1}{2}}$ and $\pi_{2j} = \pi_j - \delta n^{-\frac{1}{2}}$. Use a Taylor's expansion of $g(\pi_{ij})$ about $\pi_j \in (\pi_{1j}, \pi_{2j})$ for each group $i = 1, 2$ to derive (4.83) and (4.84).

4.4.2. Using the expression for the optimal weights that maximize the asymptotic efficiency, $\boldsymbol{W} = \boldsymbol{\Sigma}_0^{-1} \boldsymbol{G}$ in (4.95), show that the optimal weights for each of the following scales are estimated as:

Scale	$\widehat{w}_j$
Risk difference	$\left(\frac{1}{n_{1j}} + \frac{1}{n_{2j}}\right)^{-1} \frac{1}{p_j(1-p_j)}$
Log relative risk	$\left(\frac{1}{n_{1j}} + \frac{1}{n_{2j}}\right)^{-1} \frac{1}{1-p_j}$
Log odds ratio	$\left(\frac{1}{n_{1j}} + \frac{1}{n_{2j}}\right)^{-1}$

that simplify to the expressions in Table 4.8.

4.4.3. Derive the explicit expression for the test for each scale presented below:

1. Constant risk difference:

$$X^2_{A(RD)} = \frac{\left[\sum_j \widehat{V}_{0j}^{-1} (p_{1j} - p_{2j})\right]^2}{\sum_j \widehat{V}_{0j}^{-1}} \tag{4.184}$$

2. Constant log relative risk

$$X^2_{A(RR)} = \frac{\left[\sum_j p_j \widehat{V}_{0j}^{-1} (p_{1j} - p_{2j})\right]^2}{\sum_j p_j^2 \widehat{V}_{0j}^{-1}} \tag{4.185}$$

3. Constant log odds ratio

$$X^2_{A(OR)} = \frac{\left[\sum_j p_j (1-p_j) \widehat{V}_{0j}^{-1} (p_{1j} - p_{2j})\right]^2}{\sum_j [p_j (1-p_j)]^2 \widehat{V}_{0j}^{-1}} \tag{4.186}$$

4.4.4. The additional scales used in Problem 4.2.4. are also members of the Radhakrishna family. Show that the estimated optimal weights for the test for a common difference on each of these scales are as follows:

Scale	$\widehat{w}_j$
Arcsin	$\left(\frac{n_{1j}n_{2j}}{N_j}\right)\sqrt{\frac{1}{p_j\left(1-p_j\right)}}$
Probit	$\left(\frac{n_{1j}n_{2j}}{N_j}\right)\frac{\exp[-\Phi^{-1}(p_j)^2/2]}{p_j\left(1-p_j\right)\sqrt{2\pi}}$
Square root	$\left(\frac{n_{1j}n_{2j}}{N_j}\right)\frac{1}{\left(1-p_j\right)\sqrt{p_j}}$

4.5 For a scale with measure $\theta_j = g(\pi_{1j}) - g(\pi_{2j})$, we now show that the Radhakrishna test of association may also be viewed as an estimation-based test using an *MVLE* of the common parameter, but using weights estimated under H_0 rather than under H_1.

4.5.1. Since $\widehat{\theta}_j \cong g'\left(p_j\right)\left(p_{1j} - p_{2j}\right)$ show that the numerator of $X^2_{A(g)}$ in (4.98) is approximately a weighted average of the $\{\widehat{\theta}_j\}$ of the form

$$T_g \cong \sum\nolimits_j \frac{\widehat{\theta}_j}{g'\left(\pi_j\right)^2 V_{0j}} \tag{4.187}$$

4.5.2. From the δ-method show that asymptotically

$$V\left(T_g \,|\, H_0\right) \cong \sum\nolimits_j \frac{1}{g'\left(\pi_j\right)^2 V_{0j}} \tag{4.188}$$

4.5.3. Thus asymptotically the test can be expressed as

$$X^2_{A(g)} \cong \frac{\left[\sum_j \omega_{0j}\widehat{\theta}_j\right]^2}{\sum_j^2 \omega_{0j}^2 V\left(\widehat{\theta}_j \,|\, H_0\right)} \tag{4.189}$$

with weights

$$\omega_{0j} = V\left(\widehat{\theta}_j \,|\, H_0\right)^{-1} \quad \widehat{=} \quad \left[g'\left(p_j\right)^2 \widehat{V}_{0j}\right]^{-1}. \tag{4.190}$$

4.6 Now let $Z_{A(g)}$ be the standardized normal deviate test corresponding to the Radhakrishna test $X^2_{A(g)}$ in (4.98). Consider two separate tests of association on two different scales, such as Z_r for the test of common log odds ratio and Z_s for the test of common log relative risk.

4.6.1. Show that the correlation among the two standardized deviates is

$$corr(Z_r, Z_s) = \rho_{rs} = \frac{\sum_j w_{rj}w_{sj}\pi_j\left(1-\pi_j\right)\left(\frac{1}{n_{1j}} + \frac{1}{n_{2j}}\right)}{\left(V_r V_s\right)^{1/2}}, \tag{4.191}$$

where V_r and V_s are the denominators in the Radhakrishna test X_A^2 for the rth scale and sth scale, respectively.

Hint: For a vector $\boldsymbol{X}$ with covariance matrix $\boldsymbol{\Sigma}_x$, the covariance of two bilinear forms $\boldsymbol{A}'\boldsymbol{X}$ and $\boldsymbol{B}'\boldsymbol{X} = \boldsymbol{A}'\boldsymbol{\Sigma}_x\boldsymbol{B}$.

4.6.2. Consider two scales $g_s(\pi)$ and $g_r(\pi)$, where $\theta_{sj} = \theta_s \; \forall j$ for scale $g_s(\pi)$. Thus the test T_s is optimal and test T_r is not. Show that the $ARE\,(T_r, T_s \mid \theta_s) = \rho_{rs}^2$ as presented in (4.117).

4.6.3. Let Z_r and Z_s be the standardized deviates corresponding to T_r and T_s. Show that $ARE\,(T_r, T_s \mid \theta_s) = ARE\,(Z_r, Z_s \mid \theta_s)$.

4.6.4. Using the sample fractions described in (4.119), show that the sample size N factors from the above expression for ρ_{rs}^2 using weights $\widetilde{w}_{\ell j}$ presented in (4.120) to yield

$$ARE\,(T_r, T_s|\theta_s) = \frac{\left[\sum_j \widetilde{w}_{rj}\widetilde{w}_{sj}\phi_{0j}^2\right]^2}{\left[\sum_j \widetilde{w}_{rj}^2\phi_{0j}^2\right]\left[\sum_j \widetilde{w}_{sj}^2\phi_{0j}^2\right]} = \rho_{rs}^2. \tag{4.192}$$

4.6.5. To compute the $\widehat{ARE}$ for each pair of scales the covariance of the two tests involves the cross products $\widehat{w}_{rj}\widehat{w}_{sj}\widehat{V}_{0j}$ that can be simplified. Let

$$M_j = \left(\frac{1}{n_{1j}} + \frac{1}{n_{2j}}\right)^{-1} = \left(\frac{n_{1j}n_{2j}}{N_j}\right) \tag{4.193}$$

so that $\widehat{V}_{0j} = p_j(1-p_j)/M_j$. Then for the following pairs of scales show that these terms reduce as follows:

Pair of Tests	$\widehat{w}_{rj}\widehat{w}_{sj}\widehat{V}_{0j}$
log Odds ratio, Risk difference	M_j
log Odds ratio, log Relative risk	p_jM_j
log Relative risk, Risk difference	$M_j/(1-p_j)$

4.7 For the measurement error random effects model of Section 4.10.1, assume that the random error variances vary from subject to subject such that $V(\varepsilon_i) = \sigma_{\varepsilon_i}^2$ is not constant for all subjects in the population, such as where $\sigma_{\varepsilon_i}^2$ is a function of the conditional mean $E(y|v_i)$. Also assume that a consistent estimate of the error variance $\widehat{\sigma}_{\varepsilon_i}^2$ is provided for each subject.

4.7.1. Using the partitioning of variation as in (A.6), show that $V(y_i) = \sigma_{\varepsilon_i}^2 + \sigma_v^2$ and that $V(Y) = E(\sigma_{\varepsilon_i}^2) + \sigma_v^2$.

4.7.2. For a sample of N independent observations, show that

$$V\left[\sum\nolimits_i y_i\right] = NV(Y) = E\left[\sum\nolimits_i \sigma_{\varepsilon_i}^2\right] + N\sigma_v^2\,. \tag{4.194}$$

4.7.3. Then show that the moment estimator for the mixing distribution variance σ_v^2 is provided by

$$\widehat{\sigma}_v^2 = \widehat{V}(Y) - \sum\nolimits_i \widehat{\sigma}_{\varepsilon_i}^2/N \tag{4.195}$$

where

$$\widehat{V}(Y) = S_y^2 = \frac{\sum_i (y_i - \overline{y})^2}{N-1}. \tag{4.196}$$

4.7.4. Alternately, partition the sum of squares of Y about μ as in (A.4) to show that

$$S_y^2 = \frac{\sum_i (y_i - \mu)^2 - N(\overline{y} - \mu)^2}{N-1} \tag{4.197}$$

and that

$$E(S_y^2) = \frac{\sum_i \sigma_{\varepsilon_i}^2}{N} + \sigma_v^2. \tag{4.198}$$

4.7.5. Then again show that the moment estimator for σ_v^2 is provided by (4.195).

4.8 For the DerSimonian-Laird random effects model in Section 4.10.2, show the following:

4.8.1. Show that (4.149) can be expressed as

$$E\left(X_{H,C}^2\right) = \sum_j \tau_j \left(\sigma_{\widehat{\theta}_j}^2 + \sigma_\theta^2\right) - \left(\sum_j \tau_j\right) V\left(\widehat{\mu}_\theta\right). \tag{4.199}$$

4.8.2. Then derive (4.151).

4.8.3. Then derive (4.153), (4.154) and (4.155).

4.8.4. In a fixed-point iterative procedure as described in (4.157)– (4.159), at the first step, show the expressions for the weights $\widehat{\omega}_j^{(1)}$ and the estimated variance $\widehat{V}\left(\widehat{\mu}_\theta^{(1)}\right)$.

4.9 A cross-sectional occupational health study assessed the association between smoking and the prevalence of byssinosis among workers in a textile plant (Higgins and Koch, 1977). The following 2×2 tables were obtained giving the frequencies of workers with byssinosis versus not among smokers and non-smokers stratified by the length of employment:

	Smoker		*Non-Smoker*	
Stratum	a_j	n_{1j}	b_j	n_{2j}
1: <10 years	44	1587	19	1142
2: 10–20 yr	21	481	5	231
3: >20 yr.	60	1121	16	857

(reproduced with permission). In the remainder of this problem, various analyses of these data will be performed. It is preferred that you write computer programs to perform these analyses. Measures of association in terms of smokers versus non-smokers should be constructed so as to express the increase in risk, if any, associated with smoking.

4.9.1. Separately within each strata, and for the data combined into a single marginal 2×2 table with total $N = 5419$, calculate the following:

1. V_u and the Contingency (Pearson) chi-square statistic and P-value;
2. V_c and the Mantel chi-square statistic and P-value;
3. The difference in proportions, the log odds ratio and the log relative risk and estimated the variance of each; and
4. The 95% confidence limits for the difference, odds ratio and relative risk.

4.9.2. Now conduct a stratified Mantel-Haenszel analysis of odds ratios.

1. Calculate the Mantel-Haenszel estimate of the common odds ratio and the Robins et al. variance of the log odds ratio. Use these to compute asymmetric 95% confidence limits on the odds ratio.
2. Calculate the Mantel-Haenszel and Cochran stratified-adjusted tests of significance.
3. Now use the Mantel-Haenszel test value to obtain the test-inverted confidence limits for the common odds ratio.
4. Compare and comment on the difference between the marginal unadjusted analysis and the stratified-adjusted analysis of the odds ratios.

4.9.3. Now conduct a stratified Mantel-Haenszel analysis of relative risks.

1. Calculate the Mantel-Haenszel estimate of the common relative risk.
2. Now use the Mantel-Haenszel test value to obtain the test-inverted confidence limits for the common relative risk.
3. Compare and comment on the difference between the marginal unadjusted analysis and the stratified-adjusted analysis of the relative risks.

4.9.4. For each of the three scales (difference, log odds ratio, log relative risk) perform the following *MVLE* analyses over all three strata:

1. Obtain the estimate of the common parameter and its variance.
2. Calculate the 95% confidence interval on the parameter under each scale, and also the asymmetric confidence limits for the odds ratio and relative risk.
3. Compare the weights for the estimates of the common parameter on each scale.
4. Compare the *MVLE* estimates and confidence limits to the marginal unadjusted analysis.
5. Compute the relative weights (summing to 1) for the Mantel-Haenszel estimates of the common odds ratio and relative risk and compare these to the weights for the *MVLE* estimates for the log odds ratio and log relative risk, respectively.
6. Then compare the stratified adjusted estimates of the odds ratio and the relative risk and their confidence limits.

4.9.5. Evaluate the extent to which years of employment strata are associated with the smoking group and/or are a risk factor for response (byssinosis) using the appropriate 3 ×2 tables. What are the implications for the comparison of the adjusted versus unadjusted analyses?

4.9.6. Conduct the following tests of significance:

1. The 3 *df* omnibus X^2 test using the risk difference $(p_{1j} - p_{2j})$ within each stratum. Also state the null and alternative hypotheses and the result of the test.

2. An alternative is to use three separate tests (Cochran or Mantel-Haenszel), one within each stratum, at $\alpha = 0.05/3$. Interpret the tests conducted under Problem 4.9.1.1 and 2 above in terms of these Bonferroni adjusted significance levels.

3. Conduct Cochran's $K-1$ *df* test of homogeneity for each of the three scales.

4. For each of the three scales, perform the Radhakrishna test of association for the hypothesis of a common parameter H_{0A}: $\theta_j = \theta_0$ for all j versus the alternative H_1: $\theta_j = \theta \neq \theta_0$ for all j. Compare the tests on each scale in relation to the tests for homogeneity.

5. Compare the Mantel-Haenszel test obtained in 4.9.2.2 to the Radhakrishna test for log odds.

6. Compute the correlation $\widehat{\rho}_{rs}$ and $\widehat{ARE}(T_r, T_s)$ for each of the three possible pairs of scales among the risk difference, log relative risk and log odds ratio.

7. Confirm that the Gastwirth *MERT* conditions apply and then compute the scale robust *MERT* in (4.126). Compare this test to the individual Radhakrishna family tests. Describe the *ARE* of this test relative to others in the family of tests considered.

8. Likewise, compute the Wei-Lachin test of stochastic ordering using the risk difference within each stratum. State the null and alternative hypotheses. Compare to the other tests above.

4.9.7. Using a random effects model for an analysis of log odds ratios:

1. Estimate $\widehat{\sigma}^2_\theta$.

2. Compute the updated $V(\widehat{\theta}_j)$.

3. Compute the $\widehat{\omega}_j^{(1)}$.

4. Compute $\widehat{\mu}_\theta^{(1)}$.

5. Compute $V\left(\widehat{\mu}_\theta^{(1)}\right)$.

6. Compute asymmetric 95% confidence limits on θ.

7. Compare these computations under a random effects model to the previous computations under a fixed effects model.

4.10 Using the developments in Section 4.11.1,

4.10.1. Derive the expression for $\sigma^2_{\overline{T}_0}$ in (4.163).

4.10.2. Derive the expression for $\mu_{\overline{T}}$ in (4.164).

4.10.3. Derive the expression for $\sigma^2_{\widetilde{T}_0}$ in (4.166).

4.10.4. Derive the expression in (4.169).

4.11 Consider an epidemiologic study designed to compare the risks (odds) of exposure to a risk factor (E versus $\overline{E}$) on the incidence of developing a disease over a fixed period of time (D versus $\overline{D}$). Because the study will not be randomized, it is planned to conduct a stratified analysis with the following expected sample fractions and control ($\overline{E}$) group probabilities, and the specified odds ratios within each of three strata:

Stratum	*Stratum Fraction* ζ_j	*Group Fractions* ξ_{1j}	ξ_{2j}	π_{2j}	OR_j (e^{θ_j})
1	0.15	0.3	0.7	0.30	1.80
2	0.50	0.6	0.4	0.22	2.50
3	0.35	0.45	0.55	0.15	3.00

4.11.1. Compute the corresponding probabilities in the exposed group π_{1j}.

4.11.2. Compute the marginal probabilities $\pi_{1\bullet}$ and $\pi_{2\bullet}$ obtained as the weighted average of the π_{1j} and π_{2j} weighting by the expected fraction within each stratum. From these compute the sample size required to provide power of 0.90 for the unadjusted test of the difference between two proportions in the unstratified analysis with $\alpha = 0.05$, two-sided.

4.11.3. Now compute the sample size required for the stratified analysis of a common odds ratio to provide power of 0.90.

4.11.4. Suppose that the study is conducted with a total sample size of $N = 150$, what level of power is provided by the stratified analysis?

4.11.5. Alternately, for these expected probabilities within each stratum, compute the sample size required to provide power of 0.90 for the test of a common relative risk rather than a common odds ratio.

4.11.6. Explain why the required sample sizes for the test of a common odds ratio differs from that for the test of a common relative risk.

5

Case-Control and Matched Studies

The preceding chapters considered exclusively statistical methods for the assessment of the association between two independent treatment or exposure groups and the risk of a positive or negative outcome from a cross-sectional or prospective study. In this chapter we consider two generalizations. One is the case-control or retrospective study. When separate independent groups of cases and controls are constructed, many of the methods in the preceding chapters still apply. However, the odds ratio and relative risk have different interpretations in a retrospective study from those in a prospective study.

The other extension is to the matched study. Matching is often used in a case-control study to control for important factors; however, matching may be used in a prospective study as well. In either case the two exposure, treatment or case-control groups are not independent, and thus different statistical methods must be employed.

5.1 UNMATCHED CASE-CONTROL (RETROSPECTIVE) SAMPLING

Consider an unmatched case-control or retrospective study in which a sample of known cases is obtained who have the outcome disease or characteristic of interest (D) and who are to be compared with an independent sample of non-diseased controls ($\overline{D}$). For each subject in the two groups, the prior degree of exposure to the risk factor under study, classified as E and $\overline{E}$, is then determined *retrospectively*. Just the opposite is done in a prospective study where samples of exposed and non-exposed individuals (E and $\overline{E}$) are obtained and then followed prospectively to determine whether each individual develops the disease (D) or does not ($\overline{D}$). The entries in a 2×2 table of frequencies from each type of study are

Prospective

	Group		
	Exposed	*Not Exposed*	
	E	$\overline{E}$	
D	a	b	m_1
$\overline{D}$	c	d	m_2
	n_1	n_2	N

Retrospective

	Group		
	Cases	*Controls*	
	D	$\overline{D}$	
E	a	b	m_1
$\overline{E}$	c	d	m_2
	n_1	n_2	N

(5.1)

In each case the "unconditional" likelihood given samples of sizes n_1 and n_2 is a product binomial likelihood as shown in (2.52) of Chapter 2; and conditioning on all margins fixed, the conditional likelihood is a hypergeometric probability as shown in (2.57). Therefore, Fisher's exact test or the large sample contingency (Cochran) or Mantel-Haenszel 1 *df* χ^2 tests can be used with such data to test the hypothesis of association between disease (case/control) and exposure. However, there are additional considerations in describing the degree of association.

5.1.1 Odds Ratio

Since the data are obtained from retrospective sampling, the retrospective odds ratio can be distinguished from the prospective odds ratio as follows: Given retrospective samples of n_1 cases (D) and n_2 controls ($\overline{D}$), the assumed conditional probabilities are

	D	$\overline{D}$
E	ϕ_1	ϕ_2
$\overline{E}$	$1-\phi_1$	$1-\phi_2$
	1.0	1.0

(5.2)

where ϕ_1 and ϕ_2 are the retrospective *disease conditional probabilities* of exposure (E) given being a case (D) versus a control ($\overline{D}$):

$$\phi_1 = P(E|D) \text{ and } \phi_2 = P\left(E|\overline{D}\right). \tag{5.3}$$

Thus the *retrospective odds ratio* of exposure given disease is

$$OR_{retro} = \frac{\phi_1/(1-\phi_1)}{\phi_2/(1-\phi_2)} = \frac{P(E|D)\ /\ P\left(\overline{E}|D\right)}{P\left(E|\overline{D}\right)\ /\ P\left(\overline{E}|\overline{D}\right)} \mathrel{\widehat{=}} \frac{ad}{bc}, \tag{5.4}$$

where $\widehat{=}$ means "estimated as" using the cell frequencies from the retrospective 2×2 table.

However, it is of far greater interest to describe the *prospective odds ratio* (OR) of disease given exposure defined as presented in Chapter 2,

$$OR = \frac{\pi_1/(1-\pi_1)}{\pi_2/(1-\pi_2)} = \frac{P(D|E)\ /\ P(\overline{D}|E)}{P(D|\overline{E})\ /\ P(\overline{D}|\overline{E})}. \tag{5.5}$$

To obtain this quantity we must also account for the prevalence of the disease in the population from which the cases and controls arose.

$$P(D) = \delta. \tag{5.6}$$

In this case the joint and marginal probabilities of disease and exposure become

$$\begin{array}{c|c|c|c} & \begin{array}{c} Cases \\ D \end{array} & \begin{array}{c} Controls \\ \overline{D} \end{array} & \\ \hline E & P(E,D) & P(E,\overline{D}) & P(E) \\ \hline \overline{E} & P(\overline{E},D) & P(\overline{E},\overline{D}) & P(\overline{E}) \\ \hline & \delta & 1-\delta & 1.0 \end{array} \tag{5.7}$$

where, for example, $P(E,D) = P(E|D)P(D) = \phi_1\delta$, and so forth. Therefore, these probabilities are expressed as

$$\begin{array}{c|c|c|c} & \begin{array}{c} Cases \\ D \end{array} & \begin{array}{c} Controls \\ \overline{D} \end{array} & \\ \hline E & \phi_1\delta & \phi_2(1-\delta) & \phi_1\delta + \phi_2(1-\delta) \\ \hline \overline{E} & (1-\phi_1)\delta & (1-\phi_2)(1-\delta) & (1-\phi_1)\delta + (1-\phi_2)(1-\delta) \\ \hline & \delta & 1-\delta & 1.0 \end{array} \tag{5.8}$$

The prospective *exposure-conditional probabilities* of disease given exposure status, $\pi_1 = P(D|E)$ and $\pi_2 = P(D|\overline{E})$, can then be obtained as the ratio of the joint to the marginal probabilities, such as $P(D|E) = P(D,E)/P(E)$, which is equivalent to using Bayes Theorem. Thus these prospective probabilities can then be expressed in terms of the disease-conditional probabilities (ϕ_1, ϕ_2) and the prevalence δ as

$$\pi_1 = P(D|E) = \frac{P(D,E)}{P(E)} = \frac{P(E|D)P(D)}{P(E|D)P(D) + P(E|\overline{D})P(\overline{D})} \tag{5.9}$$

$$= \frac{\phi_1\delta}{\phi_1\delta + \phi_2(1-\delta)}$$

and

$$\pi_2 = P(D|\overline{E}) = \frac{(1-\phi_1)\delta}{(1-\phi_1)\delta + (1-\phi_2)(1-\delta)}. \tag{5.10}$$

It then follows that the prospective odds ratio of disease given exposure presented in (5.4) above equals the retrospective odds ratio of exposure given disease, so that

$$\begin{aligned} OR &= \frac{\pi_1/(1-\pi_1)}{\pi_2/(1-\pi_2)} = \frac{P(D|E)\ /\ P(\overline{D}|E)}{P(D|\overline{E})\ /\ P(\overline{D}|\overline{E})} \\ &= \frac{P(E|D)\ /\ P(\overline{E}|D)}{P(E|\overline{D})\ /\ P(\overline{E}|\overline{D})} = \frac{\phi_1/(1-\phi_1)}{\phi_2/(1-\phi_2)} = OR_{retro}\,. \end{aligned} \tag{5.11}$$

Even though the prevalence (probability) of the disease δ is required to estimate the prospective exposure conditional probabilities, these terms cancel from the expression for the odds ratio.

Thus all of the methods described previously for the analysis of odds ratios in single and multiple 2×2 tables for two independent groups from a cross-sectional or prospective study also apply to the analysis of a case-control study with independent groups. The only alteration is that the disease conditional probabilities (ϕ_1, ϕ_2) are substituted for the exposure conditional probabilities (π_1, π_2) in the expressions for the underlying parameters. The estimates and tests are then obtained using the sample proportions from the retrospective 2×2 table in (5.1) computed as $p_1 = a/n_1$, $p_2 = b/n_2$ and $p = m_1/N$, where $p_1 = \widehat{\phi}_1$, $p_2 = \widehat{\phi}_2$ and $p = \widehat{\phi}$.

5.1.2 Relative Risk

Now consider the relative risk. Based on a case-control study, the *retrospective relative risk* of exposure given disease versus not is

$$RR_{retro} = \frac{\phi_1}{\phi_2} = \frac{P(E|D)}{P(E|\overline{D})} \mathrel{\widehat{=}} \frac{\widehat{\phi}_1}{\widehat{\phi}_2} = \frac{a/n_1}{b/n_2}\,. \tag{5.12}$$

However, the parameter of interest is the *prospective* relative risk of disease given exposure versus not that is defined as

$$RR = \frac{\pi_1}{\pi_2} = \frac{P(D|E)}{P(D|\overline{E})} \neq RR_{retro}. \tag{5.13}$$

Thus the prospective relative risk can not be obtained from the retrospective relative risk.

However, through application of Bayes theorem, or from (5.8), Cornfield (1951) showed that the prospective probabilities π_1 and π_2 can be defined directly given knowledge (specification) of the prevalence of the disease δ. From the resulting expressions for π_1 and π_2 in (5.9) and (5.10), the prospective relative risk is

$$\begin{aligned} RR = \frac{P(D|E)}{P(D|\overline{E})} &= \left[\frac{\delta\phi_1}{\delta\phi_1 + (1-\delta)\phi_2}\right]\left[\frac{\delta(1-\phi_1) + (1-\delta)(1-\phi_2)}{\delta(1-\phi_1)}\right] \\ &= \left[\frac{\phi_1}{1-\phi_1}\right]\left[\frac{\delta(1-\phi_1) + (1-\delta)(1-\phi_2)}{\delta\phi_1 + (1-\delta)\phi_2}\right]. \end{aligned} \tag{5.14}$$

Given knowledge of the prevalence δ, this quantity can then be estimated consistently as

$$\widehat{RR} = \frac{\widehat{\pi}_1}{\widehat{\pi}_2} = \left[\frac{\delta p_1}{\delta p_1 + (1-\delta)\,p_2}\right]\left[\frac{\delta\,(1-p_1)+(1-\delta)\,(1-p_2)}{\delta\,(1-p_1)}\right]. \tag{5.15}$$

Since the prevalence of the disease is often unknown, and is usually not directly estimable from such a study, this approach is rarely practical.

Cornfield (1951), however, showed that when the disease is rare ($\delta \downarrow 0$), the value of the RR is approximately

$$RR_{\delta\downarrow 0} \doteq \left(\frac{\phi_1}{1-\phi_1}\right)\left(\frac{1-\phi_2}{\phi_2}\right) = OR_{retro}\ , \tag{5.16}$$

which is the retrospective odds ratio of exposure given disease. Thus under the assumption that the prevalence of the disease is rare ($\delta \doteq 0$), the retrospective odds ratio provides an approximate estimate of the prospective relative risk.

It can also be shown that an equivalent condition is that $\delta\,(\phi_1 - \phi_2) \downarrow 0$, or

$$RR_{\delta(\phi_1-\phi_2)\downarrow 0} \doteq OR_{retro}\ , \tag{5.17}$$

so that the approximation applies when either the prevalence becomes rare, or the difference $(\phi_1 - \phi_2)$ approaches the null hypothesis.

5.1.3 Attributable Risk

As for the precise definition of the relative risk in (5.14), other measures of association such as the risk difference or attributable risk require that explicit expressions be derived using the known or specified value of the probability of the disease (δ) in conjunction with Bayes' Theorem. As a result, the analysis of a case-control study usually focuses only on the estimation and description of the odds ratio. The odds ratio applies to any case-control study because the retrospective odds ratio equals the prospective odds ratio. In the case of a rare disease, the odds ratio can also be interpreted approximately as a relative risk.

Since the attributable risk and the population attributable risk are also a function of the relative risk, then under the rare disease assumption the odds ratio also provides an estimate of the measures of attributable risk. As described in Section 2.8.2 for a prospective study, since the population attributable risk is a direct function of the relative risk, then the point estimate and asymmetric confidence limits can be obtained from the point estimate and asymmetric confidence limits for the odds ratio.

From (2.99), under the rare disease assumption as in (5.16), the population attributable risk may be estimated as a function of the odds ratio as

$$\widehat{PAR} \doteq \frac{\alpha_1(\widehat{OR}-1)}{\alpha_1\widehat{OR}+\alpha_2}\ , \tag{5.18}$$

where the prevalence of exposure $\alpha_1 = P(E)$ is assumed to be known or specified. Using the δ-method (see Problem 5.1), it is then readily shown that asymptotically

$$V(logit\ \widehat{PAR}) \cong \left[\frac{OR}{(OR-1)}\right]^2 V[\log OR]\ , \tag{5.19}$$

where the exposure prevalence α_1 cancels from this expression. This variance is consistently estimated by substituting the retrospective estimate $\widehat{OR}$ and the estimate of the variance $\widehat{V}[\log \widehat{OR}]$ from (2.46).

Example 5.1 *Smoking and Lung Cancer*

Cornfield (1951) presents the following example to describe the accuracy of his approximation. Schrek, Baker, Ballard and Dolgoff (1950) reported a case-control study of smoking (E) and lung cancer (D). The retrospectively sampled data are

	D	$\overline{D}$
E	154	80
$\overline{E}$	46	120
	200	200

where $p_1 = \widehat{\phi}_1 = 0.77$, $p_2 = \widehat{\phi}_2 = 0.40$ and the $\widehat{OR}_{retro} = 5.0217$. Under the rare disease assumption, this provides an approximation to the prospective relative risk.

How close is this to an estimate of the relative risk RR? From a large population-based study, Dorn (1944) reported that the prevalence of lung cancer at that time was 15.5/100,000 population. Thus $\delta = P(D) = 0.155\left(10^{-3}\right)$. The joint and marginal probabilities of the exposure and disease categories are estimated to be

	D	$\overline{D}$	
E	$0.119\left(10^{-3}\right)$	0.399938	0.400057
$\overline{E}$	$0.036\left(10^{-3}\right)$	0.599907	0.599943
	$0.155\left(10^{-3}\right)$	0.999845	1.0

so that $\widehat{\pi}_1 = \widehat{P}(D|E) = (0.000119/0.400057) = 0.297\left(10^{-3}\right)$ and $\widehat{\pi}_2 = \widehat{P}\left(D|\overline{E}\right) = (0.000036/0.599943) = 0.060\left(10^{-3}\right)$. Therefore, the direct estimate of the prospective relative risk given the above value of δ is $\widehat{RR} = (\widehat{\pi}_1/\widehat{\pi}_2) = (0.297/0.060) = 4.95$. The odds ratio of 5.02 provides a relatively accurate estimate of this relative risk. The asymmetric 95% confidence limits are (3.25, 7.75) and the association is highly significant with a Mantel-Haenszel test value of 56.25 on 1 df, $p < 0.001$.

To estimate the population attributable risk, the prevalence of exposure to smoking in the population must be specified. For $\alpha_1 = 0.3$ the estimate is $\widehat{PAR} = 0.54680$, which indicates that smoking, if the sole cause of lung cancer, accounts for approximately 54% of cases of lung cancer in the population. The higher the estimated prevalence of smoking, the higher the estimate of the attributable risk. The estimated $\widehat{V}[\log \widehat{OR}] = 0.04907$ then yields an estimated $V(logit\ \widehat{PAR}) = 0.0765$. This yields symmetric confidence limits on the $logit(PAR)$, which then yield asymmetric 95% confidence limits for the PAR of (0.412, 0.675).

5.2 MATCHING

We now turn to the case of a matched prospective study and a matched case-control study. Previously a stratified analysis was employed to obtain an adjusted estimate and a test of the risk difference between the treatment or exposure groups adjusting or controlling for the effect of an intervening covariate or confounding factor. As we shall see later, a stratified analysis is equivalent to using a regression model as the basis for the adjustment. An alternative approach to control for the effects of a covariate is matching. In this case, each member of a group (either E or $\overline{E}$ in a prospective study, D or $\overline{D}$ in a retrospective study) is matched to a member of the other group with respect to the values of one or more covariates Z.

5.2.1 Frequency Matching

One of the principal types of matching is frequency or within-class matching, in which elements or members of the comparison group are sampled within separate categories of a discrete covariate class such as gender (male/female), decade of age (0–9 years, 10–19 years, etc.) so that the members of each group are matched within each category. For example, in a frequency-matched case-control study, the cases may be stratified by gender and decade of age. Then within each category, such as males in their 40s, a separate sample of controls is selected from the control population in that category (males in their 40s). In this technique, any quantitative covariates such as age are grouped.

In a frequency-matched study separate samples of cases and controls are obtained within covariate categories or strata. These samples need not be of equal size within or between strata. The objective is to have adequate numbers of subjects from each group within each stratum to provide an adequate overall comparison between groups. Thus it is only necessary that a sufficient number of exposed and non-exposed cases and controls be sampled within each stratum so as to provide an assessment of the odds ratio or relative risk in aggregate over strata.

No single overall sample from each group is obtained at random from the population, such as an unrestricted sample of cases and an unrestricted sample of controls. Thus the unadjusted marginal analysis has no simple corresponding likelihood, such as a simple product binomial. Rather, the underlying likelihood is a product binomial within strata, because separate samples are obtained within each stratum. Thus the only appropriate analysis in this case is one that is stratified by the covariate categories used in the frequency matching. Of course, the analysis may be further stratified by additional unmatched covariates other than those used as the basis for the frequency matched samples, such as by quintiles of body mass index in a study frequency matched by gender and decade of age.

Example 5.2 *Frequency Matching*
Korn (1984) describes a study of the association between a history of smoking (E versus $\overline{E}$) and bladder cancer among males (D) versus and an age frequency-matched sample of non-cancer control patients ($\overline{D}$). The study data are

	Cases		*Controls*	
Age Stratum	E	$\overline{E}$	E	$\overline{E}$
50–54	24	1	22	4
55–59	35	2	35	4
60–64	31	5	38	3
65–69	46	7	42	15
70–74	60	13	51	28
75–79	39	14	32	20
Total	235	42	220	74

(reproduced with permission). Again note that in a frequency-matched study such as this the number of matched controls sampled within a stratum need not equal the number of cases in that stratum, nor be proportional to the numbers of controls sampled in other strata.

The sampling procedure for this study was to first select all 277 cases (D) and to then stratify these cases by intervals of age. Within each age stratum, n_{2j} controls were then selected and the exposure status (E versus $\overline{E}$) was determined for all cases and controls. This design, therefore, provides six separate independent samples (strata) with an independent 2×2 table within each stratum. Because the controls were sampled separately within strata, this design does not yield a single random sample of 277 cases and 294 controls. Thus it is inappropriate to collapse the data into a single marginal 2×2 table. Rather, the proper analysis is a stratified analysis, such as a Mantel-Haenszel analysis of odds ratios.

The Mantel-Haenszel estimate of the common odds ratio is $\widehat{OR}_{MH} = 1.96$ with a Robins et al. estimated variance of $\widehat{V}[\log \widehat{OR}_{MH}] = 0.0483$. This yields a 95% $C.I.$ for the odds ratio of $(1.28, 3.02)$. The conditional Mantel-Haenszel test statistic is $X^2_{C(MH)} = 9.59$, with $p \leq 0.002$. Cochran's test of homogeneity is $X^2_{H(C)} = 4.38$ on 5 *df*, which is not statistically significant with $p \doteq 0.50$, indicating that a fixed effects model is appropriate.

5.2.2 Matched Pairs Design: Cross-Sectional or Prospective

The more common approach to matching is to construct matched pairs, or matched sets. In a cross-sectional or prospective study, matched sets are sampled, each consisting of an exposed (E) individual and one or more individuals who were not exposed ($\overline{E}$), where the members of the pair (or set) are uniquely matched with respect to a set of covariates. An example is the matched study, where a control is selected from the population of non-exposed individuals who share the same covariates as each exposed individual, such as age and gender. Matched sets may also arise naturally as when each subject serves as his/her own control and observations on each subject are obtained before ($\overline{E}$) and again after (E) exposure to a risk factor or the application of some treatment.

In some cases, the matching covariates are, in fact, unobservable quantities such as when the sample consists of a set of individuals with multiple measurements within-person. An example would be studies of matched organs within individuals (eyes, ears, kidneys, teeth, etc.), where one or more elements within each individual are exposed or treated with an experimental agent and one or more are not. Another example is a study involving multiple family members, for example, among twins, where the family members have some genes in common. A similar example is when characteristics of litter-mates are studied. In this section, the methods for the analysis of matched data are described in terms of matched pairs of individuals, but they obviously apply to these other cases as well.

Consider a sample of N matched pairs with one exposed (E) and one non-exposed ($\overline{E}$) member where the outcome of each member is designated as $(D, \overline{D})$ to denote developing the disease versus not, or having the positive outcome versus not. In this setting it would be inappropriate to treat these as two independent groups in a 2×2 table such as

	E	$\overline{E}$	
D	a	b	m_1
$\overline{D}$	c	d	m_2
	N	N	$2N$

The principal reason that methods for two independent groups do not apply is that the observations within each pair are correlated so that we do not have a sample of $2N$ independent observations.

To see this, assume that the exposed and non-exposed members of each pair are matched with respect to some continuous covariate Z. We then assume that the probability of the outcome is some function of the covariate. Let Y_1 and Y_2 denote the responses for the exposed and non-exposed members, respectively. Then for the ith pair with shared covariate value z_i, let y_{ij} designate the outcome for the jth member, $y_{ij} = 1$ if D, 0 if $\overline{D}$, with probability $E(y_{ij}) = \pi_{ij}$ for $i = 1, \ldots, N$ and $j = 1, 2$. To simplify, consider the case where the general null hypothesis is true such that the probabilities of the outcome are the same for each pair member conditional on the values z_i, that is, $\pi_{i1} = \pi_{i2}$. Denote these shared probabilities as $E(y_{ij}|z_i) = \pi(z_i)$ for the two members of the ith pair with covariate value z_i. Within the ith pair with a given value of the covariate z_i, the pair members are sampled independently so that the outcomes of the matched pairs y_{i1} and y_{i2} are *conditionally independent*. Thus $E(y_{i1}|y_{i2}, z_i) = E(y_{i1}|z_i) = \pi(z_i)$ and $E(y_{i2}|y_{i1}, z_i) = E(y_{i2}|z_i) = \pi(z_i)$. However, since the pair members are matched, then the covariance of the matched measures over all sets is

$$Cov(Y_1, Y_2) = E_Z\left[Cov(y_{i1}, y_{i2}|z_i)\right] = E_Z\left[\pi(z)^2\right] - \left(E_Z\left[\pi(z)\right]\right)^2 \neq 0, \quad (5.20)$$

so that Y_1 and Y_2 are correlated in the population of matched pairs (see Problem 5.3).

Thus the proper analysis of matched pairs must allow for the within-pair correlation. In the case of a binary outcome this requires that we treat the pair as the sampling unit and then classify each pair according to the values of the outcomes of the two pair members. Thus we have N pairs that are cross-classified as follows:

Frequencies / *Probabilities*

$$\begin{array}{cc|c|c|c} & & \multicolumn{2}{c}{\overline{E}} & \\ & & D & \overline{D} & \\ \hline E & D & e & f & n_{1\bullet} \\ \hline & \overline{D} & g & h & n_{2\bullet} \\ \hline & & n_{\bullet 1} & n_{\bullet 2} & N \end{array} \qquad \begin{array}{cc|c|c|c} & & \multicolumn{2}{c}{\overline{E}} & \\ & & D & \overline{D} & \\ \hline E & D & \pi_{11} & \pi_{12} & \pi_{1\bullet} \\ \hline & \overline{D} & \pi_{21} & \pi_{22} & \pi_{2\bullet} \\ \hline & & \pi_{\bullet 1} & \pi_{\bullet 2} & 1 \end{array} \quad (5.21)$$

Therefore, the sample frequencies are distributed as a multinomial (quadrinomial) where the likelihood of the sample is

$$P(e, f, g \mid N) = \frac{N!}{e!f!g!h!}\pi_{11}^{e}\pi_{12}^{f}\pi_{21}^{g}\pi_{22}^{h}. \quad (5.22)$$

As for the 2×2 table in Chapter 2, we use simple letters to refer to the frequencies in the table, $e = n_{11}$, $f = n_{12}$, and so on. The underlying probabilities of the four cells designate the probability of the outcomes for the exposed and non-exposed members of each pair. Thus $\pi_{11} = P(D, D)$ for the exposed (E) and non-exposed ($\overline{E}$) pair members, respectively, $\pi_{12} = P(D, \overline{D})$, $\pi_{21} = P(\overline{D}, D)$, and $\pi_{22} = P(\overline{D}, \overline{D})$.

From the margins of the table it appears that $\pi_{1\bullet}$ equals $P(D|E)$ and that $\pi_{\bullet 1}$ equals $P\left(D|\overline{E}\right)$. However, $\pi_{1\bullet}$ and $\pi_{\bullet 1}$ do not both refer to a conditional probability in the general population. Consider the case where matched pairs have been constructed on the basis of a continuous matching covariate Z. The covariate is distributed as $f_E(z) = P(z|E)$ within the exposed segment of the population, and as $f_{\overline{E}}(z) = P\left(z|\overline{E}\right)$ within the non-exposed segment. If the covariate Z is, in fact, associated with the likelihood of exposure, then these distributions will differ between the exposed and non-exposed individuals, $f_E(z) \neq f_{\overline{E}}(z)$. This, in turn, implies the following developments.

Usually we begin by selecting an exposed individual from its respective population. When the sample of exposed individuals can be viewed as a sample drawn at random from the exposed population, then $\pi_{1\bullet} = P(D|E)$ irrespective of matching, since

$$\pi_{1\bullet} = \int_z P(D|E, z)\, f_E(z)\, dz = P(D|E). \quad (5.23)$$

However, to construct a matched pair, first the covariate value z is identified for the exposed (E) member and then a control non-exposed individual $\overline{E}$ is sampled at random conditional on z. That is, the control member is sampled from the subset of the non-exposed population who share the covariate value z. Thus the distribution of the covariate Z among the sampled non-exposed individuals equals the distribution among the exposed individuals so that

$$\pi_{\bullet 1} = \int_z P\left(D|\overline{E}, z\right) f_E\left(z\right) dz = P_m\left(D|\overline{E}\right) , \tag{5.24}$$

where $P_m\left(D|\overline{E}\right)$ is used to denote the conditional probability under matching. However, since the distribution of the matching covariate differs among those exposed and those not exposed, $f_E\left(z\right) \neq f_{\overline{E}}\left(z\right)$, then this conditional probability $P_m\left(D|\overline{E}\right)$ is not the same as the conditional probability in the entire population of non-exposed individuals

$$\pi_{\bullet 1} \neq \int_z P\left(D|\overline{E}, z\right) f_{\overline{E}}\left(z\right) dz = P\left(D|\overline{E}\right). \tag{5.25}$$

Therefore, measures of association such as the relative risk or odds ratio based on the marginal probabilities from the matched or paired 2×2 table do not have an unconditional population interpretation. Rather, any such measures must be interpreted conditionally on the matching on other covariates.

5.3 TESTS OF ASSOCIATION FOR MATCHED PAIRS

Before considering measures of association for matched data, let us first consider tests of the hypothesis of association. For the paired or matched 2×2 table there are two equivalent hypotheses of interest. The first is the hypothesis of *marginal homogeneity* under matching

$$H_0\text{: } P\left(D|E\right) = P_m\left(D|\overline{E}\right) \text{ or } \pi_{1\bullet} = \pi_{\bullet 1}. \tag{5.26}$$

Since this hypothesis is stated in terms of the marginal joint probabilities from an underlying quadrinomial in (5.21), then this hypothesis implies and is implied by the hypothesis of *symmetry*

$$H_0\text{: } \pi_{12} = \pi_{21} \tag{5.27}$$

with respect to the discordant pairs.

5.3.1 Exact Test

The likelihood of the 2×2 table for matched pairs is a quadrinomial. However, we wish to conduct a test of the equality of two of the four probabilities in this likelihood. To do so we must account for or eliminate the other nuisance parameters

(π_{11}, π_{22}). This is readily done through the principle of conditioning as illustrated in Section 2.4.2 through the derivation of the conditional hypergeometric distribution. This approach requires that we identify a conditional distribution that only involves the parameters of interest (π_{12}, π_{21}).

As a Problem, it is readily shown that the conditional distribution of f given the number of discordant pairs $M = f + g$, is a simple binomial distribution

$$P(f \mid M) = \frac{M!}{f!g!}\left(\frac{\pi_{12}}{\pi_d}\right)^f \left(\frac{\pi_{21}}{\pi_d}\right)^g = B\left(f;\ M, \frac{\pi_{12}}{\pi_d}\right), \tag{5.28}$$

where $\pi_d = \pi_{12} + \pi_{21}$ is the probability of a discordant pair in which one member of the pair has a positive outcome, the other a negative outcome.

Therefore, under the null hypothesis of symmetry H_0: $\pi_{12} = \pi_{21}$, it follows that

$$P(f \mid M) = B(f;\ M, 1/2). \tag{5.29}$$

Therefore, an exact two-sided P-value is obtained as

$$p = 2 \sum_{j=0}^{\min(f,g)} B(j;\ M, 1/2), \tag{5.30}$$

which is a simple binomial test.

Example 5.3 *Small Sample*

For example consider a paired 2×2 table with discordant entries

-	2
8	-

To test the hypothesis of symmetry, the total sample size and the numbers of concordant pairs are irrelevant. Conditioning on the number of discordant pairs, $M = 10$, then the two-sided exact P-value is

$$\begin{aligned} p &= 2\left[B(0; 10, 1/2) + B(1; 10, 1/2) + B(2; 10, 1/2)\right] \\ &= 2\left[0.001 + 0.0097 + 0.0440\right] = 0.1094. \end{aligned}$$

Note that this quantity can also be computed by hand simply as

$$p = 2(0.5)^{10}\left[\binom{10}{0} + \binom{10}{1} + \binom{10}{2}\right] = 2(0.5)^{10}\left[1 + 10 + 45\right] = 0.1094.$$

5.3.2 McNemar's Large Sample Test

A large-sample test can be derived either from the normal approximation to the multinomial or from the normal approximation to the conditional binomial distribution. Here we use the former while the derivation using the latter is left to a Problem.

Let the proportion within the $ijth$ cell be $p_{ij} = n_{ij}/N$, where $E(p_{ij}) = \pi_{ij}$, $i = 1, 2;\ j = 1, 2$. From the normal approximation to the multinomial (see Example A.2 of the Appendix), the vector of cell proportions is asymptotically distributed as

$$\begin{pmatrix} p_{11} \\ p_{12} \\ p_{21} \\ p_{22} \end{pmatrix} \overset{d}{\approx} \mathcal{N}\left[\begin{bmatrix} \pi_{11} \\ \pi_{12} \\ \pi_{21} \\ \pi_{22} \end{bmatrix}, \boldsymbol{\Sigma}\right], \tag{5.31}$$

where

$$\boldsymbol{\Sigma} = \frac{1}{N}\begin{bmatrix} \pi_{11}(1-\pi_{11}) & -\pi_{11}\pi_{12} & -\pi_{11}\pi_{21} & -\pi_{11}\pi_{22} \\ -\pi_{11}\pi_{12} & \pi_{12}(1-\pi_{12}) & -\pi_{12}\pi_{21} & -\pi_{12}\pi_{22} \\ -\pi_{11}\pi_{21} & -\pi_{12}\pi_{21} & \pi_{21}(1-\pi_{21}) & -\pi_{21}\pi_{22} \\ -\pi_{11}\pi_{22} & -\pi_{12}\pi_{22} & -\pi_{21}\pi_{22} & \pi_{22}(1-\pi_{22}) \end{bmatrix} \tag{5.32}$$

Because the proportions sum to 1.0, the distribution is degenerate, meaning that the covariance matrix is singular with rank 3. However, this distribution can still be used as the basis for the evaluation of the distribution of contrasts among the proportions.

Clearly, to test the hypothesis of symmetry, H_0: $\pi_{12} = \pi_{21}$, we wish to use the test statistic based on the difference in the discordant proportions $p_{12} - p_{21}$ of the form

$$Z = \frac{p_{12} - p_{21}}{\sqrt{\widehat{V}(p_{12} - p_{21}|H_0)}} \tag{5.33}$$

with the variance of the difference evaluated under the null hypothesis. Using a linear contrast in p_{12} and p_{21}, it is readily shown (Problem 5.6) that

$$V(p_{12} - p_{21}) = \frac{(\pi_{12} + \pi_{21}) - (\pi_{12} - \pi_{21})^2}{N}. \tag{5.34}$$

Under H_0: $\pi_{12} = \pi_{21} = \pi$ where $\pi = \pi_d/2$, then

$$V(p_{12} - p_{21}|H_0) = \frac{\pi_{12} + \pi_{21}}{N} = \frac{\pi_d}{N} \mathrel{\widehat{=}} \frac{p_{12} + p_{21}}{N}. \tag{5.35}$$

Therefore,

$$Z_M = \frac{p_{12} - p_{21}}{\sqrt{(p_{12} + p_{21})/N}} = \frac{f - g}{\sqrt{f + g}} \tag{5.36}$$

is asymptotically normally distributed under H_0. This statistic provides either a one or two-sided test of H_0. For a two-sided test one can use the equivalent chi-square statistic

$$X_M^2 = \frac{(f - g)^2}{f + g} \tag{5.37}$$

which is asymptotically distributed as chi-square on 1 df. This is *McNemar's test* (McNemar, 1947).

Table 5.1 SAS PROC FREQ Analysis of Example 5.4

```
data one; input e f g h;
**** Note that the values of e and h are irrelevant;
cards;
10 80 55 10
;
title1 'Chapter 5, Example 5.4';
data two; set one;
Emember=1; Nmember=1; x=e; output;
Emember=2; Nmember =1; x=f; output;
Emember =1; Nmember =2; x=g; output;
Emember =2; Nmember =2; x=h; output;
proc freq; tables Emember* Nmember / all nopercent nocol agree;
    weight x; exact mcnem;
title2 'SAS PROC FREQ Analysis of Matched 2 x 2 tables';
run;
```

Example 5.4 *Large Sample*
Consider the paired 2×2 table with the discordant frequencies

-	80
55	-

where $M = 135$, $f = 80$, $g = 55$. McNemar's test of the hypothesis of symmetry yields $X^2_M = (25)^2/135 = 4.63$ with $p \leq 0.032$.

5.3.3 SAS PROC FREQ

The SAS PROC FREQ provides for the computation of the exact and large sample McNemar test for matched or paired 2×2 tables. Table 5.1 presents the SAS statements to conduct an analysis of the data in Example 5.4. While only the discordant observations are required to compute the exact and large sample tests, SAS requires that the complete table be provided. Thus the rows are labeled using the variable `Emember` that designates the response of the exposed member of each pair and the variable `Nmember` that designates that for the non-exposed member. Unfortunately, all of the other various tests and measures for unmatched 2×2 tables are also provided in the SAS output that can be highly misleading. Further, the measures of association for matched pairs are not presented.

5.4 MEASURES OF ASSOCIATION FOR MATCHED PAIRS

We now consider the definition and estimation of measures of association for matched pairs, starting with the odds ratio.

5.4.1 Conditional Odds Ratio

In the general unselected population, the population odds ratio is defined as in (5.11), where the conditional probabilities, such as $P(D|E)$, are defined for the general population. Under matching, however,

$$OR \neq \frac{\pi_{1\bullet}/\pi_{2\bullet}}{\pi_{\bullet 1}/\pi_{\bullet 2}} = \frac{P(D|E)/P\left(\overline{D}|E\right)}{P_m\left(D|\overline{E}\right)/P_m\left(\overline{D}|\overline{E}\right)} = OR_m\ , \tag{5.38}$$

where OR_m is the marginal odds ratio under matching. Therefore, the observed marginal odds ratio does not provide an estimate of the population odds ratio. Thus this is called the *population averaged odds ratio* because both $\pi_{1\bullet}$ and $\pi_{\bullet 1}$ are the average risks or the expectations of $P(D|E,z)$ and $P(D|\overline{E},z)$, respectively, with respect to the distribution of the matching covariate in the exposed population, $f_E(z)$.

Rather than attempting to describe the odds ratio unconditionally, let's do so conditionally with respect to the distribution of the covariate among the exposed. Conditionally, for a specific value of the covariate z, the conditional odds ratio is

$$OR_z = \frac{P(D|E,z)\,P\left(\overline{D}|\overline{E},z\right)}{P\left(D|\overline{E},z\right)P\left(\overline{D}|E,z\right)} . \tag{5.39}$$

Given z, the matched E and $\overline{E}$ pair members are sampled independently from their respective populations, and thus the pair members are conditionally independent. Therefore, for each pair with covariate value z, a 2×2 table can be constructed such as that in (5.21) with cell probabilities $\{\pi_{ij|z}\}$, row marginal probabilities $\{\pi_{i\bullet|z}\}$, and column marginal probabilities $\{\pi_{\bullet j|z}\}$ for $i = 1, 2$; $j = 1, 2$. Since the pairs are conditionally independent, then within each such table

$$\pi_{12|z} = \pi_{1\bullet|z}\pi_{\bullet 2|z} = P(D|E,z)\,P\left(\overline{D}|\overline{E},z\right) \tag{5.40}$$

and

$$\pi_{21|z} = \pi_{2\bullet|z}\pi_{\bullet 1|z} = P\left(D|\overline{E},z\right)P\left(\overline{D}|E,z\right) \tag{5.41}$$

so that the conditional marginal odds ratio is

$$OR_z = \frac{\pi_{1\bullet|z}\pi_{\bullet 2|z}}{\pi_{2\bullet|z}\pi_{\bullet 1|z}} = \frac{\pi_{12|z}}{\pi_{21|z}} . \tag{5.42}$$

Now assume that although $P(D|E,z)$ and $P\left(D|\overline{E},z\right)$ may vary with z, there is a constant conditional marginal odds ratio for all values of the matching covariate

z (Ejigou and McHugh, 1977, among others) such that

$$OR_z = OR_C \quad \forall z. \tag{5.43}$$

This implies that the population averaged discordant probabilities π_{12} and π_{21}, each averaged with respect to the distribution of the covariate in the exposed population, then satisfy

$$\pi_{21} = \int_z \pi_{21|z} f_E(z)\, dz = \int_z \pi_{2\bullet|z}\pi_{\bullet 1|z} f_E(z)\, dz \tag{5.44}$$

and from (5.42),

$$\pi_{12} = \int_z OR_z \pi_{21|z} f_E(z)\, dz = OR_c \int_z \pi_{2\bullet|z}\pi_{\bullet 1|z} f_E(z)\, dz = OR_c \pi_{21}. \tag{5.45}$$

Thus the conditional odds ratio is

$$OR_C = \frac{\pi_{12}}{\pi_{21}} \;\widehat{=}\; \frac{p_{12}}{p_{21}} = \frac{f}{g}. \tag{5.46}$$

In Chapter 6 we will show that this is quantity also arises from Cox's (1958b) conditional logit model for pair-matched data.

5.4.2 Confidence Limits for the Odds Ratio

5.4.2.1 Exact Limits Exact confidence limits for the conditional odds ratio can be obtained from the conditional binomial distribution presented in (5.28). Thus $f \sim B(f;\, M, \pi_f)$, where $\pi_f = \pi_{12}/\pi_d$, and $g \sim B(g;\, M, \pi_g)$, where $\pi_g = \pi_{21}/\pi_d$. Since $OR_c = \pi_{12}/\pi_{21}$, then

$$OR_c = \frac{\pi_f}{1-\pi_f} = \frac{1-\pi_g}{\pi_g}. \tag{5.47}$$

Let $a = \min(f, g)$. Then the upper confidence limit at level $1-\alpha$ for $\pi_a = \min(\pi_f, \pi_g)$ is that value of $\pi_{a(u)}$ such that

$$\alpha/2 = \sum_{x=0}^{a} \binom{M}{x} \pi_{a(u)}^{x} \left(1-\pi_{a(u)}\right)^{M-x}. \tag{5.48}$$

Likewise, the lower limit $\pi_{a(\ell)}$ satisfies

$$\alpha/2 = \sum_{x=a}^{M} \binom{M}{x} \pi_{a(\ell)}^{x} \left(1-\pi_{a(\ell)}\right)^{M-x}. \tag{5.49}$$

Example 5.5 *Small Sample*
Consider the above example with $f = 2$ and $M = 10$. Then using StatXact the 95% confidence limits (0.02521, 0.5561) are obtained as the solution to the equations

$$\sum_{x=3}^{10} B\left(x;\ 10, \pi_{f(\ell)}\right) = 0.025 \tag{5.50}$$

and

$$\sum_{x=0}^{2} B\left(x;\ 10, \pi_{f(u)}\right) = 0.025. \tag{5.51}$$

5.4.2.2 Large Sample Limits To obtain large sample confidence limits for the odds ratio, as for independent (unmatched) observations, it is preferable to use $\theta = \log OR_C$, which can be estimated consistently as $\widehat{\theta} = \log(f/g)$. From the normal approximation to the multinomial, then using the δ-method it can be shown that

$$V\left(\widehat{\theta}\right) = \frac{1}{N}\left(\frac{1}{\pi_{12}} + \frac{1}{\pi_{21}}\right) = \frac{\pi_d}{N\pi_{12}\pi_{21}}. \tag{5.52}$$

Because the cell proportions provide consistent estimates of the cell probabilities, the variance can be estimated consistently as

$$\widehat{V}\left(\widehat{\theta}\right) = \frac{1}{N}\left(\frac{N}{f} + \frac{N}{g}\right) = \left(\frac{1}{f} + \frac{1}{g}\right) = \frac{M}{fg}, \tag{5.53}$$

which only involves the discordant frequencies f and g, where $M = f + g$. This large sample estimated variance can then be used to construct asymmetric confidence limits on the conditional odds ratio OR_C.

5.4.3 Conditional Large Sample Test and Confidence Limits

Alternately, McNemar's test and the asymmetric confidence limits for the odds ratio can be obtained from the large sample normal approximation to the conditional distribution of f given M. Under H_1 in (5.28) we showed that $f \sim B(f;\ M, \pi_{12}/\pi_d)$, whereas under H_0 $f \sim B(f;\ M, 1/2)$. Therefore, under H_0 the large sample approximation to the binomial in terms of the proportion $p_f = f/M$ yields

$$H_0: \quad p_f \overset{d}{\approx} \mathcal{N}\left(\frac{1}{2}, \frac{1}{4M}\right) \tag{5.54}$$

and the square of the large sample Z-test equals McNemar's test in (5.37) (see Problem 5.7).

Under H_1 the large sample approximation yields

$$H_1: \quad p_f \overset{d}{\approx} \mathcal{N}\left[\pi_f, \frac{\pi_f(1-\pi_f)}{M}\right], \tag{5.55}$$

where $\pi_f = (\pi_{12}/\pi_d)$. This then provides the basis for large sample confidence limits on π_f. Although one can compute the usual symmetric confidence limits for this probability, the asymmetric limits based on the logit transformation as presented in (2.14–2.15) are preferred. Note, however, that $logit(\pi_f) = \log(\pi_{12}/\pi_{21}) = \log OR_c$. Thus in Problem 5.7 it is also shown that the confidence limits for the logit of π_f equal those for the conditional odds based on $\widehat{V}(\widehat{\theta})$ presented in (5.53).

Example 5.6 *Large Sample*
For the large sample data set in Example 5.4, the estimate of the conditional odds ratio is $\widehat{OR}_C = 80/55 = 1.45$ and the estimated log odds ratio is $\widehat{\theta} = 0.375$ with estimated $\widehat{SE}\left(\widehat{\theta}\right) = \sqrt{\frac{135}{(80)(55)}} = 0.1752$. Therefore, the asymmetric 95% confidence limits for OR_C are $\exp\left[0.375 \pm (1.96 \times 0.1752)\right] = (1.03, 2.05)$.

5.4.4 Mantel-Haenszel Analysis

Another way to approach the problem of matched pairs was proposed by Mantel and Haenszel (1959). Since each pair has two members, each sampled independently given the value of the matching covariate Z, then the members within each pair are conditionally independent. Thus they suggested that the sample of matched-pairs be considered to consist of N independent samples (strata) consisting of one member in each exposure group $(E, \overline{E})$ in a cross-sectional or prospective study. These N samples provide N independent 2×2 tables each comprising one matched pair. The ith pair then provides an unmatched 2×2 table of the form

	E	$\overline{E}$	
D	a_i	b_i	m_{1i}
$\overline{D}$	c_i	d_i	m_{2i}
	1	1	2

(5.56)

where $m_{1i} = 0, 1$ or 2. In a case-control study similar tables are constructed from each pair consisting of one member from each disease or outcome group $(D, \overline{D})$. A stratified Mantel-Haenszel analysis can then be performed over these N tables.

However, it is readily shown (Problem 5.9) that there are only four possible types of 2×2 tables, each with values a_i, $E(a_i)$, and $V_c(a_i)$ as shown in Table 5.2. It can then be shown (Problem 5.9) that the summary Mantel-Haenszel test statistic X^2_{MH} equals McNemar's test statistic X^2_M in (5.37) and that the Mantel-Haenszel estimated summary odds ratio in (4.14) is the conditional odds ratio: $\widehat{OR}_{MH} = f/g = \widehat{OR}_C$. Further, it is also readily shown that the Robins, Breslow and Greenland (1986) estimate of the variance of the log odds ratio in (4.28) equals the estimate of the large sample variance of the log conditional odds ratio presented in (5.53). Thus a Mantel-Haenszel analysis of odds ratios from the N pair matched tables provides results that are identical to those described previously for the conditional odds ratio.

Table 5.2 Mantel-Haenszel Stratified Analysis of Pair-Matched Data.

	Tables of Each Type			
#:	e	f	g	h
Table:	[1 1 / 0 0]	[1 0 / 0 1]	[0 1 / 1 0]	[0 0 / 1 1]
a_i:	1	1	0	0
$E(a_i)$:	1	1/2	1/2	0
$V_c(a_i)$:	0	1/4	1/4	0

5.4.5 Relative Risk for Matched Pairs

Like the odds ratio, the relative risk in the general population does not equal the marginal relative risk under matching. When the non-exposed controls are matched to the exposed members of each pair, as shown in Section 5.2.2, then the *population averaged relative risk* is

$$RR_A = \frac{\pi_{1\bullet}}{\pi_{\bullet 1}} = \frac{P(D|E)}{P_m(D|\overline{E})} \neq \frac{P(D|E)}{P(D|\overline{E})} = RR\,, \tag{5.57}$$

where RR_A is estimated as

$$\widehat{RR}_A = \frac{p_{1\bullet}}{p_{\bullet 1}} = \frac{p_{11}+p_{12}}{p_{11}+p_{21}} = \frac{e+f}{e+g} = \frac{n_{1\bullet}}{n_{\bullet 1}} \tag{5.58}$$

using the entries in the aggregate paired 2×2 table. In Problem 5.9, we also show that the Mantel-Haenszel estimate of the relative risk in (4.17) is obtained from the N conditionally independent pairs equals this quantity.

Exact confidence limits are not readily obtained because the parameter involves the probabilities $(\pi_{11}, \pi_{12}, \pi_{21})$. Large sample asymmetric confidence limits can be obtained using the large sample variance of $\log(\widehat{RR}_A)$. In Problem 5.6.6, using the δ-method, it is shown that

$$V\left[\log(\widehat{RR}_A)\right] = \frac{\pi_{12}+\pi_{21}}{N\pi_{1\bullet}\pi_{\bullet 1}} = \frac{\pi_d}{N\pi_{1\bullet}\pi_{\bullet 1}}\,, \tag{5.59}$$

which can be estimated consistently as

$$\widehat{V}\left[\log(\widehat{RR}_A)\right] = \frac{p_{12}+p_{21}}{Np_{1\bullet}p_{\bullet 1}} = \frac{M}{n_{1\bullet}n_{\bullet 1}}\,. \tag{5.60}$$

5.4.6 Attributable Risk for Matched Pairs

Although the population attributable risk is a function of the relative risk, it is not a function of the population averaged relative risk as estimated from a matched prospective study. Thus the PAR is rarely used to describe the results of a matched cross-sectional or prospective study.

Example 5.7 *Pregnancy and Retinopathy Progression*
So far our discussion of matched-pairs has concerned separate individuals that are matched to construct each pair. However, matched pairs may also arise from repeated measures on the same subject, among other situations. An example is provided by the study of the effects of pregnancy on progression of retinopathy (diabetic retinal disease) among women who became pregnant during the Diabetes Control and Complications Trial (DCCT, 2000). The retinopathy status of each woman was assessed before pregnancy and then again during pregnancy. Retinopathy was assessed on a 13-step ordinal scale of severity. Thus rather than a 2×2 table to describe the retinopathy level before pregnancy and that during pregnancy, a 13×13 table is required. However, the study wished only to assess whether pregnancy exacerbated the level of retinopathy represented by *any* worsening versus *any* improvement. In this case, worsening refers to any cell above the diagonal of this 13×13 table, and improvement by any cell below.

For example, if retinopathy were assessed with four steps or ordinal categories, the data would be summarized in a 4×4 table such as

		Level During Pregnancy			
		1	2	3	4
Level	1	=	w	w	w
Before	2	b	=	w	w
Pregnancy	3	b	b	=	w
	4	b	b	b	=

All entries along the diagonal refer to individuals for whom the level before pregnancy equals that after pregnancy, where pregnancy is equivalent to the risk factor exposure. Then the total number of individuals worse, say W, is the sum of the entries in all cells above the diagonal of equality, and the total number improved or better, say B, is the sum of the entries in all cells below the diagonal. These quantities are equivalent to the discordant frequencies used previously such that $f = W$ and $g = B$. Then, from multinomial distribution for the 16 cells of the table, or by conditioning on the total number of discordant observations, $M = W + B$, it can be shown that all of the above results apply.

Note that one can do likewise for any symmetric partitioning of the entries in the paired two-way table. Thus if one is only interested in worsening by two or more steps above the diagonal, only the entries in the three extreme cells in either corner of the table would be used.

The effects of pregnancy on the level of retinopathy in the DCCT are then presented in the following table.

During

Before		
	-	$W = 31$
	$B = 14$	-

Among the 77 women who became pregnant 31 (40%) had a level of retinopathy that was worse during pregnancy than before pregnancy compared to 14 (18%) whose level of retinopathy improved. Under the null hypothesis of no effect of pregnancy, the numbers worse and better are expected to be equal. The conditional odds ratio is $\widehat{OR}_C = 2.2$ with asymmetric 95% *C.I.* (1.18, 4.16). McNemar's test yields a value $X^2_M = 6.42$ with $p \leq 0.012$.

5.5 PAIR-MATCHED RETROSPECTIVE STUDY

One of the most common applications of matched pairs is in a retrospective case-control study. Here the sampling involves the selection of a case with the disease (D), the determination of the covariate values for that subject (z), the selection of a matched non-diseased control ($\overline{D}$) with the same covariate values, and then the determination of whether the case and control had previously been exposed (E) or not ($\overline{E}$) to the risk factor under study. In this setting the observations are summarized in a 2×2 table as

Frequencies

		$\overline{D}$: E	$\overline{D}$: $\overline{E}$	
D	E	e	f	$n_{1\bullet}$
	$\overline{E}$	g	h	$n_{2\bullet}$
		$n_{\bullet 1}$	$n_{\bullet 2}$	N

Probabilities

		$\overline{D}$: E	$\overline{D}$: $\overline{E}$	
D	E	ϕ_{11}	ϕ_{12}	$\phi_{1\bullet}$
	$\overline{E}$	ϕ_{21}	ϕ_{22}	$\phi_{2\bullet}$
		$\phi_{\bullet 1}$	$\phi_{\bullet 2}$	1

(5.61)

As was the case for the prospective study, the marginal probability $\phi_{1\bullet}$ under matching can be expressed as $P(E|D) = \int_z P(E|D,z) f_D(z)\,dz$ with respect to the distribution of Z among the cases. However, since the controls are selected to have covariate values that match those of the cases, then

$$\begin{aligned} \phi_{\bullet 1} &= \int_z P\left(E|\overline{D},z\right) f_D(z)\,dz = P_m\left(E|\overline{D}\right) \\ &\neq \int_z P\left(E|\overline{D},z\right) f_{\overline{D}}(z)\,dz = P\left(E|\overline{D}\right) \end{aligned} \tag{5.62}$$

analogous to (5.24) for the matched prospective study. Therefore, the marginal retrospective odds ratio under matching does not equal the retrospective odds ratio in the general population of cases and controls, and thus does not equal the prospective odds ratio.

As for the prospective study the null hypothesis of marginal homogeneity under matching H_0: $\phi_{1\bullet} = \phi_{\bullet 1}$ implies the hypothesis of symmetry H_0: $\phi_{12} = \phi_{21}$. Further, the hypothesis of symmetry of the discordant retrospective probabilities implies the null hypothesis of symmetry of the prospective discordant probabilities (Problem 5.10). Thus the conditional binomial exact test and the large sample McNemar test are again employed to test the hypothesis of no association between exposure and disease.

5.5.1 Conditional Odds Ratio

As in Section 5.4.1, conditional on the value of the matching covariate z the pair members are sampled independently such that the probability of each type of discordant pair is

$$\phi_{12|z} = \phi_{1\bullet|z}\phi_{\bullet 2|z} = P\left(E|D,z\right)P\left(\overline{E}|\overline{D},z\right) \tag{5.63}$$

and

$$\phi_{21|z} = \phi_{2\bullet|z}\phi_{\bullet 1|z} = P\left(\overline{E}|D,z\right)P\left(E|\overline{D},z\right). \tag{5.64}$$

Thus the conditional marginal retrospective odds ratio is

$$\underset{retro}{OR_z} = \frac{\phi_{1\bullet|z}\phi_{\bullet 2|z}}{\phi_{2\bullet|z}\phi_{\bullet 1|z}} = \frac{\phi_{12|z}}{\phi_{21|z}}. \tag{5.65}$$

For the unmatched retrospective study we showed in (5.11) that the retrospective odds ratio equals the prospective odds ratio. For the matched retrospective study, it can likewise be shown that the conditional retrospective odds ratio

$$\underset{retro}{OR_z} = \frac{P\left(E|D,z\right)/P\left(\overline{E}|D,z\right)}{P\left(E|\overline{D},z\right)/P\left(\overline{E}|\overline{D},z\right)} \tag{5.66}$$

equals the conditional prospective odds ratio

$$\underset{prosp}{OR_z} = \frac{P\left(D|E,z\right)/P\left(\overline{D}|E,z\right)}{P\left(D|\overline{E},z\right)/P\left(\overline{D}|\overline{E},z\right)}, \tag{5.67}$$

using the same developments as for the matched prospective study (Problem 5.10).

Also, the assumption of a common conditional retrospective odds ratio implies a common conditional prospective odds ratio

$$\underset{retro}{OR_z} = \left(\underset{retro}{OR_C} \quad \forall z\right) = \left(\underset{prosp}{OR_C} \quad \forall z\right), \tag{5.68}$$

which is estimated consistently as $\widehat{OR}_C = f/g$ with estimated large sample variance $\widehat{V}\left[\log\widehat{OR}\right] = M/(fg)$. This then provides for the computation of large sample confidence limits.

5.5.2 Relative Risks from Matched Retrospective Studies

For the unmatched retrospective study Cornfield (1951) showed that the relative risk can be approximated by the odds ratio under the rare disease assumption. This result also applies to the matched retrospective study under matching. In this case it can be shown that the relative risk for a fixed value of the matching covariate z is

$$RR_z = \frac{P(D|E,z)}{P(D|\overline{E},z)} = \frac{P(E|D,z)\,P(\overline{E}|z)}{P(\overline{E}|D,z)\,P(E|z)} \tag{5.69}$$
$$= \frac{P(E|D,z)\,P(\overline{E}|D,z)\,P(D|z) + \phi_{12|z}\,[1-P(D|z)]}{P(E|D,z)\,P(\overline{E}|D,z)\,P(D|z) + \phi_{21|z}\,[1-P(D|z)]},$$

where $\phi_{12|z}$ and $\phi_{21|z}$ are the discordant probabilities conditional on the value z and $P(D|z) = \delta_z$ is the conditional probability of the disease in the population given z. Again assuming a constant relative risk for all values of Z, then under the rare disease assumption $(\delta_z \downarrow 0)$,

$$RR_{z,\delta_z\downarrow 0} \doteq \frac{\phi_{12|z}}{\phi_{21|z}} = OR_C \tag{5.70}$$

so that the conditional odds ratio provides an approximation to the relative risk.

Example 5.8 *Estrogen Use and Endometrial Cancer*
Breslow and Day (1980) present data from Mack, Pike, Henderson, et al. (1976) for a matched case-control study of the risk of endometrial cancer among women who resided in a retirement community in Los Angeles. The data set is described in Problem 7.18 of Chapter 7. Each of the 63 cases was matched to four controls within one year of age of the case, who had the same marital status, and who entered the community at about the same time. Multiple controls were matched to each case. Here only the first matched control is employed to construct 63 matched pairs. The risk factor exposure of interest is a history of use of conjugated estrogens. The resulting table of frequencies is

		$\overline{D}$: E	$\overline{D}$: $\overline{E}$	
D	E	18	33	51
	$\overline{E}$	6	6	12
		24	39	63

for which McNemar's test is $X^2_M = 18.69$ with $p < 0.0001$. The estimate of the conditional odds ratio is $\widehat{OR}_C = 5.5$ with asymmetric 95% confidence limits (2.30, 13.13). This odds ratio provides an approximate relative risk under the rare disease assumption.

5.6 POWER FUNCTION OF McNEMAR'S TEST

5.6.1 Unconditional Power Function

Various authors have proposed methods for the evaluation of the power of the McNemar test, some unconditionally assuming only a sample of N matched pairs, others conditionally assuming that M discordant pairs are observed. Among the unconditional power function derivations, Lachin (1992b) shows that the procedure suggested by Connor (1987) and Connett, Smith and McHugh (1987) based on the underlying multinomial is the most accurate. This approach will be described for a prospectively matched study (in terms of the π_{ij}), but the same equations also apply to the matched retrospective study (in terms of the ϕ_{ij}).

The test statistic is $T = p_{12} - p_{21} = p_f - p_g$. The null hypothesis of symmetry specifies that H_0: $\pi_{12} = \pi_{21} = \pi_d/2$, where π_d is the probability of a discordant pair. Then, from (5.31) and (5.35) above, $T \overset{d}{\approx} \mathcal{N}\left(\mu_0, \sigma_0^2\right)$ with $\mu_0 = 0$ and $\sigma_0^2 = \pi_d/N$. Under the alternative hypothesis H_1: $\pi_{12} \neq \pi_{21}$, from (5.34) $T \overset{d}{\approx} \mathcal{N}\left(\mu_1, \sigma_1^2\right)$ with $\mu_1 = \pi_{12} - \pi_{21}$ and

$$\sigma_1^2 = \frac{(\pi_{12} + \pi_{21}) - (\pi_{12} - \pi_{21})^2}{N} = \frac{\pi_d - (\pi_{12} - \pi_{21})^2}{N} \,. \tag{5.71}$$

Using the general equation relating sample size to power presented in (3.18) it follows that

$$\sqrt{N}\,|\pi_{12} - \pi_{21}| = Z_{1-\alpha}\sqrt{\pi_d} + Z_{1-\beta}\sqrt{\pi_d - (\pi_{12} - \pi_{21})^2} \,, \tag{5.72}$$

from which one can determine N or solve for the standardized deviate corresponding to the level of power $Z_{1-\beta}$ (see Problem 5.11).

In order to use this expression, one only need specify π_{21} and $OR = \pi_{12}/\pi_{21}$ so that $\pi_{12} = \pi_{21} OR$. The other entries in the 2×2 table are irrelevant.

Example 5.9 *Unconditional Sample Size*

Assume $\pi_{21} = 0.125$ and that we wish to detect an odds ratio of $OR = 2$ which, in turn, implies that $\pi_{12} = 0.25$, $\pi_d = 0.375$ and $\mu_1 = 0.125$. For $\alpha = 0.05$ (two-sided) the N required to provide $\beta = 0.10$ is provided by

$$N = \left[\frac{(1.96)\sqrt{0.375} + 1.282\sqrt{(0.375) - (0.125)^2}}{0.125}\right]^2 = 248.$$

5.6.2 Conditional Power Function

The unconditional power function above is the preferred approach for determining the sample size a priori. The actual power, however, depends on the observed number of discordant pairs, $M = f + g$. In the above example, given the specified values of the discordant probabilities, then the expected number of discordant pairs is $E(M) = N\pi_d = (248)(0.375) = 93$. It is the latter quantity that determines the power of the test, as is demonstrated from the expressions for the conditional power function.

Miettinen (1968) describes the conditional power function given M in terms of the statistic $(f - g)$. However, it is simpler to do this in terms of the conditional distribution of $p_f = f/M$ described in Section 5.4.3. Under H_0 in (5.54), $p_f \overset{d}{\approx} \mathcal{N}(\mu_{0c}, \sigma_{0c}^2)$ with $\mu_{0c} = 1/2$ and $\sigma_{0c}^2 = 1/(4M)$. Conversely, under the alternative H_1 in (5.55), $p_f \overset{d}{\approx} \mathcal{N}(\mu_{1c}, \sigma_{1c}^2)$ with $\mu_{1c} = \pi_{12}/\pi_d$ and

$$\sigma_{1c}^2 = \frac{\pi_{12}\pi_{21}}{M(\pi_d)^2} \tag{5.73}$$

(see Problem 5.11). These expressions can also be presented in terms of the conditional odds ratio, say $\varphi_c = OR_c$, (see Problem 5.11.3). Since $\varphi_c = \pi_{12}/\pi_{21}$, then

$$\mu_{1c} = \frac{\pi_{12}}{\pi_d} = \frac{\pi_{12}}{\pi_{12} + \pi_{21}} = \frac{\varphi_c}{1 + \varphi_c} \tag{5.74}$$

and

$$\sigma_{1c}^2 = \frac{\pi_{12}\pi_{21}}{M(\pi_d)^2} = \frac{\varphi_c}{M(1 + \varphi_c)^2} \,. \tag{5.75}$$

Substituting into the general expression (3.20) for $Z_{1-\beta}$, then the conditional power function is provided by

$$Z_{1-\beta} = \frac{\sqrt{M}\left|\pi_{12} - \pi_{21}\right| - Z_{1-\alpha}\pi_d}{2\sqrt{\pi_{12}\pi_{21}}} \,. \tag{5.76}$$

When specified in terms of M and φ_c, we obtain

$$Z_{1-\beta} = \frac{\sqrt{M}\left|\varphi_c - 1\right| - Z_{1-\alpha}(\varphi_c + 1)}{2\sqrt{\varphi_c}} \,, \tag{5.77}$$

which provides the level of conditional power $1 - \beta$ to detect an odds ratio of φ_c with M discordant pairs.

The conditional power function should not be used to determine the sample size a priori unless the study design calls for continued sampling until the fixed number M of discordant pairs is obtained. Otherwise, there is no guarantee that the required M will be observed. Therefore, the unconditional expressions based on (5.72) should be used to determine the total sample size N to be sampled a priori. This is the value N that on average will provide the required number of discordant pairs for the specified values of π_{21} and φ_c.

Example 5.10 *Conditional Power*
For the above example where we wished to detect an odds ratio $\varphi_c = 2$, if the actual number of discordant pairs is $M = 80$ rather than the desired 93, then the standardized deviate is

$$Z_{1-\beta} = \frac{\sqrt{80}\,|2-1| - (1.96)\,(3)}{2\sqrt{2}} = 1.083\,, \tag{5.78}$$

for which the corresponding level of power is $1 - \beta \doteq 0.86$.

5.6.3 Other Approaches

Schlesselman (1982) describes an unconditional calculation based on an extension of the above expression for the conditional power function in a matched case-control study with probabilities $\{\phi_{ij}\}$ in the four cells of the table, where the $\{\phi_{ij}\}$ in the retrospective study are the counterparts to the $\{\pi_{ij}\}$ in the prospective study. His approach is appealing because the problem is parameterized in terms of the *marginal* probabilities that are often known rather than the discordant probabilities that are usually unknown. Thus his approach is based on specification of the odds ratio to be detected and the marginal population probabilities of exposure among the controls, $\phi_{\bullet 1} = P(E|\overline{D})$. However, Schlesselman's procedure assumes that the responses of the pair members are independent, or, for example, $\phi_{11} = \phi_{1\bullet}\phi_{\bullet 1}$, which only occurs when matching is ineffective.

Various authors have provided corrections to Schlesselman's calculation based on the specification of the correlation among pair members induced by matching. Lachin (1992b) describes a simpler direct calculation based on a specification of the marginal probability of exposure among the controls ($\phi_{\bullet 1}$), the conditional odds ratio to be detected ($\varphi_c = \phi_{12}/\phi_{21}$), and the correlation of exposures within pairs represented by the exposure odds ratio in the matched sample $[\omega = (\phi_{11}\phi_{22})\,/(\phi_{12}\phi_{21})]$. From these quantities the following joint relationships apply:

$$\begin{aligned}\phi_d &= \phi_{12} + \phi_{21} \\ \phi_{21} &= \phi_{12}/\varphi_c \\ \phi_{11} &= \omega\phi_{12}\phi_{21}/\phi_{22}.\end{aligned} \tag{5.79}$$

Noting that $\phi_{\bullet 1} = \phi_{11} + \phi_{21}$ and that $\phi_{22} = 1 - \phi_{\bullet 1} - \phi_{12}$, it can then be shown that $\phi_{\bullet 1}(1 - \phi_{\bullet 1})$ can be expressed as a function of ϕ_{12} alone such that

$$\frac{\omega - 1}{\varphi_c}\phi_{12}^2 + \frac{\varphi_c\phi_{\bullet 1} + (1 - \phi_{\bullet 1})}{\varphi_c}\phi_{12} - \phi_{\bullet 1}(1 - \phi_{\bullet 1}) = 0. \tag{5.80}$$

This is a standard quadratic equation of the form $a\phi_{12}^2 + b\phi_{12} + c = 0$, and the root in the interval (0,1) provides the value of ϕ_{12} that satisfies these constraints. Given ϕ_{12}, the other probabilities of the table are then obtained from the above relationships. These probabilities $\{\phi_{ij}\} \equiv \{\pi_{ij}\}$ in the four cells of the table can then be used in the unconditional sample size calculation in (5.72).

Example 5.11 *Case-Control Study*
Lachin (1992b) presents the following example: We wish to determine the total sample size for a matched case-control study that will provide unconditional power to detect a conditional odds ratio $\varphi_c = 2$ with power $1 - \beta = 0.90$ in a two-sided test at level $\alpha = 0.05$. If the probability of exposure among the controls is $\phi_{\bullet 1} = 0.3$, then Schlesselman's calculation (not shown) yields $N = 186.4$. This assumes, however, that matching is ineffective or that the exposure odds ratio is $\omega = 1$. If we assume that matching is effective, such that $\omega = 2.5$, then the root of (5.80) yields $\phi_{12} = 0.25$, which yields the following table of probabilities:

		$\overline{D}$		
		E	$\overline{E}$	
D	E	$\phi_{11} = 0.175$	$\phi_{12} = 0.250$	$\phi_{1\bullet} = 0.425$
	$\overline{E}$	$\phi_{21} = 0.125$	$\phi_{22} = 0.450$	$\phi_{2\bullet} = 0.575$
		$\phi_{\bullet 1} = 0.30$	$\phi_{\bullet 2} = 0.70$	1

The unconditional sample size calculation in (5.72) then yields $N = 248$, substantially larger than suggested by Schlesselman's procedure.

Note that the discordant probabilities are the same as those specified in Example 5.9 so that the required sample size is the same. The only difference is how the probabilities arose. In Example 5.9 the problem was uniquely specified in terms of the probabilities of the discordant pairs. In Example 5.11, these probabilities are derived indirectly, starting with the specification of one of the marginal probabilities, the conditional odds ratio and the exposure odds ratio, from which the discordant probabilities are then obtained.

5.6.4 Matching Efficiency

Various authors have considered the topic of the relative efficiency of matching versus not. Miettinen (1968; 1970) argues that "matching tends to reduce efficiency and would therefore have to be motivated by the pursuit of validity alone." He argues that the matched study has less power than an unmatched study, all other things being equal. However, his comparison and conclusions are flawed. Karon and Kupper (1981), Kupper, Karon, Kleinbaum, Morgenstern and Lewis (1981), Wacholder and Weinberg (1982) and Thomas and Greenland (1983) present further explorations of the efficiency of matching. In general, they conclude that efficient matching that provides positive within-pair correlation will provide greater power in a matched study than in an unmatched study of the same size.

5.7 STRATIFIED ANALYSIS OF PAIR-MATCHED TABLES

In some instances it may be desirable to conduct an analysis of pair-matched data using stratification or a regression model to adjust for other characteristics that may be associated either with exposure or the response. There have been few papers on stratification adjustment, in part because the conditional logistic regression model provides a convenient mechanism to conduct such an adjusted analysis. This model is described in Chapter 7. Nevertheless, it is instructive to consider a stratified analysis of K independent 2×2 tables that have been either prospectively or retrospectively matched in parallel with the methods developed for independent observations in Chapter 4. Here we only consider the case of matched pairs. In such a stratified matched analysis, we must distinguish between two different cases: pair versus member stratification.

5.7.1 Pair and Member Stratification

In *pair stratification*, the N matched pairs are characterized by pair-level covariates, where both members of the pair share the same values of the stratifying covariates. The pairs and their respective members then are grouped into K independent strata. Since all pair members have the same covariate values within each stratum, there is no covariate by exposure group association and the marginal, unadjusted odds ratio is unbiased. Thus some authors have stated that a stratified analysis in this case is unnecessary because the exposed and non-exposed groups are also balanced with respect to the pair stratification covariates. This, however, ignores the possible association between the stratifying covariates and response for which a stratification adjustment may be desirable.

The most common example of pair stratification is when the matched observations are obtained from the same individual, such as right eye versus left eye (one treated topically, the other not), or when observations are obtained before and again after some exposure as in Example 5.7. In such cases, any patient characteristic is a potential pair-stratifying covariate. With such pair-matched data, it might still be relevant to conduct an analysis stratifying on potential risk factors that may be associated with the probability of a positive outcome (D), including the matching covariates. As we shall see, such pair stratification has no effect in a Mantel-Haenszel analysis, but it does yield an adjusted *MVLE* of the common odds ratio. If, for example, a study employed before-and-after or left-and-right measurements in the same individual, and if gender were a potent risk factor, then the gender stratified-adjusted *MVLE* of the odds ratio might be preferred. In such cases, it would also be important to assess the heterogeneity of the odds ratios across strata using a test of homogeneity.

In *member stratification* the individual members of each pair are stratified with respect to the covariate characteristics of the pair members. For example, consider a study using pairs of individuals that are matched with respect to age and severity of disease, but not by gender. It might then be desirable to conduct a stratified analysis that adjusts for gender. In this case, there are four possible strata with respect to the gender of the pair members: MM, MF, FM and FF. Usually the stratified analysis is then based on those concordant pairs where the two pair members are concordant with respect to the stratifying covariates; that is, using only the MM and FF strata. This provides an estimate of the conditional odds ratio in the setting where the members are also matched for gender in addition to the other matching covariates. Clearly, this sacrifices any information provided by the other strata wherein the pair members are discordant for the stratifying covariates; that is, the MF, FM strata. While these discordant strata would exist in the general unmatched population, their exclusion here is appropriate since the frame of reference is the matched population.

5.7.2 Stratified Mantel-Haenszel Analysis

Consider either pair stratification with K separate strata defined by the covariate values, or member stratification with K strata defined from the possible concordant covariate values for the two members of each pair. Building on the construction for a single matched 2×2 table shown in Table 5.2, there are a total of $4K$ sets of tables of type e_j, f_j, g_j and h_j, $(j = 1, \ldots, K)$, as presented in Sections 5.4.4 and 5.4.5. Applying the expression for the Mantel-Haenszel estimate of the common odds ratio in (4.14) yields

$$\widehat{OR}_{C(MH)} = \frac{\sum_j f_j}{\sum_j g_j} = \frac{f_\bullet}{g_\bullet}, \tag{5.81}$$

where $f_\bullet$ and $g_\bullet$ are the numbers of discordant pairs in the aggregate matched 2×2 table. Likewise, the expression for the Robins, Breslow and Greenland (1986) variance in (4.28) is

$$\widehat{V}\left[\log\left(\widehat{OR}_{C(MH)}\right)\right] = \frac{1}{f_\bullet} + \frac{1}{g_\bullet} = \frac{M_\bullet}{f_\bullet g_\bullet}, \tag{5.82}$$

where $M_\bullet = f_\bullet + g_\bullet$. Similarly, the Mantel-Haenszel estimate of the common relative risk in (4.17) is

$$\widehat{RR}_{C(MH)} = \frac{\sum_j (e_j + f_j)}{\sum_j (e_j + g_j)} = \frac{e_\bullet + f_\bullet}{e_\bullet + g_\bullet}. \tag{5.83}$$

This relative risk estimate may be used for a matched prospective or cross-sectional study, but not a case-control study.

Applying the expression for the Mantel-Haenszel test in (4.8) to these $4K$ tables yields the Mantel-Haenszel test statistic

$$X^2_{MH} = \frac{\left[\sum_j (f_j - g_j)\right]^2}{\sum_j (f_j + g_j)} = \frac{(f_\bullet - g_\bullet)^2}{f_\bullet + g_\bullet}, \tag{5.84}$$

Which is the unadjusted McNemar test statistic. Thus the stratified-adjusted Mantel-Haenszel estimates and test are the same as the unadjusted estimates and test.

5.7.3 MVLE

As was the case for 2×2 tables of independent (unmatched) observations, the $MLVE$ of the common log odds ratio θ is obtained as the weighted combination of the stratum-specific values $\widehat{\theta}_j = \log(\widehat{OR}_{Cj})$ of the form

$$\widehat{\theta} = \sum\nolimits_j \widehat{\omega}_j \widehat{\theta}_j \tag{5.85}$$

with estimated weights obtained from the estimated variance of the estimate (5.53)

$$\widehat{\omega}_j = \frac{(f_j g_j)/M_j}{\sum_\ell (f_\ell g_\ell)/M_\ell} . \tag{5.86}$$

The variance is consistently estimated as

$$\widehat{V}\left(\widehat{\theta}\right) = \frac{1}{\sum_\ell (f_\ell g_\ell)/M_\ell} , \tag{5.87}$$

which provides asymmetric confidence limits for the adjusted odds ratio.

Likewise, the *MVLE* of the common marginal log relative risk, when such is appropriate, is obtained using $\widehat{\theta}_j = \log(\widehat{RR}_{Aj})$ with weights obtained from the variance of the estimate (5.60)

$$\widehat{\omega}_j = \frac{(n_{1\bullet j} n_{\bullet 1j})/M_j}{\sum_\ell (n_{1\bullet \ell} n_{\bullet 1\ell})/M_\ell} \tag{5.88}$$

and large sample estimated variance

$$\widehat{V}\left(\widehat{\theta}\right) = \frac{1}{\sum_\ell (n_{1\bullet \ell} n_{\bullet 1\ell})/M_\ell} . \tag{5.89}$$

These methods can be applied to either pair or member stratified analyses.

5.7.4 Tests of Homogeneity and Association

5.7.4.1 Conditional Odds Ratio The contrast or Cochran's test of homogeneity of the conditional odds ratios is readily conducted using the *MVLE* $\widehat{\theta}$ in (5.85) above, in conjunction with the inverse variances $\widehat{\tau}_j = \widehat{V}(\widehat{\theta}_j)^{-1}$ for each stratum $(j = 1, \ldots, K)$. These quantities are simply substituted in the expression for the test statistic in (4.68). Alternately, Breslow and Day (1980) suggest that a simple $K \times 2$ contingency table of the discordant frequencies n_{1j} and n_{2j} could be constructed and used as the basis for a simple contingency chi-square test as in (2.81) on $K - 1$ *df*. It is readily shown (Problem 5.12.11) that the null hypothesis of independence in the $K \times 2$ table is equivalent to the hypothesis of homogeneity of the conditional odds ratio among strata.

The asymptotically efficient test of association of a common conditional odds ratio can be derived as follows. Since the conditional odds ratio is a function of the discordant probabilities, this suggests that analogous Radhakrishna-like test statistic is of the form:

$$T = \sum\nolimits_{j=1}^{K} w_j (f_j - g_j) . \tag{5.90}$$

Now assume that $E(f_j) = N_j\pi_{12j}$, $E(g_j) = N_j\pi_{21j}$, and that $\theta_{j\ell} = g(\pi_{12j}) - g(\pi_{21j}) = \theta\ \forall j$, where $g(\pi) = \log(\pi)$. Thus by a first order Taylor's expansion, then asymptotically

$$E(T) \cong \theta \sum_{j=1}^{K} \frac{w_j N_j}{g'(\pi_{dj}/2)} = \theta \sum_{j=1}^{K} \frac{w_j N_j \pi_{dj}}{2}\ , \tag{5.91}$$

where $\pi_j = \pi_{dj}/2 \in (\pi_{12j}, \pi_{21j})$ and $\pi_{dj} = (\pi_{12j} + \pi_{21j})/2$. From (5.35), noting that $p_{12} = f/N$ and that $p_{21} = g/N$, the large sample variance of the statistic is

$$V(T \mid H_0) \cong \sum_{j=1}^{K} w_j^2 V(f_j - g_j \mid H_0) = \sum_{j=1}^{K} w_j^2 N_j \pi_{dj}. \tag{5.92}$$

Then, evaluating the asymptotic efficiency of the test, it follows that the optimal weights are $w_j = 1$ for all strata and that the asymptotically efficient test is the same as that derived as the Mantel-Haenszel test in (5.84) above.

Thus for a pair or a member stratified analysis the Radhakrishna-like asymptotically efficient test of a common odds ratio is the same as the marginal unadjusted McNemar test. This is not surprising because the Mantel-Haenszel test is a member of the Radhakrishna family.

5.7.4.2 Marginal Relative Risk Similarly, in a matched prospective study, a test of homogeneity of log relative risks is readily constructed using the *MVLE* and the variance of the estimates within strata as in (5.58) and (5.60).

Also, an asymptotically efficient test of a common relative risk can be derived as follows: The log RR is $\theta_j = g(\pi_{1\bullet j}) - g(\pi_{\bullet 1j})$, where $g(\pi) = \log(\pi)$ as above. Using the test statistic

$$T = \sum_{j=1}^{K} w_j (p_{1\bullet j} - p_{\bullet 1j})\ , \tag{5.93}$$

then, under the model with a common relative risk

$$E(T) \cong \theta \frac{\sum_{j=1}^{K} w_j}{g'(\pi_j)}\ , \tag{5.94}$$

where $\pi_j \in (\pi_{1\bullet j}, \pi_{\bullet 1j})$, and the null variance is

$$V(T \mid H_0) \cong \sum_{j=1}^{K} w_j^2 V(p_{1\bullet j} - p_{\bullet 1j} \mid H_0) = \sum_{j=1}^{K} w_j^2 \pi_{dj}/N_j \tag{5.95}$$

since $(p_{1\bullet j} - p_{\bullet 1j}) = (p_{12j} - p_{21j})$. Then it is easily shown that the asymptotically efficient test uses the optimal weights

$$w_j = \frac{\pi_j N_j}{\pi_{dj}} \mathrel{\hat{=}} \frac{(p_{1\bullet j} + p_{\bullet 1j}) N_j}{2(p_{12j} + p_{21j})}\ . \tag{5.96}$$

These derivations are left to a Problem.

Since the optimal test of a common odds ratio and the test of the common relative risks employ different statistics, the $MERT$ for the family comprising these two tests is not described. However, the correlation is readily obtained by expressing each of these asymptotically efficient tests as a linear combination of the vectors of $4K$ frequencies or proportions.

Example 5.12 *Pregnancy and Retinopathy Progression*
In the analysis of the effects of pregnancy of the level of retinopathy (see Example 5.7), it was also important to assess the influence of the effect of intensification of control of blood glucose levels on the level of retinopathy. All women who planned to, or who became pregnant, received intensified care to achieve near-normal levels of blood glucose to protect the fetus. However, other studies had shown that such intensification alone could have a transient effect on the level of retinopathy that could be confounded with the effect of pregnancy. Thus it was important to adjust for this effect in the analysis. Here the stratified analysis within the intensive treatment group is described. The analysis within the conventional treatment group is used as a Problem.

The degree of intensification of therapy was reflected by the change in the level of HbA_{1c} from before to during pregnancy, the greater the reduction the greater the intensification. The following table presents the numbers within each of four strata classified according to the decrease in HbA_{1c} during pregnancy, where a negative value represents an increase.

HbA_{1c} Decrease	N	*Better* # (%)	*Worse* # (%)	OR_c	95% $C.I.$
<0	15	5 (33)	6 (40)	1.2	0.4, 3.9
(0.0 %, 0.7%]	20	5 (25)	10 (50)	2.0	0.7, 5.9
(0.7 %, 1.3 %]	22	1 (5)	6 (27)	6.0	0.7, 50.0
>1.3 %	20	3 (15)	9 (45)	3.0	0.8, 11.1

(DCCT, 2000; reproduced with permission). The *MVLE* provides a stratified-adjusted odds ratio of 2.1 with 95% confidence limits of (1.10, 4.02), the estimate being nearly identical to the unadjusted value of 2.2 presented earlier in Example 5.7. A Wald test based on the estimated adjusted log odds ratio (0.743) and its estimated variance (0.109) yields $X^2 = (0.743)^2/(0.109) = 5.067$, which is also highly significant ($p < 0.006$). This test value is comparable to that provided by the unadjusted McNemar test in Example 5.7.

The Cochran test of homogeneity of odds ratios over strata yields $X^2 = 2.093$ on 3 df with $p \leq .56$. A simpler test that does not require the computation of the *MVLE* is to conduct a Pearson contingency test of independence for the above 4×2 table of frequencies, for which $X^2 = 2.224$ on 3 df with $p \leq 0.527$. Both tests indicate that the effect of pregnancy on the level of retinopathy does not depend on the change in the level of HbA_{1c} during pregnancy.

Example 5.13 *Member-Stratified Matched Analysis*
To illustrate a member-stratified analysis, consider the endometrial cancer data of Example 5.8. One of the covariates measured on each participant in the study is the presence or absence of a history of hypertension (H versus $\overline{H}$). Thus there are four possible strata defined by the covariate values of the pair members – $\overline{H}\overline{H}$ with 16 discordant pairs, $\overline{H}H$ with 7 pairs, $H\overline{H}$ with 11 pairs, and HH with 5 pairs. Within the concordant strata $\overline{H}\overline{H}$ and HH the numbers of discordant pairs are too small to provide a reliable analysis, $g_j = 3$ and 0 in each stratum, respectively.

With a larger sample size the same analyses as in the previous example could then be performed using the concordant strata.

5.7.5 Random Effects Model Analysis

Finally, based on the test of homogeneity, a random effects analysis of odds ratios, or relative risks when appropriate, is readily implemented as described in Section 4.10.2. The expressions previously presented apply, the only difference being the computational expressions for the log odds ratio (or relative risk) and the estimated variance within each stratum. Specifically, the test of homogeneity provides a moment estimator of the variance between strata, σ_θ^2, based on (4.155). Then the total variance of each stratum-specific estimate involves the estimation variance plus the variance between strata, $V(\widehat{\theta}_j) = V(\widehat{\theta}_j|\theta_j) + \sigma_\theta^2$, as in (4.156). This then leads to first-step reweighted estimate using the weights as in (4.157).

5.8 PROBLEMS

5.1 For an unmatched case-control study do the following:

5.1.1. Show that the prospective odds ratio in (5.5) equals the retrospective odds ratio in (5.4).

5.1.2. Show that the null hypothesis of equal prospective probabilities H_0: $P(D|E) = P(D|\overline{E})$ is equivalent to the null hypothesis of equal retrospective probabilities H_0: $P(E|D) = P(E|\overline{D})$.

5.1.3. Also show that the alternative hypothesis of unequal prospective probabilities H_1: $P(D|E) \neq P(D|\overline{E})$ is equivalent to the hypothesis of unequal retrospective probabilities H_1: $P(E|D) \neq P(E|\overline{D})$.

5.1.4. For a given prevalence $P(D) = \delta$, derive the expression for the RR presented in (5.14).

5.1.5. Under the rare disease assumption show that $RR_{\delta\downarrow 0} \doteq OR_{retro}$.

5.1.6. Also show that

$$RR_{\delta(\phi_1 - \phi_2)\downarrow 0} \doteq OR_{retro}. \tag{5.97}$$

5.1.7. For $\alpha_1 = P(E)$ known, under the rare disease assumption show that

$$logit\ (\widehat{PAR}) = \log \frac{\widehat{PAR}}{1 - \widehat{PAR}} \doteq \log \left[\alpha_1(\widehat{OR} - 1)\right]. \tag{5.98}$$

5.1.8. Derive the expression for the variance of the $logit(\widehat{PAR})$ presented in (5.19).

5.2 Consider the following hypothetical unmatched case-control study of heart failure (D) and smoking (E).

	D	$\overline{D}$
E	120	130
$\overline{E}$	80	170
	200	300

5.2.1. Conduct an appropriate test of H_0.

5.2.2. Calculate an estimate of the prospective population odds ratio with 95% confidence limits.

5.2.3. Assume the prevalence of congestive heart failure is $P(D) = 1$ in 10,000. Calculate an estimate of the prospective population relative risk and compare to the estimate based on the odds ratio.

5.2.4. Calculate an estimate of the population attributable risk with 95% confidence limits assuming that $P(E) = \alpha_1 = 0.30$.

5.3 Consider matched pairs as described in Section 5.2.2.

5.3.1. For a binary response y_{ij}, where $E(y_{ij}|z_i) = \pi(z_i)$ for both members $(j = 1, 2)$ of the ith pair, and where the responses (y_{i1}, y_{i2}) are conditionally independent, show that

$$Cov\,(Y_1, Y_2) = E_z\left[Cov(y_{i1}, y_{i2}|z)\right] \tag{5.99}$$
$$= E_z\left[E(y_{i1}|z_i)E(y_{i2}|z_i)\right] - E_z\left[E(y_{i1}|z_i)\right]E_z\left[E(y_{i2}|z_i)\right]$$

and that

$$Cov\,(Y_1, Y_2) = E_Z\left[\pi(z)^2\right] - \left(E_Z\left[\pi(z)\right]\right)^2 = V[\pi(z)] = \sigma_\pi^2 \neq 0\,. \tag{5.100}$$

5.3.2. Using the partitioning of variation as in (A.6) show that

$$V(Y_1) = V(Y_2) = E_Z\left\{\pi(z)\left[1 - \pi(z)\right]\right\} + \sigma_\pi^2\,. \tag{5.101}$$

5.3.3. Then show that

$$Corr\,(Y_1, Y_2) = \frac{\sigma_\pi^2}{E_Z\left\{\pi(z)\left[1 - \pi(z)\right]\right\} + \sigma_\pi^2}\,. \tag{5.102}$$

5.3.4. For a quantitative response assume that $y_{ij} = \mu(z_i) + \varepsilon_{ij}$, where $\mu(z_i) = E(y_{ij}|z_i)$, and where ε_{ij} are independent of μ_i with $E(\varepsilon_{ij}) = 0$, $V(\varepsilon_{ij}) = \sigma_\varepsilon^2\ \forall ij$ and $Cov(\varepsilon_{i1}, \varepsilon_{i2}) = 0$. Then show that

$$Cov\,(Y_1, Y_2) = E_z\left[Cov(y_{i1}, y_{i2}|z)\right] = E_z\left[\mu(z)^2\right] - \left(E_Z\left[\mu(z)\right]\right)^2 \tag{5.103}$$
$$= \sigma_\mu^2 \neq 0$$

and that

$$Corr\left(Y_1, Y_2\right) = \frac{E_Z\left[\mu(z)^2\right] - \left(E_Z\left[\mu(z)\right]\right)^2}{\left(E_Z\left[\mu(z)^2\right] - \left(E_Z\left[\mu(z)\right]\right)^2\right) + \sigma_\varepsilon^2} = \frac{\sigma_\mu^2}{\sigma_\mu^2 + \sigma_\varepsilon^2}. \tag{5.104}$$

5.4 Maxwell (1961) presents the following matched prospective study of recovery (D) versus not ($\overline{D}$) following treatment among matched pairs of depersonalized (E) versus non-depersonalized ($\overline{E}$) psychiatric patients

		$\overline{E}$: D	$\overline{E}$: $\overline{D}$
E	D	14	2
	$\overline{D}$	5	2

(reproduced with permission).

5.4.1. State appropriate null and alternative hypotheses (H_0, H_1) and conduct an exact test of H_0.

5.4.2. Calculate an estimate of the population odds ratio with exact 95% confidence limits.

5.5 Starting from the multinomial likelihood for a pair matched study presented in (5.22), do the following:

5.5.1. Derive the conditional binomial distribution presented in (5.28).

5.5.2. Show that asymptotically

$$p_f = f/M \overset{d}{\approx} \mathcal{N}\left(\frac{\pi_{12}}{\pi_d}, \frac{\pi_{12}\pi_{21}}{M\pi_d^2}\right). \tag{5.105}$$

5.6 For a pair-matched study, use the normal approximation to the multinomial in (5.31) to show the following:

5.6.1. Derive the expression for $V\left(p_{12} - p_{21}\right)$ in (5.34) under the alternative hypothesis.

5.6.2. Then show that under the null hypothesis, $V\left(p_{12} - p_{21}|H_0\right) = \pi_d/N$, where $\pi_d = \pi_{12} + \pi_{21} = 2\pi$.

5.6.3. Also, show that the large sample Z-test is as shown in (5.36).

5.6.4. Show that $Z \overset{d}{\approx} \mathcal{N}(0,1)$ under H_0.

5.6.5. For the conditional odds ratio defined in (5.43) and (5.46), let $\theta = \log\left(OR_C\right)$ and $\widehat{\theta} = \log(f/g)$. Then use the δ-method to derive the expression for $V\left(\widehat{\theta}\right)$ in (5.52) and the consistent estimator in (5.53).

5.6.6. Likewise, for the log population averaged relative risk $\widehat{RR}_A$ in (5.57) derive the expression for the variance in (5.59) and the estimate in (5.60).

5.7 Alternatively, one can base the test and the confidence limits on the conditional binomial distribution of f given M presented in (5.28) and derived in Problem 5.5.

5.7.1. Under the null hypothesis of symmetry H_0: $\pi_{12} = \pi_{21}$ show that $f \sim B(f; M, 1/2)$.

5.7.2. Show that this then implies that asymptotically

$$p_f \overset{d}{\approx} \mathcal{N}\left(\frac{1}{2}, \frac{1}{4M}\right). \tag{5.106}$$

5.7.3. Then show that the resulting large sample Z-test for this hypothesis is

$$Z = \frac{f/M - 1/2}{\sqrt{1/(4M)}} \tag{5.107}$$

and that this equals McNemar's Z-test in (5.36).

5.7.4. Under H_1: $\pi_{12} \neq \pi_{21}$, from the large sample distribution of p_f derived in Problem 5.5.2, and shown in Section 5.4.3, show that the large sample confidence limits based on the logit of $\pi_f = \pi_{12}/\pi_d$ equal the confidence limits for $\theta = \log(OR_C)$ based on the estimated variance derived in Problem 5.6.5 and presented in (5.53).

5.8 Now consider a larger replication of the above study in Problem 5.4 above with hypothetical frequencies.

		$\overline{E}$: D	$\overline{E}$: $\overline{D}$
E	D	135	18
	$\overline{D}$	62	24

5.8.1. State appropriate null and alternative hypotheses (H_0, H_1) and conduct a large sample test of H_0.

5.8.2. Calculate an estimate of the conditional odds ratio with asymmetric 95% confidence limits.

5.8.3. Calculate an estimate of the population averaged relative risk with asymmetric 95% confidence limits.

5.9 For the pair-matched 2×2 table, consider a Mantel-Haenszel analysis stratified by the N pairs as shown in Table 5.2 and described in Section 5.4.4.

5.9.1. Show that the a Mantel-Haenszel test statistic equals McNemar's test or that $X^2_{MH} = X^2_M$ in (5.37).

5.9.2. Show that the Mantel-Haenszel estimate of the odds ratio equals the conditional odds ratio estimate, $\widehat{OR}_{MH} = f/g = \widehat{OR}_C$.

5.9.3. Show that the Robins, Breslow and Greenland (1986) estimate of the variance of the log odds ratio in (4.28) equals that in (5.52) or that $V[\log \widehat{OR}_{MH}] = V[\log \widehat{OR}_C] \mathrel{\hat{=}} M/(fg)$.

5.9.4. Show that the Mantel-Haenszel estimate of the relative risk equals the population averaged RR_A in (5.58) or that $\widehat{RR}_{MH} = \widehat{RR}_A$.

5.10 Now consider the pair-matched retrospective study as described in Section 5.5.

5.10.1. Show that the hypothesis of symmetry with respect to the retrospective probabilities of discordant pairs implies and is implied by the hypothesis of symmetry for the prospective probabilities, that is, that $(\phi_{12} = \phi_{21}) \Longleftrightarrow (\pi_{12} = \pi_{21})$.

5.10.2. Show that the retrospective conditional odds ratio equals the prospective conditional odds ratio as described in (5.66) and (5.67).

5.10.3. Derive the relation shown in (5.69) between the prospective conditional relative risk from a pair-matched case-control study and the retrospective conditional odds ratio.

5.10.4. In a retrospective study, from (5.69) show that $RR_z \doteq OR_C$ under the rare disease assumption $(\delta \downarrow 0)$.

5.11 In a pair-matched study, from the distribution of $T = p_{12} - p_{21}$ under H_0 and H_1 in Section 5.6, do the following:

5.11.1. Derive the expression relating the total sample size and power in (5.72) unconditionally, that is, without fixing the number of discordant pairs, M.

5.11.2. From the unconditional power function of McNemar's test presented in (5.72) show that the equation for the N needed to provide power $1 - \beta$ for an α level test is given by

$$N = \left[\frac{Z_{1-\alpha}\sqrt{\pi_d} + Z_{1-\beta}\sqrt{\pi_d - (\pi_{12} - \pi_{21})^2}}{\pi_{12} - \pi_{21}} \right]^2 . \tag{5.108}$$

5.11.3. Then use the conditional distribution with p_f as the test statistic as described in Problem 5.5.2 to derive the expression relating sample size and power as

$$\sqrt{M}\,|\varphi_c - 1| = Z_{1-\alpha}\,(\varphi_c + 1) + 2Z_{1-\beta}\sqrt{\varphi_c}\,, \tag{5.109}$$

which is expressed as a function of M and φ_c alone.

5.11.4. From this, derive the expression for $Z_{1-\beta}$ in (5.77).

5.11.5. Derive the quadratic function in (5.80) given the parameters $\phi_{\bullet 1}$, ω, and φ_c.

5.11.6. For a study where $\pi_{21} = 0.25$, $\alpha = 0.05$ (two-sided), what N is required to provide 90% power to detect an odds ratio of 1.8?

5.11.7. Show that the expected number of discordant pairs with this sample size is $E(M) = 126$.

5.11.8. Suppose that because of cost considerations, the final sample size is determined to be $N = 150$ with $E(M) = 105$.

1. From (5.77), what level of power does this provide to detect an odds ratio of 1.8?
2. What odds ratio can be detected with power = 0.90 with this sample size?

5.11.9. At the end of the study, however, assume that only $M = 90$ discordant pairs are observed. What level of power does this provide to detect an odds ratio of 1.8?

5.11.10. For a study where $\phi_{\bullet 1} = 0.2$, $\omega = 1.5$, and $\varphi_c = 1.3$, determine the cell probabilities $\{\phi_{ij}\}$ using the solution to (5.80). Then determine the sample size N required to provide 90% power to detect this value of φ_c.

5.12 Now consider a pair- or member-stratified analysis with K strata.

5.12.1. Show that a stratified Mantel-Haenszel analysis with $4K$ strata yields the adjusted estimate of the odds ratio, and the estimated variance of the log odds ratio, equal those from the unadjusted marginal analysis as shown in (5.81)–(5.82).

5.12.2. Likewise, show that the stratified Mantel-Haenszel estimate of the relative risk equals that from the marginal unadjusted analysis shown in (5.83).

5.12.3. Also, show that the stratified Mantel-Haenszel test of a common odds ratio equals the marginal McNemar test as in (5.84).

5.12.4. Show that the *MVLE* of a common odds ratio uses the weights presented in (5.86) and has the variance presented in (5.87).

5.12.5. Likewise, show that the *MVLE* of a common relative risk uses the weights presented in (5.88) and has the estimated variance presented in (5.89).

5.12.6. For an asymptotically efficient test of a common odds ratio over strata, derive $E(T)$ in (5.91) with variance as shown in (5.92).

5.12.7. Then use the expressions for the asymptotic efficiency of the test as employed in Section 4.7.2 to show that the optimal weights that provide maximum efficiency for this test are $w_j = 1 \; \forall j$.

5.12.8. Then show that the resulting efficient Z-test equals the McNemar test in (5.36).

5.12.9. Similarly, for an asymptotically efficient test of a common relative risk over strata, derive $E(T)$ in (5.94) with variance as shown in (5.95).

5.12.10. Then show that the optimal weights $\{w_j\}$ that provide maximum efficiency for this test are as shown in (5.96).

5.12.11. Show that the hypothesis of homogeneity of the conditional odds ratio among strata is equivalent to the null hypothesis of independence in the $K \times 2$ contingency table.

5.13 Exercise 5.12 presents a stratified analysis of the effect of pregnancy on the level of retinopathy among women treated intensively in the Diabetes Control and Complications Trial (DCCT, 2000). For women in the conventional treatment group, the following table presents the numbers of women whose retinopathy during pregnancy was better or worse than that before pregnancy within each of four strata classified according to the decrease in HbA_{1c} during pregnancy:

HbA_{1c} Decrease	N	Better # (%)	Worse # (%)	OR_c	95% C.I.
(0.0 %, 1.7%]	23	7 (30)	8 (35)	1.1	0.4, 3.2
(1.7 %, 3.1 %]	24	2 (8)	12 (50)	6.0	1.3, 26.8
>3.1 %	23	2 (9)	19 (82)	9.5	2.2, 40.8

(reproduced with permission). The reductions in HbA_{1c} in this group were greater than those in the intensive group because the patients treated conventionally had higher levels of HbA_{1c} by about 2% during the study compared to the intensive group. Conduct the following analyses of these data.

5.13.1. In the unadjusted marginal analysis, a total of 11 patients had better and 39 worse levels of retinopathy during pregnancy than before pregnancy. Compute the conditional odds ratio and its asymmetric 95% confidence limits and conduct McNemar's test.

5.13.2. Compute the MVLE of the common log odds ratio across strata and its large sample variance. Use these to compute the 95% asymmetric confidence limits.

5.13.3. Then conduct Cochran's test of homogeneity of odds ratios over strata.

5.13.4. Also conduct a contingency chi-square test of independence for the 3×2 table of discordant frequencies.

5.13.5. Present an overall interpretation of the effect of pregnancy in the conventional treatment group on the level of retinopathy and the effect of the change in HbA_{1c} during pregnancy on this effect.

6

Applications of Maximum Likelihood and Efficient Scores

Previous chapters describe the derivation of estimators and tests for the assessment of the association between exposure or treatment groups and the risk of an outcome from either independent or matched 2×2 tables obtained from either cross-sectional, prospective and retrospective studies. Methods for the stratified analysis of multiple independent tables under fixed and random effects models are also described. All of those developments are based on "classical" statistical theory, such as least squares, asymptotic efficiency and the central limit theorem.

Although Fisher's landmark work in maximum likelihood (Fisher, 1922, 1925) predated many of the original publications, such as those of Cochran (1954a) and Mantel and Haenszel (1959), it was not until the advent of modern computers that maximum likelihood solutions to multi-parameter estimation problems became feasible. Subsequently it was shown that many of the classic methods could be derived using likelihood-based efficient scores. This chapter describes the application of this "modern" likelihood theory to develop the methods presented in the previous chapters. Subsequent chapters then use these likelihood-based methods to develop regression models and related procedures. Section A.6 of the Appendix presents a review of maximum likelihood estimation and efficient scores, and Section A.7 describes associated test statistics.

6.1 BINOMIAL

Again consider the simple binomial distribution that describes the risk of an outcome in the population. Assuming that the number of positive responses X in N

observations (trials) is distributed as binomial with $P(x) = B(x; N, \pi)$, then the likelihood is

$$L(\pi) = \binom{N}{x} \pi^x (1-\pi)^{N-x} \tag{6.1}$$

and the log likelihood is

$$\ell(\pi) = \log \binom{N}{x} + x \log \pi + (N-x) \log (1-\pi) . \tag{6.2}$$

Thus the total score is

$$U(\pi) = \frac{d\ell}{d\pi} = \frac{x}{\pi} - \frac{(N-x)}{1-\pi} = \frac{x - N\pi}{\pi(1-\pi)} . \tag{6.3}$$

When set equal to zero, the solution yields the *MLE*:

$$\widehat{\pi} = x/N = p. \tag{6.4}$$

The observed information is

$$i(\pi) = \frac{-d^2\ell}{d\pi^2} = \frac{x}{\pi^2} + \frac{(N-x)}{(1-\pi)^2} . \tag{6.5}$$

Since $E(x) = N\pi$, the expected information is

$$I(\pi) = E[i(\pi)] = \frac{E(x)}{\pi^2} + \frac{E(N-x)}{(1-\pi)^2} = \frac{N}{\pi(1-\pi)} . \tag{6.6}$$

Thus the large sample variance of the *MLE* is

$$V(\widehat{\pi}) = I(\pi)^{-1} = \frac{\pi(1-\pi)}{N} . \tag{6.7}$$

The estimated observed information is $i(\pi)_{|\pi=\widehat{\pi}} = i(\widehat{\pi})$ and the estimated expected information is $I(\pi)_{|\pi=\widehat{\pi}} = I(\widehat{\pi})$. For the binomial the two are equivalent:

$$i(\widehat{\pi}) = I(\widehat{\pi}) = \frac{N}{\widehat{\pi}(1-\widehat{\pi})} \tag{6.8}$$

and the estimated large sample variance of the *MLE* is

$$\widehat{V}(\widehat{\pi}) = I(\widehat{\pi})^{-1} = \frac{\widehat{\pi}(1-\widehat{\pi})}{N} . \tag{6.9}$$

Efron and Hinkley (1978) show that it is often better to use the observed information for computation of confidence intervals, whereas for tests of hypotheses it is customary to use the expected information. In many problems, such as this, the two estimates are identical.

Since the number of positive responses is the sum of N $i.i.d.$ Bernoulli variables, $x = \sum_{i=1}^{N} y_i$ as described in Section 2.1.2, then asymptotically $U(\pi)$ is normally distributed with mean zero and variance $I(\pi)$. Likewise, as shown in Section A.6.5, then asymptotically

$$\widehat{\pi} \overset{d}{\approx} \mathcal{N}\left[\pi, V\left(\widehat{\pi}\right)\right],$$

or more precisely, as n increases then

$$\sqrt{n}\widehat{\pi} \overset{d}{\to} \mathcal{N}\left[\pi,\ \pi\left(1-\pi\right)\right].$$

All of these results were previously obtained from basic principles in Section 2.1.2.

Then, from the invariance principle of Section A.6.8, it follows that $g\left(\widehat{\pi}\right)$ is the *MLE* of any function $g\left(\pi\right)$, under nominal regularity conditions on $g(\cdot)$. The asymptotic distribution of $g\left(\widehat{\pi}\right)$ is then provided by the δ-method and Slutsky's Theorem as illustrated in Section 2.1.3 using the logit and log(-log) transformations.

6.2 2×2 TABLE: PRODUCT BINOMIAL (UNCONDITIONALLY)

Now consider the analysis of the 2×2 table from a study involving two independent groups of subjects, either from a cross-sectional, prospective or retrospective study, unmatched. As in Chapter 2, the analysis of such a table is described from the perspective of a prospective study. Equivalent results apply to an unmatched retrospective study.

6.2.1 MLEs AND Their Asymptotic Distribution

Given two independent samples of n_1 and n_2 observations, of which we observe x_1 and x_2 positive responses, respectively, then as shown in (2.52) the likelihood is a product binomial $L(\pi_1, \pi_2) = B\left(x_1; n_1, \pi_1\right) B\left(x_2; n_2, \pi_2\right) = L_1(\pi_1) L_2(\pi_2)$ and the log likelihood is $\ell = \log L_1 + \log L_2 = \ell_1 + \ell_2$. Thus $\boldsymbol{\theta} = (\pi_1\ \pi_2)^T$ and the score vector is

$$\begin{aligned} \boldsymbol{U}(\boldsymbol{\theta}) &= \left[U\left(\boldsymbol{\theta}\right)_{\pi_1} \quad U\left(\boldsymbol{\theta}\right)_{\pi_2}\right]^T = \frac{\partial \ell}{\partial \boldsymbol{\theta}} = \left[\frac{\partial \ell}{\partial \pi_1} \quad \frac{\partial \ell}{\partial \pi_2}\right]^T \qquad (6.10)\\ &= \left[\frac{d\ell_1}{d\pi_1} \quad \frac{d\ell_2}{d\pi_2}\right]^T = \left[\frac{x_1 - n_1\pi_1}{\pi_1\left(1-\pi_1\right)} \quad \frac{x_2 - n_2\pi_2}{\pi_2\left(1-\pi_2\right)}\right]^T. \end{aligned}$$

Therefore, the *MLE* $\widehat{\boldsymbol{\theta}}$ is the vector of the *MLEs* for the two independent groups $\widehat{\boldsymbol{\theta}} = \left[\widehat{\pi}_1\ \widehat{\pi}_2\right]^T$. The Hessian matrix has diagonal elements

$$\frac{\partial^2 \ell}{\partial \pi_1^2} = \frac{d^2 \ell_1}{d\pi_1^2} = -\left[\frac{x_1}{\pi_1^2} + \frac{\left(n_1 - x_1\right)}{\left(1-\pi_1\right)^2}\right] = \boldsymbol{H}\left(\boldsymbol{\theta}\right)_{\pi_1}, \qquad (6.11)$$

$$\frac{\partial^2 \ell}{\partial \pi_2^2} = \frac{d^2 \ell_2}{d \pi_2^2} = -\left[\frac{x_2}{\pi_2^2} + \frac{(n_2 - x_2)}{(1-\pi_2)^2}\right] = \boldsymbol{H}(\boldsymbol{\theta})_{\pi_2},$$

and off-diagonal elements $\partial^2 \ell / (\partial \pi_1 \partial \pi_2) = 0$, since no terms in ℓ involve both π_1 and π_2. Thus

$$\boldsymbol{I}(\boldsymbol{\theta}) = E[\boldsymbol{i}(\boldsymbol{\theta})] = \begin{bmatrix} \dfrac{n_1}{\pi_1(1-\pi_1)} & 0 \\ 0 & \dfrac{n_2}{\pi_2(1-\pi_2)} \end{bmatrix}. \tag{6.12}$$

Then the large sample variance of $\widehat{\boldsymbol{\theta}} = [\widehat{\pi}_1 \; \widehat{\pi}_2]^T$ is

$$\mathbf{V}\left(\widehat{\boldsymbol{\theta}}\right) = \boldsymbol{I}(\boldsymbol{\theta})^{-1} = diag\left[\frac{\pi_1(1-\pi_1)}{n_1} \quad \frac{\pi_2(1-\pi_2)}{n_2}\right], \tag{6.13}$$

which is consistently estimated as

$$\widehat{\mathbf{V}}\left(\widehat{\boldsymbol{\theta}}\right) = \boldsymbol{I}\left(\widehat{\boldsymbol{\theta}}\right)^{-1} = diag\left[\frac{p_1(1-p_1)}{n_1} \quad \frac{p_2(1-p_2)}{n_2}\right]. \tag{6.14}$$

Therefore, asymptotically

$$\begin{pmatrix} p_1 \\ p_2 \end{pmatrix} \overset{d}{\approx} \mathcal{N}\left[\begin{pmatrix} \pi_1 \\ \pi_2 \end{pmatrix}, \begin{pmatrix} \dfrac{\pi_1(1-\pi_1)}{n_1} & 0 \\ 0 & \dfrac{\pi_2(1-\pi_2)}{n_2} \end{pmatrix}\right]. \tag{6.15}$$

From the invariance principle (Section A.6.8 of the Appendix), for any one-to-one function of π such as $\beta = g(\pi_1) - g(\pi_2)$, it again follows that the *MLE* is $\widehat{\beta} = g(\widehat{\pi}_1) - g(\widehat{\pi}_2)$ with large sample variance of the estimate obtained by the δ-method. Thus the *MLE* of the $\log(OR)$ is the sample log odds ratio, $\log\left(\widehat{OR}\right)$, and the *MLE* of the $\log(RR)$ is $\log(\widehat{RR})$. These estimates and the large sample variance of each are presented in Section 2.3.

6.2.2 Logit Model

Since the odds ratio plays a pivotal role in the analysis of categorical data, it is also instructive to describe a simple logit or logistic model for the association in a 2×2 table, as originally suggested by Cox (1958a). This model was first introduced in Problem 2.13. In this model the logits of the probabilities within each group can be expressed as

$$\log\left(\frac{\pi_1}{1-\pi_1}\right) = \alpha + \beta, \qquad \log\left(\frac{\pi_2}{1-\pi_2}\right) = \alpha. \tag{6.16}$$

The inverse functions that relate the probabilities to the parameters $\boldsymbol{\theta} = (\alpha\ \beta)^T$ are logistic functions of the form

$$\begin{aligned} \pi_1 &= \frac{e^{\alpha+\beta}}{1+e^{\alpha+\beta}}\,, \qquad & 1-\pi_1 &= \frac{1}{1+e^{\alpha+\beta}} \\ \pi_2 &= \frac{e^{\alpha}}{1+e^{\alpha}}\,, \qquad & 1-\pi_2 &= \frac{1}{1+e^{\alpha}} \end{aligned} \tag{6.17}$$

and the odds ratio is simply

$$\frac{\pi_1/(1-\pi_1)}{\pi_2/(1-\pi_2)} = e^{\beta}. \tag{6.18}$$

This is called a *logit link model* or a *logistic model.* Later other link functions will be introduced.

When expressed in terms of the frequencies (a, b, c, d) in the 2×2 table as in (2.24), the product binomial likelihood is

$$L(\pi_1, \pi_2) \propto \pi_1^a (1-\pi_1)^c \pi_2^b (1-\pi_2)^d \tag{6.19}$$

and the log likelihood is

$$\ell(\pi_1, \pi_2) = a \log(\pi_1) + c \log(1-\pi_1) + b \log(\pi_2) + d \log(1-\pi_2)\,. \tag{6.20}$$

Expressing the probabilities in terms of the parameters α and β and in terms of the marginal totals (n_1, n_2, m_1, m_2) in the 2×2 table, then

$$\begin{aligned} \ell(\boldsymbol{\theta}) &= a \log\left[\frac{e^{\alpha+\beta}}{1+e^{\alpha+\beta}}\right] + c \log\left[\frac{1}{1+e^{\alpha+\beta}}\right] \\ &\quad + b \log\left[\frac{e^{\alpha}}{1+e^{\alpha}}\right] + d \log\left[\frac{1}{1+e^{\alpha}}\right] \\ &= m_1\alpha + a\beta - n_1 \log\left(1+e^{\alpha+\beta}\right) - n_2 \log\left(1+e^{\alpha}\right). \end{aligned} \tag{6.21}$$

The resulting score vector has elements $\boldsymbol{U}(\boldsymbol{\theta}) = \left[U(\boldsymbol{\theta})_\alpha \ U(\boldsymbol{\theta})_\beta\right]^T$, where

$$\begin{aligned} U(\boldsymbol{\theta})_\alpha &= \frac{\partial \ell}{\partial \alpha} = m_1 - n_1 \frac{e^{\alpha+\beta}}{1+e^{\alpha+\beta}} - n_2 \frac{e^{\alpha}}{1+e^{\alpha}} \\ &= m_1 - n_1\pi_1 - n_2\pi_2 \end{aligned} \tag{6.22}$$

and

$$U(\boldsymbol{\theta})_\beta = \frac{\partial \ell}{\partial \beta} = a - n_1 \frac{e^{\alpha+\beta}}{1+e^{\alpha+\beta}} = a - n_1\pi_1. \tag{6.23}$$

Note that these quantities can be expressed interchangeably as functions of the conditional expectations (π_1, π_2) or as functions of the model parameters (α, β).

The *MLEs* $(\widehat{\alpha}, \widehat{\beta})$ are then obtained by setting each respective score equation to zero and solving for the parameter. This yields two equations in two unknowns. Taking the difference $U(\boldsymbol{\theta})_\alpha - U(\boldsymbol{\theta})_\beta$ then yields

$$\pi_2 = \frac{m_1 - a}{n_2} . \tag{6.24}$$

From (6.17), this implies that

$$\frac{e^\alpha}{(1+e^\alpha)} = \frac{b}{n_2} . \tag{6.25}$$

Solving for α then yields

$$e^{\widehat{\alpha}} = b/(n_2 - b) = b/d \tag{6.26}$$

so that the *MLE* is $\widehat{\alpha} = \log(b/d)$.

Then evaluating $U(\boldsymbol{\theta})_\beta$ in (6.23) at the value $\widehat{\alpha}$ and setting $U(\boldsymbol{\theta})_{\beta|\widehat{\alpha}} = 0$ implies that

$$a - n_1 \frac{e^{\widehat{\alpha}+\beta}}{1+e^{\widehat{\alpha}+\beta}} = 0. \tag{6.27}$$

Substituting the solution for $e^{\widehat{\alpha}}$ yields

$$e^{\widehat{\beta}} = \frac{a}{(n_1 - a)\, b/d} = \frac{ad}{bc} . \tag{6.28}$$

Therefore, the *MLE* of β is the sample log odds ratio $\widehat{\beta} = \log(ad/bc)$, which was demonstrated earlier from the invariance principle.

The elements of the Hessian matrix then are

$$\begin{aligned} \boldsymbol{H}(\boldsymbol{\theta})_\alpha &= \frac{\partial^2 \ell}{\partial \alpha^2} \\ &= -n_1 \left[\frac{e^{\alpha+\beta}}{1+e^{\alpha+\beta}} - \left(\frac{e^{\alpha+\beta}}{1+e^{\alpha+\beta}} \right)^2 \right] - n_2 \left[\frac{e^\alpha}{1+e^\alpha} - \left(\frac{e^\alpha}{1+e^\alpha} \right)^2 \right] \\ &= -n_1 [\pi_1 (1-\pi_1)] - n_2 [\pi_2 (1-\pi_2)] \\ \boldsymbol{H}(\boldsymbol{\theta})_\beta &= \frac{\partial^2 \ell}{\partial \beta^2} = -n_1 \left[\frac{e^{\alpha+\beta}}{1+e^{\alpha+\beta}} - \left(\frac{e^{\alpha+\beta}}{1+e^{\alpha+\beta}} \right)^2 \right] = -n_1 [\pi_1 (1-\pi_1)] \\ \boldsymbol{H}(\boldsymbol{\theta})_{\alpha\beta} &= \frac{\partial^2 \ell}{\partial \alpha \partial \beta} = -n_1 [\pi_1 (1-\pi_1)] . \end{aligned} \tag{6.29}$$

Thus the information matrix is

$$\begin{aligned} \boldsymbol{i}(\boldsymbol{\theta}) = \boldsymbol{I}(\boldsymbol{\theta}) &= \begin{bmatrix} n_1 [\pi_1 (1-\pi_1)] + n_2 [\pi_2 (1-\pi_2)] & n_1 [\pi_1 (1-\pi_1)] \\ n_1 [\pi_1 (1-\pi_1)] & n_1 [\pi_1 (1-\pi_1)] \end{bmatrix} \\ &= \begin{bmatrix} n_1\psi_1 + n_2\psi_2 & n_1\psi_1 \\ n_1\psi_1 & n_1\psi_1 \end{bmatrix} , \end{aligned} \tag{6.30}$$

where $\psi_i = \pi_i (1 - \pi_i)$ is the variance of the Bernoulli variates in the ith population, $i = 1, 2$. The elements of $\boldsymbol{I}(\boldsymbol{\theta})$ will be expressed as

$$\boldsymbol{I}(\boldsymbol{\theta}) = \begin{bmatrix} \boldsymbol{I}(\boldsymbol{\theta})_{\alpha} & \boldsymbol{I}(\boldsymbol{\theta})_{\alpha,\beta} \\ \boldsymbol{I}(\boldsymbol{\theta})_{\alpha,\beta}^{T} & \boldsymbol{I}(\boldsymbol{\theta})_{\beta} \end{bmatrix}. \tag{6.31}$$

The covariance matrix of the estimates then is $\boldsymbol{V}\left(\widehat{\boldsymbol{\theta}}\right) = \boldsymbol{I}(\boldsymbol{\theta})^{-1}$. The inverse of a 2×2 matrix

$$A = \begin{bmatrix} a_{11} & a_{12} \\ a_{21} & a_{22} \end{bmatrix} \tag{6.32}$$

is obtained as

$$A^{-1} = \begin{bmatrix} a_{22} & -a_{21} \\ -a_{12} & a_{11} \end{bmatrix} / |A| , \tag{6.33}$$

where the determinant is $|A| = a_{11}a_{22} - a_{12}a_{21}$.

Therefore, $|\boldsymbol{I}(\boldsymbol{\theta})| = (n_1\psi_1 + n_2\psi_2)(n_1\psi_1) - (n_1\psi_1)^2 = n_1\psi_1 n_2\psi_2$ and the inverse matrix is

$$\begin{aligned} \boldsymbol{I}(\boldsymbol{\theta})^{-1} &= \frac{1}{n_1\psi_1 n_2\psi_2} \begin{bmatrix} n_1\psi_1 & -n_1\psi_1 \\ -n_1\psi_1 & (n_1\psi_1 + n_2\psi_2) \end{bmatrix} \\ &= \begin{bmatrix} \dfrac{1}{n_2\psi_2} & -\dfrac{1}{n_2\psi_2} \\ -\dfrac{1}{n_2\psi_2} & \dfrac{n_1\psi_1 + n_2\psi_2}{n_1\psi_1 n_2\psi_2} \end{bmatrix}. \end{aligned} \tag{6.34}$$

The elements of $\boldsymbol{I}(\boldsymbol{\theta})^{-1}$ will alternately be expressed as

$$\begin{aligned} \boldsymbol{I}(\boldsymbol{\theta})^{-1} &= \begin{bmatrix} \left[\boldsymbol{I}(\boldsymbol{\theta})^{-1}\right]_{\alpha} & \left[\boldsymbol{I}(\boldsymbol{\theta})^{-1}\right]_{\alpha,\beta} \\ \left[\boldsymbol{I}(\boldsymbol{\theta})^{-1}\right]_{\alpha,\beta}^{T} & \left[\boldsymbol{I}(\boldsymbol{\theta})^{-1}\right]_{\beta} \end{bmatrix} \\ &= \begin{bmatrix} \boldsymbol{I}(\boldsymbol{\theta})^{\alpha} & \boldsymbol{I}(\boldsymbol{\theta})^{\alpha,\beta} \\ \boldsymbol{I}(\boldsymbol{\theta})^{\beta,\alpha} & \boldsymbol{I}(\boldsymbol{\theta})^{\beta} \end{bmatrix}. \end{aligned} \tag{6.35}$$

Thus the elements of the covariance matrix $\boldsymbol{V}\left(\widehat{\theta}\right)$ are

$$\begin{aligned} V(\widehat{\alpha}) &= \left[\boldsymbol{I}(\boldsymbol{\theta})^{-1}\right]_{\alpha} = \frac{1}{n_2\pi_2(1-\pi_2)} \\ V\left(\widehat{\beta}\right) &= \left[\boldsymbol{I}(\boldsymbol{\theta})^{-1}\right]_{\beta} = \frac{n_1\pi_1(1-\pi_1) + n_2\pi_2(1-\pi_2)}{n_1\pi_1(1-\pi_1)n_2\pi_2(1-\pi_2)} \\ &= \frac{1}{n_1\pi_1(1-\pi_1)} + \frac{1}{n_2\pi_2(1-\pi_2)} \\ Cov\left(\widehat{\alpha}, \widehat{\beta}\right) &= \left[\boldsymbol{I}(\boldsymbol{\theta})^{-1}\right]_{\alpha,\beta} = \frac{-1}{n_2\pi_2(1-\pi_2)}. \end{aligned} \tag{6.36}$$

Fig. 6.1 Two views of the likelihood surface in terms of the logit model parameters (α, β) for the data from the acid dependent ulcer stratum of Example 4.1.

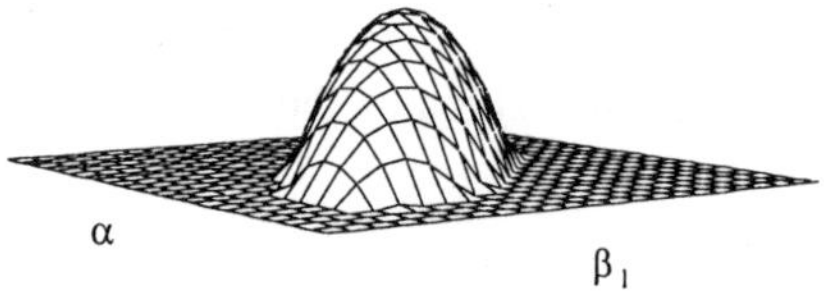

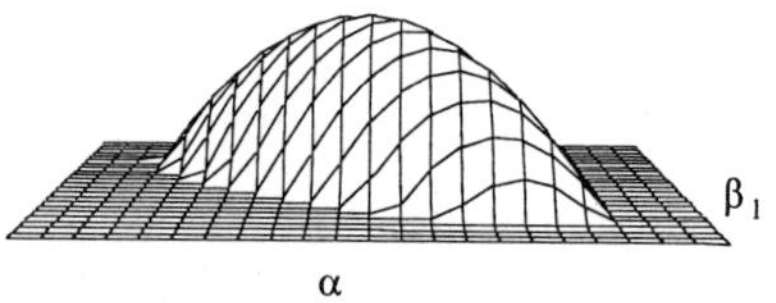

Note that the estimates are correlated even though the underlying likelihood is the product of two independent likelihoods. That is because each of these two binomial likelihoods involves the intercept α.

Also note that $\widehat{\beta} = \log\left(\widehat{OR}\right)$ and $V\left(\widehat{\beta}\right)$ is the same as Woolf's expression for the variance of the estimated log odds ratio obtained in Section 2.3.3 using the δ-method. Expressing the probabilities using the logistic function, then the estimated variance is

$$\widehat{V}\left(\widehat{\beta}\right) = \frac{\left(1 + e^{\widehat{\alpha}+\widehat{\beta}}\right)^2}{n_1 e^{\widehat{\alpha}+\widehat{\beta}}} + \frac{\left(1 + e^{\widehat{\alpha}}\right)^2}{n_2 e^{\widehat{\alpha}}} . \tag{6.37}$$

Substituting $e^{\widehat{\alpha}} = b/d$, $e^{\widehat{\beta}} = ad/bc$ and $e^{\widehat{\alpha}+\widehat{\beta}} = a/c$, it is readily shown that $\widehat{V}\left(\widehat{\beta}\right)$ equals Woolf's estimated variance of the log odds ratio. Thus large sample asymmetric confidence limits based on maximum likelihood estimation using this logit model are equivalent to those using log odds ratio in Section 2.3.3.

In Problem 6.4 we also show that the *MLE* of the log relative risk can be obtained from a model using a log link (function) in place of the logit in (6.16), with results equivalent to those presented earlier in Section 2.3.2.

Example 6.1 *Ulcer Clinical Trial*
Consider only the first stratum in the ulcer clinical trial data from Example 4.1 with $a = 16$, $n_1 = 42$, $m_1 = 36$ and $N = 89$. Figure 6.1 displays the two views of the likelihood surface as a function of the parameters of the logit model parameters for these data. The values of the parameters at the peak of the surface are the

MLEs $\widehat{\alpha} = -0.3001$ and $\widehat{\beta} = -0.1854$, the latter being the log odds ratio presented in Table 4.1. The elements of the estimated information are $I(\widehat{\alpha}) = 11.489$ and $I(\widehat{\beta}) = 5.3192$. These yield standard errors of the estimates $S.E.(\widehat{\alpha}) = 0.2950$ and $S.E.(\widehat{\beta}) = 0.4336$.

6.2.3 Tests of Significance

As described in Section A.7 of the Appendix, either a large sample Wald, likelihood ratio or score test can be applied in conjunction with maximum likelihood estimation. In the logit model for the 2×2 table we wish to test the null hypothesis H_0: $\beta = 0$ against the alternative H_1: $\beta \neq 0$. Since $\beta = 0 \Leftrightarrow \pi_1 = \pi_2$, this is equivalent to the hypothesis of no association in the 2×2 table employed previously.

6.2.3.1 Wald Test The Wald test of the log odds ratio in the logit model is defined as

$$X_W^2 = \frac{\widehat{\beta}^2}{\widehat{V}\left(\widehat{\beta}\right)} . \tag{6.38}$$

From (6.36), the variance of the estimate, that is obtained without restriction on the value of the parameter, or $\widehat{V}\left(\widehat{\beta}|H_1\right)$, is Woolf's estimated variance. Therefore, the Wald test is

$$X_W^2 = \frac{\left[\log\left(ad/bc\right)\right]^2}{\left[\frac{1}{a} + \frac{1}{b} + \frac{1}{c} + \frac{1}{d}\right]} . \tag{6.39}$$

Since the Wald test uses the variance estimated under the alternative, $\widehat{V}\left(\widehat{\beta}|H_1\right)$, this test does not equal any of the common tests for a 2×2 table presented in Chapter 2.

6.2.3.2 Likelihood Ratio Test The likelihood ratio test is based on the ratio of log likelihoods estimated with and without the parameter to be tested, β in this case, included in the model. Thus for a 2×2 table, under a logit model, the likelihood ratio test for H_0: $\beta = 0$ is provided by

$$X_{LR}^2 = 2\log[L(\widehat{\alpha}, \widehat{\beta})] - 2\log[L(\widehat{\alpha}_{|\beta=0})]. \tag{6.40}$$

From (6.21) the null log likelihood is

$$\ell(\alpha_{|\beta=0})] = \log[L(\alpha_{|\beta=0})] = m_1\alpha - N\log\left(1 + e^{\alpha}\right) \tag{6.41}$$

and the score equation is

$$U\left(\boldsymbol{\theta}\right)_{\alpha|\beta=0} = m_1 - N\frac{e^{\alpha}}{1 + e^{\alpha}} , \tag{6.42}$$

which, when set equal to zero, yields the estimate

$$\widehat{\alpha}_0 = \widehat{\alpha}_{|\beta=0} = \log(m_1/m_2). \tag{6.43}$$

Thus the likelihood ratio test statistic is

$$\begin{aligned} X^2_{LR}/2 = {} & m_1\widehat{\alpha} + a\widehat{\beta} - n_1 \log\left(1 + e^{\widehat{\alpha}+\widehat{\beta}}\right) - n_2 \log\left(1 + e^{\widehat{\alpha}}\right) \\ & - m_1\widehat{\alpha}_{|\beta=0} + N \log\left(1 + e^{\widehat{\alpha}_{|\beta=0}}\right). \end{aligned} \tag{6.44}$$

Substituting the maximum likelihood estimates under the full and the null model and simplifying then yields the expression presented earlier in (2.94); see Problem 6.2.13.

6.2.3.3 Efficient Score Test The efficient score test for β may also be derived. Since the model involves two parameters (α and β) the score test for β is a $C(\alpha)$ test, as described in Section A.7.3 of the Appendix.

By definition, the score equation for the intercept is zero when evaluated at the *MLE* for α under the null hypothesis, or $U(\boldsymbol{\theta})_{\widehat{\alpha}|\beta=0} = 0$. Then evaluating the score equation for β under the null hypothesis H_0: $\beta = \beta_0 = 0$, given the estimate $\widehat{\alpha}_0 = \widehat{\alpha}_{|\beta=0}$ of the nuisance parameter in (6.43) yields

$$U\left(\widehat{\boldsymbol{\theta}}_0\right)_\beta = a - \frac{n_1 e^{\widehat{\alpha}_0}}{1 + e^{\widehat{\alpha}_0}} = a - n_1 m_1/N = a - \widehat{E}\left(a|H_0\right). \tag{6.45}$$

This score equation can be viewed as arising from a *profile likelihood* of the form $L(\widehat{\alpha}_{|\beta=0}, \beta_0) = L(\widehat{\boldsymbol{\theta}}_0)$. Thus the restricted *MLE* of the parameter vector under the null hypothesis is $\widehat{\boldsymbol{\theta}}_0 = (\widehat{\alpha}_{|\beta=0}\ \beta_0)^T$.

To obtain the score test we also need $\boldsymbol{I}(\boldsymbol{\theta})^{-1}$ evaluated underH_0: $\beta = 0$. Under H_0, $\pi_1 = \pi_2 = \pi$, from (6.42) it follows that

$$\pi = \frac{e^\alpha}{1 + e^\alpha} \mathrel{\widehat{=}} \frac{m_1}{N}. \tag{6.46}$$

Substituting $\pi_1 = \pi_2 = \pi$ into the expressions for $\boldsymbol{I}(\boldsymbol{\theta})$ in (6.30), and evaluating at the estimate $\widehat{\pi}$, we obtain the estimated information

$$\boldsymbol{I}\left(\widehat{\boldsymbol{\theta}}_0\right) = \widehat{\pi}(1 - \widehat{\pi}) \begin{bmatrix} N & n_1 \\ n_1 & n_1 \end{bmatrix} \tag{6.47}$$

and the inverse

$$\boldsymbol{I}\left(\widehat{\boldsymbol{\theta}}_0\right)^{-1} = \left[\frac{1}{\widehat{\pi}(1 - \widehat{\pi})}\right] \left[\frac{1}{Nn_1 - n_1^2}\right] \begin{bmatrix} n_1 & -n_1 \\ -n_1 & N \end{bmatrix}. \tag{6.48}$$

As shown in Section A.7.3, the efficient score test for H_0: $\beta = 0$, given the estimate of the nuisance parameter $\widehat{\alpha}_{|\beta=0}$ evaluated under this hypothesis, is of the form

$$
\begin{aligned}
X^2 &= \left[U\left(\widehat{\boldsymbol{\theta}}_0\right)_\alpha \; U\left(\widehat{\boldsymbol{\theta}}_0\right)_\beta\right] \boldsymbol{I}\left(\widehat{\boldsymbol{\theta}}_0\right)^{-1} \left[U\left(\widehat{\boldsymbol{\theta}}_0\right)_\alpha \; U\left(\widehat{\boldsymbol{\theta}}_0\right)_\beta\right]^T \\
&= \left[0 \;\; U\left(\widehat{\boldsymbol{\theta}}_0\right)_\beta\right] \boldsymbol{I}\left(\widehat{\boldsymbol{\theta}}_0\right)^{-1} \left[0 \;\; U\left(\widehat{\boldsymbol{\theta}}_0\right)_\beta\right]^T \\
&= U\left(\widehat{\boldsymbol{\theta}}_0\right)_\beta' \left[\boldsymbol{I}\left(\widehat{\boldsymbol{\theta}}_0\right)^{-1}\right]_\beta U\left(\widehat{\boldsymbol{\theta}}_0\right)_\beta ,
\end{aligned}
\tag{6.49}
$$

where, by definition, $U\left(\widehat{\boldsymbol{\theta}}_0\right)_\alpha = \left[U(\boldsymbol{\theta})_\alpha\right]_{|\widehat{\alpha}|\beta=0} = 0$. From (6.48), the desired element of the inverse estimated information corresponding to β evaluated under H_0 is

$$
\left[\boldsymbol{I}\left(\widehat{\boldsymbol{\theta}}_0\right)^{-1}\right]_\beta = \frac{N}{\widehat{\pi}(1-\widehat{\pi})\,n_1 n_2} = \frac{N^3}{m_1 m_2 n_1 n_2} .
\tag{6.50}
$$

The resulting efficient score test is

$$
X^2 = \frac{[a - E(a)]^2}{\left(\dfrac{m_1 m_2 n_1 n_2}{N^3}\right)} = \frac{\left[a - \widehat{E}(a)\right]^2}{\widehat{V}_u(a)} ,
\tag{6.51}
$$

which equals Cochran's test for the 2×2 table.

6.3 2×2 TABLE, CONDITIONALLY

Conditioning on both margins fixed, Section 2.4.2 shows that the likelihood function for the 2×2 table is the non-central hypergeometric with parameter φ being the odds ratio. Again, it is convenient to parameterize the model in terms of the log odds ratio, $\beta = \log(\varphi)$. Then from (2.57), the log likelihood is

$$
\ell(\beta) \propto a\beta - \log\left[\sum_{i=a_\ell}^{a_u} \binom{n_1}{i}\binom{n_2}{m_1 - i} e^{i\beta}\right]
\tag{6.52}
$$

with only one parameter β. The score equation then is

$$
\begin{aligned}
U(\beta) = \frac{d\ell}{d\beta} &= a - \frac{\sum_{i=a_\ell}^{a_u} \binom{n_1}{i}\binom{n_2}{m_1-i} i e^{i\beta}}{\sum_{i=a_\ell}^{a_u} \binom{n_1}{i}\binom{n_2}{m_1-i} e^{i\beta}} \\
&= a - E(a|\beta) .
\end{aligned}
\tag{6.53}
$$

The estimating equation for the *MLE* of β is $U(\beta) = 0$ which implies that

$$
a \sum_{i=a_\ell}^{a_u} \binom{n_1}{i}\binom{n_2}{m_1 - i} e^{i\beta} = \sum_{i=a_\ell}^{a_u} \binom{n_1}{i}\binom{n_2}{m_1 - i} i e^{i\beta} ,
\tag{6.54}
$$

for which there is no closed-form solution; see Birch (1964).

However, the solution is readily obtained by an iterative procedure such as Newton-Raphson iteration described subsequently. To do so requires the expressions for the Hessian or observed information. Using a summary notation such that $C_i = \binom{n_1}{i}\binom{n_2}{m_1-i}$, the observed information is

$$i(\beta) = -\frac{d^2\ell}{d\beta^2} = \frac{\left[\sum_i C_i e^{i\beta}\right]\left[\sum_i i^2 C_i e^{i\beta}\right] - \left[\sum_i i C_i e^{i\beta}\right]^2}{\left[\sum_i C_i e^{i\beta}\right]^2} \tag{6.55}$$

$$= \frac{\sum_i i^2 C_i e^{i\beta}}{\sum_i C_i e^{i\beta}} - \left[\frac{\sum_i i C_i e^{i\beta}}{\sum_i C_i e^{i\beta}}\right]^2 .$$

Thus the observed and expected information is

$$i(\beta) = I(\beta) = E\left[a^2|\beta\right] - E(a|\beta)^2 = V(a|\beta) . \tag{6.56}$$

Although the *MLE* must be solved for iteratively, the score test for H_0: $\beta = \beta_0 = 0$ is readily obtained directly. From (6.53), then

$$U(\beta_0) = a - E(a|H_0) = a - \frac{m_1 n_1}{N} . \tag{6.57}$$

When the information in (6.56) is evaluated under the null hypothesis with $\beta_0 = 0$, the expected information reduces to

$$I(\beta_0) = V(a|\beta_0) = V_c(a) = \frac{n_1 n_2 m_1 m_2}{N^2(N-1)} \tag{6.58}$$

based on the results presented in Section 2.6.3. Therefore, the score test is

$$X^2 = \frac{U(\beta_0)^2}{I(\beta_0)} = \frac{[a - E(a)]^2}{V_c(a)} , \tag{6.59}$$

which is the Mantel-Haenszel test.

6.4 SCORE-BASED ESTIMATE

In a meta-analysis of the effects of beta blockade on post-myocardial infarction mortality, Yusuf, Peto, Lewis, et al. (1985) and Peto (1987) employed a simple estimator of the log odds ratio that can be derived as a score equation-based estimator. Generalizations of this approach are also presented by Whitehead and Whitehead (1991), based on developments in the context of sequential analysis that are also described in Whitehead (1992).

In general, consider a likelihood in a single parameter, say θ, where the likelihood is parameterized such that the null hypothesis implies H_0: $\theta = \theta_0 = 0$. Then consider a Taylor's expansion of $U(\theta)$ about the null value θ_0 so that

$$U(\theta) = U(\theta_0) + (\theta - \theta_0) U'(\theta_0) + R_2(u) \tag{6.60}$$

for $u \in (\theta, \theta_0)$ Under a sequence of local alternatives where $\theta_n = \theta_0 + \delta/\sqrt{n}$, the remainder vanishes asymptotically. Since $\theta_0 = 0$, then asymptotically

$$U(\theta) \cong U(\theta_0) + \theta U'(\theta_0). \tag{6.61}$$

Taking expectations yields

$$E[U(\theta)] \cong E[U(\theta_0)] + \theta E[U'(\theta_0)]. \tag{6.62}$$

Since $E[U(\theta)] = 0$ and $E[U'(\theta_0)] = E[H(\theta_0)] = -I(\theta_0)$ then, asymptotically

$$\theta \cong \frac{E[U(\theta_0)]}{I(\theta_0)}. \tag{6.63}$$

Therefore, the score-based estimate is

$$\widehat{\theta} = \frac{U(\theta_0)}{I(\theta_0)}, \tag{6.64}$$

which is consistent for θ under local alternatives. As will be shown in Section 6.7, this score-based estimate equals the first step estimate in a Fisher Scoring iterative solution for the maximum likelihood estimate of the parameter when the initial (starting) estimate of the parameter is the value θ_0.

From Section A.6.2 of the Appendix, the asymptotic distribution of the score given the true value of θ is $U(\theta) \overset{d}{\approx} \mathcal{N}[0, I(\theta)]$. From (6.61), asymptotically $U(\theta_0) \cong U(\theta) - \theta U'(\theta_0)$, where $U(\theta)$ converges in distribution to a normal variate, and from the law of large numbers, $-\theta U'(\theta_0) \xrightarrow{p} \theta I(\theta_0)$. Therefore, from Slutsky's Theorem (A.43), asymptotically

$$U(\theta_0) \overset{d}{\approx} \mathcal{N}[\theta I(\theta_0), I(\theta)] \tag{6.65}$$

and asymptotically

$$\widehat{\theta} \overset{d}{\approx} \mathcal{N}\left[\theta, \frac{I(\theta)}{I(\theta_0)^2}\right]. \tag{6.66}$$

Assuming that approximately $I(\theta) \doteq I(\theta_0)$, as will apply under local alternatives, then

$$\widehat{\theta} \overset{d}{\approx} \mathcal{N}\left[\theta, \frac{1}{I(\theta_0)}\right] \tag{6.67}$$

and asymptotically

$$V\left(\widehat{\theta}\right) \cong I(\theta_0)^{-1}. \tag{6.68}$$

Example 6.2 *log Odds Ratio*
Under the conditional hypergeometric distribution for the 2×2 table with fixed margins as in Section 6.3, the parameter of the likelihood is the odds ratio $\varphi = e^{\beta}$, where $\theta \equiv \beta = \log(OR)$. Using the log odds ratio, rather than the odds ratio itself, yields a value zero for the parameter under the null hypothesis ($\beta_0 = 0$) as required for the above. Therefore the score-based estimate is

$$\widehat{\beta} = \frac{U(\beta_0)}{I(\beta_0)}, \tag{6.69}$$

where the numerator is presented in (6.57) and the denominator in (6.58). Thus the score-based estimate of the log odds ratio is

$$\log\left(\widehat{OR}\right) = \widehat{\beta} = \frac{a - E(a|H_0)}{V_c(a)} \tag{6.70}$$

with estimated variance

$$V\left(\widehat{\beta}\right) \approx I(\beta_0)^{-1} = \frac{1}{V_c(a)} = \frac{N^2(N-1)}{n_1 n_2 m_1 m_2}. \tag{6.71}$$

Note that this variance estimate differs from Woolf's estimate that is also the inverse information-based estimate as shown in Section 6.2.2.

Example 6.3 *Ulcer Clinical Trial*
Again, consider only the first stratum in the ulcer clinical trial data from Example 4.1. Then $\widehat{OR} = 0.8307$ and $\log \widehat{OR} = -0.18540$ with $V\left[\log \widehat{OR}\right] = 0.1880$ based on the δ-method. The score-based estimate of $\theta = \log(OR)$ is based on $a - E(a|H_0) = -0.98876$ and $V_c(a) = 5.4033$. The ratio yields $\widehat{\theta} = -0.18299$ with large sample variance $V\left(\widehat{\theta}\right) = 1/5.4033 = 0.18507$. Both quantities are close to those obtained using the δ-method.

6.5 STRATIFIED SCORE ANALYSIS OF INDEPENDENT 2×2 TABLES

Now consider the case of a stratified analysis with K independent 2×2 tables. Day and Byar (1979) used the logit model with a product binomial likelihood for each of the K tables to show that Cochran's test of association can be derived as a $C(\alpha)$ test. However, it is somewhat simpler to show that the Mantel-Haenszel test can be obtained using the conditional hypergeometric likelihood. Here both developments are described. As a problem, it is also shown that the Day and Byar approach can be employed to yield an efficient score test under a log risk model that equals the Radhakrishna test for a common relative risk presented in Section 4.7.2.

6.5.1 Conditional Mantel-Haenszel Test and the Score Estimate

In a stratified analysis with K independent 2×2 tables, assuming a common odds ratio under H_0: $\beta_1 = \ldots = \beta_K = \beta$, the conditional hypergeometric likelihood in (2.57) involves only a single parameter $\varphi = OR = e^{\beta}$. Then the resulting likelihood is the product of the hypergeometric likelihoods for each of the K strata

$$L(\beta) = \prod_{j=1}^{K} P\left(a_j \,|\, n_{1j},\, m_{1j}, N_j,\, \beta\right). \tag{6.72}$$

Thus the total score is $U(\beta) = \sum_j U_j(\beta)$ and the total information is $I(\beta) = \sum_j I_j(\beta)$. To obtain the *MLE* requires an iterative solution; see Example 6.7 in Section 6.5. However, the estimate is readily obtained under the null hypothesis.

Under H_0: $\beta = \beta_0 = 0$, from (6.57) then

$$U(\beta_0) = \sum\nolimits_j U_j(\beta_0) = \sum\nolimits_j \left[a_j - E(a_j|H_0)\right] \tag{6.73}$$

and from (6.58)

$$I(\beta_0) = \sum\nolimits_j I_j(\beta_0) = \sum\nolimits_j V_c(a_j). \tag{6.74}$$

Therefore, the efficient score test is

$$\frac{U(\beta_0)^2}{I(\beta_0)} = \frac{\sum_j \left([a_j - E(a_j|H_0)]\right)^2}{\sum_j V_c(a_j)}, \tag{6.75}$$

which equals the Mantel-Haenszel test $X^2_{C(MH)}$ in (4.8). Therefore, the Mantel-Haenszel test is an asymptotically efficient test of H_0 against H_1: $\beta_1 = \ldots = \beta_K = \beta \neq 0$.

Also, the score-based estimate of the common log odds ratio $\beta = \log(OR)$ is

$$\widehat{\beta} = \frac{U(\beta_0)}{I(\beta_0)} = \frac{\sum_j \left[a_j - E(a_j|H_0)\right]}{\sum_j V_c(a_j)} \tag{6.76}$$

with estimated variance

$$\widehat{V}\left(\widehat{\beta}\right) = I(\beta_0)^{-1} = \frac{1}{\sum_j V_c(a_j)}. \tag{6.77}$$

Example 6.4 *Ulcer Clinical Trial*
For a stratified-adjusted analysis of the ulcer clinical trial data from Exercise 4.1 combined over strata, as shown in Table 4.3, the $MVLE$ or logit estimate of the common log odds ratio for these three 2×2 tables is $\widehat{\theta} = 0.493$, with estimated variance $\widehat{V}\left(\widehat{\theta}\right) = 0.0854$. The corresponding Mantel-Haenszel estimate of the log odds ratio is $\widehat{\theta} = 0.491$, with the Robins, Breslow and Greenland (1986) estimated variance $\widehat{V}\left(\widehat{\theta}\right) = 0.0813$. The score-based estimate involves $\sum_j (O_j - E_j) = 6.094$ and $\sum_j V_c(a_j) = 12.36$. These yield $\widehat{\theta} = 0.493$, with variance $V\left(\widehat{\theta}\right) = 0.0809$, again comparable to the $MVLE$ and Mantel-Haenszel estimates.

6.5.2 Unconditional Cochran Test as a C(α) Test

In many applications, the model involves one parameter of primary interest and one or more nuisance parameters, as illustrated by the efficient score test for the 2×2 table under a logit model in Section 6.2.3.3. In such cases, the score test for the parameter of interest is a $C(\alpha)$ test as described in Section A.7.3 of the Appendix. Another such instance is the score test for the common odds ratio using a product binomial likelihood within each stratum described by Day and Byar (1979).

Consider a logit model for the K independent 2×2 tables with a common odds ratio $OR = e^{\beta}$, where the probabilities $\{\pi_{2j}\}$ are allowed to vary over strata. Applying a logit model to each table as in Section 6.2.2, the model yields a vector of $K+1$ parameters $\boldsymbol{\theta} = (\alpha_1 \ldots \alpha_K \ \beta)^T$,where α_j refers to the intercept or log background risk for the jth table ($j = 1, \ldots, K$) and β represents the assumed common log odds ratio. Thus the $\{\alpha_j\}$ are nuisance parameters that must be jointly estimated under the null hypothesis H_0: $\beta = \beta_0 = 0$. The resulting score test for β is a $C(\alpha)$ test and it equals the Cochran test for a common odds ratio presented in Section 4.2.4. The details are derived as a problem to illustrate this approach.

For the K independent 2×2 tables from each of the K independent strata, the compound product binomial likelihood is the product of the stratum-specific likelihoods such as

$$L(\alpha_1, \ldots, \alpha_K, \beta) = \prod_{j=1}^{K} L_j(\alpha_j, \beta). \tag{6.78}$$

For the jth stratum, the likelihood is a product-binomial

$$L_j(\alpha_j, \beta) = \binom{n_{1j}}{a_j}\binom{n_{2j}}{b_j} \pi_{1j}^{a_j} (1 - \pi_{1j})^{c_j} \pi_{2j}^{b_j} (1 - \pi_{2j})^{d_j} \tag{6.79}$$

with probabilities

$$\pi_{1j} = \frac{e^{\alpha_j + \beta}}{1 + e^{\alpha_j + \beta}} \qquad \text{and} \qquad \pi_{2j} = \frac{e^{\alpha_j}}{1 + e^{\alpha_j}}, \tag{6.80}$$

where it is assumed that $\beta_1 = \ldots = \beta_K = \beta$. In this case, since $\boldsymbol{\theta}$ is a $K+1$ vector, then a score test for H_0: $\beta = \beta_0 = 0$ entails a quadratic form evaluated as

$$X^2 = \boldsymbol{U}\left(\widehat{\boldsymbol{\theta}}_0\right)' \boldsymbol{I}\left(\widehat{\boldsymbol{\theta}}_0\right)^{-1} \boldsymbol{U}\left(\widehat{\boldsymbol{\theta}}_0\right), \tag{6.81}$$

where $\boldsymbol{U}\left(\widehat{\boldsymbol{\theta}}_0\right)$ is the score vector for the $K+1$ parameters evaluated at the vector of estimates obtained under H_0 designated as $\widehat{\boldsymbol{\theta}}_0$. Thus $\widehat{\boldsymbol{\theta}}_0 = \left(\widehat{\alpha}_{1(0)} \ldots \widehat{\alpha}_{K(0)} \ 0\right)^T$, where $\widehat{\alpha}_{j(0)}$ is the *MLE* for α_j that is obtained as the solution to $\frac{\partial \ell}{\partial \alpha_j}|_{\beta=\beta_0} = \frac{\partial \ell_j}{\partial \alpha_j}|_{\beta=\beta_0}$.

However, when evaluated at the *MLEs* $\{\widehat{\alpha}_{j(0)}\}$ of the $\{\alpha_j\}$ estimated under the null hypothesis where $\beta = \beta_0$, then each respective element of the score vector is zero. Thus the score vector

$$\boldsymbol{U}\left(\widehat{\boldsymbol{\theta}}_0\right) = \left[U\left(\widehat{\boldsymbol{\theta}}_0\right)_{\alpha_1} \ldots U\left(\widehat{\boldsymbol{\theta}}_0\right)_{\alpha_K} \; U\left(\widehat{\boldsymbol{\theta}}_0\right)_{\beta}\right] = \left[0 \ldots 0 \; U\left(\widehat{\boldsymbol{\theta}}_0\right)_{\beta}\right] \tag{6.82}$$

is a single random variable augmented by zeros, where

$$U\left(\widehat{\boldsymbol{\theta}}_0\right)_{\beta} = \left[U\left(\boldsymbol{\theta}\right)_{\beta}\right] |_{\widehat{\boldsymbol{\alpha}},\beta=\beta_0} = \frac{\partial \ell}{\partial \beta}|_{\widehat{\boldsymbol{\alpha}},\beta=\beta_0} \tag{6.83}$$

and $\widehat{\boldsymbol{\alpha}} = (\widehat{\alpha}_{1(0)} \, \cdots \, \widehat{\alpha}_{K(0)})^T$. Then the quadratic form reduces to simply

$$X^2 = \left[\boldsymbol{U}\left(\widehat{\boldsymbol{\theta}}_0\right)_{\beta}\right]^2 \boldsymbol{I}\left(\widehat{\boldsymbol{\theta}}_0\right)^{\beta}, \tag{6.84}$$

where $\boldsymbol{I}\left(\widehat{\boldsymbol{\theta}}_0\right)^{\beta} = \left[\boldsymbol{I}\left(\widehat{\boldsymbol{\theta}}_0\right)^{-1}\right]_{\beta}$ is the diagonal element of the inverse information matrix corresponding to the parameter β and evaluated at $\beta = \beta_0$. Thus X^2 is distributed asymptotically as chi-square on 1 *df*.

As a problem it is readily shown that $\widehat{\alpha}_{j(0)} = \log(m_{1j}/m_{2j})$ and that the resulting $C(\alpha)$ score statistic for β is

$$U\left(\widehat{\boldsymbol{\theta}}_0\right)_{\beta_0} = \sum\nolimits_j \left(a_j - \frac{m_{1j}n_{1j}}{N_j}\right). \tag{6.85}$$

Then under H_0: $\pi_1 = \pi_2 = \pi$, from (6.30) the information matrix has elements

$$\boldsymbol{I}(\boldsymbol{\theta}_0)_{\alpha_j} = \pi_j\left(1-\pi_j\right)N_j \,\widehat{=}\, \frac{m_{1j}m_{2j}}{N_j} \tag{6.86}$$

$$\boldsymbol{I}(\boldsymbol{\theta}_0)_{\alpha_j,\beta} = \pi_j\left(1-\pi_j\right)n_{1j} \,\widehat{=}\, \frac{m_{1j}m_{2j}n_{1j}}{N_j^2}$$

$$\boldsymbol{I}(\boldsymbol{\theta}_0)_{\beta} = \sum\nolimits_j \pi_j\left(1-\pi_j\right)n_{1j} \,\widehat{=}\, \sum\nolimits_j \frac{m_{1j}m_{2j}n_{1j}}{N_j^2}$$

and $\boldsymbol{I}(\boldsymbol{\theta}_0)_{\alpha_j,\alpha_k} = 0$ for $j \neq k$ since the strata are independent. Therefore, the estimated information matrix is a patterned matrix of the form

$$\boldsymbol{I}(\widehat{\boldsymbol{\theta}}_0) = \begin{bmatrix} \boldsymbol{I}(\widehat{\boldsymbol{\theta}}_0)_{\alpha} & \boldsymbol{I}(\widehat{\boldsymbol{\theta}}_0)_{\alpha,\beta} \\ \boldsymbol{I}(\widehat{\boldsymbol{\theta}}_0)_{\alpha,\beta}^T & \boldsymbol{I}(\widehat{\boldsymbol{\theta}}_0)_{\beta} \end{bmatrix}, \tag{6.87}$$

where $\boldsymbol{I}(\widehat{\boldsymbol{\theta}}_0)_{\alpha} = diag\{I(\widehat{\boldsymbol{\theta}}_0)_{\alpha_1} \ldots I(\widehat{\boldsymbol{\theta}}_0)_{\alpha_K}\}$ is a $K \times K$ diagonal matrix.

The required element of the inverse matrix $\boldsymbol{I}(\widehat{\boldsymbol{\theta}}_0)^{\beta} = \left[\boldsymbol{I}(\widehat{\boldsymbol{\theta}}_0)^{-1}\right]_{\beta}$ is obtained using the expression for the inverse of a patterned matrix presented in (A.3) of Section A.1.2 of the Appendix. Then it is readily shown that

$$\boldsymbol{I}(\widehat{\boldsymbol{\theta}}_0)^{\beta} = \frac{1}{\sum_j \widehat{V}_u\left(a_j\right)} \tag{6.88}$$

and that

$$X^2 = \frac{\left(\sum_j \left[a_j - E\left(a_j|H_0\right)\right]\right)^2}{\sum_j \widehat{V}_u\left(a_j\right)}, \tag{6.89}$$

which is Cochran's test $X^2_{U(C)}$ in (4.12). Note that in order to compute the test one does not need to solve for the estimates of the nuisance parameters, the $\{\widehat{\alpha}_j\}$.

Gart (1985) used a similar model parameterized using the log link where $\pi_{1j} = e^{\alpha_j+\beta}$ and $\pi_{2j} = e^{\alpha_j}$, in which case $\beta = \log(RR)$. The resulting score test equals Radhakrishna's test for a common relative risk (see Problem 6.7). Gart and Tarone (1983) present general expressions for score tests for the exponential family of likelihoods, which includes the binomial with a logit link as above but not the relative risk model of Gart (1985).

6.6 MATCHED PAIRS

6.6.1 Unconditional Logit Model

Now consider a likelihood-based analysis of a sample of N pairs of observations that have been matched with respect to the values of a set of covariates, or that are intrinsically matched such as before-and-after measurements from the same subject. As in Sections 5.3 and 5.4, consider the analysis of a prospective study in which pairs of exposed and non-exposed individuals are compared with respect to the risk of developing a disease, or experiencing a positive outcome. For the jth member ($j = 1, 2$) of the ith pair ($i = 1, \ldots, N$), let the binary indicator variable X_{ij} denote the conditions to be compared where $x_{ij} = 1$ denotes the exposed (E) member and $x_{ij} = 0$ denotes non-exposed $\left(\overline{E}\right)$ member. Then let Y_{ij} denote the outcome of each pair member, where $y_{ij} = 1$ if the $ijth$ subject has a positive outcome (D), 0 if not $\left(\overline{D}\right)$. If we use the convention that $j = 1$ refers to the exposed (E) member and $j = 2$ to the non-exposed $\left(\overline{E}\right)$ member, then the observed data marginally in aggregate form the 2×2 table

		$\overline{E}$ $(x_{i2}=0)$	
		D $(y_{i2}=1)$	$\overline{D}$ $(y_{i2}=0)$
E $(x_{i1}=1)$	D $(y_{i1}=1)$	$e = n_{11}$	$f = n_{12}$
	$\overline{D}$ $(y_{i1}=0)$	$g = n_{21}$	$h = n_{22}$

(6.90)

where the underlying probabilities are $(\pi_{11}, \pi_{12}, \pi_{21}, \pi_{22})$ for the four cells of the table. The number of discordant pairs is $M = n_{12} + n_{21} = f + g$, where $E(M) = \pi_d N$, $\pi_d = \pi_{12} + \pi_{21}$ being the probability of a discordant pair.

For the ith pair a simple logistic model can be used to describe the odds ratio allowing for the underlying probabilities of the outcome to vary over pairs. Within the ith pair, since the exposed and non-exposed individuals were sampled independently given the matching covariate values, then as in Section 5.4.1 we can define the conditional probabilities of the outcome as

$$\begin{aligned} \pi_{i1} &= P\left(y_{i1}=1\,|\,x_{i1}=1\right) = P_i\left(D|E\right) = P\left(D|E,z_i\right), \text{and} \\ \pi_{i2} &= P\left(y_{i2}=1\,|\,x_{i2}=0\right) = P_i\left(D|\overline{E}\right) = P\left(D|\overline{E},z_i\right). \end{aligned} \tag{6.91}$$

In terms of the notation of Section 5.4.1, for a matched-pair with shared covariate value z, then $\pi_{i1} \equiv \pi_{1\bullet|z}$ and $\pi_{i2} \equiv \pi_{\bullet 1|z}$.

Now, as originally suggested by Cox (1958b), we can adopt a logistic model assuming a constant odds ratio for all N pairs of the form

$$\log\left(\frac{\pi_{ij}}{1-\pi_{ij}}\right) = \alpha_i + \beta x_{ij}, \qquad i=1,...,N;\ j=1,2. \tag{6.92}$$

Therefore, the log odds ratio, or difference in logits for the exposed and non-exposed members, is

$$\log\left(\frac{\pi_{i1}}{1-\pi_{i1}}\right) - \log\left(\frac{\pi_{i2}}{1-\pi_{i2}}\right) = \alpha_i + \beta x_{i1} - (\alpha_i + \beta x_{i2}) = \beta. \tag{6.93}$$

As illustrated by the Mantel-Haenszel construction for matched pairs in Section 5.4.4, within each pair the members are conditionally independent so that the likelihood is a product binomial for the ith independent pair. Using the logistic function (inverse logits) for each probability, the likelihood is

$$L_i(\alpha_i,\beta) \propto \tag{6.94}$$
$$\left(\frac{e^{\alpha_i+\beta}}{1+e^{\alpha_i+\beta}}\right)^{y_{i1}}\left(\frac{1}{1+e^{\alpha_i+\beta}}\right)^{(1-y_{i1})}\left(\frac{e^{\alpha_i}}{1+e^{\alpha_i}}\right)^{y_{i2}}\left(\frac{1}{1+e^{\alpha_i}}\right)^{(1-y_{i2})}.$$

Therefore, the total log likelihood for the complete sample of N matched pairs up to a constant is $\ell = \sum_i \log L_i = \sum_i \ell_i$, which equals

$$\ell = \sum\nolimits_i \alpha_i\left(y_{i1}+y_{i2}\right) + \sum\nolimits_i y_{i1}\beta - \sum\nolimits_i \log\left(1+e^{\alpha_i}\right) - \sum\nolimits_i \log\left(1+e^{\alpha_i+\beta}\right). \tag{6.95}$$

The resulting score equations for the parameters $\boldsymbol{\theta} = (\alpha_1\ \ldots\ \alpha_N\ \beta)^T$ are

$$U\left(\boldsymbol{\theta}\right)_{\alpha_i} = \frac{\partial\ell}{\partial\alpha_i} = \frac{\partial\ell_i}{\partial\alpha_i} = \left(y_{i1}+y_{i2}\right) - \frac{e^{\alpha_i}}{1+e^{\alpha_i}} - \frac{e^{\alpha_i+\beta}}{1+e^{\alpha_i+\beta}} \tag{6.96}$$

for $i = 1,...,N$ and

$$U\left(\boldsymbol{\theta}\right)_{\beta} = \frac{\partial\ell}{\partial\beta} = \sum\nolimits_i y_{i1} - \sum\nolimits_i \frac{e^{\alpha_i+\beta}}{1+e^{\alpha_i+\beta}}. \tag{6.97}$$

Thus in order to estimate the log odds ratio β, it is also necessary that we jointly estimate the N nuisance parameters $\{\alpha_i\}$ as well. Since the number of parameters equals $N+1$, this yields an unstable estimate of β, the bias of which increases with N (*cf.* Cox and Hinkley, 1974).

6.6.2 Conditional Logit Model

However, Cox (1958b) showed that the assumed common log odds ratio β can be estimated without simultaneously estimating the N nuisance parameters $\{\alpha_i\}$ by applying the principle of conditioning originally attributed to Fisher (*cf.* Fisher, 1956). Since the maximum likelihood estimator for a parameter is a function of the asymptotically *sufficient statistic* for that parameter, then from the log likelihood in (6.95), the number of positive responses within the ith pair, $S_i = y_{i1} + y_{i2}$, is the sufficient statistic for the parameter α_i, where $S_i = 0, 1$ or 2. In a loose sense this means that the ancillary statistic S_i captures all the information about the nuisance parameter α_i in the data and that inferences about the additional parameters in the data (the β) may be based on the conditional likelihood, after conditioning on the values of the ancillary statistics.

The conditional likelihood then is $L(\beta)_{|S} = \prod_{i=1}^{N} L(\beta)_{i|S_i}$, where the conditional likelihood for the ith pair is

$$L(\beta)_{i|S_i} = P\left(y_{i1}, y_{i2} \mid S_i\right) = \frac{P(y_{i1}, y_{i2}, S_i)}{P(S_i)} = \frac{P(y_{i1}, y_{i2})}{P(S_i)}, \tag{6.98}$$

since S_i depends explicitly on the values of (y_{i1}, y_{i2}). However, referring to the $2{\times}2$ table in (6.90), $y_{i1} = y_{i2} = 0$ for the h pairs where $S_i = 0$. Thus for these pairs, the conditional likelihood is $P\left(y_{i1} = 0,\ y_{i2} = 0 \mid S_i = 0\right) = 1$. Likewise, $y_{i1} = y_{i2} = 1$ for the e pairs where $S_i = 2$ so that $P\left(y_{i1} = 1,\ y_{i2} = 1 \mid S = 2\right) = 1$. Therefore, each concordant pair ($S_i = 0$ or 2) contributes a constant (unity) to the conditional likelihood and provides no information about β.

Conversely, there are M discordant pairs where $S_i = 1$, f of which with probability

$$P\left(y_{i1} = 1,\ y_{i2} = 0\right) = \pi_{i1}\left(1 - \pi_{i2}\right) = \left(\frac{e^{\alpha_i + \beta}}{1 + e^{\alpha_i + \beta}}\right)\left(\frac{1}{1 + e^{\alpha_i}}\right) \tag{6.99}$$

and g with probability

$$P\left(y_{i1} = 0,\ y_{i2} = 1\right) = \left(1 - \pi_{i1}\right)\pi_{i2} = \left(\frac{1}{1 + e^{\alpha_i + \beta}}\right)\left(\frac{e^{\alpha_i}}{1 + e^{\alpha_i}}\right). \tag{6.100}$$

The total probability of a discordant pair, or $P(S_i = 1)$, is the sum of these two probabilities. Therefore, the conditional likelihood is

$$L_{|S} = \prod_{i=1}^{N} \frac{P(y_{i1}, y_{i2})}{P(S_i)} = \prod_{i:S_i=1} \frac{P(y_{i1}, y_{i2})}{P(S_i = 1)} = L_{|S=1}, \tag{6.101}$$

since only the discordant pairs contribute to the likelihood.

The resulting conditional likelihood can be expressed as

$$L(\beta)_{|S=1} = \tag{6.102}$$

$$\prod_{i:S_i=1} \left[\frac{P\left(y_{i1} = 1, y_{i2} = 0\right)}{P\left(S_i = 1\right)}\right]^{y_{i1}(1-y_{i2})} \left[\frac{P\left(y_{i1} = 0, y_{i2} = 1\right)}{P\left(S_i = 1\right)}\right]^{y_{i2}(1-y_{i1})}$$

where one, but not both, of the exponents equals 1, the other 0. Then, for the f pairs where $y_{i1}(1-y_{i2})=1$, as a problem it is readily shown that

$$\frac{P(y_{i1}=1,y_{i2}=0)}{P(S_i=1)}=\frac{\pi_{i1}(1-\pi_{i2})}{\pi_{i1}(1-\pi_{i2})+\pi_{i2}(1-\pi_{i1})}=\frac{e^{\beta}}{1+e^{\beta}} \tag{6.103}$$

and for the g pairs where $y_{i2}(1-y_{i1})=1$ that

$$\frac{P(y_{i2}=1,y_{i1}=0)}{P(S_i=1)}=\frac{\pi_{i2}(1-\pi_{i1})}{\pi_{i1}(1-\pi_{i2})+\pi_{i2}(1-\pi_{i1})}=\frac{1}{1+e^{\beta}}\,. \tag{6.104}$$

Thus the conditional likelihood is

$$\begin{aligned} L_c(\beta)=L(\beta)_{|S=1} &= \prod_{i:S_i=1}\left[\frac{e^{\beta}}{1+e^{\beta}}\right]^{y_{i1}(1-y_{i2})}\left[\frac{1}{1+e^{\beta}}\right]^{y_{i2}(1-y_{i1})} \\ &= \left(\frac{e^{\beta}}{1+e^{\beta}}\right)^{f}\left(\frac{1}{1+e^{\beta}}\right)^{g}, \end{aligned} \tag{6.105}$$

which depends only on β and not on the nuisance parameters $\{\alpha_i\}$ that appear in the unconditional likelihood.

The resulting score equation for β is

$$U(\beta)=\frac{d\ell_c}{d\beta}=f-\frac{Me^{\beta}}{1+e^{\beta}}\,, \tag{6.106}$$

which, when set to zero, yields the *MLE*

$$\widehat{\beta}=\log\left(\frac{f}{g}\right). \tag{6.107}$$

The observed information is

$$i(\beta)=-\frac{d^2\log L_c(\beta)}{d\beta^2}=M\left[\frac{e^{\beta}}{(1+e^{\beta})^2}\right] \tag{6.108}$$

and the expected information is

$$I(\beta)=\frac{E(M)\,e^{\beta}}{(1+e^{\beta})^2}\,. \tag{6.109}$$

Evaluating the expected information at the *MLE* and conditioning on the observed value of M yields the estimated information

$$I\left(\widehat{\beta}\right)=\frac{fg}{M}\,. \tag{6.110}$$

Thus the large sample variance of the estimate is

$$V\left(\widehat{\beta}\right)=I\left(\widehat{\beta}\right)^{-1}=\frac{M}{fg}\,, \tag{6.111}$$

as obtained in Section 5.4.

It is also instructive to note that the conditional likelihood in (6.105) is identical to the marginal conditional binomial likelihood obtained by conditioning on the total number of discordant pairs (M) as presented earlier in (5.28) of Section 5.3.1. That conditional likelihood is

$$L(\beta)_{|M} = \binom{M}{f} \left(\frac{\pi_{12}}{\pi_d}\right)^f \left(\frac{\pi_{21}}{\pi_d}\right)^g , \tag{6.112}$$

which, apart from the constant, equals the conditional logistic regression model likelihood where

$$\frac{\pi_{12}}{\pi_d} = \frac{e^\beta}{1+e^\beta} \quad \text{and} \quad \frac{\pi_{21}}{\pi_d} = \frac{1}{1+e^\beta}. \tag{6.113}$$

6.6.3 Conditional Likelihood Ratio Test

The conditional likelihood in (6.105) then provides a likelihood ratio test for the hypothesis H_0: $\beta = 0$ of no association between exposure and the outcome in the population after matching on the matching covariates. To simplify notation, designate the conditional likelihood in (6.105) as simply $L(\beta)$. Under this conditional logit model, the null conditional likelihood has no intercept and thus is simply $L(\beta)_{|\beta=0} = -M \log(2)$. The conditional likelihood evaluated at the conditional *MLE* $\widehat{\beta}$ then yields the likelihood ratio test statistic

$$X^2_{LR} = -2 \log \left[\frac{(M/2)^M}{f^f g^g} \right], \tag{6.114}$$

which is asymptotically distributed as chi-square on 1 *df*.

6.6.4 Conditional Score Test

Now consider the score test for H_0: $\beta = 0$ in the conditional logit model. It is readily shown that

$$U(\beta)_{|\beta=0} = (f-g)/2 \tag{6.115}$$

and that

$$I(\beta)_{|\beta=0} = \frac{E(M)}{4} \,\widehat{=}\, \frac{M}{4}. \tag{6.116}$$

Therefore, the efficient score test is

$$X^2 = \frac{(f-g)^2}{M}, \tag{6.117}$$

which is McNemar's test.

Example 6.5 *Hypothetical Data*
Consider the simple example where the matched 2×2 table has discordant frequencies $f = 8$ and $g = 15$. Then the likelihood ratio test from (6.114) yields $X^2_{LR} = 2.164$, which is not significant at the 0.05 level. The score or McNemar's test yields $X^2 = 2.13$.

6.6.5 Matched Case-Control Study

Now consider a case-control study with matched pairs consisting of a case with the disease (D) and a control without the disease $(\overline{D})$, where the prior exposure status of each has been determined retrospectively (E or $\overline{E}$). In this model, being a case or control is the independent variable and having been exposed versus not is the response or dependent variable. Using a binary indicator variable $\{y_{ij}\}$ to denote exposure (E) versus not $(\overline{E})$, and an indicator variable $\{x_{ij}\}$ to denote having the disease or being a case (D) versus a control $(\overline{D})$, then the conditional probabilities associated with the ith matched pair are

$$\phi_{ij} = P\left(E|x_{ij}\right) = P\left(y_{ij} = 1|x_{ij}\right) \tag{6.118}$$

for $i = 1, \ldots, N$ and $j = 1, 2$. Note that the $\{x_{ij}\}$ and the $\{y_{ij}\}$ refer to the independent and dependent variables, respectively, under retrospective sampling.

As for the analysis of matched pairs from a prospective study in Section 6.6.2, assume a logit model for these conditional probabilities such that

$$\log\left(\frac{\phi_{ij}}{1-\phi_{ij}}\right) = \alpha_i + \beta x_{ij}. \tag{6.119}$$

Then the marginal retrospective odds ratio for the ith matched pair is

$$OR_{retro} = \frac{\phi_{i1}\left(1-\phi_{i2}\right)}{\phi_{i2}\left(1-\phi_{i1}\right)} = \frac{P_i\left(E|D\right) P_i(\overline{E}|\overline{D})}{P_i\left(\overline{E}|D\right) P_i(E|\overline{D})} = e^{\beta}. \tag{6.120}$$

Using the prior probabilities or prevalences $P(D)$ and $P(\overline{D})$ and applying Bayes theorem, as in Section 5.5, it is readily shown that the retrospective odds ratio equals the prospective odds ratio

$$OR_{retro} = OR = \frac{\pi_{i1}\left(1-\pi_{i2}\right)}{\pi_{i2}\left(1-\pi_{i1}\right)} = e^{\beta}. \tag{6.121}$$

Therefore, the results in Sections 6.6.1–6.6.4 also apply to the analysis of odds ratios in a matched case-control study. These conclusions are identical to those described in Section 5.5, where it is shown that $\widehat{\beta} = \log(\widehat{OR}) = \log\left(f/g\right)$. An example is presented in Example 5.8.

6.7 ITERATIVE MAXIMUM LIKELIHOOD

In many cases, a model is formulated in terms of a set of parameters $\boldsymbol{\theta} = (\theta_1 \ldots \theta_p)$, where the *MLE* estimating equation for $\boldsymbol{\theta}$ cannot be solved in closed form. For example, under an appropriate model, the probabilities of a C category multinomial $\{\pi_i\}$ may be a function of a smaller set of $p \leq C$ parameters. Although the p estimating equations may admit a simultaneous unique solution, meaning that the likelihood is *identifiable* in the parameters, no direct solution may exist, or may be computed tractably. In such cases, an iterative or recursive computational method is required to arrive at the solution vector. The two principal approaches are Newton-Raphson iteration and Fisher scoring, among many. Readers are referred to Thisted (1988), among others, for a comprehensive review of the features of these and other methods.

6.7.1 Newton-Raphson (or Newton's Method)

Consider the scalar parameter case and assume that we have an initial guess as to the value of the parameter that is "in the neighborhood" of the desired solution $\widehat{\theta}$. Taking a Taylor's expansion of the estimating equation $U\left(\widehat{\theta}\right) = 0$ about the starting value $\widehat{\theta}^{(0)}$, we have

$$0 = U\left(\widehat{\theta}\right) = U\left(\widehat{\theta}^{(0)}\right) + \left(\widehat{\theta} - \widehat{\theta}^{(0)}\right) U'\left(\widehat{\theta}^{(0)}\right) + R_2 \,, \tag{6.122}$$

which implies that

$$\widehat{\theta} \cong \widehat{\theta}^{(0)} - \frac{U\left(\widehat{\theta}^{(0)}\right)}{U'\left(\widehat{\theta}^{(0)}\right)} = \widehat{\theta}^{(0)} + \frac{U\left(\widehat{\theta}^{(0)}\right)}{i\left(\widehat{\theta}^{(0)}\right)} \,. \tag{6.123}$$

Because Newton's method uses the Hessian at each step, this is equivalent to using the observed information function. This equation is then applied iteratively until $\widehat{\theta}$ converges to a constant (the desired solution) using the sequence of equations

$$\widehat{\theta}^{(i+1)} = \widehat{\theta}^{(i)} - \frac{U\left(\widehat{\theta}^{(i)}\right)}{U'\left(\widehat{\theta}^{(i)}\right)} = \widehat{\theta}^{(i)} + \frac{U\left(\widehat{\theta}^{(i)}\right)}{i\left(\widehat{\theta}^{(i)}\right)} \tag{6.124}$$

for $i = 0, 1, 2, \ldots$.

For the case where $\boldsymbol{\theta}$ is a vector, the corresponding sequence of equations is characterized as

$$\widehat{\boldsymbol{\theta}}^{(i+1)} = \widehat{\boldsymbol{\theta}}^{(i)} - \left[\boldsymbol{U}'\left(\widehat{\boldsymbol{\theta}}^{(i)}\right)\right]^{-1} \boldsymbol{U}\left(\widehat{\boldsymbol{\theta}}^{(i)}\right) \tag{6.125}$$

$$= \widehat{\boldsymbol{\theta}}^{(i)} + \boldsymbol{i}\left(\widehat{\boldsymbol{\theta}}^{(i)}\right)^{-1} \boldsymbol{U}\left(\widehat{\boldsymbol{\theta}}^{(i)}\right). \tag{6.126}$$

The *MLE* is obtained when the solution $\boldsymbol{U}\left(\widehat{\boldsymbol{\theta}}^{(i+1)}\right) = 0$ by definition.

Fig. 6.2 Two steps of a Newton-Raphson iterative solution of a score estimating equation.

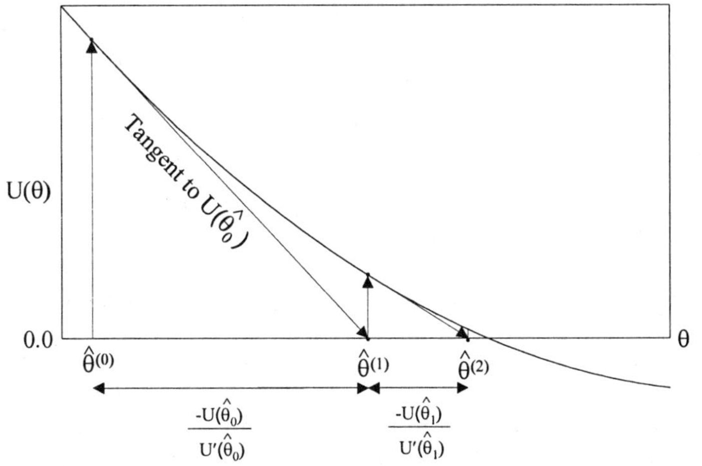

In the univariate case this can be pictured graphically, as shown in Figure 6.2, where $-U\left(\widehat{\theta}^{(i)}\right)/U'\left(\widehat{\theta}^{(i)}\right)$ is the step size and $-U\left(\widehat{\theta}^{(i)}\right)$ determines the direction of the step as shown in (6.123) above. The initial estimate is often the value expected under an appropriate null hypothesis or may be an estimate provided by a simple non-iterative moment estimator when such exists. Then the above expression is equivalent to determining the tangent to the log likelihood contour and projecting it to the abscissa to determine the next iterative estimate. This process continues until the process converges.

The Newton-Raphson iteration is quadratic convergent with respect to the rate at which $U\left(\widehat{\theta}\right) \rightarrow 0$. However, it is also sensitive to the choice of the initial or starting value.

6.7.2 Fisher Scoring (Method of Scoring)

Fisher (1935) proposed a similar procedure that he termed the *method of scoring*. Fisher scoring employs the estimated expected information to determine the step size rather than the Hessian (or observed information). Here the updated iterative estimate at the $(i+1)\,th$ step is obtained as

$$\widehat{\theta}_{i+1} = \widehat{\theta}_i + \frac{U\left(\widehat{\theta}_i\right)}{I\left(\widehat{\theta}_i\right)} \qquad i = 0, 1, 2, \ldots \tag{6.127}$$

In the multi-parameter case the vector estimate is obtained as

$$\widehat{\boldsymbol{\theta}}_{i+1} = \widehat{\boldsymbol{\theta}}_i + \left[\boldsymbol{I}\left(\widehat{\boldsymbol{\theta}}_i\right)\right]^{-1} \boldsymbol{U}\left(\widehat{\boldsymbol{\theta}}_i\right). \tag{6.128}$$

Again note that step size is determined by $I\left(\widehat{\theta}_i\right)$ and the direction by $U\left(\widehat{\theta}_i\right)$.

Scoring methods generally do not converge to a solution as quickly as does Newton-Raphson. However, the Newton-Raphson method is much more sensitive to the choice of initial values. If the initial estimate $\widehat{\theta}^{(0)}$ is far off, it is more likely to provide a divergent solution.

Example 6.6 *Recombination Fraction*
These computations are illustrated by the following well-known example from Fisher (1925) of a quadrinomial distribution where the four probabilities are each a function of a single parameter θ (*cf.* Thisted, 1988). Consider the characteristics of a crop of maize where categories of phenotypes can be defined from the combination of two characteristics: consistency (starch, A, versus sugar, $\overline{A}$) and color (green, B, versus white, $\overline{B}$), where the (A, B) phenotypes are dominant traits and the $(\overline{A}, \overline{B})$ traits are recessive. The probabilities of the four phenotypes (AB, $A\overline{B}$, $\overline{A}B$, $\overline{AB}$) of the two traits are each a function of the parameter θ of the form

$$\begin{aligned} P(AB) &= \pi_1 = (2+\theta)/4 \\ P(A\overline{B}) &= P(\overline{A}B) = \pi_2 = \pi_3 = (1-\theta)/4 \\ P(\overline{AB}) &= \pi_4 = \theta/4\,, \end{aligned} \tag{6.129}$$

where $0 < \theta < 1$. The recombination fraction is obtained as $\sqrt{\theta}$. When the genes that determine each trait are in loci on different chromosomes then the traits are independent and occur in the Mendelian ratios of 9:3:3:1 for which $\theta = 0.25$. Thus we wish to estimate θ and conduct a test of the independence null hypothesis H_0: $\theta = \theta_0 = 0.25$.

Fisher (1925) presents a sample of $N = 3839$ observations with frequencies $x_1 = \#AB = 1997$, $x_2 = \#A\overline{B} = 906$, $x_3 = \#\overline{A}B = 904$, and $x_4 = \#\overline{AB} = 32$. The likelihood function is a quadrinomial with

$$L = \frac{N!}{x_1!x_2!x_3!x_4!}\pi_1^{x_1}\pi_2^{x_2}\pi_3^{x_3}\pi_4^{x_4}. \tag{6.130}$$

Therefore,

$$\ell \propto \sum\nolimits_{j=1}^{4} x_j \log(\pi_j) = x_1 \log(2+\theta) + (x_2+x_3)\log(1-\theta) + x_4 \log(\theta) \tag{6.131}$$

and the score equation is

$$U(\theta) = \frac{x_1}{2+\theta} - \frac{x_2+x_3}{1-\theta} + \frac{x_4}{\theta}\,, \tag{6.132}$$

Table 6.1 Newton-Raphson Iterative Solution for the *MLE*

i	$\widehat{\theta}^{(i)}$	$U\left(\widehat{\theta}^{(i)}\right)$	$U'\left(\widehat{\theta}^{(i)}\right)$
0	0.05704611	−387.740680	−12340.838
1	0.02562679	376.956469	−51119.245
2	0.03300085	80.1936782	−31802.060
⋮	⋮	⋮	⋮
6	0.03571230	−0.00000000	−27519.223

which is a quadratic function in θ. Although we could solve for θ directly as the positive root of the quadratic equation, we will proceed to solve for the *MLE* of θ using iterative computations.

The Hessian is

$$U'(\theta) = -\left[\frac{x_1}{(2+\theta)^2} + \frac{x_2+x_3}{(1-\theta)^2} + \frac{x_4}{\theta^2}\right] = -\left[\frac{1997}{(2+\theta)^2} + \frac{1810}{(1-\theta)^2} + \frac{32}{\theta^2}\right] \tag{6.133}$$

and the expected information function is

$$I(\theta) = -E\left[U'(\theta)\right] \tag{6.134}$$

$$= N\left[\frac{\pi_1}{(2+\theta)^2} + \frac{\pi_2+\pi_3}{(1-\theta)^2} + \frac{\pi_4}{\theta^2}\right] = \frac{N}{4}\left[\frac{1}{2+\theta} + \frac{2}{1-\theta} + \frac{1}{\theta}\right].$$

In many problems, the iterative solution can be accelerated by starting with a good initial non-iterative estimate of the parameter, such as a moment estimator. For this model, a simple moment estimator for θ is readily obtained (see Problem 6.9). Each frequency provides a separate moment estimator of θ, for example

$$E(x_1) = \frac{N(2+\theta)}{4} \Rightarrow \widetilde{\theta}_1 = \frac{4x_1}{N} - 2. \tag{6.135}$$

Averaging these estimators from the four frequencies provides the initial moment estimator

$$\widehat{\theta}^{(0)} = (x_1 - x_2 - x_3 + x_4)/N \tag{6.136}$$

which yields $\widehat{\theta}^{(0)} = 0.05704611$ for this example.

Using $\widehat{\theta}^{(0)}$ at the initial step yields

$$U\left(\widehat{\theta}^{(0)}\right) = \frac{1997}{2.057} - \frac{(906+904)}{1-0.057} + \frac{32}{0.057} = -387.74068038 \tag{6.137}$$

Table 6.2 Fisher Scoring Iterative Solution for the *MLE*

i	$\widehat{\theta}^{(i)}$	$U\left(\widehat{\theta}^{(i)}\right)$	$I\left(\widehat{\theta}^{(i)}\right)$
0	0.05704611	−387.740680	19326.302
1	0.03698326	−33.882673	28415.310
2	0.03357908	−2.157202	29277.704
⋮	⋮	⋮	⋮
6	0.03571232	−0.00051375	29336.524
7	0.03571230	−0.00003182	29336.537
8	0.03571230	−0.00000197	29336.538

and

$$U'\left(\widehat{\theta}^{(0)}\right) = -\left[\frac{1997}{(2.057)^2} + \frac{1810}{(1-0.057)^2} + \frac{32}{(0.057)^2}\right] = -12340.84. \quad (6.138)$$

Therefore, the first step in Newton-Raphson iteration yields the revised estimate

$$\widehat{\theta}^{(1)} = \widehat{\theta}^{(0)} - \left(\frac{-387.74}{-12340.84}\right) = 0.0256268\ .$$

Table 6.1 presents the values of $\widehat{\theta}^{(i)}$ at each step of the Newton-Raphson iteration, the values of the efficient score $U\left(\widehat{\theta}^{(i)}\right)$ and the values of the Hessian $U'\left(\widehat{\theta}^{(i)}\right)$ values. Newton-Raphson required six iterations to reach a solution to eight decimal places. The *MLE* is $\widehat{\theta} = 0.03571230$ and the estimated observed information is $i\left(\widehat{\theta}\right) = -U'\left(\widehat{\theta}\right) = 27519.223$. Therefore, the estimated variance of the estimate obtained as the inverse of the estimated observed information is $\widehat{V}_i\left(\widehat{\theta}\right) = \frac{1}{27519.223} = 3.6338 \times 10^{-5}$.

Table 6.2 presents each step of the Fisher Scoring iterative procedure with the estimated information that determines the step size. Fisher scoring required eight iterations. The initial estimated information is

$$I\left(\widehat{\theta}^{(0)}\right) = \frac{3839}{4}\left[\frac{1}{2.057} + \frac{2}{1-0.057} + \frac{1}{0.057}\right] = 19326.302\ , \quad (6.139)$$

which yields the first step estimate $\widehat{\theta}^{(1)} = 0.057 + \left(\frac{-387.74}{19326.304}\right) = 0.0369833$. The remainder of the iterative calculations are shown in Table 6.2. At the final convergent iteration the estimated expected information is $I\left(\widehat{\theta}\right) = 29336.538$, which yields a slightly different estimate of the variance

$$\widehat{V}_I\left(\widehat{\theta}\right) = \frac{1}{29336.538} = 3.4087 \times 10^{-5}. \tag{6.140}$$

Now consider a test of independence of loci or H_0: $\theta = 0.25$. The Wald Test using the variance estimate $\widehat{V}_I\left(\widehat{\theta}\right) = I\left(\widehat{\theta}\right)$ obtained from the estimated information is

$$\begin{aligned} X^2 &= \left(\widehat{\theta} - 0.25\right)^2 / \widehat{V}_I\left(\widehat{\theta}\right) = \left(\widehat{\theta} - 0.25\right)^2 I\left(\widehat{\theta}\right) \\ &= (0.0357 - 0.25)^2 (29336.54) = 1347.1, \end{aligned} \tag{6.141}$$

whereas the efficient score test is

$$X^2 = U(0.25)^2 / I(0.25) = (-1397.78)^2 / 6824.89 = 286.27,$$

both of which are highly significant on 1 *df*. Even though the score test is less significant, it is preferred because it uses the variance estimated under the null rather than under the alternative hypothesis as in the Wald test. Thus the size (Type I error probability) of the Wald test may be inflated.

The likelihood ratio test is computed as $X^2 = -2\left[\log L(0.25) - \log L\left(\widehat{\theta}\right)\right]$, where the log likelihood is $\log L(\theta) \propto \sum_j x_j \log[\pi_j(\theta)] = \widetilde{\ell}(\theta)$. Evaluating each yields $\widetilde{\ell}(0.25) = -4267.62$ under the tested hypothesis and $\widetilde{\ell}\left(\widehat{\theta}\right) = \widetilde{\ell}(0.0357) = -4074.88$ globally. Therefore, the likelihood ratio test is $X^2 = 2(4267.62 - 4074.88) = 385.48$.

Example 6.7 *Conditional MLE, Ulcer Clinical Trial*

An iterative procedure such as Newton-Raphson is required to compute the *MLE* of the odds ratio from the conditional hypergeometric distribution for a 2×2 table with fixed margins, or for a set of independent tables in a stratified model with a common odds ratio. From Section 6.5.1, the score vector for K independent strata is

$$U(\beta) = \sum_{j=1}^{K} \left[a_j - \frac{\sum_{i=a_{\ell j}}^{a_{uj}} \binom{n_{1j}}{i} \binom{n_{2j}}{m_{1j}-i} i e^{i\beta}}{\sum_{i=a_{\ell j}}^{a_{uj}} \binom{n_{1j}}{i} \binom{n_{2j}}{m_{1j}-i} e^{i\beta}} \right] = \sum_{j=1}^{K} \left[a_j - E(a_j|\beta)\right] \tag{6.142}$$

and the information is

$$\begin{aligned} i(\beta) = I(\beta) &= \sum_{j=1}^{K} \frac{\left[\sum_i C_{ij} e^{i\beta}\right]\left[\sum_i i^2 C_{ij} e^{i\beta}\right] - \left[\sum_i i C_{ij} e^{i\beta}\right]^2}{\left[\sum_i C_{ij} e^{i\beta}\right]^2} \\ &= \sum_{j=1}^{K} \left[\frac{\sum_i i^2 C_{ij} e^{i\beta}}{\sum_i C_{ij} e^{i\beta}} - \frac{\left[\sum_i i C_{ij} e^{i\beta}\right]^2}{\left[\sum_i C_{ij} e^{i\beta}\right]^2} \right] \\ &= \sum_{j=1}^{K} E\left[a_j^2|\beta\right] - E(a_j|\beta)^2 = \sum_{j=1}^{K} V(a_j|\beta)\ , \end{aligned} \tag{6.143}$$

Table 6.3 Newton-Raphson Iterative Solution for the Conditional *MLE* for the Ulcer Clinical Trial Data of Example 4.1

i	$\widehat{\beta}^{(i)}$	$U\left(\widehat{\beta}^{(i)}\right)$	$U'\left(\widehat{\beta}^{(i)}\right)$
0	1.0	6.09378	−12.3594
1	1.49305	1.14934	−12.2682
2	1.58673	0.40382	−12.2314
⋮	⋮	⋮	⋮
11	1.64001	0.00021	−12.2091
12	1.64003	0.00008	−12.2090

where $C_{ij} = \binom{n_{1j}}{i}\binom{n_{2j}}{m_{1j}-i}$.

Consider the stratified analysis of the ulcer clinical trial presented in Example 4.1. Applying (6.124) recursively using these expressions provides the *MLE* and its estimated variance. The computations are shown in Table 6.3. Using the null hypothesis value as the starting point ($\beta = 0$), the solution required 12 iterations to reach a solution to 5 places. The resulting estimate of the *MLE* is $\widehat{\beta} = 0.49472$ with estimated variance $\widehat{V}(\widehat{\beta}) = 0.0819$. The estimated common odds ratio is $\widehat{OR} = 1.64$. These are close to the $MVLE$ estimates presented in Table 4.3 for which $\log(\widehat{OR}) = 0.493$ with estimated variance 0.0854 and $\widehat{OR} = 1.637$.

6.8 PROBLEMS

6.1 Consider a sample of N observations where the number of events is distributed as binomial with $P(X) = B(x; N, \pi)$ as in (6.1).

6.1.1. Show that the score equation for the binomial probability is $U(\pi)$ presented in (6.3):

6.1.2. That the MLE of π is $p = x/N$; and

6.1.3. The expected information is $I(\pi)$ presented in (6.6).

6.1.4. Show that $i(\widehat{\pi}) = I(\widehat{\pi})$ in (6.8).

6.1.5. Show that the estimated large sample variance is estimated as $\widehat{V}(\widehat{\pi}) = p(1-p)/N$.

6.2 Consider a 2×2 table with a product-binomial likelihood and assume that a logit model applies as in (6.16).

6.2.1. Show that the likelihood and log likelihood are as presented in (6.19) and (6.20), respectively.

6.2.2. Then show that $U(\boldsymbol{\theta})_\alpha = m_1 - n_1\pi_1 - n_2\pi_2$ in terms of α and β.

6.2.3. Likewise, show that $U(\boldsymbol{\theta})_\beta = a - n_1\pi_1$.

6.2.4. In addition to the above derivations, show that the score equations can be obtained using the chain rule such as

$$U(\boldsymbol{\theta})_\alpha = \frac{d\ell}{d\pi_1}\frac{\partial\pi_1}{\partial\alpha} + \frac{d\ell}{d\pi_2}\frac{d\pi_2}{d\alpha} \tag{6.144}$$
$$U(\boldsymbol{\theta})_\beta = \frac{d\ell}{d\pi_1}\frac{\partial\pi_1}{\partial\beta} .$$

6.2.5. Likewise, show that the $H(\boldsymbol{\theta})_\beta$ element of the Hessian matrix can be obtained as

$$H(\boldsymbol{\theta})_\beta = \frac{\partial}{\partial\beta}U(\boldsymbol{\theta})_\beta = \frac{\partial}{\partial\beta}\left(\frac{d\ell}{d\pi_1}\frac{\partial\pi_1}{\partial\beta}\right), \tag{6.145}$$

which is of the common form $d(uv)/d\beta$. Therefore,

$$H(\boldsymbol{\theta})_\beta = \left[\frac{d\ell}{d\pi_1}\frac{\partial^2\pi_1}{\partial\beta^2} + \frac{\partial\pi_1}{\partial\beta}\frac{\partial}{\partial\beta}\left(\frac{d\ell}{d\pi_1}\right)\right] = \left[\frac{d\ell}{d\pi_1}\frac{\partial^2\pi_1}{\partial\beta^2} + \frac{\partial\pi_1}{\partial\beta}\frac{d^2\ell}{d\pi_1^2}\frac{\partial\pi_1}{\partial\beta}\right]. \tag{6.146}$$

Solve for the other elements in like fashion.

6.2.6. Jointly solving the two score equations in two unknowns, show that the *MLEs* are $\widehat{\alpha} = \log(b/d)$ and $\widehat{\beta} = \log(ad/bc)$.

6.2.7. Show that these estimates satisfy the expressions in Problems 6.2.2 and 6.2.3.

6.2.8. Then show that the expected information matrix is $\boldsymbol{I}(\boldsymbol{\theta}) = \boldsymbol{I}(\alpha,\beta)$ as shown in (6.30)

6.2.9. Show that the large sample covariance matrix of the estimates has elements as shown in (6.36).

6.2.10. From this, show that the estimate of the variance of $\widehat{\beta}$ without restrictions, or $\widehat{V}(\widehat{\beta}|H_1)$, is as shown in (6.37) and that this equals Woolf's estimate of the variance of the log odds ratio under the alternative hypothesis in (2.46).

6.2.11. Then show that the Wald test for the null hypothesis H_0: $\beta = 0$ is as shown in (6.39).

6.2.12. For the single 2×2 table used in Problem 2.11, compute the Wald test and compare it to the Mantel and Cochran tests computed therein.

6.2.13. Evaluating the log likelihood under the null and alternative hypotheses as shown in Section 6.2.3.2, derive the expression for the likelihood ratio test in the 2×2 table presented earlier in (2.94) of Chapter 2.

6.2.14. Evaluating the score vector under the null hypothesis, show that the score equation for β is as presented in (6.45) and that the estimated information matrix is as presented in (6.47).

6.2.15. Then show that the inverse information matrix is as presented in (6.48) and that the efficient score test is equivalent to Cochran's test.

6.3 Consider a 2×2 table with fixed margins for which the likelihood function is the conditional hypergeometric probability in (2.57).

6.3.1. Derive the score equation $U(\beta)$ in (6.53) and the expression for the information in (6.55).

6.3.2. Then show that (6.56) applies.

6.3.3. Under the null hypothesis H_0: $\beta = 0$, derive the score equation $U(\beta_0)$ in (6.57) and the expression for the information in (6.58).

6.3.4. Then show that the efficient score test equals the Mantel-Haenszel test of Section 2.6.2 for a single 2×2 table as presented in (6.59).

6.3.5. Show that the score-based estimate of the log odds ratio is

$$\widehat{\beta} = \left(a - \frac{m_1 n_1}{N}\right)\left(\frac{N^2(N-1)}{m_1 m_2 n_1 n_2}\right) \tag{6.147}$$

and that the variance is as expressed in (6.71).

6.3.6. Compute this estimate, its variance, and a 95% *C.I.* for the OR for the 2×2 table used in Problem 2.11.

6.4 For a 2×2 table with a product binomial likelihood as in Section 6.2, now assume a log risk model where $\log(\pi_1) = \alpha + \beta$ and $\log(\pi_2) = \alpha$ so that $\beta = \log(\pi_1/\pi_2)$ is the log relative risk.

6.4.1. Show that

$$U(\boldsymbol{\theta})_\alpha = m_1 - \frac{c\pi_1}{1-\pi_1} - \frac{d\pi_2}{1-\pi_2} \tag{6.148}$$

in terms of α and β.

6.4.2. Likewise show that

$$U(\boldsymbol{\theta})_\beta = a - \frac{c\pi_1}{1-\pi_1}\,. \tag{6.149}$$

6.4.3. Solving jointly show that the *MLEs* are $\widehat{\alpha} = \log(b/n_2)$ and $\widehat{\beta} = \log(an_2/bn_1) = \log(\widehat{RR})$.

6.4.4. Then show that the expected information matrix is

$$\boldsymbol{I}(\boldsymbol{\theta}) = \boldsymbol{I}(\alpha, \beta) = \begin{bmatrix} n_1\psi_1 + n_2\psi_2 & n_1\psi_1 \\ n_1\psi_1 & n_1\psi_1 \end{bmatrix}, \tag{6.150}$$

where $\psi_i = \pi_i/(1-\pi_i)$, $i = 1, 2$.

6.4.5. Also show that the large sample covariance matrix of the estimates has elements

$$\mathbf{V}\left(\widehat{\alpha}, \widehat{\beta}\right) = \boldsymbol{I}(\boldsymbol{\theta})^{-1} = \begin{bmatrix} \dfrac{1-\pi_2}{n_2\pi_2} & -\dfrac{1-\pi_2}{n_2\pi_2} \\ -\dfrac{1-\pi_2}{n_2\pi_2} & \dfrac{1-\pi_1}{n_1\pi_1} + \dfrac{1-\pi_2}{n_2\pi_2} \end{bmatrix}. \tag{6.151}$$

6.4.6. Then show that the estimated variance of the log relative risk equals that presented in (2.37).

6.4.7. Evaluating the score vector under the null hypothesis H_0: $\beta = \beta_0 = 0$, show that the estimate of α under H_0 implies that $\widehat{\pi} = m_1/N$, where $\widehat{\pi}$ is the estimated common parameter in both groups.

6.4.8. Then show that the score statistic for β is

$$U(\widehat{\boldsymbol{\theta}}_0)_\beta = a - \frac{cm_1}{m_2} . \tag{6.152}$$

Note that H_0 implies that $a{:}c = m_1{:}m_2$ and that $E[U(\widehat{\boldsymbol{\theta}}_0)_\beta] = 0$.

6.4.9. Then evaluate the estimated information matrix under H_0 and show that β-element of the inverse estimated information matrix is

$$\boldsymbol{I}(\widehat{\boldsymbol{\theta}}_0)^\beta = \frac{Nm_2}{n_1 n_2 m_1} . \tag{6.153}$$

6.4.10. Show that the efficient score test is

$$X^2 = \left(a - \frac{cm_1}{m_2}\right)^2 \left(\frac{Nm_2}{n_1 n_2 m_1}\right) . \tag{6.154}$$

6.4.11. For the single 2×2 table used in Problem 2.11, compute the score test and compare it to the tests computed previously.

6.4.12. Show that the score based estimate of the common log relative risk is

$$\widehat{\beta} = \log(\widehat{RR}) = \left(a - \frac{cm_1}{m_2}\right)\left(\frac{Nm_2}{n_1 n_2 m_1}\right) \tag{6.155}$$

with estimated variance

$$\widehat{V}(\widehat{\beta}) = \frac{Nm_2}{n_1 n_2 m_1} . \tag{6.156}$$

6.4.13. For the data in Problem 2.11, compute the score-based estimate of the $\log(RR)$ and its large sample variance and compare these to the simple estimates computed in Problem 2.11.

6.5 Now consider a set of K independent 2×2 tables. Conditioning on the margins of each table as in Section 6.5.1, the model assumes a conditional hypergeometric likelihood for each table with a common odds ratio $OR_j = \varphi = e^\beta$.

6.5.1. From the likelihood in (6.52), derive the score equation $U(\beta)$ in (6.142) and the information $I(\beta)$ in (6.143).

6.5.2. Under H_0: $\beta = 0$, show that the score equation for β is as shown in (6.73) and the information is as shown in (6.74).

6.5.3. Show that the efficient score test of H_0 in (6.75) equals the Mantel-Haenszel test for stratified 2×2 tables presented in Section 4.2.3.

6.5.4. Show that the score-based stratified-adjusted estimate of the common log odds ratio in (6.76) is

$$\widehat{\beta} = \sum\nolimits_j \left(a_j - \frac{m_{1j}n_{1j}}{N_j}\right)\left(\frac{N_j^2(N_j-1)}{m_{1j}m_{2j}n_{1j}n_{2j}}\right) \tag{6.157}$$

and that

$$V\left(\widehat{\beta}\right) = \sum\nolimits_j \frac{N_j^2(N_j-1)}{m_{1j}m_{2j}n_{1j}n_{2j}}. \tag{6.158}$$

6.5.5. For the stratified 2×2 tables used in Problem 4.9, compute this estimate, its variance, and a 95% *C.I.* for the common OR.

6.6 Consider the case of K independent 2×2 tables with a compound product binomial likelihood. Section 6.2.2 presents the score equations and Hessian for a single 2×2 table using a logit model with parameters α and β. Use these with the compound product binomial likelihood for K such tables presented in (6.78) and (6.79) in terms of the parameters $\boldsymbol{\theta} = (\alpha_1 \ \dots \ \alpha_K \ \beta)^T$.

6.6.1. Show that

$$U(\boldsymbol{\theta})_{\alpha_j} = \frac{\partial \ell_j}{\partial \alpha_j} = m_{1j} - n_{1j}\frac{e^{\alpha_j+\beta}}{1+e^{\alpha_j+\beta}} - n_{2j}\frac{e^{\alpha_j}}{1+e^{\alpha_j}} \tag{6.159}$$

for $j = 1, \dots, K$; and that

$$U(\boldsymbol{\theta})_{\beta} = \frac{\partial \ell}{\partial \beta} = \sum_{j=1}^{K}\left(a_j - n_{1j}\frac{e^{\alpha_j+\beta}}{1+e^{\alpha_j+\beta}}\right). \tag{6.160}$$

6.6.2. Show that the elements of the Hessian matrix then are

$$\begin{aligned} \boldsymbol{H}(\boldsymbol{\theta})_{\alpha_j} &= \frac{\partial^2 \ell_j}{\partial \alpha_j^2} = -n_{1j}\left[\frac{e^{\alpha_j+\beta}}{1+e^{\alpha_j+\beta}} - \left(\frac{e^{\alpha_j+\beta}}{1+e^{\alpha_j+\beta}}\right)^2\right] \\ &\quad - n_{2j}\left[\frac{e^{\alpha_j}}{1+e^{\alpha_j}} - \left(\frac{e^{\alpha_j}}{1+e^{\alpha_j}}\right)^2\right] \\ \boldsymbol{H}(\boldsymbol{\theta})_{\beta} &= \frac{\partial^2 \ell}{\partial \beta^2} = -\sum_{j=1}^{K} n_{1j}\left[\frac{e^{\alpha_j+\beta}}{1+e^{\alpha_j+\beta}} - \left(\frac{e^{\alpha_j+\beta}}{1+e^{\alpha_j+\beta}}\right)^2\right] \\ \boldsymbol{H}(\boldsymbol{\theta})_{\alpha_j\beta} &= \frac{\partial^2 \ell}{\partial \alpha_j \partial \beta} = -n_{1j}\left[\frac{e^{\alpha_j+\beta}}{1+e^{\alpha_j+\beta}} - \left(\frac{e^{\alpha_j+\beta}}{1+e^{\alpha_j+\beta}}\right)^2\right]. \end{aligned} \tag{6.161}$$

6.6.3. Under the hypothesis of no association in all of the tables, H_0: $\beta = \beta_0 = 0$, show that $\widehat{\alpha}_{j(0)} = \log(m_{1j}/m_{2j})$ and that the score equation for β is as presented in (6.85).

6.6.4. Then derive the elements of the estimated information matrix as presented in (6.86).

6.6.5. Also derive the expression for $\boldsymbol{I}(\widehat{\boldsymbol{\theta}}_0)^{\beta}$ under H_0 in (6.88).

6.6.6. Show that the efficient score statistic is Cochran's test as shown in (6.89).

6.7 Similarly, assume that there is a common relative risk among the K 2×2 tables represented by a log link as in Problem 6.4 above, where $\beta = \log(RR)$.

6.7.1. Under H_0: $\beta = \beta_0 = 0$, show that the score equation for α_j is

$$U(\boldsymbol{\theta}_0)_{\alpha_j} = m_{1j} - m_{2j}\frac{e^{\alpha_j}}{1+e^{\alpha_j}}$$

and that the solution is

$$\widehat{\alpha}_{j(0)} = \log(m_{1j}/N_j).$$

6.7.2. Under H_0, show that the score statistic for β is

$$U(\widehat{\boldsymbol{\theta}}_0)_{\beta} = \sum\nolimits_j a_j - \frac{c_j m_{1j}}{m_{2j}}\,. \tag{6.162}$$

6.7.3. Then evaluate the inverse information matrix under H_0 and show that the β-element of the inverse information matrix is

$$\boldsymbol{I}(\widehat{\boldsymbol{\theta}}_0)^{\beta} = \left(\sum\nolimits_j \frac{n_{1j}n_{2j}m_{1j}}{N_j m_{2j}}\right)^{-1}, \tag{6.163}$$

where $\widehat{\pi}_j = m_{1j}/N_j$ is the estimated common probability in the jth stratum.

6.7.4. Show that the resulting score test is

$$X^2 = \frac{\left[\sum_j a_j - \dfrac{c_j m_{1j}}{m_{2j}}\right]^2}{\sum_j \dfrac{n_{1j}n_{2j}m_{1j}}{N_j m_{2j}}}\,. \tag{6.164}$$

6.7.5. For comparison, show that the above expression is equivalent to that for the Radhakrishna asymptotically fully efficient test for a common relative risk presented in (4.185) of Problem 4.4.3 based on (4.98) of Section 4.7.2. Gart (1985) presented an alternate derivation of the score test and its equivalence to the Radhakrishna test using a slightly different parameterization.

6.8 Consider Cox's (1958b) logistic model for matched pairs described in Section 6.6.1.

6.8.1. Starting with the unconditional likelihood described in Section 6.6.1, derive the expressions for the likelihood in (6.94) and log likelihood in (6.95).

6.8.2. Then derive the score equations for the stratum-specific effects $U(\boldsymbol{\theta})_{\alpha_i}$ in (6.96) and the common log odds ratio $U(\boldsymbol{\theta})_{\beta}$ in (6.97).

6.8.3. Then, conditioning on $S_i = (y_{i1} + y_{i2}) = 1$, show that the conditional likelihood in (6.102) can be expressed as

$$L(\beta)_{|S=1} = \prod_{i:S_i=1} \frac{\left[\pi_{i1}(1-\pi_{i2})\right]^{y_{i1}(1-y_{i2})}\left[\pi_{i2}(1-\pi_{i1})\right]^{y_{i2}(1-y_{i1})}}{\pi_{i1}(1-\pi_{i2})+\pi_{i2}(1-\pi_{i1})}. \tag{6.165}$$

6.8.4. Then show that the expressions in (6.103) and (6.104) hold.

6.8.5. Show that this yields $L(\beta)_{|S=1}$ as in (6.105).

6.8.6. Then derive $\ell_{(c)}$ in terms of f and $M = f + g$.

6.8.7. Derive $U(\beta)$ in (6.106) and show that $\widehat{\beta} = \log(f/g)$.

6.8.8. Then show that $i(\beta)$ is as expressed in (6.108) and $I(\beta)$ as in (6.109).

6.8.9. Then show that $V\left(\widehat{\beta}\right)$ is as expressed in (6.111).

6.8.10. Under H_0: $\beta = 0$, derive the likelihood ratio test presented in (6.114).

6.8.11. Derive the efficient score test statistic in Section 6.6.4 and show that it equals McNemar's statistic.

6.9 Consider the recombination fraction example presented in Section 6.7.1.

6.9.1. Based on the assumed probabilities, show that a simple moment estimator of the recombination fraction is $\widehat{\theta}^{(0)} = (x_1 - x_2 - x_3 + x_4)/N$.

6.9.2. Either write a computer program or use a calculator to fit the model using Newton-Raphson iteration convergent to eight decimal places. Verify the computations presented in Table 6.1

6.9.3. Do likewise, using Fisher scoring to verify the computations in Table 6.2.

6.9.4. Using the final *MLE* $\widehat{\theta}$ and estimated observed information $i\left(\widehat{\theta}\right)$ from the Newton-Raphson iteration, which provides an estimate of the variance of the estimate, compute the 95% *C.I.* for θ.

6.9.5. Likewise, using the final estimated expected information from Fisher scoring, compute a 95% *C.I.* for θ.

6.9.6. For H_0: $\theta = \theta_0 = 0.25$ compute the Wald, score, and likelihood ratio tests.

6.10 Consider the estimation of the conditional *MLE* of the log odds ratio β based on the conditional hypergeometric distribution.

6.10.1. For a single 2×2 table, use the score equation in (6.53) and the observed information in (6.55) as the basis for a computer program to use the Newton-Raphson iteration to solve for the *MLE* of the log odds ratio β and to then evaluate the estimated expected information $I(\widehat{\beta})$.

6.10.2. Apply this program to the data from Problem 2.11 and compare the point estimate of the log odds ratio and its estimated variance, and the 95% asymmetric confidence limits for the odds ratio, to those computed previously.

6.10.3. Also, generalize this program to allow for K independent strata as in Section 6.5.1 and Example 6.7.

6.10.4. Apply this program to the data from Example 4.1 to verify the computations in Table 6.3.

6.11 Now consider the maximum likelihood estimate of the common odds ratio in K independent 2×2 tables based on the compound product-binomial likelihood of Section 6.5.2 and Problem 6.6 above.

6.11.1. Use the expressions for the score equations and the Hessian matrix to write a program for Newton-Raphson iteration to jointly solve for the *MLE* of the vector $\boldsymbol{\theta}$.

6.11.2. Apply the Newton-Raphson iteration to the set of 2×2 tables in Problem 4.9 to obtain the *MLE* of the common log odds ratio and its estimated variance, and the 95% asymmetric confidence limits for the odds ratio, and compare these to those computed previously.

7

Logistic Regression Models

The previous chapter describes the analysis of 2×2 tables using a simple logit model. This model provides estimates and tests of the log odds ratio for the probability of the positive response or outcome of interest as a function of the exposure or treatment group. Now consider the simultaneous estimation of the effects of multiple covariates, perhaps some categorical and some quantitative, on the odds of the response using a logistic regression model. We first consider the logistic regression model for independent observations, as in a cross-sectional, prospective or retrospective study. The treatment herein is based on the work of Cox (1970) and others, using maximum likelihood as the basis for estimation. Later we consider the conditional logistic model for paired or clustered observations, as in matched sets, described by Breslow (1982). An excellent general reference is the book by Hosmer and Lemeshow (1989).

7.1 UNCONDITIONAL LOGISTIC REGRESSION MODEL

7.1.1 General Logistic Regression Model

In keeping with the notation in a single 2×2 table, we start with a prospective cohort or cross-sectional sample of N observations of whom m_1 have the positive response (for example, develop the disease) and m_2 do not. For each subject, the response is represented by the binary dependent variable Y, where for the ith subject $y_i = 1$ if a positive response is observed, 0 if not. We assume that the probability of the positive response for the ith subject, $\pi_i(\boldsymbol{x}_i, \boldsymbol{\theta}) = P(y_i = 1 \mid \boldsymbol{x}_i, \boldsymbol{\theta})$ can be expressed as a function of the p covariate vector $\boldsymbol{x}_i = (x_{i1} \; \ldots \; x_{ip})^T$ and parameter vector

$\boldsymbol{\theta} = (\alpha\ \beta_1\ \dots\ \beta_p)^T$. The covariates can be either quantitative characteristics or design effects to represent qualitative characteristics. We then assume that the logit of these probabilities can be expressed as a function of the p covariates, such as a linear function of the parameters $\boldsymbol{\theta}$ of the form

$$\log\left[\frac{\pi_i}{1-\pi_i}\right] = \alpha + \boldsymbol{x}_i'\boldsymbol{\beta} = \alpha + \sum\nolimits_{j=1}^{p} x_{ij}\beta_j\ , \tag{7.1}$$

where $\pi_i = \pi_i(\boldsymbol{x}_i, \boldsymbol{\theta})$ and $\boldsymbol{\beta} = (\beta_1\ \dots\ \beta_p)^T$. This implies that the probabilities can be obtained as the logistic function (inverse logit)

$$\begin{aligned} \pi_i &= \frac{e^{\alpha+\boldsymbol{x}_i'\boldsymbol{\beta}}}{1+e^{\alpha+\boldsymbol{x}_i'\boldsymbol{\beta}}} = \frac{1}{1+e^{-(\alpha+\boldsymbol{x}_i'\boldsymbol{\beta})}} \\ 1-\pi_i &= \frac{1}{1+e^{\alpha+\boldsymbol{x}_i'\boldsymbol{\beta}}}\ . \end{aligned} \tag{7.2}$$

Since the ith observation is a Bernoulli variable $\{y_i\}$, then the likelihood in terms of the probabilities $\boldsymbol{\pi} = (\pi_1\ \dots\ \pi_N)$ is

$$L(\boldsymbol{\pi}) = \prod_{i=1}^{N} \pi_i^{y_i}\,(1-\pi_i)^{1-y_i}\ . \tag{7.3}$$

Expressing the probabilities as a logistic function of the covariates and the parameters yields

$$L(\boldsymbol{\theta}) = \prod_{i=1}^{N}\left(\frac{e^{\alpha+\boldsymbol{x}_i'\boldsymbol{\beta}}}{1+e^{\alpha+\boldsymbol{x}_i'\boldsymbol{\beta}}}\right)^{y_i}\left(\frac{1}{1+e^{\alpha+\boldsymbol{x}_i'\boldsymbol{\beta}}}\right)^{1-y_i}\ . \tag{7.4}$$

Likewise, the log likelihood in the probabilities is

$$\ell(\boldsymbol{\pi}) = \sum\nolimits_{i=1}^{N} y_i \log \pi_i + \sum\nolimits_{i=1}^{N}(1-y_i)\log(1-\pi_i) \tag{7.5}$$

and that in the corresponding model parameters $\boldsymbol{\theta}$ is

$$\ell(\boldsymbol{\theta}) = \sum\nolimits_{i=1}^{N} y_i\,(\alpha+\boldsymbol{x}_i'\boldsymbol{\beta}) - \sum\nolimits_{i=1}^{N}\log\left(1+e^{\alpha+\boldsymbol{x}_i'\boldsymbol{\beta}}\right). \tag{7.6}$$

The score vector for the parameters is

$$\boldsymbol{U}(\boldsymbol{\theta}) = \left[U(\boldsymbol{\theta})_\alpha\ U(\boldsymbol{\theta})_{\beta_1}\ \dots\ U(\boldsymbol{\theta})_{\beta_p}\right]^T\ . \tag{7.7}$$

The score equation for the intercept is

$$\begin{aligned} U(\boldsymbol{\theta})_\alpha &= \frac{\partial \ell}{\partial \alpha} = \sum\nolimits_i y_i - \sum\nolimits_i \frac{e^{\alpha+\boldsymbol{x}_i'\boldsymbol{\beta}}}{1+e^{\alpha+\boldsymbol{x}_i'\boldsymbol{\beta}}} \\ &= \sum\nolimits_i (y_i - \pi_i) = \sum\nolimits_i [y_i - E(y_i|\boldsymbol{x}_i)] = m_1 - E(m_1|\boldsymbol{\theta}) \end{aligned} \tag{7.8}$$

where $E(y_i|\boldsymbol{x}_i) = \pi_i$. The score equation for the jth coefficient is

$$U(\boldsymbol{\theta})_{\beta_j} = \frac{\partial \ell}{\partial \beta_j} = \sum\nolimits_i x_{ij} \left(y_i - \frac{e^{\alpha + \boldsymbol{x}_i' \boldsymbol{\beta}}}{1 + e^{\alpha + \boldsymbol{x}_i' \boldsymbol{\beta}}} \right) \tag{7.9}$$

$$= \sum\nolimits_i x_{ij} \left[y_i - \pi_i \right] = \sum\nolimits_i x_{ij} \left[y_i - E(y_i|\boldsymbol{x}_i) \right] \;,$$

which is a weighted sum of the observed value of the response (y_i) minus that expected under the model.

Note that each score equation involves the complete parameter vector $\boldsymbol{\theta}$. The *MLE* is the vector $\widehat{\boldsymbol{\theta}} = (\widehat{\alpha}\ \widehat{\beta}_1\ \ldots\ \widehat{\beta}_p)^T$ that jointly satisfies $\boldsymbol{U}(\boldsymbol{\theta}) = \boldsymbol{0}$; or the solution obtained by setting the $p + 1$ score equations to zero and solving simultaneously. Since closed-form solutions do not exist, the solution must be obtained by an iterative procedure such as the Newton-Raphson algorithm.

Note that an important consequence of the score equation for the intercept is that setting $U(\boldsymbol{\theta})_\alpha = 0$ requires that the *MLE* for α and the estimates of the coefficients jointly satisfy $U(\widehat{\boldsymbol{\theta}})_\alpha = m_1 - \sum_i \widehat{\pi}_i = 0$ where $\widehat{\pi}_i$ is the estimated probability obtained upon substituting $\widehat{\alpha} + \boldsymbol{x}_i'\widehat{\boldsymbol{\beta}}$ into (7.2). Thus the mean estimated probability is $\widehat{\overline{\pi}} = \sum_i \widehat{\pi}_i / N = m_1/N$.

The observed information $\boldsymbol{i}(\boldsymbol{\theta}) = -\boldsymbol{H}(\boldsymbol{\theta})$ has the following elements. For $\boldsymbol{i}(\boldsymbol{\theta})_\alpha$ we require

$$\frac{\partial \pi_i}{\partial \alpha} = \frac{e^{\alpha + \boldsymbol{x}_i' \boldsymbol{\beta}}}{\left(1 + e^{\alpha + \boldsymbol{x}_i' \boldsymbol{\beta}}\right)^2} = \pi_i (1 - \pi_i) \tag{7.10}$$

so that

$$\boldsymbol{i}(\boldsymbol{\theta})_\alpha = -\frac{\partial U(\boldsymbol{\theta})_\alpha}{\partial \alpha} = \sum\nolimits_i \pi_i (1 - \pi_i)\,. \tag{7.11}$$

Likewise, it is readily shown that the remaining elements of the observed information matrix for $1 \le j < k \le p$ are

$$\boldsymbol{i}(\boldsymbol{\theta})_{\beta_j} = \frac{-\partial U(\boldsymbol{\theta})_{\beta_j}}{\partial \beta_j} = \sum\nolimits_i x_{ij} \frac{\partial \pi_i}{\partial \beta_j} = \sum\nolimits_i x_{ij} \left[\frac{x_{ij} e^{\alpha + \boldsymbol{x}_i' \boldsymbol{\beta}}}{\left(1 + e^{\alpha + \boldsymbol{x}_i' \boldsymbol{\beta}}\right)^2} \right] \tag{7.12}$$

$$= \sum\nolimits_i x_{ij}^2 \pi_i (1 - \pi_i)\,, \tag{7.13}$$

$$\boldsymbol{i}(\boldsymbol{\theta})_{\beta_j, \beta_k} = \frac{-\partial U(\boldsymbol{\theta})_{\beta_j}}{\partial \beta_k} = \sum\nolimits_i x_{ij} \frac{\partial \pi_i}{\partial \beta_k} = \sum\nolimits_i x_{ij} x_{ik} \pi_i (1 - \pi_i)\,, \tag{7.14}$$

$$\boldsymbol{i}(\boldsymbol{\theta})_{\alpha, \beta_j} = \frac{-\partial U(\boldsymbol{\theta})_\alpha}{\partial \beta_j} = \sum\nolimits_i \frac{\partial \pi_i}{\partial \beta_j} = \sum\nolimits_i x_{ij} \pi_i (1 - \pi_i)\,. \tag{7.15}$$

Since $\boldsymbol{i}(\boldsymbol{\theta})$ can be expressed in terms of the probabilities $\{\pi_i\}$ or the parameters $\{\boldsymbol{\theta}\}$, then $\boldsymbol{i}(\boldsymbol{\theta}) = \boldsymbol{I}(\boldsymbol{\theta})$ and the observed and expected information are the same for this model. Also, since $V(y_i|\boldsymbol{x}_i) = \pi_i(1-\pi_i)$ under the model, then the elements of the information matrix are weighted sums of the model based variances, such as $\boldsymbol{I}(\boldsymbol{\theta})_{\beta_j} = \sum_i x_{ij}^2 V(y_i|\boldsymbol{x}_i)$.

Using the elements of the Hessian matrix, the vector of parameter estimates can be solved using the Newton-Raphson iterative procedure, or another algorithm. Wald tests and large sample confidence limits for the individual parameters, such as the jth coefficient β_j, are readily computed using the large sample variance $\widehat{V}(\widehat{\beta}_j) = \left[\boldsymbol{I}(\widehat{\boldsymbol{\theta}})^{-1}\right]_{\beta_j}$ obtained as the corresponding diagonal element of the estimated expected information, $\boldsymbol{I}(\widehat{\boldsymbol{\theta}})$. The null and fitted likelihoods can also be used as the basis for a likelihood ratio test of the model, as described subsequently.

Section 6.2.2 shows that the estimated coefficient in a simple logit model for a 2×2 table is the *MLE* of the log odds ratio. Subsequently (see Section 7.2.1) we show that the estimated coefficients for binary covariates have the same interpretation in a multivariate logistic regression model. That is, if X_j is coded as 1 or 0, then the estimated odds ratio for this covariate is $\widehat{OR}_j = \exp(\widehat{\beta}_j)$. Confidence limits on the coefficient then yield asymmetric confidence limits on the odds ratio.

In some applications it is also of interest to obtain confidence limits for the predicted probability for an observation with covariate vector $\boldsymbol{x}$. Let $\widetilde{\boldsymbol{x}}$ denote the covariate vector augmented by the constant for the intercept $\widetilde{\boldsymbol{x}} = (1 \;//\; \boldsymbol{x})$. Then the linear predictor is

$$\widehat{\eta} = \widetilde{\boldsymbol{x}}'\widehat{\boldsymbol{\theta}}. \tag{7.16}$$

Denoting the estimated variance of the estimates as $\widehat{\boldsymbol{\Sigma}}_{\boldsymbol{\theta}} = \widehat{V}\left(\widehat{\boldsymbol{\theta}}\right) = \boldsymbol{I}\left(\widehat{\boldsymbol{\theta}}\right)^{-1}$, then the estimated variance of the linear predictor is

$$\widehat{V}\left(\widehat{\eta}\right) = \widetilde{\boldsymbol{x}}'\widehat{\boldsymbol{\Sigma}}_{\boldsymbol{\theta}}\widetilde{\boldsymbol{x}} = \widehat{\sigma}_{\widehat{\eta}}^2. \tag{7.17}$$

Therefore, the $1-\alpha$ level confidence limits on η are

$$(\widehat{\eta}_\ell, \widehat{\eta}_u) = \widehat{\eta} \pm Z_{1-\alpha/2}\widehat{\sigma}_{\widehat{\eta}} \tag{7.18}$$

and the resulting confidence limits on the true probability π are

$$\frac{e^{\widehat{\eta}_\ell}}{1+e^{\widehat{\eta}_\ell}} \le \pi \le \frac{e^{\widehat{\eta}_u}}{1+e^{\widehat{\eta}_u}}\,. \tag{7.19}$$

7.1.2 Logistic Regression and Binomial Logit Regression

Section 6.2.2 describes the logit model for the probabilities of a positive response or of developing the disease (D) in a single 2×2 table. That model is a special case of the more general model above. Consider the case where x_i is a single binary covariate for all subjects: $x_i = 1$ if E, or 0 if $\overline{E}$, for exposed versus not. Then the

Table 7.1 Incidence of Coronary Heart Disease in the Framingham Study as a Function of Blood Pressure and Cholesterol Levels

Category j	*# CHD* a_j	*# Subjects* n_j	*(Intercept, Covariates)* $(1,\ x_{1j},\ x_{2j})$
1	10	431	$(1,0,0)$
2	10	142	$(1,0,1)$
3	38	532	$(1,1,0)$
4	34	224	$(1,1,1)$

data can be summarized in a 2×2 table with frequencies.

		x_i: (1) E	x_i: (0) $\overline{E}$	
y_i	(1), D	a	b	m_1
y_i	(0), $\overline{D}$	c	d	m_2
		n_1	n_2	N

Then the logistic regression likelihood is

$$
\begin{aligned}
L &= \prod_{i=1}^{N} \pi_i^{y_i} (1-\pi_i)^{1-y_i} = \prod_{i=1}^{N} \left(\frac{e^{\alpha + x_i\beta}}{1+e^{\alpha+x_i\beta}} \right)^{y_i} \left(\frac{1}{1+e^{\alpha+x_i\beta}} \right)^{1-y_i} \\
&= \left(\frac{e^{\alpha+\beta}}{1+e^{\alpha+\beta}} \right)^{a} \left(\frac{1}{1+e^{\alpha+\beta}} \right)^{c} \left(\frac{e^{\alpha}}{1+e^{\alpha}} \right)^{b} \left(\frac{1}{1+e^{\alpha}} \right)^{d} \qquad (7.20) \\
&= \pi_1^{a} (1-\pi_1)^{c} \pi_2^{b} (1-\pi_2)^{d} \ ,
\end{aligned}
$$

which equals the product binomial logit model likelihood in (6.19) where $\pi_1 = P(D|E) = \pi_{i:x_i=1}$ and $\pi_2 = P\left(D|\overline{E}\right) = \pi_{i:x_i=0}$.

Thus for a simple qualitative covariate, such as E versus $\overline{E}$, an equivalent model can be obtained using the individual covariate values for all subjects, or using a product binomial logit model. The latter approach is also called binomial logit regression, or simply binomial regression.

Binomial regression also generalizes to a higher-order contingency table layout where the covariate vector for each subject, x_i, is now a vector of binary covariates, as illustrated in the following example.

Example 7.1 *Coronary Heart Disease in the Framingham Study*
Cornfield (1962) presents preliminary results from the Framingham Study of the association of cholesterol levels and blood pressure with the risk of developing

coronary heart disease (CHD) in the population of the town of Framingham, Massachusetts over a 10-year period. Cholesterol levels and blood pressure were measured at the beginning of the study and the cohort was then followed over time. For the purpose of illustration these predictor variables have been categorized as high versus low values of serum cholesterol (X_1: $hichol = 1$ if ≥ 200 mg/dL, 0 if <200 mg/dL) and as high versus low values of systolic blood pressure (X_2: $hisbp = 1$ if ≥ 147 mm Hg, 0 if <147 mm Hg). Thus there are four possible configurations of the covariates, and for each configuration ($j = 1, \ldots, 4$) the data consist of $(a_j, n_j, \boldsymbol{x}_j)$, where a_j is the number of patients developing CHD among the n_j subjects enrolled into the cohort (or the number at risk) for subjects sharing the covariate vector $\boldsymbol{x}_j = (x_{1j}\ x_{2j})^T$. From the sample of $N = 1329$ subjects, the data in each category of the covariates are presented in Table 7.1 (reproduced with permission).

A logistic or binomial regression model could then be fit using a statistical package such as PROC LOGISTIC in SAS with a data set containing two observations per stratum. For example, for the first stratum ($j = 1$) with two cells ($l = 1, 2$) the data are represented as

Cell (l)	x_l	y_l	F_l
1	1 0 0	1	10
2	1 0 0	0	421

(see Example 7.2). For each stratum or configuration of the covariates, the first observation specifies the number (F) with the response ($y = 1$) and the second specifies the number (F) without the response ($y = 0$). In this case the frequency F is also the weight for each term in the likelihood where

$$L = \prod_l \pi_l^{F_l y_l} (1 - \pi_l)^{F_l (1 - y_l)}, \tag{7.21}$$

$l = 1, \ldots, 8$ in this case, and

$$\ell = \sum_l F_l y_l [\log \pi_l] + \sum_l F_l (1 - y_l) \log (1 - \pi_l), \tag{7.22}$$

where $F_l y_l = a_j$ and $F_l(1 - y_l) = (n_j - a_j)$ for the jth corresponding stratum or category in Table 7.1. This is the same as the likelihood for 1329 individual observations with one record per subject.

An iterative solution is then required to fit the model. It is customary to start with an initial value of the coefficient vector evaluated under the null hypothesis, that is, using $\widehat{\boldsymbol{\theta}}^{(0)} = 0$ for the coefficients. The initial value for the intercept may be set to that for the simple logit model for a 2×2 table in (6.43): $\widehat{\alpha}^{(0)} = \log [m_1/m_2]$. For this example $m_1/(N - m_1) = 92/1237$ and $\widehat{\alpha}^{(0)} = -2.5987$. The solution converges to eight places after only six iterations, as shown in Table 7.2. Since $\boldsymbol{I}(\boldsymbol{\theta}) = \boldsymbol{i}(\boldsymbol{\theta})$ for logistic regression, then Newton-Raphson and Fisher scoring are equivalent. Thus $\boldsymbol{I}\left(\widehat{\boldsymbol{\theta}}\right)^{-1}$ at the final iteration provides the estimated covariance matrix of the coefficients.

Table 7.2 Newton-Raphson Iterative Solution of Logit Model for Data in Table 7.1

i		$\widehat{\theta}$	$U(\widehat{\theta})$	$i(\widehat{\theta})^{-1} = -H(\widehat{\theta})^{-1}$		
	$\widehat{\alpha}^{(0)}$	−2.5986560	0.0000000	0.030690	−0.026380	−0.0145450
0	$\widehat{\beta}_1^{(0)}$	0.0000000	19.665914	−0.026380	0.047753	−0.0028450
	$\widehat{\beta}_2^{(0)}$	0.0000000	18.663657	−0.014545	−0.002845	0.0586903
	$\widehat{\alpha}^{(1)}$	−3.3889060	−16.78430	0.0487293	−0.0396770	−0.0189040
1	$\widehat{\beta}_1^{(1)}$	0.8859976	−10.34558	−0.0396770	0.0541921	−0.0010600
	$\widehat{\beta}_2^{(1)}$	1.0394225	−10.46654	−0.0189040	−0.0010600	0.0416894
⋮	⋮	⋮	⋮	⋮	⋮	⋮
	$\widehat{\alpha}$	−3.6282170	−0.000000	0.0612917	−0.0515490	−0.0210170
6	$\widehat{\beta}_1$	1.0301350	−0.000000	−0.0515490	0.0678787	−0.0014840
	$\widehat{\beta}_2$	0.9168417	−0.000000	−0.0210170	−0.0014840	0.0485391

Note that this is a main effects only model. Since there are four covariate categories, the saturated model would have three coefficients in addition to the intercept, the missing term being the interaction between the two covariates. This interaction model is described in Section 7.4.

7.1.3 SAS PROCEDURES

These calculations are readily performed by any one of a number of computer programs that are available for the computation of logistic regression models. The SAS system provides four procedures that can be used for this purpose. PROC CATMOD was one of the earliest of these procedures. It implements the method of Grizzle, Starmer and Koch (1969), which fits generalizations of the binomial logit model. PROC NLIN is a general procedure for solving systems of equations that can be used to fit a family of models, including the logistic model, using iteratively reweighted least squares (IRLS). However, the user must provide the expressions for the IRLS weights and programming code is required for additional computations, such as computing $I(\widehat{\theta})$, tests of significance and confidence limits for the parameters. It is far more convenient to use PROC LOGISTIC, a general comprehensive procedure for fitting logistic regression models. A more recent program, PROC GENMOD, fits the wide family of *generalized linear models (GLMs)* described in Section A.10 of the Appendix. These models are based on distributions from the exponential family that includes as a special case the logistic regression model based on the binomial distribution. The GENMOD procedure provides some features not available through PROC LOGISTIC.

In logistic regression we wish to model the probability of either a positive or negative outcome or characteristic (D versus $\overline{D}$) by expressing the log odds (logit) of one of these probabilities as a function of the covariates. Herein we have employed the usual convention that the data analyst is interested in describing covariate effects on the odds of a positive response (D) through an assessment of the covariate effects on the log odds, say $\log(O_{\boldsymbol{x}})$, where the odds of the positive response associated with a given value of the covariate vector $\boldsymbol{x}$ is

$$O_{\boldsymbol{x}} = \frac{P(D|\boldsymbol{x})}{P(\overline{D}|\boldsymbol{x})} = \frac{\pi_{\boldsymbol{x}}}{1-\pi_{\boldsymbol{x}}} . \tag{7.23}$$

In describing the odds in this manner, the negative ($\overline{D}$) category is considered the reference category or the denominator for the calculation of the odds ratio.

The dependent variable Y that designates $\overline{D}$ and D can be coded as any unique pair of values: (0,1) or (1,2), and so on. The default convention in PROC LOGISTIC is that the highest numbered (or last) category is used as the reference category. Thus if Y is coded such that $\overline{D} = 0$ and $D = 1$, or as $\overline{D} = 1$ and $D = 2$, then the default in PROC LOGISTIC is to use the highest numbered category D as the reference category in these cases rather than $\overline{D}$ as in (7.23) above. In such cases, the DESCENDING option must be used to ensure that the program uses the category corresponding to the desired $\overline{D}$ as the reference category.

Other useful options include the COVB, which computes the estimated covariance matrix of the estimates, and RL (or risk limits), which computes the estimated odds ratios associated with each covariate.

Example 7.2 *Coronary Heart Disease in the Framingham Study*
For the above example, the SAS program in Table 7.3 would perform the logistic regression analysis. The SAS output is presented in Table 7.4 (other extraneous information deleted). The program first presents the score and likelihood ratio tests of significance of the overall model, which in this case is the null hypothesis that the two regression coefficients are zero. Each is highly significant. These tests are discussed later. This is followed by the estimates of the model parameters, standard errors, and Wald tests for the individual parameters.

In this example, the RL option provides the estimated odds ratio associated with the two binary covariates, each adjusted for the effect of the other. The effect of high versus low cholesterol, adjusting for blood pressure, is $\widehat{\beta}_1 = 1.0301$ with an estimated odds ratio of $\widehat{OR}_{H:L} = e^{1.0301} = 2.801$. This indicates that high cholesterol is associated with a 180% increase $[(2.801-1)\times 100]$ in the odds of CHD (adjusted for the effects of blood pressure). The inverse provides the decrease in odds associated with low versus high cholesterol, or $\widehat{OR}_{L:H} = e^{-1.0301} = 0.357$, which indicates a 64.3% decrease in the odds of CHD $[(1-0.357)\times 100]$. Similarly, $\widehat{\beta}_2 = 0.9168$ is the effect of high versus low systolic blood pressure adjusted for serum cholesterol which yields an odds ratio of $\widehat{OR}_{H:L} = 2.501$, indicating a 150%

Table 7.3 SAS Program for the Framingham CHD Data in Table 7.1

```
data one; input hichol hisbp chd frequncy; cards;
0 0 1 10
0 0 0 421
0 1 1 10
0 1 0 132
1 0 1 38
1 0 0 494
1 1 1 34
1 1 0 190
;
proc logistic descending;
  model chd = hichol hisbp / rl;
  weight frequncy;
run;
```

increase in the odds among females. The large sample 95% confidence limits are also presented for the odds ratio associated with each covariate.

7.1.4 Stratified 2×2 Tables

Section 6.5.2 describes the derivation of the Cochran test of association for a common odds ratio in K independent 2×2 tables as a score test in a logit model as described by Day and Byar (1979). In (6.80) the probabilities of a positive response in the jth stratum, $\pi_{1j} = P_j(D|E)$ and $\pi_{2j} = P_j\left(D|\overline{E}\right)$, are expressed as logistic functions with parameters $(\alpha_1, \ \ldots, \ \alpha_K, \ \beta)$. An equivalent model could be fit using PROC LOGISTIC using the covariate vector for the ith subject

$$\boldsymbol{x}_i = \left(z_{i2} \; z_{i3} \; \ldots \; z_{iK} \; x_i\right)^T , \tag{7.24}$$

where $z_{ij} = I\left(subject \; i \in stratum \; j\right)$ is a binary indicator variable for membership of the ith subject in the jth stratum, $j = 2, \ldots, K$, and $x_i = I(E)$ is the indicator variable for membership in the exposed group, $i = 1, \ldots, N$. The associated parameter vector is

$$\boldsymbol{\theta} = \left(\alpha \; \gamma_2 \; \gamma_3 \; \ldots \; \gamma_K \; \beta\right)^T , \tag{7.25}$$

where α is the stratum 1 effect, stratum 1 being the reference category in this case; γ_j is the difference between the jth stratum and the stratum 1 effects $(j > 1)$; and β is the stratified-adjusted effect of exposure that is assumed to be constant for all strata, that is, $\beta = \log(OR)$ as in Chapter 4. In terms of the Day-Byar parametrization in Section 6.5.2, $\alpha_1 \equiv \alpha$ and $\alpha_j \equiv \alpha + \gamma_j$.

Table 7.4 Logistic Regression Analysis of the Data in Table 7.1

The LOGISTIC Procedure
Model Fitting Information and
Testing Global Null Hypothesis BETA=0

Criterion	Intercept Only	Intercept and Covariates	Chi-Square for Covariates
-2 LOG L	668.831	632.287	36.544 with 2 DF (p=0.0001)
Score	.	.	36.823 with 2 DF (p=0.0001)

Analysis of Maximum Likelihood Estimates

Variable	DF	Parameter Estimate	Standard Error	Wald Chi-Square	Pr > Chi-Square	Odds Ratio
INTERCPT	1	-3.6282	0.2476	214.7757	0.0001	.
HICHOL	1	1.0301	0.2605	15.6334	0.0001	2.801
HISBP	1	0.9168	0.2203	17.3180	0.0001	2.501

Conditional Odds Ratios and 95% Confidence Intervals

Variable	Unit	Odds Ratio	Wald Confidence Limits Lower	Upper
HICHOL	1.0000	2.801	1.681	4.668
HISBP	1.0000	2.501	1.624	3.852

The linear predictor is then expressed as

$$\eta = \alpha + \boldsymbol{x}_i'\boldsymbol{\beta} = \alpha + \sum\nolimits_{j=2}^{K} z_{ij}\gamma_j + x_i\beta \ . \tag{7.26}$$

Therefore,

$$\begin{aligned} \pi_{11} &= \frac{e^{\alpha+\beta}}{1+e^{\alpha+\beta}} & \pi_{21} &= \frac{e^{\alpha}}{1+e^{\alpha}} & \\ \pi_{1j} &= \frac{e^{\alpha+\gamma_j+\beta}}{1+e^{\alpha+\gamma_j+\beta}} & \pi_{2j} &= \frac{e^{\alpha+\gamma_j}}{1+e^{\alpha+\gamma_j}} & 2 \le j \le K \ , \end{aligned} \tag{7.27}$$

Note that γ_j represents the log odds ratio for stratum j versus the reference stratum 1 both for those exposed, and also for those non-exposed, that is,

$$\frac{\pi_{1j}/(1-\pi_{1j})}{\pi_{11}/(1-\pi_{11})} = \frac{\pi_{2j}/(1-\pi_{2j})}{\pi_{21}/(1-\pi_{21})} = e^{\gamma_j} \tag{7.28}$$

and that the maximum likelihood estimate of the common odds ratio for exposure is $\widehat{OR}_{MLE} = e^{\widehat{\beta}}$.

Example 7.3 *Clinical Trial in Duodenal Ulcers (Continued)*
As a problem the reader is asked to conduct a logistic regression analysis of the data from the ulcer healing clinical trial presented in Example 4.1. The *MLEs* of the odds ratios, the corresponding asymmetric Wald 95% confidence limits and the Wald test *P*-values for the model effects are

Effect	$\widehat{OR}$	95% *C.I.*	$p \le$
Stratum 2 vs. 1	2.305	0.861, 6.172	0.0964
Stratum 3 vs.1	1.388	0.765, 2.519	0.2813
Drug vs. Placebo	1.653	0.939, 2.909	0.0816

Neither stratum 2 nor 3 has an effect on the overall risk that is significantly different from that of the first stratum, which is consistent with the results of the test of stratum by healing presented in Section 4.4.3. The odds of healing is increased 1.65-fold with drug treatment, which is not statistically significant. The estimated common odds ratio for drug treatment versus placebo, adjusted for stratum effects, and its confidence limits, are comparable to the Mantel-Haenszel and *MVLE* estimates presented earlier in Chapter 4, and the conditional *MLE* presented in Example 6.7 of Chapter 6.

7.1.5 Family of Binomial Regression Models

The logistic regression model is one of a family of binomial regression models for such data that employ different link functions to relate the probability π to the covariates $\boldsymbol{x}$. This family of models was first proposed by Dyke and Patterson (1952), who described a generalization of the ANOVA model for proportions. These models are also members of the family of *Generalized Linear Models* described in Section A.10 of the Appendix.

For the ith observation, let π_i be some smooth function of the covariates $\boldsymbol{x}_i$ with parameter vector $\boldsymbol{\theta} = (\alpha\ \beta_1\ \ldots\ \beta_p)^T$. In GLM notation we assume a binomial error structure with a link function $g(\cdot)$ that relates the probabilities to the covariates such that $g(\pi) = \alpha + \boldsymbol{x}_i'\boldsymbol{\beta}$. Thus $\pi = g^{-1}(\alpha + \boldsymbol{x}_i'\boldsymbol{\beta})$, where $g^{-1}(\cdot)$ is the inverse function. In logistic regression, $g(\cdot)$ is the logit and $g^{-1}(\cdot)$ is the logistic function (inverse logit). The link function $g(\cdot)$ can be any twice-differentiable one-to-one function.

The binomial likelihood and log likelihood in terms of the $\{\pi_i\}$ are given in (7.3) and (7.5). Using the chain rule, the score equations for the parameters are then obtained as

$$U(\boldsymbol{\theta})_\alpha = \frac{\partial \ell}{\partial \pi}\frac{\partial \pi}{\partial \alpha} \quad \text{and} \quad U(\boldsymbol{\theta})_{\beta_j} = \frac{\partial \ell}{\partial \pi}\frac{\partial \pi}{\partial \beta_j}, \tag{7.29}$$

where $\partial\pi/\partial\beta_j = \partial\left[g^{-1}(\alpha + \boldsymbol{x}_i'\boldsymbol{\beta})\right]/\partial\beta_j$, $j = 1, \ldots, p$. From (7.5), the resulting score equation for β_j then is of the form

$$U(\boldsymbol{\theta})_{\beta_j} = \sum\nolimits_i \left(\frac{y_i}{\pi_i} - \frac{1-y_i}{1-\pi_i}\right)\frac{\partial \pi}{\partial \beta_j} = \sum\nolimits_i \left(\frac{y_i - \pi_i}{\pi_i(1-\pi_i)}\right)\frac{\partial \pi_i}{\partial \beta_j}, \tag{7.30}$$

which is a weighted sum of standardized residuals; and likewise for $U(\boldsymbol{\theta})_\alpha$. These expressions provide the elements of the score vector $\boldsymbol{U}(\boldsymbol{\theta})$ that provide the maximum likelihood estimating equations for the parameters. Then the elements of the Hessian are obtained as

$$\begin{aligned}
\boldsymbol{H}(\boldsymbol{\theta})_\alpha &= \frac{\partial^2 \ell}{\partial \alpha^2} = \frac{\partial}{\partial \alpha}\left(\frac{\partial \ell}{\partial \pi}\frac{\partial \pi}{\partial \alpha}\right) = \frac{\partial \ell}{\partial \pi}\frac{\partial^2 \pi}{\partial \alpha^2} + \frac{\partial^2 \ell}{\partial \pi^2}\left(\frac{\partial \pi}{\partial \alpha}\right)^2 \\
\boldsymbol{H}(\boldsymbol{\theta})_{\beta_j} &= \frac{\partial^2 \ell}{\partial \beta_j^2} = \frac{\partial}{\partial \beta_j}\left(\frac{\partial \ell}{\partial \pi}\frac{\partial \pi}{\partial \beta_j}\right) = \frac{\partial \ell}{\partial \pi}\frac{\partial^2 \pi}{\partial \beta_j^2} + \frac{\partial^2 \ell}{\partial \pi^2}\left(\frac{\partial \pi}{\partial \beta_j}\right)^2 \\
\boldsymbol{H}(\boldsymbol{\theta})_{\alpha,\beta_j} &= \frac{\partial^2 \ell}{\partial \alpha \partial \beta_j} = \frac{\partial}{\partial \beta_j}\left(\frac{\partial \ell}{\partial \pi}\frac{\partial \pi}{\partial \alpha}\right) = \frac{\partial \ell}{\partial \pi}\frac{\partial^2 \pi}{\partial \alpha \partial \beta_j} + \frac{\partial^2 \ell}{\partial \pi^2}\left(\frac{\partial \pi}{\partial \alpha}\right)\left(\frac{\partial \pi}{\partial \beta_j}\right) \\
\boldsymbol{H}(\boldsymbol{\theta})_{\beta_j,\beta_k} &= \frac{\partial^2 \ell}{\partial \beta_j \partial \beta_k} = \frac{\partial}{\partial \beta_k}\left(\frac{\partial \ell}{\partial \pi}\frac{\partial \pi}{\partial \beta_j}\right) = \frac{\partial \ell}{\partial \pi}\frac{\partial^2 \pi}{\partial \beta_j \partial \beta_k} + \frac{\partial^2 \ell}{\partial \pi^2}\left(\frac{\partial \pi}{\partial \beta_j}\right)\left(\frac{\partial \pi}{\partial \beta_k}\right)
\end{aligned} \tag{7.31}$$

for $1 \le j < k \le p$, from which the observed and expected information are obtained. These expressions are special cases of the more general equations for the family of $GLMs$ described in Section A.10 of the Appendix.

As a problem, the model equations for an exponential risk model are obtained using the simple log link, and for a compound exponential model using the complimentary log-log link. Also included in this family of models is probit regression using the inverse probit link.

These models differ in the manner that the Bernoulli residuals $\{y_i - \pi_i\}$ are weighted in the score estimating equation. From (7.30), the score equation for the

jth coefficient β_j can be expressed as

$$U(\boldsymbol{\theta})_{\beta_j} = \sum\nolimits_i w_{ij} \left(y_i - \pi_i\right) \tag{7.32}$$

with weights

$$w_{ij} = \left(\frac{1}{\pi_i(1-\pi_i)}\right) \frac{\partial \pi_i}{\partial \beta_j} \,. \tag{7.33}$$

For example, in a logit link (logistic regression) model $\partial\pi/\partial\beta_j = x_{ij}\pi_i(1-\pi_i)$ and the weights are $w_i = x_{ij}$. However, in an exponential risk model with a log link, $\partial\pi/\partial\beta_j = x_{ij}\pi_i$ and the weights are $w_i = x_{ij}/(1-\pi_i)$.

7.2 INTERPRETATION OF THE LOGISTIC REGRESSION MODEL

7.2.1 Model Coefficients and Odds Ratios

In a logistic regression model, each estimated coefficient equals the estimated log odds ratio associated with a unit increase in the value of the covariate adjusted for all other effects in the model, or assuming that the values of all other covariates are held fixed. To see this, consider the effect of a single covariate, say the last X_p. The effects of all the other $p-1$ covariates in the model with fixed values $(x_1, \ldots, x_{p-1})$ are held constant, expressed as $\boldsymbol{C} = \sum_{j=1}^{p-1} x_j\beta_j$. The effect of the last covariate is $x_p\beta_p$, or simply $x\beta$. Then the model can be parameterized using $\boldsymbol{x}'\boldsymbol{\beta} = \boldsymbol{C}+\, x\beta$.

First consider a binary covariate with only two categories coded as $x = 1$ if exposed (E), 0 if not exposed $(\overline{E})$. Then the ratio of the odds for an exposed individual to those of an individual not so exposed is

$$OR_{E:\overline{E}} = \frac{O_{x=1}}{O_{x=0}} = \frac{e^{\alpha+\boldsymbol{C}+\beta}}{e^{\alpha+\boldsymbol{C}}} = e^{\beta}. \tag{7.34}$$

Therefore, the log(odds ratio)$= \beta$ for $x = 1$ versus $x = 0$, or for E versus $\overline{E}$. In some applications, however, such as in PROC CATMOD, a binary covariate is coded as a contrast using $x = 1$ if exposed (E) and $x = -1$ if not exposed $(\overline{E})$. As a problem it is readily shown that in this case $\exp(2\beta) = OR$.

Now consider a quantitative covariate. The odds ratio associated with a 1-unit difference in the value of the covariate is

$$OR_{\triangle x=1} = \frac{O_{x+1}}{O_x} = \frac{e^{\alpha+\boldsymbol{C}+\beta(x+1)}}{e^{\alpha+\boldsymbol{C}+\beta x}} = e^{\beta}. \tag{7.35}$$

Therefore, $[\triangle \log(\text{odds}) \mid \triangle\, x = 1] = \beta$. Thus for a binary covariate coded such that $\triangle x = 1$, and for any quantitative covariate, $\widehat{\beta}$ equals the log odds ratio associated with a unit change (increase) in the covariate.

Similarly, the change in the log odds associated with a K-unit difference in the value of the covariate is $[\triangle \log(\text{odds}) \mid \triangle x = K] = K\beta$ with the associated odds ratio

$$OR_{\triangle x=K} = e^{K\beta} = (OR_{\triangle x=1})^K \tag{7.36}$$

and estimated standard error

$$S.E.\left[K\widehat{\beta}\right] = |K|\, S.E.\left(\widehat{\beta}\right). \tag{7.37}$$

This also allows one to readily invert the odds ratio, or to describe the odds ratio associated with a K unit *decrease* in X using some $K < 0$. For example, if $\beta < 0$ then this implies that there is a decrease in the odds associated with an increase in X; however, in many cases it is easier to explain the association in terms of the increase in the odds associated with a decrease in X. For example, if for a binary covariate, exposure is associated with a decrease in the odds, $OR_{E:\overline{E}} < 1$, then the increase in the odds associated with non-exposure is simply $OR_{\overline{E}:E} = e^{-\beta}$. Likewise, if an increase in a quantitative covariate is associated with a decrease in the odds, then the relationship is readily inverted to describe the increase in odds associated with a decrease in the covariate. In both cases $K = -1$.

For any K, the asymmetric $1-\alpha$ confidence interval on the odds ratio associated with a K unit change in the covariate is obtained as

$$\exp\left[K\widehat{\beta} \pm Z_{1-\alpha/2}\, |K|\, \widehat{V}\left(\widehat{\beta}\right)^{1/2}\right], \tag{7.38}$$

where $\widehat{V}\left(\widehat{\beta}\right)$ is obtained from the corresponding diagonal element of the inverse estimated information matrix.

Similar developments can be used to describe the effects of covariates on the response variable after transformations of either. For example, in a log-log linear model where $\log(y)$ is regressed on $\log(x)$ then the coefficient β for $\log(x)$ is such that the proportionate change in Y associated with a c-fold change in X is c^{β}; see Problem 7.5.

Example 7.4 *DCCT Nephropathy Data*
Section 1.5 provides a description of the Diabetes Control and Complications Trial (DCCT) and presents the analysis of the cumulative incidence of the onset of microalbuminuria that is the earliest sign of diabetic renal disease (early nephropathy). The life-table analyses of cumulative incidence are described in Chapter 9. Another way to assess the effect of treatment on the risk of nephropathy is to examine the prevalence of microalbuminuria at a fixed point in time. Because the average duration of follow-up was about six years, we present analyses of the factors related to the prevalence of microalbuminuria in those subjects evaluated at six years. Further, we consider the subset of 172 patients in the secondary intervention cohort with a high-normal albumin excretion rate (AER) at baseline defined as $15 \leq AER \leq 40$. In addition to intensive versus conventional treatment group (*int*

Table 7.5 SAS Program for DCCT Nephropathy Data

```
data renal;
 input obsn micro24 int hbael duration sbp female;
yearsdm=duration/12; cards;
    1   0   1     9.63   178    104    1
    2   0   0     7.93   175    112    0
    3   1   0    11.20   126    110    1
    4   1   0    10.88   116    106    0
    5   0   0     8.22   168    110    1
    6   1   1    12.73    71    112    0
    7   0   0     8.28   107    116    1
    8   0   1     9.44    79    120    1
    9   0   0     7.44   176    120    1
   10   0   0     8.33    47    140    0
   11   1   1     9.89   135    126    1
   12   0   1    12.06   117    120    0
   13   0   1     9.01    35    128    1
   14   0   1    10.05    82    110    1
   15   1   1     9.60    70    108    1
   16   0   1    10.17   159    121    1
  ...   .   .      ...   ...    ...    .
  158   1   0    11.80   157    128    0
  159   0   1     7.00    64    126    0
  160   0   1     9.00   141    114    0
  161   0   1     8.00    46    131    0
  162   1   0    12.50    99    116    1
  163   1   0     7.10    89    114    0
  164   0   0     8.60   139    130    1
  165   0   1    12.20    76    106    1
  166   0   1    11.70   118    110    1
  167   1   0    11.30    99    122    1
  168   0   0     6.80    41    104    1
  169   0   0    10.60    82    106    1
  170   0   1     8.70   101     98    1
  171   0   0     7.90   136    126    0
  172   0   0    10.10   127    124    0
;
proc logistic descending;
  model micro24 = int hbael yearsdm sbp female
  / rl covb rsquare;
units exp=1 -1 hbael=1 yearsdm=1 sbp=1 5
  female=1 -1;
run;
```

= 1 or 0, respectively), the analysis adjusts for the level (%) of HbA_{1c} at baseline (*hbael*), the prior duration of diabetes in months (*duration*), the level of systolic blood pressure in mm Hg (*sbp*), and gender *(female)* (1 if female, 0 if male).

The HbA_{1c} is a measure of the average level of blood glucose control over the preceding 4–6 weeks. The measure employed here (*hbael*) is that obtained at the initial eligibility screening that reflects the level of control prior to participation in the trial. The underlying hypothesis is that the level of hyperglycemia and duration of diabetes together determine the risk of further progression of complications of diabetes. Here the months duration is converted to years duration of diabetes (*yearsdm*). Risk of progression of nephropathy may also be associated with increased levels of blood pressure that lead to damage to the kidneys that also may increase the risk of progression of nephropathy. These effects may also differ for men and women. Thus the objective is to obtain an adjusted assessment of the effects of intensive versus conventional treatment and also to explore the association between these baseline factors and the risk of nephropathy progression. The SAS program in Table 7.5 fits a logistic regression model using PROC LOGISTIC. The output from the program is presented in Table 7.6 (extraneous information has been deleted).

The estimated coefficient for treatment group (*int*) is $\widehat{\beta} = -1.5831$ that equals the difference in log odds, or the log odds ratio, for intensive versus conventional treatment. The large sample 95% confidence limits on the log odds ratio are obtained as $-1.5831 \pm (1.96)(0.4267) = (-2.4194, -0.7468)$. The estimated odds ratio, therefore, is $\widehat{OR}_{I:C} = e^{-1.5831} = 0.205$ with asymmetric 95% confidence limits of $\left(e^{-2.4194}, e^{-0.7468}\right) = (0.0890,\ 0.4739)$. These latter values are labeled as the Wald confidence limits because they correspond to confidence limits obtained by inverting the Wald test statistic for the parameter (described in Section 7.3.3 to follow). Thus intensive treatment results in a 79.5% reduction in the odds of having microalbuminuria after six years of treatment compared to conventional therapy.

Because the odds ratio is less than one ($\widehat{\beta} < 0$), there is a decrease in risk (odds) in the intensive therapy group. The complement of this coefficient provides the increase in odds (odds ratio) for conventional versus intensive therapy, $\widehat{OR}_{C:I} = e^{1.5831} = 4.87$ with asymmetric 95% confidence limits of $(e^{0.7468}, e^{2.4194}) = (2.11, 11.24)$. Thus conventional therapy yields a 4.87-fold increase in the odds versus intensive therapy.

Similarly, the estimated coefficient for female gender ($\widehat{\beta} = -0.0895$) yields an odds ratio of $\widehat{OR}_{F:M} = 0.410$ for females versus males. The inverse yields an estimated odds ratio for males versus females of $\widehat{OR}_{M:F} = 2.436$ with 95% *C.I.* = (1.014, 5.854).

Hbael is a quantitative covariate measured in percent, since it the percent of red cells (hemoglobin) that have been glycosylated in reaction with free glucose in blood. Thus the estimated coefficient $\widehat{\beta} = 0.5675$ is the difference in log odds (the log odds ratio) for each one percentage higher HbA_{1c}; such as the difference in the log odds for two subjects, one with a HbA_{1c} of 8 and another 9, or one a value of 10 and another 11. The 95% *C.I.* is $0.5675 \pm (1.96)(0.1449) = (0.2835, 0.8515)$.

Thus the estimated odds ratio associated with a one percent higher HbA_{1c} is $\widehat{OR} = e^{0.5675} = 1.7639$ with 95% Wald $C.I.$ of $\left(e^{0.2835}, e^{0.8515}\right) = (1.328, 2.343)$.

Similarly, the estimated coefficient for years duration ($\widehat{\beta} = 0.0096$) implies an odds ratio of 1.010 per year greater duration with 95% $C.I.$ (0.891, 1.144); and the estimated coefficient for systolic blood pressure ($\widehat{\beta} = 0.0233$) implies an odds ratio of 1.024 per mm Hg higher pressure with 95% $C.I.$ (0.983, 1.066). However, a one mm Hg difference in blood pressure is a small difference with respect to the distribution of blood pressure values with a mean of 115.6 and $S.D.$ of 11.3. Thus it is more relevant to describe the odds ratio associated with perhaps a 5 mm Hg higher pressure ($K = 5$). The associated log odds ratio is $5\widehat{\beta} = 0.1165$ with $S.E. = 5\,(0.0208) = 0.104$ and 95% $C.I.$ of (-0.087, 0.320). The corresponding odds ratio is $\widehat{OR} = e^{0.1165} = 1.123$ with 95% $C.I. = \left(e^{-0.087}, e^{0.320}\right) = (0.917,$ $1.377)$.

Using the expressions in (7.36) and (7.37) it is also possible to invert the relationship for a quantitative covariate. For example, the coefficient for the HbA_{1c} implies a reduction in risk as the HbA_{1c} is reduced. From the coefficient and its $S.E.$, for a 2 percentage reduction in HbA_{1c} ($K = -2$), the corresponding coefficient is $(-2)0.5675 = -1.135$ with $S.E. = 2(0.1449) = 0.2898$. These yield an odds ratio of 0.321 with 95% $C.I.$ (0.182, 0.567).

PROC LOGISTIC includes an option for a UNITS specification, as is shown in Table 7.5. The output in Table 7.6 provides the odds ratio associated with a one-unit change for each covariate and that associated with each of the units change specified in the UNITS statement.

Table 7.6 also presents the estimated covariance of the coefficient estimates obtained as the inverse of the estimated information matrix. This is generated by the COVB option. This covariance matrix is useful for computing the variance of an estimated probability and in other computations, such as the robust variance covariance matrix, which will be described later.

7.2.2 Partial Regression Coefficients

Any regression model has three main components, as exemplified by the GLM family described in Section A.10 of the Appendix.

First is the *structural component*, which specifies the nature of the link between the conditional expectation and the covariates, such as

$$E\left(y_i|\boldsymbol{x}_i\right) = \mu_i \text{ and } g(\mu_i) = \alpha + \boldsymbol{x}_i'\boldsymbol{\beta} \tag{7.39}$$

so that

$$E\left(y_i|\boldsymbol{x}_i\right) = g^{-1}\left(\alpha + \boldsymbol{x}_i'\boldsymbol{\beta}\right) \tag{7.40}$$

for some link function $g\left(\cdot\right)$ of the linear predictor $\eta = \alpha + \boldsymbol{x}_i'\boldsymbol{\beta}$. In ordinary multiple regression g is the identity link. In logistic regression g is the logit link. The predictor may be non-linear and, in some cases, such as *generalized additive models* (Hastie and Tibshirani, 1990), it is replaced by some smooth function of

Table 7.6 Logistic Regression Analysis of DCCT Nephropathy Data

The LOGISTIC Procedure

Model Fitting Information and Test of H0: BETA=0

Criterion	Intercept Only	Intercept and Covariates	Chi-Square for Covariates
-2 LOG L	191.215	155.336	35.879 with 5 DF (p=0.0001)
Score	.	.	33.773 with 5 DF (p=0.0001)

RSquare = 0.1883 Max-rescaled RSquare = 0.2806

Analysis of Maximum Likelihood Estimates

Variable	DF	Parameter Estimate	Standard Error	Wald Chi-Square	Pr > Chi-Square	Odds Ratio
INTERCPT	1	-8.2806	3.0714	7.2686	0.0070	.
INT	1	-1.5831	0.4267	13.7626	0.0002	0.205
HBAEL	1	0.5675	0.1449	15.3429	0.0001	1.764
YEARSDM	1	0.00961	0.0636	0.0228	0.8799	1.010
SBP	1	0.0233	0.0208	1.2579	0.2620	1.024
FEMALE	1	-0.8905	0.4473	3.9641	0.0465	0.410

Conditional Odds Ratios and 95% Confidence Intervals

Variable	Unit	Odds Ratio	Wald Confidence Limits Lower	Upper
INT	1.0000	0.205	0.089	0.474
INT	-1.0000	4.870	2.110	11.240
HBAEL	1.0000	1.764	1.328	2.343
YEARSDM	1.0000	1.010	0.891	1.144
SBP	1.0000	1.024	0.983	1.066
SBP	5.0000	1.123	0.917	1.377
FEMALE	1.0000	0.410	0.171	0.986
FEMALE	-1.0000	2.436	1.014	5.854

Estimated Covariance Matrix

Variable	INTERCPT	INT	HBAEL	YEARSDM	SBP	FEMALE
INTERCPT	9.43355	-0.0125	-0.2582	-0.0401	-0.0550	-0.1975
INT	-0.0125	0.18211	-0.0097	-0.0005	0.00032	0.02425
HBAEL	-0.2582	-0.0097	0.02100	0.00154	0.00044	-0.0158
YEARSDM	-0.0401	-0.0005	0.00154	0.00405	-0.0001	-0.0023
SBP	-0.0550	0.00032	0.00044	-0.0001	0.00043	0.00243
FEMALE	-0.1975	0.02425	-0.0158	-0.0023	0.00243	0.20005

the covariates. The estimated model then provides an estimate of the nature of the association under the assumed model. If one assumes that the logit is a linear function of the covariate, then the model provides an "optimal" estimate of the slope of the relationship. However, this does not guarantee that the estimated model is correct. Thus it is important to assess the adequacy of the model specifications using techniques such as residual analysis or the analysis of value-added plots, in addition to fitting different expressions for the predictor, such as polynomials or logarithms, and so forth, when analyzing data using these methods. Excellent references for these assessments are Pregibon (1981) and Hosmer and Lemeshow (1989), among many. Many of these model diagnostics are available through PROC LOGISTIC. Model diagnostics, however, will not be assessed herein.

Second is the *random component*, which is usually assumed to be of the form

$$y_i = \mu_i + \varepsilon_i, \qquad \varepsilon_i \perp \mu_i \tag{7.41}$$

where the random errors ε_i are assumed to be independent of the conditional expectation μ_i and where the errors have some distribution $\varepsilon_i \sim f(\varepsilon)$ with mean zero and a variance that may be some function of the expectation expressed as

$$E(\varepsilon_i) = 0, \qquad V(y_i|\mu_i) = V(\varepsilon_i|\mu_i) = \sigma_\varepsilon^2(\mu_i)\,. \tag{7.42}$$

In ordinary multiple regression we assume that the $\{\varepsilon_i\}$ are *i.i.d.* with $E(\varepsilon_i) = 0$ and constant $V(\varepsilon_i) = \sigma_\varepsilon^2$. In logistic regression we assume that the Y_i are independent (but not identically) distributed Bernoulli variables with mean $\mu_i = \pi_i$ so that $E(\varepsilon_i) = 0$ and $V(\varepsilon_i|\pi_i) = \sigma_\varepsilon^2(\pi_i) = \pi_i(1 - \pi_i)$. The random component specification, in effect, determines how each individual observation is weighted in the computation of the estimates of the model parameters.

For example, in a multiple regression model fit using ordinary least squares, constant variance is assumed such that each observation has equal weight in the computation of the sum of squared errors, the objective function to be minimized. However, if homoscedastic errors do not apply, then the model is fit using weighted least squares, where each observation is weighted inversely to its variance. Similarly, in a GLM the score equation (A.244) is a weighted sum of residuals with weights inversely proportional to the $\sigma_\varepsilon^2(\mu_i)$. In addition, the link function also affects the way each observation is weighted as shown in Section 7.1.5.

The above two components of regression models are described in every modern text on regression. However, there is a third component of importance — the covariate specification — which is rarely discussed. Any model includes a specific set of covariates $(X_1 \ \ldots \ X_p)$, where the elements are selected from a larger set $X_j \in \Omega_X$, Ω_X being the set of possible covariates in the model. If one adds or subtracts a covariate from the model, where a transformation of a measurement is considered a new covariate, then the *meaning* of the coefficients for the other covariates in the model changes, not just their values.

For example, consider an ordinary regression model with two covariates X_1 and X_2. To be mathematically precise, the structural component of the assumed model should be expressed as

$$E(y \mid x_1, x_2) = \alpha_{|\beta_1,\beta_2} + x_1\beta_{1|\alpha,\beta_2} + x_2\beta_{2|\alpha,\beta_1}. \tag{7.43}$$

If x_3 is now added to the model, then the precise model specification is that

$$\begin{aligned} E(y \mid x_1, x_2, x_3) = {} & \alpha_{|\beta_1,\beta_2,\beta_3} + x_1\beta_{1|\alpha,\beta_2,\beta_3} \\ & + x_2\beta_{2|\alpha,\beta_1,\beta_3} + x_3\beta_{3|\alpha,\beta_1,\beta_2}. \end{aligned} \tag{7.44}$$

In general, these sets of coefficients differ, so that

$$\begin{aligned} \alpha_{|\beta_1,\beta_2,\beta_3} &\neq \alpha_{|\beta_1,\beta_2} \\ \beta_{1|\alpha,\beta_2,\beta_3} &\neq \beta_{1|\alpha,\beta_2} \\ \beta_{2|\alpha,\beta_1,\beta_3} &\neq \beta_{2|\alpha,\beta_1} \;. \end{aligned} \tag{7.45}$$

Thus the value, and more importantly the meaning, of each coefficient depends explicitly on the other covariates in the model. Each coefficient is a *partial coefficient* adjusted for the other covariates in the model.

In a simple normal errors regression model, when all effects in the design matrix $\boldsymbol{x}$ are jointly orthogonal (uncorrelated) such that $\boldsymbol{x}'\boldsymbol{x}$ is a diagonal matrix, then the coefficients for each covariate are unique and are unchanged in all reduced models containing subsets of the original p covariates. If the covariates are correlated, then the coefficients in any reduced model will change from one model to the next (*cf.* Snedecor and Cochran, 1967, p. 393). In a logistic regression model, however, even when the covariates are uncorrelated, the value and the meaning of model coefficients change when variables are added to or removed from a model because the link $g(\cdot)$ is the logit, a non-linear function. Thus the meaning of each coefficient depends explicitly on the set of covariates included in the model.

To see this, consider the case of only two covariates where X_1 and X_2 are independent. The logistic model specifies that

$$E(y \mid x_1, x_2) = g^{-1}\left(\alpha_{|\beta_1,\beta_2} + x_1\beta_{1|\alpha,\beta_2} + x_2\beta_{2|\alpha,\beta_1}\right), \tag{7.46}$$

where $g(\cdot)$ is the logit and $g^{-1}(\cdot)$ is the logistic function (inverse logit). If we now drop X_2 from the model, then we obtain the reduced model

$$E(y \mid x_1) = g^{-1}\left(\alpha_{|\beta_1} + x_1\beta_{1|\alpha}\right). \tag{7.47}$$

However, from (7.46) it follows that the reduced model satisfies

$$\begin{aligned} E(y \mid x_1) &= E_{X_2}\left[E(y \mid x_1, x_2)\right] \\ &= E_{X_2}\left[g^{-1}\left(\alpha_{|\beta_1,\beta_2} + x_1\beta_{1|\alpha,\beta_2} + x_2\beta_{2|\alpha,\beta_1}\right)\right]. \end{aligned} \tag{7.48}$$

Were it not for the logistic function $g^{-1}(\cdot)$, then $E_{X_2}\left(x_2\beta_{2|\alpha,\beta_1}\right)$ might be absorbed into the intercept and $\beta_{1|\alpha} = \beta_{1|\alpha,\beta_2}$. However, with the logit link, $\beta_{1|\alpha} \neq \beta_{1|\alpha,\beta_2}$, even when X_1 and X_2 are independent; see also Problem 7.6.

7.2.3 Model Building: Stepwise Procedures

As with any regression procedure, it is possible to build a logistic regression model that meets some specified criteria for optimality or superiority for a given data set. Model building refers to an iterative process by which one explores the sensitivity of the model to each of the three components of the model specification: (1) the structural relationship represented by the expression for the predictor (linear or non-linear) and the link function; (2) the error distribution and the correlation structure of the errors, if any; and (3) the elements of the covariate vector that are included in the model.

Various diagnostic tools, such as residuals, can be used with logistic regression to assist in the specification of the form of the prediction and link functions and the error distribution. See Pregibon (1981) and the text by Hosmer and Lemeshow (1989). For example, the value-added plots can serve as a visual aid to determine whether a linear or non-linear covariate effect best fits the data. Hastie and Tibshirani (1990) describe a general approach to determine the structural elements of a model using generalized additive models. In these models, rather than specifying a form for the predictor and the link function, local "non-parametric" smoothing algorithms such as a kernel or spline smoother are applied to the relationship between X and Y to estimate $E(y|x)$ as some smooth function of X with a possibly irregular shape. However, these models are computer intensive and have the disadvantage that individual coefficients no longer describe the effects of covariates on the risk of the response in direct terms such as a partial odds ratio.

Of the three model components, covariate selection is the easiest and most commonly explored using some optimization algorithm such as stepwise model building. In stepwise models, one starts with a *full model* comprising the complete set of say p candidate covariates, including transformations of covariates such as quadratic and higher-order effects and interactions. The object is to identify a subset of $r \leq p$ covariates that yield a "best fitting" model according to some criteria. One approach is to fit all $2^p - 1$ sets of one or more covariates and to then select the model with the greatest model likelihood ratio test value, or perhaps the smallest model P-value, among many possibilities. This would be computationally tedious. As an approximation to this approach for normal errors multiple regression models, Efroymson (1960) suggested a *stepwise procedure* starting with forward covariate selection followed interchangeably with backward elimination based on some criterion. That most commonly used is a test of significance of each covariate, adjusting for the other covariates in the model. Most computer programs for regression models afford the option for some variation of stepwise model building, including PROC LOGISTIC in SAS.

In *forward selection* all remaining candidate covariates not in the model are tested to determine that one, if any, that best meets the inclusion criteria, usually a P-value $\leq \alpha_E$, the minimum significance level for entry into the model, termed the SLE in SAS. At the initial step with no variables yet added to the model, all p candidate covariates are tested for inclusion, and the most significant is added provided that its P-value is $\leq \alpha_E$. The process continues until no additional covariates meet the criteria for inclusion.

Conversely, in *backward elimination,* one starts with the full model of p candidate covariates and tests each covariate, in turn, to determine that one which has the smallest contribution to the model, such as that with the largest P-value. That covariate is then eliminated from the model provided that its P-value $\geq \alpha_S$, the maximum significance level for staying in the model, termed the SLS in SAS. The model is then refit and the process repeated until none of the covariates remaining in the model meet the criteria for elimination. Mantel (1970) suggested that backward elimination was preferred to forward selection because it required fewer models to be fit, and because it would more likely identify synergistic combinations of covariates that contributed to the model when the covariates individually (marginally) did not.

More information about such model building procedures is provided by one of the many standard texts on linear regression models. In a normal errors linear model with all p covariates, the variance of the estimates is a function of the mean square error on $N - p - 1$ *df* that also serves as the denominator in the F-tests of significance of the coefficients in the model. In this case, the variance of the coefficient estimates is reduced as the number of extraneous covariates is deleted. Also, the variance of the estimate of the predicted values (the $\widehat{y}_i$) is also reduced.

However, a feature of model building that often is not addressed in standard texts is the fact that any inferences about a model have been compromised by the application of any of the above optimization criteria to select the model. For example, any confidence limits, tests of significance or P-values obtained following a stepwise covariate selection procedure are highly biased and are virtually worthless. See, for example, Freedman (1983), Miller (1984) and Freedman and Pee (1989), among many. Thus there is a trade-off between the reduction in variance of the estimates and the potential increase in bias of the estimates associated with "stepwise" model building.

In the normal errors linear model, Freedman (1983) presents limiting expressions for the value of the model F-test when a one-stage deletion at level α_S is conducted. That is, all covariates not significant at the α_S level in the full model are deleted at once and the model is refit. For example, if one starts with $N = 100$ observations with $p = 50$ covariates and applies a one-stage deletion with $\alpha_S = 0.25$, then under the null hypothesis wherein all 50 covariates represent noise, on average 12.5 ($= p \times \alpha_S$) covariates will be selected and the F-statistic value will be about 3.95, which would have a $p \leq 0.0001$ on 12, 87 df. Thus the data-dependent selection of covariates grossly inflates the test statistics. Unfortunately, there is no simple way to correct for this inflation.

In addition, the estimates of the model coefficients following model selection are biased. Consider the simple linear model with two covariates such that $E(y \,|\, x_1, x_2) = \alpha + x_1\beta_1 + x_2\beta_2$. To simplify matters, assume that $V(\widehat{\beta}_1) = V(\widehat{\beta}_2) = \sigma^2_\beta$ and that we always select either X_1 or X_2 in the final reduced model, depending on whether $\widehat{\beta}_1 >$ or $< \widehat{\beta}_2$, respectively. Miller (1984) then presents computations of $E(\widehat{\beta}_1 | sel)$ or the expected value of $\widehat{\beta}_1$ given that X_1 was selected for inclusion in the reduced model. Even when X_1 and X_2 are uncorrelated, such that $Cov(\widehat{\beta}_1, \widehat{\beta}_2) = 0$, the following are the values of $E(\widehat{\beta}_1|sel)$ as a function of β_2 and σ^2_β when the true value of $\beta_1 = 1.0$:

	β_2				
σ^2_β	0.0	0.5	1.0	1.5	2.0
0.3	1.01	1.05	1.17	1.35	1.57
0.6	1.10	1.15	1.28	1.46	1.66

(reproduced with permission). Thus the process of covariate selection induces a positive bias in the expected value of the coefficient in the subset of models in which X_1 is the selected covariate. This bias increases as the true value of the coefficient for the competing covariate X_2 increases, and as the variance of the estimates increases. Miller also presents computations when the coefficient estimates are correlated; however, the correlation, whether positive or negative, has a negligible effect on the magnitude of the bias that is induced.

Therefore, any inferences about the values of the coefficients or the fit of the model should only be based upon the full model or on an *a priori* specified reduced model. If a reduced model is derived through a data-dependent algorithm, then it should only be characterized as a model that may work about as well as the full model as a basis for predictions, with no statement of statistical significance or importance associated with the reduced model so identified.

One technique that is useful to assess the performance of a model is *cross-validation,* whereby the fit of a model is assessed by applying it to independent observations not used to fit the model. The simplest method is *split sample validation,* where the model is fit to N_A random observations and then applied to another N_B random observations, designated as the A and B samples, respectively. Often the A and B samples are the split-halves or some other fraction of the total sample. Such split-sample validation allows one to assess the properties of a model from observations that are independent of those used to derive the model. It is often sobering to first conduct a stepwise model building exercise from two independent samples and to find that the reduced models so derived from the A and B samples are disjoint with no variables in common, even though all the selected variables may have been "nominally" significant at $p \leq 0.05$ in the reduced models.

The Coronary Drug Project Research Group (1974) presented an example in which a reduced model containing only the first 10 of 40 potential covariates were chosen by forward selection separately within split samples, all with $|Z| \geq 2.84$. Thus the "best" 10-covariate reduced models were selected separately and independently in sample A and in sample B. However, only two of the selected covariates were included in both sets of 10. This would lead one to question whether the other 16 selected covariates, 8 in each set, could merely be noise. Nevertheless, despite the discrepancy in the covariates selected in each reduced model, each model showed equal predictive power when applied to another independent sample of observations. This illustrates that, in general, there is no true "best model" for any population or set of observations. Rather, one can usually identify a group of reduced models containing different collections of covariates, where each model performs equally well according to some criterion.

There is a large literature on indices of the adequacy of reduced models, such as Mallow's C_p (Mallows, 1973) and Akaike's Information Criterion (Akaike, 1973) that assess in some sense the bias introduced in reduced models. However, these are global measures that do not help assess the actual level of significance associated with the reduced model or the selected covariates, or to assess the true importance of any individual covariates selected or not selected. There is also a growing literature on cross-validation techniques to assess model fit by eliminating k observations from the data set, fitting the model from the $N - k$ remaining observations and then applying the model to the k omitted observations. This is also termed the "leave one out" method ($k = 1$) or many-fold cross-validation ($k > 1$). An overall assessment of model fit is then obtained by averaging or aggregating over the $\binom{N}{k}$ possible combinations. Recently, Thall, Simon and Grier (1992) and Thall, Russell and Simon (1997) described such many-fold cross-validation and data splitting as a basis for reduced model selection to ensure that selected covariates are predictive when applied to independent observations. Others, such as Breiman (1992) use the bootstrap and related ideas to validate selected models. Such techniques allow one to differentiate true signal from mere noise in the data; however, they can be very computer intensive.

7.2.4 Disproportionate Sampling

Truett, Cornfield and Kannel (1967) showed that a logistic model could be used in conjunction with a linear discriminant function to provide an estimate of the posterior probability $P(D|\boldsymbol{x})$ of membership in the index population of subjects with the positive response or disease given the covariate vector $\boldsymbol{x}$. This posterior probability depends explicitly on the prior probability of the response or the prevalence of the disease in the population. In a general random sample from the population, the sample fraction with the disease provides an unbiased estimate of this prior prevalence; that is, $p_1 = m_1/N$ and $E(p_1) = P(D)$. In some cases, however, one does not have a general random sample but rather two separate disproportionate samples of m_1 of those with the response (D) and m_2 of those without ($\overline{D}$). In this case the score equation for the intercept (7.8) requires that $p_1 = \widehat{\overline{\pi}} = \sum_i \widehat{\pi}_i/N$, where $E(p_1) \neq P(D)$. Therefore, the estimated posterior probabilities obtained from the estimated model coefficients are biased.

Anderson (1972), however, shows that unbiased estimates of the posterior probabilities are readily obtained by subtracting a constant from the intercept that yields a model of the form

$$\log\left[\frac{\widehat{\pi}_i}{1-\widehat{\pi}_i}\right] = \boldsymbol{x}_i'\widehat{\boldsymbol{\theta}} - \log\left(\frac{P(\overline{D})}{P(D)}\right) . \tag{7.49}$$

When applied to the linear predictor in (7.16) this also yields unbiased estimates of the confidence limits for the posterior probabilities. This modification of the model, however, has no effect on the estimates of the coefficients, the estimated information matrix, or any of the other characteristics of the model.

7.2.5 Unmatched Case Control Study

Previously in Section 5.1.1 we showed that the retrospective odds ratio of exposure given disease derived from the unmatched retrospectively sampled case-control 2×2 table is an estimate of the population prospective odds ratio of disease given exposure. This suggests that logistic regression can be applied to the analysis of such studies to estimate the covariate effects on the retrospective odds ratio that can be interpreted as prospective odds ratios. Although cases (D) and controls $(\overline{D})$ are almost always disproportionately sampled in such studies, nevertheless, from the preceding results, allowance for the true prevalence of cases and controls serves only to add a constant to the intercept or linear predictor, affecting nothing else. See also Breslow and Day (1980).

Example 7.5 *Ischemic Heart Disease*
Dick and Stone (1973) present an unmatched case-control study comparing 146 men with ischemic heart disease (the cases) to 283 controls selected from the general population of males within the same age range as the cases (30–69 years). The study assessed the influence of hyperlipidemia (HL) versus not, smoking at least 15 cigarettes/day (SM) versus not, and hypertension $(HT$, diastolic blood pressure $\geq$ 95 mm Hg). Each covariate is coded as a binary indicator variable, 1 if yes, 0 if no. The data are

HL	*SM*	*HT*	*# Cases*	*# Controls*
0	0	0	15	82
0	0	1	10	37
0	1	0	39	81
0	1	1	23	28
1	0	0	18	16
1	0	1	7	12
1	1	0	19	19
1	1	1	15	8

(reproduced with permission). The logistic regression model estimates of the parameter, $S.E.$, odds ratio and 95% confidence limits $(OR_L$ and $OR_U)$ for each effect are

Effect	$\widehat{\beta}$	$S.E.$	OR	OR_L	OR_U
Intercept	−1.5312	0.2048			
Hyperlipidemia (HL)	1.0625	0.2316	2.894	1.838	4.556
Smoking (SM)	0.7873	0.2185	2.198	1.432	3.372
Hypertension (HT)	0.3262	0.2242	1.386	0.893	2.150

In such a model, the intercept has no direct prospective interpretation. Each of the estimated coefficients, however, provides an estimate of the log retrospective and the log prospective odds ratio for each risk factor adjusting for the other risk factors in the model. Hyperlipidemia has a nearly threefold greater odds of ischemic heart disease than not, and smoking has over a twofold greater odds than not. Hypertension produces a modest increase in the odds.

7.3 TESTS OF SIGNIFICANCE

7.3.1 Likelihood Ratio Tests

7.3.1.1 Model Test In most applications there is no reason to conduct a test of significance of the value of the intercept α that, in effect, describes the overall mean or background odds against which the effects of covariates are assessed. Rather we wish to conduct a global hypothesis test of the overall model covariate coefficients H_0: $\boldsymbol{\beta} = \mathbf{0}$ against H_1: $\boldsymbol{\beta} \neq \mathbf{0}$ for $\boldsymbol{\beta} = (\beta_1 \ \dots \ \beta_p)^T$ given α in the model. Expressed in terms of the hypothesized true likelihood, under the null hypothesis the true model is the null model or H_0: $L = L(\alpha)$ and that under the alternative is H_1: $L = L(\alpha, \boldsymbol{\beta})$. Therefore, as shown in Section A.7.2 of the Appendix, the likelihood ratio test is

$$X^2_{\boldsymbol{\beta}} = -2 \log \left(\frac{L(\widehat{\alpha})}{L\left(\widehat{\alpha}, \widehat{\boldsymbol{\beta}}\right)} \right) = -2 \log L(\widehat{\alpha}) - \left[-2 \log L\left(\widehat{\alpha}, \widehat{\boldsymbol{\beta}}\right) \right]. \tag{7.50}$$

Under H_0, $X^2_{\boldsymbol{\beta}} \sim \chi^2$ on p df. This is the termed the *Likelihood Ratio Model Chi-Square Test.*

For example, in ordinary normal errors multiple regression, from (A.137) it is readily shown that $-2 \log L(\widehat{\alpha})$ is proportional to the total SS(Y), whereas $-2 \log L\left(\widehat{\alpha}, \widehat{\boldsymbol{\beta}}\right)$ is proportional to the Error SS. Therefore, a model likelihood ratio $X^2_{\boldsymbol{\beta}}$ test in a normal errors regression model is proportional to the Regression SS.

7.3.1.2 Test of Model Components Often it is of interest to test a hypothesis concerning specific components of the coefficient vector. In this case the coefficient vector for the *full model* of p coefficients is partitioned into two subsets of q and r coefficients ($p \leq q + r$) such as $\boldsymbol{\beta} = (\boldsymbol{\beta}_q \, / / \, \boldsymbol{\beta}_r)$ and we wish to test H_0: $\boldsymbol{\beta}_r = \mathbf{0}$. Again, it should be emphasized that this hypothesis should be formally expressed as H_0: $\boldsymbol{\beta}_{r|q} = \mathbf{0}$, that is, a hypothesis with respect to the partial contribution of $\boldsymbol{\beta}_r$ to a *reduced model* that contains only the subset $\boldsymbol{\beta}_q$. In this case, the null and alternative hypotheses H_{0r}: $\boldsymbol{\beta}_{r|q} = \mathbf{0}$ and H_{1r}: $\boldsymbol{\beta}_{r|q} \neq \mathbf{0}$ imply

$$\begin{aligned} H_{0r} &: \ L = L(\alpha, \boldsymbol{\beta}_q) \\ H_{1r} &: \ L = L\left(\alpha, \boldsymbol{\beta}_{q|r}, \boldsymbol{\beta}_{r|q}\right) = L(\alpha, \boldsymbol{\beta}). \end{aligned} \tag{7.51}$$

Then as shown in Section A.7.2.2 of the Appendix, the likelihood ratio test is

$$X^2_{\boldsymbol{\beta}_{r|q}} = -2\log\left(\frac{L\left(\widehat{\alpha},\widehat{\boldsymbol{\beta}}_q\right)}{L\left(\widehat{\alpha},\widehat{\boldsymbol{\beta}}_{q|r},\widehat{\boldsymbol{\beta}}_{r|q}\right)}\right) \tag{7.52}$$

$$= \left[-2\log L\left(\widehat{\alpha},\widehat{\boldsymbol{\beta}}_q\right)\right] - \left[-2\log L\left(\widehat{\alpha},\widehat{\boldsymbol{\beta}}\right)\right],$$

which equals the change in $-2\log L$ when $\boldsymbol{\beta}_r$ is added to, or dropped from, the model. Under the reduced model null hypothesis, $X^2_{\boldsymbol{\beta}_{r|q}} \sim \chi^2$ on r df.

In order to compute this statistic both the full and reduced models must be fit. The test can then be obtained as the difference between the likelihood ratio model chi-square statistics $X^2_{\boldsymbol{\beta}_{r|q}} = X^2_{\boldsymbol{\beta}} - X^2_{\boldsymbol{\beta}_q}$, the $-2\log L(\widehat{\alpha})$ canceling from the expressions for each model X^2 to yield (7.52) above.

Likewise, a likelihood ratio test can be computed for an individual component of the parameter vector by fitting the model with and then without that variable included in the model. To conduct this test for each component of the model requires using PROC LOGISTIC to fit p sub-models with each variable, in turn, eliminated from the full model. However, the SAS procedure GENMOD will compute these likelihood ratio tests directly through its TYPE3 option. See Section 7.3.4.

7.3.2 Efficient Scores Test

7.3.2.1 Model Test As shown in Section A.7.3 of the Appendix, the efficient scores test for the model null hypothesis H_0: $\boldsymbol{\beta} = \mathbf{0}$ is based on the score vector $\boldsymbol{U}(\boldsymbol{\theta})$ and the expected information $\boldsymbol{I}(\boldsymbol{\theta})$, each estimated under the tested hypothesis, designated as $\boldsymbol{U}\left(\widehat{\boldsymbol{\theta}}_0\right)$ and $\boldsymbol{I}\left(\widehat{\boldsymbol{\theta}}_0\right)$, respectively. Since $\boldsymbol{\theta} = (\alpha \,//\, \boldsymbol{\beta})$, and the tested hypothesis specifies that $\boldsymbol{\beta} = \mathbf{0}$, then the *MLE* estimated under this hypothesis is the vector $\widehat{\boldsymbol{\theta}}_0 = (\widehat{\alpha}_0 \,//\, \mathbf{0})$. Thus we need only obtain the *MLE* of the intercept under this hypothesis, designated as $\widehat{\alpha}_0$.

Again, let $m_1 = \#(y_i = 1) = \sum_i y_i$ and $m_2 = \#(y_i = 0) = \sum_i (1 - y_i) = N - m_1$. Evaluating (7.8) under H_0: $\boldsymbol{\beta} = 0$ yields

$$[U(\boldsymbol{\theta})_\alpha]_{|\boldsymbol{\beta}=\mathbf{0}} = m_1 - \sum_{i=1}^{N} \frac{e^\alpha}{1+e^\alpha} = m_1 - N\overline{\pi} = 0 \tag{7.53}$$

such that

$$\widehat{\overline{\pi}} = \frac{m_1}{N} \tag{7.54}$$

and

$$\widehat{\alpha}_0 = \log\left(\frac{\widehat{\overline{\pi}}}{1-\widehat{\overline{\pi}}}\right) = \log\left(\frac{m_1}{m_2}\right). \tag{7.55}$$

By definition, the resulting score for the intercept evaluated at the *MLE* is

$$U(\widehat{\boldsymbol{\theta}}_0)_\alpha = [U(\boldsymbol{\theta})_\alpha]_{|\widehat{\alpha}_0, \boldsymbol{\beta}=\mathbf{0}} = 0. \tag{7.56}$$

Given the estimate $\widehat{\alpha}_0$, the score equation for each coefficient as in (7.9) is then evaluated under the tested hypothesis. The score for the jth coefficient is then

$$\begin{aligned} U(\widehat{\boldsymbol{\theta}}_0)_{\beta_j} &= \left[U(\boldsymbol{\theta})_{\beta_j}\right]_{|\widehat{\alpha}_0, \boldsymbol{\beta}=\mathbf{0}} = \sum\nolimits_i x_{ij}\left[y_i - \frac{e^{\widehat{\alpha}_0}}{1+e^{\widehat{\alpha}_0}}\right] \\ &= \sum\nolimits_i x_{ij}\left(y_i - \widehat{\overline{\pi}}\right) = \sum\nolimits_i x_{ij}\left(y_i - \frac{m_1}{N}\right) = m_1\left[\overline{x}_{j(1)} - \overline{x}_j\right], \end{aligned} \tag{7.57}$$

where $\overline{x}_{j(1)}$ is the mean of X_j among the m_1 subjects with a positive response $(y_j = 1)$ and $\overline{x}_j$ is the mean in the total sample of N observations. Thus the score for the jth coefficient is proportional to the observed mean of the covariate among those with the response minus that expected under the null hypothesis. If the covariate is associated with the risk of the response then we expect that $\overline{x}_{j(1)} \neq \overline{x}_j$. For example, if the risk of the response increases as age increases, then we would expect the mean age among those with the response to be greater than the mean age in the total population. On the other hand, if there is no association we expect $\overline{x}_{j(1)}$ to be approximately equal to $\overline{x}_j$. Therefore, the total score vector evaluated under H_0 is

$$\begin{aligned} \boldsymbol{U}\left(\widehat{\boldsymbol{\theta}}_0\right) &= m_1\left[0 \;\left(\overline{x}_{1(1)} - \overline{x}_1\right) \;\ldots\; \left(\overline{x}_{p(1)} - \overline{x}_p\right)\right]^T \\ &= m_1\left[0 \,\|\, \left(\overline{\boldsymbol{x}}_{(1)} - \overline{\boldsymbol{x}}\right)^T\right]^T. \end{aligned} \tag{7.58}$$

The information matrix estimated under H_0, $\boldsymbol{I}\left(\widehat{\boldsymbol{\theta}}_0\right) = \boldsymbol{I}(\boldsymbol{\theta})_{|\widehat{\alpha}_0, \boldsymbol{\beta}=\mathbf{0}}$, then has elements obtained from (7.11)–(7.15) that are

$$\begin{aligned} \boldsymbol{I}\left(\widehat{\boldsymbol{\theta}}_0\right)_\alpha &= \sum\nolimits_i \left[\frac{e^\alpha}{(1+e^\alpha)^2}\right] = N\overline{\pi}\left(1-\overline{\pi}\right) \\ \boldsymbol{I}\left(\widehat{\boldsymbol{\theta}}_0\right)_{\alpha,\beta_j} &= \sum\nolimits_i x_{ij}\frac{e^\alpha}{(1+e^\alpha)^2} = N\overline{x}_j\overline{\pi}\left(1-\overline{\pi}\right) \\ \boldsymbol{I}\left(\widehat{\boldsymbol{\theta}}_0\right)_{\beta_j,\beta_k} &= \overline{\pi}\left(1-\overline{\pi}\right)\sum\nolimits_i x_{ij}x_{ik} \\ \boldsymbol{I}\left(\widehat{\boldsymbol{\theta}}_0\right)_{\beta_j} &= \overline{\pi}\left(1-\overline{\pi}\right)\sum\nolimits_i x_{ij}^2. \end{aligned} \tag{7.59}$$

Therefore, the estimated information matrix under H_0 is

$$\boldsymbol{I}\left(\widehat{\boldsymbol{\theta}}_0\right) = \widehat{\overline{\pi}}\left(1-\widehat{\overline{\pi}}\right)\left(\boldsymbol{x}'\boldsymbol{x}\right), \tag{7.60}$$

where $\boldsymbol{x}$ is the $N \times (p+1)$ design matrix, the first column being the unit vector.

For given $\boldsymbol{\theta}_0$, since $\boldsymbol{U}\left(\widehat{\boldsymbol{\theta}}_0\right) \overset{d}{\approx} \mathcal{N}\left[0,\ \boldsymbol{I}\left(\boldsymbol{\theta}_0\right)\right]$, then the model score test is the quadratic form

$$X_p^2 = \boldsymbol{U}\left(\widehat{\boldsymbol{\theta}}_0\right)' \boldsymbol{I}\left(\widehat{\boldsymbol{\theta}}_0\right)^{-1} \boldsymbol{U}\left(\widehat{\boldsymbol{\theta}}_0\right) \tag{7.61}$$
$$= m_1^2 \left[0 \parallel \left(\overline{\boldsymbol{x}}_{(1)} - \overline{\boldsymbol{x}}\right)^T\right] \frac{\left(\boldsymbol{x}'\boldsymbol{x}\right)^{-1}}{\widehat{\overline{\pi}}\left(1 - \widehat{\overline{\pi}}\right)} \left[0 \parallel \left(\overline{\boldsymbol{x}}_{(1)} - \overline{\boldsymbol{x}}\right)^T\right]^T$$

which is asymptotically distributed as chi-square on p *df* under the model null hypothesis H_0: $\boldsymbol{\beta} = \mathbf{0}$.

Asymptotically, a score test is fully efficient as shown in the Appendix, Section A.7.3. Computationally the test can be obtained without actually fitting the model, that is, without obtaining the *MLEs* of the parameter vector and the associated estimated information. Thus the score test can be computed for a degenerate model or one for which the Newton-Raphson iteration does not converge. However, when a convergent solution exists, the Likelihood Ratio Test, in general, is preferred. When the model is degenerate, then it implies either that there is a linear dependency among the covariates, or that a covariate has nearly constant values among those with a positive or negative response. In such cases the model should be refit after the suspect covariate(s) have been removed, in which case the score test from the degenerate model becomes irrelevant.

7.3.2.2 Test of Model Components Similarly, it would be possible to conduct a score test of specific components of the coefficient vector. However, the score test in this instance offers few advantages over the likelihood ratio test, especially considering that the latter is readily obtained by fitting the full and reduced models of interest. Likewise, score tests of the individual components of the parameter vector are rarely used, because this would require evaluation of the score vector and the information matrix under the null hypothesis for each covariate, in turn.

7.3.3 Wald Tests

As described in Section A.7.1 of the Appendix, the Wald model test is obtained as a quadratic form in the parameter estimates and the estimated covariance matrix obtained from the inverse of the estimated information. See (A.147). Although the Wald test is readily computed, the likelihood ratio test, in general, is preferred.

Similarly, a Wald test can be computed for the parameters of a sub-model or for individual coefficients in the model. To test the hypothesis H_0: $\beta_j = 0$ for the jth coefficient, the Wald test is readily computed as

$$X_j^2 = \frac{\widehat{\beta}_j^2}{\widehat{V}(\widehat{\beta}_j)}, \tag{7.62}$$

where $\widehat{V}(\widehat{\beta}_j) = \left[\boldsymbol{I}(\widehat{\boldsymbol{\theta}})^{-1}\right]_{\beta_j}$ is the corresponding diagonal element of the inverse information matrix. Since Wald tests are based on the variance of the coefficient estimate obtained under the alternative hypothesis, then the significance of a Wald test can be ascertained from the 95% confidence limits. Thus these limits are also referred to as Wald confidence limits.

For example, for the case-control study in Example 7.5, the 95% confidence limits for the odds ratio associated with hyperlipidemia and with smoking each do not include 1.0, and thus the corresponding Wald test for each are significant at the 0.05 significance level (two-sided). However, that for hypertension does include 1.0 and thus the Wald test is not significant.

Hauck and Donner (1977) and Vaeth (1985) have shown that the Wald test may be highly anomalous in special, albeit unrealistic, cases. In particular, for fixed values of all the coefficients except the last $(\beta_1, \ldots, \beta_{p-1})$, then as the value of β_p increases away from the null, the Wald test may, in fact, decrease approaching zero. Mantel (1987) shows that the problem is related to the fact that the Wald test employs the variance estimated under the model (under the alternative) rather than under the null hypothesis. Thus the score or likelihood ratio tests of the individual parameters or for a submodel are also preferred over the Wald test.

Example 7.6 *DCCT Nephropathy Data*
Table 7.6 of Example 7.4 presents the likelihood ratio and score tests for the overall model. The likelihood ratio test is the difference between the null (intercept only) and full model values of $-2\log(L)$, or $191.215 - 155.336 = 35.879$ on 5 *df*, $p < 0.0001$.

The model score test is based on the score vector and inverse information matrix evaluated under the null hypothesis. For these data, the score vector evaluated from (7.58) is

$$\boldsymbol{U}\left(\widehat{\boldsymbol{\theta}}_0\right) = \boldsymbol{U}\left[\widehat{\alpha}, \boldsymbol{\beta}\right]_{|\boldsymbol{\beta}=0} = (0 \quad -10.73 \quad 30.01 \quad -7.133 \quad 78.33 \quad -4.047)^T$$

and the estimated information evaluated under the null hypothesis from (7.60) is $\boldsymbol{I}(\widehat{\boldsymbol{\theta}}_0) =$

31.744186	16.425771	294.00468	299.33968	3692.6609	14.395619
16.425771	16.425771	151.85902	154.53759	1907.235	7.3823688
294.00468	151.85902	2791.717	2753.4845	34128.414	138.08536
299.33968	154.53759	2753.4845	3174.566	34875.341	138.92695
3692.6609	1907.235	34128.414	34875.341	433238.01	1623.9366
14.395619	7.3823688	138.08536	138.92695	1623.9366	14.395619

The resulting model score test from (7.61) is 33.773 on 5 *df* with $p < 0.0001$.

Table 7.6 presents the Wald tests for each coefficient in the model. However, these tests employ the estimated variance of the coefficient estimates obtained under the general alternative hypothesis rather than under the covariate coefficient-specific null hypothesis. Thus likelihood ratio tests of the coefficients are preferred to these Wald tests. However, PROC LOGISTIC does not compute these likelihood ratio tests for the individual coefficients. Rather, when using PROC LOGISTIC it is necessary to compute these by hand by successively fitting a model without each covariate, in turn, which is compared to the full model with all covariates. For example, to compute the likelihood ratio test for the covariate *int*, the model without *int* is fit yielding a model likelihood ratio test $X^2 = 20.298$ on 4 *df* (not shown). Compared to the full model likelihood ratio test in Table 7.6, the resulting likelihood ratio test for *int* is 35.88 − 20.30 = 15.58 with $p < 0.0001$ on 1 df.

Alternately, PROC GENMOD can be used to fit the model and to directly compute the likelihood ratio test for each covariate.

7.3.4 Type III Tests in SAS PROC GENMOD

As described in Section 7.1.5 above, the logistic regression model is one of a family of possible regression models that could be fit to such data. The SAS procedure GENMOD can be used to fit various members of this family. The family of generalized linear models ($GLMs$) is described in Section A.10 of the Appendix. Also, PROC GENMOD will readily compute the likelihood ratio tests for the individual coefficients in the model, called Type III tests of covariate effects. This terminology is used to refer to the test of the partial effect of a covariate in the SAS PROC GLM that fits normal errors regression models with design (class) effects and quantitative covariate effects.

To fit a logistic regression model, as opposed to other possible models in the *GLM* family, the model uses the logit link in conjunction with a binomial variance, which equals the Bernoulli variance since each observation is a separate individual. GENMOD also provides a TYPE 3 option which specifies that the likelihood ratio tests be computed in addition to the Wald tests.

In addition, for models with categorical covariates, GENMOD allows a *class* statement that automatically generates the appropriate number of binary contrasts in the design matrix. For example, for the stratified analysis of the ulcer clinical trial data in Example 7.3, the statement "`class stratum;`" can be used with the model statement "`model=stratum group;`" that will automatically include the 2 *df* stratum effect in the model.

In GLM terminology, the deviance equals the difference in $-2\log(L)$ for the present model versus a model that fits the data perfectly (see Section A.10.3 of the Appendix). For a logistic regression model with a binary response, but not binomial regression, the saturated model log likelihood is zero so that in this special case $deviance = -2\log(L)$. The scaled deviance and Pearson chi-square are also described in the Appendix.

Example 7.7 *DCCT Nephropathy Data*

The following SAS code fits a logistic regression model to the DCCT data using PROC GENMOD.

```
proc genmod data = renal;
class int female;
model micro24 = int hbael yearsdm sbp female
  / dist=binomial link=logit type3;
title2 'logistic regression model fit through GENMOD';
run;
```

The resulting output from the fitted model is presented in Table 7.7 (extraneous information deleted).

In Table 7.7 the `class` statement provides the estimates of the regression coefficients for each category versus the reference category, zero in these instances. The estimates and Wald tests are followed by the likelihood ratio (`type3`) tests for each coefficient. Although the Wald test is occasionally greater than the likelihood ratio test, in general the latter has a type I error probability that is closer to the desired level and is more efficient.

7.3.5 Robust Inferences

All regression models that are based on a full likelihood specification, such as the GLM family in general, and the logistic regression model in particular, explicitly assume that the variance–covariance structure specified by the model is correct and applies to the population from which the observations were obtained. In the logistic model, conditional on the covariate vector $\boldsymbol{x}$, the assumption is that $V(y|\boldsymbol{x}) = \pi(\boldsymbol{x})[1-\pi(\boldsymbol{x})]$, where $E(y|\boldsymbol{x}) = \pi(\boldsymbol{x})$ is the conditional expectation (probability). In some cases, however, the variation in the data may be greater or less than that assumed by the model; that is, there may be over- or under-dispersion. In this case, the properties of confidence limits and test statistics can be improved by using an estimate of the variance–covariance matrix of the estimates that is robust to departures from the model variance assumptions.

One simple approach is to fit an over- (under-)dispersed GLM regression model by adopting a quasi-likelihood (see Section A.10.4 of the Appendix) with an additional dispersion parameter. This approach is widely used in Poisson regression models of count data and thus is described in Chapter 8. However, it could also be employed with logistic (binomial) regression. As described in Section A.10.3 of the Appendix, the ratio of the Pearson chi-square to its degrees of freedom is an estimate of the degree of departure from the model variance assumption, a value substantially greater than one indicating over-dispersion, substantially less indicating under-dispersion. When over-dispersion exists, the model under-estimates the degree of variation in the data and thus the variance–covariance matrix of the coefficients is under-estimated so that Wald tests are inflated and confidence intervals are too narrow. The opposite occurs with under-dispersion.

For example, in the above analysis of the DCCT nephropathy data in Table 7.7, the $chi\text{-}square/df = 1.098$ which suggests that the actual variance–covariance matrix of the estimates may be about 10% greater than that estimated from the model

Table 7.7 Logistic Regression Analysis of DCCT Data Using PROC GENMOD

The GENMOD Procedure

Criteria For Assessing Goodness Of Fit

Criterion	DF	Value	Value/DF
Deviance	166	155.3362	0.9358
Scaled Deviance	166	155.3362	0.9358
Pearson Chi-Square	166	182.2750	1.0980
Scaled Pearson X2	166	182.2750	1.0980
Log Likelihood	.	-77.6681	.

Analysis Of Parameter Estimates

Parameter		DF	Estimate	Std Err	ChiSquare	Pr>Chi
INTERCEPT		1	-10.7543	3.0731	12.2463	0.0005
INT	0	1	1.5831	0.4267	13.7626	0.0002
INT	1	0	0.0000	0.0000	.	.
HBAEL		1	0.5675	0.1449	15.3429	0.0001
YEARSDM		1	0.0096	0.0636	0.0228	0.8799
SBP		1	0.0233	0.0208	1.2579	0.2620
FEMALE	0	1	0.8905	0.4473	3.9641	0.0465
FEMALE	1	0	0.0000	0.0000	.	.
SCALE		0	1.0000	0.0000	.	.

NOTE: The scale parameter was held fixed.

LR Statistics For Type 3 Analysis

Source	DF	ChiSquare	Pr>Chi
INT	1	15.5803	0.0001
HBAEL	1	17.6701	0.0001
YEARSDM	1	0.0229	0.8798
SBP	1	1.2726	0.2593
FEMALE	1	4.1625	0.0413

based on the inverse estimated information matrix. This degree of over-dispersion is well within the range one might expect under random variation with 95% confidence when the model assumptions actually apply, $1 \pm 0.215 = 2.77/\sqrt{166}$ (see Section A.10.3). However, if substantial departures are suggested, and there are no obvious deficiencies or errors in the model specification, then this approach can be used to estimate the over-dispersion scale parameter and to adjust the estimates of the variances of the model coefficients. Section 8.3.2 describes the application of this approach to Poisson regression using PROC GENMOD.

Another approach to adjust for departures from the model variance assumption is to employ the *robust information sandwich estimator* described in the Section A.9 of the Appendix. The information sandwich provides a consistent estimate of the variance of the estimates for any model where the first moment specification (the structural component) is correct, but where the second moment specification (the error variance–covariance structure) may not be correctly specified.

Let $\boldsymbol{X}$ denote the $n \times (p+1)$ design matrix where the ith row is the covariate vector $\boldsymbol{x}_i^T$ for the ith subject augmented by the constant (unity) for the intercept. Also, let $\boldsymbol{\Gamma} = diag[\pi_i(1-\pi_i)]$, where $\pi_i = E(y_i|\boldsymbol{x}_i)$ is the conditional probability expressed as a logistic function of the covariates. Then, from (7.11–7.15), the expected information can be expressed as

$$\boldsymbol{I}(\boldsymbol{\theta}) = \boldsymbol{X}'\boldsymbol{\Gamma}\boldsymbol{X}. \tag{7.63}$$

This is the covariance matrix of the score vector when the specified logistic model is assumed to be correct.

However, when the model-specified covariance matrix is not correct, let $\boldsymbol{\Sigma}_\varepsilon = diag[E(y_i - \pi_i)^2]$ refer to the true covariance matrix of the errors. Then, from (7.8–7.9), it follows that the true covariance matrix of the score vector is

$$\boldsymbol{J}(\boldsymbol{\theta}) = \sum_i E[\boldsymbol{U}_i(\boldsymbol{\theta})\boldsymbol{U}_i(\boldsymbol{\theta})'] = \boldsymbol{X}'\boldsymbol{\Sigma}_\varepsilon\boldsymbol{X}. \tag{7.64}$$

From (A.221) of the Appendix, the expression for the robust information sandwich covariance matrix of the coefficient estimates is then

$$\boldsymbol{\Sigma}_R(\widehat{\boldsymbol{\theta}}) = \boldsymbol{I}(\boldsymbol{\theta})^{-1}\boldsymbol{J}(\boldsymbol{\theta})\boldsymbol{I}(\boldsymbol{\theta})^{-1} = (\boldsymbol{X}'\boldsymbol{\Gamma}\boldsymbol{X})^{-1}(\boldsymbol{X}'\boldsymbol{\Sigma}_\varepsilon\boldsymbol{X})(\boldsymbol{X}'\boldsymbol{\Gamma}\boldsymbol{X})^{-1}. \tag{7.65}$$

This matrix can be consistently estimated as

$$\widehat{\boldsymbol{\Sigma}}_R(\widehat{\boldsymbol{\theta}}) = \boldsymbol{I}(\widehat{\boldsymbol{\theta}})^{-1}\widehat{\boldsymbol{J}}(\widehat{\boldsymbol{\theta}})\boldsymbol{I}(\widehat{\boldsymbol{\theta}})^{-1}, \tag{7.66}$$

where $\boldsymbol{I}(\widehat{\boldsymbol{\theta}}) = \boldsymbol{X}'\widehat{\boldsymbol{\Gamma}}\boldsymbol{X}$, $\widehat{\boldsymbol{\Gamma}} = diag[\widehat{\pi}_i(1-\widehat{\pi}_i)]$, $\widehat{\boldsymbol{J}}(\widehat{\boldsymbol{\theta}}) = \sum_i[\boldsymbol{U}_i(\widehat{\boldsymbol{\theta}})\boldsymbol{U}_i(\widehat{\boldsymbol{\theta}})'] = \left(\boldsymbol{X}'\widehat{\boldsymbol{\Sigma}}_\varepsilon\boldsymbol{X}\right)$ and $\widehat{\boldsymbol{\Sigma}}_\varepsilon = diag[(y_i - \widehat{\pi}_i)^2]$. The "bread" of the information sandwich estimate is the model estimated inverse information, or the estimated covariance matrix of the coefficient estimates. The "meat" of the sandwich is the *empirical* estimate of the observed information based on the empirical estimate of the error variance of the observations.

This robust estimate then can be used to construct confidence intervals and to compute Wald tests of significance. A robust score test of the model, or of model components, can also be constructed based on the model estimated under the tested hypothesis. The model score test addresses the null hypothesis H_0: $\boldsymbol{\beta} = \mathbf{0}$. The empirical estimate of the covariance matrix of the score vector is then obtained as

$$\widehat{\boldsymbol{J}}(\widehat{\boldsymbol{\theta}}_0) = \sum_i \boldsymbol{U}_i(\widehat{\boldsymbol{\theta}}_0)\boldsymbol{U}_i(\widehat{\boldsymbol{\theta}}_0)' = \mathbf{X}'\widehat{\Sigma}_{\varepsilon 0}\mathbf{X}, \tag{7.67}$$

where

$$\boldsymbol{U}_i(\widehat{\boldsymbol{\theta}}_0) = [U_i(\widehat{\boldsymbol{\theta}}_0)_\alpha \parallel \boldsymbol{U}_i(\widehat{\boldsymbol{\theta}}_0)_{\boldsymbol{\beta}}{}^T]^T \tag{7.68}$$

is the score vector with the parameters estimated ($\widehat{\alpha}_0$) or evaluated ($\boldsymbol{\beta}$) under the tested hypothesis as in Section 7.3.2. From (7.8) and (7.53) it follows that

$$U_i(\widehat{\boldsymbol{\theta}}_0)_\alpha = y_i - \overline{\pi}, \tag{7.69}$$

and from (7.9) and (7.57) then

$$U_i(\widehat{\boldsymbol{\theta}}_0)_{\beta_j} = x_{ij}(y_i - \overline{\pi}) \tag{7.70}$$

for the jth coefficient. Then the total score vector $\boldsymbol{U}(\widehat{\boldsymbol{\theta}}_0)$ evaluated under the tested hypothesis is as presented in (7.58). Note that while the total score $U(\widehat{\boldsymbol{\theta}})_\alpha = 0$, the individual terms in (7.68) are not all zero. The robust model score test then is

$$X^2 = \boldsymbol{U}(\widehat{\boldsymbol{\theta}}_0)'\widehat{\boldsymbol{J}}(\widehat{\boldsymbol{\theta}}_0)^{-1}\boldsymbol{U}(\widehat{\boldsymbol{\theta}}_0), \tag{7.71}$$

which is asymptotically distributed as chi-square on p df.

Example 7.8 *DCCT Nephropathy Data*
The analysis of the DCCT prevalence of microalbuminuria in Table 7.6 contains the estimated covariance matrix of the coefficient estimates, $\widehat{\Sigma}(\widehat{\boldsymbol{\theta}}) = \boldsymbol{I}(\widehat{\boldsymbol{\theta}})^{-1}$. These and the estimated coefficients can be output to a data set by PROC LOGISTIC using the OUTEST option. The program will also create a data set with the model estimated probabilities, the $\{\widehat{\pi}_i\}$. Using a routine written in IML, additional computations may then be performed. The matrix of score vector outer products is $\widehat{\boldsymbol{J}}(\widehat{\boldsymbol{\theta}}) = \sum_i [\boldsymbol{U}_i(\widehat{\boldsymbol{\theta}})\boldsymbol{U}_i(\widehat{\boldsymbol{\theta}})'] = \mathbf{X}'\widehat{\Sigma}_\varepsilon\mathbf{X} =$

24.555087	9.1094437	237.04689	235.26848	2906.3223	9.5713568
9.1094437	9.1094437	88.132987	90.047355	1058.265	3.8137591
237.04689	88.132987	2351.74	2254.7264	27920.217	98.190859
235.26848	90.047355	2254.7264	2513.004	27998.372	92.930592
2906.3223	1058.265	27920.217	27998.372	347070.83	1105.9189
9.5713568	3.8137591	98.190859	92.930592	1105.9189	9.5713568

This yields the robust information sandwich estimate $\widehat{\Sigma}_R(\widehat{\boldsymbol{\theta}}) =$

10.664905	0.1689527	-0.267092	-0.05978	-0.063752	-0.374126
0.1689527	0.1965141	-0.02177	0.0012571	-0.000466	0.0515362
-0.267092	-0.02177	0.0243438	0.0021586	0.0002403	-0.017867
-0.05978	0.0012571	0.0021586	0.0041104	0.0000171	-0.001376
-0.063752	-0.000466	0.0002403	0.0000171	0.0005139	0.0039347
-0.374126	0.0515362	-0.017867	-0.001376	0.0039347	0.2162162

This can be used to compute large sample 95% confidence intervals for the model estimates and a Wald test for significance of the individual coefficients. The following are the vectors of estimated standard errors for the respective coefficients in the model, the resulting upper and lower 95% confidence limits for the odds ratios, the Wald test chi-square value and the *P*-value

		95% Confidence Limits			
Effect	$S.E.(\widehat{\theta})$	*Lower*	*Upper*	*Wald* X^2	$p \leq$
Intercept	3.2657	−14.6810	−1.8800	6.429	0.0113
Intensive group	0.4433	−2.4520	−0.7143	12.750	0.0004
HbA_{1c}	0.1560	0.2617	0.8733	13.230	0.0003
Years duration	0.0641	−0.1160	0.1353	0.022	0.8809
Systolic B.P.	0.0227	−0.0211	0.0677	1.055	0.3043
Female	0.4650	−1.8019	0.0208	3.668	0.0555

For some of the coefficients, particularly intensive treatment group and HbA_{1c}, the robust standard errors are slightly larger than those estimated from the model presented in Table 7.6, so that the confidence limits for the odds ratios are slightly wider and the Wald tests slightly smaller.

The above computations are also provided by PROC GENMOD. This is described in Section 8.4.2 in the context of Poisson regression.

This robust variance estimate can also be used to compute an overall model Wald test of H_0: $\boldsymbol{\beta} = \mathbf{0}$. The value of the test is $X^2 = 21.34$ on 5 *df* with $p \leq 0.0007$. However, it is preferable that a score test be used that employs a robust estimate of the variance under the tested hypothesis. In this case, the score vector $\boldsymbol{U}(\widehat{\boldsymbol{\theta}}_0)$ evaluated under H_0 from (7.58) is presented in Example 7.6 and the robust estimate of the covariance (information) matrix from (7.67) is $\widehat{\boldsymbol{J}}(\widehat{\boldsymbol{\theta}}_0) =$

31.744186	10.934694	309.35822	295.69036	3732.7345	12.325311
10.934694	10.934694	105.21842	103.05156	1263.9045	4.4315846
309.35822	105.21842	3092.0635	2845.4577	36218.088	126.01731
295.69036	103.05156	2845.4577	3058.0105	34860.286	113.10115
3732.7345	1263.9045	36218.088	34860.286	442423.16	1407.7262
12.325311	4.4315846	126.01731	113.10115	1407.7262	12.325311

The resulting robust score test is $X^2 = 33.64$ on 5 *df* with $p \leq 0.0001$.

Overall, for this model, inferences based on the robust information sandwich are nearly identical to those based on the simple model-based estimates of the expected information. There is every indication that the model-based inferences are indeed appropriate. However, an example is presented in Chapter 8 where this is not the case.

7.3.6 Power and Sample Size

The power of a Wald test for the overall model or for an individual covariate in a logistic regression model analysis is a function of the non-centrality parameter of the test statistic. For a test of the coefficients in a model with a specific coefficient vector X, Whittemore (1981) shows that the non-centrality parameter of the Wald test is a function of the moment generating function of the joint multivariate distribution of X. Because the expression for this mgf is, in general, unknown, it is not practical to describe the power function for a set of covariates, especially including one or more quantitative covariates.

However, the power of the Wald test is readily obtained when all covariates are categorical (*cf.* Rochon, 1989). In this case, the binomial logit model of Section 7.1.2 specifies that the probability of the response within each sub-population or cell characterized by the covariate vector $\boldsymbol{x}$ is the logistic function $\pi = e^{\alpha+\boldsymbol{x}'\boldsymbol{\beta}}/(1+e^{\alpha+\boldsymbol{x}'\boldsymbol{\beta}})$ with coefficients $\boldsymbol{\beta} = (\beta_1 \ \ldots \ \beta_p)^T$ and intercept α. Assume that the covariate vector generates K distinct cells. For example, for S binary covariates, $K = 2^S$ and $p \leq K$. This yields the $K \times (p+1)$ design matrix $\boldsymbol{X}$ where the ith row is the covariate vector $\boldsymbol{x}_i^T$ for the ith cell augmented by the constant (unity) for the intercept. Then assume that the ith cell has expected sample size $E(n_i) = N\zeta_i$, N being the total sample size and $\{\zeta_i\}$ being the sample fractions. Within the ith cell, the observed proportion with the index characteristic is p_i. Since $V\left[\log\left(\frac{p}{1-p}\right)\right] = [n\pi(1-\pi)]^{-1}$ the parameters can be estimated through weighted least squares such that

$$\widehat{\boldsymbol{\theta}} = (\mathbf{X}'\Omega^{-1}\mathbf{X})^{-1}\mathbf{X}'\Omega^{-1}Y \tag{7.72}$$

and

$$V(\widehat{\boldsymbol{\theta}}) = \boldsymbol{\Sigma}_{\widehat{\theta}} = (\mathbf{X}'\Omega^{-1}\mathbf{X})^{-1}/N \ , \tag{7.73}$$

where $\Omega = diag\{1/[\zeta_i\pi_i(1-\pi_i)]\}$ and $\Omega^{-1} = diag[\zeta_i\pi_i(1-\pi_i)]$. Note that Ω is directly obtained as a function of $\boldsymbol{x}_i$ and $\boldsymbol{\theta}$, $i = 1, \ldots, K$.

From Section A.7.1 of the Appendix, the Wald test for a linear hypothesis of the form H_0: $\boldsymbol{C}'\boldsymbol{\theta} = 0$ for a $r \times (p+1)$ matrix $\boldsymbol{C}'$ of row rank r is the form

$$X^2 = \widehat{\boldsymbol{\theta}}'\boldsymbol{C}(\boldsymbol{C}'\boldsymbol{\Sigma}_{\widehat{\theta}}\boldsymbol{C})^{-1}\boldsymbol{C}'\widehat{\boldsymbol{\theta}} \tag{7.74}$$

on r df. Thus the non-centrality parameter or expected value of the test statistic under the alternative hypothesis is

$$\begin{aligned} \psi^2 &= \boldsymbol{\theta}'\boldsymbol{C}(\boldsymbol{C}'\boldsymbol{\Sigma}_{\widehat{\theta}}\boldsymbol{C})^{-1}\boldsymbol{C}'\boldsymbol{\theta} \\ &= N\boldsymbol{\theta}'\boldsymbol{C}(\boldsymbol{C}'(\mathbf{X}'\Omega^{-1}\mathbf{X})^{-1}\boldsymbol{C})^{-1}\boldsymbol{C}'\boldsymbol{\theta} = NK^2 \ , \end{aligned} \tag{7.75}$$

where K^2 is the non-centrality factor. Since ψ^2 is a function of $\boldsymbol{C}$, $\boldsymbol{X}$ and $\boldsymbol{\theta}$, then sample size and power can readily be determined from the non-central chi-square distribution as described in Section 3.4.

Finally, in the event that over-dispersion is thought to apply, the above is readily modified. All that is required is to inflate the values of the covariance matrix $\Sigma_{\widehat{\theta}}$ by the desired factor that would correspond to the value of the over-dispersion scale parameter in an over-dispersed quasi likelihood model. Alternately, if power is being computed based on observed results, then the model may be fit using the information sandwich estimate of the variances that may then be used to compute the non-central factor and the power of the test.

Example 7.9 *Stratified Analysis*
Consider the determination of the sample size for a study comparing two treatments within three strata with the design matrix

$$\boldsymbol{X} = \begin{bmatrix} 1 & 0 & 0 & 0 \\ 1 & 0 & 0 & 1 \\ 1 & 1 & 0 & 0 \\ 1 & 1 & 0 & 1 \\ 1 & 0 & 1 & 0 \\ 1 & 0 & 1 & 1 \end{bmatrix}$$

where the corresponding coefficients $\boldsymbol{\theta} = (\alpha\, \beta_1\, \beta_2\, \beta_3)^T$ represent the effects for the intercept, stratum 2 versus 1, stratum 3 versus 1, and the active (1) versus the control (0) treatment, respectively. Assume that subjects are assigned at random to each treatment such that there are equal sample sizes for each treatment group within each stratum. Also, assume that the three strata comprise 15%, 50% and 35% of the total population, respectively. Thus the expected cell fractions are specified to be $\{\zeta_i\} = (0.075, 0.075, 0.25, 0.25, 0.175, 0.175)$. Under the alternative hypothesis we specify that the probabilities for each cell are $\{\pi_i\} = (0.75, 0.923, 0.70, 0.882, 0.65, 0.770)$ for which the associated parameter vector is $\boldsymbol{\theta} = (1.2412, -0.3158, -0.7680, 0.9390)^T$.

These values of $\boldsymbol{\theta}$ correspond to a model where the adjusted (common) odds ratio for active treatment versus control is $e^{\beta_3} = e^{0.9390} = 2.557$. This, in fact, is a simplification because the actual odds ratios for the active versus control treatment within each stratum are 4.0, 3.2 and 1.8, respectively. Thus sample size and power are evaluated under a fixed effects model with some heterogeneity, which provides a conservative assessment relative to the model with a common odds ratio for all strata.

Substituting the values of $\{\pi_i\}$ and $\{\zeta_i\}$ into (7.73) yields the covariance matrix of the estimates. For the test of H_0: $\beta_3 = 0$, with vector $\boldsymbol{C}' = (0\ 0\ 0\ 1)$, the resulting value of the non-centrality factor for the 1 *df* test is $K^2 = 0.03484$. For a 1 *df* Wald chi-square test at $\alpha = 0.05$ with a critical value of 3.84146, the SAS function CNONCT(3.84146, 1, 0.1) yields the non-centrality parameter $\psi^2(0.05, 0.10, 1) = 10.5074$ that provides power $= 0.90$. Thus the total N required to provide 90% power for this test is $N = 10.5074/0.03484 = 302$ (rounded up from 301.6).

Conversely, for the same model, a sample size of only $N = 200$ provides a non-centrality parameter $\psi^2(0.05, \beta, 1) = NK^2 = (200 \times 0.0348427) = 6.969$. The SAS function PROBCHI(3.841, 1, 6.969) then yields $\beta = 0.248$ and power of 0.752.

For this and similar examples, one approach to specifying the model under the alternative is to first specify the $\{\pi_i\}$ in terms of the odds ratios under the alternative hypothesis that one wishes to detect. For the above example, the values of $\{\pi_i\}$ were obtained by first specifying the values in the control group within each stratum, (0.75, 0.70 and 0.65), and then specifying the odds ratio within each stratum (4.0, 3.2 and 1.8), which then yields the probabilities in the active treatment group within each stratum (0.923, 0.882 and 0.770). The parameter vector $\boldsymbol{\theta}$ was then obtained by generating a set of cell frequencies summing to a large number, say 10,000, the frequencies being proportional to the corresponding probabilities, and then fitting the logistic regression model to obtain the corresponding values of $\boldsymbol{\theta}$.

Alternatively, Rochon (1989) first uses the vector of probabilities $\{\pi_i\}$ and cell sample fractions $\{\zeta_i\}$ to define the covariance matrix Ω. The design matrix for the assumed model is then employed in (7.72) to obtain the expected vector of parameters $\boldsymbol{\theta}$. These correspond to the "first step" estimates in an iteratively reweighted computation and thus are not precisely correct because the covariance matrix Ω is based on the specified alternative model in the $\{\pi_i\}$, not the model in $\boldsymbol{\theta}$, that is, using the vector $\boldsymbol{\pi_\theta} = 1/\left[1 - \exp(\boldsymbol{x}'\boldsymbol{\theta})\right]$. However, the iteratively reweighted computation with Ω defined using the elements of $\boldsymbol{\pi_\theta}$ will yield parameters $\boldsymbol{\theta}$ and a sample size similar to those described above.

7.4 INTERACTIONS

Thus far, all of the models we have considered contain main effects only. In many instances, however, models that include interactions between covariates provide a better description of the risk relationships. In the simplest sense, an interaction implies that the effect of one covariate depends on the value of another covariate and *vice versa*. Interactions are assessed in logistic regression models by adding an additional covariate that consists of the product of two or more covariates. In PROC LOGISTIC a new variable must be defined in the data step such as `x12=x1*x2` where `x1*x2` is the product of the values for X_1 and X_2. The variable `x12` is then used in the model statement. Other programs, such as PROC GENMOD in SAS, will automatically fit an interaction in a model statement such as `model y=x1 x2 x1*x2`. Interactions involving categorical variables are also readily fit using a CLASS statement in PROC GENMOD.

First, consider the simple linear regression model with an identity link, for example, where Y is a quantitative variable and $E(y|x)$ is the conditional expectation for Y given the value of the covariate vector $\boldsymbol{x}$. The resulting model can be parameterized as

$$E(y|\boldsymbol{x}) = \alpha + x_1\beta_1 + x_2\beta_2 + x_1x_2\beta_{12} \,, \tag{7.76}$$

where $x_1x_2\beta_{12}$ is the interaction term with coefficient β_{12}. When $\beta_{12} = 0$ then the simple main-effects only model applies. However, when $\beta_{12} \neq 0$, we have, in effect, two alternate representations of the same model:

$$\begin{aligned} E(y|\boldsymbol{x}) &= \alpha + x_1\widetilde{\beta}_{1|x_2} + x_2\beta_2, & \widetilde{\beta}_{1|x_2} &= \beta_1 + x_2\beta_{12} \\ &= \alpha + x_1\beta_1 + x_2\widetilde{\beta}_{2|x_1}, & \widetilde{\beta}_{2|x_1} &= \beta_2 + x_1\beta_{12}. \end{aligned} \tag{7.77}$$

Thus the coefficient (slope) of the effect for one covariate depends on the value of the other covariate and *vice versa*, and the two covariates interact.

Note that in this case, the main effects β_1 and β_2 do not describe the complete effect of each covariate in the model. Rather the complete effect of X_1 is represented by the values of β_1 and β_{12} and the complete effect of X_2 by the values of β_2 and β_{12}. Thus the tests of the main effects alone do not provide tests of the complete effects of X_1 and X_2. Rather, a 2 *df* test of H_0: $\beta_1 = \beta_{12} = 0$ is required to test the overall effect of X_1, and likewise a 2 *df* test of H_0: $\beta_2 = \beta_{12} = 0$ is required to test the overall effect of X_2. These tests are not automatically computed by any standard program, but some programs, including LOGISTIC, provide a TEST option to conduct multiple degree of freedom Wald tests of hypotheses.

In simple regression, the coefficients are interpreted in terms of the change in the expected value of the mean given a unit change in the covariates. In logistic regression, these effects are interpreted in terms of the change in the log odds or the odds ratio.

7.4.1 Qualitative-Qualitative Covariate Interaction

The simplest case is where the two covariates X_1 and X_2 are both binary. As in Example 7.1, let $O_{\boldsymbol{x}}$ designate the odds of the positive outcome for a subject with covariate values $\boldsymbol{x}$, that includes the x_1x_2 interaction term. Then

$$O_{\boldsymbol{x}} = \exp(\alpha + \boldsymbol{x}'\boldsymbol{\beta}) = \exp\left(\alpha + x_1\beta_1 + x_2\beta_2 + x_1x_2\beta_{12}\right). \tag{7.78}$$

There are only four categories of subjects in the population, each with odds given by

Category	x_1	x_2	x_1x_2	$O_{\boldsymbol{x}}$
1	0	0	0	e^{α}
2	0	1	0	$e^{\alpha+\beta_2}$
3	1	0	0	$e^{\alpha+\beta_1}$
4	1	1	1	$e^{\alpha+\beta_1+\beta_2+\beta_{12}}$

. (7.79)

Note that if there is no interaction term, $\beta_{12} = 0$, then in the last cell the effects are additive in the exponent, or multiplicative.

In this case, it is instructive to consider the odds ratio of each category versus the first, designated as $OR_{(x_1,x_2):(0,0)}$, as presented in the cells of the following table

$$
\begin{array}{ccccc}
 & & \multicolumn{2}{c}{OR_{(x_1,x_2):(0,0)}} & \\
 & & \multicolumn{2}{c}{x_2} & \\
 & & 0 & 1 & OR_{(x_1,1):(x_1,0)} \\
\cline{3-4}
x_1 & 0 & \multicolumn{1}{|c|}{1.0} & \multicolumn{1}{c|}{e^{\beta_2}} & e^{\beta_2} \\
\cline{3-4}
 & 1 & \multicolumn{1}{|c|}{e^{\beta_1}} & \multicolumn{1}{c|}{e^{\beta_1+\beta_2+\beta_{12}}} & e^{\beta_2+\beta_{12}} \\
\cline{3-4}
OR_{(1,x_2):(0,x_2)} & & e^{\beta_1} & e^{\beta_1+\beta_{12}} &
\end{array}
\quad . \qquad (7.80)
$$

The odds ratios associated with each covariate, conditional on the value of the other, are obtained by the ratios within rows and within columns. These are given along the row and column margins of the table. The variance of each log odds ratio is then obtained from the variance of the corresponding linear combination of the estimated coefficients. For example, $V\left[\log \widehat{OR}_{(1,1):(0,1)}\right] = V(\widehat{\beta}_1 + \widehat{\beta}_{12}) = V(\widehat{\beta}_1) + V(\widehat{\beta}_{12}) + 2Cov(\widehat{\beta}_1, \widehat{\beta}_{12})$. From these, confidence limits and Wald tests may then be computed.

Example 7.10 *Coronary Heart Disease in the Framingham Study*
Using the data from Example 7.1, a logistic model with an interaction yields the following model coefficients

	No Interaction			*Interaction*		
Effect	$\widehat{\beta}$	$\widehat{OR}$	$p \leq$	$\widehat{\beta}$	$\widehat{OR}$	$p \leq$
X_1: *HiChol*	1.0301	2.801	0.0001	1.1751	3.238	0.0012
X_2: *HiSBP*	0.9168	2.501	0.0001	1.1598	3.189	0.0114
X_1X_2	–	–	–	−0.3155	0.729	0.5459

(7.81)

For comparison, the no-interaction model from Table 7.4 is also presented. The interaction effect is not statistically significant, indicating that the additive exponential model applies. Nevertheless, it is instructive to describe the interaction model.

To assess the overall significance of each effect, a 2 *df* test could be obtained using the following statements as part of the SAS PROC LOGISTIC analysis:

```
test hichol=interact=0;
test hisbp=interact=0;
```

The resulting test of the effect of high cholesterol yields a 2 *df* chi-square test value of $X^2 = 15.7577$ with $p \leq 0.0004$, and the test for the effect of high blood pressure yields $X^2 = 17.7191$ with $p \leq 0.0001$. In such instances, the 1 *df* tests computed by the program and presented above should not be used as tests of the overall effect. Rather, they are tests of the simple effect of each covariate when added to the interaction. In rare instances, these simple 1 *df* effects may not be significant, in which case the simple effect may not add significantly to the interaction effect.

From this interaction model we can construct a table of odds ratios such as that in (7.80) as follows

$OR_{(x_1,x_2):(0,0)}$

		x_2: *HiSBP* 0	1	$OR_{(x_1,1):(x_1,0)}$
x_1: *HiChol*	0	1.0	3.189	3.189
	1	3.238	7.527	2.325
	$OR_{(1,x_2):(0,x_2)}$	3.238	2.360	

(7.82)

Thus the odds ratio associated with high versus low cholesterol is 3.238 for those with low blood pressure, and 2.360 for those with high blood pressure. Likewise, the odds ratio associated with high versus low blood pressure is 3.189 for those with low cholesterol, and 2.325 for those with high cholesterol. Note that the odds ratios associated with the main effects in the interaction model (3.189 and 3.238) are those for each covariate, where the value of the other covariate is zero.

For comparison, the following table provides the odds ratios using the no-interaction model

$OR_{(x_1,x_2):(0,0)}$

		x_2: *HiSBP* 0	1	$OR_{(x_1,1):(x_1,0)}$
x_1: *HiChol*	0	1.0	2.501	2.501
	1	2.801	7.007	2.501
	$OR_{(1,x_2):(0,x_2)}$	2.801	2.801	

(7.83)

In this case the odds ratios for one covariate are constant for the values of the other covariate. Also, the effects of the covariates are multiplicative or exponentially additive where $OR_{(1,1):(0,0)} = e^{x_1\beta_1 + x_2\beta_2} = (2.501 \times 2.801) = 7.007$. Thus the test of the interaction effect can be viewed as a test of the difference between the odds ratios in the (1,1) cell, or between the 7.007 under the no-interaction model versus the 7.527 under the interaction model.

Example 7.11 *Ulcer Clinical Trial: Stratum by Group Interaction*
Table 4.4 presents a SAS program for the analysis of the Ulcer Clinical Trial data of Example 4.1 stratified by the ulcer type. The following statements when added to this program would fit a logistic regression model that includes a treatment by stratum interaction:

```
data three; set two;
z2=0; if k=2 then z2=1;  iz2=i*z2;
z3=0; if k=3 then z3=1;  iz3=i*z3
proc logistic descending;
 model j = i z2 z3 iz2 iz3 / rl;
weight f;
test iz2=iz3=0;
run;
```

The covariates z2 and z3 represent the effects of stratum 2 and 3, respectively, versus the reference stratum 1, and iz2 and iz3 represent the treatment group by covariate interaction. The fitted model is of the form

$$\begin{aligned}\log(\widehat{O}_{\boldsymbol{x}}) &= \widehat{\alpha} + i\widehat{\beta}_1 + z_2\widehat{\beta}_2 + z_3\widehat{\beta}_3 + (iz_2)\widehat{\beta}_{12} + (iz_3)\widehat{\beta}_{13} \\ &= -0.3001 - i(0.1854) + z_2(0.077) - z_3(0.2595) \\ &\quad + (iz_2)(1.5072) + (iz_3)(1.1869),\end{aligned} \tag{7.84}$$

where β_2 and β_3 are equivalent to the γ_2 and γ_3 of Section 7.1.4.

In this model, all covariates are binary so that $e^{\widehat{\alpha}} = 0.7407$ is the estimated odds of healing within the category of the population with all covariates equal to 0, or for the control group of stratum 1. Then, $e^{\widehat{\alpha}+\widehat{\beta}_1} = 0.6154$ is the odds of healing in the drug-treated group within the first stratum. Their ratio is the estimate of the odds ratio for the drug vs. placebo within this stratum, $\widehat{OR}_1 = 0.831$. Likewise, the odds within each group within each stratum, and the odds ratio within each stratum, are obtained from the fitted model as

Stratum	z_2	z_3	*Placebo Odds (i=0)*	*Drug Odds (i=1)*	*Drug:Placebo Odds Ratio*
1	0	0	0.741	0.615	0.831
2	1	0	0.800	3.000	3.750
3	0	1	0.571	1.556	2.722

(7.85)

These model-based estimates of the treatment group odds ratios within each stratum are identical to those within each stratum as shown in Table 4.1. Converting these odds back to the corresponding proportions yields the observed proportions within each cell of the data structure. Such models are called *saturated* because they provide an exact fit to the observed data and no other parameters can be added to the model without the addition of more covariates to the data structure.

Clearly, if the two interaction terms equal zero, then the model reduces to the stratified-adjusted model of Section 7.1.4 that provides an estimate of the common odds ratio for treatment group within each stratum. Thus a test of no-interaction on 2 *df* provides a test of homogeneity comparable to those described in Sections 4.6.1 and 4.6.2. A Wald test of H_0: $\beta_{12} = \beta_{13} = 0$ is provided by the test option in PROC LOGISTIC. The resulting test yields $X^2 = 4.5803$ with $p \leq 0.1013$, which is equivalent to the value of the Cochran test of homogeneity presented in Example 4.10.

Alternatively, the likelihood ratio test can be computed as the difference in the model X^2 tests for the interaction versus no-interaction model to yield $X^2 = 11.226 - 6.587 = 4.639$ with $p \leq 0.0983$ on 2 df. This test can also be obtained from PROC GENMOD using a model with class effects for group and stratum and with the `type3` option.

7.4.2 Interactions with a Quantitative Covariate

Similar considerations apply to the interpretation of a model with covariate effects between a qualitative and a quantitative covariate, or between two quantitative covariates. These cases are illustrated in the following examples.

Example 7.12 *DCCT Nephropathy: Treatment by Duration Interaction*
A model fit to the DCCT nephropathy data of Example 7.4 with interactions between treatment and duration of diabetes, and also between the HbA_{1c} and level of systolic blood pressure, yields the following

	Effect	$\widehat{\beta}$	$\widehat{OR}$	$p \leq$
X_1:	*Intensive Treatment*	−2.4410	0.087	0.0697
X_2:	*HbA_{1c}*	3.7478	42.428	0.0182
X_3:	*Years Duration of DM*	−0.0103	0.990	0.8997
X_4:	*Systolic Blood Pressure*	0.2752	1.317	0.0297
X_5:	*Female*	−0.9505	0.387	0.0386
X_1X_3:	*Group×Duration*	0.0772	1.080	0.5636
X_2X_4:	*HbA_{1c}×SBP*	−0.0272	0.973	0.0431

(7.86)

Note that the model does not include an interaction with gender so that the female effect can be interpreted as a true main effect. The other covariates each involve an interaction and thus the combined effect of each covariate is contained in part in the main effect and in part in the interaction term.

The treatment group by duration interaction is not statistically significant. Nevertheless, it is instructive to describe the interpretation of such an interaction between a qualitative and a quantitative covariate. Here, one must consider a specific value (or values) of the quantitative covariate. For example, consider the odds of microalbuminuria in each treatment group for subjects with nine and 10 years duration of diabetes:

Group	*Duraton*	$\widehat{O}_{\boldsymbol{x}}$	
1 (Intensive)	*9*	$Ce^{\widehat{\beta}_1+9(\widehat{\beta}_3+\widehat{\beta}_{13})}$	$= C(0.06213)$
0 (Conv.)	*9*	$Ce^{9\widehat{\beta}_3}$	$= C(0.91147)$
1 (Intensive)	*10*	$Ce^{\widehat{\beta}_1+10(\widehat{\beta}_3+\widehat{\beta}_{13})}$	$= C(0.05984)$
0 (Conv.)	*10*	$Ce^{10\widehat{\beta}_3}$	$= C(0.90213)$

, (7.87)

where the constant $C = \exp(x_2\widehat{\beta}_2 + x_4\widehat{\beta}_4 + x_5\widehat{\beta}_5 + x_2x_4\widehat{\beta}_{24})$ for any fixed values of the other covariates. This yields odds ratios for intensive versus conventional therapy for those with nine years duration of $\widehat{OR}_{I:C|D=9} = (0.06213/0.91147) = 0.0682 = \exp(\widehat{\beta}_1 + 9\widehat{\beta}_{13})$; and for those with 10 years duration of $\widehat{OR}_{I:C|D=10} = 0.0663 = \exp(\widehat{\beta}_1 + 10\widehat{\beta}_{13})$. Likewise, within the intensive therapy group, the odds ratio associated with one year longer duration, such as 10 versus nine years, is $\widehat{OR}_{D+1|I} = (0.05984/0.06213) = 0.963 = \exp(\widehat{\beta}_3 + \widehat{\beta}_{13})$ and $\widehat{OR}_{D+1|I} = \exp(\widehat{\beta}_3) = 0.990$ as shown above.

Similar calculations for other values of duration would allow the inspection of the odds ratio for intensive versus conventional therapy over a range of values of duration. Such computations are described in the next example.

Note that neither interaction involves the covariates in the other interaction. Thus the effects of covariates $X_1 - X_4$ each involve only one other covariate. If a covariate is involved in multiple interactions, then the effect of that covariate depends on the values of all other covariates included in any of these interactions. This complicates the interpretation of the results, but using simple generalizations of the above, it is possible to describe the effects of each covariate as a function of the values of the other covariates in higher dimensions.

Example 7.13 *DCCT Nephropathy: HbA$_{1c}$ By Blood Pressure Interaction*
In the above model, the Wald test for the interaction between HbA$_{1c}$ and systolic blood pressure is statistically significant at the 0.05 level. This indicates that the effect of one variable depends explicitly on the value of the other. Since the biological hypothesis is that blood sugar levels, as represented by the HbA$_{1c}$ are the underlying causal mechanism, it is more relevant to examine the effect of HbA$_{1c}$ over a range of values of blood pressure than vice versa. For a given value of systolic blood pressure (say S) then the odds ratio associated with a one unit (%) increase in HbA$_{1c}$ (H) is provided by

$$\widehat{OR}_{H+1|S} = \exp\left[\widehat{\beta}_2 + x_4\widehat{\beta}_{24}\right]. \tag{7.88}$$

This function can be plotted over a range of values of blood pressure (X_4) as shown in Figure 7.1. For each value of blood pressure (x_4) the 95% asymmetric confidence limits for this odds ratio can be computed from those based on the linear combination of the estimates where

$$\widehat{V}(\widehat{\beta}_2 + x_4\widehat{\beta}_{24}) = \widehat{V}(\widehat{\beta}_2) + x_4^2\widehat{V}(\widehat{\beta}_{24}) + 2x_4\widehat{Cov}(\widehat{\beta}_2, \widehat{\beta}_{24}). \tag{7.89}$$

The variances and covariances (not shown) can be obtained from the COVB option in PROC LOGISTIC.

As the level of baseline blood pressure increases, the odds ratio associated with a unit increase in the HbA$_{1c}$ decreases. For a moderately low blood pressure, such as 110 mm Hg, the odds ratio for a unit increase in HbA$_{1c}$ is 2.13; for an average blood pressure of about 116, the HbA$_{1c}$ odds ratio is 1.81, close to that in the no-interaction model in Table 7.6; and for a moderately high blood pressure, such as 120 mm Hg, the HbA$_{1c}$ odds ratio is 1.62.

Fig. 7.1 Odds ratio per unit higher value of HbA_{1c} (%) as a function of the level of systolic blood pressure in mm/Hg.

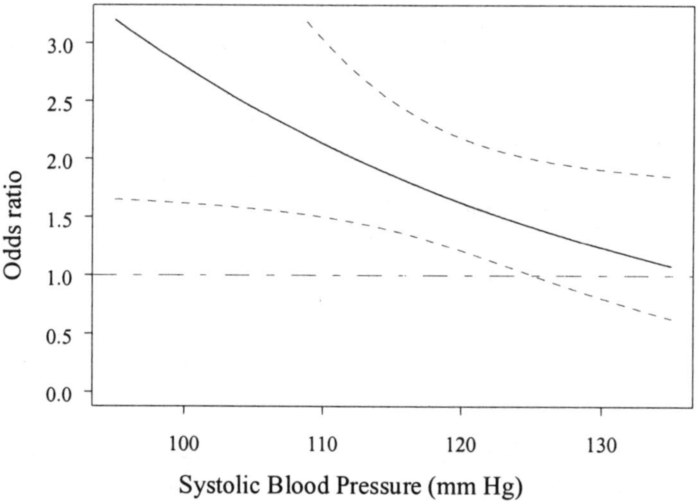

7.5 MEASURES OF THE STRENGTH OF ASSOCIATION

There are three central objectives in the analysis of data. The first is to describe the *nature* of the association among the independent variables or covariates and the dependent variable. In a logistic regression model this is expressed by the coefficient estimates $\widehat{\boldsymbol{\beta}}$ or the corresponding odds ratios and their confidence limits. The second is to assess the *statistical significance* of the association by conducting a test of a relevant hypothesis concerning the values of the coefficients, such as the test of the model null hypothesis H_0: $\boldsymbol{\beta} = \mathbf{0}$. The third is to describe the *strength* of the association using some measure of the fraction of variation in the dependent variable that is explained by the covariates through the estimated model relationship.

7.5.1 Squared Error Loss

Section A.8.2 of the Appendix describes the use of the squared error loss function to quantify the fraction of squared error loss explained, $\rho^2_{\varepsilon^2}$. In a logistic regression model, where y_i is an indicator variable for the response (D) versus not ($\overline{D}$), this is expressed as a measure of the explained variation in risk, $\rho^2_{\varepsilon^2} = \rho^2_{risk}$ (Korn and Simon, 1991). Unconditionally, with no covariate effects, or under the null model, $E(y_i) = P(D) = \overline{\pi}$ and $V(y) = \overline{\pi}(1 - \overline{\pi})$. Alternatively, the logistic regression model specifies that the probability of the response conditional on covariates is $E(y_i \mid \boldsymbol{x}_i) = P(D \mid \boldsymbol{x}_i) = \pi_i(\boldsymbol{x}_i) = \pi_i$, where $\pi_i = \left[1 + e^{-\boldsymbol{x}_i\boldsymbol{\theta}}\right]^{-1}$. Then

$$V[E(y|\boldsymbol{x})] = E(\pi_i - \overline{\pi})^2 \;\widehat{=}\; \frac{1}{N}\sum\nolimits_i \left(\widehat{\pi}_i - \widehat{\overline{\pi}}\right)^2, \tag{7.90}$$

where $\widehat{\overline{\pi}} = m_1/N$. Therefore, the estimated explained risk with squared error loss from (A.183) is

$$\widehat{\rho}^2_{\varepsilon^2} = \widehat{\rho}^2_{risk} = \frac{\widehat{V}\left[\widehat{E}(y|\boldsymbol{x})\right]}{\widehat{V}(y)} = \frac{\frac{1}{N}\sum_i \left(\widehat{\pi}_i - \widehat{\overline{\pi}}\right)^2}{\widehat{\overline{\pi}}\left(1 - \widehat{\overline{\pi}}\right)}. \tag{7.91}$$

When the logistic regression model fits perfectly, then $\widehat{\pi}_i = 1.0$ for all subjects where $y_i = 1$ (or for whom the response occurred), and $\widehat{\pi}_i = 0$ for all subjects where $y_i = 0$. In this case it is readily shown that $\widehat{\rho}^2_{risk} = 1.0$. Conversely, when the model explains none of the variation in risk, then $\widehat{\pi}_i = \overline{\pi}$ for all subjects, irrespective of the covariate values and $\widehat{\rho}^2_{risk} = 0$; see Problem 7.14.

As described in Section A.8.2 of the Appendix, the fraction of squared error loss explained by the model is also estimated by the fraction of residual variation. In logistic regression it is readily shown that $R^2_{\varepsilon^2,resid} = \widehat{\rho}^2_{risk}$ above.

7.5.2 Entropy Loss

Although squared error loss pervades statistical practice, other loss functions have special meaning in some situations. The entropy loss in logistic regression and the analysis of categorical data is one such instance, as described by Efron (1978).

Consider a discrete random variable Y with probability mass function $P(y)$, $0 \leq P(y) \leq 1$ for $y \in S$, the sample space of Y. The certainty or uncertainty with which one can predict an individual observation depends on the higher moments of $P(y)$ or the degree of dispersion in the data. One measure of this predictive uncertainty or dispersion is the entropy of the distribution of Y (Shannon, 1948) defined as

$$H(Y) = -\sum\nolimits_{y \in S} P(y)\log[P(y)] = -E(\log[P(y)]). \tag{7.92}$$

For example, assume that the sample space consists of the set of K integers $y \in S = (1, \ldots, K)$ with equal probability $P(y) = 1/K$ for $y \in S$. Then the entropy for this distribution is

$$H(Y) = -\sum\nolimits_{i=1}^{K} \left(\frac{1}{K}\right)\log\left(\frac{1}{K}\right) = -\log\left(\frac{1}{K}\right) = \log(K). \tag{7.93}$$

In fact, this is the maximum possible entropy $H(Y)$ for a discrete random variable with K categories. Thus for this distribution there is the least possible certainty with which one can predict any single observation drawn at random compared to any other member of the family of K-element discrete distributions.

Conversely, if the sample space consists of a single integer, say $y = b$ where $P(y = b) = 1$ and $P(y) = 0$ for $y \neq b$, then the entropy for this point mass distribution is $H(Y) = 1 \log(1) = 0$. This is the minimum entropy possible for any distribution, discrete or continuous.

Now consider a Bernoulli random variable such as the indicator variable for the response in a logistic regression model. Here Y is an indicator variable with sample space $S = (0, 1)$ and where $E(Y) = P(Y = 1) = \pi$. Then the entropy for this distribution is

$$\begin{aligned} H(Y) &= -\sum\nolimits_{y=0}^{1} \left[y\pi \log \pi + (1-y)(1-\pi) \log (1-\pi) \right] \\ &= -\left[\pi \log(\pi) + (1-\pi) \log(1-\pi) \right] \\ &= -E(\log [p(y)]). \end{aligned} \tag{7.94}$$

This suggests that an *entropy loss function* can be defined for any one random observation as

$$\mathcal{L}_E(y, \widehat{\pi}) = -\left[y \log(\widehat{\pi}) + (1-y) \log(1-\widehat{\pi}) \right] \tag{7.95}$$

for any prediction function for the Bernoulli probability of that observation based on covariates such that $\widehat{\pi} = P(D \mid \boldsymbol{x})$ (see Problem 7.14.3). Then for a population of observations, each with a predicted or estimated probability $\widehat{\pi}_i = P(D \mid \boldsymbol{x}_i)$, the expected loss is the entropy for the corresponding Bernoulli probabilities

$$D_E(\boldsymbol{x}) = E\left[\mathcal{L}_E(y_i, \widehat{\pi}_i)\right] = -\left[\pi_i \log(\pi_i) + (1-\pi_i) \log(1-\pi_i)\right] = H(y|\pi_i). \tag{7.96}$$

Now under the null model with no covariates, all observations are identically distributed as Bernoulli with probability $\overline{\pi}$, and the expected entropy is

$$D_{E0} = E\left[\mathcal{L}_E(y, \overline{\pi})\right] = -\left[\overline{\pi} \log(\overline{\pi}) + (1-\overline{\pi}) \log(1-\overline{\pi})\right] = H(y|\overline{\pi}). \tag{7.97}$$

Thus the expected fraction of entropy loss explained, ρ_E^2, is obtained by substituting these quantities into (A.173).

This fraction of entropy explained can then be estimated as

$$\widehat{\rho_E^2} = \frac{\widehat{D}_{E0} - \widehat{D}_E(\boldsymbol{x})}{\widehat{D}_{E0}}, \tag{7.98}$$

using the sample estimate of the entropy loss based on the estimated logistic function with covariates

$$\widehat{D}_E(\boldsymbol{x}) = -\frac{1}{N}\left[\sum\nolimits_i \widehat{\pi}_i \log(\widehat{\pi}_i) + (1-\widehat{\pi}_i) \log(1-\widehat{\pi}_i)\right], \tag{7.99}$$

and that unconditionally without covariates

$$\widehat{D}_{E0} = -\left[\widehat{\overline{\pi}} \log\left(\widehat{\overline{\pi}}\right) + \left(1-\widehat{\overline{\pi}}\right) \log\left(1-\widehat{\overline{\pi}}\right)\right]. \tag{7.100}$$

Therefore,

$$\widehat{\rho}_E^2 = 1 - \frac{\frac{1}{N}\sum_i \left[\widehat{\pi}_i \log\left(\widehat{\pi}_i\right) + \left(1-\widehat{\pi}_i\right)\log\left(1-\widehat{\pi}_i\right)\right]}{\widehat{\overline{\pi}}\log\left(\widehat{\overline{\pi}}\right) + \left(1-\widehat{\overline{\pi}}\right)\log\left(1-\widehat{\overline{\pi}}\right)} . \tag{7.101}$$

Alternately, the fraction of explained entropy loss can be estimated from the fraction of residual variation explained in terms of the entropy loss function, which from (A.192) is

$$\begin{aligned} R^2_{E,resid} &= \frac{\sum_i \mathcal{L}_E\left(y_i, \widehat{\overline{\pi}}\right) - \sum_i \mathcal{L}_E\left(y_i, \widehat{\pi}_i\right)}{\sum_i \mathcal{L}_E\left(y_i, \widehat{\overline{\pi}}\right)} \\ &= 1 - \frac{\sum_i \left[y_i \log\left(\widehat{\pi}_i\right) + \left(1-y_i\right)\log\left(1-\widehat{\pi}_i\right)\right]}{\sum_i \left[y_i \log\left(\widehat{\overline{\pi}}\right) + \left(1-y_i\right)\log\left(1-\widehat{\overline{\pi}}\right)\right]} . \end{aligned} \tag{7.102}$$

In Problem 7.14 it is then shown that these estimates are identical $\widehat{\rho}_E^2 = R^2_{E,resid}$ based on the fact that the score equation for the intercept $U(\boldsymbol{\theta})_\alpha$ implies that $\sum_i y_i = \sum_i \widehat{\pi}_i = N\overline{\pi}$, and that for each coefficient $U(\boldsymbol{\theta})_{\beta_j}$ implies that $\sum_i y_i x_{ij} = \sum_i \widehat{\pi}_i x_{ij}$, $j = 1, \ldots, p$.

The latter expression for $R^2_{E,resid}$ is especially meaningful in logistic regression since the two expressions in (7.102) refer to logistic regression log likelihood functions, such that

$$R^2_{E,resid} = 1 - \left[\frac{\log L\left(\widehat{\alpha}, \widehat{\boldsymbol{\beta}}\right)}{\log L\left(\widehat{\alpha}\right)}\right] = \frac{X^2_{LR}}{-2\log L\left(\widehat{\alpha}\right)} = R^2_\ell \tag{7.103}$$

where X^2_{LR} is the model likelihood ratio chi-square test value on p df.

The latter quantity R^2_ℓ has a natural interpretation as the fraction of $-2\log L$ explained by the covariates in a logistic regression model. However, the interpretation as a fraction of entropy loss explained by the model is applicable only to regression models with a Bernoulli response, such as the logistic regression model. Nevertheless, the concept of explained log likelihood has become widely used in other regression models, such as Poisson regression, since the null model $-2\log L\left(\widehat{\alpha}\right)$ for any model is analogous to the unconditional $SS(Y)$. Further, as shown in Section A.8.3 of the Appendix, any model based on a likelihood with distribution $f(y|x)$ implies a loss function equal to $-\log[f(y|x)]$ that then yields a fraction of explained loss equal to R^2_ℓ.

Another measure of explained variation described in Section A.8.4 of the Appendix is Madalla's R^2_{LR}, which is also a function of the model likelihood ratio statistic. However, this measure differs from the entropy R^2_E. From (7.103), stated more generally in terms of a null model with $\boldsymbol{\theta} = \boldsymbol{\theta}_0$, then

$$R^2_E = 1 - \frac{\log L\left(\widehat{\boldsymbol{\theta}}\right)}{\log L\left(\boldsymbol{\theta}_0\right)} \tag{7.104}$$

$$\neq R^2_{LR} = 1 - \exp[-X^2_{LR}/N] . \tag{7.105}$$

Also, Nagelkerke (1991) shows that the likelihood ratio explained variation $R^2_{LR} < 1$ for models based on discrete probability distributions, such as the binomial or Bernoulli in logistic regression, where $\max(R^2_{LR}) = 1 - L(\boldsymbol{\theta}_0)^{2/N}$. Thus he suggested that the R^2_{LR} be rescaled so as to achieve a value of 1.0 when R^2_{LR} is at its maximum. Thus the maximum rescaled R^2_{LR} is computed as

$$\max\ rescaled\ R^2_{LR} = \frac{R^2_{LR}}{\max(R^2_{LR})} = \frac{1 - \exp\left[-\frac{X^2_{LR}}{N}\right]}{1 - L(\boldsymbol{\theta}_0)^{2/N}} = \frac{1 - \left[\frac{L(\boldsymbol{\theta}_0)}{L\left(\widehat{\boldsymbol{\theta}}\right)}\right]^{2/N}}{1 - L(\boldsymbol{\theta}_0)^{2/N}}. \tag{7.106}$$

These R^2_{LR} measures are generated using the RSQUARE option in PROC LOGISTIC. However, the program does not compute the other R^2 measures based on squared error loss, $\widehat{\rho}^2_{risk} = R^2_{\varepsilon^2,resid}$, or on the entropy, $\widehat{\rho}^2_E = R^2_{E,resid}$.

Example 7.14 *DCCT Nephropathy Data*
To evaluate $\widehat{\rho}^2_{risk} = R^2_{\varepsilon^2,resid}$ for the model presented in Table 7.6, it is necessary that the estimated probabilities for all observations be output into a separate data set and that the numerator sum of squares be calculated, such as using PROC UNIVARIATE. By definition, from the score estimating equation for the intercept, $\sum_i \widehat{\pi}_i = 42 = m_1$ which is the number of responses, so that $\widehat{\overline{\pi}} = 42/172 = 0.244186$. From the computed estimated probabilities of all observations, $\sum_i \left(\widehat{\pi}_i - \widehat{\overline{\pi}}\right)^2 = 6.662041$. Therefore,

$$R^2_{\varepsilon^2,resid} = \frac{6.662041}{(172)(0.244186)(1 - 0.244186)} = 0.20987.$$

For these data, $-2\log L(\widehat{\alpha}) = -2\log L(\boldsymbol{\theta}_0) = 191.215$ and the model likelihood ratio chi-square test is $X^2_{LR} = 35.879$. Thus the explained entropy loss or negative log likelihood is $R^2_E = 35.879/191.215 = 0.18764$. Similarly, Madalla's likelihood ratio based measure is $R^2_{LR} = 1 - \exp[-35.879/172] = 0.18828$. This is the value of *RSquare* computed using the RSQUARE option as shown in Table 7.6. The program also presents the *Max-rescaled RSquare* = 0.2806. The latter is inflated relative to the other measures and has no obvious interpretation in terms of a loss function. The $R^2_{\varepsilon^2,resid}$ and R^2_E are clearly preferable.

7.6 CONDITIONAL LOGISTIC REGRESSION MODEL FOR MATCHED STUDIES

7.6.1 Conditional Logistic Model

Section 6.6.2 introduces the conditional logit model for the analysis of the 2×2 table from a prospective sample of matched pairs of exposed and non-exposed individuals. Section 6.6.5 also shows that this model can be applied to the analysis of a retrospective sample of matched pairs of cases and controls. Those developments apply to the special case of 1:1 or pair matching with a single binary covariate (exposure or case group).

Breslow (1982) describes the general conditional logistic regression model to assess the effects of covariates on the odds of the response in studies that involve matched sets of multiple exposed and non-exposed individuals in a prospective study, or multiple cases and controls in a retrospective study. This model allows for a vector of covariates for each subject, and also for each pair. See the discussion of pair and member stratification in Section 5.7.1. The model is described herein in terms of a prospective study, but may also be applied to a retrospective study.

Consider the case of N matched sets of size n_i for the ith set, $i = 1, \ldots, N$ The jth member of the ith set has covariate vector $\boldsymbol{x}_{ij} = (x_{ij1} \; \ldots \; x_{ijp})^T$ for $i = 1, \ldots, N$; $j = 1, \ldots, n_i$. For each member one or more of these covariates describe the exposure status of that member, either qualitatively such as exposed versus not, or quantitatively such as the dose and the duration of exposure. The other covariates may then represent pair characteristics that are shared by all members of the matched set, or characteristics that are unique to each member of each set. See the discussion of pair and member stratification in Section 5.7. Then for the $ijth$ member let y_{ij} be an indicator variable representing whether the member experienced or developed the positive response of interest $(D : y_{ij} = 1)$ or not $(\overline{D} : y_{ij} = 0)$. We then adopt a logistic model which specifies that the probability of the response π_{ij} for the $ijth$ member is a logistic function of the covariates $\boldsymbol{x}_{ij}$. The model allows each matched set to have a unique risk of the response represented by a set-specific intercept α_i, and then assumes that the covariates have a common effect on the odds of the response over all matched sets represented by the coefficient vector $\boldsymbol{\beta} = (\beta_1 \; \ldots \; \beta_p)^T$. Thus the model is of the form

$$P_i\left(D|\boldsymbol{x}_{ij}\right) = \pi_{ij} = \pi_i(\boldsymbol{x}_{ij}) = \frac{e^{\alpha_i + \boldsymbol{x}'_{ij}\boldsymbol{\beta}}}{1 + e^{\alpha_i + \boldsymbol{x}'_{ij}\boldsymbol{\beta}}} \tag{7.107}$$

and

$$\log\left[\frac{\pi_{ij}}{1 - \pi_{ij}}\right] = \alpha_i + \boldsymbol{x}'_{ij}\boldsymbol{\beta}. \tag{7.108}$$

Now consider two members from any matched set (the ith), one with covariate vector $\boldsymbol{x}$, the other $\boldsymbol{x}_0$. Then the odds ratio for these two members is

$$\frac{\pi_i\left(\boldsymbol{x}\right) / \left(1 - \pi_i\left(\boldsymbol{x}\right)\right)}{\pi_i\left(\boldsymbol{x}_0\right) / \left(1 - \pi_i\left(\boldsymbol{x}_0\right)\right)} = \frac{e^{\alpha_i + \boldsymbol{x}'\boldsymbol{\beta}}}{e^{\alpha_i + \boldsymbol{x}'_0\boldsymbol{\beta}}} = \exp\left[\left(\boldsymbol{x} - \boldsymbol{x}_0\right)'\boldsymbol{\beta}\right]. \tag{7.109}$$

Thus the odds ratio for two members of any other set with covariate values differing by the amount $\boldsymbol{x} - \boldsymbol{x}_0$ is assumed to be constant for all matched sets regardless of the value of α_i.

Within the ith set the n_i members are conditionally independent. Thus the unconditional likelihood in terms of the parameters $\boldsymbol{\alpha} = (\alpha_1 \; \ldots \; \alpha_N)^T$ and $\boldsymbol{\beta}$ is of the form

$$L(\boldsymbol{\alpha}, \boldsymbol{\beta}) = \prod_{i=1}^{N} \prod_{j=1}^{n_i} \pi_i \left(\boldsymbol{x}_{ij}\right)^{y_{ij}} \left[1 - \pi_i \left(\boldsymbol{x}_{ij}\right)\right]^{1-y_{ij}} , \tag{7.110}$$

which yields the log likelihood

$$\begin{aligned} \ell(\boldsymbol{\alpha}, \boldsymbol{\beta}) &= \sum\nolimits_{i=1}^{N} \sum\nolimits_{j=1}^{n_i} y_{ij} \log\left[\pi_i \left(\boldsymbol{x}_{ij}\right)\right] + \left(1 - y_{ij}\right) \log\left[1 - \pi_i \left(\boldsymbol{x}_{ij}\right)\right] \qquad (7.111) \\ &= \sum\nolimits_i \sum\nolimits_j \left[y_{ij} \log\left(\frac{e^{\alpha_i + \boldsymbol{x}'_{ij}\boldsymbol{\beta}}}{1 + e^{\alpha_i + \boldsymbol{x}'_{ij}\boldsymbol{\beta}}}\right) - \left(1 - y_{ij}\right) \log\left(1 + e^{\alpha_i + \boldsymbol{x}'_{ij}\boldsymbol{\beta}}\right) \right] \\ &= \sum\nolimits_i m_{1i}\alpha_i + \sum\nolimits_i \sum\nolimits_j \left[y_{ij}\boldsymbol{x}'_{ij}\boldsymbol{\beta} - \log\left(1 + e^{\alpha_i + \boldsymbol{x}'_{ij}\boldsymbol{\beta}}\right) \right] \end{aligned}$$

where m_{1i} denotes the number of members with a positive response ($y_{ij} = 1$) and m_{2i} those without ($y_{ij} = 0$) in the ith matched set. As was the case with a single covariate to represent exposed versus not as in Section 6.6, to obtain an estimate of the covariate vector $\boldsymbol{\beta}$ requires that we also solve for the N nuisance parameters $\{\alpha_i\}$. However, this can be avoided by conditioning on the number of positive responses m_{1i} within the ith matched set that is the sufficient statistic for the nuisance parameter α_i for that set.

To do so, it is convenient to order the members in the ith set such that the first m_{1i} subjects are those with a response. Now, the conditional probability of interest for the ith matched set is the probability of m_{1i} responses with covariates $\boldsymbol{x}_1 \ldots \boldsymbol{x}_{m_{1i}}$ and m_{2i} non-responses with covariates $\boldsymbol{x}_{(m_{1i}+1)} \cdots \boldsymbol{x}_{n_i}$ given m_{1i} responses out of n_i members, or

$$P\left[\left\{\left(m_{1i} \ ; \boldsymbol{x}_1 \ldots \boldsymbol{x}_{m_{1i}}\right) \cap \left(m_{2i} \ ; \boldsymbol{x}_{(m_{1i}+1)} \cdots \boldsymbol{x}_{n_i}\right)\right\} \mid m_{1i}, n_i\right] . \tag{7.112}$$

Thus the conditional likelihood for the ith matched set is

$$L_{i|m_{1i},n_i} = \frac{\prod\limits_{j=1}^{m_{1i}} \pi_i \left(\boldsymbol{x}_{ij}\right) \prod\limits_{j=m_{1i}+1}^{n_i} \left[1 - \pi_i \left(\boldsymbol{x}_{ij}\right)\right]}{\sum\limits_{\ell=1}^{\binom{n_i}{m_{1i}}} \prod\limits_{j(\ell)=1}^{m_{1i}} \pi_i \left(\boldsymbol{x}_{ij(\ell)}\right) \prod\limits_{j(\ell)=m_{1i}+1}^{n_i} \left[1 - \pi_i \left(\boldsymbol{x}_{ij(\ell)}\right)\right]} . \tag{7.113}$$

The denominator is the sum over all possible combinations of m_{1i} of the n_i subjects with the response and m_{2i} subjects without. Within the ℓth combination $j(\ell)$ then refers to the original index of the member in the jth position.

Substituting the logistic function for $\pi_i \left(\boldsymbol{x}_{ij}\right)$ it is readily shown that

$$L_{i|m_{1i},n_i} = \frac{\prod\limits_{j=1}^{m_{1i}} e^{\boldsymbol{x}'_{ij}\boldsymbol{\beta}}}{\sum\limits_{\ell=1}^{\binom{n_i}{m_{1i}}} \prod\limits_{j(\ell)=1}^{m_{1i}} e^{\boldsymbol{x}'_{ij(\ell)}\boldsymbol{\beta}}} . \tag{7.114}$$

Therefore, for the sample of N matched sets, the conditional likelihood is

$$L_{(c)}(\boldsymbol{\beta}) = \prod\nolimits_i L_{i|m_{1i},n_i} \tag{7.115}$$

and the log likelihood is

$$\ell_{(c)}(\boldsymbol{\beta}) = \sum\nolimits_i \ell_{i|m_{1i},n_i}(\boldsymbol{\beta}), \tag{7.116}$$

where

$$\begin{aligned} \ell_{i|m_{1i},n_i}(\boldsymbol{\beta}) &= \sum\nolimits_{j=1}^{m_{1i}} \boldsymbol{x}_{ij}'\boldsymbol{\beta} - \log\left[\sum_{\ell=1}^{\binom{n_i}{m_{1i}}} \prod_{j(\ell)=1}^{m_{1i}} e^{\boldsymbol{x}_{ij(\ell)}'\boldsymbol{\beta}}\right] \\ &= \boldsymbol{s}_i'\boldsymbol{\beta} - \log\left[\sum_{\ell=1}^{\binom{n_i}{m_{1i}}} e^{\boldsymbol{s}_{i(\ell)}'\boldsymbol{\beta}}\right] \end{aligned} \tag{7.117}$$

where $\boldsymbol{s}_i = \sum_{j=1}^{m_{1i}} \boldsymbol{x}_{ij}$ for the ith set and $\boldsymbol{s}_{i(\ell)} = \sum_{j(\ell)=1}^{m_{1i}} \boldsymbol{x}_{ij(\ell)}$ for the ℓth combination of m_{1i} subjects within that set. The resulting score vector is $\boldsymbol{U}(\boldsymbol{\beta}) = [U(\boldsymbol{\beta})_{\beta_1} \ \ldots \ U(\boldsymbol{\beta})_{\beta_p}]^T$, where the equation for the kth element of the coefficient vector is

$$U(\boldsymbol{\beta})_{\beta_k} = \sum\nolimits_{i=1}^{N} \frac{\partial \ell_{i|m_{1i},n_i}(\boldsymbol{\beta})}{\partial \beta_k} = \sum\nolimits_{i=1}^{N} \left[s_{ik} - \frac{\sum_\ell s_{i(\ell)k} e^{\boldsymbol{s}_{i(\ell)}'\boldsymbol{\beta}}}{\sum_\ell e^{\boldsymbol{s}_{i(\ell)}'\boldsymbol{\beta}}} \right], \tag{7.118}$$

where s_{ik} and $s_{i(\ell)k}$ denote the kth element of $\boldsymbol{s}_i$ and $\boldsymbol{s}_{i(\ell)}$, respectively, and $\sum_\ell$ denotes $\sum_{\ell=1}^{\binom{n_i}{m_{1i}}}$. The expected information matrix $\boldsymbol{I}(\boldsymbol{\beta})$ then has diagonal elements $(1 \le k \le p)$

$$\boldsymbol{I}(\boldsymbol{\beta})_{\beta_k} = \sum\nolimits_{i=1}^{N} \left[\left(\frac{\sum_\ell s_{i(\ell)k}^2 e^{\boldsymbol{s}_{i(\ell)}'\boldsymbol{\beta}}}{\sum_\ell e^{\boldsymbol{s}_{i(\ell)}'\boldsymbol{\beta}}} \right) - \left(\frac{\sum_\ell s_{i(\ell)k} e^{\boldsymbol{s}_{i(\ell)}'\boldsymbol{\beta}}}{\sum_\ell e^{\boldsymbol{s}_{i(\ell)}'\boldsymbol{\beta}}} \right)^2 \right] \tag{7.119}$$

and off-diagonal elements $(1 \le k < u \le p)$

$$\begin{aligned} \boldsymbol{I}(\boldsymbol{\beta})_{\beta_k,\beta_u} &= \sum\nolimits_{i=1}^{N} \left(\frac{\sum_\ell s_{i(\ell)k} s_{i(\ell)u} e^{\boldsymbol{s}_{i(\ell)}'\boldsymbol{\beta}}}{\sum_\ell e^{\boldsymbol{s}_{i(\ell)}'\boldsymbol{\beta}}} \right) \\ &\quad - \sum\nolimits_{i=1}^{N} \left(\frac{\sum_\ell s_{i(\ell)k} e^{\boldsymbol{s}_{i(\ell)}'\boldsymbol{\beta}}}{\sum_\ell e^{\boldsymbol{s}_{i(\ell)}'\boldsymbol{\beta}}} \right) \times \left(\frac{\sum_\ell s_{i(\ell)u} e^{\boldsymbol{s}_{i(\ell)}'\boldsymbol{\beta}}}{\sum_\ell e^{\boldsymbol{s}_{i(\ell)}'\boldsymbol{\beta}}} \right). \end{aligned} \tag{7.120}$$

Note that since the coefficient estimates $\widehat{\boldsymbol{\beta}}$ are derived from a conditional likelihood, with no intercept, then this model describes the relative odds as a function of the covariates $(\boldsymbol{x}_{ij})$ not the absolute probabilities or risks $\{\pi_i(\boldsymbol{x}_{ij})\}$. To model the absolute risks, we would have to use the full unconditional likelihood (7.111), in which we must also estimate the nuisance parameters $(\alpha_1, \ldots, \alpha_N)$.

7.6.2 Special Case: 1:1 Matching

For the special case of matched pairs or 1:1 matching then $n_i = 2$, and $m_{1i} = 0, 1,$or 2; where m_{1i} is equivalent to the ancillary statistic S_i employed in Section 6.6 that describes the simple conditional logit model for the matched 2×2 table. As was the case in Section 6.6, it is readily shown that only the discordant pairs contribute to the likelihood. If we adopt the convention that the first member of each discordant pair is that with the response $(y_{i1} = 1, D)$ and the second is that without $(y_{i2} = 0, \overline{D})$, then it is readily shown that the conditional likelihood reduces to

$$L_{(c)}(\boldsymbol{\beta}) = \prod\nolimits_{i:S_i=1} L_{i|S_i=1} = \prod\nolimits_{i=1}^{n_d} \frac{e^{\boldsymbol{x}'_{i1}\boldsymbol{\beta}}}{e^{\boldsymbol{x}'_{i1}\boldsymbol{\beta}} + e^{\boldsymbol{x}'_{i2}\boldsymbol{\beta}}} \tag{7.121}$$

$$= \prod\nolimits_{i=1}^{n_d} \frac{e^{(\boldsymbol{x}_{i1}-\boldsymbol{x}_{i2})'\boldsymbol{\beta}}}{1 + e^{(\boldsymbol{x}_{i1}-\boldsymbol{x}_{i2})'\boldsymbol{\beta}}},$$

where n_d is the number of discordant pairs, $n_d = M = f + g$ in (6.90) in the model for a simple matched 2×2 table.

The latter expression is equivalent to the log likelihood of an ordinary logistic model for n_d observations, all of which have a positive response, with no intercept in the model and with covariate values $\boldsymbol{z}_i = \boldsymbol{x}_{i1} - \boldsymbol{x}_{i2}$ for the ith pair. Therefore, PROC LOGISTIC can be used in this case using a data set with n_d observations, each with the dependent variable coded as a response $(y_i = 1)$ and with covariate vector $\boldsymbol{z}_i$. Then a model is fit with no intercept.

7.6.3 Matched Retrospective Study

Now consider a matched retrospective case-control study with N matched sets consisting of n_i members of whom m_{1i} are cases $(y_{ij} = 1)$ and m_{2i} are controls $(y_{ij} = 0)$. Each member of each set then has an associated covariate vector $\boldsymbol{x}_{ij}$, some elements of which may represent prior exposure variables. As described by Breslow, Day, Halvorsen, Prentice and Sabai (1978), we then wish to model the posterior (retrospective) probabilities for cases and controls

$$\phi_{i1}(\boldsymbol{x}) = P_i(\boldsymbol{x}|y = 1) \qquad \text{and} \qquad \phi_{i2}(\boldsymbol{x}) = P_i(\boldsymbol{x}|y = 0) \tag{7.122}$$

described in Section 6.6.5. We again apply a conditioning argument to the ith matched set to assess the conditional probability as in (7.112) of m_{1i} covariate vectors $\boldsymbol{x}_{i1} \ldots \boldsymbol{x}_{im_{1i}}$ among m_{1i} cases and m_{2i} covariate vectors $\boldsymbol{x}_{i(m_{1i}+1)} \cdots \boldsymbol{x}_{in_i}$ among m_{2i} controls given m_{1i} cases among the matched set of n_i subjects. For the ith matched set this yields the conditional likelihood

$$\widetilde{L}_{i|m_{1i},n_i} = \frac{\prod_{j=i}^{m_{1i}} \phi_i(\boldsymbol{x}_{ij}) \prod_{j=m_{1i}+1}^{n_i} [1 - \phi_i(\boldsymbol{x}_{ij})]}{\sum_{\ell=1}^{\binom{n_i}{m_{1i}}} \prod_{j(\ell)=1}^{m_{1i}} \phi_i(\boldsymbol{x}_{ij(\ell)}) \prod_{j(\ell)=m_{1i}+1}^{n_i} \left[1 - \phi_i(\boldsymbol{x}_{ij(\ell)})\right]}. \tag{7.123}$$

However, as was the case for a single 2×2 table in Section 6.6.5, each of the retrospective probabilities ϕ_{ij} then can be expressed as a function of the prospective probabilities π_{ij}. Substituting into (7.123) it is readily shown that

$$\widetilde{L}_{i|m_{1i},n_i} = L_{i|m_{1i},n_i} \tag{7.124}$$

as presented in (7.113) and (7.114). Therefore, the conditional logistic regression model can be applied to a matched retrospective study to describe the effects of covariates on the prospective odds of the disease, or of being a case.

7.6.4 Fitting the General Conditional Logistic Regression Model: The Conditional PH Model

Before the conditional logistic regression model was developed, Seigel and Greenhouse (1973) showed that the ordinary logistic regression model for unmatched data could be applied, approximately, to a matched study to obtain an estimate of the odds ratio. They showed that the unmatched model provides an estimate of β that is biased, if anything, toward 0. However, the variances of the estimates obtained from the ordinary model in this case are incorrect. Therefore, PROC LOGISTIC should not be used to analyze matched data, except in the special case of matched pairs with 1:1 matching described above. Rather, a program for fitting the discrete proportional hazards model for tied event times described in Section 9.4.6.2, such as PROC PHREG in SAS, can be used to fit the conditional logistic model.

In survival analysis (see Chapter 9), the dependent variable for the jth subject is the time t_j at which either an event is observed to occur (an event time) or at which the subject was last observed and was still event free (a right censoring time). The distinction between an event versus censoring time is provided by an event indicator $\delta_j = I(\text{event at } t_j) = 1$ if an event, 0 if censored for the jth subject.

Now, consider a set of N intervals of time. During the ith interval assume that n_i subjects are followed (at risk) of whom m_{1i} subjects experience the outcome event and m_{2i} subjects survive the interval event-free. Then conditioning on m_{1i} events and m_{2i} non-events during the ith interval, the conditional likelihood is as shown in (9.104) of Chapter 9. Thus the conditional logistic model for matched sets has a likelihood function that is of identical form to that of a stratified conditional logistic PH model with ties. Therefore, the model can be fit to matched sets using PROC PHREG with the DISCRETE option and with each matched set being a unique stratum.

Example 7.15 *Low Birth Weight*
Hosmer and Lemeshow (1989) present data from a study of factors associated with the risk of giving birth to a low-weight infant. Table 7.8 presents a subset of the data (presented with permission). These data are also presented in the SAS (1997) manual and on-line documentation that describes the use of the program PROC PHREG to fit the conditional logistic model for matched sets.

Sets of pregnant women were matched according to the year of age, the set sizes ranging from 3 to 18 women. The risk factors considered (in addition to age) were

Table 7.8 SAS Program for Conditional Logistic Regression Analysis of Low Birth Weight Data Using PROC PHREG

```
data one;
 input matchset casen controln lwt smoke ht ui case time;
cards;
   1   1   0   130   0   0   0   1   1
   1   0   1   110   0   0   0   0   2
   1   0   2   112   0   0   0   0   2
   1   0   3   135   1   0   0   0   2
   1   0   4   135   1   0   0   0   2
   1   0   5   170   0   0   0   0   2
   1   0   6    95   0   0   0   0   2
   2   1   0   130   1   0   1   1   1
   2   2   0   110   1   0   0   1   1
   2   3   0   120   1   0   0   1   1
   2   4   0   120   0   0   0   1   1
   2   5   0   142   0   1   0   1   1
   2   0   1   103   0   0   0   0   2
   2   0   2   122   1   0   0   0   2
   2   0   3   113   0   0   0   0   2
   2   0   4   113   0   0   0   0   2
   2   0   5   119   0   0   0   0   2
   2   0   6   119   0   0   0   0   2
   2   0   7   120   1   0   0   0   2
   .   .   .    .
   .   .   .    .
   .   .   .    .
  17   1   0   105   1   0   0   1   1
  17   0   1   121   0   0   0   0   2
  17   0   2   132   0   0   0   0   2
  17   0   3   134   1   0   0   0   2
  17   0   4   170   0   0   0   0   2
  17   0   5   186   0   0   0   0   2
;
proc sort; by matchset time;
proc phreg;
   model time*case(0) = lwt smoke ht ui / ties = discrete;
   strata matchset;
run;
```

the mother's weight (in lb) at her last menstrual period (*lwt*), an indicator variable for whether or not the mother smoked (*smoke*, 1=yes, 0=no), whether she had hypertension (*ht*, 1=yes, 0=no), and whether there was evidence of uterine irritation (*ui*, 1=yes, 0=no).

The data in Table 7.8 are organized by matched sets indicated by the variable *matchset,* each set comprising all women with the same age (16 years for *matchset* 1, 17 for *matchset* 2, etc.). This includes the number of each case and control (*casen* and *controln*) within each matched set, the covariate values, the indicator variable for the low birth weight response (*case*, 1=yes, 0=no), and a variable *time* ($= 2 - case$) that equals 1 if a case, 2 if control. This data set is available from the www (see Preface). The complete data set, not arranged by matched set, is available from the above references.

Table 7.9 then presents the results of the conditional logistic regression analysis of these data, extraneous material deleted. There are 17 matched sets with a total of 174 mothers of whom 54 (69%) gave birth to a low birth-weight infant. These were age matched to 120 controls who gave birth to a normal birth-weight infant. The conditional logistic model yields a model likelihood ratio test of 17.961 on 4 *df* with $p \leq 0.0013$. The parameter estimates equal the estimated log odds ratio and the risk ratio is the estimated odds ratio. Although this is a matched case-control study, these risk factors are estimates of the prospective odds ratio. Women who smoked and who had hypertension had significantly ($p \leq 0.05$) increased odds of having a low-birth-weight infant. Women with uterine irritation had an estimated 2.4-fold increased odds, but this was not statistically significant. The odds of a low-birth-weight infant decreased by 1.5% per lb higher maternal body weight. When expressed in terms of a 10 lb increase in body weight, there is an estimated odds ratio of 0.86 or a 14% risk reduction. Alternately, when expressed in terms of a 10 lb decrease, there is an estimated $OR = 1.162$ or a 16.2% increase in risk.

7.6.5 Robust Inference

For the case of 1:1 matching, where the ordinary logistic regression model can be used to fit the conditional logistic regression model, the information sandwich may be used as the basis for robust inferences as described for the unconditional logistic regression model. In the more general case where the proportional hazards regression model must be used to fit the conditional logistic model, the information sandwich for the Cox model can be employed, as described by Lin and Wei (1989) and Lin (1994). These computations are also provided by the SAS PROC PHREG, as described in Section 9.4.9.

7.6.6 Explained Variation

For the unconditional logistic regression model, based on the expression for the log-likelihood in (7.111) it follows that the fraction of explained log likelihood equals the fraction of explained entropy, or $R^2_E = R^2_\ell$. However, in the conditional logistic

Table 7.9 Conditional Logistic Regression Analysis of Low Birth Weight Data Using PROC PHREG

The PHREG Procedure
Summary of the Number of Event and Censored Values

Stratum	MATCHSET	Total	Event	Censored	Percent Censored
1	1	7	1	6	85.71
2	2	12	5	7	58.33
3	3	10	2	8	80.00
4	4	16	3	13	81.25
5	5	18	8	10	55.56
6	6	12	5	7	58.33
7	7	13	2	11	84.62
8	8	13	5	8	61.54
9	9	13	5	8	61.54
10	10	15	6	9	60.00
11	11	8	4	4	50.00
12	12	3	2	1	33.33
13	13	9	2	7	77.78
14	14	7	1	6	85.71
15	15	7	1	6	85.71
16	16	5	1	4	80.00
17	17	6	1	5	83.33
Total		174	54	120	68.97

Testing Global Null Hypothesis: BETA=0

Criterion	Without Covariates	With Covariates	Model Chi-Square
-2 LOG L	159.069	141.108	17.961 with 4 DF (p=0.0013)
Score	.	.	17.315 with 4 DF (p=0.0017)
Wald	.	.	15.558 with 4 DF (p=0.0037)

Analysis of Maximum Likelihood Estimates

Variable	DF	Parameter Estimate	Standard Error	Wald Chi-Square	Pr > Chi-Square	Risk Ratio
LWT	1	-0.014985	0.00706	4.50021	0.0339	0.985
SMOKE	1	0.808047	0.36797	4.82216	0.0281	2.244
HT	1	1.751430	0.73932	5.61199	0.0178	5.763
UI	1	0.883410	0.48032	3.38266	0.0659	2.419

model, based on the expression for the log-likelihood in (7.116), the fraction of explained log-likelihood, R_ℓ^2, though still valid, does not have a direct interpretation as a fraction of explained entropy loss, except in the case of 1:1 matching. Also, since the conditional logistic model likelihood equals that of the Cox proportional hazards model, measures of explained variation for the PH model may also be applied to the conditional logistic model. In a simulation study, Schemper (1992) showed that Madalla's likelihood ratio measure R_{LR}^2 (see Section 9.4.8) provides a simple approximation to more precise measures of explained variation for the PH model. In the conditional logistic model stratified by matched set, this measure is computed as

$$R_{LR}^2 = 1 - \exp\left[\frac{-X_{LR}^2}{\sum_{i=1}^N n_i}\right], \tag{7.125}$$

where the sample size in this expression is actually the sum total number of members.

In the above example, the fraction of explained log likelihood is

$$R_\ell^2 = 17.961/159.069 = 0.129; \tag{7.126}$$

whereas Madalla's $R_{LR}^2 = 1 - \exp\left[\frac{-17.961}{174}\right] = 0.098$.

7.7 PROBLEMS

7.1 Consider a logistic regression model with two covariates and $\boldsymbol{\theta} = (\alpha\ \beta_1\ \beta_2)^T$. Show that:

7.1.1. The likelihood is as expressed in (7.4);

7.1.2. $\log L(\boldsymbol{\theta})$ is as expressed in (7.5);

7.1.3. $U(\boldsymbol{\theta})_\alpha = m_1 - \sum_i \pi_i$;

7.1.4. $U(\boldsymbol{\theta})_{\beta_j} = \sum_i x_{ij}(y_i - \pi_i)$ for $j = 1, 2$;

7.1.5. The observed information matrix $\boldsymbol{i}(\boldsymbol{\theta})$ with elements in (7.11)–(7.15)

7.2 Use a logistic regression model to conduct an adjusted analysis over strata for the stratified 2×2 tables in Chapter 4. Compare the logistic model estimated *MLEs* of the odds ratios, their asymmetric confidence limits, the likelihood ratio and Wald tests to the results obtained previously using the *MVLE* and Mantel-Haenszel procedures for:

7.2.1. The Ulcer Clinical Trial data presented in Example 4.1 and with results shown in Example 7.3;

7.2.2. The Religion and Mortality data of Example 4.6;

7.2.3. The Pre-eclampsia data of Example 4.24;

7.2.4. The Smoking and Byssinosis data of Problem 4.9.

7.3 Thisted (1988, p.194) presents the results of a study conducted to assess the influence of the route of administration (X_2: 1=local, 0=general) of the anesthetic

Table 7.10 Allergic Reactions in Lumbar Surgery as a Function of Route of Administration and Gender

Category j	*# Allergic Reactions* a_j	*# Surgeries* n_j	*(Intercept, Route, Gender)* $(1,\ x_{1j},\ x_{2j})$
1	76	$20,959$	$(1,0,0)$
2	132	$12,937$	$(1,0,1)$
3	15	$6,826$	$(1,1,0)$
4	29	$4,191$	$(1,1,1)$

chymopapain and gender (X_2: 1=female, 0=male) on the risk of an anaphylactic allergic reaction during lumbar disc surgery. From a sample of $N = 44,913$ surgical procedures, the data in each category of the covariates are presented in Table 7.10 (reproduced with permission).

7.3.1. Conduct a logistic regression analysis of the influence of route of administration on the odds of a reaction with no adjustment.

7.3.2. Then conduct a logistic regression analysis that is also adjusted for gender. Include the COVB option in this analysis.

7.3.3. Fit an additional model that includes an interaction between route of administration and gender. The Wald or likelihood ratio test of the interaction effect is equivalent to the hypothesis of homogeneity among strata described in Section 4.6.

7.3.4. From the model in Problem 7.3.2, based on the estimated coefficients and the estimated covariance matrix of the estimates, compute the estimated probability of an allergic reaction and its 95% confidence limits from (7.18) for each of the following sub-populations:

1. Female with local anesthesia;
2. Male with local anesthesia;
3. Female with general anesthesia;
4. Male with general anesthesia.

Note that these probabilities may also be obtained from PROC LOGISTIC using the `ppred` option.

7.4 Use the GLM construction for binomial models described in Section 7.1.5 and the expressions (7.29)–(7.31) based on the chain rule. Then, for a two-covariate model as in Problem 7.1, derive the likelihood, log likelihood, the score estimating equations, the elements of the Hessian, and the observed and expected information matrices for:

7.4.1. An exponential risk model using the log link: $\log(\pi_i) = \alpha + \boldsymbol{x}_i'\boldsymbol{\beta}$;

7.4.2. A compound exponential risk model using the complimentary log-log link: $\log\left[-\log(\pi_i)\right] = \alpha + \boldsymbol{x}_i'\boldsymbol{\beta}$;

7.4.3. A probit regression model using the inverse probit link: $\Phi(\pi_i) = \alpha + \boldsymbol{x}_i'\boldsymbol{\beta}$, where $\Phi(\cdot)$ is the cumulative normal distribution function, the inverse function for which is the probit, or $\pi_i = \Phi^{-1}(\alpha + \boldsymbol{x}_i'\boldsymbol{\beta})$.

7.4.4. Use PROC GENMOD to conduct an analysis of the data from Problem 4.9 using each of the above links.

7.5 Consider a model of the form $\log(y) = \alpha + \log(x)\beta$ where the coefficient equals the change in $\log(Y)$ per unit change in $\log(X)$.

7.5.1. Show that a c-fold difference in X, such as $x_1 = cx_0$ is associated with a c^β-fold change in Y. Thus the percent change in Y per c-fold change in X is $100(c^\beta - 1)$ for $c > 0$.

7.5.2. Suppose that a model fit using covariate $\log(X)$ yields $\widehat{\beta} = -2.1413$ with $S.E.(\widehat{\beta}) = 0.9618$. Show that this implies that there is a 25.3% increase in the odds per 10% reduction in X, with 95% $C.I.$ (2.74, 52.8%).

7.6 Consider a regression model with two binary covariates (X_1, X_2) and a structural relationship as expressed in (7.46) for some smooth link function $g(\cdot)$ with independent errors having mean zero. Then for the bivariate model we assume that $E(y \,|\, x_1, x_2) = g^{-1}(\alpha + x_1\beta_1 + x_2\beta_2)$. For simplicity we assume that X_1 and X_2 are statistically independent and that X_2 has probabilities $P(x_2 = 1) = \phi$ and $P(x_2 = 0) = 1 - \phi$. Now consider the values of the coefficients when X_2 is dropped from the model to yield $E(y|x_1) = g^{-1}(\widetilde{\alpha} + x_1\widetilde{\beta}_1)$. Note that this notation is simpler than that used in Section 7.2.2.

7.6.1. Show that the coefficients in the reduced model can be obtained as

$$\begin{aligned} E(y \,|\, x_1) = g^{-1}(\widetilde{\alpha} + x_1\widetilde{\beta}_1) &= E_{x_2}\left[E(y \,|\, x_1, x_2)\right] \\ &= (1 - \phi)g^{-1}(\alpha + x_1\beta_1) + \phi g^{-1}(\alpha + x_1\beta_1 + \beta_2). \end{aligned} \tag{7.127}$$

7.6.2. Let $g(\cdot)$ be the identity function. Show that $\widetilde{\alpha} = \alpha + \phi\beta_2$ and $\widetilde{\beta}_1 = \beta_1$.

7.6.3. Let $g(\cdot)$ be the log function, $g^{-1}(\cdot) = \exp(\cdot)$. Show that $\widetilde{\alpha} = \alpha + \log(1 - \phi + \phi e^{\beta_2})$ and $\widetilde{\beta}_1 = \beta_1$.

7.6.4. Let $g(\cdot)$ be the logit function and $g^{-1}(\cdot)$ the inverse logit. Show that

$$\begin{aligned} \widetilde{\alpha} &= \log \frac{\pi_0}{1 - \pi_0}, \\ \pi_0 &= E(y \,|\, x_1 = 0) = \frac{1 - \phi}{1 + e^{-\alpha}} + \frac{\phi}{1 + e^{-(\alpha + \beta_2)}} \end{aligned} \tag{7.128}$$

and that

$$\begin{aligned} \widetilde{\beta}_1 &= \log \frac{\pi_1}{1 - \pi_1} - \widetilde{\alpha}, \\ \pi_1 &= E(y \,|\, x_1 = 1) = \frac{1 - \phi}{1 + e^{-(\alpha + \beta_1)}} + \frac{\phi}{1 + e^{-(\alpha + \beta_1 + \beta_2)}}, \end{aligned} \tag{7.129}$$

where $\widetilde{\beta}_1$, in general, does not equal β_1. Note that $e^a/(1 + e^a) = (1 + e^{-a})^{-1}$.

7.6.5. Assume a logit model with $\alpha = 0.2$, $\beta_1 = 0.5$ and $\beta_2 = 1.0$, where $\phi = 0.4$. If X_2 is dropped from the model, show that $\widetilde{\alpha} = 0.5637$ and $\widetilde{\beta}_1 = 0.4777$.

7.6.6. Let $g(\cdot)$ be the complimentary log-log function with the inverse link function $g^{-1}(\cdot) = \exp[-\exp(\cdot)]$. Show that

$$\begin{aligned} \widetilde{\alpha} &= \log[-\log(\pi_0)], \\ \pi_0 &= E(y|x_1 = 0) = e^{-e^{\alpha}} \left[1 - \phi + \phi e^{-e^{\alpha} e^{\beta_2 - 1}}\right] \end{aligned} \tag{7.130}$$

and that

$$\begin{aligned} \widetilde{\beta}_1 &= \log[-\log(\pi_0)] - \widetilde{\alpha}, \\ \pi_1 &= E(y|x_1 = 1) = e^{-e^{\alpha+\beta_1}} \left[1 - \phi + \phi e^{-e^{\alpha+\beta_1} e^{\beta_2 - 1}}\right], \end{aligned} \tag{7.131}$$

where $\widetilde{\beta}_1$, in general, does not equal β_1.

7.7 For the logistic regression model, show that under H_0: $\boldsymbol{\beta} = \mathbf{0}$,

7.7.1. $\boldsymbol{U}\left[\widehat{\alpha}, \boldsymbol{\beta}\right]^T_{|\boldsymbol{\beta}=\mathbf{0}}$ is as presented in (7.58);

7.7.2. $\boldsymbol{I}\left(\widehat{\alpha}, \boldsymbol{\beta}\right)_{|\boldsymbol{\beta}=\mathbf{0}}$ is as presented in (7.60);

7.7.3. The efficient score test for H_0 is as presented in (7.61).

7.7.4. Verify the computation of the score vector in Example 7.6 under H_0, the corresponding information matrix and the score test value.

7.8 Now consider the log-risk model with a log link as in Problem 7.4.1. Generalizing the results of Problem 6.4, for a model with p covariates $\boldsymbol{x}$, intercept α and coefficient vector $\boldsymbol{\beta}$, derive the expression for the model score test under H_0: $\boldsymbol{\beta} = \mathbf{0}$.

7.9 Direct Adjustment (Gastwirth and Greenhouse, 1995): In model-based direct adjustment, the model obtained from one sample is applied to a different sample of subjects. For example, a logistic model might be obtained from a sample of M males and then applied to a sample of N females to determine whether the number of responses among females differs from what would be expected if the male risk function also applied to females. In the sample of females, the observed number of responses is $O = \sum_{i=1}^{N} y_i$. Then the expected number of responses is estimated as $\widehat{E} = \sum_{i=1}^{N} \widehat{\pi}_i$, where $\widehat{\pi}_i$ is provided by the model derived from the sample of males. Since the model coefficients are obtained from an independent sample, then

$$V(O - \widehat{E}) = V\left(\sum\nolimits_{i=1}^{N} y_i\right) + V\left(\sum\nolimits_{i=1}^{N} \widehat{\pi}_i\right) = V(O) + V(\widehat{E}), \tag{7.132}$$

which requires $V\left(\widehat{\pi}_i\right)$. Now do the following:

7.9.1. Show that $V(O) = \sum_i \pi_i(1 - \pi_i)$, where $\pi_i = E(y_i)$.

7.9.2. Under the hypothesis that the male model also applies to females, show that the male model provides an estimate of $V(O)$ as

$$\widehat{V}(O) = \sum\nolimits_{i=1}^{N} \widehat{\pi}_i(1 - \widehat{\pi}_i). \tag{7.133}$$

7.9.3. Then show that

$$V\left[\widehat{E}\right] = \sum\nolimits_i V\left(\widehat{\pi}_i\right). \tag{7.134}$$

7.9.4. The estimated probabilities are based on the linear predictor for the ith subject, $\widehat{\eta}_i = \widetilde{\boldsymbol{x}}_i'\boldsymbol{\theta} = \alpha + \sum_{j=1}^{p} x_{ij}\widehat{\beta}_j$ in the notation of (7.16). Use the δ-method to show that

$$\begin{aligned}\widehat{V}\left(\widehat{\pi}_i\right) &\cong \left(\frac{d\widehat{\pi}_i}{d\widehat{\eta}_i}\right)^2 \widehat{V}\left(\widehat{\eta}_i\right) = \frac{\left(e^{-\widehat{\eta}_i}\right)^2}{\left(1+e^{-\widehat{\eta}_i}\right)^4}\widetilde{\boldsymbol{x}}_i'\widehat{\boldsymbol{\Sigma}}_{\boldsymbol{\theta}}\widetilde{\boldsymbol{x}}_i \\ &= \widehat{\pi}_i^2\left(1-\widehat{\pi}_i\right)^2\widetilde{\boldsymbol{x}}_i'\widehat{\boldsymbol{\Sigma}}_{\boldsymbol{\theta}}\widetilde{\boldsymbol{x}}_i.\end{aligned} \tag{7.135}$$

7.9.5. Assume that a new anesthetic drug has been developed and that a new study is conducted that shows the following numbers of allergic reactions with local and general anesthesia among males and females in the same format as presented for the drug chymopapain in Table 7.10:

Stratum j	*# Allergic Reactions* a_j	*# Surgeries* n_j	*(Intercept, Route, Gender)* $(1,\ x_{1j},\ x_{2j})$
1	53	18,497	(1,0,0)
2	85	11,286	(1,0,1)
3	19	7,404	(1,1,0)
4	34	5,369	(1,1,1)

Apply the logistic regression model obtained from the study of chymopapain obtained in Problem 7.3, with its estimated covariance matrix, to these data obtained for the new anesthetic drug. Estimate $O - E$ and its variance under the null hypothesis that the risk of allergic reactions with chymopapain also applies to the new agent. Compute a Wald test of this hypothesis.

7.10 Use the DCCT nephropathy data of Table 7.5 that can be obtained from the www (see Preface).

7.10.1. Verify the computation of the matrices $\widehat{\boldsymbol{J}}(\widehat{\boldsymbol{\theta}})$ and $\widehat{\boldsymbol{\Sigma}}_R(\widehat{\boldsymbol{\theta}})$ presented in Example 7.8, the robust confidence limits and Walt tests for each of the model parameters.

7.10.2. Verify the computation of the matrices $\widehat{\boldsymbol{J}}(\widehat{\boldsymbol{\theta}}_0)$, and $\widehat{\boldsymbol{\Sigma}}_R(\widehat{\boldsymbol{\theta}}_0)$ and the computation of the robust efficient score test of the model.

7.11 Consider the determination of sample size and power for a logistic regression model using the study design in Problem 4.11. Although the odds ratios may differ among strata, we desire the sample size for a stratified-adjusted analysis with an overall treatment group odds ratio. Assume that all tests are conducted at the 0.05 level.

7.11.1. Determine the corresponding logistic regression model parameters by generating a large data set with cell frequencies proportional to the projected sample fractions, and with numbers of positive responses proportional to the response probabilities within each cell.

7.11.2. Then compute the covariance matrix Ω of the Bernoulli variables and $V(\widehat{\boldsymbol{\theta}})$ as in (7.73).

7.11.3. Now, consider a Wald test of the adjusted group effect on 1 *df* based on a contrast vector $C = (0\ 0\ 0\ 1)^T$. Compute the non-centrality factor K^2 as in (7.75).

7.11.4. Determine the sample size required to provide 90% power for this 1 *df* Wald test.

7.11.5. Likewise, consider the Wald model test on 3 *df* of the null hypothesis H_0: $\boldsymbol{\beta} = \mathbf{0}$. Here the matrix $\boldsymbol{C}$ is a 4×3 matrix of rank 3 where $\boldsymbol{C} = (0\ 0\ 0$ // 1 0 0 // 0 1 0 // 0 0 1) and "//" denotes concatenation of row vectors. Compute the non-centrality factor K^2 as in (7.75).

7.11.6. Determine the sample size required to provide 90% power for this 3 *df* Wald model test.

7.11.7. Suppose that a sample size of $N = 220$ is proposed. What level of power does this provide for the 1 *df* test of treatment group?

7.11.8. What level of power does this provide for the 3 *df* model test?

7.12 Consider the case-control study of ischemic heart disease that is presented in Example 7.5. The effect of hypertension, adjusting for smoking and hyperlipidemia, is an odds ratio of 1.386 that is not statistically significant at the 0.05 level based on the Wald test (i.e., on the 95% confidence intervals).

7.12.1. Treating the sample frequencies within each cell as fixed, and treating the intercept and coefficients for smoking and hyperlipidemia as fixed (equal to the estimates shown in the example), what level of power was there to detect an odds ratio for hypertension versus not of 1.5 in the 1 *df* Wald test?

7.12.2. What sample size would be required to provide 90% power to detect an odds ratio for hypertension of 2.0?

7.13 Consider models with an interaction between two covariates as in Section 7.4.

7.13.1. From the interaction model in (7.84) for the ulcer clinical trial data in Example 4.1, compute the odds and odds ratios for the cells of the table in (7.85) for drug versus placebo within each stratum.

7.13.2. Fit a similar stratum by group interaction model to:

1. The Religion and Mortality data of Example 4.6;
2. The Pre-eclampsia data of Example 4.24;
3. The Smoking and Byssinosis data of Problem 4.9;
4. The allergic reactions data of Table 7.10, treating route of administration as the exposure factor and gender as the stratification factor.

7.13.3. For each, compute the odds and odds ratios for the principal exposure or treatment group versus not within each stratum.

7.13.4. For each, compute the Wald and likelihood ratio tests of homogeneity among strata.

7.13.5. Using the DCCT interaction model presented in Example 7.12, compute the odds ratio for intensive versus conventional treatment over the range 3–14 years duration of diabetes that corresponds approximately to the 5th to 95th percentiles of the distribution of duration. Display the estimates in a figure such as Figure 7.1.

7.13.6. For this model, also compute the odds ratio per unit increase in blood pressure over the range of HbA_{1c} values of 7–12 %, also corresponding approximately to the 5th to 95th percentiles. Display the estimates in a figure.

7.13.7. Consider a model with two covariates (X_1, X_2) where $X_1 = \log(Z)$ and there is an interaction between X_1 and X_2. Show that the odds ratio per c-fold change in Z given the value of X_2 is obtained as

$$\widehat{OR} = c^{\widehat{\beta}_1 + x_2 \widehat{\beta}_{12}} . \tag{7.136}$$

7.13.8. Using the DCCT data, fit an interaction model as in Example 7.12 using $X_2 = \log(HbA_{1c})$.

7.13.9. Conduct a 2 *df* test of the effect of HbA_{1c}.

7.13.10. Compute the odds ratio per 10% reduction in HbA_{1c}, and its 95% confidence limits, over the range of blood pressure values and display the estimates in a figure.

7.14 For a logistic model, using expressions for the measures of explained variation, do the following:

7.14.1. Show that the model fits perfectly when $\widehat{\pi}_i y_i + (1 - \widehat{\pi}_i)(1 - y_i) = 1$ for all $i = 1, \ldots, N$. In this case, since $\widehat{\overline{\pi}} = m_1/N$, also show that $\widehat{\rho}^2_{risk} = 1$.

7.14.2. Conversely, show that the model explains none of the variation in risk when $\widehat{\pi}_i = \widehat{\overline{\pi}}$ for all subjects, irrespective of the covariate values, and that in this case $\widehat{\rho}^2_{risk} = 0$.

7.14.3. Noting that $\sum_i y_i = \sum_i \widehat{\pi}_i$, show that the following equality applies:

$$\sum\nolimits_i y_i \log(\widehat{\pi}_i) + (1 - y_i) \log(1 - \widehat{\pi}_i) = \sum\nolimits_i \widehat{\pi}_i \log(\widehat{\pi}_i) + (1 - \widehat{\pi}_i) \log(1 - \widehat{\pi}_i).$$

7.14.4. From (A.193), show that the fraction of explained residual variation with squared error loss in a logistic regression model is

$$R^2_{\varepsilon^2, resid} = \frac{\sum_i \left(\widehat{\pi}_i - \widehat{\overline{\pi}}\right)^2}{\sum_i \left(y_i - \widehat{\overline{\pi}}\right)^2} \tag{7.137}$$

where $\widehat{y}_i = \widehat{\pi}_i$ and $\overline{y} = \overline{\pi}$.

7.14.5. Noting that $\widehat{\overline{\pi}} = \sum_i y_i/N$, show that $R^2_{resid} = \widehat{\rho}^2_{risk}$ in (7.91)

7.14.6. Show that $R^2_{E,resid}$ in (7.102) equals $\widehat{\rho}^2_E$ in (7.101).

7.14.7. Show that $R^2_{E,resid}$ in (7.102) equals R^2_ℓ in (7.103).

7.14.8. For a single 2×2 table with a single binary covariate for treatment group, based on Problem 6.2.13, show that the entropy R^2 equals

$$R_E^2 = \frac{\log\left[\dfrac{a^a b^b c^c d^d N^N}{n_1^{n_1} n_2^{n_2} m_1^{m_1} m_2^{m_2}}\right]}{\log\left[\dfrac{N^N}{m_1^{m_1} m_2^{m_2}}\right]} . \tag{7.138}$$

7.14.9. Among the many measures of association in a 2×2 table, Goodman and Kruskal (1972) suggested the uncertainty coefficient that is based on the ability to predict column membership (the response) from knowledge of the row membership (the exposure group). For the 2×2 table, this coefficient can be expressed as

$$U(C|R) = \frac{\sum_{i=1}^{2} n_i \log \frac{n_i}{N} + \sum_{j=1}^{2} m_j \log \frac{m_j}{N} - \sum_{i=1}^{2}\sum_{j=1}^{2} n_{ij} \log \frac{n_{ij}}{N}}{\sum_{j=1}^{2} m_j \log \frac{m_j}{N}} , \tag{7.139}$$

where the $\{n_{ij}\}$, $\{n_i\}$, and $\{m_j\}$ are the cell frequencies, row (group) totals and column (response) totals, respectively. Show that $R_E^2 = U(C|R)$.

7.15 Collett (1991) presents the data given in Table 7.11 (reproduced with permission) from a study of factors associated with local metastases to the lymph nodes (nodal involvement) in subjects with prostate cancer, indicated by the variable *nodalinv* (=1 if yes, 0 if no). The risk factors considered are age at diagnosis (*age*), the level of acid phosphatase in serum in King-Armstrong units (*acid*) that is a measure of the degree of tissue damage, positive versus negative signs of nodal involvement on x-ray (*xray*=1 if positive, 0 if not), size of the tumor based on rectal examination (*size*=1 if large, 0 if not), and the histologic grade of severity of the lesion based on a tissue biopsy (*grade*=1 if serious, 0 if not). The variable *acid* is log transformed (*lacd*). This data set is also included in the SAS (1995) text on logistic regression and is available from the www (see Preface). Conduct the following analyses of these data:

7.15.1. Use PROC FREQ of *nodalinv*(xray size grade)* to obtain an unadjusted estimate of the odds ratio and the associated logit confidence limits associated with each of the binary covariates.

7.15.2. Conduct a logistic regression analysis of the odds of nodal involvement using PROC LOGISTIC with *age lacd xray size grade* as covariates. Interpret all coefficients in terms of the associated odds ratio with their asymmetric confidence limits.

7.15.3. For *age*, also compute $\triangle \log(\text{odds})$ per five years lower age and the associated confidence limits.

7.15.4. Since *lacd* is log(*acid*), describe the effect of acid phosphatase on the risk of nodal involvement in terms of the % change in odds per 10% larger acid phosphatase, with the associated 95% confidence limits (see Problem 7.5.1).

7.15.5. Compare the unadjusted and adjusted odds ratio for *xray size* and *grade*.

Table 7.11 SAS Program to Input Prostate Cancer Data from Collett (1991)

```
data prostate; input case age acid xray size grade nodalinv @@;
lacd=log(acid);
datalines;
1  66  .48 0 0 0 0  2  68  .56 0 0 0 0  3  66  .50 0 0 0 0
4  56  .52 0 0 0 0  5  58  .50 0 0 0 0  6  60  .49 0 0 0 0
7  65  .46 1 0 0 0  8  60  .62 1 0 0 0  9  50  .56 0 0 1 1
10 49  .55 1 0 0 0  11 61  .62 0 0 0 0  12 58  .71 0 0 0 0
13 51  .65 0 0 0 0  14 67  .67 1 0 1 1  15 67  .47 0 0 1 0
16 51  .49 0 0 0 0  17 56  .50 0 0 1 0  18 60  .78 0 0 0 0
19 52  .83 0 0 0 0  20 56  .98 0 0 0 0  21 67  .52 0 0 0 0
22 63  .75 0 0 0 0  23 59  .99 0 0 1 1  24 64 1.87 0 0 0 0
25 61 1.36 1 0 0 1  26 56  .82 0 0 0 1  27 64  .40 0 1 1 0
28 61  .50 0 1 0 0  29 64  .50 0 1 1 0  30 63  .40 0 1 0 0
31 52  .55 0 1 1 0  32 66  .59 0 1 1 0  33 58  .48 1 1 0 1
34 57  .51 1 1 1 1  35 65  .49 0 1 0 1  36 65  .48 0 1 1 0
37 59  .63 1 1 1 0  38 61 1.02 0 1 0 0  39 53  .76 0 1 0 0
40 67  .95 0 1 0 0  41 53  .66 0 1 1 0  42 65  .84 1 1 1 1
43 50  .81 1 1 1 1  44 60  .76 1 1 1 1  45 45  .70 0 1 1 1
46 56  .78 1 1 1 1  47 46  .70 0 1 0 1  48 67  .67 0 1 0 1
49 63  .82 0 1 0 1  50 57  .67 0 1 1 1  51 51  .72 1 1 0 1
52 64  .89 1 1 0 1  53 68 1.26 1 1 1 1
;
```

7.15.6. Conduct a logistic regression analysis using PROC GENMOD to compute likelihood ratio tests for each covariate in the model. Compare these to the Wald tests.

7.15.7. Compute R^2_{resid}, R^2_E, and R^2_{LR} for these data. To obtain the corrected sum of squares of the predicted probabilities, use the following statements after the PROC LOGISTIC model statement:

```
output out=two pred=pred;
proc univariate data=two; var pred;
```

7.15.8. Use PROC GENMOD to compute the robust information sandwich estimator of the covariance matrix of the estimates and the associated robust 95% confidence limits and the robust Wald tests for each covariate (see Section 8.4.2). Compare these to those obtained under the assumed model.

7.15.9. Write a program to verify the computation of the robust information sandwich estimate of the covariance matrix of the estimates.

7.15.10. Write a program to evaluate the score vector $U(\widehat{\theta}_0)$ under the model null hypothesis H_0: $\boldsymbol{\beta} = \mathbf{0}$, and to also compute the robust information sandwich estimate of the covariance matrix under this null hypothesis. Use these to compute the robust model score test that equals $X^2 = 15.151$ on 5 *df* with $p \leq 0.0098$. Compare this to the likelihood ratio and score tests obtained under the model.

7.15.11. Fit a model with pairwise interactions between *age* and *lacd*, and between *size* and *grade*. Even though the interaction between *age* and *lacd* is not significant at the 0.05 level, provide an interpretation of both interaction effects as follows.

1. Compute the odds ratio for serious versus not grade of cancer as a function of the size category of the tumor. Compute the variance of each log odds ratio and then the asymmetric 95% confidence limits for each odds ratio.
2. Compute the odds ratio for large versus not size of the tumor as a function of the grade category or the tumor. Compute the variance of each log odds ratio and then the asymmetric 95% confidence limits for each odds ratio.
3. Compute the odds ratio associated with a 10% higher level of the acid phosphatase as a function of the age of the patient, using at least ages 50, 55, 60 and 65, and for each compute the 95% asymmetric confidence limits.
4. Compute the odds ratio associated with a 5 year higher age as a function of the level of log acid phosphatase, using at least the values of acid phosphatase of 0.5, 0.65, 0.8 and 1.0, and for each compute the 95% asymmetric confidence limits.

7.15.12. Now fit a model with interactions between *lacd* and *grade*, and between *size* and *grade*.

1. Compute the odds ratio for serious versus not grade of cancer as a function of the size category of the tumor and the level of log acid phosphatase, using at least the above specified values.
2. Compute the variance of each log odds ratio and then the asymmetric 95% confidence limits for each odds ratio.
3. Use a model with the `test` option to compute a multiple degree of freedom test of the overall effect of *lacd, grade* and *size.*

7.16 Consider the logistic regression model for matched sets in Section 7.5.

7.16.1. Derive the likelihood $L(\boldsymbol{\alpha}, \boldsymbol{\beta})$ in (7.110).

7.16.2. Then conditioning on the m_{1i} responses with respective covariate values as in (7.112), derive the conditional logistic likelihood for matched sets in (7.113).

7.16.3. Show that the likelihood reduces to (7.114).

7.16.4. Derive the expression for the score equation for β_k presented in (7.118).

7.16.5. Derive the expressions for the elements of the information matrix presented in (7.119) and (7.120).

7.16.6. For the special case of 1:1 matching, show that the conditional likelihood reduces to (7.121).

7.17 For the matched retrospective study described in Section 7.6.3, show that the retrospective conditional likelihood in (7.123) equals the prospective conditional likelihood in (7.113).

7.18 Breslow and Day (1980) present data from a matched case-control study with four controls per case from a study of the risk of endometrial cancer among women who resided in a retirement community in Los Angeles (Mack et al., 1976).

These data are available from the www (see Preface) and are used with permission. Each of the 63 cases was matched to four controls within one year of age of the case, who had the same marital status, and who entered the community at about the same time. The variables in the data set are:

caseset: the matched set number.

case: 1 if case, 0 if control, in sets of 5 (1+4).

age: in years (a matching variable).

gbdx: history of gallbladder disease (1=yes, 0=no).

hypert: history of hypertension, (1=yes, 0=no).

obese: (1=yes, 0=no, 9=unknown). Breslow and Day combine the no and unknown categories in their analyses.

estrogen: history of estrogen use, (1=yes, 0=no).

dose: dose of conjugated estrogen use: 0=none, 1 = 0.3; 2 = (0.301-0.624); 3 = 0.625; 4 = (0.626-1.249); 5 = 1.25; 6 = (1.26-2.50); 9 = unknown.

dur: duration of conjugated estrogen use in months, values >96 are truncated at 96, 99=unknown.

nonest: history of use of non-estrogen drug use (1=yes, 0=no).

Note that conjugated estrogen use is a subset of all estrogen use. Also the variables for dose and duration are defined only for *conjugated* estrogen use. Thus, for some analyses below an additional variable *conjest* must be defined as 0 for $dose = 0$, 1 otherwise. Those subjects with an unknown dose are assumed to have used conjugated estrogens.

Also, in the Breslow and Day (1980) analyses of these data, the *dose* categories 3-6 were combined, and those with unknown dose treated as missing. This yields a variable with four categories – none (0), low (1), middle (2) and high (3).

From these data, perform the following analyses: In each case, state the model and describe the interpretation of the regression coefficients.

7.18.1. Fit the conditional logistic model using PROC PHREG to assess the effect of estrogen use on the odds of having cancer with no other covariates (unadjusted) and then adjusted for other covariates (*gbdx, hypert, obese, nonest*).

7.18.2. Assess the effect of the dose of conjugated estrogen on risk unadjusted and then adjusted for the other four covariates. Do so using the four groupings of dose described above (none, low, middle high). Since this is a categorical variable, use a 3 *df* test and 3 odds ratios for the low, middle and high dose categories versus none. This can be done using three indicator variables.

7.18.3. Now consider the effect of duration of estrogen use including both *conjest* and *dur* in the model because we wish to describe the dose effect only among the estrogen users

1. Fit the model with both covariates unadjusted and then adjusted for other covariates.

2. To show the effects of both variables simultaneously; explain the covariate effects for a non-user, and then for a user with increasing doses of conjugated estrogen use.

7.18.4. The analyses in Problem 7.18.2 compare each dose level to the non-users. Now perform an analysis comparing the high and middle dose to the low dose. This can be done with a new class variable where the 0 category is the low dose, or using three binary variables for each of the other categories versus the low dose (including the nonusers).

7.18.5. Now select only the first control matched to each case, and set up matched pairs. Do the analyses above using only these 1:1 matched pairs in PROC LOGISTIC. This is suggested only to demonstrate use of the technique.

8

Analysis of Count Data

Thus far we have considered only the risk of a single outcome such as a positive or negative response, D or $\overline{D}$, over an assumed fixed period of exposure for each subject. In many settings, however, we are interested in the risk of possibly recurrent events during some period of exposure. Although the exposure period may vary from subject to subject, the frequency of events is usually summarized as a rate or number per year of exposure, or per 100 patient years, and so on.

Common examples include the rate of seizures per year among subjects with epilepsy; the rate of infections per year among school children; the rate of hospitalizations per year among patients with cardiovascular disease; and the rate of episodes of hypoglycemia per year among patients with diabetes, among many. Although time is the unit of exposure in these examples, it need not always be the case. Other examples are the rate of infections per hospitalization, cavities per set of teeth, defects or abnormalities per item, and so on.

This chapter describes methods for the analysis of such data analogous to those presented in previous chapters. Since the majority of these applications occur in the setting of a prospective study, we omit discussion of methods for the analysis of retrospective studies.

8.1 EVENT RATES AND THE HOMOGENEOUS POISSON MODEL

8.1.1 Poisson Process

Count data arises when we observe d_j events during t_j units of exposure (months, years, etc.) for the jth in a sample of N subjects ($j = 1, ..., N$). Although the actual event times may also be observed, here we focus only on the number of events, not their times. We then assume that the sequence of events and the aggregate count for each subject is generated by a *Poisson process*. The Poisson process is described in any general reference on stochastic processes such as Cox and Miller (1965), Karlin and Taylor (1975), and Ross (1983), among many. The Poisson process specifies that the underlying risk or rate of events over time is characterized by the *intensity* $\alpha(t)$, possibly varying in t. The intensity of the process is the instantaneous probability of the event among those at risk at time t. Now, let $\mathrm{N}(t)$ denote the *counting process,* which is the cumulative number of events observed over the interval $(0, t]$. Then one of the features of the Poisson process with intensity $\alpha(t)$ is that probability of any number of events, say $d = m$, in an interval $(t, t+s]$ is given by the Poisson probability

$$P\{[\mathrm{N}(t+s) - \mathrm{N}(t)] = m\} = \frac{e^{-\Lambda_{(t,t+s)}} \left(\Lambda_{(t,t+s)}\right)^m}{m!}, \qquad t > 0, s > 0 \quad (8.1)$$

with rate parameter

$$\Lambda_{(t,t+s)} = \int_t^{t+s} \alpha(u)\, du \quad (8.2)$$

equal to the *cumulative intensity* over the interval. This cumulative intensity or rate parameter can also be expressed as $\Lambda_{(t,t+s)} = \lambda_{(t,t+s)} s$, where the *mean or linearized rate* is obtained by the mean value theorem as

$$\lambda_{(t,t+s)} = \frac{\Lambda_{(t,t+s)}}{s}. \quad (8.3)$$

Therefore, the probability of events over any interval of time can be obtained from the Poisson distribution with rate parameter $\theta = \Lambda_{(t,t+s)} = \lambda_{(t,t+s)} s$.

8.1.2 Doubly Homogeneous Poisson Model

In the most general case, the Poisson process may be time-heterogeneous where the intensity $\alpha(t)$ varies over time, also called a non-homogeneous Poisson process. In the simplest case, the intensity of the process may be a fixed constant $\alpha(t) = \lambda$ for $\forall t$, in which case the process is homogeneous over time. In this *homogeneous Poisson process* the number of events d in an interval of length t is distributed as *Poisson*(λt) or

$$P(d; t, \lambda) = \frac{e^{-\lambda t} (\lambda t)^d}{d!}, \qquad d \geq 0, t > 0. \quad (8.4)$$

Herein we also make the additional assumption that this constant intensity applies to all subjects in the population, what we call the *doubly homogeneous Poisson model*. Under this model, the likelihood function for a sample of N independent observations is

$$L(\lambda)=\prod_{j=1}^{N}\frac{e^{-\lambda t_j}\left(\lambda t_j\right)^{d_j}}{d_j!} \tag{8.5}$$

given the set of exposure times $\{t_1,...,t_N\}$, that is, conditioning on the exposure times as fixed constants. The log likelihood is

$$\ell(\lambda)=\sum\nolimits_{j=1}^{N}\left[(-\lambda t_j)+d_j\log\left(t_j\right)+d_j\log\left(\lambda\right)-\log\left(d_j!\right)\right]. \tag{8.6}$$

Then the efficient score is

$$U\left(\lambda\right)=\frac{\partial\ell}{\partial\lambda}=\frac{1}{\lambda}\sum\nolimits_j\left[d_j-\lambda t_j\right]=\frac{1}{\lambda}\sum\nolimits_j\left[d_j-E(d_j|\lambda,t_j)\right], \tag{8.7}$$

where d_j and $E(d_j|\lambda,t_j)$ are the observed and expected number of events, respectively. The maximum likelihood estimating equation $U\left(\lambda\right)=0$ then implies that $\sum_j d_j=\sum_j \lambda t_j$ so that the MLE is

$$\widehat{\lambda}=\frac{\sum_j d_j}{\sum_j t_j}=\frac{D}{T}, \tag{8.8}$$

which equals the total number of events (D) divided by the total period of exposure (T). This estimator is called the *crude rate*, the *linearized rate* estimate or the *total time on test* estimate, the latter term arising in reliability theory.

This estimator can also be expressed as a weighted *mean rate*. Let the rate for the jth subject be denoted as $r_j=d_j/t_j$. Under the constant intensity assumption, $E\left(r_j\right)=\lambda$ for all subjects. Then it is readily shown that the MLE in (8.8) can be expressed as a weighted mean of the r_j, weighted by the t_j

$$\widehat{\lambda}=\overline{r}=\frac{\sum_j t_j r_j}{\sum_j t_j}=\frac{D}{T}. \tag{8.9}$$

The variance of the estimate can be easily derived as follows. Under the Poisson assumption $E\left(d_j\right)=V\left(d_j\right)=\lambda t_j$, the expression for the MLE then implies that

$$\begin{aligned} V\left(\widehat{\lambda}\right)&=\left(\frac{1}{\sum_j t_j}\right)^2\sum\nolimits_j V\left(d_j\right)=\frac{\lambda}{\sum_j t_j}=\frac{\lambda}{T} \\ &=\frac{\lambda^2}{\sum_j \lambda t_j}=\frac{\lambda^2}{\sum_j E\left(d_j\right)}=\frac{\lambda^2}{E\left(D\right)}. \end{aligned} \tag{8.10}$$

The latter expression shows that the variance of the estimate is λ^2 divided by the expected number of events. Thus for given λ, the variance is inversely proportional to the total exposure time and also to the total expected number of events.

This expression for the variance can also be derived as the inverse of Fisher's Information. The observed information is

$$i(\lambda) = \frac{-dU(\lambda)}{d\lambda} = \frac{\sum_j d_j}{\lambda^2} = \frac{D}{\lambda^2} \tag{8.11}$$

and the expected information $I(\lambda) = E[i(\lambda)]$ is

$$I(\lambda) = E\left[\frac{\sum_j d_j}{\lambda^2}\right] = \frac{E(D)}{\lambda^2} = \frac{T}{\lambda}. \tag{8.12}$$

Thus the total information in the data is proportional to the total expected number of events $E(D)$ conditional on the $\{t_j\}$. Then the variance of the estimate is as shown in (8.10) and is consistently estimated as

$$\widehat{V}\left(\widehat{\lambda}\right) = \frac{\widehat{\lambda}^2}{D} = \frac{\widehat{\lambda}}{T}. \tag{8.13}$$

Asymptotically, since $\widehat{\lambda}$ is the MLE, then it follows that

$$\left(\widehat{\lambda} - \lambda\right) \overset{d}{\approx} \mathcal{N}\left[0, V\left(\widehat{\lambda}\right)\right]. \tag{8.14}$$

This result can also be derived from the central limit theorem conditioning on the $\{t_j\}$ since $\widehat{\lambda}$ is the weighted mean rate. Although this provides large sample tests of significance and confidence limits, since λ is positive a more accurate approximation is provided by $\widehat{\eta} = \log\left(\widehat{\lambda}\right)$. From the δ-method it is readily shown that

$$V(\widehat{\eta}) = \left(\frac{1}{\lambda}\right)^2 V\left(\widehat{\lambda}\right) = \frac{1}{\sum_j E(d_j)} = \frac{1}{E(D)} \tag{8.15}$$

so that the estimated variance is $\widehat{V}(\widehat{\eta}) = D^{-1}$. This then yields asymmetric confidence limits for the assumed constant intensity.

8.1.3 Relative Risks

Now assume that we have independent samples of n_1 and n_2 subjects drawn at random from two separate populations, such as exposed versus not or treated versus control. We then wish to assess the relative risk of events between the two groups, including possible multiple or recurrent events. Under a doubly homogeneous Poisson model, this relative risk is described as simply the ratio of the assumed constant intensities

$$RR = \lambda_1/\lambda_2 \,\widehat{=}\, \widehat{\lambda}_1/\widehat{\lambda}_2 \tag{8.16}$$

where $\widehat{\lambda}_i = D_i/T_i$ is the estimated rate in the ith group of n_i subjects with $D_i = \sum_{j=1}^{n_i} d_{ij}$ and $T_i = \sum_{j=1}^{n_i} t_{ij}$, $i = 1, 2$. Thus we can use $\log\left(\widehat{RR}\right)$ as the basis for confidence intervals or use $\widehat{\lambda}_1 - \widehat{\lambda}_2$ for a statistical test.

Let $\eta_i = \log(\lambda_i)$ and $\theta = \log(RR) = \eta_1 - \eta_2$. Then

$$\widehat{\theta} = \log\left(\widehat{RR}\right) = \widehat{\eta}_1 - \widehat{\eta}_2 , \tag{8.17}$$

where $\widehat{\eta}_i = \log\left(\widehat{\lambda}_i\right)$, $i = 1, 2$. The variance of the estimate is

$$V\left(\widehat{\theta}\right) = V(\widehat{\eta}_1) + V(\widehat{\eta}_2) = \frac{1}{E(D_1)} + \frac{1}{E(D_2)} , \tag{8.18}$$

which is estimated as

$$\widehat{V}\left(\widehat{\theta}\right) = \frac{1}{D_1} + \frac{1}{D_2} = \frac{D_\bullet}{D_1 D_2} , \tag{8.19}$$

where $D_\bullet = D_1 + D_2$. This provides large sample confidence limits for θ and asymmetric confidence limits for the relative risk.

An efficient large sample test of H_0: $\lambda_1 = \lambda_2 = \lambda$ is then obtained as

$$Z = \frac{\widehat{\lambda}_1 - \widehat{\lambda}_2}{\sqrt{\widehat{V}\left(\widehat{\lambda}_1 - \widehat{\lambda}_2 | H_0\right)}} , \tag{8.20}$$

using an estimate of the variance under the null hypothesis. This null variance is defined as

$$V\left(\widehat{\lambda}_1 - \widehat{\lambda}_2 | H_0\right) = V\left(\widehat{\lambda}_1 | H_0\right) + V\left(\widehat{\lambda}_2 | H_0\right) . \tag{8.21}$$

From (8.10)

$$V\left(\widehat{\lambda}_i | H_0\right) = \frac{\lambda^2}{E(D_i)_{|\lambda_i = \lambda}} = \frac{\lambda}{T_i} . \tag{8.22}$$

Under H_0 the MLE of the assumed common rate is

$$\widehat{\lambda} = \frac{D_1 + D_2}{T_1 + T_2} = \frac{D_\bullet}{T_\bullet} . \tag{8.23}$$

Therefore, the null variance is estimated as

$$\widehat{V}\left(\widehat{\lambda}_1 - \widehat{\lambda}_2 | H_0\right) = \widehat{\lambda}\left[\frac{1}{T_1} + \frac{1}{T_2}\right] = \frac{D_\bullet}{T_1 T_2} . \tag{8.24}$$

The resulting large sample test statistic is

$$Z = \frac{(\widehat{\lambda}_1 - \widehat{\lambda}_2)}{\sqrt{D_\bullet / T_1 T_2}} , \tag{8.25}$$

which is asymptotically distributed as standard normal under H_0. In a problem it is also shown that this test is the efficient score test for the effect of a single binary covariate for treatment group in a Poisson regression model.

An asymptotically equivalent large sample test can be obtained as a test of H_0: $\theta = \log(RR) = 0$ using

$$Z = \frac{\widehat{\theta}}{\sqrt{\widehat{V}\left(\widehat{\theta}|H_0\right)}}, \tag{8.26}$$

where the null variance is estimated as

$$\widehat{V}\left(\widehat{\theta}|H_0\right) = \frac{1}{\overline{\lambda}}\left[\frac{1}{T_1} + \frac{1}{T_2}\right]. \tag{8.27}$$

Using a Taylor's expansion about λ it is readily shown that the two tests are asymptotically equivalent.

Example 8.1 *Hypoglycemia in the DCCT*
In the Diabetes Control and Complications Trial (DCCT, see Section 1.5), the major potential adverse effect of intensive diabetes therapy with then available technology was an episode of hypoglycemia where the blood glucose level falls too low, at which point the patient becomes dizzy, disoriented and may pass out. Here we use the data from the secondary cohort of 715 patients with more advanced and longer duration diabetes. The following is a summary of the overall incidence of severe hypoglycemia in the two treatment groups within this cohort:

	Intensive $i = 1$	*Conventional* $i = 2$	*Total*
n_i	363	352	715
Events (D_i)	1723	543	2266
Exposure time in years (T_i)	2598.5	2480.2	5078.7
$\widehat{\lambda}_i$ per patient year (PY)	0.6631	0.2189	0.4462
$\widehat{\lambda}_i$ per 100 PY	66.3	21.9	44.6

The crude rates are conveniently expressed as a number per 100 patient years. This is sometimes described as the percent of patients per year. Such a description is appropriate if there are no recurrent events, such as in survival analysis. Here, however, some patients experienced as many as 20 or more episodes of severe hypoglycemia in which case it is inappropriate to describe these rates in terms of a percent of patients per year.

The ratio of these crude rates yields $\widehat{RR} = 3.029$ and $\widehat{\theta} = \log(\widehat{RR}) = 1.108$ with estimated variance $\widehat{V}\left(\widehat{\theta}\right) = \frac{1}{1723} + \frac{1}{543} = 0.002422$ and $S.E.\left(\widehat{\theta}\right) = 0.0492$. This yields asymmetric 95% confidence limits on RR of (2.75, 3.335). Under the doubly homogeneous Poisson model, the large sample test of H_0 yields

$$Z = \frac{0.6631 - 0.2189}{\sqrt{\frac{2266}{(2598.5 \times 2480.2)}}} = 23.689 ,$$

which is highly statistically significant.

8.1.4 Violations of the Homogeneous Poisson Assumptions

All of the above is based on the doubly homogeneous Poisson assumptions that the intensity of the process is constant over time, and that it is the same for all subjects in the population. These assumptions may not apply in practice. Violation of the first assumption is difficult, if not impossible, to assess with count data when the exact times of the events are not recorded. However, when the event times are known, a time-varying intensity is easily assessed or allowed for in a multiplicative intensity model that is a generalization of the Cox proportional hazards model for recurrent event times (see Section 9.6). The second assumption specifies that for given exposure time t, the mean number of events in the population is $E(d) = \lambda t$ and that the variance of the number of events also is $V(d) = \lambda t$ as is characteristic of the Poisson distribution. In any population, violation of either the homogeneous mean assumption or of the mean:variance relationship leads to over- (under-)dispersion in the data where $V(d) > (<) \lambda t$.

Cochran (1954a) suggested that the homogeneous Poisson process assumption be tested using a simple chi-square goodness of fit test. Under the assumption of a common intensity for all subjects, then $E(d_j) = V(d_j) = \lambda t_j$ and the test is provided by

$$X^2 = \sum\nolimits_{j=1}^{N} \frac{(d_j - \widehat{\lambda} t_j)^2}{\widehat{\lambda} t_j} , \tag{8.28}$$

which is distributed as chi-square on $N-1$ *df* under the hypothesis of homogeneity. An alternate approach is to assess the degree of over- (under-)dispersion in a Poisson regression model, as shown in Section 8.4.

Example 8.2 *Hypoglycemia in the DCCT*
Applying the above to each DCCT treatment group, separately, yields chi-square values of $X^2 = 1737.12$ in the conventional treatment group on 351 *df* ($p \leq 0.001$) and 3394.22 in the intensive group on 362 *df* ($p \leq 0.001$). In each case there is a strong indication that the simple homogeneous Poisson model does not apply. These tests can also be obtained from a Poisson regression model as described in Example 8.4.

8.2 OVER-DISPERSED POISSON MODEL

One way to allow for this extra variation in risks is to assume that the over dispersion arises because of mixtures of subjects with different characteristics that, in turn, determine the intensity or risk of events in the population. This implies that the intensity for a subject is a function of covariate values and that all subjects with a given covariate vector $\boldsymbol{x}$ share a common intensity $\lambda(\boldsymbol{x})$. These relationships can then be assessed using a Poisson regression model.

Another approach is to assume a random-effects over-dispersed Poisson model in which we assume that each subject has a unique intensity that is drawn from a distribution of intensities. We first consider this over-dispersed model, followed by use of the Poisson regression model.

8.2.1 Two-Stage Random Effects Model

The doubly homogeneous Poisson model assumes that the mean and variance of the number of events for the jth subject with exposure t_j is $E(d_j|t_j) = V(d_j|t_j) = \lambda t_j$. Over- (under-)dispersion arises when these relationships are violated. In this case it is more realistic to describe the process in terms of a two-stage random effects model, where each subject has a unique intensity λ that determines the mean and variance for that subject, and where the subject-specific intensities are then drawn from some distribution in the population. That is, we assume that

$$E(d|\lambda, t) = V(d|\lambda, t) = \lambda t \tag{8.29}$$
$$\lambda \sim G(\mu_\lambda, \sigma_\lambda^2)$$

for some "mixing" distribution G, where $E(\lambda) = \mu_\lambda$ is now the overall mean rate in the population and $V(\lambda) = \sigma_\lambda^2$ is the *over-dispersion variance component.* Throughout we also assume that λ is independent of t and we condition on the exposure times $\{t_j\}$ as fixed quantities.

Then unconditionally for the jth subject,

$$E(d_j) = E_\lambda E[d_j|\lambda_j, t_j] = \mu_\lambda t_j. \tag{8.30}$$

The rate of events for each subject then provides an estimate of the subject specific rate

$$\widehat{\lambda}_j = r_j = d_j/t_j\ , \tag{8.31}$$

where

$$E\left[\widehat{\lambda}_j|\lambda_j\right] = E\left[\frac{d_j}{t_j}|\lambda_j\right] = \frac{\lambda_j t_j}{t_j} = \lambda_j \tag{8.32}$$
$$V\left[\widehat{\lambda}_j|\lambda_j\right] = V\left[\frac{d_j}{t_j}|\lambda_j\right] = \frac{1}{t_j^2}V(d_j) = \frac{\lambda_j t_j}{t_j^2} = \frac{\lambda_j}{t_j}\ .$$

Note that this model is directly analogous to the simple measurement error model described in Section 4.10.1. Here the subject-specific number of events is equivalent to the subject-specific cholesterol value ($d_j \equiv y_j$), the subject-specific intensity is equivalent to the subject-specific true cholesterol value ($\lambda_j \equiv v_j$), the mean intensity is equivalent to the mean cholesterol ($\mu_\lambda \equiv \mu$), and the variance of the intensity between subjects is equivalent to the variance of the true cholesterol measurements ($\sigma_\lambda^2 \equiv \sigma_\mu^2$).

Also note that a dispersion variance component value of $\sigma_\lambda^2 = 0$ implies that there is no over-dispersion, in which case G is degenerate and $\lambda_j = \mu_\lambda$ for all subjects. In this case the homogenous Poisson model applies.

Therefore, the objective is to estimate the dispersion variance component σ_λ^2, and if non-zero, to then estimate μ_λ and $V(\widehat{\mu}_\lambda)$ allowing for over-dispersion. Various authors have proposed methods that are based on a parametric specification of the mixing distribution $G(\mu_\lambda, \sigma_\lambda^2)$, such as the gamma mixing distribution or the log normal, among others. Here we take a different approach wherein an estimator of the variance component is obtained using distribution-free or robust methods that do not require specification of the specific form of the mixing distribution.

Consider the weighted mean rate in (8.9). Then

$$E(\overline{r}) = \frac{\sum_j E(d_j)}{\sum_j t_j} = \frac{\mu_\lambda \sum_j t_j}{\sum_j t_j} = \mu_\lambda \tag{8.33}$$

and the crude rate $\overline{r}$ provides an unbiased estimate of μ_λ under the random effects model. We now require an estimate of $V(\overline{r})$. Unconditionally, for the jth subject, the variance of d_j can be expressed as

$$\begin{aligned} V(d_j) &= V[E(d_j|\lambda_j t_j)] + E[V(d_j|\lambda_j t_j)] \\ &= V[\lambda_j t_j] + E[\lambda_j t_j] \\ &= t_j^2 \sigma_\lambda^2 + \mu_\lambda t_j \,, \end{aligned} \tag{8.34}$$

which reflects both the variance of the intensities between subjects and the Poisson variation within subjects. Thus

$$V(\overline{r}) = \frac{\sum_j V(d_j)}{\left(\sum_j t_j\right)^2} = \frac{\sum_j t_j^2 \sigma_\lambda^2 + \mu_\lambda \sum_j t_j}{\left(\sum_j t_j\right)^2} . \tag{8.35}$$

The two-stage model described in Section 4.10 can then be employed to obtain a moment estimator for the dispersion parameter, σ_λ^2, based on the expected value of a sum of squares that involves σ_λ^2. Since $\widehat{\mu}_\lambda = \overline{r}$ is a weighted mean of the $\{r_j\}$ with weights $\{t_j\}$, consider the weighted SSE. From (8.31) and (8.34),

$$E\left[\sum\nolimits_j t_j (r_j - \mu_\lambda)^2\right] = \sum\nolimits_j t_j V(r_j) = \sum\nolimits_j \frac{V(d_j)}{t_j} = \left(\sum\nolimits_j t_j \sigma_\lambda^2\right) + N\mu_\lambda . \tag{8.36}$$

Solving for σ_λ^2 we obtain the moment estimator

$$\widehat{\sigma}_\lambda^2 = \max\left[0, \frac{\sum_j t_j (r_j - \widehat{\mu}_\lambda)^2 - N\widehat{\mu}_\lambda}{\sum_j t_j}\right] . \tag{8.37}$$

Substituting into the expression for $V(\overline{r})$ in (8.35), we obtain the estimated variance of the estimated mean intensity

$$\widehat{V}(\overline{r}) = \frac{\sum_j t_j^2 \widehat{\sigma}_\lambda^2 + \overline{r}\sum_j t_j}{\left(\sum_j t_j\right)^2}. \tag{8.38}$$

For confidence intervals it is customary that we use the log scale. Again let $\eta = \log(\mu_\lambda)$ and $\widehat{\eta} = \log\left(\widehat{\overline{r}}\right)$, where

$$V(\widehat{\eta}) = \frac{V(\overline{r})}{\mu_\lambda^2} \widehat{=} \frac{\widehat{V}(\overline{r})}{\overline{r}^2}. \tag{8.39}$$

Under weak conditions, using the Liapunov Central Limit Theorem (Section A.2.1 of the Appendix) it can be shown that $\sum_j d_j$ is asymptotically normally distributed so that asymptotically

$$\overline{r} \overset{d}{\approx} \mathcal{N}\left[\mu_\lambda, V(\overline{r})\right] \tag{8.40}$$

and

$$\widehat{\eta} = \log(\overline{r}) \overset{d}{\approx} \mathcal{N}\left[\log(\mu_\lambda), V(\overline{r})/\mu_\lambda^2\right]. \tag{8.41}$$

Therefore, asymmetric confidence limits on the mean rate are obtained from the symmetric limits on the log mean rate using $\widehat{\eta} \pm Z_{1-\alpha/2}\sqrt{\widehat{V}(\widehat{\eta})}$.

A fixed-point iterative method can then be used to obtain jointly convergent estimates of the mean rate μ_λ and the over-dispersion variance component σ_λ^2. Let the initial values for the iteration be the values of the crude rate $\widehat{\mu}_\lambda^{(0)} = \overline{r}$ from (8.9) and the variance estimate $\widehat{\sigma}_\lambda^{2(0)}$ from (8.37). Then using the fixed-point method, we can alternatively compute updated estimates of the mean and the variance component until both converge. The updated estimate of the mean at the $(\ell+1)th$ iteration $(\ell = 0, 1, 2, \ldots)$ can be obtained as an inverse variance weighted estimate

$$\widehat{\mu}_\lambda^{(\ell+1)} = \overline{r}^{(\ell+1)} = \frac{\sum_j \widehat{\tau}_j^{(\ell)} r_j}{\sum_j \widehat{\tau}_j^{(\ell)}}, \tag{8.42}$$

where $\widehat{\tau}_j^{(\ell)} = [\widehat{V}(r_j)^{(\ell)}]^{-1}$ and where, from (8.34),

$$\widehat{V}(r_j)^{(\ell)} = \frac{t_j^2 \widehat{\sigma}_\lambda^{2(\ell)} + \widehat{\mu}_\lambda^{(\ell)} t_j}{t_j^2} = \widehat{\sigma}_\lambda^{2(\ell)} + \frac{\widehat{\mu}_\lambda^{(\ell)}}{t_j}. \tag{8.43}$$

Then the updated estimate of the variance component $\widehat{\sigma}_\lambda^{2(\ell+1)}$ is obtained upon substitution of $\widehat{\mu}_\lambda^{(\ell+1)}$ into the moment estimating equation (8.37). The process continues until both the mean and variance estimates converge to constants.

8.2.2 Relative Risks

Again, consider the case of samples of n_1 and n_2 observations from two populations where within each we assume a two-stage model with mixing distributions $G_1\left(\mu_{\lambda_1}, \sigma^2_{\lambda_1}\right)$ and $G_2\left(\mu_{\lambda_2}, \sigma^2_{\lambda_2}\right)$, respectively. We then wish to estimate the relative risk, or its logarithm $\theta = \log(RR) = \eta_1 - \eta_2$, where $\eta_i = \log\left(\mu_{\lambda_i}\right)$ and $\widehat{\theta} = \widehat{\eta}_1 - \widehat{\eta}_2$. In this case, we first obtain an estimate of the dispersion variance components $\widehat{\sigma}^2_{\lambda_1}$ and $\widehat{\sigma}^2_{\lambda_2}$ separately within each group. These are then employed in (8.38) to obtain estimates of the variances $\widehat{V}(\overline{r}_i)$ and in (8.39) to obtain estimates of the $\widehat{V}(\widehat{\eta}_i)$, $i = 1, 2$. The estimated variance of the estimated log(RR) then is $V\left(\widehat{\theta}\right) = V(\widehat{\eta}_1) + V(\widehat{\eta}_2)$, which can be used as the basis for a large sample confidence interval calculation.

Now consider a test of the null hypothesis H_0: $\mu_{\lambda_1} = \mu_{\lambda_2}$. Ideally we wish to construct a test using the variance under this null hypothesis. However, to do so we must also allow for the dispersion variance components to differ between groups, that is, assuming $\sigma^2_{\lambda_1} \neq \sigma^2_{\lambda_2}$. Thus we first estimate μ_λ from the pooled sample with both groups combined. We then estimate $\sigma^2_{\lambda_i}$ separately for the ith group but using $\widehat{\mu}_\lambda$ in the moment estimating equation (8.37). We would then substitute $\widehat{\mu}_\lambda$ and $\widehat{\sigma}^2_{\lambda_i}$ in (8.38) to obtain an estimate of the $\widehat{V}(\overline{r}_i|H_0)$ separately for each group ($i = 1, 2$). Then an asymptotically efficient test of H_0 is provided by

$$Z = \frac{(\overline{r}_1 - \overline{r}_2)}{\sqrt{\widehat{V}(\overline{r}_1|H_0) + \widehat{V}(\overline{r}_2|H_0)}}, \tag{8.44}$$

which is asymptotically normally distributed. Note that this test is also asymptotically efficient when there is homogeneity of dispersion parameters ($\sigma^2_{\lambda_1} = \sigma^2_{\lambda_2}$) because the estimates within each group each provide a consistent estimate of the common parameter in this case.

Example 8.3 *Hypoglycemia in the DCCT (continued)*
Table 8.1 presents a partial listing of the data from the DCCT that also includes covariate information (discussed later). For each subject this shows the treatment group assignment (*grp*: $i = 1$ for intensive, 0 for conventional), the number of events ($d_i = D$), the years of exposure or follow-up ($t_i = T$), and the corresponding rate of events per year ($r_i = rate$). The following is a summary of the over-dispersed analysis within the two groups:

	Intensive $i=1$	*Conventional* $i=2$
$\sum_{j=1}^{n_i} t_{ij}\left(r_{ij} - \widehat{\mu}_{\lambda_i}\right)^2$	2250.64	380.309
$\widehat{\sigma}^2_{\lambda_i}$	0.77351	0.12226
$\sum_{ij} t^2_{ij}$	18315.17	19488.77
$\widehat{V}(\overline{r}_i)$	0.0024878	0.0004523
$\widehat{V}(\widehat{\eta}_i)$	0.0056582	0.0094363

Table 8.1 Episodes of Severe Hypoglycemia for a Subset of Subjects in the DCCT

GRP	D	T	RATE	INSULIN	DUR	FEM	ADU	BCV	HBAE	HXC
1	5	9.50582	0.52599	0.33557	110	0	1	0.01	9.74	0
0	9	9.47296	0.95007	0.71197	100	1	1	0.04	8.90	0
1	0	9.45106	0.00000	0.55487	119	1	1	0.01	9.99	0
1	6	9.24025	0.64933	0.58361	111	0	1	0.01	6.94	0
0	3	9.37714	0.31993	0.40984	99	1	1	0.01	10.94	0
0	1	9.37988	0.10661	0.48409	44	0	1	0.18	9.56	0
1	1	9.43190	0.10602	0.57143	180	0	1	0.01	9.67	0
0	0	8.09035	0.00000	0.34783	60	1	1	0.04	12.75	0
1	0	7.88775	0.00000	0.75472	112	0	1	0.04	8.57	0
1	3	7.76728	0.38624	1.01587	161	0	1	0.03	7.14	0
0	1	7.76454	0.12879	0.36111	157	0	1	0.06	10.02	0
0	0	7.61944	0.00000	0.57143	104	1	1	0.03	6.93	0
1	1	7.35113	0.13603	0.59829	152	1	1	0.03	8.80	0
0	1	6.99795	0.14290	0.66598	56	0	1	0.03	6.76	0
1	0	6.89938	0.00000	0.67747	150	1	1	0.03	7.75	0
1	0	6.83094	0.00000	0.60102	11	0	1	0.11	10.99	0
0	0	6.47775	0.00000	0.62208	72	0	1	0.03	8.23	0
0	0	6.02053	0.00000	1.22172	99	1	0	0.03	12.50	0
0	1	6.25873	0.15978	0.71778	79	1	1	0.03	11.74	0
0	0	6.17112	0.00000	0.88183	136	1	1	0.03	9.30	0
0	0	5.86995	0.00000	0.42159	117	1	1	0.03	8.90	0
0	0	5.73854	0.00000	0.67518	128	1	1	0.03	10.40	0
1	0	5.56331	0.00000	0.79498	129	0	1	0.03	7.30	0
0	0	5.52498	0.00000	0.40040	56	0	1	0.27	8.00	0
1	15	5.19918	2.88507	0.34169	176	0	1	0.03	8.60	0
1	14	5.31143	2.63582	0.54701	149	1	1	0.03	8.10	0
0	1	5.12799	0.19501	0.67935	57	0	1	0.04	7.10	1
1	3	5.04860	0.59422	0.70661	155	1	1	0.03	8.50	0
0	0	9.62081	0.00000	0.33635	50	0	1	0.20	10.65	0
0	8	9.62081	0.83153	0.81505	176	1	1	0.01	7.44	0
0	6	9.48665	0.63247	1.04405	84	1	0	0.01	8.96	0
1	27	9.48665	2.84610	1.53061	66	1	0	0.01	9.56	0
1	2	2.59274	0.77138	0.41176	157	0	1	0.07	9.11	0
0	0	9.39083	0.00000	0.51988	124	1	1	0.09	10.23	0
1	17	9.34155	1.81983	0.76823	71	0	1	0.12	7.01	0
0	0	9.31964	0.00000	0.46642	48	1	1	0.05	12.68	0
1	0	9.29500	0.00000	0.72488	59	0	0	0.01	9.35	1
1	2	8.01916	0.24940	0.45147	121	0	1	0.03	9.69	0
0	0	7.85763	0.00000	0.48544	30	0	1	0.03	8.98	0

etc.

These yield an estimated variance $\widehat{V}\left(\widehat{\theta}\right) = 0.0056582 + 0.0094363 = 0.015095$ and $\widehat{SE}\left(\widehat{\theta}\right) = \sqrt{0.015095} = 0.12286$. The resulting asymmetric 95% confidence limits on RR are (2.381, 3.853) slightly wider than the homogeneous model confidence limits presented in Example 8.1. Using this $S.E.$ computed under the alternative yields a Wald Z-test of $Z = 1.10814/0.12286 = 9.01954$, still highly statistically significant.

In the combined sample, as shown in Example 8.1, the mean rate is $\widehat{\mu}_\lambda = 0.44618$. This can then be used in (8.37) to compute the estimate of the dispersion variance component within each group under the null hypothesis of no difference in rates between groups, designated as $\widehat{\sigma}^2_{\lambda_{i(0)}}$. When both quantities are employed in (8.38) this yields estimates of the variances of the rates within each group under the null hypothesis, $\widehat{V}\left(\overline{r}_i|H_0\right)$. The resulting values for the dispersion variance components and the variances of the rates are:

	Intensive $i = 1$	*Conventional* $i = 2$
$\widehat{\sigma}^2_{\lambda_{i(0)}}$	0.85086	0.14165
$\widehat{V}\left(\overline{r}_i\|H_0\right)$	0.0026276	0.0006016

Then the test of the difference between groups allowing for over-dispersion is

$$Z = \frac{0.6631 - 0.2189}{\sqrt{0.0026276 + 0.0006016}} = 7.816 ,$$

which is again highly significant.

8.2.3 Stratified-Adjusted Analyses

As was the case for the analysis of odds ratios and relative risks of single binary events, in some cases it is desirable to conduct an analysis that is stratified or adjusted for other patient covariates. Such analyses are readily performed by adapting the methods described in Chapter 4 to the analysis of crude rates, either under a homogeneous Poisson model, or under an over-dispersed Poisson model with an additional parameter (the over-dispersion variance component) estimated separately within each group and within each stratum. These methods include the $MVLE$ of an assumed common relative risk over strata and its large sample variance, the Radhakrishna-like asymptotically fully efficient test, the Cochran test of homogeneity of relative risks over stratum and the random-effects stratified-adjusted estimate. These developments are left as Problems.

8.3 POISSON REGRESSION MODEL

8.3.1 Homogeneous Poisson Regression Model

Another mechanism by which over-dispersion may arise in the aggregate sample is when the population consists of a mixture of sub-populations characterized by different covariate values that, in turn, are associated with different intensities or risks. Thus the population consists of a mixture of subjects with different covariate values that implies a mixture of some subjects with inherently high risks and some subjects with inherently low risks. One way to account for such variation in risks is through a Poisson regression model in which the conditional expectation (the intensity) is modeled as a function of covariates (Frome, 1983). Since the Poisson distribution is a member of the exponential family, the Poisson regression model is a member of the family of generalized linear models (*GLMs*) described in Section A.10 of the Appendix.

Assume that each patient ($i = 1, \ldots, N$) has an underlying risk or intensity that is a function of a covariate vector $\boldsymbol{x}_i = (x_{i1} \; \ldots \; x_{ip})^T$. Because the rate parameter $\lambda(\boldsymbol{x}_i)$ must be positive, it is natural to use a log-linear model with a log link which is the canonical link for the Poisson rate parameter. Thus

$$\log[\lambda(\boldsymbol{x}_i)] = (\alpha + x_{i1}\beta_1 + \ldots + x_{ip}\beta_p) = \alpha + \boldsymbol{x}_i'\boldsymbol{\beta} = \eta_i \qquad (8.45)$$
$$\lambda(\boldsymbol{x}_i) = e^{\alpha + \boldsymbol{x}_i'\boldsymbol{\beta}}$$

with parameters $\boldsymbol{\theta} = (\alpha \parallel \boldsymbol{\beta}^T)^T$, where $\boldsymbol{\beta} = (\beta_1 \; \ldots \; \beta_p)^T$. For the jth covariate, β_j is the log relative risk per unit change in X_j. Then for the ith patient with d_i events over t_i years of exposure, the expected number of events is

$$E(d_i|\boldsymbol{x}_i, t_i) = \lambda(\boldsymbol{x}_i)\, t_i = \exp[\alpha + \boldsymbol{x}_i'\boldsymbol{\beta} + \log(t_i)]. \qquad (8.46)$$

Here the $\log(t_i)$ is an *offset* so that each patient has an implied intercept $\widetilde{\alpha}_i = \alpha + \log(t_i)$ given the fixed exposure time t_i.

Then assume that

$$d_i \sim \text{Poisson}[\lambda(\boldsymbol{x}_i)\, t_i] = P(d_i; \lambda(\boldsymbol{x}_i), t_i) \qquad (8.47)$$

or that $E(d_i|\lambda(\boldsymbol{x}_i), t_i) = V(d_i|\lambda(\boldsymbol{x}_i), t_i) = \lambda(\boldsymbol{x}_i)t_i$. Alternately, this is equivalent to specifying that $d_i = \lambda(\boldsymbol{x}_i)\, t_i + \varepsilon_i$, where the errors are distributed with mean zero and variance $\lambda(\boldsymbol{x}_i)t_i$ independently of t_i. Therefore, the likelihood is the product of Poisson probabilities of the form

$$L(\boldsymbol{\theta}) = \prod_{i=1}^{N} \frac{e^{-\lambda(\boldsymbol{x}_i)t_i}\,[\lambda(\boldsymbol{x}_i)\, t_i]^{d_i}}{d_i!} \qquad (8.48)$$

with log likelihood

$$\ell(\boldsymbol{\theta}) = \sum\nolimits_{i=1}^{N} [-\lambda(\boldsymbol{x}_i)\, t_i + d_i \log[\lambda(\boldsymbol{x}_i)\, t_i] - \log(d_i!)]$$
$$= \sum\nolimits_{i=1}^{N} \left[-e^{\alpha + \boldsymbol{x}_i'\boldsymbol{\beta}} t_i + d_i[\alpha + \boldsymbol{x}_i'\boldsymbol{\beta} + \log(t_i)] - \log(d_i!)\right].$$

The score vector is $\boldsymbol{U}(\boldsymbol{\theta}) = \left[U(\boldsymbol{\theta})_\alpha \; U(\boldsymbol{\theta})_{\beta_1} \; \ldots \; U(\boldsymbol{\theta})_{\beta_p}\right]^T$ with elements

$$U(\boldsymbol{\theta})_\alpha = \sum\nolimits_{i=1}^{N} \left[d_i - e^{\alpha + \boldsymbol{x}_i'\boldsymbol{\beta}} t_i\right] = \sum\nolimits_{i=1}^{N} \left[d_i - \lambda(\boldsymbol{x}_i) t_i\right] \tag{8.49}$$

and

$$U(\boldsymbol{\theta})_{\beta_j} = \sum\nolimits_{i=1}^{N} x_{ij} \left[d_i - e^{\alpha + \boldsymbol{x}_i'\boldsymbol{\beta}} t_i\right] = \sum\nolimits_{i=1}^{N} x_{ij} \left[d_i - \lambda(\boldsymbol{x}_i) t_i\right] \tag{8.50}$$

for $j = 1, \ldots, p$. Therefore, $\boldsymbol{U}(\boldsymbol{\theta})_{\boldsymbol{\beta}} = \sum_{i=1}^{N} \boldsymbol{x}_i \left[d_i - \lambda(\boldsymbol{x}_i) t_i\right]$

From $U(\boldsymbol{\theta})_\alpha$ it follows that $D = \sum_i d_i = \sum_i \widehat{\lambda}_i t_i$. Also, for the jth covariate, given fixed values of the other covariates and a fixed exposure time t, then β_j equals the difference in the log intensity (log risk) per unit difference in the value of the covariate X_j, and e^{β_j} equals the log relative intensity (risk) per unit difference.

It is then readily shown that the observed and expected information matrices have elements

$$\begin{aligned}
\boldsymbol{i}(\boldsymbol{\theta})_\alpha &= \boldsymbol{I}(\boldsymbol{\theta})_\alpha = \sum\nolimits_i e^{\alpha + \boldsymbol{x}_i'\boldsymbol{\beta}} t_i = \sum\nolimits_i \lambda_i(\boldsymbol{x}_i) t_i \\
\boldsymbol{i}(\boldsymbol{\theta})_{\alpha,\beta_j} &= \boldsymbol{I}(\boldsymbol{\theta})_{\alpha,\beta_j} = \sum\nolimits_i x_{ij} t_i e^{\alpha + \boldsymbol{x}_i'\boldsymbol{\beta}} = \sum\nolimits_i x_{ij} \lambda_i(\boldsymbol{x}_i) t_i \\
\boldsymbol{i}(\boldsymbol{\theta})_{\beta_j} &= \boldsymbol{I}(\boldsymbol{\theta})_{\beta_j} = \sum\nolimits_i x_{ij}^2 t_i e^{\alpha + \boldsymbol{x}_i'\boldsymbol{\beta}} = \sum\nolimits_i x_{ij}^2 \lambda_i(\boldsymbol{x}_i) t_i \\
\boldsymbol{i}(\boldsymbol{\theta})_{\beta_j,\beta_k} &= \boldsymbol{I}(\boldsymbol{\theta})_{\beta_j,\beta_k} = \sum\nolimits_i x_{ij} x_{ik} t_i e^{\alpha + \boldsymbol{x}_i'\boldsymbol{\beta}} = \sum\nolimits_i x_{ij} x_{ik} \lambda_i(\boldsymbol{x}_i) t_i
\end{aligned} \tag{8.51}$$

for $1 \leq j < k \leq p$. Therefore, $i\left(\widehat{\boldsymbol{\theta}}\right) = \boldsymbol{I}\left(\widehat{\boldsymbol{\theta}}\right)$ with elements obtained by substituting $\widehat{\boldsymbol{\theta}} = (\widehat{\alpha} \; \widehat{\beta}_1 \; \ldots \; \widehat{\beta}_p)^T$ into the above expressions. As for any member of the exponential (GLM) family, the elements of the information are weighted sums of the model-based conditional variances, such as where $\boldsymbol{I}(\boldsymbol{\theta})_{\beta_j} = \sum_i x_{ij}^2 V(d_i|\boldsymbol{x}_i, t_i)$.

The estimated information then provides likelihood ratio tests, Wald tests and efficient score tests. As a problem we show that the score test for the coefficient of a single binary covariate for treatment group equals the simple Z-test in (8.26) derived under the homogeneous Poisson model.

The model is readily fit to a data set using the SAS procedure for generalized linear models, PROC GENMOD, as illustrated in the following example.

Example 8.4 *Hypoglycemia in the DCCT (continued)*
We now conduct a Poisson regression analysis of the incidence of episodes of severe hypoglycemia in the DCCT adjusting for the other covariates in the data set. These include *insulin*: the baseline daily insulin dose in units per kg body weight; *duration*: the number of months duration of diabetes on entry into the study in the range (12–180); *female*: an indicator variable for female (1) versus male (0); *adult*: an indicator variable for adult (1, ≥ 18 years) versus adolescent on entry; *bcval5*: the level of c-peptide on entry in pico-moles/mL; *hbael*: the percent glycosylated hemoglobin (HbA_{1c} %); and *hxcoma*: an indicator variable for a history of coma and/or seizure prior to entry (1=yes, 0=no).

Table 8.2 PROC GENMOD Program for Analysis of the DCCT Data

```
data one; set DCCT;
lnyears = log(T);

proc genmod;
  model nevents =
  / dist = poisson link = log  offset = lnyears;
TITLE1 'Poisson regression models of risk of hypoglycemia';
title2 'null model';

proc genmod; class group;
  model nevents = group
  / dist = poisson link = log  offset = lnyears;
title2 'unadjusted treatment group effect';

proc genmod; class group;
  model nevents = group insulin duration female
                  adult bcval5 hbael hxcoma
  / dist = poisson link = log  offset = lnyears covb;
title2 'covariate adjusted treatment group effect';
```

The HbA_{1c} is a measure of the overall level of blood glucose control. The lower the level of HbA_{1c}, the greater will be the risk of hypoglycemia that occurs when the blood glucose falls below the levels required to maintain consciousness. The basal c-peptide is a measure of the residual endogenous insulin secreted by the pancreas; the higher the value, the less demand there is for external (exogenous) insulin. A small level of endogenous insulin (c-peptide) may be more protective against hypoglycemia than virtually no endogenous insulin. The c-peptide level diminishes to zero over time. The total insulin dose may reflect past efforts to maintain low blood glucose values or may reflect the need for exogenous insulin to compensate for deficiencies in endogenous insulin. Patients with multiple prior episodes of hypoglycemia were excluded from entry, but a small fraction reported a history of one or two prior such episodes. The objective is to explore the effects of these covariates on the risk of hypoglycemia and to obtain an assessment of the treatment group effect after adjusting for these possible risk factors.

Table 8.2 presents the program statements using PROC GENMOD to fit a Poisson regression model to these data. Here the variable $nevents = d_j$. The results of the analyses are presented in Tables 8.3–8.6, with some extraneous information deleted.

The first model presented in Table 8.3 is the null or intercept only model. This null model yields an estimated intercept of $\widehat{\alpha} = -0.807 = \log(0.4462) = \log(\widehat{\lambda})$ where $\widehat{\lambda}$ is the overall crude rate computed as in (8.23). Also, the $S.E.(\widehat{\alpha}) =$

Table 8.3 PROC GENMOD Poisson Regression Analysis of the Risk of Hypoglycemia in the DCCT: The Null Model

```
                        null model
                   The GENMOD Procedure

          Criteria For Assessing Goodness Of Fit

  Criterion              DF         Value       Value/DF

  Deviance              714     4520.8419         6.3317
  Scaled Deviance       714     4520.8419         6.3317
  Pearson Chi-Square    714     6457.7296         9.0444
  Scaled Pearson X2     714     6457.7296         9.0444
  Log Likelihood          .      479.1508              .

             Analysis Of Parameter Estimates

Parameter    DF    Estimate     Std Err   ChiSquare  Pr>Chi

INTERCEPT     1     -0.8070      0.0210   1475.8849  0.0001
SCALE         0      1.0000      0.0000           .       .

NOTE:  The scale parameter was held fixed.
```

$1/\sqrt{D_{\bullet}} = 0.0210$. The log likelihood value is $\log[L(\alpha)] = 479.1508$, with a corresponding $-2\log[L(\alpha)] = -958.3$. This is not the complete log likelihood since the computation does not include the negative constant $-3371.1 = \sum_i \log(d_i!)$ (computed separately). This constant would cancel from any likelihood ratio test computation comparing this model to any other model. For this model, the $Deviance = 4520.84$, which is obtained as the difference between the $-2\log[L(\alpha)]$ and that of a model that fits the data perfectly (see Section A.10.3 of the Appendix), where the log likelihood for the latter is $\log[L(\lambda_1, \ldots, \lambda_N)] = 2739.57$, also ignoring the large negative constant (computed separately). Thus the $Deviance = -2(479.1508 - 2739.57) = 4520.84$.

Since PROC GENMOD fits the family of $GLMs$ using quasi-likelihood, it can allow for a dispersion or scale parameter, as described in Section A.10.4. Since no over-dispersion parameter is included in the models in this example, then in each model the scale parameter is fixed at the value 1.0. However, the Pearson Chi-square is $X^2 = 6457.73$ on 714 df. This is the same as Cochran's variance test of the hypothesis of no-overdispersion described in Section 8.1.4. The associated P-value is $p \leq 0.0001$ which indicates substantial extra-variation. Nevertheless,

Table 8.4 PROC GENMOD Poisson Regression Analysis of the Risk of Hypoglycemia in the DCCT: Unadjusted Treatment Group Effect

```
            Unadjusted treatment group effect
                  The GENMOD Procedure

           Criteria For Assessing Goodness Of Fit

      Criterion              DF          Value        Value/DF

      Deviance              713      3928.7828          5.5102
      Scaled Deviance       713      3928.7828          5.5102
      Pearson Chi-Square    713      5131.3432          7.1968
      Scaled Pearson X2     713      5131.3432          7.1968
      Log Likelihood          .       775.1804               .

              Analysis Of Parameter Estimates

Parameter          DF    Estimate      Std Err   ChiSquare  Pr>Chi

INTERCEPT           1     -1.5190       0.0429   1252.8961  0.0001
GROUP       Int     1      1.1081       0.0492    507.0072  0.0001
GROUP       Conv    0      0.0000       0.0000           .       .
SCALE               0      1.0000       0.0000           .       .

 NOTE:  The scale parameter was held fixed.
```

models assuming homogeneous Poisson variation are presented here for illustration. Models allowing for over-dispersion are presented in Section 8.4 to follow.

The second model presented in Table 8.4 then includes the effect of treatment group alone. Since the model includes an indicator variable for intensive versus conventional therapy, then $\widehat{\alpha} = \log(\widehat{\lambda}_{conv}) = \log(0.2189) = -1.5190$; and $\widehat{\beta} = \log(\widehat{RR}) = \log(\widehat{\lambda}_{int}/\widehat{\lambda}_{conv}) = \log(3.029) = 1.1081$. Also, $S.E.(\widehat{\beta}) = \sqrt{\frac{1}{D_1} + \frac{1}{D_2}} = 0.0492$. All of these quantities equal those obtained from the homogeneous Poisson model analysis presented in Example 8.1. The program also computes the Wald test for each coefficient, which for the group effect yields a chi-square value of 507.0072 that corresponds to a Z-value of 22.517. This model, when contrasted to the *Deviance* of the null model, also yields a likelihood ratio chi-square test of $4520.84 - 3928.78 = 592.06$, which corresponds to a Z-value of 24.33. These values are similar to the Z-value computed earlier in Example 8.1 under homogeneous Poisson model assumptions.

The third model presented in Table 8.5 describes the effect of treatment group in addition to those of other covariates. The likelihood ratio test for the model is

Table 8.5 PROC GENMOD Poisson Regression Analysis of the Risk of Hypoglycemia in the DCCT: Adjusted For Other Covariate Effects

The GENMOD Procedure

Criteria For Assessing Goodness Of Fit

Criterion	DF	Value	Value/DF
Deviance	706	3707.7027	5.2517
Scaled Deviance	706	3707.7027	5.2517
Pearson Chi-Square	706	4792.7878	6.7887
Scaled Pearson X2	706	4792.7878	6.7887
Log Likelihood	.	885.7204	.

Analysis Of Parameter Estimates

Parameter		DF	Estimate	Std Err	ChiSquare	Pr>Chi
INTERCEPT		1	-0.9568	0.2174	19.3765	0.0001
GROUP	Int	1	1.0845	0.0493	483.9072	0.0001
GROUP	Conv	0	0.0000	0.0000	.	.
INSULIN		1	0.0051	0.0995	0.0026	0.9593
DURATION		1	0.0015	0.0006	6.7882	0.0092
FEMALE		1	0.1794	0.0424	17.9276	0.0001
ADULT		1	-0.5980	0.0656	83.1309	0.0001
BCVAL5		1	-0.5283	0.3630	2.1175	0.1456
HBAEL		1	-0.0335	0.0151	4.8868	0.0271
HXCOMA		1	0.6010	0.0685	77.0919	0.0001
SCALE		0	1.0000	0.0000	.	.

NOTE: The scale parameter was held fixed.

obtained as the difference in the deviances versus the null model, so that $X^2_{LR} = 4520.84 - 3707.7027 = 813.14$ on $714 - 706 = 8$ *df*, which is highly significant. In this model, the adjusted $\widehat{\beta} = \log(\widehat{RR}) = 1.0845$ corresponds to an estimated $\widehat{RR} = 2.96$ adjusted for other covariates, virtually unchanged from the unadjusted analysis. Each of the other covariates, other than the levels of insulin and c-peptide (*bcval5*) were statistically significantly associated with the risk of hypoglycemia (based on Wald tests).

As described in Section 7.3.4, PROC GENMOD will also compute likelihood ratio tests for each of the covariates in the model using the TYPE3 option. Such tests could have been computed here as well.

Table 8.6 PROC GENMOD Poisson Regression Analysis of the Risk of Hypoglycemia in the DCCT: Estimated Covariance Matrix in the Adjusted Model

Estimated Covariance Matrix

Parameter	INTERCEPT	GROUP Int	INSULIN	DURATION	FEMALE
INTERCEPT	0.04725	-0.001773	-0.01189	-0.000051	-0.000241
GROUP Int	-0.001773	0.002431	0.0000531	-5.444E-9	0.0000214
INSULIN	-0.01189	0.0000531	0.009890	-8.33E-7	-0.000107
DURATION	-0.000051	-5.444E-9	-8.33E-7	3.1565E-7	-2.211E-6
FEMALE	-0.000241	0.0000214	-0.000107	-2.211E-6	0.001795
ADULT	-0.006857	0.0000547	0.003462	-6.594E-6	0.000089
BCVAL5	-0.02945	0.0004454	0.006824	0.0000956	-0.000535
HBAEL	-0.002572	-0.00002	0.0001494	2.0192E-6	-0.000048
HXCOMA	-0.001927	-0.000127	-0.000069	2.6715E-6	0.000013

Parameter	ADULT	BCVAL5	HBAEL	HXCOMA
INTERCEPT	-0.006857	-0.02945	-0.002572	-0.001927
GROUP Int	0.0000547	0.0004454	-0.00002	-0.000127
INSULIN	0.003462	0.006824	0.0001494	-0.000069
DURATION	-6.594E-6	0.0000956	2.0192E-6	2.6715E-6
FEMALE	0.000089	-0.000535	-0.000048	0.000013
ADULT	0.004301	-0.001021	0.0001582	0.0000196
BCVAL5	-0.001021	0.13179	0.0008153	0.002335
HBAEL	0.0001582	0.0008153	0.0002291	0.0001239
HXCOMA	0.0000196	0.002335	0.0001239	0.004686

The covariance of the estimates is also presented in Table 8.6. These allow estimation of the variance of the estimated rates for individual subjects with specific covariate values. For example, for the first patient in the listing in Table 8.1, the linear predictor is $\widehat{\eta}_i = \widehat{\alpha} + \boldsymbol{x}_i'\widehat{\boldsymbol{\beta}} = -0.63870$, which yields an estimated rate per year of $\widehat{\lambda}(\boldsymbol{x}_i) = \exp(-0.6387) = 0.52798$. Given the $t_i = 9.50582$ years of follow-up, this yields a predicted number of events of $\widehat{\lambda}(\boldsymbol{x}_i)t_i = 5.0189$. Using the above covariance matrix, the $S.E.$ of the linear predictor is 0.0537, which yields 95% confidence limits of (–0.7441, –0.5333). This yields 95% limits on the predicted number of events of (4.5169, 5.5766). The linear predictor $\widehat{\eta}_i$ and its $S.E.$ may also be obtained from PROC GENMOD using the statements:

```
%global _disk_ ; %let _disk_=on ;
%global _print_; %let _print_=off ;
PROC GENMOD;
MAKE 'OBSTATS' OUT=PRED ;
model nevents = group insulin duration female
                adult bcval5 hbael hxcoma
/ dist = poisson link = log offset = lnyears covb OBSTATS;
PROC PRINT DATA=PRED;
```

8.3.2 Explained Variation

Using the developments in Section A.8.1 of the Appendix, the explained fraction of squared error loss from a Poisson regression model is readily shown to be

$$\widehat{\rho}^2_{\varepsilon^2} = \frac{\widehat{V}\left[E(d|\boldsymbol{x},t)\right]}{\widehat{V}(d|t)} = \frac{\sum_i \left[\widehat{\lambda}\left(\boldsymbol{x}_i\right) t_i - \widehat{\lambda} t_i\right]^2}{\sum_i \left[d_i - \widehat{\lambda} t_i\right]^2} = R^2_{\varepsilon^2, resid}\,, \tag{8.52}$$

which equals the fraction of explained residual variation. In this expression, $\widehat{\lambda} = \overline{r}$ is the crude rate estimate in the combined sample that also equals $e^{\widehat{\alpha}}$ from the null model. The derivations are left to a Problem.

Alternately, the fraction of $-$log-likelihood explained R^2_ℓ can be used as described in Section A.8.3 of the Appendix, or Maddala's measure R^2_{LR} based on the likelihood ratio statistic as in Section A.8.4 of the Appendix.

Example 8.5 *Hypoglycemia in the DCCT (continued)*
Using the output of the data set PRED, with additional calculations, it is then possible to compute the sums of squares in the numerator and denominator of (8.52). The resulting value of the $R^2_{\varepsilon^2, resid} = (3139.217/21856.4) = 0.14363$.

Alternately, the Madalla's likelihood ratio measure yields $R = R^2_{LR} = 1 - \exp(-813.14/715) = 0.6793$, which is clearly substantially different from the fraction of squared error loss explained by the model.

The other possible measure of explained variation is the fraction of explained negative log likelihood. As described in Example 8.4, PROC GENMOD only computes the log likelihood and the deviance up to an additive constant. Thus it is necessary to compute the complete log likelihood using a separate program based on the output data set (PRED) described above. The complete log likelihoods are $\log[L(\alpha)] = -2891.94$ and $\log[L(\alpha, \boldsymbol{\beta})] = -2485.37$, for which $R^2_\ell = (2891.94 - 2485.37)/2891.94 = 0.14059$, which is rather close to the value of the $R^2_{\varepsilon^2, resid}$.

For illustration, the value of the ratio based on the kernel of the log likelihoods using the values presented in Tables 8.3 and 8.5, which ignore the constant, is $(885.7204 - 479.1508)/479.1508 = 0.84852$. This is clearly incorrect.

Alternately, one could use the fraction of the explained deviance as a measure of explained variation, in which case the constant is irrelevant because it cancels in the computation of the deviance. This yields $R^2_D = (4520.8419 - 3707.7027)/4520.8419 = 0.17986$, which is more in line with the value of R^2_ℓ.

8.3.3 Applications of Poisson Regression

The above example illustrates the application of Poisson regression to the analysis of subject or unit specific count data. Count data also arises in other contexts such as the analysis of rates of events within subpopulations or vital statistics. An example is given in Problem 8.6. Another example is presented by Gail (1978). Poisson regression has also been applied to the analysis of grouped survival data by Holford (1980) and Laird and Oliver (1981). In each instance the data structure consists of a count of the number of events associated with a measure of exposure, such as population size, within subsets of the population, where each subset is characterized by covariate values. The risk (rate) within each sub-population is then expressed as a function of the covariate values.

8.4 OVER-DISPERSED AND ROBUST POISSON REGRESSION

The ordinary Poisson regression model, and the above analyses of the DCCT, all assume that there is no over-dispersion after allowing for covariates. That is, the model assumes that the observed number of events for all subjects with a given covariate vector $\boldsymbol{x}$ and fixed time t, or the conditional distribution $d|\boldsymbol{x}$, is distributed as Poisson with parameter $\lambda(\boldsymbol{x})$. However, even after adjusting for covariates, the conditional distribution of $d|\boldsymbol{x}$ may be over-dispersed. There is a strong suggestion that this also applies to the analysis of hypoglycemia in the DCCT. The null model Pearson Chi-square value shown in Table 8.3, which is equivalent to the Cochran variance test, is $X^2 = 6457.73$ on 714 df with $p \leq 0.0001$ indicating a significant degree of extra-variation. Another indication is that the $SE\left(\widehat{\beta}\right)$ for the effect of treatment group in the above covariate-adjusted regression model (0.0493) is still too small compared to the variance of the $\log(\widehat{RR})$ estimated from the two-stage over-dispersed Poisson model (0.12286). Here we consider generalizations that allow for various kinds of model misspecification such as over-dispersion.

8.4.1 Quasi-Likelihood Over-Dispersed Poisson Regression

Many methods have been described for fitting over-dispersed Poisson regression models. Breslow (1984, 1990) employed the method of moments to estimate the over-dispersion variance component that is assumed to be homoscedastic for all values of the covariates. He then used quasi-likelihood to estimate the regression coefficients. Moore (1986) also discusses this approach. Others adopt a specific form for the mixing distribution, such as the Gaussian by Dean and Lawless (1989). An alternate approach is to adopt a quasi-likelihood that incorporates a scale or variance inflation dispersion parameter as described in Section A.10.3 of the Appendix that is also implemented in PROC GENMOD.

The over-dispersed quasi-likelihood Poisson errors regression model adopts the simplifying assumption that

$$V(d_i|\boldsymbol{x}_i) = \phi E(d_i|\boldsymbol{x}_i, t_i) = \phi\lambda(\boldsymbol{x}_i)t_i \tag{8.53}$$

for scale parameter $\phi \neq 1$, or that the true conditional variance is ϕ times the expected variance (the mean) under a homogeneous Poisson model. As described in Section A.10.3 of the Appendix, both the deviance and the Pearson chi-square tests of goodness of fit are asymptotically distributed as chi-square on $N-p-1$ *df* when the model assumptions are correct. Thus a simple moment estimator of ϕ is provided by either $\widehat{\phi}$ = *Deviance/df* or by $\widehat{\phi}$ = *Pearson* χ^2/df. Since the modes of convergence for the deviance and the Pearson statistics differ for some models (see McCullagh and Nelder, 1989) it is generally recommended that the Pearson statistic be used. When the model assumptions are correct, the 95% tolerance limits of the distribution of X^2/df are $1 \pm 2.77/\sqrt{df} = 1 \pm 0.104$ for the covariate-adjusted Poisson regression model in Table 8.5 on 706 *df*. The *Pearson* $X^2/df = 6.7887$ thus indicates some degree of over-dispersion. This provides the basis for specifying a starting value for an iterative procedure by which PROC GENMOD obtains an estimate of ϕ.

PROC GENMOD uses a different parametrization where the scale parameter is $scale = \sqrt{\phi}$. In the model statement we then use the option "SCALE=$scale_0$ PSCALE" where $scale_0 = \sqrt{\phi_0}$ is the starting value for the solution of $\widehat{\phi}$. Therefore, in this example, $scale_0 = \sqrt{6.7887} = 2.606$ is the starting value. The solution $\widehat{\phi}$ is that value such that the *scaled Pearson chi-square* for the fitted model equals

$$\frac{Pearson\ X^2/\ df}{\widehat{\phi}} = 1.0. \tag{8.54}$$

All of the elements of $V\left(\widehat{\beta}\right)$ are then multiplied by $\widehat{\phi}$, and all the $S.E.\left(\widehat{\beta}\right)$ by $\sqrt{\widehat{\phi}}$. The $\widehat{\beta}$ themselves are unchanged.

The corresponding option to estimate the scale parameter using *Deviance/df* is DSCALE.

Example 8.6 *Hypoglycemia in the DCCT (continued)*
Table 8.7 presents the over-dispersed covariate adjusted model using the options in the model statement as follows:

```
/ dist = poisson link = log offset = logyears
  SCALE=2.606 PSCALE;
```

The scale parameter is estimated to be $\sqrt{\phi} = 2.6055$ as above and the standard errors of all the coefficients have been inflated by this amount. Thus the $S.E.(group) = (2.6055 \times 0.0493) = 0.1285$, which is comparable to that estimated from the random effects model (0.1229) without adjustment for covariates.

Table 8.7 PROC GENMOD Over-dispersed Poisson Regression Analysis of the DCCT Data Using *Pearson/df*

The GENMOD Procedure

Criteria For Assessing Goodness Of Fit

Criterion	DF	Value	Value/DF
Deviance	706	3707.7027	5.2517
Scaled Deviance	706	546.1619	0.7736
Pearson Chi-Square	706	4792.7878	6.7887
Scaled Pearson X2	706	706.0000	1.0000
Log Likelihood	.	130.4707	.

Analysis Of Parameter Estimates

Parameter		DF	Estimate	Std Err	ChiSquare	Pr>Chi
INTERCEPT		1	-0.9568	0.5664	2.8542	0.0911
GROUP	Int	1	1.0845	0.1285	71.2818	0.0001
GROUP	Conv	0	0.0000	0.0000	.	.
INSULIN		1	0.0051	0.2591	0.0004	0.9844
DURATION		1	0.0015	0.0015	0.9999	0.3173
FEMALE		1	0.1794	0.1104	2.6408	0.1042
ADULT		1	-0.5980	0.1709	12.2456	0.0005
BCVAL5		1	-0.5283	0.9459	0.3119	0.5765
HBAEL		1	-0.0335	0.0394	0.7198	0.3962
HXCOMA		1	0.6010	0.1783	11.3560	0.0008
SCALE		0	2.6055	0.0000	.	.

NOTE: The scale parameter was estimated by the square root of Pearson's Chi-Squared/DOF.

8.4.2 Robust Inference Using the Information Sandwich

Essentially, over-dispersion arises when the variance of the random errors is misspecified. Another approach, therefore, is to base inferences on the robust information sandwich estimate of the variance/covariance matrix of the estimates that is based on the empirical estimate of the observed information. From the devel opments in Section A.9 of the Appendix, it is readily shown that the empirical estimate of the covariance matrix of the Poisson model score vector is the matrix

$$\widehat{\boldsymbol{J}}(\widehat{\boldsymbol{\theta}}) = \sum\nolimits_i \boldsymbol{U}(\widehat{\boldsymbol{\theta}})\boldsymbol{U}(\widehat{\boldsymbol{\theta}})^T = \boldsymbol{X}'\widehat{\boldsymbol{\Sigma}}_\varepsilon \boldsymbol{X} \,, \tag{8.55}$$

where $\boldsymbol{x}$ is the $n \times (p+1)$ design matrix and $\widehat{\boldsymbol{\Sigma}}_\varepsilon = diag\left[\left(d_i - \widehat{\lambda}(\boldsymbol{x}_i)t_i\right)^2\right]$. When the scores, and thus the covariance matrix, are estimated under a specified null hypothesis, this matrix can be used as the basis for a robust efficient score test as described in the Appendix.

The robust information sandwich estimate of the covariance matrix of the errors is obtained as

$$\widehat{\boldsymbol{\Sigma}}_R(\widehat{\boldsymbol{\theta}}) = \boldsymbol{I}(\widehat{\boldsymbol{\theta}})'\widehat{\boldsymbol{J}}(\widehat{\boldsymbol{\theta}})^{-1}\boldsymbol{I}(\widehat{\boldsymbol{\theta}}), \tag{8.56}$$

where from (8.51) $\boldsymbol{I}(\widehat{\boldsymbol{\theta}}) = \boldsymbol{X}'\widehat{\boldsymbol{\Gamma}}\boldsymbol{X}$ and $\widehat{\boldsymbol{\Gamma}} = diag[\widehat{\lambda}(\boldsymbol{x}_i)t_i]$. This robust covariance matrix could be used as the basis for the computation of Wald tests and confidence limits.

Such computations can be performed using a supplemental SAS PROC IML program as described in Section 7.3.5 for a logistic regression model. Alternately, computation of the robust information sandwich and associated confidence limits and Wald tests, but not the robust score test, can be obtained from PROC GENMOD.

One feature of PROC GENMOD is the option to conduct an analysis of correlated observations using generalized estimating equations (*GEE*), as described in Section A.10.6 of the Appendix. *GEE* models employ the information sandwich to estimate the covariance matrix of the estimated coefficients when the observations may be correlated, such as where repeated measures are obtained on the same subject. This is designated by the REPEATED statement. When there is only a single observation per subject, as herein, then the REPEATED statement may still be used in which case the robust information sandwich above is computed.

Example 8.7 *Hypoglycemia in the DCCT (continued)*
The following SAS statements would be used for the robust analysis of the DCCT hypoglycemia data employed earlier in Table 8.2:

```
proc genmod; class group patid;
 model nevents = group insulin duration female
 adult bcval5 hbael hxcoma
 / dist = poisson link = log offset = logyears;
 repeated subject=patid / type=unstr covb;
title2 'covariate adjusted treatment group effect';
```

Because there is only one observation per subject, indicated by the class effect `patid` in this case, then basically the `type=unstr` option specifies that the covariance matrix of the scores be estimated empirically. The robust (empirical) estimate of the covariance matrix of the coefficient estimates is presented in Table

Table 8.8 PROC GENMOD Robust Poisson Regression Analysis of the DCCT Data: Robust Information Sandwich Covariance Estimate

The GENMOD Procedure

Covariance Matrix (Empirical)
Covariances are Above the Diagonal and Correlations are Below

PRM1 (intercep	PRM2 group	PRM4 insulin	PRM5 duration	PRM6 female	PRM7 adult
0.32773	-0.01447	-0.09258	-0.000257	0.005107	-0.05755
-0.20799	0.01478	-0.000606	5.1525E-6	-0.001735	0.002650
-0.52964	-0.01607	0.09303	-0.000043	-0.007640	0.03841
-0.32059	0.03030	-0.10143	1.9571E-6	-7.34E-6	-0.00007
0.07366	-0.11785	-0.20681	-0.04332	0.01467	-0.004818
-0.47736	0.10353	0.59801	-0.23659	-0.18889	0.04435
-0.50214	0.04150	0.17092	0.47185	0.10377	-0.03604
-0.73454	0.07888	-0.02292	0.19724	-0.02465	2.4554E-6
-0.20426	-0.01986	0.004646	0.23973	0.06898	0.009142

PRM8 bcval5	PRM9 hbael	PRM10 hxcoma)
-0.28247	-0.01729	-0.02070
0.004957	0.0003942	-0.000427
0.05123	-0.000287	0.0002503
0.0006486	0.0000113	0.0000594
0.01235	-0.000123	0.001479
-0.007458	2.1256E-8	0.0003409
0.96557	0.01293	0.04096
0.32006	0.001690	0.0007505
0.23545	0.10312	0.03135

8.8, and the robust analysis of the parameter estimates is presented in Table 8.9, extraneous information deleted. The coefficient estimates are unchanged from those shown previously because the score vectors are identical to those employed in the previous models, there being only one observation per subject. The robust information sandwich variance/covariance estimates are similar to those obtained from the quasi-likelihood analysis with a single scale or dispersion parameter.

Table 8.9 PROC GENMOD Robust Poisson Regression Analysis of the DCCT Data: Robust Analysis of Parameter Estimates

Analysis Of GEE Parameter Estimates
Empirical Standard Error Estimates

			Empirical 95% Confidence Limits			
Parameter	Estimate	Std Err	Lower	Upper	Z	Pr>|z|
INTERCEPT	-0.9568	0.5725	-2.0789	0.1652	-1.671	0.0946
GROUP - IntI	1.0845	0.1216	0.8463	1.3228	8.9216	0.0000
GROUP - Conv	0.0000	0.0000	0.0000	0.0000	0.0000	0.0000
INSULIN	0.0051	0.3050	-0.5927	0.6029	0.0166	0.9867
DURATION	0.0015	0.0014	-0.0013	0.0042	1.0463	0.2954
FEMALE	0.1794	0.1211	-0.0580	0.4168	1.4811	0.1386
ADULT	-0.5980	0.2106	-1.0107	-0.1852	-2.839	0.0045
BCVAL5	-0.5283	0.9826	-2.4542	1.3976	-.5376	0.5908
HBAEL	-0.0335	0.0411	-0.1140	0.0471	-.8139	0.4157
HXCOMA	0.6010	0.1770	0.2540	0.9480	3.3947	0.0007
Scale	2.5891	.	.	.	.	.

NOTE: The scale parameter for GEE estimation was computed as the square root of the normalized Pearson's chi-square.

8.5 POWER AND SAMPLE SIZE FOR POISSON MODELS

In Problem 3.5. of Chapter 3 we derived the expressions for the power of the simple Z-test of the difference in the crude mean rates between two groups in (8.25). If extra-variation is observed or is anticipated, the power function of the test in the over-dispersed model is readily obtained given a specification of the over-dispersion variance components within each group, see Problem 8.7.

In the Poisson regression model, the power of the Wald test with one or more qualitative covariates is also readily obtained along the lines of that for the Wald test in logistic regression, as described in Section 7.3.6. The resulting equations are also derived in Problem 8.7.

The power of a Wald test in an over-dispersed Poisson regression model may likewise be obtained using $V\left(\widehat{\boldsymbol{\theta}}\right) = \phi\Sigma_{\boldsymbol{\theta}}$, where ϕ is the specified over-dispersion parameter and $\Sigma_{\boldsymbol{\theta}}$ is the model-based covariance matrix of the estimates. Alternately, for power computations for a test based on an observed data set, the robust estimate of the covariance matrix of the estimates may be used.

8.6 CONDITIONAL POISSON REGRESSION FOR MATCHED SETS

As in Chapter 7, now consider the assessment of covariate effects on the intensity or rate parameter for individual subjects sampled in matched sets. For example, in a prospective study of the number of episodes of some condition (hypoglycemia in diabetes, epileptic seizures, etc.) subjects may be matched with respect to one or more covariates. As in Chapter 7 the data consist of N matched sets of size n_i for the ith set. The jth member of the ith set has covariate vector $\boldsymbol{x}_{ij} = (x_{ij1} \, \ldots \, x_{ijp})^T$ and the count variable d_{ij} designates the number of events experienced during exposure time t_{ij} for $i = 1, \ldots, N$; $j = 1, \ldots, n_i$. The Poisson model then specifies that the rate λ_{ij} for the $ijth$ member is a function of the covariates $\boldsymbol{x}_{ij}$ through the link function $g(\lambda_{ij}) = \alpha_i + \boldsymbol{x}'_{ij}\boldsymbol{\beta}$. The model allows each matched set to have a unique risk of the outcome represented by a set-specific intercept α_i, and then assumes that the covariates have a common effect on the odds of the outcome over all matched sets represented by the coefficient vector $\boldsymbol{\beta} = (\beta_1 \, \ldots \, \beta_p)^T$.

The unconditional Poisson regression model with parameters $\boldsymbol{\alpha} = (\alpha_1 \, \ldots \, \alpha_N)^T$ and the vector of coefficients $\boldsymbol{\beta}$ then is of the form

$$L(\boldsymbol{\alpha}, \boldsymbol{\beta}) = \prod_{i=1}^{N} \prod_{j=1}^{n_i} \frac{e^{-\lambda(\boldsymbol{x}_{ij})t_{ij}} \left[\lambda\left(\boldsymbol{x}_{ij}\right) t_{ij}\right]^{d_{ij}}}{d_{ij}!}, \tag{8.57}$$

where $\lambda\left(\boldsymbol{x}_{ij}\right) = \exp(\alpha_i + \boldsymbol{x}'_{ij}\boldsymbol{\beta})$. To fit this model, we must solve for the N nuisance parameters $\{\alpha_i\}$ in order to obtain an estimate of the covariate vector $\boldsymbol{\beta}$. Again, however, this can be avoided by conditioning on the appropriate sufficient statistic within each matched set, which in this case is the total number of events within the ith set, $D_i = \sum_j d_{ij}$.

Within the ith matched set, the conditional probability of $\{d_{i1}, \ldots, d_{in_i}\}$ events among the n_i members with covariates $\{\boldsymbol{x}_1, \ldots, \boldsymbol{x}_{n_i}\}$, respectively, is

$$P\left[\{(d_{i1}, \boldsymbol{x}_1), \ldots, (d_{in_i}, \boldsymbol{x}_{in_i})\} \mid D_i,\, n_i\right]. \tag{8.58}$$

It is then readily shown that the conditional likelihood is

$$L_{(c)}(\boldsymbol{\beta}) = \prod_{i=1}^{N} L_{i|D_i,n_i} = \prod_{i=1}^{N} \frac{\prod_{j=1}^{n_i} \left[e^{\boldsymbol{x}'_{ij}\boldsymbol{\beta} + \log(t_{ij})}\right]^{d_{ij}}}{\sum_{\ell=1}^{n_i!} \prod_{j(\ell)=1}^{n_i} \left[e^{\boldsymbol{x}'_{ij(\ell)}\boldsymbol{\beta} + \log(t_{ij(\ell)})}\right]^{d_{ij}}} \tag{8.59}$$

(see Problem 8.8). Standard software is not available to fit this model; however, it could be fit using the general PROC NLIN.

For the special case of matched pairs, this conditional likelihood reduces to

$$\begin{aligned} L_{(c)}(\boldsymbol{\beta}) &= \prod_{i=1}^{N} \left(1 + \left[e^{(\boldsymbol{x}_{i1} - \boldsymbol{x}_{i2})\boldsymbol{\beta}} \left(\frac{t_{i1}}{t_{i2}}\right)\right]^{d_{i2} - d_{i1}}\right)^{-1} \\ &= \prod_{i=1}^{N} \left(1 + e^{-\Delta\eta}\right)^{-1}, \end{aligned} \tag{8.60}$$

where

$$\Delta\eta = (d_{i1} - d_{i2}) \left[(\boldsymbol{x}_{i1} - \boldsymbol{x}_{i2}) \boldsymbol{\beta} + \log \left(\frac{t_{i1}}{t_{i2}} \right) \right]. \tag{8.61}$$

The latter is a logistic regression model with no intercept, covariate vector $(d_{i1} - d_{i2})(\boldsymbol{x}_{i1} - \boldsymbol{x}_{i2})$ and offset $(d_{i1} - d_{i2}) \log (t_{i1}/t_{i2})$; and where the binary dependent variable $y_i = 1$ for all pairs.

In either case, the score equations and information matrix are readily obtained.

8.7 PROBLEMS

8.1 Consider the doubly homogeneous Poisson process described in Section 8.1.2.

8.1.1. Show that the crude rate in (8.9) under a doubly homogeneous Poisson model can be expressed a weighted least squares (inverse variance weighted) estimate of the form

$$\overline{r} = \frac{\sum_j \widehat{\tau}_j r_j}{\sum_j \widehat{\tau}_j}, \tag{8.62}$$

where $\widehat{\tau}_j = \widehat{V}(r_j)^{-1}$.

8.1.2. Then show that its estimated variance equals $\widehat{V}(\overline{r}) = \left[\sum_j \widehat{\tau}_j\right]^{-1} = \widehat{V}(\widehat{\lambda})$ in (8.10).

8.2 Now consider the robust information sandwich estimate of the variance as described in Section A.8 of the Appendix.

8.2.1. Based on the score equation in (8.7), show that the empirical estimate of the observed information is

$$\widehat{J}(\widehat{\lambda}) = \frac{\sum_{j=1}^{N} (d_j - \widehat{\lambda} t_j)^2}{\widehat{\lambda}^2} \tag{8.63}$$

8.2.2. Also show that the robust estimate of the variance of the estimate is

$$\widehat{V}_R(\widehat{\lambda}) = \frac{\sum_{j=1}^{N} (d_j - \widehat{\lambda} t_j)^2}{T^2}. \tag{8.64}$$

8.2.3. Under the doubly homogeneous assumptions show that

$$E\left[\widehat{V}_R(\widehat{\lambda})\right] = \frac{\lambda}{T} = I(\lambda)^{-1}. \tag{8.65}$$

8.3 Consider a random effects over-dispersed model as in Section 8.2.

8.3.1. To facilitate computation of the variance component, show that the moment estimator of the dispersion variance component presented in (8.37) can also be expressed as

$$\widehat{\sigma}_{\lambda}^2 = \max\left[0, \frac{\sum_j \frac{(d_j - \widehat{\mu}_\lambda t_j)^2}{t_j} - n\widehat{\mu}_\lambda}{\sum_j t_j}\right]. \tag{8.66}$$

8.3.2. Also show that this can be expressed as a weighted mean of variables a_j of the form

$$\widehat{\sigma}_{\lambda}^2 = \frac{\sum_j t_j a_j}{\sum_j t_j}, \qquad a_j = \frac{1}{t_j}\left[\frac{(d_j - \widehat{\mu}_\lambda t_j)^2}{t_j} - \widehat{\mu}_\lambda\right]. \tag{8.67}$$

8.4 Gail, Santner and Brown (1980) present data on the numbers of tumor recurrences over a period of 122 days among 23 female rats treated with retinoid versus 25 untreated controls. The following are the numbers of tumor recurrences for the rats in each group:

Retinoid: 1, 0, 2, 1, 4, 3, 6, 1, 1, 5, 2, 1, 5, 2, 3, 4, 5, 5, 1, 2, 6, 0, 1
Control: 7, 11, 9, 2, 9, 4, 6, 7, 6, 1, 13, 2, 1, 10, 4, 5, 11, 11, 9, 12, 1, 3, 1, 3, 3

(reproduced with permission).

8.4.1. Assuming a doubly homogeneous model, compute the estimate of the Poisson rate (per day) for tumor recurrence in each group $(\widehat{\lambda}_1, \widehat{\lambda}_2)$ and in the combined sample of 48 rats $(\widehat{\lambda})$.

8.4.2. Also compute the large sample variance of the log rate and the asymmetric 95% confidence limits for the rate in each group and in the combined sample.

8.4.3. Compute the relative risk and the asymmetric 95% confidence limits based on the estimated variance of the $\log(\widehat{RR})$.

8.4.4. Compute the estimated variance of the difference in rates under the null hypothesis $\widehat{V}(\widehat{\lambda}_1 - \widehat{\lambda}_2 | H_0)$ and the simple Z-test of the difference in rates between groups.

8.4.5. Compute Cochran's variance test of the hypothesis of no over-dispersion separately within each treatment group.

8.4.6. Using (8.64), compute the information sandwich robust estimate of the variance of the estimated rate within each group $\widehat{V}_R(\widehat{\lambda}_i)$, $i = 1, 2$.

8.4.7. Use this with (8.15) to compute the robust estimate of the variance of the $\log(\widehat{\lambda}_1)$ and $\log(\widehat{\lambda}_2)$.

8.4.8. Then compute the robust estimate of the variance of the $\log(\widehat{RR})$ and the robust 95% confidence limits for the RR.

8.4.9. Alternately, adopt a two-stage model as in Section 8.2. Using (8.37), compute the moment estimate of the over-dispersion variance $\widehat{\sigma}_{\lambda_i}^2$ for each group $(i = 1, 2)$.

8.4.10. Use this to compute an over-dispersion estimate of the variance of the crude rate within each group as in (8.38) and of the log rate as in (8.39). Use these to compute the over-dispersed estimate of the variance of the $\log(\widehat{RR})$. Compare these to those obtained using the robust estimate of the variance.

8.5 Now consider the case of two independent groups where the observations are divided into K independent strata.

8.5.1. Assume that the rates in the control group, $\{\lambda_{2j}\}$ vary among strata but that there is a common log relative risk among the K strata, $\theta_j = \log(\lambda_{1j}/\lambda_{2j}) = \theta$, $j = 1, \ldots, K$. Show that the $MVLE$ of the common $\log(RR)$ is provided by

$$\widehat{\theta} = \frac{\sum_{j=1}^{K} \left(\frac{D_{1j}D_{2j}}{D_{\bullet j}}\right) \log(\widehat{\lambda}_{1j}/\widehat{\lambda}_{2j})}{\sum_{\ell=1}^{K} \left(\frac{D_{1\ell}D_{2\ell}}{D_{\bullet \ell}}\right)} \tag{8.68}$$

and its large sample variance is estimated as

$$\widehat{V}(\widehat{\theta}) = \left[\sum_{\ell=1}^{K} \left(\frac{D_{1\ell}D_{2\ell}}{D_{\bullet \ell}}\right)\right]^{-1}. \tag{8.69}$$

8.5.2. Using the developments in Section 4.7.2, derive the expression for the maximally efficient asymptotic test of the joint null hypothesis H_0: $\theta_j = 0$ for all K strata against the alternative that a common log relative risk applies over all strata H_1: $\theta_j = \theta \neq 0$. Show that the test is of the form

$$X_A^2 = \frac{\left[\sum_{j=1}^{K} \left(\frac{T_{1j}T_{2j}}{T_{\bullet j}}\right) (\widehat{\lambda}_{1j} - \widehat{\lambda}_{2j})\right]^2}{\sum_{j=1}^{K} \left(\frac{T_{1j}T_{2j}}{T_{\bullet j}}\right) \widehat{\lambda}_j}. \tag{8.70}$$

8.5.3. Show that Cochran's test of the hypothesis of homogeneity of relative risks among strata, analogous to the Cochran test for independent 2×2 tables in Section 4.10.2, is given by

$$X_{H,C}^2 = \sum_{j=1}^{K} \widehat{\tau}_j (\widehat{\theta}_j - \widehat{\theta})^2 ,$$

where $\widehat{\tau}_j = D_{1j}D_{2j}/D_{\bullet j}$, and where $X_{H,C}^2$ is distributed as chi-square on $K-1$ *df* under the hypothesis of homogeneity H_0: $\theta_j = \theta \, \forall j$.

8.5.4. Now adopt a two-stage random effects model across strata. The random effects model estimate of the variance component between strata $\widehat{\sigma}_\theta^2$ is obtained directly from (4.155) using the above expression for $\widehat{\tau}_j$. From this, derive the resulting updated (one-step) estimate of the mean $\log(RR)$ over strata and its variance.

8.6 Table 8.10 presents data from Frome and Checkoway (1985) giving the numbers of cases of non-melanoma skin cancer and the population size within the Dallas/Fort Worth versus Minneapolis/St. Paul metropolitan areas, stratified by age groups (reproduced with permission). The hypothesis is that those in the Dallas

Table 8.10 Numbers of Cases of Skin Cancer and Population Size, Stratified by Age, in Dallas/Fort Worth and Minneapolis/St. Paul, from Frome and Checkoway (1985).

Age	*Dallas/Fort Worth*		*Minneapolis/St. Paul*		
Stratum	d_1	T_1	d_2	T_2	$\widehat{RR}$
15–24	4	181,343	1	172,675	3.81
25–34	38	146,207	16	123,065	2.00
35–44	119	121,374	30	96,216	3.14
45–54	221	111,353	71	92,051	2.57
55–64	259	83,004	102	72,159	2.21
65–74	310	55,932	130	54,722	2.33
75–84	226	29,007	133	32,185	1.89
85+	65	7,538	40	8,328	1.80
Total	1242	735758	523	651401	2.10

area would have a higher rate of such cancer because of greater exposure to the sun.

8.6.1. Based on the aggregate data for all age strata combined, compute the unadjusted relative risk for location (Dallas versus Minneapolis), the $\log(\widehat{RR})$, its estimated variance and the asymmetric 95% confidence limits for RR.

8.6.2. Compute the unadjusted Z-test of the hypothesis of equal rates in the two metropolitan areas.

8.6.3. Compute the $MVLE$ of the stratified-adjusted common log relative risk, its variance, and the asymmetric confidence limits for the common relative risk.

8.6.4. Compute the stratified-adjusted efficient test of a common relative risk and compare the results to the unadjusted test.

8.6.5. Compute Cochran's test of homogeneity and the estimate of the random effects model variance component.

8.6.6. Then compute the one step estimate of the mean log relative risk and its variance under the random effects model. Also compute the asymmetric confidence limits for the RR. Compare these to the $MVLE$ estimate and its limits.

8.6.7. Alternately, these data could be analyzed using Poisson regression. Fit a Poisson regression model with an effect for location and binary indicator variables for 7 of the 8 strata. Compare the model-based estimate of the relative risk for location to the $MVLE$ above.

8.7 Show that the power function of the Z-test for the difference in rates between two groups in (8.26) under the doubly homogeneous Poisson model is a function of the total patient years of exposure of the form

$$\lambda_1 - \lambda_2 = Z_{1-\alpha}\sqrt{\lambda\left(\frac{T_\bullet}{T_1 T_2}\right)} + Z_{1-\beta}\sqrt{\frac{\lambda_1}{T_1} + \frac{\lambda_2}{T_2}} \tag{8.71}$$

$$= Z_{1-\alpha}\sqrt{\lambda^2\left(\frac{E(D_\bullet)}{E(D_1)E(D_2)}\right)} + Z_{1-\beta}\sqrt{\frac{\lambda_1^2}{E(D_1)} + \frac{\lambda_2^2}{E(D_2)}},$$

which is also a function of the expected number of events in each group.

8.7.1. Generalize this expression to allow for an over-dispersed random effects model with variance components $\sigma_{\lambda 1}^2$ and $\sigma_{\lambda 2}^2$ in the two groups that describes the power of the Z-test in (8.44).

8.7.2. Assume that we wished to detect a difference between groups of $\lambda_1 = 0.25$/year versus $\lambda_2 = 0.15$/year with two equal sized patient groups, that is, $T_1 = T_2 = T_\bullet/2$. At $\alpha = 0.05$ two-sided,

1. How many total patient years are required to provide 90% power assuming a doubly homogeneous model?
2. If we recruit 200 patients per group, how many years must each patient be followed?
3. Assume that an over-dispersed model applies with variance component $\sigma_\lambda^2 = 0.06$ within each of the two groups. How many patient years are required to provide 90% power to detect the same difference in the mean rates?
4. If we recruit 200 patients per group, how many years must each patient be followed?

8.7.3. Now consider a Poisson regression model with grouped or discrete covariates analogous to the logistic regression model with discrete covariates. In this case, the developments in Section 7.3.6 can be used to derive the sample size for, or power of, a Wald test of coefficient effects. Denoting the expected fraction of total exposure time in the ith cell as $\zeta_i = E(T_i/T)$, where $T = \sum_i T_i$, show that $\Omega = diag\{[\zeta_i \lambda(\boldsymbol{x}_i)]^{-1}\}$ where the expected number of events in the ith cell with covariate vector $\boldsymbol{x}_i$ is $E(D_i) = \zeta_i T \lambda(\boldsymbol{x}_i)$. The remainder of the developments in Section 7.3.6 then apply.

8.7.4. Consider a model to assess the effect of a treatment or exposure group within three strata with a design matrix as presented in Example 7.9. Assume that the fractions of patient years of exposure within each of the six cells are $\{\zeta_i\} = (0.08, 0.12, 0.30, 0.20, 0.125, 0.175)$. The assumed intensities per year in each cell are $\{\lambda_i\} = (0.15, 0.28, 0.12, 0.27, 0.20, 0.42)$. Generate a large data set with 10,000 patient years and numbers of events within each cell proportional to the assumed rates to determine the corresponding Poisson model parameter vector $\boldsymbol{\theta}$.

8.7.5. Use the values of $\{\lambda_i\}$ and $\{\zeta_i\}$ to compute Ω and substitute this into (7.73) to obtain the covariance matrix of the estimates. For the test of H_0: $\beta_3 = 0$, with vector $\boldsymbol{C}' = (0\ 0\ 0\ 1)$, determine the value of the non-centrality factor K^2 for the 1 df test of the group effect. Then, determine the total patient years T required to provide 90% power for this test.

8.7.6. Conversely, for the same model, what level of power is provided by a total exposure of only $T = 200$ years?

8.8 Consider a Poisson Regression Model with a single binary covariate for treatment group ($x = 1 = E$, $0 = \overline{E}$).

8.8.1. Assuming $\beta = 0$, show that $U(\alpha) = 0$ implies that

$$e^{\alpha} = \frac{\sum_i d_i}{\sum_i t_i} = \frac{D}{T}. \tag{8.72}$$

8.8.2. Derive the elements of $I(\widehat{\theta}_0)$.

8.8.3. Show that the score test for β equals the Z-test presented in (8.26).

8.9 Derive the expressions for the proportion of explained variation in the Poisson regression model.

8.9.1. $\rho^2_{\varepsilon^2}$ assuming squared error loss;

8.9.2. R^2_{resid} and show that this equals $\widehat{\rho}^2_{\varepsilon^2}$ in (8.52).

8.10 Fleming and Harrington (1991) present the results of a randomized clinical trial of the effects of gamma interferon versus placebo on the incidence of serious infections among children with chronic granulotomous disease (CGD). For each subject the number of infections experienced and the total duration of follow-up are presented. In Chapter 9 an analysis is presented using the actual event times. Here we consider only the number of events experienced by each subject. The count data set is available from the www (see Preface). The data set includes the patient *id*, number of severe infections experienced (*nevents*), the number of days of follow-up (*futime*) and the following covariates: Z_1: treatment group: interferon (1) versus placebo (0); Z_2: Inheritance pattern: X-linked (1) versus autosomal recessive (2); Z_3: Age (years); Z_4: Height (cm); Z_5: Weight (kg); Z_6: Corticosteroid use on entry: yes (1) versus no (2); Z_7: Antibiotic use on entry: yes (1) versus no (2); Z_8: Gender: male (1) versus female (2); and Z_9: Type of hospital: NIH (1), other US (2), Amsterdam (3), other European (4). Use these data to conduct the following analyses.

8.10.1. Compute the crude rates of infection per day in each treatment group and the asymmetric confidence limits on each using the log(rate) under the homogeneous Poisson assumptions.

8.10.2. Also compute the asymmetric confidence limits for the relative risk and the Z-test for the difference between groups under the homogeneous Poisson assumptions.

8.10.3. Within each group compute Cochran's variance test of the hypothesis of no over-dispersion.

8.10.4. Within each group, adopt an over-dispersed Poisson model and use the moment equation to compute an estimate of the mixing distribution variance. Use this to revise the computations for the variance and confidence limits on the rates within each group and the relative risk.

8.10.5. Compute the estimate of the over-dispersion variance under the null hypothesis of equal intensities in each group and use these estimates to compute a Z-test to test the difference between groups allowing for over-dispersion.

8.10.6. Now assume that the over-dispersion can be accounted for by inclusion of covariate effects in a Poisson regression model. Assess the relative risk for interferon versus placebo in a Poisson regression model and also describe the model and the covariate effects:

1. unadjusted;
2. adjusted for the other covariates. Note that Z_9 should be used as a class effect to compare the four hospital types.

8.10.7. Compare the model-based results to those obtained above.

8.10.8. If the relative risks for treatment group in Problem 8.10.6.1 versus 2 differ, perform additional analyses as needed to help explain why.

8.10.9. Based on the analysis in Problem 8.10.3, and based on the value of $Pearson\ X^2/df$ in Problem 8.10.6, is there evidence that an over-dispersed model is appropriate or not?

8.10.10. Refit the model using an over-dispersed quasi-likelihood model using the PSCALE option.

8.10.11. Refit the model using the information sandwich empirical estimates of the covariances using PROC GENMOD.

8.10.12. Write a program to compute the score vector and the information sandwich estimate of the observed information under the model null hypothesis. Compute the robust model score test.

8.11 Starting from the full unconditional likelihood for Poisson count data from matched pairs in (8.57)

8.11.1. Derive the score equation for the intercept α_i for the ith matched set and show that the MLE is a function of the sufficient statistic $D_i = \sum_{j=1}^{n_i} d_{ij}$.

8.11.2. Then show that the conditional likelihood from (8.58) for the ith matched set is

$$L_{i|D_i,n_i} = \frac{\prod_{j=1}^{n_i} \frac{e^{-\lambda(\boldsymbol{x}_{ij})t_{ij}}\left[\lambda(\boldsymbol{x}_{ij})\,t_{ij}\right]^{d_{ij}}}{d_{ij}!}}{\sum_{\ell=1}^{n_i!}\prod_{j(\ell)=1}^{n_i} \frac{e^{-\lambda\left(\boldsymbol{x}_{ij(\ell)}\right)t_{ij(\ell)}}\left[\lambda\left(\boldsymbol{x}_{ij(\ell)}\right)t_{ij(\ell)}\right]^{d_{ij}}}{d_{ij(\ell)}!}}, \tag{8.73}$$

which yields (8.59).

8.11.3. For the case of matched pairs, show that (8.60) and (8.61) result.

8.11.4. For the model in (8.58), derive the elements of the score vector for $\boldsymbol{\beta}$ and the elements of the expected information matrix.

9

Analysis of Event-Time Data

Up to this point we have considered the description of the risk of an event without consideration of the *time* at which events occur in individuals in the population. One of the major advances in biostatistics has been the development of statistical methods for the analysis of event-time data, commonly known as *survival analysis*, a topic that has dominated statistical methodological research and biostatistical practice for over three decades. There are many excellent texts on survival analysis that cover this vast field in detail. Although originally motivated by the analysis of survival time or mortality data, these methods may be used to describe the distribution of time to any single event of interest. These methods have also been generalized to the analysis of event-times of possibly recurrent events in a subject. Many of the methods of survival or event-time analysis are generalizations of the methods developed in previous chapters.

There are many excellent general references on survival analysis. Kalbfleisch and Prentice (1980), Elandt-Johnson and Johnson (1980), Lawless (1982), and Cox and Oakes (1984), among others, give a precise review of the early methods from a classical perspective. Lee (1992), Collett (1994) and Marubini and Valsecchi (1995), among others, present thorough descriptions of the application of these methods. Fleming and Harrington (1991) and Andersen, Borgan, Gill and Keiding (1993) present a rigorous description of the theory of martingales and counting processes and the derivation of more general methods based on these theories. Here I present a description of the principal methods for the analysis of event-time data and their application.

9.1 INTRODUCTION TO SURVIVAL ANALYSIS

9.1.1 Hazard and Survival Function

In survival or time-to-event analysis, each individual subject is followed up to some time t_i at which time either an event is observed to occur or follow-up is curtailed without observation of an event. Thus we also observe an indicator variable δ_i that designates whether an event was observed to occur ($\delta_i = 1$) or not ($\delta_i = 0$). In the latter case, the event time is *right censored* because the actual event time is greater than the observed time of exposure. Thus for each subject two values (t_i, δ_i) are observed that are realizations of two separate processes, the event process T and the censoring process C. Under the assumption of *censoring at random*, these two processes are assumed to be statistically independent. In this case, censored event times are considered unbiased, meaning that the same stochastic event process is assumed to apply to all observations, those censored and those not.

When the time of an event is observed to an instant of time, then event times have a right continuous, monotonically increasing distribution function $F(t) = P(T \leq t), t > 0$, with a corresponding event probability density function $f(t) = dF(t)/dt$. The complement of the *cdf* is the right continuous, monotonically decreasing *survival distribution* $S(t) = 1 - F(t) = P(T > t)$. Some authors define the survival function as the left continuous $P(T \geq t)$. The distinction is irrelevant for a continuous distribution, but non-trivial when the distribution is discrete. Herein I define $S(t)$ throughout as the probability of surviving *beyond* time t so that the survival function can be the defined as the complement of the cumulative distribution function for both continuous and discrete distributions.

A pivotal and informative quantity is the *hazard function* that describes the instantaneous probability of the event among those still at risk, or those still free of the event. For a continuous distribution the hazard function at time t is defined as

$$\lambda(t) = \lim_{\Delta t \downarrow 0} \frac{P(t < T \leq t + \Delta t)}{P(T > t)} = \frac{f(t)}{S(t)} = \frac{-d \log[S(t)]}{dt} . \tag{9.1}$$

Thus the density can also be obtained as $f(t) = \lambda(t)S(t)$. The *cumulative hazard function* at time t then is

$$\Lambda(t) = \int_0^t \lambda(u)du = \int_0^t -d \log S(u) = -\log S(t), \tag{9.2}$$

from which it follows that

$$S(t) = \exp[-\Lambda(t)]. \tag{9.3}$$

Numerous parametric survival distributions and corresponding hazard functions and event probability distributions can be used to describe the experience in a population. The simplest instance is the *exponential* survival distribution, in which it is assumed that the hazard function is constant over time, or $\lambda(t) = \lambda \ \forall t$, in which case $\Lambda(t) = \lambda t$, $S(t) = e^{-\lambda t}$, $F(t) = 1 - e^{-\lambda t}$, and $f(t) = \lambda e^{-\lambda t}$. Other parametric functions are explored as Problems.

In general, no one parametric distribution will apply to every population. Thus common methods of survival analysis are non-parametric or distribution free. These include the Kaplan-Meier estimate of the underlying survival distribution, the family of generalized Mantel-Haenszel tests for the comparison of the survival distributions of two or more groups, and the proportional hazards model for the assessment of the effects of covariates on the risk or hazard function of events over time.

9.1.2 Censoring at Random

If all subjects were followed until the time of the event such as mortality or failure, then it would be a simple matter to estimate the parameters of a parametric model from the corresponding likelihood. With right censored observations, however, the full likelihood is a function of both the failure time distribution and the distribution of the right censored times. Assume that the event time distribution has parameter θ, where the probability density function, hazard function and survival distribution are designated as $f(t;\theta)$, $\lambda(t;\theta)$ and $S(t;\theta)$, respectively. Also, assume that the censored observations have a cumulative distribution function $G(t;\phi)$ and probability density function $g(t;\phi)$. The assumption of random censoring is equivalent to the assumption that the two sets of parameters θ and ϕ are distinct.

For any event-time and censoring distributions, the likelihood of a sample of N observations $\{t_i, \delta_i\}$ under random censoring is

$$L(\theta,\phi) = \prod_{i=1}^{N} \{f(t_i;\theta)\,[1 - G(t_i;\phi]\}^{\delta_i} \{g(t_i;\phi)S(t_i;\theta)\}^{1-\delta_i} \,, \tag{9.4}$$

where an individual with the event at time t_i is also not censored by that time, and an individual censored at time t_i has also survived to that time. Thus the likelihood can be factored as

$$\begin{aligned} L(\theta,\phi) &= \prod_{i=1}^{N} f(t_i;\theta)^{\delta_i} S(t_i;\theta)^{1-\delta_i} \prod_{i=1}^{N} [1 - G(t_i;\phi]^{\delta_i} g(t_i;\phi)^{1-\delta_i} \\ &= L(\theta)L(\phi) \propto L(\theta). \end{aligned} \tag{9.5}$$

Thus the full likelihood is proportional to the likelihood associated with the event time distribution $L(\theta)$, which implies that inferences regarding the event-time distribution may be obtained without the need to simultaneously specify or consider the censoring distribution.

When the data consist of observations from two or more groups or strata of subjects, the assumption of censoring at random then specifies that the observations within each group or strata are censored at random, but the random mechanisms and the extent of censoring may differ between groups. The resulting likelihood is then the product of the group or stratum specific event-time likelihoods.

9.1.3 Kaplan-Meier Estimator

Kaplan and Meier (1958) describe a *product limit estimator* of the underlying survival distribution without assuming any particular parametric form. Under random censoring, the likelihood of the sample is

$$L \propto \prod_{i=1}^{N} f(t_i)^{\delta_i} S(t_i)^{1-\delta_i} = \prod_{i=1}^{N} \lambda(t_i)^{\delta_i} S(t_i) \,. \tag{9.6}$$

Thus individuals who experience the event contribute the term $f(t)$ to the likelihood at the event time, whereas censored observations contribute the term $S(t)$ at the time of censoring.

Consider a sample of N observations (t_i, δ_i), $i = 1, \ldots, N$, in which events are observed at J distinct event times $t_{(1)} < t_{(2)} < \ldots < t_{(J)}$ and that the underlying survival distribution is a step function with points of discontinuity at these event times. The left-hand limit $F(t^-) = \lim_{\epsilon \downarrow 0} F(t - \epsilon) = \lim_{u \uparrow t} F(u)$ designates the value of the *cdf* immediately prior to a "jump" or discontinuity at time t. Then the probability of an event at time $t_{(j)}$ is $f(t_{(j)}) = F(t_{(j)}) - F(t_{(j)}^-)$. Thus $f(t_{(j)})$ also equals the size of the drop in the survival function $f(t_{(j)}) = S(t_{(j)}^-) - S(t_{(j)})$, where $S(t_{(j)}^-) = P(T \geq t_{(j)})$ is the probability of being at risk at $t_{(j)}$ or survival up to $t_{(j)}$ and $S(t_{(j)}) = P(T > t_{(j)})$ is the probability of surviving beyond t_j. Since $f(t_{(j)})$ is the probability of survival to t_j and an event at t_j, then

$$f(t_{(j)}) = P\left[event\ at\ t_{(j)} \mid T \geq t_{(j)}\right] P[T \geq t_{(j)}], \tag{9.7}$$

which can be expressed as $f(t_{(j)}) = \pi_j S(t_{(j)}^-)$, where the conditional probability of an event at $t_{(j)}$ is

$$\pi_j = \lim_{\Delta t \downarrow 0} P[t_{(j)}^- < T \leq t_{(j)}^- + \Delta t \mid T > t_{(j)}^-] \tag{9.8}$$

and its complement is the continuation probability

$$1 - \pi_j = P[T > t_{(j)} \mid T > t_{(j)}^-] = P[T > t_{(j)} \mid T \geq t_{(j)}] \,. \tag{9.9}$$

Since $S(t)$ is a decreasing step function, then $f(t_{(j)}) = \pi_j S(t_{(j)}^-) = \pi_j S(t_{(j-1)})$.

By construction the observed event times $\{t_i, \delta_i = 1\} \in \{t_{(1)}, \ldots, t_{(J)}\}$. Let $n_j = \#\{t_i \geq t_{(j)}\}$ denote the number of subjects at risk, or still under observation, at time $t_{(j)}$. Then let $d_j = \#\{t_i = t_{(j)}, \delta_i = 1\}$ denote the number of events observed at time $t_{(j)}$ among the n_j subjects at risk at time $t_{(j)}$. Also, let w_j denote the number of observations that are right censored at times after the jth event time but prior to the $(j+1)th$ time, or $w_j = \#\{t_i, \delta_i = 0\} \in [t_{(j)}, t_{(j+1)})$, so that $n_{j+1} = n_j - d_j - w_j$. Thus observations that are right censored at event time $t_{(j)}$ are considered to be at risk at that time and are removed from the risk set immediately thereafter. Then the likelihood can be expressed as

$$L(\pi_1, \ldots, \pi_N) \propto \prod_{j=1}^{J} \pi_j^{d_j} S(t_{(j-1)})^{d_j} S(t_{(j)})^{w_j} \,, \tag{9.10}$$

where $t_{(0)} = 0$ and $S(t_{(0)}) = 1$.

An essential property of an event time process is that in order to survive event-free beyond time $t_{(j)}$ requires that the subject survive event-free beyond $t_{(j-1)}$, and $t_{(j-2)}$, etc. Thus

$$\begin{aligned} S(t_{(j)}) &= P(T > t_{(j)}) = P(T > t_{(j)} \mid T > t_{(j-1)})P(T > t_{(j-1)}) \\ &= P(T > t_{(j)} \mid T > t_{(j-1)})P(T > t_{(j-1)} \mid T > t_{(j-2)})P(T > t_{(j-2)}), \end{aligned} \tag{9.11}$$

and so on. Note, however, that

$$P(T > t_{(j)} \mid T > t_{(j-1)}) = P(T > t_{(j)} \mid T \geq t_{(j)}) = (1 - \pi_j) \tag{9.12}$$

so that

$$\begin{aligned} S(t_{(j)}) &= (1 - \pi_j)\, S(t_{(j-1)}) \\ &= (1 - \pi_j)\,(1 - \pi_{j-1}) \cdots (1 - \pi_2)\,(1 - \pi_1)\,. \end{aligned} \tag{9.13}$$

After expanding the likelihood in this manner and simplifying (see Problem 9.8), it follows that

$$L(\pi_1, \ldots, \pi_J) \propto \prod_{j=1}^{J} \pi_j^{d_j} (1 - \pi_j)^{n_j - d_j} \tag{9.14}$$

with log likelihood

$$\ell(\pi_1, \ldots, \pi_J) = \sum\nolimits_{j=1}^{J} \{d_j \log(\pi_j) + (n_j - d_j) \log(1 - \pi_j)\}\,. \tag{9.15}$$

It then follows that the maximum likelihood estimates of the event probabilities are provided by the solution to the estimating equations

$$\frac{\partial \ell(\pi_1, \ldots, \pi_J)}{\partial \pi_j} = \frac{d_j}{\pi_j} - \frac{n_j - d_j}{1 - \pi_j} = 0; \qquad 1 \leq j \leq J\,, \tag{9.16}$$

from which the MLE of the jth conditional probability is

$$\begin{aligned} \widehat{\pi}_j &= \frac{d_j}{n_j} = p_j \\ 1 - \widehat{\pi}_j &= \frac{n_j - d_j}{n_j} = q_j\,. \end{aligned} \tag{9.17}$$

Thus Kaplan and Meier (1958) showed that the generalized maximum likelihood estimate of the underlying survival function is provided by

$$\widehat{S}(t) = \prod_{j=1}^{J} \left(\frac{n_j - d_j}{n_j} \right)^{I[t_{(j)} \leq t]} = \prod_{j=1}^{J} q_j^{I[t_{(j)} \leq t]} = \prod_{j: t_{(j)} \leq t} q_j \tag{9.18}$$

for values $0 \leq t \leq t_{(J)}$ and is undefined for values $t > t_{(J)}$.

Under random censoring, it can then be shown using classical methods (Peterson, 1977), or using counting processes (Aalen, 1978, Gill, 1980), that $\lim_{n\to\infty} \widehat{S}(t) \xrightarrow{p} S(t)$ whatever the form of the underlying distribution. Clearly $\widehat{S}(t)$ is a step function that describes the probability of surviving beyond time t. Its complement is the *cumulative incidence function*, which describes the probability of the event occurring up to time t. As the number of observations increases, and thus the number of events also increases, the number of steps increases and the intervals between steps become infinitesimally small, thus converging to the true survival distribution, whatever its form.

The estimator is obtained as a product of successive survival proportions, each being the conditional probability of surviving beyond an instant of time given survival up to that point in time. Thus the assumption of censoring at random implies that the survival experience of those at risk at any event time $t_{(j)}$ applies to all observations in the population, including those censored prior to $t_{(j)}$; or that the subset of n_j observations at risk at $t_{(j)}$ is a random subset of the initial cohort who actually survive to $t_{(j)}$. Alternately, random censoring implies that the estimate of the survival experience beyond any point in time is not biased by prior censored observations, or that the censored observations are subject to the same hazard function as those who continue follow-up.

These successive conditional event proportions $\{p_j\}$ clearly are not statistically independent. However, it is readily shown that these successive proportions are indeed uncorrelated; see Problem 9.8.4. Because the vector of proportions is asymptotically distributed as multivariate normal, this implies that these proportions are asymptotically conditionally independent given the past sequence of events and censored observations, that is, conditionally on the numbers at risk at the times of the preceding events.

Because the survival function estimate is a product of probabilities, it is more convenient to use the log survival function to obtain expressions for the variance of the estimate, where

$$\log[\widehat{S}(t)] = \sum_{j=1}^{J} I(t_{(j)} \leq t) \log[q_j] = \sum_{j:t_{(j)} \leq t} \log[q_j]. \tag{9.19}$$

Using the δ-method, the variance is

$$V\left(\log[\widehat{S}(t)]\right) = \sum_{j:t_{(j)} \leq t} \frac{\pi_j}{n_j(1-\pi_j)} \tag{9.20}$$

as is readily demonstrated in a Problem. Substituting $\widehat{\pi}_j$ from (9.17) yields the estimated variance

$$\widehat{V}\left(\log[\widehat{S}(t)]\right) = \sum_{j:t_{(j)} \leq t} \frac{d_j}{n_j(n_j - d_j)} . \tag{9.21}$$

Again, applying the δ-method yields Greenwood's (1926) variance estimate of the survival probability at any time $t \leq t_{(J)}$, which is expressed as

$$\widehat{V}[\widehat{S}(t)] = \widehat{S}(t)^2 \left[\sum\nolimits_{j:t_{(j)} \leq t} \frac{d_j}{n_j(n_j - d_j)}\right]. \tag{9.22}$$

Greenwood's estimator can be used to provide symmetric confidence bands for the survival function. However, these confidence limits are not bounded by (0,1). Asymmetric confidence limits so bounded are obtained using the complimentary log-log transformation, with corresponding variance again obtained by the δ-method as

$$\widehat{V}\left[\log\left(-\log[\widehat{S}(t)]\right)\right] = \frac{1}{\left(\log[\widehat{S}(t)]\right)^2} \sum\nolimits_{j:t_{(j)} \leq t} \frac{d_j}{n_j(n_j - d_j)}. \tag{9.23}$$

This yields asymmetric $1 - \alpha$ level confidence limits for $S(t)$ of the form

$$\widehat{S}(t)^{\exp\left[\pm Z_{1-\alpha/2}\sqrt{\widehat{V}\left[\log\left(-\log[\widehat{S}(t)]\right)\right]}\right]} \tag{9.24}$$

(*cf.* Collett, 1994). Alternately, asymmetric confidence limits could be obtained using the logit of the survival probability; see Problem 9.8.9.

Finally, note that since $S(t) = \exp[-\Lambda(t)]$, then the estimate of the survival function at an event time also provides an estimate of the cumulative hazard as simply $\widehat{\Lambda}(t_{(j)}) = -\log[\widehat{S}(t_{(j)})]$. From (9.21) the estimated variance is simply

$$\widehat{V}\left[\widehat{\Lambda}(t_{(j)})\right] = \sum\nolimits_{\ell=1}^{j} \frac{d_\ell}{n_\ell(n_\ell - d_\ell)}. \tag{9.25}$$

In what follows we designate the Kaplan-Meier estimate of the survival and cumulative hazard function as $\widehat{S}_{KM}(t)$ and $\widehat{\Lambda}_{KM}(t)$, to distinguish them from the Nelson-Aalen estimates, which are based on estimation of the cumulative hazard function directly.

9.1.4 Estimation of the Hazard Function

When event times are observed continuously over time, one approach to estimation of the hazard function is to adopt a piecewise exponential model where we assume that the hazard function is piecewise constant between the successive event times, or $\lambda(t) = \lambda_{(j)}$ for $t \in (t_{(j-1)}, t_j]$ for the jth interval. The probability of the event over a small interval of time is $1 - \exp\left[-\lambda_{(j)}(t_j - t_{(j-1)})\right] \doteq \lambda_{(j)}(t_j - t_{(j-1)})$ since $1 - e^{-\varepsilon} \doteq \varepsilon$ for small $\varepsilon \downarrow 0$. This probability is intuitively estimated by p_j in (9.17), which is simply the number of events divided by the number at risk at time $t_{(j)}$. Thus the hazard function over the interval prior to the jth event is estimated as

$$\widehat{\lambda}_{(j)} = \frac{p_j}{(t_{(j)} - t_{(j-1)})}. \tag{9.26}$$

Then the cumulative hazard up to that time $t_{(j)}$ is estimated as

$$\widehat{\Lambda}_{NA}(t_{(j)}) = \sum\nolimits_{\ell=1}^{j} \widehat{\lambda}_{(\ell)}(t_{(\ell)} - t_{(\ell-1)}) = \sum\nolimits_{\ell=1}^{j} p_\ell. \tag{9.27}$$

As described in Section 9.6, this latter expression is the *Nelson-Aalen estimator* $\widehat{\Lambda}_{NA}(t_{(j)})$ after Nelson (1969, 1972) and Aalen (1978), who provided estimates of the intensity of a counting process. The Nelson-Aalen estimate of the cumulative hazard function then yields the Nelson-Aalen estimate of the survival function at time $t_{(j)}$ as

$$\widehat{S}_{NA}(t_{(j)}) = \exp\left[-\widehat{\Lambda}_{NA}(t_{(j)})\right], \tag{9.28}$$

which was also proposed by Altschuler (1970).

Different expressions apply to discrete time observations, such as for the actuarial lifetable described in Section 9.2.1. These are presented in many standard texts.

Since the successive proportions are uncorrelated, then it readily follows that the estimated variances are

$$\widehat{V}(\widehat{\lambda}_{(j)}) = \frac{p_j(1-p_j)}{n_j(t_{(j)} - t_{(j-1)})^2}, \tag{9.29}$$

$$\widehat{V}\left[\widehat{\Lambda}_{NA}(t_{(j)})\right] = \sum\nolimits_{\ell=1}^{j} \frac{p_\ell(1-p_\ell)}{n_\ell}, \tag{9.30}$$

and

$$\widehat{V}\left[\widehat{S}_{NA}(t_{(j)})\right] = \widehat{S}_{NA}(t_{(j)})^2 \sum\nolimits_{\ell=1}^{j} \frac{p_\ell(1-p_\ell)}{n_\ell}. \tag{9.31}$$

From 9.28, asymmetric confidence limits on the survival probabilities can be obtained from those on the estimated cumulative hazard $\widehat{\Lambda}_t$. Also, since

$$\widehat{V}\left[\log\left(-\log[\widehat{S}_{NA}(t)]\right)\right] = \widehat{V}\left[\widehat{\Lambda}_{NA}(t)\right]\left(\log[\widehat{S}_{NA}(t)]\right)^{-2}, \tag{9.32}$$

then asymmetric confidence limits for $S(t)$ may be obtained from (9.24) upon substituting the corresponding Nelson-Aalen estimates.

In Problem 9.8.11 it is shown that as $N \to \infty$, the Nelson-Aalen and Kaplan-Meier estimators of the survival function are asymptotically equivalent, and that the Nelson-Aalen and Kaplan-Meier estimators of the variance of the cumulative hazard are also asymptotically equivalent. Later, in Section 9.6.2 we show that the Nelson-Aalen estimators can be applied to the intensity of a counting process for recurrent events. Smoothed estimates of the intensity described therein may also be applied to the Nelson-Aalen estimate of the hazard function for survival data.

9.1.5 Comparison of Survival Probabilities for Two Groups

In many instances we wish to compare the event-time distributions between two independent groups of subjects. We now consider the comparison of two groups at a specific point in time using the Kaplan-Meier estimate. In section 9.3 we describe tests for the difference between hazard functions and survival curves over time.

The Greenwood variance estimator provides a large sample confidence interval for the difference between the survival probabilities of the two groups at any point in time ($t > 0$), say $\widehat{S}_1(t) - \widehat{S}_2(t)$, with $\widehat{V}\left[\widehat{S}_1(t) - \widehat{S}_2(t)\right] = \widehat{V}\left[\widehat{S}_1(t)\right] + \widehat{V}\left[\widehat{S}_2(t)\right]$, where $\widehat{S}_i(t)$ and $\widehat{V}\left[\widehat{S}_i(t)\right]$ are obtained from the observations in each group separately ($i = 1, 2$). Alternately, asymmetric confidence limits on the ratio of the survival probabilities at time t can be obtained using the log transformation since $\log \widehat{S}_1(t) - \log \widehat{S}_2(t) = \log[\widehat{S}_1(t)/\widehat{S}_2(t)]$ with estimated variance $\widehat{V}\left[\log \widehat{S}_1(t)\right] + \widehat{V}\left[\log \widehat{S}_2(t)\right]$ from (9.21).

Likewise, asymmetric confidence limits on the ratio of the cumulative hazards can be obtained using the complimentary log-log transformation since

$$\log\{-\log \widehat{S}_1(t)\} - \log\{-\log \widehat{S}_2(t)\} = \log[\widehat{\Lambda}_1(t)/\widehat{\Lambda}_2(t)] \tag{9.33}$$

with estimated variance $\widehat{V}\left[\log\{-\log \widehat{S}_1(t)\}\right] + \widehat{V}\left[\log\{-\log \widehat{S}_2(t)\}\right]$ from (9.23). Asymmetric confidence limits may also be obtained for the survival odds ratio at time t using the difference between the logits of the survival probabilities.

These developments also provide a large sample test of the difference between the survival probabilities of two groups at a specific point in time, say t^*. Under the null hypothesis H_0: $S_1(t) = S_2(t)$, the estimated survival function from the combined groups, say $\widehat{S}_\bullet(t)$, provides a consistent estimator of the common survival function. This pooled estimate is computed from (9.18) with estimated conditional event probabilities $\widehat{\pi}_{\bullet j} = (d_{1j} + d_{2j})/(n_{1j} + n_{2j})$ for $1 \le j \le J$. Then for the ith group ($i = 1, 2$), from (9.22) the estimated variance of $S_i(t^*)$ at the specified time t^* under the null hypothesis is expressed as

$$\widehat{V}_0\left[\widehat{S}_i(t^*)\right] = \widehat{S}_\bullet(t^*)^2 \left[\sum\nolimits_{j:t_{(j)} \le t^*} \frac{\widehat{\pi}_{\bullet j}}{n_{ij}(1 - \widehat{\pi}_{\bullet j})}\right], \tag{9.34}$$

where n_{ij} is the number at risk in the ith group at the jth event time. Thus a large sample test of equality of the survival probabilities at time t^* is provided by

$$X^2 = \frac{\left[\widehat{S}_1(t^*) - \widehat{S}_2(t^*)\right]^2}{\widehat{V}_0\left[\widehat{S}_1(t^*)\right] + \widehat{V}_0\left[\widehat{S}_2(t^*)\right]}, \tag{9.35}$$

where asymptotically X^2 is distributed as chi-square on 1 *df*.

An alternate approach is to simply compute the test for two proportions ignoring the presence of censored observations and variable durations of follow-up. We then wish to test H_0: $\pi_1 = \pi_2 = \overline{\pi}$, where π_i denotes the aggregate probability of an event among the n_i observations in the ith group with the set of exposure times $\{t_{i1}, \dots, t_{in_i}\}$. This assumes that $E(\delta_{ij}) = \overline{\pi}$ for all subjects ($j = 1, \dots, n_i$; $i = 1, 2$), where δ_{ij} is the Bernoulli variable for the occurrence of the event versus being censored. However, in a staggered-entry, prolonged follow-up trial of total duration T_S, each patient has a potential exposure time defined as $E_{ij} = T_S - r_{ij}$; where r_{ij} is the time of entry into the trial, $0 < r_{ij} \leq T_R$, where all subjects are assumed to be followed until the common termination time T_S. Thus the planned exposure time for each subject is such that $T_S - T_R \leq E_{ij} \leq T_S$. However, some subjects may be randomly censored at time c_{ij} for other than administrative reasons. Thus the actual exposure time for the $(ij)th$ subject is $t_{ij} = \min[c_{ij}, E_{ij}]$, where $0 \leq t_{ij} \leq T_S$. Thus each subject has a different implied probability of the event $E(\delta_{ij}|t_{ij}) = \pi_{ij}$ given the exposure time t_{ij}. Clearly, if the survival function $S_i(t)$ is declining sharply over the range of the t_{ij}, then the variance of the simple proportions test will be affected.

Since the sample proportion in the ith group is $p_i = \sum_{j=1}^{n_i} \delta_{ij}/n_i$, then,

$$E(p_i) = \frac{\sum_j E(\delta_{ij}|t_{ij})}{n_i}, \tag{9.36}$$

where $E(\delta_{ij}|t_{ij}) = F_i(t_{ij})$. Thus

$$V_c(p_i) = \frac{\sum_j V(\delta_{ij}|t_{ij})}{n_i^2} = \frac{\sum_{ij} F_i(t_{ij}) S_i(t_{ij})}{n_i^2}. \tag{9.37}$$

In Problem 9.9 it is then shown that

$$V_c(p_i) \leq V(p_i) = \pi_i(1 - \pi_i)/n_i \tag{9.38}$$

with equality when $F(t_{ij})$ is nearly constant for all t_{ij} (Lachin, Lan, et al., 1992). Therefore, the ordinary proportions test in (2.80) is flawed in an extended follow-up study with differential exposure among the subjects because the ordinary binomial variance is incorrect. This test, therefore, should not be used.

Example 9.1 *Squamous Cell Carcinoma*
Lagakos (1978) presents data on the time to spread of disease among patients with squamous cell carcinoma who received either of two treatments – A (Experimental, group 1) versus B (Control, group 2). Here we consider only the subset of patients who were non-ambulatory on entry into the study. The complete data set is described later in Example 9.6 and can be obtained from the www (see Preface, reproduced with permission). Patients were assigned treatment B versus A in a 3:1 allocation. In treatment group A 12 of 23 patients (52%) failed versus 40 of the 62 patients (65%) in treatment group B.

Table 9.1 presents the times at which spread of the disease occurred along with the numbers still at risk at each event time and the numbers of events at that time. From this, the conditional probabilities of survival beyond each point in time are computed (q_j) as is the estimate of the survival probabilities.

Table 9.1 Numbers at Risk (n_j) in Both Groups Individually and Combined at Each Distinct Event Time $t_{(j)}$, Numbers of Events (d_j) at That Time, and Number Lost-to-Follow-up or Right Censored (w_j) During the Interval $[t_{(j)}, t_{(j+1)})$

$t_{(j)}$	n_j	d_j	w_j	n_{1j}	d_{1j}	w_{1j}	n_{2j}	d_{2j}	w_{2j}
1	85	1	5	23	0	2	62	1	3
2	79	2	3	21	0	0	58	2	3
3	74	2	1	21	0	0	53	2	1
4	71	1	5	21	0	4	50	1	1
5	65	5	1	17	1	1	48	4	0
6	59	1	1	15	0	0	44	1	1
7	57	1	1	15	0	0	42	1	1
8	55	6	0	15	1	0	40	5	0
9	49	2	0	14	0	0	35	2	0
10	47	1	1	14	0	0	33	1	1
11	45	7	1	14	2	0	31	5	1
12	37	3	2	12	2	0	25	1	2
13	32	2	2	10	1	1	22	1	1
14	28	0	2	8	0	2	20	0	0
15	26	0	1	6	0	0	20	0	1
16	25	1	0	6	0	0	19	1	0
17	24	1	1	6	0	0	18	1	1
18	22	0	2	6	0	0	16	0	2
19	20	2	0	6	1	0	14	1	0
20	18	1	0	5	0	0	13	1	0
21	17	4	1	5	1	0	12	3	1
22	12	2	1	4	0	0	8	2	1
26	9	1	0	4	0	0	5	1	0
28	8	0	1	4	0	0	4	0	1
29	7	1	0	4	0	0	3	1	0
30	6	1	1	4	0	1	2	1	0
34	4	1	0	3	1	0	1	0	0
55	3	1	0	2	0	0	1	1	0
84	2	1	0	2	1	0	0	0	0
88	1	1	0	1	1	0	0	0	0

Table 9.2 Within the Combined Sample, Time of Each Event $t_{(j)}$, Number at Risk (n_j), and Number of Events at That Time (d_j), Estimated Continuation Probability of Survival Beyond That Time (q_j), Kaplan-Meier Estimated Probability $\widehat{S}(t_{(j)})$, and Greenwood Estimate of the $S.E.$, Nelson-Aalen Estimate $\widehat{\lambda}_j$ of the Hazard over the Interval $(t_{(j)}, t_{(j-1)}]$ and Its $S.E.$

$t_{(j)}$	$n_{(j)}$	d_j	q_j	$\widehat{S}(t_{(j)})$	$S.E.[\widehat{S}(t_{(j)})]$	$\widehat{\lambda}_j$	$S.E.[\widehat{\lambda}_j]$
1	85	1	0.988	0.988	0.012	0.012	0.012
2	79	2	0.975	0.963	0.021	0.025	0.018
3	74	2	0.973	0.937	0.027	0.027	0.019
4	71	1	0.986	0.924	0.030	0.014	0.014
5	65	5	0.923	0.852	0.041	0.077	0.033
6	59	1	0.983	0.838	0.043	0.017	0.017
7	57	1	0.982	0.824	0.045	0.018	0.017
8	55	6	0.891	0.734	0.053	0.109	0.042
9	49	2	0.959	0.704	0.055	0.041	0.028
10	47	1	0.979	0.689	0.056	0.021	0.021
11	45	7	0.844	0.582	0.060	0.156	0.054
12	37	3	0.919	0.535	0.061	0.081	0.045
13	32	2	0.937	0.501	0.061	0.062	0.043
16	25	1	0.960	0.481	0.062	0.013	0.013
17	24	1	0.958	0.461	0.063	0.042	0.041
19	20	2	0.900	0.415	0.064	0.050	0.034
20	18	1	0.944	0.392	0.065	0.056	0.054
21	17	4	0.765	0.300	0.064	0.235	0.103
22	12	2	0.833	0.250	0.062	0.167	0.108
26	9	1	0.889	0.222	0.061	0.028	0.026
29	7	1	0.857	0.190	0.060	0.048	0.044
30	6	1	0.833	0.159	0.058	0.167	0.152
34	4	1	0.750	0.119	0.055	0.062	0.054
55	3	1	0.667	0.079	0.049	0.016	0.013
84	2	1	0.500	0.040	0.037	0.017	0.012
88	1	1	0.000	0.000	.	0.250	0.000

From these entries the estimated conditional event probabilities (p_j) and continuation probabilities (p_j) are obtained, from which the estimated survival curves are obtained for both groups combined along with the Greenwood estimated standard error and the piecewise constant hazard function and its standard error. These are presented in Table 9.2 for the combined group.

Figure 9.1 then presents a plot of the estimated event-free survival function and the asymmetric 95% confidence limits obtained using the standard error of the log(-log) survival function.

Fig. 9.1 The Kaplan-Meier estimated probability of remaining free of spread of cancer and the 95% confidence limits at each point in time based on the log(-log) survival probabilities.

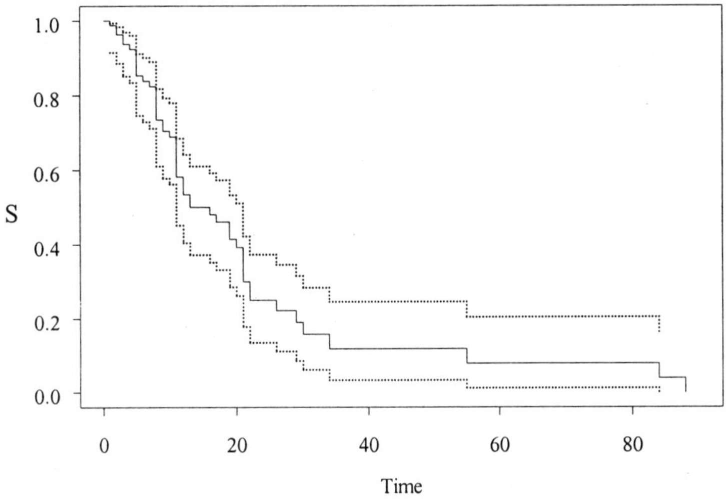

Table 9.3 presents the estimated survival and hazard functions for each group separately.

Figure 9.2 then presents a plot of the estimated event-free survival function in each group.

There is a small increase in the probability of remaining free of spread of cancer, or a small reduction in risk, with the experimental drug treatment. The apparent widening of the difference in survival functions beyond 20 weeks is highly unreliable because of the small numbers at risk in each group, especially the control group, as indicated by the large standard errors for $\widehat{S}(t)$ in Table 9.3.

Suppose that the objective of the study was to compare the "survival" probabilities of remaining free of spread of disease beyond 24 weeks so that the event-free probabilities at this fixed point in time are the only quantities of interest. The estimated probabilities in each group are $\widehat{S}_1(24) = 0.376$ and $\widehat{S}_2(24) = 0.208$ with a difference of 0.169 with standard error 0.153. Using (9.34) the estimated variances under the null hypothesis are $\widehat{V}_0\left[\widehat{S}_1(24)\right] = .0129$ and $\widehat{V}_0\left[\widehat{S}_2(24)\right] = 0.00556$, which yields a pointwise chi-square test of the difference in remaining free of spread of the disease for 24 weeks with value $X^2 = 1.540$ and $p \leq 0.22$.

In addition to the difference in event-free probabilities, other useful measures include the log ratio of event-free probabilities, or the difference in log event-free proportions, is $\log\left[\widehat{S}_1(24)/\widehat{S}_2(24)\right] = 0.594$ with estimated standard error 0.490.

Table 9.3 Within Each Group, Number at Risk (n_{ij}) and Number of Events (d_{ij}) at Each Event Time $t_{(j)}$, Estimated Continuation Probability of Survival Beyond That Time (q_{ij}), Kaplan-Meier Estimated Probability $\widehat{S}_i(t_{(j)})$, and Greenwood Estimate of the $S.E.$, Nelson-Aalen Estimate $\widehat{\lambda}_{ij}$ of the Hazard over the Interval Since the Last Event and Its $S.E.$

Control Group

$t_{(j)}$	n_{1j}	d_{1j}	q_{1j}	$\widehat{S}_1(t_{(j)})$	$S.E.[\widehat{S}_1(t_{(j)})]$	$\widehat{\lambda}_{1j}$	$S.E.[\widehat{\lambda}_{1j}]$
5	17	1	0.941	0.941	0.057	0.012	0.011
8	15	1	0.933	0.878	0.081	0.022	0.021
11	14	2	0.857	0.754	0.107	0.048	0.031
12	12	2	0.833	0.627	0.121	0.167	0.108
13	10	1	0.900	0.566	0.124	0.100	0.095
19	6	1	0.833	0.471	0.134	0.028	0.025
21	5	1	0.800	0.376	0.136	0.100	0.089
34	3	1	0.667	0.251	0.137	0.026	0.021
84	2	1	0.500	0.125	0.112	0.010	0.007
88	1	1	0.000	0.000	.	0.250	0.000

Experimentally Treated Group

$t_{(j)}$	n_{2j}	d_{2j}	q_{2j}	$\widehat{S}_2(t_{(j)})$	$S.E.[\widehat{S}_2(t_{(j)})]$	$\widehat{\lambda}_{2j}$	$S.E.[\widehat{\lambda}_{2j}]$
1	62	1	0.984	0.984	0.016	0.016	0.016
2	58	2	0.966	0.950	0.028	0.034	0.024
3	53	2	0.962	0.914	0.037	0.038	0.026
4	50	1	0.980	0.896	0.040	0.020	0.020
5	48	4	0.917	0.821	0.051	0.083	0.040
6	44	1	0.977	0.802	0.054	0.023	0.022
7	42	1	0.976	0.783	0.056	0.024	0.024
8	40	5	0.875	0.685	0.064	0.125	0.052
9	35	2	0.943	0.646	0.066	0.057	0.039
10	33	1	0.970	0.627	0.067	0.030	0.030
11	31	5	0.839	0.526	0.070	0.161	0.066
12	25	1	0.960	0.505	0.070	0.040	0.039
13	22	1	0.955	0.482	0.070	0.045	0.044
16	19	1	0.947	0.456	0.071	0.018	0.017
17	18	1	0.944	0.431	0.071	0.056	0.054
19	14	1	0.929	0.400	0.073	0.036	0.034
20	13	1	0.923	0.369	0.073	0.077	0.074
21	12	3	0.750	0.277	0.072	0.250	0.125
22	8	2	0.750	0.208	0.069	0.250	0.153
26	5	1	0.800	0.166	0.066	0.050	0.045
29	3	1	0.667	0.111	0.063	0.111	0.091
30	2	1	0.500	0.055	0.050	0.500	0.354
55	1	1	0.000	0.000	.	0.040	0.000

Fig. 9.2 The Kaplan-Meier estimated probability of remaining free of spread of cancer in the experimentally treated and the control groups.

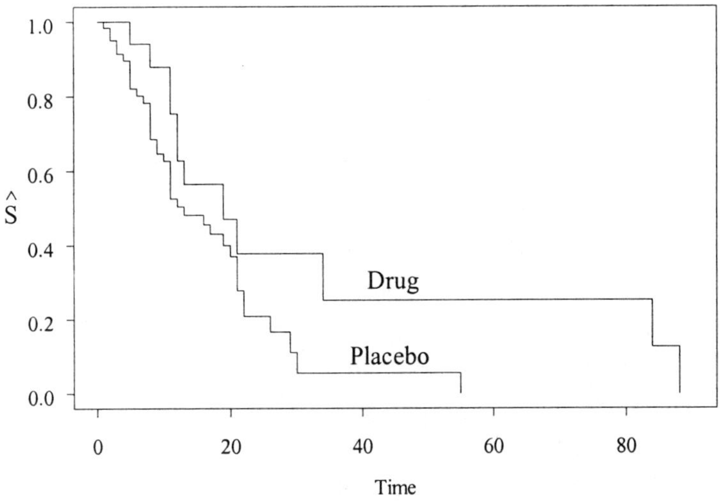

Thus the likelihood of survival beyond 24 weeks is $1.81 = \exp(0.594)$ times greater in group 1 than in group 2.

The log(-log) of the event-free proportions, or the log cumulative hazards in each group are $\log\left[\widehat{\Lambda}_1(24)\right] = -0.0234$ and $\log\left[\widehat{\Lambda}_2(24)\right] = 0.452$, so that the difference, which equals the log cumulative hazard ratio, is -0.475 with estimated standard error 0.426. Under a proportional hazards model where the ratio of the hazards is assumed constant over time, the estimated hazard ratio over the 24 weeks of study is $\exp(-.475) = 0.622$, indicating that the hazard of the event is approximately 38% lower in group 1 than in group 2.

Another useful summary is the survival odds ratio. The log odds of remaining event free in the experimental drug group is $\log\left[\widehat{S}_1(24)/\widehat{F}_1(24)\right] = -0.505$ and in the control group is $\log\left[\widehat{S}_2(24)/\widehat{F}_2(24)\right] = -1.338$. The log event-free odds ratio then is 0.834 with estimated standard error 0.715. This indicates that the odds of survival beyond 24 weeks is estimated to be $\exp(0.834) = 2.30$ times greater in group 1 than in group 2. The $S.E.$ of this survival odds ratio could then be estimated using the results of Problem 9.8.9.

9.2 LIFETABLE CONSTRUCTION

Although the Kaplan-Meier estimate is derived assuming a discontinuous distribution, it provides an estimate of the survival probabilities for a continuous survival distribution at the times that events are observed to occur. This assumes that both the time of the event and the time of censoring are observed up to the instant of time for each subject. Usually, however, the closest one comes to continuous observation is when the events are recorded to the day in a long-term follow-up study, or to the hour in a short term follow-up study. In this case tied event times may be observed and some observations will have censored times that are tied with the event times. The usual convention is that any censored times that are also tied with an event time are included as being at risk at that time and are then censored or removed from the risk set immediately after the event time. This convention was employed in the analysis of Example 9.1. Modifications are necessary, however, when the event times are either grouped or are truly discrete.

9.2.1 Discrete Distributions: Actuarial Lifetable

In many studies the exact time of the events are unknown. Rather the events are known to have occurred within some interval of time. For example, in a population lifetable, mortality is recorded up to the year of age, not the day. In this case the data are observed in *discrete or grouped time*. Likewise, in a follow-up study, events may be recorded up to the year of follow-up. In such cases, it is common to employ an *actuarial lifetable*, first described in the context of a follow-up study by Cutler and Ederer (1958). This is often termed simply the *lifetable method*. This method is appropriate when the same interval of time applies to all subjects. Herein, we consider factors that commonly arise in a discrete time follow-up study.

In the more general case, observations may be known to have occurred only within intervals of time that vary from subject to subject. Such observations are observed in *interval time* and censored observations are *interval censored*. See Section 9.4.6.5. Readers are referred to standard survival analysis texts for a treatment of interval censored data.

In an actuarial lifetable, it is assumed that the events occur within fixed intervals of time and that censoring of follow-up also occurs within intervals of time. This can be viewed as converting continuous time observations into *grouped time*. Here we assume that the event or censoring is observed to have occurred within one of a contiguous sequence of intervals $A_j = (\tau_{j-1}, \tau_j]$. Within the jth interval, let r_j be the number still at risk at the beginning of the interval, that is, event-free and under follow-up at τ_j, where $r_1 = N$ is the size of the cohort at the beginning of the study. Also, let d_j be the number of events that occur during the jth interval; and let w_j be the number of *study exits* or randomly censored (unbiased) observations during the interval. The ordinary actuarial estimate assumes that exits are known to be event-free (e.g. alive) at the time of exit. Usually, the exact time of exit is not known and thus we assume that the exits on average were followed for half the interval. Then the likelihood for the jth interval is

$$L(\pi_j) = \pi_j^{d_j}(1-\pi_j)^{w_j/2}(1-\pi_j)^{r_{j+1}}\ , \tag{9.39}$$

where $r_{j+1} = r_j - d_j - w_j$ (*cf.* Elandt-Johnson and Johnson, 1980). In Problem 9.8.3 it is shown that the resulting actuarial estimate of the conditional probability of experiencing an event during the interval is

$$p_j = \widehat{\pi}_j = \frac{d_j}{n_j} \tag{9.40}$$

with denominator

$$n_j = r_j - w_j/2 \tag{9.41}$$

that is the estimated *units or extent of exposure* for the jth interval and is equivalent to the number at risk in the Kaplan-Meier estimate. The resulting actuarial estimator of the grouped survival function then is obtained as

$$\widehat{S}[\tau_j] = \prod_{\ell=1}^{j} q_\ell = \prod_{\ell=1}^{j} \frac{n_\ell - d_\ell}{n_\ell}\ , \tag{9.42}$$

where the estimated continuation probability is $q_j = 1 - p_j$. The actuarial estimate $\widehat{S}[\tau_j]$ thus provides an estimate of the probability of surviving *beyond* the jth interval, or $\widehat{P}(T > \tau_j)$. The complement $1 - \widehat{S}[\tau_j]$ provides an estimate of the cumulative probability of the event up *through* the ith interval, or $\widehat{P}(T \le \tau_j)$.

The estimate is of the same form as the Kaplan-Meier estimate, and thus the expressions for the variance of $\widehat{S}[\tau_j]$, its log, and so forth, are similar to those presented in Section 9.1.3 upon substituting τ_j for t_j. Estimates of the hazard function within each interval and the cumulative hazard function for the actuarial lifetable are described in many general texts (*cf.* Marubini and Valsecchi, 1995).

9.2.2 Modified Kaplan-Meier Estimator

The actuarial estimate assumes that all subjects who exit during an interval are *known* to be event free at the time of exit. In many instances this may not apply in which case the actuarial lifetable estimate does not apply. For example, in many studies in order to determine whether a non-fatal event has occurred during an interval requires that an examination be performed at the end of the interval. In this case, let w'_j designate the number of such observations who exit the study during the jth interval. In fact, some of these patients could have had the event, but because the examination or test was not performed at the time of exit, their event status is unknown at the time of exit from the study.

Since the w'_j exits during the interval were last known to be event free at the beginning of the interval, then the exposure time during the interval is $n_j = r_j - w'_j$. Thus the conditional event probability in (9.40) becomes the ratio of the number of events observed in the interval to the number examined during the interval. This implies that such exits are included up to and including the last interval where status was observed and that they are dropped from the denominator for the interval in which they exited. This adjustment was proposed by Koch, McCanless and Ward (1984) and by Peto (1984), among others.

When all subjects are scheduled to be evaluated at fixed times, at least approximately, then the lifetable estimate is calculated using the Kaplan-Meier method as in Section 9.1.3 based on the numbers of events and the number at risk (examined) for each successive study evaluation. The hazard function within successive intervals would then be computed using the piecewise exponential model of Section 9.1.4. These methods are illustrated subsequently by Example 9.2 .

9.2.3 Competing Risks

An additional consideration arises when some of the censored observations are *informative* with respect to the time of the index event of interest. The most direct instance of informative censoring is the case of a *competing risk,* where each individual is at risk of an absorbing or terminal event other than the index event of interest, such as death caused by cancer in a long study of cardiovascular mortality. Numerous methods have been developed for examination of cause-specific risks including the multiple decrement lifetable (*cf.* Elandt-Johnson and Johnson, 1980), the cause-specific hazard function (*cf.* Kalbfleisch and Prentice, 1980), the cause-specific sub-distribution function (*cf.* Gray, 1988; Gaynor, Fener, Tan, et al., 1993), and the conditional probability function (Pepe, 1991; Pepe and Mori, 1993), among many. More difficult problems arise when the censoring time is predictive of the ultimate failure time, but not in a deterministic manner as for a competing risk. These methods are beyond the scope of this text. However, it is important that the impact of competing risks on the analysis of event-time data be appreciated.

Consider the case where our primary objective is to describe the risk of an index event in the presence of competing risks. For example, we may wish to describe the risk of end-stage renal disease where death because of non-renal causes such as accident or cancer is a competing risk. Let T_I refer to the time of the index event, and T_C refer to the time of a competing risk event. As described by Prentice, Kalbfleisch, Peterson, et al. (1978), treating the competing risk events as unbiased censored observations provides an estimate of the *cause-specific hazard function*

$$\lambda_I(t) = \lim_{\Delta t \downarrow 0} \frac{P(t < T_I \leq t + \Delta t)}{P[(T_I \wedge T_C) > t]} , \tag{9.43}$$

where $T_I \wedge T_C = \min(T_I, T_C)$. This is the probability of the index event at time t among those at risk at that time, meaning having "survived" both the index event and the competing risk event. This is commonly called *censoring on death* because the most common application is where the index event is non-fatal, such as hospitalization, healing, remission, and so on, but where death is a competing risk. A generalized Mantel-Haenszel test then is readily constructed to test the equality of the cause-specific hazard functions for the index event in two or more groups, or a proportional hazards regression model applied to assess covariate effects on the cause-specific hazard function (see Chapter 7 of Kalbfleisch and Prentice, 1980).

In this case, however, the associated estimate of the corresponding cause-specific survival function, such as a Kaplan-Meier estimate, has no logical or statistical interpretation. For example, consider an analysis of the time to progression of renal disease where deaths from other causes are competing risks. If one treats the competing risks (non-renal deaths) as ordinary censored observations, that is, censors on such deaths, then the estimated renal progression "survival" function provides an estimate of the probability of remaining free of progression of renal disease *in a population where there are no deaths from other causes*. Thus the population to which the estimate applies is non-existent. See Peto (1984), Gaynor, Fener, Tan, et al. (1993), Pepe and Mori (1993) and Mauribini and Valsecchi (1995) for other examples and further discussion.

In such instances, a much more useful quantity is the *sub-distribution function*, which is the estimate of the *cumulative incidence* of the index event in a population subject to other absorbing or terminal events. The following simple adjustment provides an estimate of this function. Consider the case of a simple proportion in a cohort of patients exposed for a common unit of time in which there are no exits for any reason other than a competing risk. As before, let r be the number of patients entering the cohort and d be the number of events observed in a fixed unit of time. Then let e be the number of competing risk events observed, that is, events that preclude the possibility of observing the index event. As described by Cornfield (1957), in the parlance of competing risk events we can choose to estimate either of two underlying probabilities of the index event: the *net (pure) rate* as $d/(r-e)$ or the *crude (mixed) rate* as d/r. The net or pure rate estimates the probability of the event that would be observed if the competing risk event could be eliminated in the population, and thus is purely hypothetical. Conversely, the mixed rate estimates the probability of the event, or its complement, in a population at risk of the index event and also other competing risk events, analogous to the sub-distribution function.

Now consider a discrete time lifetable. Let r_j be the number entering the jth interval (that is, at risk at the beginning of the interval); and during the jth interval let d_j be the number of index events, e_j be the number of competing risk events, and w_j be the number of unbiased randomly censored observations. To motivate a simple adjustment for the presence of competing risk events, consider the following example with three time intervals ($j = 1$, 2 or 3) and no exits for any reason other than a competing risk. Of the 100 patients entered, $\sum_j d_j = 30$ had the event and $\sum_j e_j = 20$ exited because of a competing risk event as follows:

j	r_j	d_j	e_j
1	100	5	5
2	90	10	10
3	70	15	5
Total		30	20

Because no patients are censored at random, the crude (mixed) proportion who survive the index event beyond the third interval is simply $(100 - 30)/100 = 0.70$. It is reasonable, therefore, to expect that a lifetable estimator over these three intervals should provide the same estimate because no patients were censored at random. The usual product-limit estimator, however, yields the estimate $(95/100)(80/90)(55/70) = 0.6635$, which is clearly incorrect. Alternately, if competing risk exits are *not* treated as censored, but rather are retained in the lifetable throughout, then the product-limit estimator of the proportion survivors after the last interval is $(95/100)(85/95)(70/85) = 0.70$, the desired result. For such cases, in Problem 9.6.7 it is shown that this adjustment, in general, provides the proper estimate when there are no exits other than those caused by the competing risk. This approach is also discussed by Lagakos, Lim and Robins (1990).

In the more general case, the sub-density function of the index event, or the probability of the index event at time t, is the probability of survival to time t and experiencing the index event at t, or

$$f_I(t) = \lambda_I(t)S_{I,C}(t), \tag{9.44}$$

where $S_{I,C}(t) = P[(T_I \wedge T_C) > t]$ is the probability of surviving both the index and competing risk events at time t and where $\lambda_I(t)$ is the cause-specific hazard for the index event at that time. Thus the cumulative incidence or the sub-distribution function of the index event is the cumulative probability

$$F_I(t) = \Pr[T_I \leq t] = \int_0^t \lambda_I(u)S_{I,C}(u)du. \tag{9.45}$$

This function may be estimated by

$$\widehat{F}_I(t) = \prod_{j:t_{(j)} \leq t} \left(\frac{d_{Ij}}{n_j}\right) \widehat{S}_{I,C}(t_{(j)}), \tag{9.46}$$

where d_{Ij} is the number of index events at the jth event time $t_{(j)}$ among the n_j subjects still at risk of either the index or competing risk events, and $\widehat{S}_{I,C}(t)$ is the Kaplan-Meier estimate of the probability of surviving both the index and competing risk event. Gaynor, Fener, Tan, et al. (1993) describe a generalization of the Greenwood estimate of the variance of this function; see also Marubini and Valsecchi (1995).

Gray (1988) then presents a test for the difference between the index event sub-distribution functions for two or more groups that is a generalization of the tests described in Section 9.3. Pepe (1991) also describes a test of equality of these and other related functions. Lunn and McNeil (1995) show that the Cox proportional hazards model described in Section 9.4 may be used to assess covariate effects on the cause-specific hazard functions of the index and competing risk events simultaneously.

It can be shown that if the number of randomly censored observations during the interval between event times is small relative to the number at risk, then the simple adjustment of not censoring on competing risks in a simple Kaplan-Meier calculation will yield estimates of the complement of the sub-distribution function, $1 - \widehat{F}_I(t)$, and standard errors that are close to those provided by the above precise computations. Lagakos, Lim and Robins (1990) called this computation the *intent-to-treat lifetable.* The large sample properties of this estimate, and of test statistics comparing two or more groups, relative to the more precise computations based on (9.46) have not been explored.

Finally, it should be noted that another approach is to eliminate the problem of competing risks by using a composite outcome consisting of the time to either the index or the competing risk event. In this case, inference is based on the quantity $S_{I,C}(t)$ and its associated hazard function. For example, one might use the time to progression of renal disease, transplant or death from any cause as a composite outcome where the observed time for each subject is the time of the first of these events to occur. Likewise, one might use all-cause mortality as an outcome rather than cause-specific mortality. Apart from the statistical simplifications, for clinical trials in cardiology, Fleiss, Bigger, McDermot, et al. (1990) have argued that the combined outcome of fatal or non-fatal myocardial infarction (MI) is indeed more relevant than is non-fatal MI where fatal MI is a competing risk. Their arguments also apply to an analysis of all cause mortality versus cause-specific mortality. This approach is only feasible, however, when the component events all represent worsening of disease. In other cases, where the index event represents improvement, but some patients may worsen and ultimately die, there is no alternative other than to evaluate the cause-specific hazard, its sub-distribution function or other relevant quantities. In the remainder, we assume that no competing risks apply.

Example 9.2 *Nephropathy in the DCCT*

Figures 1.1 and 1.2 present the cumulative incidence of progression from normal albumin excretion ($\leq$ 40 mg/24 h) to microalbuminuria (>40 mg/24 h) among subjects treated intensively or by conventional therapy in the Diabetes Control and Complications Trial (1993). The cumulative incidence was estimated as $1 - \widehat{S}(t)$, where $\widehat{S}(t)$ is the estimate of the probability of remaining free of (or surviving) microalbuminuria. The presence or absence of microalbuminuria was assessed annually by a four-hour timed collection of serum and urine. This is an example in which the event is observed in discrete time because a procedure must be performed in order to discern whether or not the index event has occurred. Thus we used the approach described in Section 9.2.2, whereby individuals were considered to be right censored immediately after the last annual collection when still free of microalbuminuria and the actuarial adjustment was not applied.

The actual annual visits were not conducted at intervals of exactly 365 days. Rather, the visit times varied about the annual target visit dates, some being late, some early, by as much as a month or more. Thus the renal evaluations conducted during the interval 0–18 months were considered year 1 evaluations, those during 19–30 months as year 2 evaluations, and so on. During the study, only eight subjects were lost to follow-up for various reasons that were considered to constitute random censoring events, for example, moved away, and 11 patients died. Because of the

Table 9.4 Number at Risk (n) and Number of Events (d) at each Annual Evaluation in the Intensive and Conventional Treatment Groups of the Secondary Intervention Cohort of the DCCT

Visit	*Intensive*		*Conventional*	
t_j	n_{1j}	d_{1j}	n_{2j}	d_{2j}
1	325	18	316	19
2	307	15	297	22
3	290	9	275	17
4	281	7	258	16
5	274	7	242	10
6	204	7	167	5
7	129	2	96	3
8	73	2	52	4
9	57	2	33	3

small number of deaths, the simple adjustment of not censoring on death was employed rather than a more complicated analysis of the sub-distribution function allowing for death as a competing risk. Because patients were recruited over a six-year period, the follow-up times ranged from 4–9 years, averaging 6.5 years. Patients who were followed up to the termination of the study in the spring of 1993 who had not developed microalbuminuria were then *administratively censored* at that time.

Table 9.4 presents the numbers at risk in the ith group at the jth annual evaluation (n_{ij}) and the number of patients who had microalbuminuria for the first time at that visit (d_{ij}) among the 715 patients in the secondary intervention cohort of the study (see Section 1.5). The number of patients censored immediately after the jth evaluation in the ith group is $w'_{ij} = n_{i(j+1)} - n_{ij} + d_{ij}$. In this case, the number at risk is the number of annual renal evaluations conducted so that the numbers censored are transparent. Only two subjects were lost to follow-up (censored) during the first four years.

Table 9.5 then presents the resulting estimates of the survival function, the corresponding cumulative incidence function displayed in Figure 1.2 of Chapter 1, and the Greenwood estimated standard errors.

In both treatment groups there is a steady progression in the cumulative incidence of microalbuminuria. However, beyond two years of treatment, the risk in the intensive treatment group is substantially less than that in the conventional group, estimated to be 26% after nine years versus 42%.

Table 9.5 Estimated Survival and Cumulative Incidence Functions for Developing Microalbuminuria at Each Annual Evaluation in the Intensive and Conventional Treatment Groups of the Secondary Intervention Cohort of the DCCT

Visit	*Intensive*			*Conventional*		
t_j	$\widehat{S}_{1j}$	$\widehat{F}_{1j}$	$S.E.$	$\widehat{S}_{2j}$	$\widehat{F}_{2j}$	$S.E.$
1	0.945	0.055	0.013	0.940	0.060	0.013
2	0.898	0.102	0.017	0.870	0.130	0.019
3	0.871	0.129	0.019	0.916	0.184	0.022
4	0.849	0.151	0.020	0.766	0.234	0.024
5	0.827	0.173	0.021	0.734	0.266	0.025
6	0.799	0.201	0.023	0.712	0.288	0.026
7	0.786	0.214	0.024	0.690	0.310	0.028
8	0.765	0.235	0.028	0.637	0.363	0.036
9	0.738	0.262	0.033	0.579	0.421	0.046

9.2.4 SAS PROC LIFETEST: Survival Estimation

Some of the above analyses are provided by standard programs such as the SAS procedure LIFETEST. For continuous observations these programs compute the Kaplan-Meier estimate and its standard errors directly. For the squamous cell carcinoma data in Example 9.1, the SAS procedure LIFETEST with the following specification provides the estimates of the survival function and its standard errors within each group as shown in Table 9.3:

```
proc lifetest; time time*delta(0); strata group;
```

where the variable `time` is the survival/censoring time and censoring is indicated by the variable `delta=0`.

However, with grouped or discrete data one must apply such programs with care because some programs such as LIFETEST employ different conventions from those employed herein. Throughout we take the view that the survival function is $S(t) = P(T > t)$, whereas LIFETEST defines the survival function as $S(t) = P(T \geq t)$. Thus for grouped time data, the jth interval is defined herein as $A_j = (\tau_{j-1}, \tau_j]$, whereas LIFETEST defines this interval as $A_j = [\tau_{j-1}, \tau_j)$, closed on the left rather than the right. Thus the user must be careful when using this or another program to construct the basic table of numbers at risk and numbers of events within an interval, such as for the DCCT Nephropathy data in Example 9.2. In such cases, its is recommended that one should carefully construct this basic table before using a standard program. However, the output of the program will still appear somewhat different because of the manner in which the intervals are defined.

Example 9.3 *Actuarial Lifetable in PROC LIFETEST*
For example, the following is a hypothetical set of data from a study where survival was grouped into two year intervals. The number entering the interval (r_j), the number of deaths during the interval (d_j), the number lost-to-follow-up and known to still be alive during the interval (w_j), the units of exposure (n_j) from (9.41), the estimated probability of survival *beyond* each interval $\widehat{S}(\tau_j)$ and its estimated standard error are:

$(\tau_{j-1}, \tau_j]$	r_j	d_j	w_j	n_j	$\widehat{S}(\tau_j)$	S.E.
(0, 2]	305	18	45	282.5	0.936	0.015
(2, 4]	242	12	22	231.0	0.888	0.019
(4, 6]	208	6	9	203.5	0.861	0.022

These data would be analyzed using the SAS procedure LIFETEST, using the following statements:

```
Data one; input time freqn delta @@; cards;
2 18 1   2  45 0
4 12 1   4  22 0
6  6 1   6   9 0
8  0 1   8 193 0
;
PROC LIFETEST METHOD=act WIDTH=2;
     TIME time*delta(0); FREQ freqn;
```

Note the addition of an extra interval with the number censored after the study was concluded.

The SAS output would then include the following (extraneous material deleted):

Interval [Lower,	Upper)	Number Failed	Number Censored	Effective Sample Size	Survival	Survival Standard Error
0	2	0	0	305.0	1.0000	0
2	4	18	45	282.5	1.0000	0
4	6	12	22	231.0	0.9363	0.0145
6	8	6	9	203.5	0.8876	0.0194
8	.	0	193	96.5	0.8615	0.0216

Because of the manner in which the intervals and the survival function are defined, the survival function estimates are offset by one interval compared to those computed directly from the data. Thus the last extra interval is required in order to obtain the estimates of survival beyond six years using LIFETEST.

Thus for other than continuous observations, the user should take care to properly construct the appropriate time variables for the censored and non-censored observations to ensure that the lifetable estimates are computed properly.

9.3 FAMILY OF WEIGHTED MANTEL-HAENSZEL TESTS

9.3.1 Weighted Mantel-Haenszel Test

The Kaplan-Meier and actuarial estimates provide a description of the survival experience in a cohort over time. However, the simple test for the difference between groups in proportions surviving beyond a specific point in time in (9.35) leaves much to be desired because it does not provide a test of equality of the complete survival curves, or equivalently, the hazard functions, for two or more cohorts over time. Thus another major advance was provided by Mantel (1966), when he reasoned that the Mantel-Haenszel test for a common odds ratio over independent strata also provides a test for differences between groups with respect to the pattern of event times, as represented by the underlying hazard function over time. Although the conditional proportions of events among those at risk within a cohort over time are not statistically independent, they are still uncorrelated and are asymptotically conditionally independent. Thus Mantel suggested that the Mantel-Haenszel test statistic applied to survival times should converge in distribution to the central 1 *df* chi-square distribution. A general formal proof of Mantel's conjecture was later provided through the application of the theory of martingales for counting processes by Aalen (1978) and Gill (1980); see also Harrington and Fleming (1982).

In Chapter 4 we showed that the Mantel-Haenszel test for multiple 2×2 tables may be differentially weighted over tables, the weights providing a test that is asymptotically fully efficient under a specific model, or efficient against a specific alternative hypothesis. So also, various authors have shown that a *weighted Mantel-Haenszel test* using a specific set of weights is asymptotically fully efficient against a specific alternative.

Let $t_{(j)}$ denote the jth ordered event time from the combined sample of two independent groups of sizes n_1 and n_2 initially at risk at time zero, $j = 1, \ldots, J$, where J is the number of distinct event times, $J \leq D$, D being the total number of subjects with an event. Then, at the jth event time $t_{(j)}$ a 2×2 table can be constructed from the n_{ij} subjects still under follow-up and at risk of the event in the ith group ($i = 1, 2$). In the notation of stratified 2×2 tables introduced in Section 4.2.1, let a_j denote the number of events observed at $t_{(j)}$ in group 1 and let m_{1j} denote the total number of events in both groups at time $t_{(j)}$. Then, the expected value of a_j under the null hypothesis H_0: $S_1(t) = S_2(t)\ \forall t$, or equivalently H_0: $\lambda_1(t) = \lambda_2(t)\ \forall t$, is estimated to be $\widehat{E}(a_j|H_0) = n_{1j}m_{1j}/N_j$, where $N_j = n_{1j} + n_{2j}$ is the total number at risk at that time. The weighted Mantel-Haenszel test is then of the form

$$X^2_{WMH} = \frac{\left(\sum_{j=1}^{J} w(t_{(j)})\left[a_j - E(a_j|H_0)\right]\right)^2}{\sum_{j=1}^{J} w(t_{(j)})^2 V_c(a_j|H_0)} \tag{9.47}$$

where $w(t_{(j)}) \neq 0$ for some j, and the conditional hypergeometric expectation $E(a_j|H_0)$ and variance $V_c(a_j|H_0)$ are as presented in Section 4.2.3,

$$E(a_j|H_0) = \frac{n_{1j}m_{1j}}{N_j}, \quad \text{and} \quad V_c(a_j|H_0) = \frac{n_{1j}n_{2j}m_{1j}m_{2j}}{N_j^2(N_j-1)} . \tag{9.48}$$

Asymptotically, the statistic is distributed as χ^2 on 1 df under H_0, and clearly, the size of the test asymptotically will be the desired level α regardless of the chosen weights.

This test is a weighted generalization of the Mantel-Haenszel test analogous to the Radhakrishna weighted Cochran-Mantel-Haenszel test presented in Section 4.7. In principle, the asymptotically most powerful test against a specific alternative could be derived by maximizing the asymptotic efficiency. In this setting, however, for continuous event times, as $N \to \infty$ then $J \to \infty$ and the sums becomes integrals. Schoenfeld (1981) presents the expression for the non-centrality parameter of the test with no ties and shows that, in general, the optimal weights for a weighted Mantel-Haenszel test are of the form

$$w(t) \propto \log[\lambda_1(t)/\lambda_2(t)], \tag{9.49}$$

proportional to the log hazard ratio over time under the alternative hypothesis. In cases where the alternative can be expressed in the form

$$\log[\lambda_1(t)/\lambda_2(t)] \cong g[F_0(t)] \tag{9.50}$$

or as some function of the cumulative event distribution function under the null hypothesis, Schoenfeld (1981) shows that the optimal weights are

$$w\,(t) = g[\widehat{F}(t)], \tag{9.51}$$

where $\widehat{F}(t) = 1 - \widehat{S}(t)$ and $\widehat{S}(t)$ is the Kaplan-Meier estimate of the survival distribution for the combined sample. These expressions allow the determination of the optimal weights against a specific alternative, or the alternative for which a specific set of weights is fully efficient.

9.3.2 Mantel-logrank Test

The test with unit weights $w\,(t) = 1\ \forall t$ yields the original unweighted Mantel-Haenszel test for multiple 2×2 tables applied to lifetables as suggested by Mantel (1966). When there are no tied event times ($J = D$), then $m_{1j} = 1$ for $j = 1, \ldots, D$ and the test is equivalent to a rank-based test statistic using a permutational variance derived by Peto and Peto (1972), which they termed the *logrank test.* However, when there are tied event times, the Peto-Peto logrank test statistic differs from the Mantel-Haenszel test statistic, and its variance is greater. The Mantel-Haenszel test, therefore, is widely referred to as the logrank test, even though the test of Peto and Peto does not accommodate tied observations in the same manner.

Cox (1972) also showed that the Mantel-Haenszel or logrank test (without ties) could also be derived as an efficient score test in a proportional hazards regression model with a single binary covariate to represent treatment group (see Problem 9.14). Thus the test is also called the *Mantel-Cox test.*

Based on the work of Cox (1972), Peto and Peto (1972), and Prentice (1978), among many, the Mantel-logrank test with $w(t) = 1$ is asymptotically fully efficient against a *proportional hazards* or *Lehman alternative,* which specifies that

$$S_1(t) = S_2(t)^{\phi} \tag{9.52}$$

for some real constant ϕ. Thus

$$\exp\left[-\Lambda_1(t)\right] = \exp\left[-\phi\Lambda_2(t)\right] \tag{9.53}$$

and it follows that

$$\begin{aligned} d\Lambda_1(t) &= \phi d\Lambda_2(t) \\ \lambda_1(t) &= \phi\lambda_2(t) \end{aligned} \tag{9.54}$$

or that the hazard functions are proportional over time. From (9.49), therefore, the test with constant or unit weights $w(t) = 1$ is fully efficient against this alternative.

Under this alternative, the constant ϕ is the hazard ratio over time, which is a general measure of the instantaneous relative risk between the two groups. As is shown below, an estimate of the relative risk and its variance is conveniently provided by the Cox proportional hazards model. A location shift of an exponential or a Weibull distribution also satisfy this alternative, among others.

9.3.3 Modified Wilcoxon Test

Forms of the weighted Mantel-Haenszel test have also been shown to be equivalent to a modified Wilcoxon rank test for censored event times. The *Peto-Peto-Prentice Wilcoxon test* uses weights equal to the Kaplan-Meier estimate of the survival function for the combined sample, $w(t) = \widehat{S}(t)$, after asymptotically equivalent tests derived independently by Peto and Peto (1972) and Prentice (1978). This test provides greater weight to the early events, the weights diminishing as $\widehat{S}(t)$ declines. A similar test, the *Gehan Modified Wilcoxon test* (Gehan, 1965), was shown by Mantel (1967) to be a weighted Mantel-Haenszel test using the weights $w(t_{(j)}) = n_j$ at the jth event time. Although Gehan's Wilcoxon test will be fully efficient against some alternative, just as any set of weights provides an efficient test against some alternative, the precise alternative for which it will be asymptotically efficient cannot be described in general terms, because its weights depend on both the event time distribution and the censoring time distribution (Prentice and Marek, 1979).

The Peto-Peto-Prentice Wilcoxon test, however, is asymptotically fully efficient against a *proportional odds alternative* of the form

$$\frac{F_1(t)}{1 - F_1(t)} = \left[\frac{F_2(t)}{1 - F_2(t)}\right]\varphi\,, \tag{9.55}$$

which specifies that the odds of the event are proportional over time, or that the event odds ratio φ is constant over time. Note that this model also specifies that the survival odds ratio is $1/\varphi$. This is also called a *logistic alternative* because this relationship is satisfied under a location shift of a logistic probability distribution (see Problem 9.4.2). Under this model, Bennett (1983), among others, shows that the Prentice (1978) Wilcoxon test can be obtained as an efficient score test of the effect of a binary covariate for treatment group. Also, using Schoenfeld's result in (9.51), in Problem 9.10 it is shown that the optimal weights under the proportional odds model are $w(t) = \widehat{S}(t)$.

9.3.4 G^ρ Family of Tests

To allow for a family of alternatives, Harrington and Fleming (1982) proposed that the Mantel-Haenszel test be weighted by a power of the estimated survival function $w(t) = \widehat{S}(t)^\rho$ at time t. Because they used G^ρ to denote the resulting test statistic for given ρ, the family of tests has come to be known as the G^ρ family of tests. A related family of tests was proposed by Tarone and Ware (1977), using weights $w(t_{(j)}) = n_j^\rho$ based on the total number at risk at the jth event time. In each family, the value of ρ determines the weight assigned to events at different times and thus the alternative for which the test is asymptotically efficient. For $\rho = 0$ in the G^ρ or Tarone-Ware family, all events are weighted equally, yielding the original unweighted Mantel-logrank test. When $\rho = 1$, the resulting G^ρ test is the Peto-Peto-Prentice Wilcoxon test, whereas the Tarone-Ware test is Gehan's test. The Tarone-Ware weights, however, reflect both the survival and the censoring distributions. Thus the Harrington-Fleming formulation is preferred.

The relative efficiency of various members of these families of tests have been explored, focusing principally on the relative efficiency of the logrank and Wilcoxon tests under specific alternatives. Tarone and Ware (1977) explored the efficiency of the Mantel-logrank test versus the Gehan-Wilcoxon test by simulation. Peto and Peto (1972) derived both the logrank and a modified Wilcoxon test based on the theory of optimal linear rank tests of Hájek and Śidak (1967). Prentice (1978) also showed that the logrank and a modified Wilcoxon test could be obtained as asymptotically efficient linear rank tests using scores derived from an accelerated failure time model. Prentice and Marek (1979) and Mehrotra, Michalek and Mihalko (1982) showed that these linear rank tests are algebraically equivalent to a weighted Mantel-Haenszel test with censored observations. Harrington and Fleming (1982) also explored in detail the relative efficiency of members of the G^ρ family of tests based on the Aalen-Gill martingale representation of the underlying counting process, where these tests are also equivalent to the weighted Mantel-Haenszel tests shown in (9.47); see Section 9.6.3.

Clearly, if the distributions from the two groups satisfy proportional hazards then they cannot also satisfy proportional survival odds, and vice versa. Further, the efficiency of a test with specified weights, or a specified value of ρ, will depend on the extent to which the data support the particular alternative hypothesis for which that test is most efficient. Thus if one applies a test with logrank (constant) weights and the hazard functions are not proportional, there will be a loss of power. Lagakos and Schoenfeld (1984) describe the loss in power for the Mantel or logrank test when the hazard ratios are not constant and, in fact, may cross. Conversely, if Wilcoxon weights are used and the hazards are proportional, then likewise there will be a loss in power. These cases are analogous to the loss of efficiency incurred in the test for multiple independent 2×2 tables described in Section 4.8, when the test for one scale is chosen and the true alternative is that there is a constant difference over strata on a different scale.

As was the case for multiple 2×2 tables, it is inappropriate to conduct multiple tests and then cite that one with the smallest P-value. Rather, in order to preserve the size of the test, the specific test, and the specific set of weights, must be specified *a priori*. Thus one risks a loss of power or efficiency if the chosen test is sub-optimal. As for the analysis of multiple 2×2 tables described in Section 4.9, one safeguard against this loss of power, while preserving the size of the test, is to employ the Gastwirth (1985) maximin efficient robust test (*MERT*), which is obtained as a combination of the extreme pair in the family of tests. Apart from the fact that the weighted Mantel-Haenszel tests use the conditional hypergeometric variance, whereas the Radhakrishna tests use the unconditional product-binomial variance, the *MERT* in the two cases is identical; see Gastwirth (1985). Thus the developments in Section 4.9.2 also apply to the computation and description of the *MERT* for the G^ρ family of tests.

Finally, one may also conduct a stratified-adjusted test of the difference between the lifetables of two (or more) groups. If the total sample of N subjects are divided among K independent strata ($k = 1, \dots, K$), then the stratified adjusted weighted Mantel-Haenszel test is of the form

$$X^2_{\rho A} = \frac{\left(\sum_{k=1}^{K} \sum_{j=1}^{J_k} w_k(t_{(j,k)}) \left[a_{(j,k)} - E(a_{(j,k)}|H_0)\right]\right)^2}{\sum_{k=1}^{K} \sum_{j=1}^{J} w_k(t_{(j,k)})^2 V_c(a_{(j,k)}|H_0)}, \tag{9.56}$$

where $t_{(j,k)}$ denotes the jth of the J_k event times observed among the n_k subjects within the kth stratum, $w_k(t_{(j,k)})$ is the weight within the kth stratum at time $t_{(j,k)}$, $a_{(j,k)}$ is the observed number of events in group 1 at that time, $E(a_{(j,k)}|H_0)$ its expected value with respect to the observations from the kth stratum at risk at that time, and $V_c(a_{(j,k)}|H_0)$ its conditional variance. For the G^ρ family, $w_k(t_{(j,k)}) = \widehat{S}_k(t_{(j,k)})^\rho$ based on the estimated survival function within the kth stratum. Breslow (1975) presents a justification for the stratified logrank test ($\rho = 0$) under a proportional hazards model; see also Problem 9.14.5. However, the stratified Gehan and Peto-Peto-Prentice Wilcoxon tests with censored observations have not been formally studied.

9.3.5 Measures of Association

For the analysis of simple and stratified 2×2 tables, various measures of association were employed including the odds ratio and relative risk. Mantel (1966) suggested that the Mantel-Haenszel stratified-adjusted estimate of the common odds ratio be used to summarize the differences between two survival curves, each event time representing a stratum. As shown above, however, the Mantel or logrank test is asymptotically efficient under a proportional hazards alternative. Thus Anderson and Bernstein (1985) show that the Mantel-Haenszel stratified-adjusted estimate of the common relative risk expressed in (4.17) provides an estimate of the assumed common hazard ratio over time.

Peto, Pike, Armitage, et al. (1976) suggested that a score-based estimate of the hazard ratio (relative risk) be obtained as $\widehat{\phi} = e^{\widehat{\beta}_{RR}}$, where $\widehat{\beta}_{RR} = S_L/\widehat{V}(S_L)$ with estimated variance $\widehat{V}(\beta_{RR}) = \widehat{V}(S_L)^{-1}$, when the Mantel or logrank test in (9.47) with unit weights ($w(t) = 1$) is expressed as $X_L^2 = S_L^2/\widehat{V}(S_L)$. This estimate is derived as a first-order approximation to the MLE under the Cox proportional hazards model described subsequently, in like manner to the derivation of the Peto estimate of the odds ratio for a 2×2 table described in Section 6.4 (see Problem 9.14.6). This provides an estimate of the nature of the treatment group difference to which the logrank test is optimal.

Similarly, the Peto-Peto-Prentice Wilcoxon test may be obtained as an efficient score test under a proportional odds model (Bennett, 1983). This suggests that an estimate of the log odds ratio of the cumulative event probability, or the negative log of the survival odds ratio, may be obtained as $\widehat{\beta}_{OR} = S_{PW}/\widehat{V}(S_{PW})$ with estimated variance $\widehat{V}(\beta_{OR}) = \widehat{V}(S_{PW})^{-1}$ using the Peto-Peto-Prentice Wilcoxon statistic S_{PW} and its estimated variance $\widehat{V}(S_{PW})$.

Example 9.4 *Squamous Cell Carcinoma (continued)*
For the data from Lagakos (1978) presented in Example 9.1, the commonly used weighted Mantel-Haenszel test statistics are

Test	*Weights*	*Statistic*	*Variance*	X^2	$p \leq$
Logrank	1.0	-6.082	9.67277	3.824	0.0506
Gehan Wilcoxon	n_j	-246.0	21523.4	2.812	0.0936
PPP Wilcoxon	$\widehat{S}(t_{(j)})$	-3.447	3.90095	3.046	0.0810

Since the data appear to satisfy the proportional hazards assumption to a greater degree than the proportional survival odds assumption, the logrank test yields a higher value for the test and a smaller *P*-value than does either the Gehan test or the Peto-Peto-Prentice (PPP) Wilcoxon test. Of course, one must prespecify which of these tests will be used *a priori* or the size of the test will be inflated.

An alternative is to use Gastwirth's maximin combination of the logrank and PPP Wilcoxon tests. The covariance between the two test statistics is 5.63501, which yields a correlation of 0.91735. Then, using the expression presented in (4.126), the maximin efficient robust test (MERT) test value is $X_m^2 = 3.572$ with $p \leq 0.059$.

Although not statistically significant, these tests provide a test of differences between the complete survival curves of the two groups, rather than a test of the difference between groups at a particular point in time, as in Example 9.1.

Based on the logrank test, the Peto, Pike, Armitage, et al. (1976) estimate of the log hazard ratio or relative risk is $\widehat{\beta}_{RR} = -0.629$ with estimated variance $\widehat{V}(\widehat{\beta}_{RR}) = 0.1034$, which yields an estimate of the relative risk of 0.533 with asymmetric 95% confidence limits (0.284, 1.001). A Cox proportional hazards regression model fit to these data with a single covariate for treatment effect yields a regression coefficient of -0.677 with $S.E. = 0.364$, and with an estimated relative risk of 0.508 and confidence limits of (0.249, 1.036), close to those provided by the score-based estimate. This quantity corresponds to the effect tested by the logrank test.

Likewise, for the PPP Wilcoxon test, the score-based estimate of the log cumulative event odds ratio is $\widehat{\beta}_{OR} = -0.884$ with estimated variance $\widehat{V}(\widehat{\beta}_{OR}) = 0.2564$, which yields an estimate of the cumulative event odds ratio of 0.413 with asymmetric 95% confidence limits (0.153, 1.115). The corresponding estimate of the survival odds ratio is 1/0.413=2.42 with 95% confidence limits (0.897, 6.53). This quantity corresponds to the effect tested by the Peto-Peto-Wilcoxon test.

Example 9.5 *Nephropathy in the DCCT (continued)*
For the analysis of the cumulative incidence of developing microalbuminuria in the Secondary Intervention cohort of the DCCT presented in Example 9.2, the study protocol specified *a priori* that the Mantel-logrank test would be used in the primary analysis to assess the difference between groups. This test yields $X^2 = 9.126$ with $p \leq 0.0026$. For comparison, the Peto-Peto-Prentice Wilcoxon test yields $X^2 = 8.452$ with $p \leq 0.0037$. For this example, the two tests are nearly equivalent.

The relative risk estimated from a Cox proportional hazards model is 1.62 with 95% confidence limits of (1.18, 2.22). This corresponds to a 38% reduction in the risk of progression to microalbuminuria among patients in the intensive versus conventional treatment groups. An additional calculation that adjusts for the log of the albumin excretion rate at baseline yields a 42.5% reduction in risk with intensive therapy.

9.3.6 SAS PROC LIFETEST: Tests of Significance

For the squamous cell carcinoma data set of Example 9.1 with continuous (or nearly so) survival times, the statements shown in section 9.2.4 would be used to conduct an analysis using the SAS procedure LIFETEST. In this program, the independent treatment groups are referred to as `STRATA`, not as a `CLASS` variable as in other SAS procedures. In addition to estimates of the survival and hazard functions, the program computes the following tests of significance of the difference between groups (strata):

```
        Test of Equality over Strata
                                    Pr >
        Test   Chi-Square  DF  Chi-Square
    Log-Rank     3.8245     1    0.0505
    Wilcoxon     2.8116     1    0.0936
   -2Log(LR)     3.8869     1    0.0487
```

The log-rank test value is the Mantel-logrank test and the Wilcoxon test is the Gehan Wilcoxon test, not the preferred Peto-Peto-Prentice Wilcoxon test presented in Example 9.4. The third test is a likelihood ratio test of the difference between groups based on the assumption of an underlying exponential distribution in each group (Lawless, 1982).

PROC LIFETEST also provides a TEST statement that conducts either a logrank or a Gehan-Wilcoxon scores linear rank test that can be used to assess the association between a quantitative covariate and the event times. These tests are approximately equivalent to the value of the Wald test of the covariate when used in either a proportional hazards or a proportional odds model, respectively. These tests could also be used with a binary covariate representing treatment group. The tests, however, are based on the asymptotic permutational variance of the linear rank test and not the hypergeometric variance, as employed in the family of weighted Mantel-Haenszel tests. For the squamous cell carcinoma data, the logrank scores chi-square test value for the effect of treatment group is 3.579 with $p \leq 0.0585$, and the Wilcoxon scores test value is 3.120 with $p \leq 0.0773$. These tests do not correspond to the members of the G^{ρ} family of weighted-Mantel-Haenszel tests, which, in general, are preferred.

9.4 PROPORTIONAL HAZARDS MODELS

The logistic and Poisson regression models of previous chapters assess the effects of covariates on the risk of the occurrence of an outcome or event, or multiple events, respectively, over a fixed period of time without consideration of the exact times at which events occurred, or the precise interval of time during which the event(s) occurred. The Cox (1972) Proportional Hazards (PH) regression model provides for the assessment of covariate effects on the risk of an event over time, where some of the subjects may not have experienced the event during the period of study, or may have a censored event time. The most common example is the time to death and the time of exposure (censoring) if a subject is still alive at the time of analysis. The Multiplicative Intensity Model of Aalen (1978), and Andersen and Gill (1992) allows the assessment of covariate effects on risk in the more general situation where subjects may experience multiple events during the period of exposure, or may experience none, in which case the time to the first event is also censored. These multiple events may be recurrent events of the same type, such as successive hospitalizations, episodes of infection, epileptic seizures, and so forth. These multiple events, however, may be times to different events, such as the time to the development of retinopathy, nephropathy or neuropathy among subjects with diabetes.

9.4.1 Cox's Proportional Hazards Models

The assumption of proportional hazards is defined such that

$$\lambda(t) = \phi\lambda_0(t) \qquad \text{and} \qquad S(t) = S_0(t)^{\phi} \tag{9.57}$$

for some real constant of proportionality ϕ. Because the hazard function is the instantaneous risk of the event over time, the hazard ratio ϕ is a measure of relative risk. Now let $\boldsymbol{x}$ be a vector of covariates measured at baseline $(t = 0)$. Later we generalize to allow for time-varying (time-dependent) covariates. A proportional hazards regression model is one where

$$\phi = h(\boldsymbol{x}, \boldsymbol{\beta}) \tag{9.58}$$

for some smooth function $h(\boldsymbol{x}, \boldsymbol{\beta})$. Since the constant of proportionality must be positive, it is convenient to adopt a *multiplicative risk model*

$$h(\boldsymbol{x}, \boldsymbol{\beta}) = e^{\boldsymbol{x}'\boldsymbol{\beta}} = e^{\sum_{j=1}^{p} x_j\beta_j} = \prod_{j=1}^{p} e^{x_j\beta_j}, \tag{9.59}$$

where the covariate effects on risk (λ) are multiplicative. The coefficient for the jth covariate is such that

$$\beta_j = \log(\phi) \text{ per } [\Delta X_j = 1] = \Delta \log[\lambda(t)] \text{ per } [\Delta X_j = 1] \tag{9.60}$$

and e^{β_j} is the log relative hazard (relative risk) per unit change in the covariate.

Such regression models can be derived parametrically, by assuming a specified form for $\lambda_0(t)$ and thus $\lambda(t|\boldsymbol{x})$. For example, Feigl and Zelen (1965) describe a multiplicative exponential regression model when the hazard is assumed to be constant over time, and Pike (1966) and Peto and Lee (1983) describe a multiplicative Weibull regression model when the hazard function can be characterized by a Weibull distribution over time.

To avoid the need to specify the shape of the hazard function, or a specific distribution, Cox (1972) proposed a *semi-parametric* proportional hazards model with multiplicative risks such that

$$\lambda(t|\boldsymbol{x}_i) = \lambda_0(t)\, e^{\boldsymbol{x}'_i\boldsymbol{\beta}} \qquad \text{and} \qquad \phi = e^{\boldsymbol{x}'_i\boldsymbol{\beta}} \quad \forall t, \tag{9.61}$$

where $\lambda_0(t)$ is an arbitrary background or nuisance hazard function interpreted as the hazard for an individual with covariate vector $\boldsymbol{x} = \boldsymbol{0}$. In this case the cumulative hazard is

$$\Lambda(t|\boldsymbol{x}_i) = \Lambda_0(t)\, e^{\boldsymbol{x}'_i\boldsymbol{\beta}} = \int_0^{t_i} \lambda_0(u)\, e^{\boldsymbol{x}'_i\boldsymbol{\beta}} du \tag{9.62}$$

with corresponding survival function

$$S(t|\boldsymbol{x}_i) = S_0(t)^{e^{\boldsymbol{x}_i'\boldsymbol{\beta}}}. \tag{9.63}$$

The background hazard function $\lambda_0(t)$ determines the shape of the background survival function $S_0(t)$ that is then expanded or shrunk through the effects of the covariates.

Assume that the observations are observed in continuous time with no tied event times. Then let $t_{(1)} < t_{(2)} < \ldots < t_{(D)}$ refer to the D unique event times. To derive estimates of the coefficients, Cox (1972) employed a *partial likelihood* motivated heuristically as follows.

Let $\boldsymbol{x}_{(j)}$ be the covariate vector for the patient who experiences the event at the jth event time $t_{(j)}$. Then the probability of that patient experiencing the event at time $t_{(j)}$, given the patient is still at risk, is

$$\lambda\left(t_{(j)}|\boldsymbol{x}_{(j)}\right) = \lambda_0\left(t_{(j)}\right) e^{\boldsymbol{x}_{(j)}'\boldsymbol{\beta}}. \tag{9.64}$$

Let $R\left(t_{(j)}\right)$ denote the *risk set*

$$R\left(t_{(j)}\right) = \{\ell : t_\ell \geq t_{(j)}\} \tag{9.65}$$

consisting of the indices of all subjects in the original cohort still at risk at time $t_{(j)}$. Then the total probability of an event occurring at $t_{(j)}$ among all subjects in the risk set is

$$\sum_{\ell \in R\left(t_{(j)}\right)} \lambda\left(t_{(j)}|\boldsymbol{x}_\ell\right) = \sum_{\ell \in R\left(t_{(j)}\right)} \lambda_0\left(t_{(j)}\right) e^{\boldsymbol{x}_\ell'\boldsymbol{\beta}}. \tag{9.66}$$

Therefore, the conditional probability of an event occurring at $t_{(j)}$ in a subject with covariate vector $\boldsymbol{x}_{(j)}$ at that time is

$$\frac{\lambda\left(t_{(j)}|\boldsymbol{x}_{(j)}\right)}{\sum_{\ell \in R\left(t_{(j)}\right)} \lambda\left(t_{(j)}|\boldsymbol{x}_\ell\right)} = \frac{e^{\boldsymbol{x}_{(j)}'\boldsymbol{\beta}}}{\sum_{\ell \in R\left(t_{(j)}\right)} e^{\boldsymbol{x}_\ell'\boldsymbol{\beta}}}. \tag{9.67}$$

This leads to a likelihood:

$$\widetilde{L}(\boldsymbol{\beta}) = \prod_{j=1}^{D} \frac{e^{\boldsymbol{x}_{(j)}'\boldsymbol{\beta}}}{\sum_{\ell \in R\left(t_{(j)}\right)} e^{\boldsymbol{x}_\ell'\boldsymbol{\beta}}}, \tag{9.68}$$

where only times at which events occur contribute to the likelihood. In fact, however, as is shown below, this is not a complete likelihood but rather is a partial likelihood, the complete likelihood involving other terms. Thus the partial likelihood is designated as $\widetilde{L}(\boldsymbol{\beta})$.

This likelihood can also be expressed in terms of individuals, rather than event times. Using the joint observations (δ_i, t_i), where δ_i is the indicator variable that denotes either the event or right censoring at time t_i, $i = 1, \ldots, N$, Cox's partial likelihood is

$$\widetilde{L}(\boldsymbol{\beta}) = \prod_{i=1}^{N} \left[\frac{e^{\boldsymbol{x}_i'\boldsymbol{\beta}}}{\sum_{\ell \in R(t_i)} e^{\boldsymbol{x}_\ell'\boldsymbol{\beta}}} \right]^{\delta_i}, \tag{9.69}$$

where only those individuals with an event ($\delta_i = 1$) contribute to the likelihood. The partial likelihood can also be expressed as

$$\widetilde{L}(\boldsymbol{\beta}) = \prod_{i=1}^{N} \left[\frac{e^{\boldsymbol{x}_i'\boldsymbol{\beta}}}{\sum_{\ell=1}^{N} Y_\ell(t_i)\, e^{\boldsymbol{x}_\ell'\boldsymbol{\beta}}} \right]^{\delta_i} = \frac{\exp\left[\sum_{i=1}^{N} \delta_i \boldsymbol{x}_i'\boldsymbol{\beta}\right]}{\prod_{i=1}^{N} \left[\sum_{\ell=1}^{N} Y_\ell(t_i)\, e^{\boldsymbol{x}_\ell'\boldsymbol{\beta}}\right]^{\delta_i}}, \tag{9.70}$$

where $Y(t)$ refers to a time-dependent at risk indicator variable to denote being at risk at time t. In the analysis of survival times, or the time of a single event,

$$Y_i(u) = \left\{ \begin{matrix} 1 : u \leq t_i \\ 0 : u > t_i \end{matrix} \right\}, \qquad u \geq 0. \tag{9.71}$$

To see why Cox's likelihood is a partial likelihood, it is useful to examine the full model and its partition into two partial likelihoods as described by (Johansen, 1983). From (9.57) and (9.61), the full likelihood under random censoring in (9.6) is

$$\begin{aligned} L[\boldsymbol{\beta}, \lambda_0(t)] &= \prod_{i=1}^{N} \lambda(t_i|\boldsymbol{x}_i)^{\delta_i} S(t_i|\boldsymbol{x}_i) \qquad (9.72) \\ &= \prod_{i=1}^{N} \left[\lambda_0(t_i)\, e^{\boldsymbol{x}_i'\boldsymbol{\beta}}\right]^{\delta_i} \left[\exp\{-\int_0^{t_i} \lambda_0(u)\, e^{\boldsymbol{x}_i'\boldsymbol{\beta}} du\}\right] \\ &= \prod_{i=1}^{N} \left[\lambda_0(X_i)\, e^{\boldsymbol{x}_i'\boldsymbol{\beta}}\right]^{\delta_i} \prod_{i=1}^{N} \left[\exp\{-\int_0^{\infty} Y_i(u)\, \lambda_0(u)\, e^{\boldsymbol{x}_i'\boldsymbol{\beta}} du\}\right] \end{aligned}$$

in terms of the at risk indicator process $Y(u)$ for each subject. Then let

$$B(u) = \sum_{\ell=1}^{N} Y_\ell(u)\, e^{\boldsymbol{x}_\ell'\boldsymbol{\beta}}, \qquad u \geq 0. \tag{9.73}$$

Multiplying and dividing by $B(u)$, and rearranging terms, the likelihood becomes

$$\begin{aligned} L[\boldsymbol{\beta}, \lambda_0(t)] &= \prod_{i=1}^{N} \left[\frac{e^{\boldsymbol{x}_i'\boldsymbol{\beta}}}{B(t_i)}\right]^{\delta_i} \prod_{i=1}^{N} [B(t_i)\, \lambda_0(t_i)]^{\delta_i} \exp\left[-\int_0^{\infty} B(u)\, \lambda_0(u)\, du\right] \qquad (9.74) \\ &= \widetilde{L}_{(1)}(\boldsymbol{\beta})\, \widetilde{L}_{(2)}[\boldsymbol{\beta}, \lambda_0(t)], \end{aligned}$$

where $\widetilde{L}_1(\boldsymbol{\beta})$ is Cox's partial likelihood $\widetilde{L}(\boldsymbol{\beta})$ in (9.70) and $\widetilde{L}_{(2)}[\boldsymbol{\beta}, \lambda_0(t)]$ is the remaining partial likelihood involving both the coefficients $\boldsymbol{\beta}$ and the baseline hazard function $\lambda_0(t)$.

The technical difficulty is to demonstrate that a solution to the estimating equation based on the partial likelihood $\widetilde{L}_{(1)}(\boldsymbol{\beta})$ provides estimates with the same properties as those of an MLE that maximizes a full likelihood. Cox (1972) simply applied the theory of maximum likelihood to the partial likelihood to provide estimates of the coefficients and to describe their asymptotic distribution, as though the associated score vector and information matrix were obtained from a full likelihood. However, this was received with some skepticism. Kalbfleisch and Prentice (1973) showed that Cox's partial likelihood could be obtained as a marginal likelihood when there were no tied observations. Cox (1975) provided a justification for the validity of inferences using a conditional likelihood argument. Tsiatis (1981) and others, under specific assumptions showed that the partial likelihood score vector is asymptotically normally distributed with mean zero and covariance matrix equal to the partial likelihood information function, and thus that the partial maximum likelihood estimates were also asymptotically normally distributed like those based on a full model. However, a rigorous justification was also provided using counting process methods by Andersen and Gill (1982) in the context of the more general multiplicative intensity model that generalizes the Cox model to recurrent events; see also Gill (1984).

There are many generalizations of the basic proportional hazards model. A few of the most important are now described.

9.4.2 Stratified Models

Assume that the sample of N observations is divided among S mutually exclusive strata, and that there is a stratum-specific background hazard within each stratum. If we also assume that the covariate effects on the relative risks are homogeneous over strata, then for the hth stratum the stratified PH model specifies that

$$\lambda_h(t|\boldsymbol{x}) = \lambda_{0h}(t)\, e^{\boldsymbol{x}'\boldsymbol{\beta}} \qquad h = 1, \ldots, S, \tag{9.75}$$

where $\lambda_{0h}(t)$ may vary over strata. For example, if we wished to assess the effects of age and weight on the risk of death, it might be appropriate to consider a model stratified by gender, where the background hazards would be allowed to differ for men and women, but where the effects of age and weight on relative risks are assumed to be the same for men and women.

A more general model arises when the covariate effects are also assumed to be heterogeneous over strata, in which case the model specifies that

$$\lambda_h(t|\boldsymbol{x}) = \lambda_{0h}(t)\, e^{\boldsymbol{x}'\boldsymbol{\beta}_h} \qquad h = 1, \ldots, S, \tag{9.76}$$

where the $\boldsymbol{\beta}_h$ also differ among strata. Thus the background hazards differ among strata as well as the constant of proportionality for each covariate. In this more general case, the stratified PH model partial likelihood is

$$\widetilde{L}_i(\boldsymbol{\beta}_1, \ldots, \boldsymbol{\beta}_S) = \prod_{h=1}^{S} \prod_{i=1}^{N_h} \left[\frac{e^{\boldsymbol{x}'_{i(h)}\boldsymbol{\beta}_h}}{\sum_{\ell \in R_h(t_{i(h)})} e^{\boldsymbol{x}'_{\ell(h)}\boldsymbol{\beta}_h}} \right]^{\delta_{i(h)}}, \tag{9.77}$$

where $i(h)$ denotes the ith subject within the hth stratum and where the risk set is defined only among the N_h observations in the hth stratum: $R_h\left(t_{i(h)}\right) = \{\ell \in stratum\ h : t_{\ell(h)} \geq t_{i(h)}\}$.

This stratified model implies that there is an interaction between the stratification variable and the background hazard function and with the other covariates that enter into the model through $\boldsymbol{x}'\boldsymbol{\beta}_h$. When the coefficients are homogeneous over stratum as in (9.75), then it is assumed that $\boldsymbol{\beta}_h = \boldsymbol{\beta}$ for all strata, which yields a slight simplification of the partial likelihood. In this case, there is an interaction between stratum and the background hazard only. Thall and Lachin (1986) describe these and other forms of covariate interactions in the PH model.

Stratified models are useful when the proportional hazards assumption does not apply in the aggregate population, but it does apply within strata. For example, when the relative risks differ within intervals of time, so that the proportional hazards assumption does not apply in general, it may still apply within separate intervals of time, such as $0 < t \leq 1$, $1 < t \leq 2$, and so forth. In this case a stratified model as in (9.77) could allow for proportional hazards within time strata with differences in covariate effects on risk among the different time strata.

9.4.3 Time-Dependent Covariates

In some instances the covariate process may vary over time, in which case the covariate is termed time-dependent. Since the hazard function is the instantaneous probability of the event at time t given exposure to time t, or being at risk at time t, any aspect of the history of the covariate process up to time t may be incorporated into the model to assess the effect of the covariate on the relative risk of the event over time. Technically, a time-dependent covariate is assumed to be observable up to, but not including, t itself, or $X(t) \in \Im_{t-}$, which represents the history of the process up to the instant before time t, designated as t^-. Then the time-dependent PH model is of the form

$$\lambda\left[t|\boldsymbol{x}(t)\right] = \lambda_0(t)\, e^{\boldsymbol{x}(t)'\boldsymbol{\beta}}, \tag{9.78}$$

where the jth coefficient now represents

$$\beta_j = \log(\phi) \text{ per } \left[\Delta X_j(t) = 1\right] = \Delta \log\left\{\lambda\left[t|X_j(t)\right]\right\} \text{ per } \left[\Delta X_j = 1\right]. \tag{9.79}$$

For a non-time-dependent covariate, the value over time is the same as that at baseline as employed in (9.68). With a time-dependent covariate vector, the basic partial likelihood then becomes

$$\widetilde{L}(\boldsymbol{\beta}) = \prod_{i=1}^{N} \left[\frac{e^{\boldsymbol{x}_i(t_i)'\boldsymbol{\beta}}}{\sum_{\ell \in R(t_i)} e^{\boldsymbol{x}_\ell(t_i)'\boldsymbol{\beta}}} \right]^{\delta_i} . \tag{9.80}$$

To fit this model then requires that the time-dependent covariates be evaluated for all subjects at risk at each event time.

9.4.4 Fitting the Model

Consider the non-stratified model with a possibly time-dependent covariate vector with no tied event times for which the partial likelihood is (9.80). Then the jth element of the score vector $\boldsymbol{U}(\boldsymbol{\beta})$ corresponding to coefficient β_j is

$$U(\boldsymbol{\beta})_{\beta_j} = \frac{\partial \log \widetilde{L}(\boldsymbol{\beta})}{\partial \beta_j} = \sum\nolimits_{i=1}^{N} \delta_i \left[x_{ij}(t_i) - \overline{x}_j(t_i, \boldsymbol{\beta}) \right], \tag{9.81}$$

where $E\left[x_{ij}(t_i) \mid \boldsymbol{\beta}\right] = \overline{x}_j(t_i, \boldsymbol{\beta})$ and

$$\overline{x}_j(t_i, \boldsymbol{\beta}) = \left[\frac{\sum_{\ell \in R(t_i)} x_{\ell j}(t_i) e^{\boldsymbol{x}_\ell(t_i)'\boldsymbol{\beta}}}{\sum_{\ell \in R(t_i)} e^{\boldsymbol{x}_\ell(t_i)'\boldsymbol{\beta}}} \right] = \left[\frac{\sum_{\ell=1}^{N} Y_\ell(t_i) x_{\ell j}(t_i) e^{\boldsymbol{x}_\ell(t_i)'\boldsymbol{\beta}}}{\sum_{\ell=1}^{N} Y_\ell(t_i) e^{\boldsymbol{x}_\ell(t_i)'\boldsymbol{\beta}}} \right] \tag{9.82}$$

is the weighted average of the covariate among all those at risk at time t_i. Thus the score equation is of the form $\sum$(observed - expected) expressed in terms of the covariate values for the ith subject who experiences the event at time t_i. The contribution of each subject to the total score is also known as the Schoenfeld (1982) or score residual.

The corresponding information matrix, where $\boldsymbol{i}(\boldsymbol{\beta}) = \boldsymbol{I}(\boldsymbol{\beta})$, then has elements

$$\boldsymbol{I}(\boldsymbol{\beta})_{jk} = E\left[\frac{-\partial \log \widetilde{L}(\boldsymbol{\beta})}{\partial \beta_j \partial \beta_k} \right] = \sum\nolimits_{i=1}^{N} \delta_i \left[C_{ijk}(t_i, \boldsymbol{\beta}) - \overline{x}_j(t_i, \boldsymbol{\beta}) \overline{x}_k(t_i, \boldsymbol{\beta}) \right] \tag{9.83}$$

with

$$C_{ijk}(t_i, \boldsymbol{\beta}) = \left[\frac{\sum_{\ell \in R(t_i)} x_{\ell j}(t_i) x_{\ell k}(t_i) e^{\boldsymbol{x}_\ell(t_i)'\boldsymbol{\beta}}}{\sum_{\ell \in R(t_i)} e^{\boldsymbol{x}'_\ell(t_i)\boldsymbol{\beta}}} \right] \tag{9.84}$$

$$= \left[\frac{\sum_{\ell=1}^{N} Y_\ell(t_i) x_{\ell j}(t_i) x_{\ell k}(t_i) e^{\boldsymbol{x}_\ell(t_i)'\boldsymbol{\beta}}}{\sum_{\ell=1}^{N} Y_\ell(t_i) e^{\boldsymbol{x}_\ell(t_i)'\boldsymbol{\beta}}} \right]$$

for $1 \leq j \leq k \leq p$. The model is then fit using the Newton-Raphson iteration.

In a stratified model with homogeneous covariate effects among strata ($\boldsymbol{\beta}_h = \boldsymbol{\beta}\ \forall h$) as in (9.75), the score equation for the jth coefficient is simply $U(\boldsymbol{\beta})_{\beta_j} = \sum_{h=1}^{S} U_h(\boldsymbol{\beta})_{\beta_j}$, where the $U_h(\boldsymbol{\beta})_{\beta_j}$ is the score equation for the jth covariate evaluated with respect to those subjects within the hth stratum only, $h = 1, \ldots, S$. Likewise, using summations over subjects within strata, the covariate means within the hth strata $\{\overline{x}_{j(h)}(t_{i(h)}, \boldsymbol{\beta})\}$ are computed with respect to the risk set comprising subjects within that stratum $R_h(t_{i(h)})$. The information matrix then has elements $\boldsymbol{I}(\boldsymbol{\beta})_{jk} = \sum_{h=1}^{S} \boldsymbol{I}_h(\boldsymbol{\beta})_{jk}$, where $\boldsymbol{I}_h(\boldsymbol{\beta})$ is computed from the subjects in the hth stratum only.

In a stratified model with stratum-specific covariate effects $\boldsymbol{\beta}_h$ for $h = 1, \ldots, S$ as in (9.76), the score vector contains Sp elements

$$\boldsymbol{U}(\boldsymbol{\beta}_1, \ldots, \boldsymbol{\beta}_S) = \left[\boldsymbol{U}_1(\boldsymbol{\beta}_1)^T \parallel \ldots \parallel \boldsymbol{U}_S(\boldsymbol{\beta}_S)^T\right]^T, \tag{9.85}$$

where $\boldsymbol{U}_h(\boldsymbol{\beta}_h)$ is computed as in (9.81) among the subjects within the hth stratum. Since the strata are independent, the information matrix is a $Sp \times Sp$ block diagonal matrix of the form

$$\boldsymbol{I}(\boldsymbol{\beta}_1, \ldots, \boldsymbol{\beta}_S) = blockdiag\left[\boldsymbol{I}_1(\boldsymbol{\beta}_1) \ \ldots \ \boldsymbol{I}_S(\boldsymbol{\beta}_S)\right]. \tag{9.86}$$

The information matrix for the hth stratum $\boldsymbol{I}_h(\boldsymbol{\beta}_h)$ has elements $\boldsymbol{I}_h(\boldsymbol{\beta}_h)_{jk}$ evaluated with respect to the subjects within that stratum.

It is also straightforward to further generalize the above to the case with stratum-specific effects for some covariates and homogeneous effects for others.

Although the model is based on a partial likelihood, the three general families of tests can be employed as for any other likelihood-based model: the Wald test, efficient scores test or the likelihood ratio test. As a problem it is readily shown that the Mantel-logrank test is the efficient scores test for the treatment group effect in the Cox model using the Peto-Breslow adjustment for ties (see Section 9.4.6.4).

9.4.5 Robust Inference

Inferences using the Cox model are dependent on the proportional hazards assumption that may not apply. In this case, one may still be interested in the Cox model coefficients as estimates of log *average* hazard ratios over time, even though a constant hazards ratio does not apply. The model-based standard errors and tests of significance in this case are possibly biased because of the departure from the proportional hazards assumption.

Gail, Wieand and Piantidosi (1984) and Lagakos (1988), among others, have also shown that the estimates of the model coefficients in a Cox model are biased when important covariates are omitted from the model, even when the covariates are independent of each other. Whereas the coefficients in a simple exponential model are not biased by the omission of an independent relevant covariate, as shown in Problem 7.6.3, those in the Cox model are biased because of the presence of censoring, and because the estimates are obtained from a partial likelihood rather than the full likelihood. Thus model-based inferences are also biased when an important covariate is omitted from the Cox model.

The Cox model also assumes that the same background hazard function applies to all subjects in the population. A generalization of this assumption leads to a *frailty model,* in which it is assumed that the background hazard function has some distribution between subjects in the population (Clayton and Cuzick, 1985; see also Andersen, Borgan, Gill and Keiding, 1993). This is similar to a regression model such as logistic or Poisson regression, where it is assumed that the intercept in the model is distributed randomly in the population. Whereas the robust sandwich estimator will account for overdispersion in a logistic or Poisson regression model, it does not account for frailty in the Cox model. The reason is that under a frailty model, the coefficients are distinct from those in the marginal Cox model.

In cases of model misspecification (other than frailty), proper estimates of the covariance matrix of the coefficient estimates, and related confidence intervals and statistical tests can be provided by the robust information sandwich (see Section A.9 of the Appendix). The score vector for the ith subject is

$$\boldsymbol{U}_i(\boldsymbol{\beta}) = \frac{\partial \log \widetilde{L}_i(\boldsymbol{\beta})}{\partial \boldsymbol{\beta}} = \delta_i \left[\boldsymbol{x}_i(t_i) - \overline{\boldsymbol{x}}(t_i, \boldsymbol{\beta})\right], \tag{9.87}$$

where

$$\overline{\boldsymbol{x}}(t_i, \boldsymbol{\beta}) = \left[\frac{\sum_{j=1}^{N} Y_j(t_i)\boldsymbol{x}_j(t_i)e^{\boldsymbol{x}_j(t_i)'\boldsymbol{\beta}}}{\sum_{j=1}^{N} Y_j(t_i)e^{\boldsymbol{x}_j(t_i)'\boldsymbol{\beta}}}\right] = \left[\frac{\boldsymbol{S}_1(\boldsymbol{\beta}, t_i)}{\boldsymbol{S}_0(\boldsymbol{\beta}, t_i)}\right] = \widehat{E}\left[\boldsymbol{x}(t_i) \mid \boldsymbol{\beta}\right]. \tag{9.88}$$

However, because the proportional hazards model is based on a partial rather than a full likelihood, the score vectors are not independently and identically distributed with expectation zero. Thus Lin and Wei (1989; see also Lin 1994), describe a modification to the information sandwich that provides a consistent estimate of the covariance matrix of the coefficient estimates under model misspecification. Lin and Wei (1989) show that

$$\boldsymbol{W}_i(\boldsymbol{\beta}) = \boldsymbol{U}_i(\boldsymbol{\beta}) - \widehat{E}\left[\boldsymbol{U}_i(\boldsymbol{\beta})\right] \tag{9.89}$$

$$\widehat{E}\left[\boldsymbol{U}_i(\boldsymbol{\beta})\right] = \sum_{\ell=1}^{N} \frac{\delta_\ell Y_i(t_\ell) e^{\boldsymbol{x}_i(t_\ell)'\boldsymbol{\beta}}}{\boldsymbol{S}_0(\boldsymbol{\beta}, t_\ell)} \left(\boldsymbol{x}_i(t_\ell) - \overline{\boldsymbol{x}}(t_\ell, \boldsymbol{\beta})\right),$$

where $\widehat{E}\left[\boldsymbol{U}_i(\boldsymbol{\beta})\right]$ is a weighted average of the scores of all subjects with events prior to time t_i with respect to the covariate value for the ith subject at that time. Then, the $\{\boldsymbol{W}_i(\boldsymbol{\beta})\}$ are $i.i.d.$

Thus the robust estimator of the covariance matrix of the score vector is

$$\widehat{V}\left[\boldsymbol{U}(\boldsymbol{\beta})\right] = \widehat{\boldsymbol{J}}\left(\widehat{\boldsymbol{\beta}}\right) = \sum_{i=1}^{N} \boldsymbol{W}_i\left(\widehat{\boldsymbol{\beta}}\right) \boldsymbol{W}_i\left(\widehat{\boldsymbol{\beta}}\right)' \tag{9.90}$$

and the robust estimate of the covariance matrix of the coefficient vector is

$$\widehat{\boldsymbol{\Sigma}}_R\left(\widehat{\boldsymbol{\beta}}\right) = \boldsymbol{I}\left(\widehat{\boldsymbol{\beta}}\right)^{-1} \widehat{\boldsymbol{J}}\left(\widehat{\boldsymbol{\beta}}\right) \boldsymbol{I}\left(\widehat{\boldsymbol{\beta}}\right)^{-1}. \tag{9.91}$$

This provides robust confidence limits for the parameter estimates and robust Wald and score tests for the parameters. By simulation, Lin and Wei (1989) show that the model-based Wald and score tests may have markedly inflated size when important covariates have been inadvertently omitted from the model, whereas the robust estimates retain their nominal size.

Wei, Lin and Weissfeld (1989) and Lee, Wei and Amato (1992) generalized the Cox model to the analysis of multiple event times. They used a generalization of the information sandwich to provide estimates of the covariance matrix of the coefficient estimates for each of the event times simultaneously. These generalizations are provided by PROC PHREG. The program does not compute the robust estimate of the covariance matrix of the coefficients directly; however, it can be computed from by an additional routine as follows: Useful measures of the influence of an observation on the fitted model are the leave-one-out estimates of the regression coefficients, called the DFBETA$_i$ $= \Delta\widehat{\boldsymbol{\beta}}_i = \widehat{\boldsymbol{\beta}} - \widehat{\boldsymbol{\beta}}_{(i)}$, where $\widehat{\boldsymbol{\beta}}_{(i)}$ is the vector of coefficient estimates obtained by deleting the ith observation when fitting the model. Cain and Lange (1984) showed that

$$\Delta\widehat{\boldsymbol{\beta}}_i = \boldsymbol{I}\left(\widehat{\boldsymbol{\beta}}\right)^{-1} \boldsymbol{W}_i\left(\widehat{\boldsymbol{\beta}}\right). \tag{9.92}$$

Thus the robust covariance matrix can be obtained from the DFBETA statistics as

$$\widehat{\boldsymbol{\Sigma}}_R\left(\widehat{\boldsymbol{\beta}}\right) = \sum\nolimits_{i=1}^{N} \left(\Delta\widehat{\boldsymbol{\beta}}_i\right)' \left(\Delta\widehat{\boldsymbol{\beta}}_i\right) \tag{9.93}$$

and the variance of the scores as

$$\widehat{V}\left[\boldsymbol{U}(\boldsymbol{\beta})\right] = \widehat{\boldsymbol{J}}\left(\widehat{\boldsymbol{\beta}}\right) = \boldsymbol{I}\left(\widehat{\boldsymbol{\beta}}\right) \widehat{\boldsymbol{\Sigma}}_R\left(\widehat{\boldsymbol{\beta}}\right) \boldsymbol{I}\left(\widehat{\boldsymbol{\beta}}\right). \tag{9.94}$$

A robust model score test of H_0: $\boldsymbol{\beta} = \boldsymbol{\beta}_0 = \mathbf{0}$ may be computed by evaluating the score vector $\boldsymbol{U}_i(\boldsymbol{\beta}_0)$, the centered scores $\boldsymbol{W}_i(\boldsymbol{\beta}_0)$, and the matrix $\boldsymbol{J}(\boldsymbol{\beta}_0)$ under the null hypothesis. From (9.89) the centered score vector for the ith subject evaluated under H_0 is

$$\boldsymbol{W}_i(\boldsymbol{\beta}_0) = \delta_i\left[\boldsymbol{x}_i(t_i) - \overline{\boldsymbol{x}}(t_i)\right] - \sum\nolimits_{\ell=1}^{N} \frac{\delta_\ell Y_i(t_\ell)}{n(t_\ell)} \left(\boldsymbol{x}_i(t_\ell) - \overline{\boldsymbol{x}}(t_\ell)\right), \tag{9.95}$$

where $\overline{\boldsymbol{x}}(t)$ is the unweighted mean vector of covariate values among the $n(t)$ subjects at risk at time t. The robust covariance matrix under H_0 is then obtained as in (9.90) by $\widehat{\boldsymbol{J}}(\boldsymbol{\beta}_0) = \sum_{i=1}^{N} \boldsymbol{W}_i(\boldsymbol{\beta}_0)\, \boldsymbol{W}_i(\boldsymbol{\beta}_0)'$. Using the total score vector $\boldsymbol{W}(\boldsymbol{\beta}_0) = \sum_i \boldsymbol{W}_i(\boldsymbol{\beta}_0)$, the robust model score test is then provided by $X^2 = \boldsymbol{W}(\boldsymbol{\beta}_0)'\, \widehat{\boldsymbol{J}}(\boldsymbol{\beta}_0)^{-1}\, \boldsymbol{W}(\boldsymbol{\beta}_0)$, which asymptotically is distributed as chi-square on p df under H_0. Similarly robust score tests may be obtained for the individual parameters of the model, although the computations are more tedious; see Section A.7.3. of the Appendix.

9.4.6 Adjustments for Tied Observations

The above presentation of the Cox PH model assumes that no two subjects experience the event at the same time, or that there are no tied event times, such as where multiple subjects die on the same study day. One crude and unsatisfactory option in this instance is to arbitrarily (by chance) break the ties and order them. With one or two ties, and a large sample size, this will have a trivial effect on the analysis. However, when there are many ties, or ties arise because of the manner in which the observations are recorded, the analysis should include an appropriate adjustment for ties.

9.4.6.1 Discrete and Grouped Failure Time Data The simplest structure for a model that allows tied event times is to assume that events can only occur at fixed times $\tau_1 < \tau_2 < \ldots < \tau_K$. However, this is unrealistic because almost always events occur continuously over time but may only be *observed* within grouped intervals of time, where the jth interval includes all times $A_j = [\tau_{j-1}, \tau_j)$. For example, in many studies, an examination or procedure must be performed to determine whether an event has occurred, in which case events during the jth interval will only be observed to have occurred on the examination conducted at τ_j. This structure assumes that *fixed intervals* apply to all subjects, or that all subjects are examined at the fixed times $\{\tau_j\}$. When the observation times vary from subject to subject, then the observations are *interval censored*, in which case different models will apply (see below). In some cases, however, a fixed interval model will apply approximately to interval censored data.

Prentice and Gloeckler (1978) describe the generalization of the proportional hazards model to such discrete or grouped time data. The continuous time proportional hazards model specifies that the survival probabilities satisfy the relationship $S(\tau_j|\boldsymbol{x}_i) = S_0(\tau_j)^{\exp(\boldsymbol{x}_i'\boldsymbol{\beta})}$ at any time τ_j, where $S_0(\tau_j)$ is the background survival function for a subject with covariate vector $\boldsymbol{x} = \mathbf{0}$. In the grouped time model,

$$S_0(\tau_j) = \exp\left[-\int_0^{\tau_j} \lambda_0(s)ds\right], \tag{9.96}$$

however, no information is provided regarding the form of the underlying hazard function $\lambda_0(\tau_j)$. In either the grouped or discrete time model, the background conditional probability of a subject with covariate vector $\boldsymbol{x} = \mathbf{0}$ surviving interval A_j is

$$\varphi_{0j} = \frac{S_0(\tau_j)}{S_0(\tau_{j-1})} = \exp\left[-\int_{\tau_{j-1}}^{\tau_j} \lambda_0(s)ds\right] \tag{9.97}$$

and the conditional probability that the event is observed at τ_j is $1 - \varphi_{0j}$. Note that φ_{0j} is analogous to the continuation probability $1 - \pi_j$ in (9.9) in the Kaplan-Meier construction. Under the proportional hazards model, it follows that $\varphi_{j|\boldsymbol{x}} = \varphi_{0j}{}^{\exp(\boldsymbol{x}'\boldsymbol{\beta})}$.

To construct the likelihood, let a_i denote the final interval during which the ith subject was observed to be at risk, $a_i \in (1, \ldots, K+1)$, where the last possible interval is $A_{K+1} = [\tau_K, \tau_{K+1})$ with $\tau_{K+1} = \infty$; and let δ_i be the indicator variable to denote event ($\delta_i = 1$) or censoring ($\delta_i = 0$). Any subject censored during the jth interval ($a_i = j$, $\delta_i = 0$) is assumed not to be at risk during that interval such that any observations event free and under follow-up at the end of the study are right censored after τ_K. Thus only censored observations have values $a_i = K + 1$. From (9.6) the likelihood function for a sample of N observations under a random censoring model is

$$L(\boldsymbol{\theta}) = \prod_{i=1}^{N} \prod_{j=1}^{K+1} \left[I_{\{a_i=j\}} \left(1 - \varphi_{0j}^{\exp(\boldsymbol{x}_i'\boldsymbol{\beta})} \right)^{\delta_i} \right] \left[I_{\{a_i<j\}} \left(\varphi_{0j}^{\exp(\boldsymbol{x}_i'\boldsymbol{\beta})} \right) \right] \tag{9.98}$$
$$= \prod_{i=1}^{N} \left(1 - \varphi_{0(a_i)}^{\exp(\boldsymbol{x}_i'\boldsymbol{\beta})} \right)^{\delta_i} \prod_{j=1}^{a_i-1} \left(\varphi_{0j}^{\exp(\boldsymbol{x}_i'\boldsymbol{\beta})} \right)$$

for $\boldsymbol{\theta} = (\varphi_{01} \ \dots \ \varphi_{0K} \ \beta_1 \ \dots \ \beta_p)^T$.

Because the $\{\varphi_{0j}\}$ are probabilities, Prentice and Gloeckler (1978) use the complimentary log(-log) link such that

$$\gamma_j = \log[-\log(\varphi_{0j})], \qquad \varphi_{0j} = e^{-e^{\gamma_j}} \tag{9.99}$$

for $1 < j < K$. Substitution into the above yields estimating equations for the parameters $\boldsymbol{\theta} = (\gamma_1 \ \dots \ \gamma_K \ \beta_1 \ \dots \ \beta_p)^T$.

Note that the estimation of the relative risk coefficients requires joint estimation of the nuisance parameters ($\gamma_1 \ \dots \ \gamma_K$). For finite samples, as the number of such parameters (intervals) increases, the bias of the coefficient estimates also increases (*cf.* Cox and Hinkley, 1974).

The resulting model is then fit by maximum likelihood estimation. Whitehead (1989) shows that this model can be fit using PROC GENMOD or a program for generalized linear models that allows for a binomial distribution with a complimentary log(-log) link as described in Section 7.1.5. An example is also provided in the SAS (1997) description of PROC LOGISTIC.

9.4.6.2 Cox's Adjustment for Ties In practice, even when the event-time distribution is continuous, ties may be caused by coarsening of the time measures, such as measuring time to the day or the week. For such cases, adjustments to the Cox PH model are employed to allow for ties. When there are tied event times, Cox (1972) suggested a discrete logistic model conditional on covariates x for some time interval of length dt of the form

$$\frac{\lambda(t|\boldsymbol{x})\,dt}{1 - \lambda(t|\boldsymbol{x})\,dt} = \frac{\lambda_0(t)\,dt}{1 - \lambda_0(t)\,dt} e^{\boldsymbol{x}'\boldsymbol{\beta}}. \tag{9.100}$$

Therefore, $\lim dt \downarrow 0$ yields the PH model for continuous time in (9.61) since $\lim_{dt\downarrow 0} [1 - \lambda(t)\,dt] \rightarrow 1.0$. In Problem 9.14.12 it is also shown that the PH model results from a grouped time logistic model, as above, where $K \rightarrow \infty$. If we let

$$\alpha = \log\left[\frac{\lambda_0(t)\,dt}{1 - \lambda_0(t)\,dt}\right], \tag{9.101}$$

then

$$\frac{\lambda(t|\boldsymbol{x})\,dt}{1-\lambda(t|\boldsymbol{x})\,dt}=e^{\alpha+\boldsymbol{x}'\boldsymbol{\beta}}. \tag{9.102}$$

Therefore, we have a logistic model of the form

$$\lambda(t|\boldsymbol{x})\,dt=\frac{e^{\alpha+\boldsymbol{x}'\boldsymbol{\beta}}}{1+e^{\alpha+\boldsymbol{x}'\boldsymbol{\beta}}}=\psi(\boldsymbol{x}). \tag{9.103}$$

Now, consider a set of N small but finite intervals of time. During the jth interval assume that n_j subjects are followed (at risk) of whom $d_j=m_{1j}$ subjects experience the outcome event and $n_j-d_j=m_{2j}$ subjects survive the interval event-free, where m_{1j} and m_{2j} are the numbers with and without the event in the notation of matched sampling used in Section 7.6. Let $\boldsymbol{x}_{jk}$ denote the covariate vector for the kth subject to have the event at time $t_{(j)}$. Then conditioning on m_{1j} events and m_{2j} non-events during the jth interval, the conditional likelihood is

$$L(\boldsymbol{\beta})_{j|m_{1j},n_j}=\frac{\prod_{k=1}^{m_{1j}}\psi(\boldsymbol{x}_{jk})\prod_{k=m_{1j}+1}^{n_j}\left[1-\psi(\boldsymbol{x}_{jk})\right]}{\sum_{\ell=1}^{\binom{n_j}{m_{1j}}}\prod_{k(\ell)=1}^{m_{1j}}\psi\left(\boldsymbol{x}_{jk(\ell)}\right)\prod_{k(\ell)=m_{1j}+1}^{n_j}\left[1-\psi\left(\boldsymbol{x}_{jk(\ell)}\right)\right]} \tag{9.104}$$

$$=\frac{\prod_{k=1}^{m_{1j}}e^{\boldsymbol{x}'_{jk}\boldsymbol{\beta}}}{\sum_{\ell=1}^{\binom{n_j}{m_{1j}}}\prod_{k(\ell)=1}^{m_{1j}}e^{\boldsymbol{x}'_{jk(\ell)}\boldsymbol{\beta}}}.$$

Because the successive intervals are conditionally independent, the total conditional likelihood is

$$L_c(\boldsymbol{\beta})=\prod_{j=1}^{N}L(\boldsymbol{\beta})_{j|m_{1j},n_j}. \tag{9.105}$$

This likelihood is equivalent to that of the conditional logistic regression model for matched sets of Section 7.6.1. The score equations and the expressions for the information function are presented in (7.118)–(7.120).

This model is appropriate when there is some natural grouping of the event times. Thus this model is appropriate for the instance described in Example 9.2, where the occurrence of the event can only be ascertained at fixed times during the study. One could also use the Prentice-Gloeckler grouped time model for such data.

9.4.6.3 Kalbfleisch-Prentice Marginal Model Kalbfleisch and Prentice (1973; see also Kalbfleisch and Prentice 1980), showed that Cox's partial likelihood with no ties could also be expressed as a marginal likelihood obtained by integrating out the background hazard function. For tied observations, they provide an expression for the corresponding marginal likelihood that is somewhat more computationally intensive than Cox's logistic likelihood.

9.4.6.4 Peto-Breslow Adjustment for Ties A computationally simple approach was suggested by Peto (1972) and Breslow (1974) as an approximation to a precise model allowing for ties. Breslow (1974) showed that this provides an approximation to the marginal likelihood of Kalbfleisch and Prentice (1973) adjusting for ties. Again, let $\{t_{(j)}\}$ denote the set of J distinct event times among the total of D patients who experience the event, $J < D$. The events are assumed to occur in continuous time but some ties are observed because of rounding or grouping of the times into small intervals, such that the number of tied event times is small relative to the total number of events D. Then let d_j denote the number of events observed among those at risk at time $t_{(j)}$. Generalizing (9.68), the approximate likelihood is

$$\widetilde{L}_{PB}(\boldsymbol{\beta}) = \prod_{j=1}^{J} \frac{\exp\left(\sum_{k=1}^{d_j} \boldsymbol{x}'_{jk}\boldsymbol{\beta}\right)}{\left(\sum_{\ell \in R(t_j)} e^{\boldsymbol{x}'_\ell \boldsymbol{\beta}}\right)^{d_j}}, \tag{9.106}$$

where $\boldsymbol{x}_{jk}$ denotes the covariate vector for the kth subject to experience the event at time $t_{(j)}$, $1 \leq k \leq d_j$; and where the d_j subjects with the event at $t_{(j)}$ are included in the risk set at that time. This likelihood can also be expressed as a product over individuals as in (9.69). Thus the above equations (9.81)–(9.84) and the expressions for the robust information matrix in Section 9.4.5 also apply to the Peto-Breslow approximate likelihood with tied observations.

9.4.6.5 Interval Censoring Ties may also be caused by interval censoring or grouping. Rather than observing an event time, we only observe the boundaries of an interval of time within which an event occurred. For an individual known to experience the event we then know that $t_i \in (a_i, b_i]$, where the subject was known to be at risk and event free at time a_i and to have had the event at some time up to time b_i. Observations that are right censored then have the associated interval $t_i \in (a_i, \infty)$. Turnbull (1976) describes a method for estimation of the underlying hazard and survival functions for such data. Finklestein (1986) describes a generalization of the proportional hazards model for such data. Younes and Lachin (1997) describe a family of models that includes the proportional hazards and proportional odds models for interval censored observations, and for mixtures of simple right-censored and interval censored observations. All of these methods are computer intensive and will not be considered further.

9.4.7 Model Assumptions

Consider a model with a single possibly time-dependent covariate $X(t)$. The principal assumption of the proportional hazards model is that the effect of $X(t)$ on the background hazard function over time is described by the constant of proportionality $\phi = e^{x(t)\beta}$. To test this assumption, Cox (1972) proposed a test of H_0: $\phi[t|x(t)] = e^{x(t)\beta}\ \forall t$ of a constant hazard ratio over time against the alternative H_1: $\phi[t|x(t)] = h(t) \neq e^{x(t)\beta}$ that the relative hazard is a monotonically increasing or decreasing function of time $h(t)$ such as where $h(t) = \log(t)$. This specific alternative implies that the true model includes an interaction term between the covariate and $h(t)$ such as $x(t)\beta_1 + x(t)h(t)\beta_2$. Note that no term for $h(t)$ is required in the exponent because this term, if added, would be absorbed into the background hazard function $\lambda_0(t)$. A test of H_0: $\beta_2 = 0$ then provides a test of the PH assumption for that covariate in the model against the specific alternative that the hazard ratio $\phi\left[t|x(t)\right]$ is assumed to be log linear in $h(t)$.

However, this is a test of a specific mode of departure from the PH model assumption. Numerous authors have proposed alternate assessments of this assumption, many based on graphical assessments. Lin and Wei (1991) present a review of these methods. Among the simplest is the following. Assume that we wish to assess the proportionality assumption for covariate Z when added to the covariate vector $\boldsymbol{X}$, where $\phi(t|z, \boldsymbol{x}) = \exp(z\gamma + \boldsymbol{x}'\boldsymbol{\beta})$. Then using the complimentary log(-log) transformation of the survival function yields

$$\log(-\log[S(t|z, \boldsymbol{x})]) = z\gamma + \boldsymbol{x}'\boldsymbol{\beta} + \log(-\log[S_0(t|z, \boldsymbol{x})]). \tag{9.107}$$

This implies that if Z is used to construct separate strata ($h = 1, \ldots, S$), then for fixed values of the other covariates, say $\widetilde{\boldsymbol{x}}$,

$$\log(-\log[\widehat{S}_h(t|\widetilde{\boldsymbol{x}})]) = \widetilde{\boldsymbol{x}}\widehat{\boldsymbol{\beta}} + \log(-\log[\widehat{S}_{0h}(t|\widetilde{\boldsymbol{x}})]). \tag{9.108}$$

Thus if the hazards are indeed proportional for different values of Z, now represented by strata, then the background hazard functions within strata should be proportional. In this case, plots of the functions $\log(-\log[\widehat{S}_{0h}(t|\widetilde{\boldsymbol{x}})])$ versus t or $\log(t)$ should have a constant difference, approximately, over time.

Various types of residual diagnostics have also been proposed that may be used to identify influential observations and to detect departures from the proportional hazards assumption. A summary of these methods is provided by Fleming and Harrington (1991). Many of these methods are also inherently graphical.

Lin, Wei and Zing (1993) propose tests of model assumptions that can be used with the martingale residuals of Therneau, Grambsch and Fleming (1990). A somewhat simpler test of the PH model assumption was also described by Lin (1991). Without showing the details, the basic idea is related to the properties of efficiently weighted Mantel-Haenszel tests where a unit weight for the difference $(O - E)$ at each event time is asymptotically efficient against a proportional hazards alternative. Thus if a different set of weights yields a test with a significantly greater value than the test with unit weights, this is an indication that the proportional hazards assumption does not apply. Lin (1991), therefore, considers the difference between the weighted sum of the score vectors of the form

$$T = \sum\nolimits_i \left[\boldsymbol{U}_i(\boldsymbol{\beta}) - w(t_i)\boldsymbol{U}_i(\boldsymbol{\beta})\right] \tag{9.109}$$

for some weight $w(t_i) \neq 1$. To test the proportional hazards assumption against a proportional odds assumption, the test would employ $w(t_i) = \widehat{S}(t_i)$ as would be used in a Peto-Peto-Prentice Wilcoxon test. Lin (1991) then describes a test of significance for the PH assumption based on this statistic and provides a program for its computation. A SAS macro is also available.

9.4.8 Explained Variation

Since the PH model is semi-parametric and is based on a partial likelihood, no simple, intuitively appealing measure of explained variation arises naturally. Schemper (1990) proposed a measure (his V_2), which is defined from the ratio of the weighted sum of squares of the deviations of the empirical survival function for each individual over time with respect to the Cox model fitted survival function under the null ($\boldsymbol{\beta} = \mathbf{0}$) and alternative ($\boldsymbol{\beta} = \widehat{\boldsymbol{\beta}}$) hypotheses. O'Quigley, Flandre and Reiner (1999) show that this measure is a Korn-Simon-like measure of explained variation (see Section A.8 of the Appendix) in terms of survival probabilities that are weighted by the increments in the empirical cumulative distribution function of the event times.

Schemper's measure, however, is bounded above by some constant less than 1 when the model fits perfectly. Also, it is based on the Cox model estimated survival function rather than the hazard function on which the model is based. Computation requires estimation of the background survival function $S_0(t)$, from which one obtains an estimate of the conditional survival function as $\widehat{S}(t|\boldsymbol{x}) = \widehat{S}_0(t)^{\exp[\boldsymbol{x}'\widehat{\boldsymbol{\beta}}]}$. Kalbfleisch and Prentice (1980) describe a generalization of the Kaplan-Meier estimate of the background survival function $S_0(t)$, which is obtained by maximizing the non-parametric profile likelihood for the hazard probabilities at the event times given the PH model estimated coefficients $\widehat{\boldsymbol{\beta}}$. This estimate requires iterative solution when there are tied event times. Simplifications and approximations can also be obtained (see Collett, 1994). Breslow (1974) also describes a non-iterative estimate that is based on an underlying piecewise exponential model.

Korn and Simon (1990) also defined a measure of explained variation based on the survival times. To allow for censored observations, they use the expected square deviation of the survival time from its expectation.

Alternately, Kent and O'Quigley (1988) describe a measure of explained variation based on the Kullback-Leibler measure of distance or information gained. Their method is derived using a Weibull regression model that is a fully parametric proportional hazards model. They then show how the measure may be computed for a Cox PH model, which they denoted as $\widetilde{\rho}_W$. An S-plus macro (Koq) is available from $Statlib$.

O'Quigley and Flandre (1994) proposed a measure that is based on the sum of the squared scores or Schoenfeld (1982) residuals in (9.81) for each subject under the full versus the null model. This measure can be applied to models with time-dependent covariates, but like the Schemper and Kent-O'Quigley measures, requires additional computations.

Kent and O'Quigley (1988) also suggested a simple measure of explained variation analogous to Helland's (1987) $\rho^2_{\varepsilon^2}$ presented in (A.187) of the Appendix. Given a vector of estimated coefficients $\widehat{\boldsymbol{\beta}}$ from the PH model with covariate vector

$\boldsymbol{X}$, they suggested that an approximate measure of explained variation is

$$R^2_{\varepsilon^2} = \frac{\widehat{\boldsymbol{\beta}}'\widehat{\Sigma}_{\boldsymbol{x}}\widehat{\boldsymbol{\beta}}}{\widehat{\boldsymbol{\beta}}'\widehat{\Sigma}_{\boldsymbol{x}}\widehat{\boldsymbol{\beta}} + \sigma^2_\varepsilon}, \tag{9.110}$$

where $\widehat{\Sigma}_{\boldsymbol{x}}$ is the empirical estimate of the covariance matrix of the covariate vector $\boldsymbol{X}$. When the survival times are distributed as Weibull, then the covariate effects can be derived from an accelerated failure time model in $\log(T)$, where the errors are distributed as a Gumbel distribution (see Problem 9.3). In this case, $\sigma^2_\varepsilon = 1.645$. However, because the PH model is distribution-free, Kent and O'Quigley suggested using $\sigma^2_\varepsilon = 1$ in (9.110), yielding their $R^2_{W,A}$. They showed that the latter provides an adequate approximation to the more precise measure based on the proportion of information gained $\widetilde{\rho}_W$. In the two examples presented the measure based on the estimated proportion information gain ($\widetilde{\rho}^2_W$) equaled 0.56 and 0.13, whereas the approximation $R^2_{W,A}$ equaled 0.59 and 0.13, respectively.

In a widely used computer program (superceded by PROC PHREG), Harrell (1986) suggested that the proportion of explained log partial likelihood (equivalent to the estimated entropy R^2, but not based on an entropy loss function) be used as a measure of explained variation. However, Schemper (1990) and Kent and O'Quigley (1988) have shown that this measure grossly underestimates the proportion of explained variation estimated from their respective measures.

Schemper (1992) conducted a simulation to assess the accuracy of simple approximations to his measure and concluded that Madalla's likelihood ratio R^2_{LR} in (A.203) provided a satisfactory approximation to his V_2, where the N is the total sample size, not the number of events. Over a range of settings, the median difference $V_2 - R^2_{LR}$ ranged from -0.033 to 0.003, indicating a slight negative bias. R^2_{LR} also provides a rough approximation to the measures of Kent and O'Quigley (1988). In the two examples presented, their measure ($\widetilde{\rho}^2_W$) equaled 0.56 and 0.13, respectively, whereas the approximation R^2_{LR} yields values of 0.49 and 0.042, respectively.

Schemper and Stare (1996) presented an extensive assessment by simulation of the properties of the Schemper, Korn-Simon and Kent-O'Quigley measures. All but the latter were highly sensitive to the degree of censoring, whereas $\widetilde{\rho}^2_W$ and its approximation $R^2_{W,A}$ were largely unaffected.

Some of these measures may also be used to assess the partial R^2 or partial variation explained by individual covariates or sets of covariates, adjusted for other factors in the model. The simple approximation R^2_{LR} would use the likelihood ratio chi-square statistic for the contribution of the covariate(s) to the full model. Kent and O'Quigley's measure of the proportion of variation explained by a covariate, say the first $j = 1$, is computed as

$$R^2_{W,A(1)} = \frac{\widehat{\beta}_1'\widehat{\Sigma}_{11.2}\widehat{\beta}_1}{\widehat{\beta}_1'\widehat{\Sigma}_{11.2}\widehat{\beta}_1 + 1}, \tag{9.111}$$

where $\widehat{\Sigma}_{11.2} = \widehat{\Sigma}_{11} - \widehat{\Sigma}_{12}\widehat{\Sigma}_{22}^{-1}\widehat{\Sigma}_{21}$ is the conditional variance $\widehat{V}(x_1|\boldsymbol{x}_2)$ for $\boldsymbol{X}_2 = (X_2, \ldots, X_p)$. Schemper's V_2, however, is a measure of the variation in the survival probabilities that is explained by the full model. It does not describe the contribution of the individual covariates.

All of the above measures apply to models with baseline (fixed) covariates. For models with time-dependent covariates, Schemper's V_2 and the approximation R^2_{LR} may be applied. However, it is not clear whether the Kent-O'Quigley approximation $R^2_{W,A}$ would apply in this case.

For a stratified PH model, there is no suitable measure of the proportion of variation explained by the stratification levels because the background hazards for each strata do not enter into the model directly.

9.4.9 SAS PROC PHREG

SAS PROC PHREG provides many of the above computations. The syntax of the model specification is similar to that of LIFETEST using a statement of the form

```
model time*censor(0)=x1 x2 / covb corrb risklimits;
```

to indicate that the hazard function is modified proportionately by the effects of the covariates `x1` and `x2`. The program will provide likelihood ratio and score tests of the model and Wald tests of the coefficients. The `covb` and `corrb` options print the model-based estimates of the covariance matrix of the coefficient estimates and their correlation. The `risklimits` option prints the estimated hazard ratio per unit increase in the value of the covariate obtained as $\exp(\beta)$ and the corresponding 95% confidence limits. The program does not provide type III or likelihood ratio tests; however, these can be computed by hand by fitting the appropriate nested models.

To allow for ties, the program provides four options of the form `ties=exact` or `Breslow` or `Efron` or `Discrete`. The `exact` option fits the exact marginal likelihood (see Kalbfleisch and Prentice, 1973, 1980), whereas the `Breslow` option fits the Peto-Breslow approximation and the `Efron` option fits an approximation attributed to Efron (1977). The `Discrete` option fits the Cox discrete logistic model, which is more appropriate when the times are coarsely grouped or are discrete. The others are more appropriate when the times may be tied because of rounding of the event times, such as to the day or week. The `Breslow` option is the default.

The program provides a `strata` option to fit a stratified model as described in Section 9.4.2. It also allows the use of programming statements to define covariate values, depending on the values of the strata effects or on time. Thus it accommodates time-dependent covariates of any form.

Later versions of the program allow a counting process data structure that provides extensions of the Cox model to the analysis of multiple events as described by Wei, Lin and Weissfeld (1989); to recurrent events as described by Prentice, Williams and Peterson (1981); and also implements the multiplicative intensity model for recurrent events described in Section 9.6. The program does not compute the Lin and Wei (1989) robust information sandwich directly. However, this may be obtained by generating an output data set and then employing a separate program, as described in Example 44.8 of the SAS PROC PHREG documentation (SAS, 1997) Thus the information sandwich for a single event-time analysis and the robust score test may also be computed in this manner. Many of the program features are illustrated in the following examples.

Example 9.6 *Squamous Cell Carcinoma (continued)*
The analyses in Example 9.1 were restricted to the subset of patients who were non-ambulatory. The complete data set includes the age of the patient and the indicator variable for performance status (`perfstat`: 0 if ambulatory, 1 if not). Treatment is indicated by the value of `group`: 0 if B, 1 if A. The complete data set includes 194 patients of whom 127 showed spread of disease during follow-up and 67 had right censored event times. The data set is available from the www (see Preface, reproduced with permission).

An overall unadjusted assessment of the treatment group effect would employ the SAS statements:

```
proc phreg; model time*delta(0) = group / risklimits;
```

which would produce the following output:

```
            Testing Global Null Hypothesis: BETA=0
                  Without          With
Criterion       Covariates      Covariates      Model Chi-Square
-2 LOG L          1085.464        1084.090      1.374 with 1 DF (p=0.2411)
Score                 .               .         1.319 with 1 DF (p=0.2507)
Wald                  .               .         1.313 with 1 DF (p=0.2519)

           Analysis of Maximum Likelihood Estimates
                       Parameter  Standard       Wald          Pr >
Variable   DF          Estimate   Error        Chi-Square   Chi-Square
GROUP       1         -0.250804   0.21889       1.31282       0.2519

Risk       95% Confidence Limits
Ratio         Lower         Upper
0.778         0.507         1.195
```

This shows that the overall hazard or risk ratio for treatment $A{:}B$ is 0.778, which is not statistically significant with $p \leq 0.25$ by the likelihood ratio test.

In a model fit using only the two covariates (not shown), age is not statistically significant but the effect of performance status is highly significant ($p \leq 0.0045$ by a Wald test). Those who are not ambulatory have a risk 1.7 times greater than those who are ambulatory. The likelihood ratio model chi-square value is 8.350 on 2 df.

An additional model was fit using the following statements to provide an estimate of the treatment group effect adjusted for the effects of age and performance status:

```
proc phreg; model time*delta(0) = age perfstat group
   / risklimits;
```

This yields the following results:

```
          Testing Global Null Hypothesis: BETA=0
                Without      With
Criterion    Covariates  Covariates  Model Chi-Square
-2 LOG L       1085.464    1073.951  11.513 with 3 DF (p=0.0093)
Score                .           .   11.766 with 3 DF (p=0.0082)
Wald                 .           .   11.565 with 3 DF (p=0.0090)
```

```
         Analysis of Maximum Likelihood Estimates
                  Parameter  Standard      Wald         Pr >
Variable  DF      Estimate   Error      Chi-Square  Chi-Square
AGE        1    -0.010664    0.00933     1.30737       0.2529
PERFSTAT   1     0.595259    0.19015     9.80008       0.0017
GROUP      1    -0.387619    0.22524     2.96159       0.0853
```

```
Risk     95% Confidence Limits
Ratio        Lower         Upper
0.989        0.971         1.008
1.814        1.249         2.633
0.679        0.436         1.055
```

The likelihood ratio test for the addition of treatment group, computed by hand is $X^2 = 11.513 - 8.35 = 3.163$ with $p \leq 0.0753$, slightly more significant than the Wald test for treatment group.

The proportion of variation in the empirical survival functions explained by the model using Schemper's $V_2 = 0.055$. The approximate variation explained by the model is $R^2_{LR} = 1 - \exp(-11.513/194) = 0.058$. Based on the likelihood ratio test for the effect of treatment, the approximate proportion of variation explained by treatment group is $R^2_{LR} = 1 - \exp(-3.163/194) = 0.016$. The Kent and O'Quigley (1988) approximation to their measure of explained information gained by the full model is $R^2_{W,A} = 0.105$, and that explained by treatment group is 0.027. Because these measures are not affected by censoring, they are somewhat larger than the Schemper and R^2_{LR} measures which are reduced by censoring.

An additional model was fit with pairwise interactions among the three covariates in the linear exponent. The Wald test of the interaction between treatment group and performance status had a value $X^2 = 3.327$ with $p \leq 0.0681$, which suggests heterogeneity of treatment group effects between the performance status groups. Thus an additional model was fit, dropping the other non-significant interactions with a treatment group effect nested within the levels of performance status. Among those who were ambulatory (`perfstat=0`) the estimated relative risk for treatment group $A{:}B$ is 0.996 with $p \leq 0.99$. However, among those who were not ambulatory (`perfstat=1`), the estimated relative risk is 0.436 with $p \leq 0.0175$.

Since the treatment effect appears to depend on the value of performance status, a final model was fit that was stratified by performance status and which had nested effects for age and treatment group within strata using the likelihood (9.77). This model allows the background hazard function and the covariate effects to differ between performance status strata.

```
          Testing Global Null Hypothesis: BETA=0
                    Without        With
Criterion       Covariates   Covariates    Model Chi-Square
-2 LOG L           918.810      912.486    6.324 with 4 DF (p=0.1762)
Score                    .            .    5.861 with 4 DF (p=0.2097)
Wald                     .            .    5.751 with 4 DF (p=0.2185)

          Analysis of Maximum Likelihood Estimates
                    Parameter  Standard        Wald          Pr >
Variable  DF        Estimate      Error   Chi-Square    Chi-Square
AGE0       1       -0.006504    0.01229      0.27995        0.5967
AGE1       1       -0.021947    0.01491      2.16759        0.1409
GROUP0     1        0.037793    0.29517      0.01639        0.8981
GROUP1     1       -0.770036    0.37309      4.25977        0.0390

Risk      95% Confidence Limits
Ratio         Lower        Upper
0.994         0.970        1.018
0.978         0.950        1.007
1.039         0.582        1.852
0.463         0.223        0.962
```

This analysis shows that the treatment group effect is nominally significant at the 0.05 level within the subgroup of patients who are not ambulatory with an estimated relative risk of 0.463 with $p \leq 0.039$, whereas there is no evidence of any treatment effect within the non-ambulatory subgroup with an estimated relative risk of 1.04.

Unfortunately, it is not possible to conduct a direct test of the difference between the stratified model and that which is not stratified by performance status because the models are not nested in the coefficient parameters and thus the likelihoods are not comparable. Thall and Lachin (1986) describe a visual assessment of the hypothesis of homogeneous background hazards within strata.

If the protocol had specified that the primary analysis was the comparison of the event-free or cumulative incidence curves, then the principal analysis would be the logrank test supplemented by an estimate of the relative hazard (relative risk) using either the Mantel-Haenszel estimate, the score-based estimate, or the estimate obtained from the Cox proportional hazards model. The latter is fully efficient under the proportional hazards model and is a consistent estimate of the time-averaged relative hazard when the proportional hazards model does not apply. In the latter case, however, the model-based confidence limits are too narrow and thus limits based on the robust information sandwich estimate of the covariance matrix are preferred.

For this study, the logrank test applied to the complete sample of 194 patients yields a Mantel-logrank test value of $X^2 = 1.372$ with $p \leq 0.25$. The Peto score-based estimate of the relative risk is $\widehat{RR} = 0.783$ with 95% confidence limits (0.52, 1.18). These are close to the Cox model-based estimates from the first of the above models.

For this study, however, it is also important to note the suggestion of heterogeneity of the treatment effect within strata defined by the performance status of the patients. In this case, especially if the objective of the study were to conduct exploratory analyses of these relationships, it would be most appropriate to report the final model that is stratified by ambulatory status with stratum-specific covariate effects.

Example 9.7 *Robust Information Sandwich*
For the no-interaction model with treatment group adjusted for age and performance status, the COVB option provides the following model-based estimate of the covariance matrix of the estimates obtained as the inverse of the information matrix:

	AGE	PERFSTAT	GROUP
AGE	0.0000869908	-.0001999462	0.0001887417
PERFSTAT	-.0001999462	0.0361561494	-.0085786551
GROUP	0.0001887417	-.0085786551	0.0507323757

Using the vector of DFBETA values for each subject, the robust estimate of the covariance matrix is computed to be

	AGE	PERFSTAT	GROUP
AGE	0.0000809359	-0.0003168883	0.0002510116
PERFSTAT	-0.0003168883	0.0342840251	0.0017102024
GROUP	0.0002510116	0.0017102024	0.0516267618

This matrix yields Wald tests and confidence limits for the model coefficients that are similar to the model based estimates.

The similarity of the two covariance matrices indicates that the proportional hazards model specification appears to be appropriate for these data.

Using a separate program, the centered score vector from (9.95) evaluated under the model null hypothesis H_0: $\boldsymbol{\beta} = \mathbf{0}$ is

$$\boldsymbol{W}(\boldsymbol{\beta}_0) = [\ -82.50022 \quad 14.11157 \quad -5.600537\ ]$$

and the robust estimate of the covariance matrix is

$$\widehat{J}(\beta_0) = \begin{bmatrix} 9592.8705 & -9.248928 & 27.70094 \\ -9.248928 & 26.649602 & 8.3026425 \\ 27.70094 & 8.3026425 & 8.3026425 \end{bmatrix}.$$

The resulting score test is $X^2 = 13.155$ on 3 *df* with $p \leq 0.0014$. This test value is comparable to the likelihood ratio test ($X^2 = 11.513$) and efficient score test ($X^2 = 11.766$) based on the assumed model.

Example 9.8 *Testing the Proportional Hazards Assumption*
For the same model used in the preceding example, Lin's (1991) procedure was applied using an estimating equation derived from weighted scores as in (9.109), using the Kaplan-Meier estimated survival function from the combined sample as the weights. The coefficient estimates and their estimated standard errors derived from the weighted score equation, the difference of the weighted estimates from the Cox model estimates, and the standard error of the difference were

	Weighted		Difference	
Parameter	Estimate	Std. Error	Estimate	Std. Error
GROUP	-0.48504	0.251705	-0.09742	0.112351
AGE	-0.01812	0.010092	-0.00746	0.003855
PERFSTAT	0.704442	0.203207	0.109183	0.071672

The difference in the estimates of the coefficient for age is nearly twice its standard error ($Z = -0.00746/0.003855 = -1.935$), suggesting that the proportional hazards assumption for age may not be applicable. Using the vector of differences and its associated estimated covariance matrix (not shown) as the basis for an overall test of the three covariate proportional hazards model versus a non-proportional hazards model yields a Wald test $X^2 = 6.36$ with 3 *df* ($p \leq 0.0954$), which is not statistically significant.

An additional analysis can be conducted using Cox's method, in which a separate model is fit containing an interaction between each covariate, in turn, with $\log(t)$. The Wald test of the interaction effect for each coefficient did not approach significance, $p \leq 0.27$ for age, 0.15 for performance status and 0.41 for treatment group. Cox's test, however, is designed to detect a monotone shift in the hazard ratio over time, and not a general alternative.

Since Lin's analysis suggests some departure from proportional hazards for the effect of age, age was divided into two strata and a model fit for the other two covariates stratified by age. Figure 9.3 presents the plots of the log(-log($\widehat{S}_0(t)$)) within age strata. Noting that the failure probability is directly proportional to the complimentary log-log function of the survival probability, this plot suggests (weakly) that those in the lower age stratum tend to have higher risk during the middle of the study, but equivalent risk otherwise. This non-monotonic departure from the proportional hazards assumption was not detected by Cox's interaction test but was suggested by Lin's test. Overall, however, there does not appear to be

Fig. 9.3 The Cox-model estimated background log(-log) survival probabilities, or the log cumulative hazard function, within the low (L) and upper (H) halves of the distribution of age, including treatment group and performance status in the model.

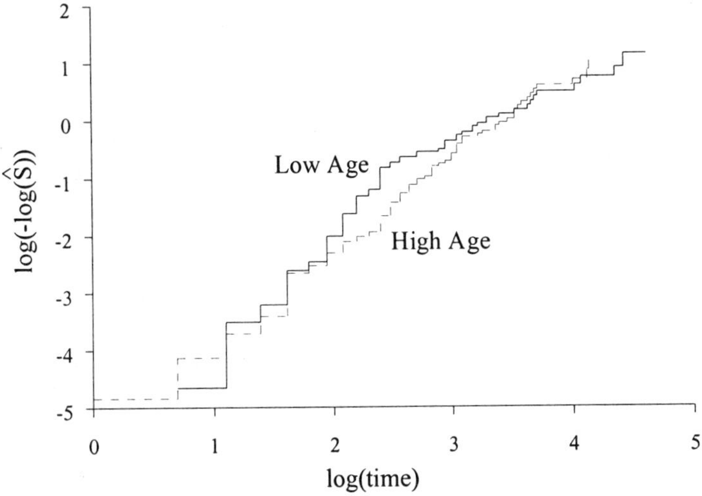

a major departure from the proportional hazards assumption for this or the other covariates.

Example 9.9 *DCCT Time-Dependent HbA_{1c} and Nephropathy*
Example 9.2 describes the cumulative incidence of developing microalbuminuria among subjects in the Secondary Intervention Cohort of the DCCT. Such analyses established the effectiveness of intensive therapy aimed at near-normal levels of blood glucose at reducing the risk of progression of diabetic eye, kidney and nerve disease. However, it was also important to describe the relationship between the risk of progression and the level of blood glucose control achieved, as measured by the percent of total hemoglobin that was glycosylated, or the % HbA_{1c}. The median of the current average HbA_{1c} over the period of follow-up was 8.9% among patients in the conventional treatment group (upper and lower quartiles of 9.9% and 8%), compared to a median of 7.0% (extreme quartiles of 6.5% and 7.6%) in the intensive group. The questions were whether the risk gradients for the level of HbA_{1c} were the dominant determinants of the risk of progression of complications, whether the risk gradients were different between the groups, and whether this difference between groups in the level of HbA_{1c} accounted for the difference in risk of complications. These questions were addressed by the DCCT Research Group (1995). An additional paper (DCCT, 1996) showed that the there is no

threshold of glycemia (HbA_{1c}) above the normal range below which there is no further reduction in risk.

To address these questions, the DCCT Research Group (1995) describes Cox PH model analyses of the effects of the log of the current mean HbA_{1c}, as a time-dependent covariate, on the relative risk of developing nephropathy (microalbuminuria) in the Diabetes Control and Complications Trial. To fit the model, for each subject a vector (array) of mean HbA_{1c} values was computed as of the time of each annual visit, the values `MHBA1-MHBA9` for the nine years of study. Then the following SAS statements were used to fit the model separately within each treatment group:

```
proc phreg; by group;
 model time*flag(0)= lmhba / ties=discrete alpha=0.05 rl;
 array mhba (9) mhba1-mhba9;
 do j=1 to 9;
   if time eq j then lmhba=log(mhba(j));
 end;
```

Note that the observation for each subject includes the array of current mean HbA_{1c} values and additional programming statements are used to define the appropriate covariate value for each subject at each event (annual visit) time. Also, note that the array must include the values of the covariate at all event (visit) times up to the last visit for each subject. Thus a subject with the event at year 5, or one with the last evaluation at year 5, would have the vector of values defined for years 1–5 and missing for years 6–9.

The following results were obtained from the analysis of the 316 patients in the conventional treatment group of the Secondary Intervention cohort.

```
        Testing Global Null Hypothesis: BETA=0
                  Without        With
Criterion  Covariates   Covariates   Model Chi-Square
-2 LOG L    715.917      698.106    17.811 with 1 DF (p=0.0001)
Score          .            .       17.646 with 1 DF (p=0.0001)
Wald           .            .       17.364 with 1 DF (p=0.0001)

        Analysis of Maximum Likelihood Estimates
                   Parameter  Standard      Wald         Pr >
Variable   DF     Estimate     Error    Chi-Square   Chi-Square
LMHBA       1     3.170502   0.76086     17.36408     0.0001
```

Because the log HbA_{1c} was used in the model, it is inappropriate to simply interpret $e^{\beta} = e^{3.17} = 23.8$ as a risk ratio per unit change in the HbA_{1c}, although this is the estimate of the risk ratio per unit change in log(HbA_{1c}). Rather, it is more informative to consider the change in risk associated with a proportionate change in HbA_{1c}. Since $\beta = \log \phi$ per $\Delta \log X(t) = 1$, it can be shown (see Problem 9.14.11) that $100(c^{\beta} - 1)$ represents the *percentage* change in risk per a c-fold change in $X(t) = \mathrm{HbA}_{1c}$ $(c > 0)$. Therefore, the estimated coefficient $\widehat{\beta} = 3.17$ represent a 35.3% increase in the risk of developing nephropathy per 10% higher value of the current mean HbA_{1c} at any point in time ($c = 1.1$), or a 28.4% decrease in risk per 10% lower HbA_{1c} ($c = 0.9$). Using the 95% confidence limits for β yields 95% confidence limits for the risk reduction per 10% lower HbA_{1c} of (16.2, 38.8%).

Among the measures of explained variation described in Section 9.4.8, only the crude approximate measure R^2_{LR} may be readily applied because the model included time-dependent covariates. Based on the likelihood ratio chi-square test, the log of the current mean HbA_{1c} explains $100(1 - \exp(-17.811/316)) = 5.48\%$ of the variation in risk. In DCCT (1995), such measures were used to describe the relative importance of different covariates, not as an absolute measure of explained variation.

In the analyses of the total conventional group, stratified by primary and secondary cohort, and also adjusting for the baseline level of the log albumin excretion rate, the estimated coefficient $\widehat{\beta} = 2.834$ corresponds to 25.8% risk reduction per 10% lower HbA_{1c}, with similar risk reductions in the primary and secondary intervention cohorts (DCCT, 1995). Nearly equivalent results were obtained in the intensive treatment group, $\widehat{\beta} = 2.639$. Thus virtually all of the difference between the treatment groups in the risk of developing microalbuminuria was attributable to the differences in the level of glycemia as represented by the HbA_{1c}.

9.5 EVALUATION OF SAMPLE SIZE AND POWER

9.5.1 Exponential Survival

In general, a distribution-free test such as the Mantel-logrank test is used for the analysis of survival (event-time) data from two groups. In principle, the power of any such test can be assessed against any particular alternative hypothesis with hazard functions that differ in some way over time. It is substantially simpler, however, to consider the power of such tests assuming some simple parametric model. The Mantel-logrank test is the most commonly used test in this setting, which is asymptotically fully efficient against a proportional hazards or Lehmann alternative. The simplest parametric form of this model is the exponential model with constant hazard rates λ_1 and λ_2 over time in each group.

Asymptotically, the sample estimate of the log hazard rate $\log(\widehat{\lambda})$ is distributed as $\mathcal{N}[\log(\lambda),\ E(D|\lambda)^{-1}]$. Thus the power of the test depends on the expected total number of events $E(D|\lambda)$ to be observed during the study. Here $E(D|\lambda) = NE(\delta|\lambda)$, where δ is a binary variable representing observation of the event ($\delta = 1$) versus right censoring of the event time ($\delta = 0$); and $E(\delta|\lambda)$ is the probability that the event will be observed as a function of λ and the total exposure of the cohort (patient years of follow-up). The test statistic then is $T = \log(\widehat{\lambda}_1/\widehat{\lambda}_2)$. Under H_0: $\lambda_1 = \lambda_2 = \lambda$ and the statistic has expectation $\mu_0 = \log(\lambda_1/\lambda_2) = 0$, while under H_1, $\mu_1 = [\log(\lambda_1) - \log(\lambda_2)]$. As in Section 3.3.1, let ζ_i refer to the expected sample fraction expected in the ith group ($i = 1, 2$) where $E(n_i) = N\zeta_i$. Then the variance of the test statistic under the alternative hypothesis is

$$V(T|H_1) = \sigma_1^2 = \frac{1}{N}\left[\frac{1}{\zeta_1 E(\delta|\lambda_1)} + \frac{1}{\zeta_2 E(\delta|\lambda_2)}\right] \tag{9.112}$$

and under the null hypothesis is

$$V(T|H_0) = \sigma_0^2 = \frac{1}{N}\left[\frac{1}{E(\delta|\lambda)}\left(\frac{1}{\zeta_1\zeta_2}\right)\right]. \tag{9.113}$$

The basic equation relating sample size N and power $Z_{1-\beta}$ is

$$\sqrt{N}\,|\log(\lambda_1) - \log(\lambda_2)| = Z_{1-\alpha}\left[\frac{1}{E(\delta|\lambda)\zeta_1\zeta_2}\right]^{1/2} + Z_{1-\beta}\left[\frac{1}{\zeta_1 E(\delta|\lambda_1)} + \frac{1}{\zeta_2 E(\delta|\lambda_2)}\right]^{1/2}, \tag{9.114}$$

where $\lambda = \zeta_1\lambda_1 + \zeta_2\lambda_2$ analogously to (3.34) in the test for proportions.

Lachin (1981) and Lachin and Foulkes (1986) present a similar expression for the case where the test statistic is the difference in estimated hazard rates, $T = \widehat{\lambda}_1 - \widehat{\lambda}_2$. Freedman (1982) shows that these expressions can also be derived from the null and alternative distributions of the Mantel-logrank statistic. As cited by Lachin and Foulkes (1986), computations of sample size and power using the difference in hazards are conservative relative to those using the log hazard ratio in that the former yields larger required sample sizes, lower computed power for the same values of λ_1 and λ_2. As the difference in hazards approaches zero, the difference in the two methods also approaches zero. In some respects the use of the difference in hazards would be preferred because, in fact, the Mantel-Haenszel (logrank) test statistic can be expressed as the weighted sum of the difference in the estimated hazards between groups (see Section 9.6.3). Herein, however, we use the log hazard ratio because generalizations also apply to the Cox PH model.

In the simplest case of a study with no censoring of event times, $E(\delta|\lambda) = 1$ and the event times of all subjects are observed ($N = D$). In this case, the total number of events D and power $Z_{1-\beta}$ are obtained as the solutions to

$$\sqrt{D}\,|\log(\lambda_1) - \log(\lambda_2)| = \frac{Z_{1-\alpha} + Z_{1-\beta}}{\sqrt{\zeta_1\zeta_2}}. \tag{9.115}$$

Thus the total number of events D required to provide power $1 - \beta$ to detect a specified hazard ratio in a test at level α is

$$D = \frac{(Z_{1-\alpha} + Z_{1-\beta})^2}{\zeta_1\zeta_2\left[\log(\lambda_1/\lambda_2)\right]^2} \tag{9.116}$$

(George and Desu, 1974; Schoenfeld, 1981). Usually, however, there are censored event times because of administrative curtailment of follow-up (administrative censoring) or because of random losses to follow-up.

To allow for administrative censoring, let T_S designate the maximum total length of study. In the simplest case, each subject is followed for T_S years of exposure and $E(\delta|\lambda) = 1 - e^{-\lambda T_S}$. Typically, however, patients enter a study during a period of recruitment of T_R years and are then followed for a maximum total duration of T_S years ($T_S \geq T_R$) so that the first patient entered is followed for T_S years and the last patient for $T_S - T_R$ years. In a study with uniform entry over the recruitment interval of T_R years and with no random losses to follow-up, it is readily shown (see Problem 9.1.5) that

$$E(\delta|\lambda) = \left[1 - \frac{e^{-\lambda(T_S - T_R)} - e^{-\lambda T_S}}{\lambda T_R}\right] \tag{9.117}$$

(Lachin, 1981). Substitution into (9.114) and solving for N yields the sample size needed to provide power $1 - \beta$ in a test at level α to detect a given hazard ratio, say $N(\lambda)$.

Rubenstein, Gail and Santner (1981) and Lachin and Foulkes (1986) present generalizations that allow for randomly censored observations because of loss to follow-up. Let γ_i be an indicator variable that denotes loss to follow-up at random during the study prior to the planned end at T_S. If we assume that times of loss are exponentially distributed with constant hazard rate η over time, then for a study with uniform entry the probability of the event is

$$E(\delta|\lambda, \eta) = \frac{\lambda}{\lambda + \eta}\left[1 - \frac{e^{-(\lambda+\eta)(T_S - T_R)} - e^{-(\lambda+\eta)T_S}}{(\lambda + \eta)T_R}\right] \tag{9.118}$$

and the probability of loss to follow-up is

$$E(\gamma|\lambda, \eta) = \frac{\eta}{\lambda} E(\delta|\lambda, \eta) \tag{9.119}$$

(see Problem 9.1.6). When $\eta_1 = \eta_2$ for the two groups, then the equation relating sample size and power is obtained by substituting $E(\delta|\lambda, \eta)$ in (9.114) evaluated at λ_1, λ_1, and λ. When $\eta_1 \neq \eta_2$, then the probability of the event in each group under H_0 will differ. In this case, the general equation in (9.114) is modified to employ the term

$$Z_{1-\alpha}\left[\frac{1}{\zeta_1 E(\delta|\lambda, \eta_1)} + \frac{1}{\zeta_2 E(\delta|\lambda, \eta_2)}\right]^{1/2}. \tag{9.120}$$

In the more general case, where losses to follow-up may not be exponentially distributed, Lachin and Foulkes (1986) show that the sample size with losses that follow any distribution $G(t)$ can be obtained approximately as

$$N[\lambda, G(t)] \doteq N(\lambda) \frac{E(\delta|\lambda)}{E[\delta|\lambda, G(t)]}, \tag{9.121}$$

where $E[\delta|\lambda, G(t)]$ is the probability of an event during the study. Thus random losses to follow-up, whatever their distribution, that result in a 10% reduction in the probability of the event being observed require that the sample size needed with no losses to follow-up, $N(\lambda)$, be increased by 11.1%.

For cases where the exponential model is known not to apply, Lakatos (1988) describes the assessment sample size based on the power function of a weighted Mantel-Haenszel test against an arbitrarily specified alternative hypothesis. In this procedure one specifies the hazard rates in the control and experimental groups over intervals of time along with other projected features of the study, such as the proportion censored or lost to follow-up within each interval. This allows the assessment of power under any alternative, including cases where the hazards may not be proportional, or may even cross, and where the pattern of random censoring is uneven over time and may differ between groups. Wallenstein and Wittes (1993) also describe the power of the Mantel-logrank test in an analysis where the hazards need not be constant nor proportional over time, although the Lakatos procedure is more general.

9.5.2 Cox's Proportional Hazards Model

As with the logistic regression model, the power function of a test for the vector of coefficients in the Cox PH model is a function of the joint distribution of the vector of covariates among those with the event and those not at each event time, which are unknown. However, one can assess the power of a Wald or score test in a PH model that is stratified by other factors. Under an exponential model, Lachin and Foulkes (1986) describe the relationship between sample size and power for a test of the difference in hazards for two groups that is stratified by other factors. Since the logrank test is the score test in a Cox PH model, these methods also apply, approximately, to a test of the difference between groups in a Cox PH model with other binary covariates that define a set of strata.

Schoenfeld (1983) showed that the probability of events in each of two groups under the assumption of constant proportional hazards over strata can be obtained if one has information on the survival function over time in the control group. Let $E(\delta)$ denote the resulting probability of an event in the total sample. Then the sample size required to provide power $1-\beta$ in a test at level α to detect a given hazard ratio (RR) is provided as $N = D/E(\delta)$, where D is the total number of events required from (9.116) substituting RR for λ_1/λ_2. Many, such as Palta and Amini (1985), describe generalizations of this approach.

Example 9.10 *Lupus Nephritis: A Study*

Lewis, Hunsicker, Lan, et al. (1992) describe the results of a clinical trial of plasmapheresis (plasma filtration and exchange) plus standard immunosuppressive therapy versus standard therapy alone in the treatment of severe lupus nephritis. The sample size for this study was based on two previous studies in which the survival function was log-linear with constant hazard approximately $\lambda = 0.3$ yielding median survival of 2.31 years. Since lupus is a rare disease, recruitment was expected to be difficult. Thus initial calculations determined the sample size required to provide 90% power to detect either a 40 or 50% risk reduction (relative hazards of 0.6 and

0.5) with a one-sided test at level $\alpha = 0.05$. From (9.116), the total number of events required are 131.3 for a 40% risk reduction and 71.3 for a 50% reduction, assuming no right censoring. In practice these would be rounded up to 132 and 72 events, respectively.

The study was planned to recruit patients over a period of $T_R = 4$ years with a total study duration of $T_S = 6$ years. With no adjustments for losses to follow-up, to detect a 40% risk reduction, a total $N = 222$ (221.9) subjects would be required. Among those in the control group with hazard $\lambda_2 = 0.3$, the probability of the event is $E(\delta|\lambda_2 = 0.3) = 0.6804$, which yields 75.5 expected events. In the plasmapheresis group with hazard $\lambda_1 = 0.18$, the event probability is $E(\delta|\lambda_1 = 0.18) = 0.5027$ with 55.8 expected events. To detect a 50% risk reduction, the total sample size required is $N = 127.3$. In the control group, the event probability is unchanged, yielding 43.3 expected events. In the plasmapheresis group with hazard $\lambda_1 = 0.15$, the event probability is $E(\delta|\lambda_1 = 0.15) = 0.4429$ with 28.2 expected events.

In each case, the total expected number of events under the alternative hypothesis is approximately equal to the total number of events required in a study with no censoring. Thus the sample size is the number to be placed at risk such as to yield the required expected number of events in the presence of censoring. The greater the average period of exposure, the larger the expected number of events and the smaller the required sample size. For example, if all patients could be recruited over a period of only six months ($T_R = 0.5$) in a six year study, then the numbers of patients required to detect a 40% risk reduction is 178.2, substantially less than the 222 required with a four year recruitment period. With a six month recruitment period, the average duration of follow-up is 5.75 years, yielding a total of 131 expected events among the 179 patients. With a four year recruitment period, the average duration of follow-up is four years, again yielding 131 expected events among the 222 patients.

Because lupus is a life-threatening disease, it was considered unlikely that patients would be lost to follow-up at a rate greater than 0.05 per year. An additional calculation allowing for losses using (9.114) with a loss hazard rate of $\eta_1 = \eta_2 = \eta = 0.05$ in each group in (9.118) yields the following:

λ_1	N	$E(\gamma\|\lambda_1, \eta)$	$E(\delta\|\lambda_1, \eta)$	$E(\delta\|\lambda, \eta)$	D_1	D_2
0.18	241.6	0.128	0.460	0.553	55.5	75.9
0.15	138.8	0.135	0.404	0.532	28.0	43.6

In the control group, for both values of λ_1, about 11% of the patients would be lost to follow-up ($E(\gamma|\lambda_2, \eta) = 0.105$); and the probability of observing an event is reduced to $E(\delta|\lambda_2, \eta) = 0.628$ because of these losses to follow-up. With a 40% risk reduction ($\lambda_1 = 0.18$), about 13% of the patients in the experimental group, $E(\gamma|\lambda_1, \eta)$, would be lost to follow-up; and likewise the probability of an event is reduced to $E(\delta|\lambda_1, \eta) = 0.46$. Substituting these probabilities into (9.114), the total sample size is increased to $N = 241.6$ compared to $N = 221.9$ with no losses to follow-up. The total expected number of events, however, is still approximately $D = 131$ as before. The combined probability of the event with no losses, 0.59 = (0.680+0.503)/2), is 8.7% greater than that with losses (0.54). Thus the simple approximation in (9.121) indicates that the sample size with losses should be approximately $N = 1.087 \times 221.9 = 241.3$. Similarly, to detect a 50% risk reduction ($\lambda_1 = 0.15$), a total sample size of $N = 138.8$ is required.

Based on such calculations, the final sample size was determined to be $N = 125$ recruited over four years with a total duration of six years. Allowing for a loss to follow-up hazard rate of 0.05 per year, this provides power of 0.872 to detect a 50% reduction in the hazard rate relative to the control group hazard of 0.3 per year with a one-sided test at the 0.05 level. This sample size provides power of 0.712 to detect a 40% reduction in the hazard, and only 0.553 to detect a one-third reduction. Were it not for the fact that lupus nephritis is a very rare disease, a larger sample size would have been desirable, one that would provide perhaps a minimum of 80% power to detect a one-third risk reduction. Eventually the trial was terminated early because of lack of evidence of a beneficial effect of plasmapheresis on disease progression (Lachin, Lan, et al., 1992).

9.6 ANALYSIS OF RECURRENT EVENTS: THE MULTIPLICATIVE INTENSITY MODEL

In many prospective studies, the exact times of possibly recurrent events are recorded for each subject over a period of exposure that may vary among subjects because of staggered subject entry or random loss to follow-up. For example, such events may include successive hospitalizations, epileptic seizures, infections, accidents or any other transient, non-absorbing or non-fatal event. For each subject, the sequence of times of the event constitutes a simple point process observed continuously over time (or in *continuous time*). From such data it is desired to estimate the underlying hazard function over time and to test the difference between groups in these functions. In the more general case, we wish to assess the effects of covariates on the aggregate risk of events over time.

One of the major advances in event-time analysis is the application of stochastic models for such counting processes to the analysis of survival and recurrent event-time data. Excellent general references are the texts by Fleming and Harrington (1991) and by Andersen, Borgan, Gill and Keiding (1993). This approach generalizes methods for the analysis of survival data to the analysis of recurrent event data, including tests for differences between two groups of subjects and the generalized proportional hazards regression model for a counting process. A rigorous treatment is beyond the scope of this text. In the following I provide a brief description of the extensions to the analysis of recurrent events, with only a brief introduction to martingales.

9.6.1 Counting Process Formulation

We begin with a cohort of N subjects, where each subject is observed over a subset of the interval $(0, \tau]$. The *counting process* for the ith subject, designated as $\mathrm{N}_i(s)$, is the number of events experienced by that subject up to study time $s \leq \tau$, for $i = 1, \ldots, N$. The aggregate counting process for the cohort is $\mathrm{N}(s) = \sum_{i=1}^{N} \mathrm{N}_i(s)$, $s > 0$.

To indicate when events occur in each subject, let $d\mathrm{N}_i(s)$ and $d\mathrm{N}(s)$ designate the *point processes* corresponding to the "jumps" in the counting process for the ith subject or the aggregate process, respectively. Using a Stieltjes integral, then

$$\mathrm{N}(t) = \int_0^t d\mathrm{N}(s). \tag{9.122}$$

For a continuous time process, where it is assumed that no subject experiences multiple events at any instant and that no two subjects experience an event at the same instant, then $d\mathrm{N}(s) = 1$ or 0, depending on whether or not an event occurred at time s, $0 < s \leq \tau$. However, such strict continuity is not necessary. To allow for tied event times, such as multiple recurrent events in a subject on the same day, or two or more subjects with events on the same day, the point process may be defined as

$$\Delta\mathrm{N}(s) = \mathrm{N}(s) - \mathrm{N}(s^-), \tag{9.123}$$

where $\mathrm{N}(s^-) = \lim_{\Delta s \downarrow 0} \mathrm{N}(s - \Delta s)$ is the left-hand limit at s^-, or the value of the process at the instant before time s.

To designate the individuals at risk of an event at any time s, for the ith subject let $Y_i(s)$ designate the at risk process where $Y_i(s) = 1$ if the subject is alive and at risk (under observation) at time s; 0 otherwise. The total number of subjects in the cohort at risk of an event at time s is $Y(s) = \sum_{i=1}^{N} Y_i(s)$. The at risk indicator $Y_i(s)$ is a generalization of the simple right censoring at risk function employed in (9.71) with the proportional hazards model, where for the ith subject followed to time t_i, $Y_i(s) = 1$ for $s \leq t_i$, 0 for $s > t_i$. In the analysis of recurrent events, however, $Y_i(s)$ may vary over time such that $Y_i(s) = 1$ over intervals of time while a subject is under surveillance and at risk and $Y_i(s) = 0$ over intervals of time while a subject is lost to follow-up or is not at risk. In this setting $Y_i(s)$ may switch back and forth between the values 0,1 over time. By construction, a subject is not at risk of a recurrent event over the duration of an event. For example, in an analysis of the risk of recurrent hospitalizations, $Y_i(s) = 0$ over those intervals of time while a subject is hospitalized and is not at risk of another hospitalization.

We also assume that the risk process $Y_i(s)$ is statistically independent of the counting process $\mathrm{N}_i(s)$ for the ith subject, and likewise for the aggregate processes in the cohort. This is a generalization of the assumption of censoring at random that was employed in the analysis of survival data.

Note that the above formulation is easily adapted to the case of calendar time rather than study time. In the case of calendar time, time $s = 0$ is simply the earliest calendar time for the start of observation among all subjects, and time $s = \tau$ is the latest calendar time for the end of observation among all subjects. The $\mathrm{N}_i(s)$, $Y_i(s)$ and $d\mathrm{N}_i(s)$ processes are then defined in calendar rather than in study time.

Now let $\mathcal{F}_s$ designate the history or the *filtration* of the aggregate processes for all N subjects up to time s $\{\mathrm{N}_i(u), Y_i(u); i = 1, \ldots, N; 0 < u \leq s\}$. Likewise, $\mathcal{F}_{s-}$ designates the history up to the instant prior to s. Then in continuous time the conditional probability *(intensity)* of an event in the ith subject at time s is

$$\alpha_i(s)ds = E\left[d\mathrm{N}_i(s)|\mathcal{F}_{s-}\right] = \lambda(s)Y_i(s)ds \qquad i = 1, ..., N, \tag{9.124}$$

where $\lambda(s)$ is the *underlying intensity or rate function* that applies to all the subjects in the cohort. The hazard function for a single event in continuous time is now generalized to the case of possibly recurrent events as

$$\begin{aligned} \lambda\left(s\right) &= \lim_{\Delta s \downarrow 0} \frac{\Pr\left[\left\{\mathrm{N}_i\left(s + \Delta s\right) - \mathrm{N}_i\left(s\right) = 1\right\}|\mathcal{F}_{s-}\right]}{\Delta s} \\ &= \lim_{\Delta s \downarrow 0} \frac{\Pr\left[d\mathrm{N}_i\left(s\right)|\mathcal{F}_{s-}\right]}{\Delta s} \end{aligned} \tag{9.125}$$

for the ith subject. Likewise, for the ith subject, the *cumulative intensity function* of the counting process $\mathrm{N}_i\left(t\right)$ is defined as

$$\mathcal{A}_i(t) = E\left[\mathrm{N}_i(t)|\mathcal{F}_t\right] = \int_0^t \alpha(s)ds = \int_0^t \lambda(s)Y_i(s)ds\,, \qquad i = 1, ..., N. \tag{9.126}$$

The intensity $\alpha_i(t)$ and cumulative intensity $\mathcal{A}_i(t)$ are interpreted as the point and cumulative expected numbers of events per subject at time t. Note that the intensity of the point process $\alpha_i(s)$ is distinguished from the underlying intensity $\lambda(s)$ in the population, in that the former allows for the extent of exposure over time within individual subjects.

Similar results apply to the aggregate counting process in the aggregate cohort. These results are now presented using Stieltjes integrals to allow for observations in continuous or discrete time. The intensity function for the aggregate point process is then defined by

$$d\mathcal{A}(s) = E\left[d\mathrm{N}(s)|\mathcal{F}_{s-}\right] = Y(s)d\Lambda(s), \tag{9.127}$$

where, in continuous time, $d\Lambda(s) = \lambda(s)ds$, and in discrete time $d\Lambda(s) = \Delta\Lambda(s) = [\Lambda(s) - \Lambda(s^-)]$. Thus the cumulative intensity function for the aggregate counting process is

$$\mathcal{A}(t) = E\left[\mathrm{N}(t)|\mathcal{F}_t\right] = \int_0^t d\mathcal{A}(s). \tag{9.128}$$

In continuous time, $d\mathcal{A}(s) = \alpha(s)ds = \lambda(s)Y(s)ds$. In discrete time $d\mathcal{A}(s) = Y(s)\Delta\Lambda(s)$ and the integral may be replaced by a sum.

Since $\mathrm{N}(t)$ is a non-decreasing process in time, it is termed a *submartingale.* The cumulative intensity $\mathcal{A}(t)$ is referred to as the *compensator* of the counting process $\mathrm{N}(t)$. Since $\mathcal{A}(t)$ is the conditional expectation of the counting process given the past history of the process $\mathcal{F}_t$, then the difference $\mathcal{M}(t) = \mathrm{N}(t) - \mathcal{A}(t)$ is a *martingale* which satisfies the essential property that

$$E\left[\mathcal{M}(t)|\mathcal{F}_s\right] = \mathcal{M}(s), \qquad s < t \tag{9.129}$$

or, equivalently, that

$$E\left[\mathcal{M}(t) - \mathcal{M}(s)|\mathcal{F}_s\right] = 0, \qquad s < t. \tag{9.130}$$

Likewise, the change in $\mathcal{M}(t)$ at any instant in time $d\mathcal{M}(t) = d\mathrm{N}(t) - d\mathcal{A}(t)$ is a martingale with respect to the history of the process $\mathcal{F}_t$. Note that $\mathcal{M}(t)$ is of the form (Observed - Expected) and thus represents the random noise associated with the observed counting process $\mathrm{N}(t)$ over time.

The theory of martingales then provides powerful tools that establish the large sample distribution of regular functions of martingales. The variance of the counting or point process is provided by the *variation process* and the large sample distribution is described by the martingale central limit theorem.

9.6.2 Nelson-Aalen Estimator

We now wish to estimate the underlying intensity $\lambda(t)$ for the cohort of N individuals followed up to some time τ. Aalen (1978) presented the following estimators of the intensity for possibly recurrent events as a generalization of the estimators of the intensity for survival data first proposed by Nelson (1972). Let $\{t_{(1)} < t_{(2)} < \cdots < t_{(J)}\}$ refer to the sequence of J distinct event times in the combined cohort of size N, where $d\mathrm{N}(s) > 0$ for $s \in [t_{(1)}, \cdots, t_{(J)}]$ and $= 0$ otherwise. We again designate the total number of subjects in the cohort at risk at the jth event time as $Y(t)$. If we assume that the intensity is discrete with probability mass at the observed event times, then from (9.127), a simple moment estimator of the jump in the cumulative hazard at time $t_{(j)}$ is

$$d\widehat{\Lambda}\left(t_{(j)}\right) = \frac{d\mathrm{N}\left(t_{(j)}\right)}{Y\left(t_{(j)}\right)}, \qquad 1 \le j \le J, \tag{9.131}$$

where $d\widehat{\Lambda}(s) = 0$ for values $s \notin [t_{(1)}, \cdots, t_{(J)}]$. The corresponding *Nelson-Aalen estimator* of the cumulative intensity then is

$$\widehat{\Lambda}(t) = \int_0^t d\widehat{\Lambda}(s) = \int_0^t \frac{d\mathrm{N}(s)}{Y(s)} = \sum\nolimits_{j:t_{(j)} \le t} \frac{d\mathrm{N}(t_{(j)})}{Y(t_{(j)})} \tag{9.132}$$

for $0 < s \le \tau$. Allowing for tied event times, the variance of the estimate is

$$V\left[\widehat{\Lambda}(t)\right] = \int_0^t \frac{\left[Y(s) - \Delta\mathrm{N}(s)\right] d\mathrm{N}(s)}{Y(s)^3}, \tag{9.133}$$

which is estimated consistently as

$$\widehat{V}\left[\widehat{\Lambda}(t)\right] = \sum\nolimits_{j:t_{(j)}\leq t} \frac{d\mathrm{N}(t_{(j)})\left[Y(t_{(j)}) - d\mathrm{N}(t_{(j)})\right]}{Y(t_{(j)})^3} . \tag{9.134}$$

For the analysis of survival data, or the time to the first or a single event, in the notation of Section 9.1, $d_j = d\mathrm{N}(t_{(j)})$, $n_j = Y(t_{(j)})$ and $p_j = d\mathrm{N}(t_{(j)})/Y(t_{(j)})$. In this case, the cumulative intensity equals the cumulative hazard and the estimate in (9.132) equals that in (9.27). Likewise, the estimate of the variance in (9.134) equals that in (9.30).

Clearly the estimated intensity $d\widehat{\Lambda}(t)$ is a step function that can be smoothed by standard methods. The simplest approach is to use linear interpolation between successive event times such that

$$d\widehat{\Lambda}(s) = \frac{s - t_{(j)}}{t_{(j+1)} - t_{(j)}}\left[d\widehat{\Lambda}(t_{(j+1)}) - d\widehat{\Lambda}(t_{(j)})\right] + d\widehat{\Lambda}(t_{(j)}), \quad t_{(j)} < s < t_{(j+1)} , \tag{9.135}$$

where $\widehat{\Lambda}(s)$ has successive steps at $t_{(j)}$ and $t_{(j+1)}$. This is equivalent to the simple histogram method of density estimation (*cf.* Bean and Tsokos, 1980).

Ramlau-Hansen (1983a, 1983b), and others, have described kernel functions that yield a smoothed estimator of the intensity $\lambda(s)$. In general, the kernel estimator of the intensity is defined as

$$\widehat{\lambda}^*(s) = \frac{1}{b}\int_0^\infty K\left(\frac{s-t}{b}\right) d\widehat{\Lambda}(t) \tag{9.136}$$

with bandwidth b, where $K(u)$ is a smoothing kernel function defined for $|u| \leq 1$, 0 otherwise, and where $\int_{-1}^{1} K(u)du = 1$. Computationally, the smoothed estimator is obtained as

$$\widehat{\lambda}^*(s) = \frac{1}{b}\sum\nolimits_{j=1}^{J} K\left(\frac{s - t_{(j)}}{b}\right)\left[\frac{d\mathrm{N}(t_{(j)})}{Y(t_{(j)})}\right] . \tag{9.137}$$

Andersen, Borgan, Gill and Keiding (1993) describe an adjustment for values that are within b units of the left boundary ($t = 0$) that may also be applied to values approaching the right boundary ($t = t_{(J)}$).

Ramlau-Hansen (1983a) also describes a consistent estimator of the variance of this smoothed estimator. Allowing for ties, the estimated variance, say $\widehat{\sigma}^2(s) = \widehat{V}\left[\widehat{\lambda}^*(s)\right]$, is provided as

$$\widehat{\sigma}^2(s) = \frac{1}{b^2}\sum\nolimits_{j=1}^{J}\left[K\left(\frac{s - t_{(j)}}{b}\right)\right]^2\left[\frac{d\mathrm{N}(t_{(j)})\left[Y(t_{(j)}) - d\mathrm{N}(t_{(j)})\right]}{Y(t_{(j)})^3}\right] \tag{9.138}$$

(Andersen, Borgan, Gill and Keiding, 1993). Therefore, a point-wise large-sample $1-\alpha$ confidence band is provided by $\widehat{\lambda}^*(s) \pm Z_{1-\alpha/2}\widehat{\sigma}^2(s)$.

Basically, $\widehat{\lambda}^*(s)$ is a weighted average of the intensity in the neighborhood of time s using weights defined by the kernel function. If $K(u)$ is the uniform distribution, then $\widehat{\lambda}^*(s)$ is the unweighted mean of the intensity over the region $s \pm b/2$. Usually, however, a kernel function gives weights that decline as the distance $s-t$ increases. Ramlau-Hansen (1983b) also derives a family of optimal smoothing kernels for counting process intensities. If the true intensity can be expressed as a polynomial of degree r, Ramlau-Hansen shows how to derive the optimal smoothing kernel in the sense that it minimizes the variance of the estimator of the vth derivative of the cumulative intensity function, for specified r and v. For a polynomial of degree 1, that is, a linear function in the immediate region, the optimal smoothing function for $v = 1$, that is, for the intensity itself, is the Epanechnikov (1969) kernel

$$K(u) = 0.75(1-u^2), \qquad |u| \leq 1. \tag{9.139}$$

However, as is the general case in non-parametric density estimation, the choice of the smoothing kernel is not nearly as important as the choice of the bandwidth. Although iterative procedures have been proposed for determining the optimal bandwidth, they are computationally tedious; see Andersen, Borgan, Gill and Keiding (1993). Often, therefore, the bandwidth is determined by trial and error, starting with a bandwidth that is too wide, and then successively reducing it to the point where further reductions in the bandwidth yield an overly noisy estimate. Andersen and Rasmussen (1986) present an example of kernel smoothed intensity function estimates.

9.6.3 Aalen-Gill Test Statistics

Now consider the case where there are two groups of subjects and we wish to test the equality of the intensities

$$H_0 : \lambda_1(s) = \lambda_2(s) = \lambda(s) \qquad 0 < s \leq \tau\ , \tag{9.140}$$

where $\lambda(s) = d\Lambda(s)/ds$ is the assumed common intensity. This hypothesis can be tested using a family of non-parametric tests that are generalizations of the weighted Mantel-Haenszel tests for lifetables which includes the logrank test and the Gehan Wilcoxon test as special cases. These are also called "Aalen-Gill" tests, based on the pioneering work of Aalen (1978) and Gill (1980). Andersen, Borgan, Gill and Keiding (1982) and Harrington and Fleming (1982) also present further results. The properties of these tests are established using the theory of martingales for counting processes. Lan and Lachin (1995) present a heuristic introduction to martingales and their application to rank tests for survival data using the analogy of a video game originally described by Lan and Wittes (1985).

Let $\mathrm{N}_i(s)$ and $Y_i(s)$ denote the aggregate counting and at risk processes summed over all n_i members of the ith group, $i = 1, 2$; and let $\mathrm{N}_\bullet(s) = \mathrm{N}_1(s) + \mathrm{N}_2(s)$ and

$Y_\bullet(s) = Y_1(s) + Y_2(s)$ refer to the aggregate processes for both groups combined, all evaluated over $0 < s \leq \tau$. Then, let $K(s)$ be a *predictable process* meaning that it is a function of the past history $\mathcal{F}_{s-}$, or is $\mathcal{F}_{s-}$-measurable, where $K(s)$ defines the weighting process for different test statistics.

The Aalen-Gill test statistic is defined as the *stochastic integral*

$$T_{AG} = \int_0^\tau K(s)\frac{Y_1(s)Y_2(s)}{Y_\bullet(s)}\left[\frac{d\mathcal{M}_1(s)}{Y_1(s)} - \frac{d\mathcal{M}_2(s)}{Y_2(s)}\right], \tag{9.141}$$

where

$$d\mathcal{M}_i(s) = d\mathrm{N}_i(s) - d\mathcal{A}_i(s) \tag{9.142}$$

is a martingale with the compensator $d\mathcal{A}_i(s) = Y_i(s)d\Lambda(s)$ defined in terms of the assumed underlying common intensity under H_0. Because $K(s)$, $Y_1(s)$ and $Y_2(s)$ are predictable processes, then the test statistic is the difference between two *stochastic integrals* each of which is a *martingale transform*, and thus is also a martingale. Expanding, the terms $d\Lambda(s)$ cancel to yield

$$T_{AG} = \int_0^\tau K(s)\frac{Y_1(s)Y_2(s)}{Y_\bullet(s)}\left[\frac{d\mathrm{N}_1(s)}{Y_1(s)} - \frac{d\mathrm{N}_2(s)}{Y_2(s)}\right]. \tag{9.143}$$

Since the Stieltjes integral can be expressed as a sum over the observed event times at which $d\mathrm{N}_\bullet(s) > 0$, then from (9.132),

$$T_{AG} = \int_0^\tau K(s)\frac{Y_1(s)Y_2(s)}{Y_\bullet(s)}\left[d\widehat{\Lambda}_1(s) - d\widehat{\Lambda}_2(s)\right], \tag{9.144}$$

which is a weighted sum of the differences between the estimated intensities of the two groups.

From (9.143) the test statistic can also be expressed as

$$\begin{aligned} T_{AG} &= \int_0^\tau \frac{K(s)}{Y_\bullet(s)}\left[d\mathrm{N}_1(s)Y_2(s) - d\mathrm{N}_2(s)Y_1(s)\right] \\ &= \int_0^\tau K(s)\left[d\mathrm{N}_1(s) - \frac{Y_1(s)d\mathrm{N}_\bullet(s)}{Y_\bullet(s)}\right]. \end{aligned} \tag{9.145}$$

From (9.132), the estimate of the common underlying intensity under H_0 is

$$d\widehat{\Lambda}(s) = \left\{\begin{array}{ll} \dfrac{d\mathrm{N}_\bullet(s)}{Y_\bullet(s)}, & s \in \left[t_{(1)}, \ldots, t_{(J)}\right] \\ 0 & \text{otherwise} \end{array}\right\}, \tag{9.146}$$

which is obtained from the combined sample. Thus

$$T_{AG} = \int_0^\tau K(s)\left[d\mathrm{N}_1(s) - \widehat{E}\left[d\mathrm{N}_1(s)|H_0, \mathcal{F}_{s-}\right]\right], \tag{9.147}$$

where the conditional expectation is

$$\widehat{E}\left[dN_1(s)|H_0,\mathcal{F}_{s-}\right] = Y_1(s)d\widehat{\Lambda}(s). \tag{9.148}$$

Expressing the Stieltjes integral as a sum over the observed event times $\{t_{(j)}\}$ yields

$$T_{AG} = \sum\nolimits_{j=1}^{J} K(t_{(j)})\left[dN_1(t_{(j)}) - \frac{Y_1(t_{(j)})dN_\bullet(t_{(j)})}{Y_\bullet(t_{(j)})}\right]. \tag{9.149}$$

Allowing for ties (*cf.* Fleming and Harrington, 1991; Andersen, Borgan, Gill and Keiding, 1993), the variance of the test statistic is consistently estimated as

$$\begin{aligned}\widehat{\sigma}^2 &= \int_0^{\tau} \frac{K(s)^2 Y_1(s)Y_2(s)}{Y_\bullet(s)^2}\left(\frac{Y_\bullet(s)-\Delta N_\bullet(s)}{Y_\bullet(s)-1}\right)dN_\bullet(s) \\ &= \sum\nolimits_{j=1}^{J} \frac{K(t_{(j)})^2 Y_1(t_{(j)})Y_2(t_{(j)})}{Y_\bullet(t_{(j)})^2}\left(\frac{Y_\bullet(t_{(j)})-\Delta N_\bullet(t_{(j)})}{Y_\bullet(t_{(j)})-1}\right)\Delta N_\bullet(t_{(j)}),\end{aligned} \tag{9.150}$$

where $\Delta N_\bullet(t_{(j)}) = \left[N_\bullet(t_{(j)}) - N_\bullet(t_{(j)}^-)\right]$. Then from the martingale central limit theorem, asymptotically $X_{AG}^2 = (T_{AG}^2/\widehat{\sigma}^2) \overset{d}{\approx} \chi^2$ on 1 *df* under H_0.

In terms of the notation of a 2×2 table at each event time, then $a_j \equiv dN_1(t_{(j)})$, $n_{1j} \equiv Y_1(t_{(j)})$, $m_{1j} \equiv \Delta N_\bullet(t_{(j)})$, $N_j \equiv Y_\bullet(t_{(j)})$, and $w(t_{(j)}) \equiv K(t_{(j)})$. Substituting into the above, the Aalen-Gill test statistic for either the time to a single event, or for recurrent events, is equivalent to the weighted Mantel-Haenszel test in (9.47).

The predictable weight process $K(s)$ then determines the specific form of the test. The logrank or Mantel statistic employs $K(s) = 1$, for $0 < s \leq \tau$ and the Gehan-type modified Wilcoxon statistic employs $K(s) = Y_\bullet(s)$. For recurrent events, the Harrington-Fleming G^ρ family of tests may also be defined, but in terms of the weight processes $K(s) = \exp\left[-\rho\widehat{\Lambda}(s)\right]$ using the estimated cumulative hazard rather than the estimated survival function. Again $\rho = 0$ yields a logrank test, $\rho = 1$ a modified Wilcoxon-like test.

Example 9.11 *DCCT Hypoglycemia Incidence*
Examples in Chapter 8 describe the rate of severe hypoglycemia in the intensive and conventional treatment groups of the Diabetes Control and Complications Trial. Those results describe the overall incidence rate of possibly recurrent events over time, and the relative risk between groups, based only on the total number of events experienced by each subject in the cohort. Here we expand those analyses to incorporate information based on the time of each event. A detailed assessment of the time-varying incidence of severe hypoglycemia over time is presented by the DCCT (1997). The DCCT hypoglycemia data for the secondary intervention cohort used herein is available from the www (see Preface).

Among the 715 patients in the secondary cohort, a total of 2266 episodes of severe hypoglycemia occurred at a total of 1565 distinct event times during the total of 9.4 years of follow-up. The total number at risk ranged from 715 at the

Fig. 9.4 Kernel-smoothed estimates of the intensity function expressed as a rate per 100 patient years within the intensive and conventional treatment groups in the secondary cohort of the DCCT, using a band width of nine months. The 95% confidence bands are also presented for each group.

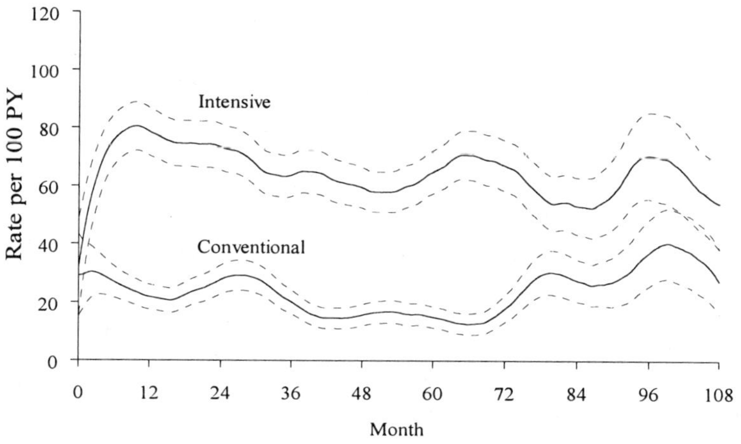

start of treatment, 713 at year 1 (365 days), 711 at year 3, 705 at year 5, 469 at year 7 and 173 at year 9. Figure 9.4 presents the kernel-smoothed estimates of the intensity (rate per 100 patient years) using a bandwidth of nine months. This shows that as intensive therapy was implemented over the first three months of treatment, the risk in the intensive group rose to a level about three times that in the conventional group. This excess risk persisted over time. The estimates beyond five years are increasingly unstable because of the declining numbers at risk.

The computation of the Aalen-Gill statistic requires that essentially a 2×2 table be constructed for each of the 1565 times at which one or more events occurred. The test using logrank (unit) weights yields $X^2 = 562.2$ on 1 df, whereas the test using Gehan-Wilcoxon weights yields $X^2 = 564.5$, both being markedly significant.

9.6.4 Multiplicative Intensity Model

The Cox (1972) proportional hazards regression model was adapted to the problem of modeling the risks of recurrent events by Prentice, Williams and Peterson (1981) and Kay (1982), among others, using a partial likelihood argument. However, using the martingale theory for counting processes, Andersen and Gill (1982) describe a generalization of the PH model that can be applied to any continuous time point process, including multiple recurrent events.

Consider the combined cohort of N subjects. For the ith subject with covariate vector $x_i(s)$, possibly time-dependent for $0 < s \leq \tau$, the intensity is assumed to be of the form

$$\alpha_i(s) = Y_i(s)\lambda_0(s)e^{\boldsymbol{x}_i(s)'\boldsymbol{\beta}}; \qquad 0 < s \leq \tau, \quad i = 1, \ldots, N, \tag{9.151}$$

where $\lambda_0(s)$ is the background intensity of possibly recurrent events over time and $\alpha_i(s)$ is defined for those points in time when the subject is at risk of the event, $Y_i(s) = 1$. As before, let $d\mathrm{N}_i(s)$ designate whether the ith individual experienced an event at time s; $d\mathrm{N}_i(s) = 1$ if yes, 0 otherwise, and denote the successive event times among all subjects as $t_{(j)}$, $1 \leq j \leq J$. The conventional partial likelihood allowing for time-dependent covariates can then be written as

$$\widetilde{L}(\beta) = \prod_{i=1}^{N}\prod_{j=1}^{J}\left[\frac{Y_i(t_{(j)})e^{\boldsymbol{x}_i(t_{(j)})'\boldsymbol{\beta}}}{\sum_{k=1}^{N} Y_k(t_{(j)})e^{\boldsymbol{x}_k(t_{(j)})'\boldsymbol{\beta}}}\right]^{d\mathrm{N}_i(t_{(j)})}, \tag{9.152}$$

which is an obvious generalization of Cox's partial likelihood for survival data in (9.69) where the "risk set" of individuals at risk at each event time is explicitly defined in the denominator by the $Y_k(t_{(j)})$. The principal distinction is that subjects are removed from the risk set immediately following the event in the analysis of survival data, whereas in this model, individuals may be at risk over different periods of time and may be retained in the risk set following each event, designated by the values of $Y_i(s)$.

As for the Cox model in Section 9.4.4, the kth element of the score vector $\boldsymbol{U}(\boldsymbol{\beta})$ corresponding to coefficient β_k is

$$U(\boldsymbol{\beta})_{\beta_k} = \sum\nolimits_{i=1}^{N}\sum\nolimits_{j=1}^{J} d\mathrm{N}_i(t_{(j)})\left[x_{ik}(t_{(j)}) - \overline{x}_k(t_{(j)}, \boldsymbol{\beta})\right], \tag{9.153}$$

where

$$\overline{x}_k(t_{(j)}, \boldsymbol{\beta}) = \left[\frac{\sum_{\ell=1}^{N} Y_\ell(t_{(j)})x_{\ell k}(t_{(j)})e^{\boldsymbol{x}_\ell(t_{(j)})'\boldsymbol{\beta}}}{\sum_{\ell=1}^{N} Y_\ell(t_{(j)})e^{\boldsymbol{x}_\ell(t_{(j)})'\boldsymbol{\beta}}}\right] \tag{9.154}$$

is again the weighted average of the covariate among all those at risk at time $t_{(j)}$. The information matrix then has elements

$$\boldsymbol{I}_{km}(\boldsymbol{\beta}) = \sum\nolimits_{i=1}^{N}\sum\nolimits_{j=1}^{J} d\mathrm{N}_i(t_{(j)})\left[C_{ikm}(t_{(j)}, \boldsymbol{\beta}) - \overline{x}_k(t_{(j)}, \boldsymbol{\beta})\overline{x}_m(t_{(j)}, \boldsymbol{\beta})\right] \tag{9.155}$$

with

$$C_{ikm}(t_{(j)}, \boldsymbol{\beta}) = \left[\frac{\sum_{\ell=1}^{N} Y_\ell(t_{(j)})x_{\ell k}(t_{(j)})x_{\ell m}(t_{(j)})e^{\boldsymbol{x}_\ell(t_{(j)})'\boldsymbol{\beta}}}{\sum_{\ell=1}^{N} Y_\ell(t_{(j)})e^{\boldsymbol{x}_\ell(t_{(j)})'\boldsymbol{\beta}}}\right] \tag{9.156}$$

for $1 \leq k \leq m \leq p$.

As a problem it is shown that the score test for the coefficient of a binary covariate for each of two treatment groups reduces to the Aalen-Gill generalized Mantel or logrank test.

Andersen and Gill (1982) show that the usual properties of large sample estimators and score tests are then provided by the theory of martingales for counting processes. Gill (1984) provides a readable account of these derivations with an introduction to the martingale central limit theorem. The basic result is the demonstration that the total score vector $\boldsymbol{U}(\boldsymbol{\beta})$ is a martingale. Even though the score vectors for the individual subjects are not $i.i.d.$ with mean zero, then from the martingale central limit theorem it can be shown that asymptotically

$$\widehat{\boldsymbol{\beta}} \overset{d}{\approx} \mathcal{N}[\boldsymbol{\beta},\ \boldsymbol{I}(\boldsymbol{\beta})^{-1}], \tag{9.157}$$

where $\boldsymbol{I}(\boldsymbol{\beta})$ is the expected information matrix. Large sample inferences can then be obtained as for the Cox model.

There has been no exploration of measures of explained variation for use with the multiplicative intensity model for recurrent events. For lack of an alternative, the proportion of explained log likelihood, analogous to the entropy R^2 in logistic regression, may be used to assess the relative importance of alternate models or of different covariates. This measure likely under-estimates the proportion of explained variation with respect to other possible metrics.

Example 9.12 *DCCT Hypoglycemia Risk*
Analyses were also conducted to assess the relationship between the level of HbA_{1c} as a time-dependent covariate and the risk of severe hypoglycemia over time, including the risk of recurrent events (DCCT, 1997). In both the intensive and conventional treatment groups, the dominant predictor of future hypoglycemic episodes was a history of past hypoglycemia as represented by an additional time-dependent covariate giving the number of prior episodes since entry into the trial. The analyses employed the multiplicative intensity model using the SAS PROC PHREG with "counting process" input. For example, the data for a single subject are:

(start	stop]	*Event*	$\log(HbA_{1c})$	*# Prior*
0	34	0	2.399712	0
34	66	0	2.198335	0
⋮		⋮	⋮	⋮
145	162	1	1.947338	0
162	189	1	1.947338	1
189	194	0	1.947338	2
⋮		⋮	⋮	⋮
600	608	1	1.891605	2
608	630	0	1.891605	3
⋮		⋮	⋮	⋮
3421	3451	0	1.960095	27

Each interval is defined by a start and stop time, closed on the right. A new interval is also constructed whenever an event occurs, in which case the value of the stop time equals the event time. A new interval is also constructed whenever the value of a time-dependent covariate changes, where the start time indicates when the change becomes effective.

In this example, the patient starts at day 0 with a log HbA_{1c} of 2.399712 and with no prior episodes of hypoglycemia. On day 34 a new blood sample was drawn for which the log HbA_{1c} value was 2.198335. An updated HbA_{1c} value was obtained on day 66. As of day 145 a new log HbA_{1c} value of 1.947338 was obtained. The patient then experienced the first episode of severe hypoglycemia on day 162, followed by a second episode on day 189. On day 194 an updated value of HbA_{1c} was then obtained. On day 608 the patient experienced a third episode, and so on. The last updated HbA_{1c} was obtained on day 3421 and the patient completed the study on day 3451, having experienced a total of 27 episodes of severe hypoglycemia during the study.

The number of prior episodes during the study is a second time-dependent covariate. It is incremented beginning with the interval immediately following that in which an event occurs. If a patient experienced more than one event on the same day, then the additional events were included in the data set as though they had occurred on successive days.

Among the 715 intensive group patients in the Secondary cohort, the following SAS statements would fit a quadratic model in the log HbA_{1c}:

```
proc phreg;
strata phase2;
model (start,stop)*hypo(0) = lhba1c lhba1c2 nprior;
```

where lhba1c2 is the square of lhba1c, the log(HbA_{1c}). The model is stratified by the phase in which the subject entered the study because after the first year of recruitment (Phase 2) the eligibility criteria were changed to exclude patients with a prior history of severe hypoglycemia. The principal output of the program, including the model estimates of the coefficients, is presented in Table 9.6. Note that the data set contained 32,413 records based on the data from the 715 patients who experienced a total of 1720 episodes of severe hypoglycemia.

Since a quadratic effect for the log HbA_{1c} was included in the model, it can then be shown that the percentage reduction in risk for a fixed 10% reduction in the HbA_{1c} at an HbA_{1c} value of x is computed as $100(0.9^{\eta} - 1)$, where $\eta = \beta_1 + \beta_2 \log(0.9x^2)$. Thus from the estimated coefficients, a 10% reduction from an HbA_{1c} of 10 yields a 37.9% reduction in risk, whereas a 10% reduction from an HbA_{1c} of 8 yields a 23.6% reduction in risk. Because the coefficient for the quadratic effect is negative, the risk reduction per 10% lower HbA_{1c} declines as the reference value of the HbA_{1c} declines.

As a crude measure of explained variation for the model, the proportion of log likelihood explained by these three covariates (see Section A.8.3 in the Appendix) is provided by $R^2_\ell = 974.037/17520.386 = 0.056$. In analyses of the complete cohort (DCCT, 1997), the risk relationships were stronger among patients in the conventional than in the intensive treatment groups, explaining a higher fraction

Table 9.6 Multiplicative Intensity Model of Recurrent Hypoglycemia with a Quadratic Effect for log Hba_1c

Summary of the Number of Event and Censored Values

Stratum	PHASE2	Total	Event	Censored	Percent Censored
1	0	21038	1015	20023	95.18
2	1	11375	705	10670	93.80
Total		32413	1720	30693	94.69

Testing Global Null Hypothesis: BETA=0

Criterion	Without Covariates	With Covariates	Model Chi-Square
-2 LOG L	17520.386	16546.349	974.037 with 3 DF (p=0.0001)
Score	.	.	1628.964 with 3 DF (p=0.0001)
Wald	.	.	1278.827 with 3 DF (p=0.0001)

Analysis of Maximum Likelihood Estimates

Variable	DF	Parameter Estimate	Standard Error	Wald Chi-Square	Pr > Chi-Square
LHBA1C	1	13.557041	3.94309	11.82104	0.0006
LHBA1C2	1	-3.943712	1.00977	15.25346	0.0001
NPRIOR	1	0.088835	0.00269	1090	0.0001

of the log likelihood. Further, the risk gradients were different between the two groups, such that the difference in the level of glycemia alone was not the principal determinant of the difference in risk of severe hypoglycemia between the two groups.

9.7 PROBLEMS

9.1 Exponential Model. Assume a simple exponential survival model with constant hazard $\lambda(t) = \lambda$, $t > 0$, where $S(t) = e^{-\lambda t}$.

9.1.1. In a cohort of size N, none of whom have censored survival (event or death) times (that is, all N event times are observed), derive the maximum likelihood estimator of λ and its asymptotic variance as

$$\widehat{\lambda} = \frac{N}{\sum_{j=1}^{N} t_j}, \qquad V(\widehat{\lambda}) = \frac{\lambda^2}{N}, \tag{9.158}$$

in terms of the observed event times t_j $(j = 1, \ldots, N)$.

9.1.2. In a cohort of size N, now let there be D event (deaths) observed and the remainder $N - D$ right censored at random. Let δ_j be the binary indicator variable that denotes whether the jth subject's event time was observed ($\delta_j = 1$) or was right censored ($\delta_j = 0$), where $D = \sum_j \delta_j$. Using the likelihood in (9.6), derive the maximum likelihood estimator and variance as

$$\widehat{\lambda} = \frac{D}{\sum_{j=1}^{D} t_j + \sum_{j=D+1}^{N} t_j^+}, \qquad V(\widehat{\lambda}) = \frac{\lambda^2}{E(D)} = \frac{\lambda^2}{NE(\delta)}, \tag{9.159}$$

where t_j $(j = 1, \dots, D)$ are the observed event times and t_j^+ $(j = D+1, \dots, N)$ are the right censored event times. Thus asymptotically $\widehat{\lambda} \sim \mathcal{N}[\lambda,\ \lambda^2/E(D)]$.

9.1.3. Under random censoring, use the delta method to show that the estimated log survival distribution at a given time t, $\log[\widehat{S}(t)]$, is asymptotically normally distributed with expectation $-\lambda t$ and variance $(\lambda t)^2/E(D)$.

9.1.4. Also show that the estimated survival distribution at a given time t, $\widehat{S}(t)$, is asymptotically normally distributed with expectation $e^{-\lambda t}$ and variance $e^{-2\lambda t}(\lambda t)^2/E(D)$.

9.1.5. Now consider the case of staggered entry into the trial where we assume subjects enter the study uniformly over the interval 0 to T_R in calendar time, meaning that the entry times, say r_j are distributed as uniform over $(0, T_R]$. Also assume that all subjects are followed to time $T_S > T_R$, T_S being the total study duration. Then any subjects event free at the end of the study are administratively censored with a censored event time of $t_j^+ = T_S - r_j$. Allowing only for administrative censoring (that is, no other censoring during the study), derive the expression for $E(\delta|\lambda)$ presented in (9.117).

9.1.6. Now assume that losses to follow-up occur randomly over time with a constant hazard rate η. Show that the probability of the event in the study is given by the expression in (9.118) and that the probability of being lost to follow-up is given by (9.119).

9.1.7. Consider the case of two populations with constant hazards λ_i and corresponding survival functions $S_i(t)$; $i = 1, 2$. The constant of proportionality of the hazards is $\theta = \lambda_1/\lambda_2$. Show that $S_1(t) = S_2(t)^\theta$.

9.1.8. Let $t_{i\alpha}$ be the time at which α is the remaining proportion of survivors in the ith population, or $S_i(t_{i\alpha}) = \alpha$, $0 < \alpha < 1$, $i = 1, 2$. Show that $t_{1\alpha}/t_{2\alpha} = 1/\theta$ for all $\alpha \in (0,\ 1)$. Thus a hazard ratio of $\theta = 2$ implies that it will take half the time for subjects in population 1 to reach the α fraction of survivors than those in population 2; or $\theta = 1/3$ implies that it will take 3 times as long. Thus when θ is the hazard ratio, $1/\theta$ is the factor by which the event times are either accelerated or decelerated.

9.2 Weibull Model. The Weibull distribution with rate parameter μ and shape parameter γ has a hazard function $\lambda(t) = \mu\gamma(t)^{\gamma-1}$, $\mu > 0$, $\gamma > 0$, where $\gamma = 1$ equals the exponential distribution.

9.2.1. Show that the Weibull survival distribution is expressed as

$$S(t) = \exp(-\mu t^\gamma) \tag{9.160}$$

and the death density as

$$f(t) = \mu\gamma(t)^{\gamma-1}\exp(-\mu t^{\gamma}). \tag{9.161}$$

9.2.2. For a random sample of N observations containing D event times $\{t_j\}$ and $N-D$ censored event times $\{t_j^+\}$, show that the likelihood function in the parameters $\boldsymbol{\theta} = (\mu\ \gamma)^T$ is

$$L(\boldsymbol{\theta}) = \prod_{j=1}^{N} \left[\mu\gamma(t_j)^{\gamma-1}\right]^{\delta_j} \exp(-\mu t_j^{\gamma}). \tag{9.162}$$

From the log-likelihood and its derivatives, show that the maximum likelihood estimating equations for μ and γ are

$$\begin{aligned} U(\boldsymbol{\theta})_\mu &= \frac{D}{\mu} - \sum\nolimits_j t_j^{\gamma} \\ U(\boldsymbol{\theta})_\gamma &= \frac{D}{\gamma} + \sum\nolimits_j \delta_j \log(t_j) - \mu \sum\nolimits_j t_j^{\gamma}\log(t_j). \end{aligned} \tag{9.163}$$

Given an estimate of γ, the MLE of the rate parameter is $\widehat{\mu} = D/\left(\sum_j t_j^{\widehat{\gamma}}\right)$. What does this suggest as to a starting value for μ and γ in the Newton-Raphson iteration?

9.2.3. Then show that the expected information matrix for μ and γ has elements:

$$\begin{aligned} \boldsymbol{I}(\boldsymbol{\theta})_\mu &= \frac{D}{\mu^2} \\ \boldsymbol{I}(\boldsymbol{\theta})_\gamma &= \frac{D}{\gamma^2} + \mu \sum\nolimits_j t_j^{\gamma}\left[\log(t_j)\right]^2 \\ \boldsymbol{I}(\boldsymbol{\theta})_{\mu,\gamma} &= \sum\nolimits_j t_j^{\gamma}\log(t_j). \end{aligned} \tag{9.164}$$

9.2.4. Use the δ-method to show that the variance of the log estimated survival distribution at time t is

$$V\left(\log[\widehat{S}(t)]\right) = t^{2\gamma}\sigma_\mu^2 + \mu^2 t^{2\gamma}\left[\log(t)\right]^2 \sigma_\gamma^2 + \mu t^{2\gamma}\log(t)\sigma_{\mu\gamma}\,, \tag{9.165}$$

where the variances $(\sigma_\mu^2, \sigma_\gamma^2)$ and covariance $(\sigma_{\mu\gamma})$ of the parameter estimates are provided by the inverse information matrix.

9.2.5. Now let the rate parameter be a function of a vector of covariates $\boldsymbol{x}$ of the form $\mu = \exp(\alpha + \boldsymbol{x}'\boldsymbol{\beta})$. Show that the score equations for the elements $(\alpha, \boldsymbol{\beta})$ are

$$\begin{aligned} U(\boldsymbol{\theta})_\alpha &= \sum\nolimits_j \left(\delta_j - e^{\alpha + \boldsymbol{x}_j'\boldsymbol{\beta}} t_j^{\gamma}\right) = D - \sum\nolimits_j \mu_j t_j^{\gamma}\,, \\ U(\boldsymbol{\theta})_\gamma &= \frac{D}{\gamma} + \sum\nolimits_j \delta_j \log(t_j) - \sum\nolimits_j \mu_j t_j^{\gamma}\log(t_j)\,, \end{aligned} \tag{9.166}$$

and

$$U(\boldsymbol{\theta})_{\beta_k} = \sum\nolimits_j x_{jk}\left(\delta_j - e^{\alpha + \boldsymbol{x}_j'\boldsymbol{\beta}} t_j^{\gamma}\right) = \sum\nolimits_j x_{jk}\left(\delta_j - \mu_j t_j^{\gamma}\right)$$

for $1 \leq k \leq p$.

9.2.6. Show that the elements of the information matrix are

$$\begin{aligned}
\boldsymbol{I}(\boldsymbol{\theta})_{\alpha} &= \sum\nolimits_j \mu_j t_j^{\gamma}\ , & \boldsymbol{I}(\boldsymbol{\theta})_{\gamma} &= \frac{D}{\gamma^2} + \sum\nolimits_j \mu_j t_j^{\gamma}\left[\log(t_j)^2\right]\ , \\
\boldsymbol{I}(\boldsymbol{\theta})_{\alpha\gamma} &= \sum\nolimits_j \mu_j t_j^{\gamma}\log(t_j)\ , & \boldsymbol{I}(\boldsymbol{\theta})_{\beta_k} &= \sum\nolimits_j x_{jk}^2 \mu_j t_j^{\gamma}\ , \\
\boldsymbol{I}(\boldsymbol{\theta})_{\alpha\beta_k} &= \sum\nolimits_j x_{jk}\mu_j t_j^{\gamma}\ , & \boldsymbol{I}(\boldsymbol{\theta})_{\gamma\beta_k} &= \sum\nolimits_j x_{jk}\mu_j t_j^{\gamma}\log(t_j)\ , \\
\boldsymbol{I}(\boldsymbol{\theta})_{\beta_k\beta_m} &= \sum\nolimits_j x_{jk}x_{jm}\mu_j t_j^{\gamma}\ ,
\end{aligned} \tag{9.167}$$

for $1 \leq k < m \leq p$.

9.2.7. Show that e^{β_j} equals the hazard ratio per unit increase in the jth covariate x_j. Thus the Weibull model with this parameterization is a parametric proportional hazards model.

9.2.8. In this model show that the survival function for an individual with covariate vector $\boldsymbol{x}$ is

$$S(t|\boldsymbol{x}) = \exp\left[-e^{(\alpha+\boldsymbol{x}'\boldsymbol{\beta})}t^{\gamma}\right] = \exp\left[-\exp\left(\gamma\log(t) + \alpha + \boldsymbol{x}'\boldsymbol{\beta}\right)\right]. \tag{9.168}$$

9.2.9. Then show that

$$\log(-\log[\widehat{S}(t|\boldsymbol{x})]) = \widehat{\gamma}\log(t) + \widehat{\alpha} + \boldsymbol{x}'\widehat{\boldsymbol{\beta}} \tag{9.169}$$

and that $V\left[\log(-\log[\widehat{S}(t|\boldsymbol{x})])\right] = \boldsymbol{H}'\boldsymbol{\Sigma}\boldsymbol{H}$, where

$$\boldsymbol{H} = [\log(t) \quad 1 \quad \boldsymbol{x}']^T. \tag{9.170}$$

The covariance matrix of the parameter estimates, Σ, is provided by the inverse information and is partitioned as

$$\boldsymbol{\Sigma} = \begin{bmatrix} \sigma_{\gamma}^2 & \sigma_{\gamma\alpha} & \Sigma_{\gamma\boldsymbol{\beta}} \\ \sigma_{\alpha\gamma} & \sigma_{\alpha}^2 & \Sigma_{\alpha\boldsymbol{\beta}} \\ \Sigma_{\boldsymbol{\beta}\gamma} & \Sigma_{\boldsymbol{\beta}\alpha} & \Sigma_{\boldsymbol{\beta}} \end{bmatrix}, \tag{9.171}$$

where $\Sigma_{\boldsymbol{\beta}}$ is a $p \times p$ matrix.

9.3 Accelerated Failure Time Model. The accelerated failure time model assumes that the time to any fraction of survivors is accelerated or decelerated by some function of a covariate vector $\boldsymbol{X}$ with coefficient vector $\widetilde{\boldsymbol{\beta}}$, usually of the form $\exp\left[-(\widetilde{\alpha} + \boldsymbol{x}'\widetilde{\boldsymbol{\beta}})/\sigma\right]$ where σ is a scale parameter. That is, for a specified survival function $S_0(t)$, then the survival function for a patient with covariate vector $\boldsymbol{x}$ is of the form

$$S(t|\boldsymbol{x}) = S_0(e^{-(\widetilde{\alpha}+\boldsymbol{x}'\widetilde{\boldsymbol{\beta}})/\sigma}t) = S_0(\widetilde{t}_{\boldsymbol{x}}) \tag{9.172}$$

where $\widetilde{t}_{\boldsymbol{x}} = e^{-(\widetilde{\alpha}+\boldsymbol{x}'\widetilde{\boldsymbol{\beta}})/\sigma}t$ is the transformed accelerated failure time determined by the covariates vector value $\boldsymbol{x}$.

9.3.1. Noting that $\Lambda(t|\boldsymbol{x}) = \Lambda_0(e^{-(\widetilde{\alpha}+\boldsymbol{x}'\widetilde{\boldsymbol{\beta}})/\sigma}t)$, show that the hazard function is of the form

$$\lambda(t|\boldsymbol{x}) = e^{-(\widetilde{\alpha}+\boldsymbol{x}'\widetilde{\boldsymbol{\beta}})/\sigma}\lambda_0(e^{-(\widetilde{\alpha}+\boldsymbol{x}'\widetilde{\boldsymbol{\beta}})/\sigma}t). \tag{9.173}$$

Hint: Use Leibniz' rule for the derivative of an integral. Thus

$$f(t|\boldsymbol{x}) = e^{-(\widetilde{\alpha}+\boldsymbol{x}'\widetilde{\boldsymbol{\beta}})/\sigma}\lambda_0(e^{-(\widetilde{\alpha}+\boldsymbol{x}'\widetilde{\boldsymbol{\beta}})/\sigma}t)S_0(e^{-(\widetilde{\alpha}+\boldsymbol{x}'\widetilde{\boldsymbol{\beta}})/\sigma}t).$$

9.3.2. Now, let $y/\sigma = \log(t)$ so that $t = \exp(y/z)$. Given any underlying distribution with hazard function $\lambda_0(t)$ and corresponding survival function $S_0(t)$, show that the conditional distribution of Y given $\boldsymbol{x}$ is

$$f(y|\boldsymbol{x}) = e^{[y-(\widetilde{\alpha}+\boldsymbol{x}'\widetilde{\boldsymbol{\beta}})]/\sigma}\lambda_0(e^{[y-(\widetilde{\alpha}+\boldsymbol{x}'\widetilde{\boldsymbol{\beta}})]/\sigma})S_0(e^{[y-(\widetilde{\alpha}+\boldsymbol{x}'\widetilde{\boldsymbol{\beta}})]/\sigma})(\frac{1}{\sigma}). \tag{9.174}$$

Then for $\varepsilon = \left[y - (\widetilde{\alpha} + \boldsymbol{x}'\widetilde{\boldsymbol{\beta}})\right]/\sigma$ it follows that

$$f(\varepsilon) = e^{\varepsilon}\lambda_0(\varepsilon)S_0(\varepsilon). \tag{9.175}$$

Thus we can adopt a linear model of the form $y_i = \widetilde{\alpha} + \boldsymbol{x}'\widetilde{\boldsymbol{\beta}} + \varepsilon_i\sigma$ with error distribution $f(\varepsilon)$.

9.3.3. Since $P(t_i > t) = P[\log(t_i) > \log(t)]$, then show that

$$S(t|\boldsymbol{x}_i) = P\left[\varepsilon_i > \frac{y_i - (\widetilde{\alpha} + \boldsymbol{x}_i'\widetilde{\boldsymbol{\beta}})}{\sigma}\right] \tag{9.176}$$

evaluated with respect to the distribution of the errors that in turn can be obtained from the assumed underlying survival distribution $S_0(t)$. These developments can be used to obtain an accelerated failure time model for a given parametric distribution $S_0(t)$.

9.3.4. Weibull Accelerated Failure Time Model. In a standard Weibull model with $\mu = 1$ then $\lambda(t) = \gamma(t)^{\gamma-1}$. Adopting the accelerated failure time transformation as in (9.172) with the acceleration factor $-(\widetilde{\alpha} + \boldsymbol{x}'\widetilde{\boldsymbol{\beta}})$ for a subject with covariate vector $\boldsymbol{x}$, show that

$$\lambda(t|\boldsymbol{x}) = e^{-(\widetilde{\alpha}+\boldsymbol{x}'\widetilde{\boldsymbol{\beta}})\gamma}\gamma(t)^{\gamma-1} \tag{9.177}$$

and that

$$S(t|\boldsymbol{x}) = \exp\left[-e^{-\gamma(\widetilde{\alpha}+\boldsymbol{x}'\widetilde{\boldsymbol{\beta}})}t^{\gamma}\right] = \exp\left(-\exp\left[\log(t) - (\widetilde{\alpha} + \boldsymbol{x}'\widetilde{\boldsymbol{\beta}})\right]^{\gamma}\right). \tag{9.178}$$

Substituting $\gamma = 1/\sigma$ it follows that

$$S(t|\boldsymbol{x}) = \exp\left(-\exp\left[\frac{\log(t) - (\widetilde{\alpha} + \boldsymbol{x}'\widetilde{\boldsymbol{\beta}})}{\sigma}\right]\right). \tag{9.179}$$

This is the parameterization used by the SAS procedure LIFEREG that fits a Weibull accelerated failure time model in which the covariates determine the expected value of Y and σ is the scale parameter.

Compared to (9.168) it follows that the Weibull proportional hazards parameters in Problem 9.2.5 can be obtained from the accelerated failure time model as $\alpha = -\widetilde{\alpha}/\sigma$, $\beta_j = -\widetilde{\beta}_j/\sigma$, and $\gamma = 1/\sigma$. The SAS procedure LIFEREG provides estimates of $\widetilde{\alpha}$, $\widetilde{\boldsymbol{\beta}}$ and σ and of the large sample covariance matrix of the estimates. From these, the estimates of α, $\boldsymbol{\beta}$ and γ are readily derived, and their covariance matrix obtained via the δ-method.

9.3.5. The Weibull accelerated failure time model in $\log(t)$ may also be derived as follows: Let the rate parameter be expressed as a function of a covariate vector $\boldsymbol{X}$ as $\mu = \exp\left[-(\widetilde{\alpha} + \boldsymbol{x}'\widetilde{\boldsymbol{\beta}})\gamma\right]$ where $\gamma = \sigma^{-1}$. Then show that the distribution of $y = \log(t)$ can be expressed as

$$f(y|\boldsymbol{x}) = \frac{1}{\sigma}\exp\left[\frac{y - (\widetilde{\alpha} + \boldsymbol{x}'\widetilde{\boldsymbol{\beta}})}{\sigma} - \exp\left(\frac{y - (\widetilde{\alpha} + \boldsymbol{x}'\widetilde{\boldsymbol{\beta}})}{\sigma}\right)\right]. \tag{9.180}$$

9.3.6. Now let $\varepsilon = \left[y - E(y|\boldsymbol{x})\right]/\sigma$ be the scaled residual where $E(y|\boldsymbol{x}) = \widetilde{\alpha} + \boldsymbol{x}'\widetilde{\boldsymbol{\beta}}$. Show that the density of ε is an extreme value (Gumbel) distribution with density $f(\varepsilon) = \exp\left[\varepsilon - e^{\varepsilon}\right]$. Thus the log-linear model is of the form $y = \widetilde{\alpha} + \boldsymbol{x}'\widetilde{\boldsymbol{\beta}} - \sigma\varepsilon$ with Gumbel distributed errors.

9.3.7. Now let $w = e^{\varepsilon}$. Show that the density function for w is the unit exponential, $f(w) = e^{-w}$.

9.3.8. The survival function is defined as $S(t) = P(T > t) = P[\log(T) > \log(t)]$. Since e^{ε} is distributed as the unit exponential, show that the survival function for a subject with covariate vector $\boldsymbol{x}$ equals that presented in (9.178).

9.4 Log-logistic Model. Consider the log-logistic distribution with rate parameter μ and shape parameter γ, where the hazard function is

$$\lambda(t) = \frac{\mu\gamma(t)^{\gamma-1}}{[1 + \mu t^{\gamma}]}. \tag{9.181}$$

9.4.1. Show that the survival function is $S(t) = [1 + \mu t^{\gamma}]^{-1}$.

9.4.2. Show that the survival distributions for two groups with the same shape parameter (γ) but with different rate parameters $\mu_1 \neq \mu_2$ have proportional failure and survival odds as in (9.55) with failure (event) odds ratio μ_1/μ_2 and survival odds ratio μ_2/μ_1.

9.4.3. To generalize the log-logistic proportional odds model as a function of a covariate vector $\boldsymbol{x}$, let $\mu = \exp(\alpha + \boldsymbol{x}'\boldsymbol{\beta})$. Show that

$$S(t|\boldsymbol{x}) = \left(1 + e^{\alpha + \boldsymbol{x}'\boldsymbol{\beta}} t^{\gamma}\right)^{-1}. \tag{9.182}$$

9.4.4. Show that the coefficient e^{β_j} equals the failure odds ratio per unit increase in the jth covariate x_j and that $e^{-\beta_j}$ equals the survival odds ratio.

9.4.5. Now, let $Y = \log(T)$. Show that the density of the distribution of $y = \log(t)$ is

$$f(y) = \frac{\mu e^{y\gamma}\gamma}{(1 + \mu e^{y\gamma})^2} \tag{9.183}$$

and the survival function is

$$S(y) = (1 + \mu e^{y\gamma})^{-1} . \tag{9.184}$$

9.4.6. Then for an individual with covariate vector $\boldsymbol{x}$, show that the conditional distribution is

$$f(y|\boldsymbol{x}) = \frac{\exp\left[\frac{y - (\tilde{\alpha} + \boldsymbol{x}'\tilde{\boldsymbol{\beta}})}{\sigma}\right]\frac{1}{\sigma}}{\left[1 + \exp\left(\frac{y - (\tilde{\alpha} + \boldsymbol{x}'\tilde{\boldsymbol{\beta}})}{\sigma}\right)\right]^2} = \frac{\exp\left[\frac{y - E(y|\boldsymbol{x})}{\sigma}\right]\frac{1}{\sigma}}{\left[1 + \exp\left(\frac{y - E(y|\boldsymbol{x})}{\sigma}\right)\right]^2} . \tag{9.185}$$

where the proportional odds model parameters are now expressed as

$$\mu = \exp\left[-E(y|\boldsymbol{x})\gamma\right] = \exp\left[-(\tilde{\alpha} + \boldsymbol{x}'\tilde{\boldsymbol{\beta}})/\sigma\right] \tag{9.186}$$

with $\gamma = \sigma^{-1}$ so that $\alpha = -\tilde{\alpha}/\sigma$, $\boldsymbol{\beta} = -\tilde{\boldsymbol{\beta}}/\sigma$. Therefore, the proportional failure or survival odds model parameters can be obtained from the accelerated failure time model.

9.4.7. Then show that the survival function in $\log t$ can be expressed as

$$S(y|\boldsymbol{x}) = \left(1 + \exp\left[\frac{y - (\tilde{\alpha} + \boldsymbol{x}'\tilde{\boldsymbol{\beta}})}{\sigma}\right]\right)^{-1} = \left(1 + \exp\left[\frac{y - E(y|\boldsymbol{x})}{\sigma}\right]\right)^{-1} . \tag{9.187}$$

9.4.8. Show that the density of $\varepsilon = \left[y - E(y|\boldsymbol{x})\right]/\sigma$ is the logistic distribution with density

$$f(\varepsilon) = \frac{e^{\varepsilon}}{[1 + e^{\varepsilon}]^2} . \tag{9.188}$$

Thus the errors in the log-linear model are distributed as logistic and the hazard function in (9.181) corresponds to a log-logistic density.

9.5 Consider two populations with arbitrary hazard functions $\lambda_1(t)$ and $\lambda_2(t)$, respectively.

9.5.1. Show that the hazard ratio and the survival odds ratio cannot be proportional simultaneously under the same survival model. That is,

a. If $\lambda_1(t)/\lambda_2(t) = \phi \; \forall t$, then $O_1(t)/O_2(t) \neq \mathit{constant} \; \forall t$; and

b. If $O_1(t)/O_2(t) = \varphi \; \forall t$, then $\lambda_1(t)/\lambda_2(t) \neq \mathit{constant} \; \forall t$ where $O_i(t) = S_i(t)/[1 - S_i(t)]$, $i = 1, 2$.

9.5.2. Show that two populations with Weibull survival distributions with rate parameters μ_1 and μ_2, where $\mu_1 \neq \mu_2$ but with the same shape parameter $\gamma_1 = \gamma_2 = \gamma$ satisfy (a).

9.5.3. Show that two populations with logistic survival distributions with rate parameters μ_1 and μ_2, where $\mu_1 \neq \mu_2$ but with the same shape parameter $\gamma_1 = \gamma_2 = \gamma$ satisfy (b).

b

9.6 Let T_a, and T_b be two independent random variables corresponding to two separate causes of death, each with corresponding hazard functions $\lambda_a(t)$, and $\lambda_b(t)$, respectively. A subject will die from only one cause, and the observed death time for a subject is $t^* = \min(t_a, t_b)$, where t_a and t_b are the potential event times for that subject.

9.6.1. Show that t^* has the hazard function

$$\lambda^*(t^*) = \lambda_a(t^*) + \lambda_b(t^*). \tag{9.189}$$

Note that this generalizes to any number of competing causes of death.

9.6.2. Consider the case where one cause of death is the index event of interest with hazard function $\lambda_I(t)$, and the other is a competing risk with hazard $\lambda_C(t)$. Suppose each is exponentially distributed with constant cause-specific hazards λ_I and λ_C. From (9.44) show that the sub-distribution of the index event has density

$$f_I(t) = \lambda_I \exp\left[-\left(\lambda_I + \lambda_C\right) t\right]. \tag{9.190}$$

9.6.3. Now consider two populations with constant cause specific hazards (λ_{I1}, λ_{C1}) in the first population and (λ_{I2}, λ_{C2}) in the second. Show that the ratio of the ratio of the sub-distribution densities is

$$\frac{f_{I1}(t)}{f_{I2}(t)} = \frac{\lambda_{I1}}{\lambda_{I2}} \exp\left[\left(\lambda_{I2} - \lambda_{I1} + \lambda_{C2} - \lambda_{C1}\right) t\right]. \tag{9.191}$$

9.6.4. Show that if the cause-specific index hazards are the same in each population, $\lambda_{I1} = \lambda_{I2}$, but the competing risk hazards differ, then this index event density ratio is increasing (decreasing) in t when $\lambda_{C1} < (>) \; \lambda_{C2}$.

9.6.5. Use numerical examples to show that when the cause-specific index hazards differ, then depending on the values of the competing risk hazards, the above density ratio may remain constant over time, may be increasing in t, or decreasing in t. Also show that in the latter two cases the density ratio over time may cross from < 1 to > 1 or vice versa.

9.6.6. For $\lambda_{I1} = 4$, $\lambda_{I2} = 2$, $\lambda_{C1} = 3$, $\lambda_{C2} = 1$, show that the index density ratio is decreasing in t and that the ratio equals 1 at $t = -\frac{1}{2}\log(1/2)$.

9.6.7. Section 9.2.3 presents an example involving competing risks where inclusion of competing risk exits in the risk set throughout yields a correct estimate of the survival probability. Show that this adjustment, in general, is appropriate when there are no exits other than those caused by the competing risk. For the jth interval, the extent of exposure or number at risk becomes $n_j = r_j + \sum_{\ell=1}^{j-1} e_\ell = N - \sum_{\ell=1}^{j-1} d_\ell$. Then show that the adjusted product-limit estimator at $t_{(j)}$ is

$$\widehat{S}(t_{(j)}) = \frac{N - \sum_{\ell=1}^{j} d_\ell}{N} . \tag{9.192}$$

9.7 Use the following set times of remission (in weeks) of leukemia subjects treated with 6-MP from Freireich, Gehan, Frei, et al. (1963), with censored survival times designated as t^+:

$$6^+ \ 6\ 6\ 6\ 7\ 9^+ \ 10^+ \ 10\ 11^+ \ 13\ 16\ 17^+ \ 19^+ \ 20^+ \ 22\ 23\ 25^+ \ 32^+ \ 32^+ \ 34^+ \ 35^+$$

9.7.1. Assuming exponential survival, estimate λ and the variance of the estimate. Use $\widehat{\lambda}$ to estimate the survival function $S(t)$ and its 95% asymmetric confidence bands based on the estimated variance of $\log[\widehat{S}(t)]$.

9.7.2. Alternately, assuming a Weibull distribution, use Newton-Raphson iteration or SAS PROC LIFEREG to compute the maximum likelihood estimates of the parameters (μ, γ) and the estimated covariance matrix of the estimates. Note that PROC LIFEREG uses the extreme value representation in Problem 9.3 to provide estimates of α and σ^2 in a no-covariate model. Use these to compute the estimated hazard function, the estimated survival function, and the asymmetric 95% confidence bands.

9.7.3. Likewise, assuming a log-logistic distribution, use Newton-Raphson iteration, or SAS PROC LIFEREG to compute the maximum likelihood estimates of the parameters (μ, γ) and the estimated covariance matrix of the estimates. Use these to compute the estimated hazard function, the estimated survival function, and the asymmetric 95% confidence bands.

9.7.4. Compute the Kaplan-Meier product limit estimate of the survival function and its asymmetric 95% confidence band.

9.7.5. Is there any substantial difference among these estimates of the survival distribution and its confidence bands?

9.8 Consider the case of a discrete survival distribution with lumps of probability mass of the event f_j at the jth discrete event time $t_{(j)}$, $1 \le j \le J$. Then the event probability π_j at the jth event time is defined as

$$\pi_j = \frac{f_j}{\sum_{\ell \ge j} f_\ell} . \tag{9.193}$$

9.8.1. Show that

$$S(t_{(j)}) = \prod_{\ell \leq j} [1 - \pi_\ell] = 1 - \sum_{\ell \leq j} f_\ell. \tag{9.194}$$

9.8.2. At the jth discrete event time $t_{(j)}$, let n_j be the number at risk, and let d_j be the number of events observed, $d_j > 0$. Also let w_j be the number of observations censored during the interval $(t_{(j)}, t_{(j+1)}]$ between the jth and $(j+1)th$ event times. Then $n_{j+1} = n_j - d_j - w_j$. Using the product representation of the survival function in (9.194), show that

$$\prod_{j=1}^{J} S\left(t_{(j-1)}\right)^{d_j} S\left(t_{(j)}\right)^{w_j} = \prod_{j=1}^{J} (1 - \pi_j)^{n_j - d_j}, \tag{9.195}$$

thus deriving the simplification in (9.14) from the likelihood in (9.10).

9.8.3. Using similar steps, show that the modified likelihood in (9.39) yields the actuarial estimate $\widehat{\pi}_j = p_j$ in (9.40).

9.8.4. The product-limit or actuarial estimator of the survivor function can be expressed as $\widehat{S}(t_{(j)}) = \prod_{\ell=1}^{j} q_\ell$, as shown in (9.18) and (9.42), respectively, where the estimate of the continuation probability is $q_j = 1 - p_j$ and $p_j = d_j/n_j$. Consider the estimates p_j and p_{j+1} at two successive event times $t_{(j)}$ and $t_{(j+1)}$. Despite the clear dependence of the denominator n_{j+1} of p_{j+1} on the elements of p_j, show that p_j and p_{j+1} are uncorrelated. Do likewise for any pair p_j and p_k for $1 \leq j < k \leq J$. Thus asymptotically the large sample distribution of the vector of proportions $\boldsymbol{p} = (p_1 \dots p_J)^T$ is asymptotically distributed as multivariate normal with expectation $\boldsymbol{\pi} = (\pi_1 \dots \pi_J)^T$ and covariance matrix $\boldsymbol{\Sigma}$ that is diagonal with elements $\sigma_j^2 = \pi_j(1 - \pi_j)/n_j$.

9.8.5. Then using the δ-method, derive the variance of $\log[\widehat{S}(t_{(j)})]$ presented in (9.20).

9.8.6. Again use the δ-method to derive Greenwood's expression for the variance of $\widehat{S}(t_{(j)})$ shown in (9.22). This yields the large sample *S.E.* of $\widehat{S}(t)$ and a confidence interval which is symmetric about $\widehat{S}(t)$, but not bounded by 0 and 1.

9.8.7. To obtain a bounded confidence interval, derive the expression for the variance of the $\log(-\log[\widehat{S}(t)])$ in (9.23).

9.8.8. Noting that $-\log[\widehat{S}(t)] = \widehat{\Lambda}(t)$, derive the asymmetric confidence limits presented in (9.24).

9.8.9. As an alternative to the complimentary log-log transformation, show that the variance of the logit of the survival probability at time t is

$$\widehat{V}\left[\log\left(\frac{\widehat{S}(t)}{1 - \widehat{S}(t)}\right)\right] = \left(\frac{1}{1 - \widehat{S}(t)}\right)^2 \widehat{V}\left(\log\left[\widehat{S}(t)\right]\right). \tag{9.196}$$

This also yields asymmetric confidence limits on the survival odds ratio at time t. Then use the logistic function to obtain the expression for the asymmetric confidence limits on the survival probability $S(t)$.

9.8.10. Show that the Kaplan-Meier estimate of the survival function estimate implies an estimate of the hazard function at time $t_{(j)}$, which is obtained as

$$\widehat{\lambda}_{KM,j} = \frac{-\log q_j}{t_{(j)} - t_{(j-1)}} = \frac{-\log(1-p_j)}{t_{(j)} - t_{(j-1)}} . \tag{9.197}$$

This is also called Peterson's estimate (Peterson, 1977).

9.8.11. Asymptotically as $J \to \infty$ and $p_j \downarrow 0$, show that $\widehat{\lambda}_{KM,j} \cong \widehat{\lambda}_j$ in (9.26) and that the Kaplan-Meier estimate of the survival function $\widehat{S}(t)$ in (9.18) is approximately equal to the Nelson-Aalen estimate $\widehat{S}_{NA}(t)$ in (9.28). *Hint*: Use $\Lambda_{NA}(t_{(j)})$ and note that $\log(1-\epsilon) \doteq \epsilon$ for $\epsilon \downarrow 0$.

9.9 Consider a sample of N observations with an indicator variable δ_j that denotes whether the observed time t_j is an observed event time ($\delta_j = 1$) or censoring time ($\delta_j = 0$). The overall proportion of events then is $p = \sum_j \delta_j / N$. The naive binomial variance is $V(p) = \pi(1-\pi)/N$, where $\pi = E(p)$.

9.9.1. Condition on the set of exposure times $(t_1, \ldots, t_N)$ as fixed quantities. Then, given the underlying cumulative distribution and survival functions $F(t)$ and $S(t)$, respectively, show that

$$E(p) = \pi = \frac{\sum_j E(\delta_j | t_j)}{N} = \frac{\sum_j F(t_j)}{N} . \tag{9.198}$$

9.9.2. Also show that the variance of p is

$$V_c(p) = \frac{\sum_j V(\delta_j | t_j)}{N} = \frac{\sum_j F(t_j) S(t_j)}{N} . \tag{9.199}$$

9.9.3. Show that this correct variance $V_c(p) \leq V(p)$ where $V(p)$ is the naive binomial variance. *Hint*: Note that $V(\delta) = E[V(\delta|t)] + V[E(\delta|t)]$.

9.10 The proportional odds model assumes that the survival odds over time for a subject with covariate vector $\boldsymbol{x}$ is proportional to the background survival odds. Bennett (1983), among others, describes the proportional odds model where the survival function is of the form

$$S(t|\boldsymbol{x}) = \left[1 + e^{\alpha_0(t) + \boldsymbol{x}'\boldsymbol{\beta}}\right]^{-1} , \tag{9.200}$$

where $\alpha_0(t)$ is some function of time that is expressed as a function of additional parameters.

9.10.1. Show that the survival odds for two subjects with covariate values $\boldsymbol{x}_1$ and $\boldsymbol{x}_2$ are proportional over time.

9.10.2. Show that the cumulative distribution is a logistic function of $\alpha_0(t) + \boldsymbol{x}'\boldsymbol{\beta}$.

9.10.3. Show that the hazard ratio for two subjects with covariate values $\boldsymbol{x}_1$ and $\boldsymbol{x}_2$ is

$$\frac{\lambda(t|\boldsymbol{x}_2)}{\lambda(t|\boldsymbol{x}_1)} = \frac{F(t|\boldsymbol{x}_2)}{F(t|\boldsymbol{x}_1)} . \tag{9.201}$$

9.10.4. Now consider a single binary indicator variable to represent treatment group, $x = (0, 1)$, with coefficient β. Then let

$$H(\beta) = \log \frac{\lambda(t|x=1)}{\lambda(t|x=0)} \, . \tag{9.202}$$

Under the null hypothesis of no treatment group difference, H_0: $\beta = \beta_0 = 0$, use a Taylor's expansion of $H(\beta)$ about $H(\beta_0)$ to show that

$$H(\beta) \cong (\beta - \beta_0)H'(\beta_0). \tag{9.203}$$

9.10.5. Then show that

$$H'(\beta) = 1 - F(t|x). \tag{9.204}$$

Evaluating $H'(\beta)$ under the null hypothesis at $\beta = \beta_0$, it follows that

$$\log \frac{\lambda(t|x=1)}{\lambda(t|x=0)} \cong (\beta - \beta_0)[1 - F_0(t)] = g[F_0(t)], \tag{9.205}$$

which is a function of $F_0(t)$, thus satisfying Schoenfeld's (1981) condition in (9.50).

9.10.6. Using (9.51), show that the asymptotically efficient test against the alternative of proportional odds over time for two groups is the weighted Mantel-Haenszel test in (9.47) with weights $w(t) = \widehat{S}(t)$ based on the Kaplan-Meier estimate of the survival function in the combined sample. This is the Peto-Peto-Prentice Wilcoxon test.

9.11 The following table presents the data from Freireich, Gehan, Frei, et al. (1963) with the times of remission in weeks of leukemia subjects treated with placebo

1 1 2 2 3 4 4 5 5 8 8 8 8 11 11 12 12 15 17 22 23

where no event times were censored. We now wish to compare the disease-free survival (remission) times between the group of patients treated with 6-MP presented in Problem 9.7 with those treated with placebo.

9.11.1. Order the distinct observed death times (ignoring ties or multiplicities) as $(t_{(1)}, \ldots, t_{(J)})$ for the J distinct times ($J \leq D = \#$ events) in the combined sample.

9.11.2. Compute the number at risk n_{ij} in the ith group ($i = 1$ for 6-MP, 2 for placebo) and the number of events d_{ij} at the jth remission time $t_{(j)}$, $j = 1, \ldots, J$. Construct a table with columns $t_{(j)}$, d_j, N_j, d_{1j}, n_{1j}, d_{2j}, n_{2j}. Note that for each event time we can construct a 2×2 table of treatment (1 or 2) versus event (remission or not) with N_j being the total sample size for that table.

9.11.3. From these data, calculate the Kaplan-Meier estimator of the survivor distribution and its standard errors for each of the treatment groups.

9.11.4. Compute the $S.E.$ of the $\log(-\log[\widehat{S}(t)])$ and the 95% asymmetric confidence limits for each survivor function.

Table 9.7 Number Entering Each Year of Follow-Up, Numbers Withdrawn Alive During That Year, and Number of Deaths During the Year for the Tolbutamide and Placebo Groups of the UGDP

	Placebo			*Tolbutamide*		
Year	n_{1j}	w_{1j}	d_{1j}	n_{2j}	w_{2j}	d_{2j}
1	205	0	0	204	0	0
2	205	0	5	204	0	5
3	200	0	4	199	0	5
4	196	4	4	194	5	5
5	188	23	4	184	24	5
6	161	43	3	155	41	4
7	115	50	1	110	47	5
8	64	36	0	58	33	1

9.11.5. Likewise, using the expressions in Section 9.1.4, compute the piecewise constant (linearized) hazard function estimate $\widehat{\lambda}_{(j)}$, the Nelson-Aalen estimate of the cumulative hazard $\widehat{\Lambda}_{NA}(t)$ and the Nelson-Aalen estimate of the survivor function $\widehat{S}_{NA}(t)$ for each group.

9.11.6. At each event time, compute $\widehat{V}[\widehat{\Lambda}_{NA}(t_{(j)})]$ and the corresponding $S.E.$ From these, for each group compute the asymmetric 95% confidence limits on $S(t)$ obtained via confidence limits on $\Lambda(t)$.

9.11.7. Compute the Mantel-logrank, Gehan-Wilcoxon and the Peto-Peto-Prentice Wilcoxon tests of equality of the hazard functions. In practice, the specific test to be used would be specified *a priori*. From examination of the linearized hazard function estimates, explain why the logrank test yields the larger test value.

9.11.8. As described in Section 9.3.5, compute the Peto approximate estimate of the relative risk and its 95% confidence limits, and also for the survival odds.

9.11.9. Using the methods described in Section 4.9.2, compute the correlation between the Mantel-logrank and the Peto-Peto-Prentice Wilcoxon tests, the MERT combination of these two tests, and the estimated maximin efficiency of the test.

9.12 Table 9.7 presents data showing the incidence of cardiovascular mortality among subjects treated with tolbutamide versus placebo over eight years of treatment and follow-up in the University Group Diabetes Project (UDGP, 1970). This is an example of grouped-time survival data.

9.12.1. Compute the actuarial adjusted number of units of exposure at risk n_j, n_{1j}, and n_{2j} during each interval, and from these calculate the actuarial estimator of the survivor function for each treatment group and their standard errors using Greenwood's equation.

9.12.2. Compute the logrank test, the Gehan-Wilcoxon test, the Peto-Peto-Prentice Wilcoxon test and the MERT to compare the two groups. In practice the test to be used should be specified *a priori.*

9.12.3. Compare the results of these tests based on the nature of the differences between groups with respect to the survival distributions and the censoring distributions.

9.13 Table 9.8 presents observations from a hypothetical time-to-event study in discrete time. Here we studied an experimental drug versus placebo for treatment of bone cancer in an extremity among an elderly population. Patients were x-rayed every two months after treatment was started to see if the cancer had gone into remission (that is, the subject had healed). The outcome for each subject was categorized as either healed at a designated evaluation visit at week 2, 4, or 6; the subject died during the stated month (and prior x-rays showed that the subject had not yet healed); the limb was removed surgically during the stated month because the cancer had spread; the subject dropped out (withdrew) during the stated month; or the subject had not gone into remission (healed) after completing the six months of follow-up. For each subject, the month of the last x-ray is also given.

9.13.1. Using the standard intervals (0–2] months, (2–4] months, and (4–6] months; construct a table that shows the timing of losses to follow-up in relation to the time of the last examination. As described in Section 9.2.3, construct a worktable that shows the numbers evaluated at each time, the number healed and the numbers "censored" between evaluations.

9.13.2. *Censoring on Death/Surgery*: Treat death, surgery and lost-to-follow-up as censored at random immediately following the last x-ray evaluation. From this, present a work table with the numbers of events and the adjusted numbers at risk separately for each group and for both groups combined. Within each treatment group, compute the modified Kaplan-Meier "survival" or "disease duration" function and the linearized estimate of the hazard function. Also compute the Mantel-logrank and Peto-Peto-Prentice Wilcoxon tests. Note that this survival function estimates the probability of continuing to have cancer in a population where no one dies or has limb amputation. The hazard function is an estimate of the cause-specific hazard function that also forms the basis for the statistical tests.

9.13.3. *Competing Risk Adjustment*: Treat death and surgery as competing risks that are not censored during the study but are retained in the adjusted numbers at risk. Also treat lost-to-follow-up as censoring at random. From this, present a work table with the numbers of events and the adjusted numbers at risk separately for each group. Use this table to calculate the modified Kaplan-Meier intent-to-treat survival function, including the standard errors of the survival function, separately for each treatment group. Also compute the Mantel-logrank and Peto-Peto-Prentice Wilcoxon tests.

9.13.4. *Sub-Distribution Function*: Now use a combined outcome of healing, death or surgery and treat losses-to-follow-up as censored after the last x-ray evaluation. Compute the estimate of the survival function for the combined outcome, des-

Table 9.8 Hypothetical Data from Clinical Trial of Drug Treatment for Bone Cancer

Drug Group			*Placebo Group*		
Status	*Month*	*Last x-ray*	*Status*	*Month*	*Last x-ray*
Healed	2		Healed	2	
Healed	2		Healed	2	
Healed	2		Lost	1	0
Healed	2		Lost	1	0
Healed	2		Lost	2	0
Healed	2		Surgery	2	0
Lost	1	0	Surgery	2	0
Lost	2	0	Death	1	0
Surgery	1	0	Healed	4	
Healed	4	0	Healed	4	
Healed	4		Lost	3	2
Healed	4		Lost	4	2
Healed	4		Surgery	4	2
Death	3	2	Surgery	3	2
Lost	3	2	Surgery	4	2
Lost	4	2	Healed	6	
Lost	4	2	Healed	6	
Surgery	3	2	Lost	5	4
Surgery	4	2	Lost	5	4
Healed	6		Lost	5	4
Healed	6		Surgery	6	4
Lost	5	4	Surgery	5	4
Lost	5	4	Death	5	4
Surgery	6	4	Not Healed	6	6
Not Healed	6	6			
Not Healed	6	6			
Not Healed	6	6			
Not Healed	6	6			

ignated as $\widehat{S}_{I,C}(t)$ in Section 9.2.3. Then calculate the cumulative sub-distribution function for healing $\widehat{F}_I(t)$ using (9.46). Note that the estimates of $\widehat{F}_I(t)$ are approximately equal to $1 - \widehat{S}(t)$ in Problem 9.13.3 above. In the drug treated group, the values of $\widehat{F}_I(t)$ at 2, 4, and 6 months are 0.231, 0.413 and 0.531, respectively, whereas the estimates of $\widehat{S}(t)$ in Problem 9.3.3 are 0.769, 0.588 and 0.481, respectively. For a definitive analysis, the sub-distribution function calculation is preferred. In this case, the standard errors would be calculated using the results of Gaynor, Fener, Tan, et al. (1993) and the tests of significance computed as described by Gray (1988). These computations are described by Marubini and Valsecchi (1995).

9.14 Consider the case of J distinct event times $t_{(1)} < t_{(2)} < \ldots < t_{(J)}$, where allowing for ties, the total number events at $t_{(j)}$ is $d_j \geq 1$ for all j, $1 \leq j \leq J$; where $d_j > 1$ indicates tied event times.

9.14.1. Using the Peto-Breslow approximate likelihood allowing for ties in (9.106), show that the kth element of the score vector $\boldsymbol{U}(\boldsymbol{\beta})$ corresponding to coefficient β_k is

$$U(\boldsymbol{\beta})_{\beta_k} = \frac{\partial \log \widetilde{L}_{PB}(\boldsymbol{\beta})}{\partial \beta_k} = \sum\nolimits_{j=1}^{J} s_{jk} - \overline{x}_k(t_j, \boldsymbol{\beta}), \tag{9.206}$$

where $s_{jk} = \sum_{j=1}^{d_j} x_{jk}$ for the d_j subjects with the event at time t_j; and where $\overline{x}_k(t_j, \boldsymbol{\beta})$ is as shown in (9.82) with the subscript i changed to j to designate the jth event time.

9.14.2. Then show that the information matrix has elements

$$\boldsymbol{I}(\boldsymbol{\beta})_{km} = E\left[\frac{-\partial \log \widetilde{L}_{PB}(\boldsymbol{\beta})}{\partial \beta_k \partial \beta_m}\right] = \sum\nolimits_{j=1}^{J} \left[C_{jkm}(t_j, \boldsymbol{\beta}) - \overline{x}_k(t_j, \boldsymbol{\beta})\overline{x}_m(t_j, \boldsymbol{\beta})\right] \tag{9.207}$$

with elements $C_{jkm}(t_j, \boldsymbol{\beta})$ as shown in (9.84) evaluated at the jth event time t_j for $1 \leq k \leq m \leq p$.

9.14.3. Now consider that we wish to use a Cox proportional hazards model with a solitary binary covariate $x_i = 1$ or 0 for $i = 1, \ldots, N$, such as where X designates treatment or exposure group. Using the Peto-Breslow approximate likelihood allowing for ties, show that the score equation for β can be expressed as

$$U(\beta) = \sum\nolimits_{j=1}^{J} d_{1j} - \frac{m_{1j} n_{1j} e^{\beta}}{n_{1j} e^{\beta} + n_{2j}}, \tag{9.208}$$

where at the jth event time $t_{(j)}$, d_{ij} is the number of events among the n_{ij} subjects at risk in the ith group and m_{1j} is the total number of events in both groups combined, $m_{1j} = d_j = d_{1j} + d_{2j}$. Thus under the null hypothesis H_0: $\beta = \beta_0 = 0$, the score is

$$U(\beta_0) = \sum\nolimits_{j=1}^{J} d_{1j} - \frac{m_{1j} n_{1j}}{N_j} = \sum\nolimits_{j=1}^{J} d_{1j} - E(d_{1j}|H_0), \tag{9.209}$$

where $N_j = n_{1j} + n_{2j}$.

9.14.4. Show that the expected information then is

$$I(\beta) = \sum\nolimits_{j=1}^{J} \frac{m_{1j} n_{1j} e^{\beta} n_{2j}}{(n_{1j} e^{\beta} + n_{2j})^2}, \tag{9.210}$$

which, when evaluated under the null hypothesis, yields

$$I(\beta_0) = \sum\nolimits_{j=1}^{J} \frac{m_{1j} n_{1j} n_{2j}}{N_j^2} . \tag{9.211}$$

9.14.5. Then show that the score test equals the Mantel-logrank test with no ties. For the case of ties, show that the efficient score test variance $I(\beta_0)$ differs from the Mantel-logrank test variance in (9.47) by the factor $m_{2j}/(N_j - 1)$, where $m_{2j} = N_j - m_{1j}$.

9.14.6. Using the score equation for the PH model with no ties, use a Taylor's expansion about the null value as described in Section 6.4 to show that the Peto score based estimate of the log relative risk equals that described in Section 9.3.5 based on the logrank test.

9.14.7. Now consider the case of K independent strata, where it is assumed that there is a constant ratio of the hazards of the two groups over time within all strata. This entails the stratified likelihood in (9.77), where $\beta_h = \beta$ for all strata but where the background hazard functions within each stratum may be different. With no ties, generalize the above results to show that the efficient stratified-adjusted score test equals the stratified-adjusted Mantel-logrank test presented in (9.56).

9.14.8. To allow for ties, now use the Cox conditional likelihood in (9.105). Show that the efficient scores test for H_0: $\beta = 0$ equals the Mantel-logrank test in (9.47) with $w_j = 1$. In this derivation, note that the score equation and the information are functions of the mean and variance, respectively, of a hypergeometric random variable, from which the result follows.

9.14.9. Now consider a model with multiple covariates. If one of the covariates, say X_1, is a binary indicator variable for group membership, 1 if experimental, 0 if control, then show that the hazard ratio $\phi = \lambda_e(t)/\lambda_c(t) = e^{\beta_1}$.

9.14.10. Alternately, let X_1 be a quantitative variable. Show that $e^{\beta_1} = \lambda(t|x+1)/\lambda(t|x)$ is the relative hazard per unit increase in the value of the covariate, the values of other covariates held constant.

9.14.11. Now let X_1 be the log of a quantitative variable, $X_1 = \log(Z_1)$. Consider two subjects, say i and j, with values x_{i1} and x_{j1} (z_{i1} and z_{j1}). Express the relative hazards for these two subjects in terms of the value of β_1, the values of other covariates held constant. Show that if $z_{i1}/z_{j1} = c$ for $c > 0$ then the relative hazard for these two subjects is c^{β_1}. Thus the relative percentage change in the hazard is $100(c^{\beta_1} - 1)$ per c-fold change in Z.

9.14.12. The discrete or grouped time model of Section 9.4.6.2 assumes that a logistic model applies to the jth interval such that

$$\frac{\pi_{j|\boldsymbol{x}}}{1 - \pi_{j|\boldsymbol{x}}} = \frac{\pi_{0j}}{1 - \pi_{0j}} e^{\boldsymbol{x}'\boldsymbol{\beta}} \tag{9.212}$$

for $1 < j < K$, where $\pi_{0j} = \lambda_0(t)dt$ and $\pi_{j|\boldsymbol{x}} = \pi_{0j}^{\exp(\boldsymbol{x}'\boldsymbol{\beta})}$. Show that as $K \to \infty$, such that the $\tau_j \downarrow \tau_{j-1}\ \forall j$, then this model approaches the continuous time PH model in (9.61). *Hint*: Use L'Hospital's rule.

9.15 Consider planning a two-group clinical trial to test the hypothesis H_0: $\lambda_1 = \lambda_2$ under exponential model assumptions at $\alpha = 0.05$, two-sided. We desire 90% power to detect a specific hazard ratio $\phi = \lambda_1/\lambda_2$ for a specified hazard rate in the control group λ_2.

9.15.1. Assume no censoring other than administrative censoring with uniform entry over the interval $(0, T_R]$ and total study duration $T_S > T_R$. Consider each of the following designs:

Design	λ_2	ϕ	T_R	T_S
1	0.1	2	3	5
2	0.1	2	5	5
3	0.2	2	3	5
4	0.1	2.5	3	5

Assuming equal allocations to the two groups, compute the required total sample N, the expected proportion of events within each group $E(\delta|\lambda_1)$ and $E(\delta|\lambda_2)$, and that under the null hypothesis $E(\delta|\lambda)$.

9.15.2. What is the implication of the difference between designs 1 and 2 ($T_R = 3$ vs. 5 years)?

9.15.3. What is the implication of the difference between designs 1 and 3 ($\lambda_2 = 0.1$ vs. 0.2)?

9.15.4. What is the implication of the difference between designs 1 and 4 ($\phi = 2$ vs. 2.5)?

9.15.5. Now assume random censoring with an exponential hazard rate $\eta = 0.05$ within each of the two groups. For each of the four designs, compute the above quantities plus the expected proportions lost to follow-up within each group $E(\xi|\lambda_1, \eta)$ and $E(\xi|\lambda_2, \eta)$ and that under the null hypothesis $E(\xi|\lambda, \eta)$.

9.15.6. What impact do these losses have on the relative comparisons in Problems 9.15.2, 9.15.3 and 9.15.4?

9.16 Table 9.9 presents data from Prentice (1973) for subjects with lung cancer treated by a standard or by a test method. The covariates are Z_1: performance status with nine levels that is treated as a quantitative covariate; Z_2: time since diagnosis; Z_3: age (years) at diagnosis; Z_4: any prior treatment (1=yes, 0=no); and Z_5: study drug (1=test, 0=standard). Survival times are recorded to the nearest day since entry into the study and *delta* = I(death).

The principal objectives of this analysis are to compare the study drug treatment groups (test vs. standard) to assess the impact of other covariates on survival, and to assess the differences between treatment groups adjusting for these other covariates.

9.16.1. Compute and plot the Kaplan-Meier estimated survival functions and the Nelson-Aalen estimates of the cumulative hazard functions for the two drug groups (Z_5).

9.16.2. Compute the unadjusted logrank test comparing the drug groups and an estimate of the average relative risk (hazard ratio) computed using the Peto estimate along with its 95% confidence limits.

Table 9.9 Prentice (1973) Lung Cancer Data

ID	t_i	δ_i	Z_1	Z_2	Z_3	Z_4	Z_5	ID	t_i	δ_i	Z_1	Z_2	Z_3	Z_4	Z_5
16	999	1	90	12	54	1	1	8	110	1	80	29	68	0	0
21	991	1	70	7	50	1	1	10	100	0	70	6	70	0	0
24	587	1	60	3	58	0	1	18	87	0	80	3	48	0	1
29	457	1	90	2	64	0	1	7	82	1	40	10	69	1	0
2	411	1	70	5	64	1	0	1	72	1	60	7	69	0	0
25	389	1	90	2	62	0	1	33	44	1	60	13	70	1	1
28	357	1	70	13	58	0	1	11	42	1	60	4	81	0	0
9	314	1	50	18	43	0	0	26	33	1	30	6	64	0	1
34	283	1	90	2	51	0	1	32	30	1	70	11	63	0	1
20	242	1	50	1	70	0	1	27	25	1	20	36	63	0	1
19	231	0	50	8	52	1	1	14	25	0	80	9	52	1	0
3	228	1	60	3	38	0	0	35	15	1	50	13	40	1	1
30	201	1	80	28	52	1	1	15	11	1	70	11	48	1	0
13	141	1	30	4	63	0	0	6	10	1	10	5	49	0	0
4	126	1	60	9	63	1	0	12	8	1	40	58	63	1	0
5	118	1	70	11	65	1	0	23	1	1	20	21	65	1	1
17	112	1	80	6	60	0	1	31	1	1	50	7	35	0	1
22	111	1	70	3	62	0	1								

9.16.3. Compute an estimate of the average relative risk (hazard ratio) from an unadjusted PH model and its 95% confidence limits.

9.16.4. Compute a measure of explained variation using Madalla's R^2_{LR} in (A.203) and also using $R^2_{\varepsilon^2}$ in (9.110) with $\sigma^2_\varepsilon = 1$; i.e., using the Kent and O'Quigley $R^2_{W,A}$.

9.16.5. The covariate Z_4 is binary. Consider the model where Z_4 is used to define strata and where the covariates within each strata are $Z = (Z_1, Z_2, Z_3, Z_5)$. Compute $\log\left[-\log[\widehat{S}_j(t)]\right]$ plots for the two strata ($j = 1, 2$) to assess whether the PH assumption for Z_4 is appropriate.

9.16.6. Do likewise for Z_5 used to define strata and for (Z_1, Z_2, Z_3, Z_4).

9.16.7. Fit a PH model with an interaction between Z_4 and $\log(t)$ to assess whether a monotone departure in $\log t$ from the PH assumption exists for this covariate.

9.16.8. Likewise, fit a PH model with an interaction between Z_5 and $\log(t)$ to assess whether a monotone departure from the PH assumption exists for the differences between drug treatment groups. Since there are no substantial departures from the PH model assumptions for these two covariates, there is no need to adopt remedial measures.

9.16.9. Now fit a model with the adjusting covariates only (Z_1, Z_2, Z_3, Z_4). Use $R^2_{\varepsilon^2}$ in (9.110) to provide a measure of explained variation for the model.

9.16.10. For an "average" individual with covariate vector equal to the mean values for these adjusting covariates $x = \overline{x}$ (including the binary covariate Z_4), describe the estimated survival function $\widehat{S}(t|\overline{x})$.

9.16.11. The coefficients are negative for (Z_1, Z_3) and positive for (Z_2, Z_4). Thus consider a high-risk subject with covariate values equal to the 25, 75, 75 and 25th percentiles for $(Z_1,\ Z_2,\ Z_3,\ Z_4)$, respectively; and a low-risk subject with values equal to the 75, 25, 25 and 75th percentiles. Compute and plot the estimated survival function $\widehat{S}(t|x)$ for each of these two patients.

9.16.12. Now fit a model with treatment group adjusting for the other covariates $(Z_1,\ Z_2,\ Z_3,\ Z_4)$. Compute the likelihood ratio test of the treatment group effect from the change in the model chi-square compared to the covariate only model. Use Madalla's approximate R^2_{LR} based on this chi-square test value to describe the partial variation explained by treatment group. Also compute the $R^2_{W,A}$ in (9.111) to describe the treatment group effect.

9.16.13. Then fit a model with a simple interaction between treatment group and each of the other covariates to assess the presence of interactions in the regression coefficients.

9.16.14. Also fit a model stratifying on the value of Z_4 with no interactions with other covariates but with separate treatment group coefficients within each Z_4 stratum. Conduct a Wald test of the difference between the treatment group coefficients within the two strata. Because there is no indication of a treatment group by covariate (or by stratum) interaction, the no-interaction adjusted model appears adequate.

9.16.15. Using the no-interaction model, compute the information sandwich robust estimate of the covariance matrix of the coefficient estimates. Also compute the robust model score test for the model on 5 *df.* Compared to the model-based covariance of the estimates and score test, does it appear that there is gross model misspecification?

9.17 Byar (1985) presents the data from the Veterans Administration Cooperative Urologic Research Group (VACURG) study of the treatment of prostate cancer with high-, versus low-dose estrogen versus placebo. These data are available from StatLib. Analyses of these data were presented by Byar (1985) and Thall and Lachin (1986), among others. For this Problem, some manipulations are required, the SAS program for which is available from the www (see Preface). Here we employ only the data from the high- versus low-dose groups. The variables in this data set are: *pat* = patient number (not a covariate); *trt* = 1 (high dose), 0 (low dose) estrogen; *died* = 1 (died), 0 (survived, censored); *mosfu* = number of months of follow-up to death or censoring; *history* = 1 (yes), 0 (no) for history of cardiovascular disease; *age* in years; and *size* (of tumor) = 1 (large), 0 (small). The objective is to examine covariate effects on the risk of all-cause mortality, including cardiovascular mortality, for which a history of cardiovascular disease and age are the dominant risk factors. Perform the following analyses of these data:

9.17.1. Compute a Kaplan-Meier estimate of the survival function within each group and conduct a logrank test of the difference between groups. Also compute the score-based estimate of the unadjusted relative risk and its 95% confidence limits.

9.17.2. Fit a PH model to obtain an unadjusted estimate of the relative risk and its 95% confidence limits.

9.17.3. Fit a PH model with *trt history age size* as covariates to obtain an adjusted estimate of the relative risk for treatment group. Interpret each coefficient in terms of the hazard ratio (relative risk) with asymmetric confidence limits.

9.17.4. For *age*, also compute the relative risk per ten years of age and the associated confidence limits.

9.17.5. Also, fit a model without *trt* and compute a likelihood ratio test of the effect of treatment.

9.17.6. State your conclusion as to the effects of high-dose estrogens on all-cause mortality.

9.17.7. Compute the robust information sandwich estimator of the covariance matrix of the estimates and the associated robust 95% confidence limits and the robust Wald tests for each covariate. Compare these to those obtained under the assumed model.

9.18 Problem 8.10 describes the data presented by Fleming and Harrington (1991) on the rate of serious infections among children in a clinical trial of interferon versus placebo. There, only the total number of infections experienced by each child were considered. Here, we consider the actual times of infection and the underlying intensity functions over time. The more complete data set includes one record per interval of time per subject. Each interval is designated by a start time (T_2) and a stop time (T_1), where T_1 is either the time of an event designated by $d = 1$ or the time of curtailment of follow-up or right censoring ($d = 0$). The record number for each subject is designated by the variable s. The complete event time data are also available from the www (see Preface).

9.18.1. Separately within each treatment group, compute the Nelson-Aalen estimate of the cumulative intensity function over time.

9.18.2. Also, compute the linearized estimate of the underlying intensity function, or the kernel-smoothed estimate using Epanechnikov's kernel.

9.18.3. Compute the Aalen-Gill logrank test of the difference between intensities using all recurrent events.

9.18.4. Fit the multiplicative intensity model with treatment group as the only covariate to obtain an unadjusted estimate of the relative risk and its confidence limits. Compare the Wald and likelihood ratio tests for the treatment group effect to the logrank Aalen-Gill test.

9.18.5. Fit a more complete model to assess the treatment group effect adjusted for the other covariates. Note that three separate binary indicator variables must be defined to reflect the effect of hospital type. Compare the adjusted to the unadjusted assessment of the treatment effect.

9.18.6. Fit an additional model that includes an interaction between treatment group and weight and an interaction between group and height. Describe the effect of treatment as a function of height and weight.

Appendix
Statistical Theory

A.1 INTRODUCTION

This appendix contains all of the mathematical statistical developments employed in the text. It is intended to be used hand-in-hand with the chapters in the main body of the book. In keeping with the general tone of the text, these developments are presented in an informal but precise manner. Readers who desire a more rigorous development of these results, such as the required regularity conditions, are referred to one of the many excellent texts on theoretical statistics such as those by Bickel and Doksum (1977), Cox and Hinkley (1974), Cramer (1946), Kendall and Stuart (1979), Lehmann (1983 and 1986), and Rao (1973), among many.

A.1.1 Notation

Throughout this text, the following notation conventions are employed.

$E(Y)$ and $V(Y)$ denote the mean and variance of the random variable Y.

N denotes the fixed total sample size.

n denotes an increasing sequence of observations of sample size n as $n \to \infty$.

The notation $\xrightarrow{p}$ refers to convergence in probability.

The notation $\xrightarrow{d}$ refers to convergence in distribution, or in law.

The symbol $\widehat{=}$ means "estimated as" as in $\mu \widehat{=} \overline{y}$.

The symbol $\sim$ means "is distributed as."

The symbol $\cong$ means "is asymptotically equal to."

The symbol $\doteq$ means "is approximately equal to."

The symbol $\overset{d}{\approx}$ means "is asymptotically distributed as."

The symbol $|$ is used as a conditioning operator as in $E(y|x)$.

The symbol ; is used to designate the assumed parameters of a distribution, as in $f(y;\theta)$ to designate the distribution of Y with parameter θ.

A.1.2 Matrices

A review of matrix algebra will not be provided. Selected results will be presented as needed. The following conventions are employed.

All vectors are column vectors, such that $\boldsymbol{\theta} = (\theta_1 \ \ldots \ \theta_p)^T$, the "^{T}" being the transpose operator. Alternately, $\boldsymbol{\theta}'$ is used to designate the row vector transpose of $\boldsymbol{\theta}$.

The symbol $\|$ indicates the concatenation by columns of two matrices or vectors. For example, if $\boldsymbol{A}$ is a $p\times 1$ vector and $\boldsymbol{B}$ is a $p\times 2$ vector, then $\boldsymbol{C} = (\boldsymbol{A} \parallel \boldsymbol{B})$ is $p \times 3$.

The symbol $//$ indicates the concatenation by rows of two matrices. For the above example, then $\boldsymbol{C} = (\boldsymbol{A}^T // \boldsymbol{B}^T)^T$ is obtained as the transpose of the concatenation by rows of the two matrices $\boldsymbol{A}^T$ and $\boldsymbol{B}^T$. Alternately, $\boldsymbol{D} = (\boldsymbol{A} // \boldsymbol{A})$ is a $2p \times 1$ vector and $\boldsymbol{E} = (\boldsymbol{B} // \boldsymbol{B})$ is a $2p \times 2$ matrix.

The $(jk)th$ element of the inverse of a matrix $\left[\boldsymbol{A}^{-1}\right]_{jk}$ is also expressed as $\boldsymbol{A}^{jk}$.

Likewise, let the $p \times p$ square matrix $\boldsymbol{A}$ be partitioned as

$$\boldsymbol{A} = \begin{bmatrix} \boldsymbol{A}_{rr} & \boldsymbol{A}_{rs} \\ \boldsymbol{A}_{sr} & \boldsymbol{A}_{ss} \end{bmatrix}, \tag{A.1}$$

with diagonal square matrices $\boldsymbol{A}_{rr}$ and $\boldsymbol{A}_{ss}$ of dimensions r and s, respectively, $r + s = p$. Then the partitioned inverse matrix is expressed as

$$\boldsymbol{A}^{-1} = \begin{bmatrix} \boldsymbol{A}^{rr} & \boldsymbol{A}^{rs} \\ \boldsymbol{A}^{sr} & \boldsymbol{A}^{ss} \end{bmatrix} \tag{A.2}$$

where, for example, $\boldsymbol{A}^{rr}$ refers to the elements of $\boldsymbol{A}^{-1}$ in the positions corresponding to the $r \times r$ submatrix, or $\left[\boldsymbol{A}^{-1}\right]_{rr}$. Thus $\boldsymbol{A}^{rr} \neq \boldsymbol{A}_{rr}^{-1}$. For such a matrix, the upper left submatrix of the inverse is obtained as

$$\boldsymbol{A}^{rr} = \left[\boldsymbol{A}_{rr} - \boldsymbol{A}_{rs}\boldsymbol{A}_{ss}^{-1}\boldsymbol{A}_{sr}\right]^{-1}. \tag{A.3}$$

A.1.3 Partition of Variation

A useful result is the principle of *partitioning of sums of squares* that forms the basis for the Analysis of Variance. Given a set of constants $\{a_i\}$ and values $\{z_i\}$ such that $\sum_i y_i = \sum_i z_i = n\overline{y}$, then the total sum of squares of Y can be partitioned as

$$\sum_i a_i (y_i - \overline{y})^2 = \sum_i a_i (y_i - z_i)^2 + \sum_i a_i (z_i - \overline{y})^2 \tag{A.4}$$

whenever it can be shown that $\sum_i a_i z_i (y_i - z_i) = 0$. The values z_i may be random variables or constants.

A similar result provides the *partitioning of variation (Mean Square Error)* for an estimate $\widehat{\theta}$ of a parameter θ. When the estimate has expectation $E(\widehat{\theta}) \neq \theta$, then the mean square error of the estimate can be partitioned as

$$\begin{aligned} E\left(\widehat{\theta} - \theta\right)^2 &= E\left[\widehat{\theta} - E(\widehat{\theta})\right]^2 + E\left[E(\widehat{\theta}) - \theta\right]^2 \\ &= V(\widehat{\theta}) + \left[Bias(\widehat{\theta})\right]^2 . \end{aligned} \tag{A.5}$$

A related expression is the well known result

$$V(Y) = E_X\left[V(Y|X)\right] + V_X\left[E(Y|X)\right] \tag{A.6}$$

expressed in terms of the conditional moments of Y given the value of another variable X, integrated with respect to the distribution of X.

A.2 CENTRAL LIMIT THEOREM AND THE LAW OF LARGE NUMBERS

A.2.1 Univariate Case

The Central Limit Theorem and the Law of Large Numbers, when used in conjunction with other powerful theorems, such as Slutsky's Theorem, provides the mechanism to derive the large sample or asymptotic distribution of many statistics, and virtually all of those employed in this text.

Let $\{y_1, y_2, \ldots, y_N\}$ refer to a sample of N independent and identically distributed (*i.i.d.*) observations of the random variable Y drawn at random from a specified distribution, or

$$y_i \sim f(y; \mu, \sigma^2)\ \forall i \tag{A.7}$$

for some density or probability mass distribution $f(y; .)$ with mean μ and variance σ^2. Let $\overline{y}$ be the sample mean, $\overline{y} = \sum_{i=1}^N y_i/N$. Then the *Central Limit Theorem* is often expressed as

$$\overline{y} \overset{d}{\approx} \mathcal{N}(\mu, \sigma^2/N) \tag{A.8}$$

to indicate that the asymptotic or large sample distribution of the sample mean based on a large sample of size N is normal with expectation $E(\overline{y}) = \mu$ and variance $V(\overline{y}) = \sigma^2/N$. This is a casual notation that is fine in practice and thus is used throughout the text. The *large sample variance*, σ^2/N, of the sample mean $\overline{y}$ may then be used as the basis for the computation of a large sample confidence interval or a statistical test. In the limit, however, this distribution is degenerate because $\lim_{n\to\infty} \sigma^2/n = 0$. Thus a more statistically precise description is as follows:

Let $\{y_n\} = \{y_1 \ \ldots \ y_n\}$ be a sequence of n $i.i.d.$ observations drawn from some distribution $f(y;.)$ with mean μ and finite variance σ^2. Let $\overline{y}_n$ be the sample mean computed from the first n observations. The *Weak Law of Large Numbers* specifies that the limiting value of the sample mean is the expected value of the random variable, often designated as the expectation of the ith observation in a sample; or that as $n \to \infty$ then

$$\overline{y}_n \xrightarrow{p} E(Y) = E(y_i) = \mu. \tag{A.9}$$

Thus the sample mean provides a *consistent estimator* of (converges in probability to) the mean of the distribution from which the observations arose.

Now let $S_n = (y_1 + \cdots + y_n)$ be a sequence of partial sums of n $i.i.d.$ observations drawn from this distribution, where $\overline{y}_n = S_n/n$. The *Central Limit Theorem* then asserts that a sequence of partial sums computed as $n \to \infty$, satisfies

$$\lim_{n\to\infty} P\left[\frac{S_n - n\mu}{\sqrt{n}\sigma} \leq z\right] = \Phi(z), \tag{A.10}$$

where $\Phi(z)$ is the standard normal cumulative distribution function. Thus

$$\lim_{n\to\infty} P\left[\frac{\overline{y}_n - \mu}{\sigma/\sqrt{n}} \leq z\right] = \Phi(z), \tag{A.11}$$

where $\overline{y}_n$ refers to the corresponding sequence of sample means as $n \to \infty$. Then, from (A.10)

$$\lim_{n\to\infty} P\left[\frac{1}{\sqrt{n}}\left(\frac{S_n - n\mu}{\sigma}\right) \leq z\right] = \Phi(z) \tag{A.12}$$

and from (A.11)

$$\lim_{n\to\infty} P\left[\sqrt{n}\left(\frac{\overline{y}_n - \mu}{\sigma}\right) \leq z\right] = \Phi(z). \tag{A.13}$$

Using $\mathcal{N}(.,.)$ to denote the normal distribution, the above expressions in terms of S_n and $\overline{y}_n$, respectively, can be expressed as

$$\frac{1}{\sqrt{n}}(S_n - n\mu) \xrightarrow{d} \mathcal{N}(0,\ \sigma^2) \tag{A.14}$$

and

$$\sqrt{n}\,(\overline{y}_n - \mu) \xrightarrow{d} \mathcal{N}(0,\ \sigma^2) \tag{A.15}$$

to indicate convergence in distribution or in law to the normal distribution. These lead to the equivalent expressions

$$\frac{(\overline{y}_n - \mu)}{\sigma/\sqrt{n}} \xrightarrow{d} \mathcal{N}(0,\ 1) \tag{A.16}$$

or as

$$\frac{(\overline{y}_n - \mu)^2}{\sigma^2/n} \xrightarrow{d} \chi^2_{(1)}\ , \tag{A.17}$$

where $\chi^2_{(1)}$ refers to the central chi-square distribution on 1 *df*.

The expression in (A.15) implies that $P\left[(\overline{y}_n - \mu) > \epsilon\right] \xrightarrow{p} 0$ for any value ϵ which indicates that $\overline{y}$ is a *consistent* estimator of the population mean μ regardless of the distribution $f(y;.)$ from which the observations are assumed drawn at random. A formal proof of the above results, generally known as the *Lindberg-Levy Central Limit Theorem,* requires regularity conditions that essentially assert that the sequence of random variables $\{y_n\}$ has bounded variation, which is virtually always the case for the statistics considered herein.

The *Liapunov Central Limit Theorem* (*cf.* Rao, 1973, p. 127) provides a generalization to the case of a sample of independent, though not identically distributed, random variables.

Expression (A.15) also demonstrates that $\overline{y}_n$ is a $\sqrt{n}$*-consistent* estimator of μ. Any estimator $\widehat{\theta}$ of a parameter θ is said to be $\sqrt{n}$-consistent if $\sqrt{n}(\widehat{\theta} - \theta)$ converges in distribution to a non-degenerate probability distribution (*cf.* Lehmann, 1998). Such estimators converge to the true value at the rate of $n^{-\frac{1}{2}}$.

Example A.1 *Simple Proportion*
Consider the simple proportion of positive (+) responses from a sample of N *i.i.d.* Bernoulli observations where $y_i = 1$ if +, 0 if −, where $E(y_i) = P(y_i = 1) = \pi$ and $V(y_i) = E(y_i - \pi)^2 = \sigma^2 = \pi(1-\pi)$. Then the sample proportion can be expressed as the mean of the $\{y_i\}$, or $p_n = \overline{y}_n = \sum_{i=1}^n y_i/n$ as $n \to \infty$, where $\sum_{i=1}^n y_i$ equals the number of + responses in the sample. Thus the Law of Large Numbers provides that the limiting value of the proportion is the probability, or that $p_n \xrightarrow{p} E(y_i) = \pi$, and thus p is a consistent estimator for π. The Central Limit Theorem further provides that

$$\sqrt{n}\,(p_n - \pi) \xrightarrow{d} \mathcal{N}\left[0,\ \pi(1-\pi)\right], \tag{A.18}$$

and asymptotically the large sample distribution of p with total sample size N is

$$p \overset{d}{\approx} \mathcal{N}\left[\pi,\ \pi(1-\pi)/N\right]. \tag{A.19}$$

A.2.2 Multivariate Case

Now consider the case where the observation for the *ith* subject is a p-vector of measurements $\boldsymbol{y}_i = (y_{i1}\ y_{i2}\ \cdots\ y_{ip})^T$, for $i = 1, \ldots, n$, drawn from some distribution $f(\boldsymbol{y};.)$ with mean vector $\boldsymbol{\mu} = (\mu_1 \ \ldots \ \mu_p)^T$ and covariance matrix $\boldsymbol{\Sigma}$. Let $\overline{\boldsymbol{y}}_n$ refer to a sequence of sample means computed as $n \to \infty$. Then, analogously to (A.9), the Weak Law of Large Numbers specifies that the limiting value of the sample mean vector is the mean vector

$$\overline{\boldsymbol{y}}_n \xrightarrow{p} \boldsymbol{\mu} \tag{A.20}$$

and analogously to (A.13), the Central Limit Theorem specifies that

$$\lim_{n\to\infty} P\left[\sqrt{n}\boldsymbol{\Sigma}^{-1/2}(\overline{\boldsymbol{y}}_n - \boldsymbol{\mu}) \leq z\right] = \boldsymbol{\Phi}_p(z), \tag{A.21}$$

where $\boldsymbol{\Phi}_p(z)$ refers to the standard p-variate normal cumulative distribution function. Using $\mathcal{N}(\boldsymbol{\mu},\ \boldsymbol{\Sigma})$ to denote the p-variate normal distribution with mean $\boldsymbol{\mu}$ and variance $\boldsymbol{\Sigma}$, this can be expressed as

$$\sqrt{n}\boldsymbol{\Sigma}^{-1/2}(\overline{\boldsymbol{y}}_n - \boldsymbol{\mu}) \xrightarrow{d} \mathcal{N}(\boldsymbol{0},\ \boldsymbol{I}_p), \tag{A.22}$$

where $\boldsymbol{I}_p$ is the identity matrix of order p. Thus

$$\sqrt{n}(\overline{\boldsymbol{y}}_n - \boldsymbol{\mu}) \xrightarrow{d} \mathcal{N}(\boldsymbol{0},\ \boldsymbol{\Sigma}), \tag{A.23}$$

and for a large sample of size N, the large sample distribution of the mean vector is asymptotically (approximately) distributed as

$$\overline{\boldsymbol{y}} \overset{d}{\approx} \mathcal{N}(\boldsymbol{\mu},\ \boldsymbol{\Sigma}/N). \tag{A.24}$$

Further, from (A.23) the quadratic form

$$n(\overline{\boldsymbol{y}}_n - \boldsymbol{\mu})^T\boldsymbol{\Sigma}^{-1}(\overline{\boldsymbol{y}}_n - \boldsymbol{\mu}) \xrightarrow{d} \chi^2_{(p)}\ , \tag{A.25}$$

where $\chi^2_{(p)}$ is the central chi-square distribution on p df.

Example A.2 *Multinomial Distribution*
Consider the trinomial distribution of the frequencies within each of three mutually exclusive categories from a sample of N $i.i.d$ observations. In this case, the ith observation is a vector $\boldsymbol{y}_i = (y_{i1}\ y_{i2}\ y_{i3})$, where y_{ij} is a Bernoulli variable that denotes whether the ith observation falls in the jth category; $y_{ij} = 1$ if jth category, 0 if otherwise. Since y_{ij} for the jth category is a Bernoulli variable, then $E(y_{ij}) = P(y_{ij} = 1) = \pi_j$ and $V(y_{ij}) = \sigma_j^2 = \pi_j(1 - \pi_j)$. It also follows that for any two categories $j \neq k$, $Cov(y_{ij},\ y_{ik}) = \sigma_{jk} = E(y_{ij}y_{ik}) - E(y_{ij})E(y_{ik}) = -\pi_j\pi_k$, since by construction, $y_{ij}y_{ik} = 0$. Thus the $\boldsymbol{y}_i$ have mean vector $\boldsymbol{\mu} = \boldsymbol{\pi} = (\pi_1\ \pi_2\ \pi_3)^T$ and covariance matrix

$$\boldsymbol{\Sigma}(\boldsymbol{\pi}) = \begin{bmatrix} \pi_1(1-\pi_1) & -\pi_1\pi_2 & -\pi_1\pi_3 \\ -\pi_1\pi_2 & \pi_2(1-\pi_2) & -\pi_2\pi_3 \\ -\pi_1\pi_3 & -\pi_2\pi_3 & \pi_3(1-\pi_3) \end{bmatrix}. \tag{A.26}$$

The covariance matrix is often denoted as $\boldsymbol{\Sigma}(\boldsymbol{\pi})$ because it depends explicitly on the mean vector $\boldsymbol{\pi}$. Since the $\{y_{ij}\}$ are subject to the linear constraint that $\sum_{j=1}^{3} y_{ij} = 1$, then the covariance matrix is singular with rank 2.

The vector of sample proportions can be expressed as the mean vector of the $\{\boldsymbol{y}_i\}$, or as $\boldsymbol{p} = (p_1\ p_2\ p_3)^T = \overline{\boldsymbol{y}} = \sum_{i=1}^{n} \boldsymbol{y}_i/N$, where $\sum_{i=1}^{n} \boldsymbol{y}_i$ is now the vector of frequencies within the three categories. Thus from (A.24), for a large sample of size N, $\boldsymbol{p}$ is asymptotically normally distributed with mean vector $\boldsymbol{\pi}$ and covariance matrix $\boldsymbol{\Sigma}(\boldsymbol{\pi})/N$. Because these proportions are likewise subject to the linear constraint $\sum_{j=1}^{3} p_j = 1$, the asymptotic distribution is degenerate with covariance matrix of rank 2. However, this poses no problems in practice because we need only characterize the distribution of any two of the three proportions.

A.3 DELTA METHOD

A.3.1 Univariate Case

A common problem in statistics is to derive the large sample moments, principally the expectation and variance, of a transformation of a statistic, including non-linear transformations. These expressions are readily obtained by use of the *δ-method*.

Let t be any statistic for which the first two central moments are known, $E(t) = \mu$ and $V(t) = \sigma^2$. We desire the moments of a transformation $y = g(t)$ for some function $g(\cdot)$ which is assumed to be twice differentiable, with derivatives designated as $g'(\cdot)$ and $g''(\cdot)$. A first-order *Taylor's Series expansion* of $g(t)$ about μ is:

$$g(t) = g(\mu) + g'(\mu)(t - \mu) + R_2(a). \tag{A.27}$$

From the mean value theorem, the remainder to second order is

$$R_2(a) = (t - \mu)^2 g''(a)/2 \tag{A.28}$$

for some value a contained in the interval (t, μ). If the remainder vanishes under specified conditions, such as asymptotically, then

$$g(t) \cong g(\mu) + g'(\mu)(t - \mu) \tag{A.29}$$

so that

$$E(y) = \mu_y = E\left[g(t)\right] \cong g(\mu) + g'(\mu)E(t - \mu) = g(\mu) \tag{A.30}$$

and

$$\begin{aligned} V(y) &= E(y - \mu_y)^2 \cong E\left[g(t) - g(\mu)\right]^2 \\ &= E\left[g'(\mu)(t - \mu)\right]^2 = \left[g'(\mu)\right]^2 V(t). \end{aligned} \tag{A.31}$$

Herein, we frequently consider the moments of a transformation of a statistic t that is a consistent estimate of μ. In such cases, since $t \xrightarrow{p} \mu$, then the remainder in (A.28) vanishes asymptotically, or $R_2(a) \xrightarrow{p} 0$, and the above results apply to any transformation of t. Furthermore, if $\widehat{V}(t)$ is a consistent estimator of $V(t)$, then it follows from Slutsky's Convergence Theorem (A.45, below), that $\widehat{V}(y) = [g'(t)]^2 \widehat{V}(t)$ is a consistent estimator of $V(y)$.

Example A.3 *log(p)*
For example, consider the moments of the natural log of the simple proportion p for which $\mu = \pi$ and $\sigma^2 = \pi(1-\pi)/N$. The Taylor's expansion yields

$$\log(p) = \log(\pi) + \frac{d\log(\pi)}{d\pi}(p-\pi) + R_2(a), \tag{A.32}$$

where the remainder for some value $a \in (p, \pi)$ is $R_2(a) = (p-\pi)^2 g''(a)/2$. Since p is consistent for π, $p \xrightarrow{p} \pi$, then asymptotically $R_2(a) \to 0$ and thus

$$E[\log(p)] \cong \log(\pi) \tag{A.33}$$

and

$$V[\log(p)] \cong \left[\frac{d\log(\pi)}{d\pi}\right]^2 V(p) = \left[\frac{1}{\pi^2}\right]\frac{\pi(1-\pi)}{N} = \frac{1-\pi}{\pi N}. \tag{A.34}$$

A.3.2 Multivariate Case

Now consider a transformation of a p-vector $\boldsymbol{T} = (t_1 \ \ldots \ t_p)^T$ of statistics with mean vector $\boldsymbol{\mu}$ and covariance matrix $\boldsymbol{\Sigma}_T$. Assume that $\boldsymbol{Y} = (y_1 \ \ldots \ y_m)^T = \boldsymbol{G}(\boldsymbol{T}) = [g_1(\boldsymbol{T}) \ \ldots \ g_m(\boldsymbol{T})]^T$, $m \leq p$, where the kth transformation $g_k(\boldsymbol{T})$ is a twice-differentiable function of $\boldsymbol{T}$. Applying a first-order Taylor's series, as in (A.27), and assuming that the vector of remainders $\boldsymbol{R}_2(\boldsymbol{A})$ vanishes for values $\boldsymbol{A} \in (\boldsymbol{T}, \boldsymbol{\mu})$, yields

$$\begin{aligned} E(\boldsymbol{Y}) &= \boldsymbol{\mu}_Y \cong \boldsymbol{G}(\boldsymbol{T}) \\ V(\boldsymbol{Y}) &= \boldsymbol{\Sigma}_Y \cong \boldsymbol{H}(\boldsymbol{\mu})'\boldsymbol{\Sigma}_T\boldsymbol{H}(\boldsymbol{\mu}), \end{aligned} \tag{A.35}$$

where $\boldsymbol{H}(\boldsymbol{\mu})$ is a $p \times m$ matrix with elements

$$\boldsymbol{H}(\boldsymbol{\mu}) = \begin{bmatrix} \partial g_1(\boldsymbol{T})/\partial \boldsymbol{T} \\ \vdots \\ \partial g_m(\boldsymbol{T})/\partial \boldsymbol{T} \end{bmatrix}^T_{\Big|\boldsymbol{T}=\boldsymbol{\mu}} = \begin{bmatrix} \partial g_1(\boldsymbol{T})/\partial t_1 & \cdots & \partial g_1(\boldsymbol{T})/\partial t_p \\ & \vdots & \\ \partial g_m(\boldsymbol{T})/\partial t_1 & \cdots & \partial g_m(\boldsymbol{T})/\partial t_p \end{bmatrix}^T_{\Big|\boldsymbol{T}=\boldsymbol{\mu}} \tag{A.36}$$

evaluated at $\boldsymbol{T} = \boldsymbol{\mu}$.

When $\boldsymbol{T}$ is a jointly consistent estimator for $\boldsymbol{\mu}$, then (A.35) provides the first two moments of the asymptotic distribution of $\boldsymbol{Y}$. Further, from Slutsky's Theorem (A.45, below) if $\widehat{\boldsymbol{\Sigma}}_T$ is consistent for $\boldsymbol{\Sigma}_T$, then

$$\widehat{\boldsymbol{\Sigma}}_Y = \widehat{\boldsymbol{H}}(\boldsymbol{T})'\widehat{\boldsymbol{\Sigma}}_T\widehat{\boldsymbol{H}}(\boldsymbol{T}) = \widehat{\boldsymbol{H}}'\widehat{\boldsymbol{\Sigma}}_T\widehat{\boldsymbol{H}} \tag{A.37}$$

is a consistent estimator of $\boldsymbol{\Sigma}_Y$.

Example A.4 *Multinomial Generalized Logits*
For example, consider the case of a trinomial where we wish to estimate the mean and variance of the vector of log odds (*logits*) of the second category versus the first, $\log(p_2/p_1)$, and also the third category versus the first, $\log(p_3/p_1)$. Thus $\boldsymbol{p} = (p_1\ p_2\ p_3)^T$ has mean vector $\boldsymbol{\pi} = (\pi_1\ \pi_2\ \pi_3)^T$ and covariance matrix

$$\boldsymbol{\Sigma}(\boldsymbol{\pi}) = \frac{1}{N}\begin{bmatrix} \pi_1(1-\pi_1) & -\pi_1\pi_2 & -\pi_1\pi_3 \\ -\pi_1\pi_2 & \pi_2(1-\pi_2) & -\pi_2\pi_3 \\ -\pi_1\pi_3 & -\pi_2\pi_3 & \pi_3(1-\pi_3) \end{bmatrix}. \tag{A.38}$$

The transformation is $\boldsymbol{Y} = \boldsymbol{G}(\boldsymbol{p}) = [g_1(\boldsymbol{p})\ g_2(\boldsymbol{p})]^T$, where $g_1(\boldsymbol{p}) = \log(p_2/p_1)$ and $g_2(\boldsymbol{p}) = \log(p_3/p_1)$. Asymptotically, from (A.35),

$$E(\boldsymbol{Y}) = \boldsymbol{\mu}_Y = [\log(\pi_2/\pi_1)\ \log(\pi_3/\pi_1)]^T. \tag{A.39}$$

To obtain the asymptotic variance requires the matrix of derivatives, which are

$$\begin{aligned} \boldsymbol{H}(\boldsymbol{\pi}) &= \begin{bmatrix} \partial g_1(\boldsymbol{\pi})/\partial\pi_1 & \partial g_1(\boldsymbol{\pi})/\partial\pi_2 & \partial g_1(\boldsymbol{\pi})/\partial\pi_3 \\ \partial g_2(\boldsymbol{\pi})/\partial\pi_1 & \partial g_2(\boldsymbol{\pi})/\partial\pi_2 & \partial g_2(\boldsymbol{\pi})/\partial\pi_3 \end{bmatrix}^T \\ &= \begin{bmatrix} -1/\pi_1 & 1/\pi_2 & 0 \\ -1/\pi_1 & 0 & 1/\pi_3 \end{bmatrix}^T. \end{aligned} \tag{A.40}$$

Thus

$$\boldsymbol{\Sigma}_Y = \boldsymbol{H}'\boldsymbol{\Sigma}(\boldsymbol{\pi})\boldsymbol{H} = \frac{1}{N}\begin{bmatrix} 1/\pi_1 + 1/\pi_2 & 1/\pi_1 \\ 1/\pi_1 & 1/\pi_1 + 1/\pi_3 \end{bmatrix} \tag{A.41}$$

provides the asymptotic covariance matrix of the two logits.

A.4 SLUTSKY'S CONVERGENCE THEOREM

Slutsky's Theorem (*cf.* Cramer, 1946; Serfling, 1980) is a multifaceted result which can be used to establish the convergence in distribution and/or the convergence in probability (consistency) of multidimensional transformations of a vector of statistics. For the purposes herein, I shall present these as two results, rather than as a single theorem. The theorem is then used in conjunction with the delta method to obtain the asymptotic distribution of transformations of statistics.

A.4.1 Convergence in Distribution

The most common application of Slutsky's Theorem concerns the asymptotic distribution of a linear combination of two sequences of statistics, one of which converges in probability to a constant, the other of which converges in distribution to a specified distribution. In this text we are only concerned with functions of statistics that are asymptotically normally distributed, for which the theorem is so described.

The result, however, applies more generally to statistics that follow any specified distribution. The theorem also readily generalizes to more than two such statistics.

Let t_n be a sequence of statistics that converges in distribution to a normal distribution as $n \to \infty$ such that

$$\sqrt{n}\,(t_n - \mu) \xrightarrow{d} \mathcal{N}(0,\ \sigma^2), \tag{A.42}$$

where the variance σ^2 may be a function of the expectation μ. Also, let r_n be a sequence of statistics that converges in probability to a constant ρ, expressed as $r_n \xrightarrow{p} \rho$. Then

$$\sqrt{n}\,(r_n + t_n) \xrightarrow{d} \mathcal{N}(\rho + \mu,\ \sigma^2) \tag{A.43}$$

and

$$\sqrt{n}\,(r_n t_n) \xrightarrow{d} \mathcal{N}(\rho\mu,\ \rho^2\sigma^2). \tag{A.44}$$

An example is provided below.

A.4.2 Convergence in Probability

Consider a set of statistics, each of which in sequence converges in probability to known quantities, such as $a_n \xrightarrow{p} \alpha$, $b_n \xrightarrow{p} \beta$, and $c_n \xrightarrow{p} \gamma$. For any continuous function $R(\cdot)$, then

$$R(a_n, b_n, c_n) \xrightarrow{p} R(\alpha, \beta, \gamma). \tag{A.45}$$

This result will be used to demonstrate the consistency of various "plug-in" estimators of parameters.

A.4.3 Convergence in Distribution of Transformations

The δ-method and Slutsky's Theorem together provide powerful tools that can be used to derive the asymptotic distribution of statistics which are obtained as transformations of basic statistics that are known to be asymptotically normally distributed, such as transformations of means or proportions. The δ-method provides the expressions for the mean and variance of the statistic under a linear or non-linear transformation. Slutsky's Theorem then provides the proof of the asymptotic convergence to a normal, or multivariate normal, distribution. This result is stated herein for the univariate case, the multivariate case follows similarly (*cf.* Rao, 1973, p. 385; or Bickel and Doksum, 1977, pp. 461–462). Under certain conditions, similar results can be used to demonstrate convergence to other distributions. However, herein we only consider the application to transformations of statistics that converge in distribution to a normal distribution.

As above, let $\sqrt{n}(t_n - \mu)$ be a sequence of statistics which converges in distribution to the normal distribution with mean μ and variance σ^2 as in (A.42). Also, let $g(t)$ be a single variable function with derivative $g'(\cdot)$ that is continuous at μ. Then $g(t)$ converges in distribution to

$$\sqrt{n}\left[g(t_n) - g(\mu)\right] \xrightarrow{d} \mathcal{N}\left(0, \ \left[g'(\mu)\right]^2 \sigma^2\right). \qquad \text{(A.46)}$$

Thus for large N, the approximate large sample distribution of $g(t)$ is

$$g(t) \stackrel{d}{\approx} \mathcal{N}\left(g(\mu), \ \frac{\left[g'(\mu)\right]^2 \sigma^2}{N}\right). \qquad \text{(A.47)}$$

Example A.5 *log(p)*

To illustrate the application of this result and its derivation using Slutsky's Theorem, consider the asymptotic distribution of log(p). Applying the Taylor's expansion (A.27), then from (A.32) asymptotically

$$\sqrt{n}\left[\log(p) - \log(\pi)\right] = \sqrt{n}\frac{(p-\pi)}{\pi} - \sqrt{n}\frac{(p-\pi)^2}{2a^2}, \qquad \text{(A.48)}$$

where the last term is $\sqrt{n}R_2(a)$. Since p is asymptotically normally distributed, then the first term on the r.h.s. is likewise asymptotically normally distributed.

To evaluate the remainder asymptotically, let $\{p_n\}$ denote a sequence of values as $n \to \infty$. In Example A.1 we saw that p_n is a sample mean of n Bernoulli variables, and thus is a $\sqrt{n}$-consistent estimator of π so that $(p_n - \pi)^2 \to 0$ faster than $n^{-\frac{1}{2}}$, and thus $\sqrt{n}R_2(a) \xrightarrow{p} 0$. Therefore, asymptotically $\sqrt{n}\left[\log(p) - \log(\pi)\right]$ is the sum of two random variables, one converging in distribution to the normal, the other converging in probability to a constant (zero). From Slutsky's Convergence in Distribution Theorem (A.43) it follows that

$$\sqrt{n}\left[\log(p) - \log(\pi)\right] \xrightarrow{d} \mathcal{N}\left[0, \ \left(\frac{d\log(\pi)}{d\pi}\right)^2 \pi(1-\pi)\right] \qquad \text{(A.49)}$$

and for large N asymptotically

$$\log(p) \stackrel{d}{\approx} \mathcal{N}\left(\log(\pi), \frac{1-\pi}{N\pi}\right). \qquad \text{(A.50)}$$

The large sample variance of log(p) is

$$V\left[\log(p)\right] \cong \frac{1-\pi}{N\pi}, \qquad \text{(A.51)}$$

which can be estimated by "plugging in" the estimate p for π to obtain

$$\widehat{V}\left[\log(p)\right] = \frac{1-p}{Np}. \qquad \text{(A.52)}$$

Then from Slutsky's Convergence in Probability Theorem (A.45), since $p_n \xrightarrow{p} \pi$ it follows that (A.52) $\xrightarrow{p}$ (A.51), which proves the consistency of the large sample estimate of $V[\log(p)]$.

From the asymptotic distribution of $\log(p)$ in (A.50) and the consistency of the estimate of the variance using the "plug in" approach in (A.52), again using Slutsky's Convergence in Distribution Theorem (A.44) it follows that asymptotically

$$\frac{\log(p) - \log(\pi)}{\sqrt{\widehat{V}[\log(p)]}} \overset{d}{\approx} N(0,1). \tag{A.53}$$

Thus the asymptotic coverage probability of $(1-\alpha)$ level confidence limits based on the estimated large sample variance in (A.52) is approximately the desired level $1-\alpha$.

Example A.6 *Multinomial Logits*

For the multinomial logits in Example A.4, it follows from the above theorems that the sample logits $\boldsymbol{Y} = [\log(p_2/p_1)\ \log(p_3/p_1)]^T$ are distributed as bivariate normal with expectation vector $\boldsymbol{\mu}_Y = [\log(\pi_2/\pi_1)\ \log(\pi_3/\pi_1)]^T$ and with variance as presented in (A.41). Substituting the elements $\boldsymbol{p} = (p_1\ p_2\ p_3)^T$ for $\boldsymbol{\pi} = (\pi_1\ \pi_2\ \pi_3)^T$ yields the estimated large sample covariance matrix

$$\widehat{\Sigma}_Y = \frac{1}{n}\begin{bmatrix} 1/p_1 + 1/p_2 & 1/p_1 \\ 1/p_1 & 1/p_1 + 1/p_3 \end{bmatrix}. \tag{A.54}$$

Since $\boldsymbol{p}$ is jointly consistent for $\boldsymbol{\pi}$ then from Slutsky's Convergence in Probability Theorem (A.45), $\widehat{\boldsymbol{\Sigma}}_Y$ is a consistent estimate of $\boldsymbol{\Sigma}_Y$. This estimate may also be obtained by evaluating

$$\widehat{\boldsymbol{\Sigma}}_Y = \widehat{\boldsymbol{H}}'\widehat{\Sigma}(\boldsymbol{p})\widehat{\boldsymbol{H}}, \tag{A.55}$$

in which $\widehat{\Sigma}(\boldsymbol{p}) = \Sigma(\boldsymbol{\pi})_{|\boldsymbol{\pi}=\boldsymbol{p}}$ and in which $\widehat{\boldsymbol{H}} = \boldsymbol{H}(\boldsymbol{p}) = \boldsymbol{H}(\boldsymbol{\pi})_{|\boldsymbol{\pi}=\boldsymbol{p}}$ as in (A.37).

A.5 LEAST SQUARES ESTIMATION

A.5.1 Ordinary Least Squares (OLS)

Ordinary least squares is best known as the method for the derivation of the parameter estimates in the simple linear regression model. In general, OLS can be described as a method for the estimation of the conditional expectation of the dependent variable Y within the context of some model as a function of a covariate vector with value x, which may represent a vector of p covariates, plus a constant (1) for the intercept. The values of x are considered fixed quantities; that is, we condition on the observed values of x. In the simple linear multiple regression model, the conditional expectation is expressed as $E(y|x) = \boldsymbol{x}'\boldsymbol{\theta}$ as a function of a vector of parameters $\boldsymbol{\theta}$. This specification is termed the *structural component* of the regression model, which in this instance is a linear function of the covariates. Given $\boldsymbol{x}$, it is then assumed that the random errors $\varepsilon = y - E(y|\boldsymbol{x})$ are statistically independent and identically distributed ($i.i.d.$) with mean zero and common variance σ_ε^2. This specification is termed the *random component* of the regression model. The structural and random components together specify that $y = E(y|\boldsymbol{x}) + \varepsilon$, which then specifies that $V(y|\boldsymbol{x}) = \sigma_\varepsilon^2$ conditionally on $\boldsymbol{x}$.

For a sample of N observations, let $\boldsymbol{Y} = (y_1 \ \ldots \ y_N)^T$ refer to the column vector of dependent variable values, and $\mathbf{X}$ refer to the $N \times (p+1)$ vector of covariate values for the N observations. The ith row of $\mathbf{X}$ consists of the vector of $(p+1)$ covariate values for the ith observation, $\boldsymbol{x}_i^T = (1 \ x_{i1} \ \ldots \ x_{ip})$. For the ith observation, the random error is $\varepsilon_i = y_i - E(y|\boldsymbol{x}_i)$, and for the sample of N observations the vector of random errors is $\boldsymbol{\varepsilon} = (\varepsilon_1 \ldots \varepsilon_N)^T$, where, as stated above, it is assumed that $E(\boldsymbol{\varepsilon}) = \mathbf{0}$ and $Cov(\boldsymbol{\varepsilon}) = \boldsymbol{\Sigma}_{\boldsymbol{\varepsilon}} = \sigma_\varepsilon^2 \boldsymbol{I}_N$, with $\boldsymbol{I}_N$ being the identity matrix of dimension N. In vector notation, the linear model then can be expressed as

$$\boldsymbol{Y} = \boldsymbol{X\theta} + \boldsymbol{\varepsilon} \ , \tag{A.56}$$

where $\boldsymbol{\theta} = (\alpha \,//\, \boldsymbol{\beta})$, $\boldsymbol{\beta} = (\beta_1 \ \ldots \ \beta_p)^T$, α being the intercept. Note that throughout the Appendix the intercept is implicit in the expression $\boldsymbol{x}'\boldsymbol{\theta}$ whereas elsewhere in the text we use the explicit notation $\alpha + \boldsymbol{x}'\boldsymbol{\beta}$.

We then desire an estimate of the coefficient vector $\boldsymbol{\theta}$ that satisfies some desirable statistical properties. In this setting squared error loss suggests choosing the vector $\boldsymbol{\theta}$ so as to minimize the sums of squares of errors $\sum_j \varepsilon_j^2 = \boldsymbol{\varepsilon}'\boldsymbol{\varepsilon} = SSE$. Thus the estimate satisfies

$$\min_{\widehat{\boldsymbol{\theta}}} \left(\mathbf{Y} - \mathbf{X}\widehat{\boldsymbol{\theta}}\right)' \left(\mathbf{Y} - \mathbf{X}\widehat{\boldsymbol{\theta}}\right) = \min_{\widehat{\boldsymbol{\theta}}} \left(\mathbf{Y}'\mathbf{Y} - 2\mathbf{Y}'\mathbf{X}\widehat{\boldsymbol{\theta}} + \widehat{\boldsymbol{\theta}}'\mathbf{X}'\mathbf{X}\widehat{\boldsymbol{\theta}}\right). \tag{A.57}$$

Using the calculus of maxima/minima the solution is obtained by setting the vector of first derivatives equal to $\mathbf{0}$ and solving for $\widehat{\boldsymbol{\theta}}$. For this purpose we require the derivatives of the bilinear and quadratic forms with respect to $\widehat{\boldsymbol{\theta}}$ which are as follows:

$$\frac{\partial \mathbf{X}\widehat{\boldsymbol{\theta}}}{\partial \widehat{\boldsymbol{\theta}}} = \mathbf{X}, \qquad \frac{\partial \widehat{\boldsymbol{\theta}}'\mathbf{A}\boldsymbol{\theta}}{\partial \boldsymbol{\theta}} = 2\widehat{\boldsymbol{\theta}}'\mathbf{A}. \tag{A.58}$$

Thus

$$\frac{\partial SSE}{\partial \widehat{\boldsymbol{\theta}}} = \frac{\partial \left(\mathbf{Y}'\mathbf{Y} - 2\mathbf{Y}'\mathbf{X}\widehat{\boldsymbol{\theta}} + \widehat{\boldsymbol{\theta}}'\mathbf{X}'\mathbf{X}\widehat{\boldsymbol{\theta}}\right)}{\partial \widehat{\boldsymbol{\theta}}} = -2\mathbf{Y}'\mathbf{X} + 2\widehat{\boldsymbol{\theta}}'\mathbf{X}'\mathbf{X}, \tag{A.59}$$

and the matrix of second derivatives $(2\mathbf{X}'\mathbf{X})$ is positive definite, provided it is of full rank. Thus setting the vector of first derivatives equal to $\mathbf{0}$ yields the OLS estimating equation

$$\widehat{\boldsymbol{\theta}}'\mathbf{X}'\mathbf{X} - \mathbf{Y}'\mathbf{X} = \mathbf{0} \ , \tag{A.60}$$

for which the SSE is minimized with respect to the choice of $\widehat{\boldsymbol{\theta}}$. Solving for $\widehat{\boldsymbol{\theta}}$ yields the OLS estimate

$$\widehat{\boldsymbol{\theta}} = (\mathbf{X}'\mathbf{X})^{-1}\mathbf{X}'\mathbf{Y}. \tag{A.61}$$

A unique solution vector for $\widehat{\boldsymbol{\theta}}$ is obtain provided that $\boldsymbol{X}'\boldsymbol{X}$ is of full rank and the inverse exists. Another basic result from the algebra of matrices is that the rank of $\boldsymbol{X}'\boldsymbol{X}$ is the minimum of the row and column rank of $\boldsymbol{X}$. Since $\boldsymbol{X}$ is a $n \times (p+1)$ matrix, then $\boldsymbol{X}'\boldsymbol{X}$ will be positive definite of full rank $(p+1)$ unless there is a linear dependency or degeneracy among the columns of the $\boldsymbol{X}$ matrix, which would require that one covariate in the design matrix $\boldsymbol{X}$ be a linear combination of the others. If such a degeneracy exists, then there is no unique solution vector $\widehat{\boldsymbol{\theta}}$ that satisfies the OLS estimating equation. In this case, one among the many solutions is obtained by using a generalized inverse. Throughout, however, we assume that $\boldsymbol{X}'\boldsymbol{X}$ is of full rank.

A.5.2 Gauss-Markov Theorem

The properties of least squares estimators are provided by the *Gauss-Markov Theorem.* From the assumption of $i.i.d.$ homoscedastic errors (with common variance), the following properties are readily derived (*cf.* Rao, 1973).

The least squares estimates of the coefficients are unbiased, since

$$E(\widehat{\boldsymbol{\theta}}) = (\mathbf{X}'\mathbf{X})^{-1}\mathbf{X}'E(\mathbf{Y}) = (\mathbf{X}'\mathbf{X})^{-1}\mathbf{X}'\mathbf{X}\boldsymbol{\theta} = \boldsymbol{\theta}. \tag{A.62}$$

To obtain the variance of the estimates, note that the solution $\widehat{\boldsymbol{\theta}}$ is a linear combination of the $\boldsymbol{Y}$ vector of the form $\widehat{\boldsymbol{\theta}} = \boldsymbol{H}'\boldsymbol{Y}$ where $\boldsymbol{H}' = (\mathbf{X}'\mathbf{X})^{-1}\mathbf{X}'$. Since $V(\mathbf{Y}) = V(\varepsilon) = \sigma_\varepsilon^2 \boldsymbol{I}_N$, then

$$\begin{aligned} V(\widehat{\boldsymbol{\theta}}) &= \boldsymbol{H}'\left[V(\mathbf{Y})\right]\boldsymbol{H} = (\mathbf{X}'\mathbf{X})^{-1}\mathbf{X}'\left[\sigma_\varepsilon^2 \boldsymbol{I}_N\right]\mathbf{X}(\mathbf{X}'\mathbf{X})^{-1} \\ &= (\mathbf{X}'\mathbf{X})^{-1}\sigma_\varepsilon^2. \end{aligned} \tag{A.63}$$

Since

$$\widehat{\sigma}_\varepsilon^2 = MSE = \frac{\left(\mathbf{Y} - \mathbf{X}\widehat{\boldsymbol{\theta}}\right)'\left(\mathbf{Y} - \mathbf{X}\widehat{\boldsymbol{\theta}}\right)}{N - p - 1} \tag{A.64}$$

provides a consistent estimator of σ_ε^2, then a consistent estimator of the covariance matrix of the estimates is provided by

$$\widehat{V}\left(\widehat{\boldsymbol{\theta}}\right) = (\mathbf{X}'\mathbf{X})^{-1}\widehat{\sigma}_\varepsilon^2. \tag{A.65}$$

Since the least squares estimator is that linear function of the observations which is both unbiased and which minimizes the SSE, or which has the smallest variance among all possible unbiased linear estimators, then this is a *best linear unbiased estimator (BLUE)*. Again, this result only requires the assumption that the $\{y_i\}$ are independent and that the random errors are $i.i.d.$ with mean zero and constant variance σ_ε^2 conditional on the covariates $\{\boldsymbol{x}_i\}$. No further assumptions are necessary.

In addition, if it is assumed that the $\{\varepsilon_i\}$ are normally distributed, then the F distribution can be used to characterize the distribution of a ratio of independent sums of squares as the basis for parametric tests with finite samples. However, the normal errors assumption is not necessary for a large sample inference. Since the $\{y_i\}$ are independent with constant conditional variance σ_ε^2 given $\{\boldsymbol{x}_i\}$, then from (A.62) and the Liapunov Central Limit Theorem, asymptotically

$$\left(\widehat{\boldsymbol{\theta}} - \boldsymbol{\theta}\right) \overset{d}{\approx} \mathcal{N}\left[\boldsymbol{0}, V\left(\widehat{\boldsymbol{\theta}}\right)\right]. \tag{A.66}$$

This provides the basis for large sample Wald tests and confidence limits for the elements of $\boldsymbol{\theta}$.

A.5.3 Weighted Least Squares (WLS)

These developments may also be generalized to the case where the random errors have expectation zero but are not $i.i.d.$, meaning that $V(\boldsymbol{\varepsilon}) = \boldsymbol{\Sigma_\varepsilon} \neq \sigma_\varepsilon^2 \boldsymbol{I}_N$. Thus there may be heteroscedasticity of the error variances such that $V(\varepsilon_i) \neq V(\varepsilon_j)$ for some two observations $i \neq j$, and/or the errors may be correlated such that $Cov(\varepsilon_i, \varepsilon_j) \neq 0$. As with ordinary least squares, we start with the assumption of a simple linear multiple regression model such that the conditional expectation is expressed as $E(Y|\boldsymbol{x}) = \boldsymbol{x}'\boldsymbol{\theta}$. Thus the model specifies that $\boldsymbol{Y} = \boldsymbol{X\theta} + \boldsymbol{\varepsilon}$, where $E(\boldsymbol{\varepsilon}) = \boldsymbol{0}$, but where $V(\boldsymbol{\varepsilon}) = \boldsymbol{\Sigma_\varepsilon}$. When $\boldsymbol{\Sigma_\varepsilon}$ is of full rank, then by a transformation using the root of the inverse, $\boldsymbol{\Sigma_\varepsilon}^{-1/2}$, we have

$$\boldsymbol{\Sigma_\varepsilon}^{-1/2}\mathbf{Y} = \boldsymbol{\Sigma_\varepsilon}^{-1/2}\mathbf{X}\boldsymbol{\theta} + \boldsymbol{\Sigma_\varepsilon}^{-1/2}\boldsymbol{\varepsilon}\ , \tag{A.67}$$

which can be expressed as

$$\widetilde{\mathbf{Y}} = \widetilde{\mathbf{X}}\boldsymbol{\theta} + \widetilde{\boldsymbol{\varepsilon}}\ , \tag{A.68}$$

where

$$V(\widetilde{\boldsymbol{\varepsilon}}) = \boldsymbol{\Sigma_\varepsilon}^{-1/2}\boldsymbol{\Sigma_\varepsilon}\boldsymbol{\Sigma_\varepsilon}^{-1/2} = \boldsymbol{I}_N. \tag{A.69}$$

Thus the transformed random errors $\widetilde{\boldsymbol{\varepsilon}}$ satisfy the assumptions of the OLS estimators so that

$$\widehat{\boldsymbol{\theta}} = \left(\widetilde{\mathbf{X}}'\widetilde{\mathbf{X}}\right)^{-1}\widetilde{\mathbf{X}}'\widetilde{\mathbf{Y}} = \left(\mathbf{X}'\boldsymbol{\Sigma_\varepsilon}^{-1}\mathbf{X}\right)^{-1}\left(\mathbf{X}'\boldsymbol{\Sigma_\varepsilon}^{-1}\mathbf{Y}\right). \tag{A.70}$$

As in (A.62)–(A.63), it follows that the WLS estimates are unbiased with covariance matrix

$$V\left(\widehat{\boldsymbol{\theta}}\right) = \left(\widetilde{\mathbf{X}}'\widetilde{\mathbf{X}}\right)^{-1} = \left(\mathbf{X}'\boldsymbol{\Sigma}_{\boldsymbol{\varepsilon}}^{-1}\mathbf{X}\right)^{-1}, \tag{A.71}$$

and that the estimates are asymptotically normally distributed as in (A.66). Further, if a consistent estimator $\widehat{\boldsymbol{\Sigma}}_{\boldsymbol{\varepsilon}}$ of $\boldsymbol{\Sigma}_{\boldsymbol{\varepsilon}}$ is available, then the covariance matrix of the estimates can be consistently estimated as

$$\widehat{V}\left(\widehat{\boldsymbol{\theta}}\right) = \left(\mathbf{X}'\widehat{\boldsymbol{\Sigma}}_{\boldsymbol{\varepsilon}}^{-1}\mathbf{X}\right)^{-1}, \tag{A.72}$$

which provides the basis for large sample confidence intervals and tests of significance. In these expressions, the matrix $\boldsymbol{\Sigma}_{\boldsymbol{\varepsilon}}^{-1}$ is termed the *weight matrix* since the solution vector $\widehat{\boldsymbol{\theta}}$ minimizes the weighted SSE computed as

$$SSE = \left(\widetilde{\mathbf{Y}} - \widetilde{\mathbf{X}}\widehat{\boldsymbol{\theta}}\right)'\left(\widetilde{\mathbf{Y}} - \widetilde{\mathbf{X}}\widehat{\boldsymbol{\theta}}\right) = \left(\mathbf{Y} - \mathbf{X}\widehat{\boldsymbol{\theta}}\right)'\boldsymbol{\Sigma}_{\boldsymbol{\varepsilon}}^{-1}\left(\mathbf{Y} - \mathbf{X}\widehat{\boldsymbol{\theta}}\right). \tag{A.73}$$

Example A.7 *Heteroscedasticity*

One common application of weighted least squares is the case of heteroscedastic variances. One instance is where y_i is the mean of n_i measurements for the ith observation, where n_i varies among the observations. Thus

$$\boldsymbol{\Sigma}_{\boldsymbol{\varepsilon}} = \sigma_\varepsilon^2[diag(n_1^{-1} \cdots n_N^{-1})]. \tag{A.74}$$

Example A.8 *Correlated Observations*

Another common application is to the case of clustered sampling, such as repeated or clustered measures on the same subject, or measures on the members of a family. In this case the sampling unit is the cluster and the observations within a cluster are correlated. Let n_j refer to the number of members within the jth cluster and let Σ_j denote the $n_j \times n_j$ matrix of the variances and covariances among the n_j observations within the jth cluster. Then

$$\boldsymbol{\Sigma}_{\boldsymbol{\varepsilon}} = blockdiag[\Sigma_1 \cdots \Sigma_N] = \begin{bmatrix} \Sigma_1 & \dots & 0 \\ \vdots & \dots & \vdots \\ 0 & \dots & \Sigma_N \end{bmatrix}, \tag{A.75}$$

where N is now the total number of clusters. The within-cluster variances and covariances may be specified on the basis of the statistical properties of the sampling procedure, or may be estimated from the data.

For example, for n_j repeated measures on the same subject, one may assume an *exchangeable correlation* structure such that there is constant correlation, say ρ, among measures within the subject (cluster). Given an estimate of the common correlation, $\widehat{\rho}$, and estimates of the variances of each repeated measure, say $\widehat{\sigma}_\ell^2$, then the covariance between the first and second repeated measures, for example, is $\widehat{\rho\sigma_1\sigma_2}$. From these the estimated covariance matrix could be obtained for the set of repeated measures for the jth subject.

A.5.4 Iteratively Reweighted Least Squares (IRLS)

In some instances, the covariance matrix of the random errors is a function of the estimated conditional expectations, such that $\boldsymbol{\Sigma}_{\boldsymbol{\varepsilon}} = \boldsymbol{G}(\boldsymbol{\theta})$ for some $N \times N$ matrix with elements a function of the coefficient vector $\boldsymbol{\theta}$. In this case, the weight matrix depends on the values of the coefficients. In general, an iterative procedure is required as follows: Given some initial estimates of the coefficients, say $\widehat{\boldsymbol{\theta}}_0$, one computes the weight matrix, say $\boldsymbol{G}_0 = \boldsymbol{G}(\widehat{\boldsymbol{\theta}}_0)$, and then computes the first-step estimates as

$$\widehat{\theta}_1 = \left(\mathbf{X}'G_0^{-1}\mathbf{X}\right)^{-1}\left(\mathbf{X}'G_0^{-1}\mathbf{Y}\right). \tag{A.76}$$

Then the weight matrix is updated as $\boldsymbol{G}_1 = \boldsymbol{G}(\widehat{\boldsymbol{\theta}}_1)$, and then the second-step estimate $\widehat{\boldsymbol{\theta}}_2$ obtained. The iterative process continues until the coefficient estimates converge to a constant vector, or equivalently until some objective function such as the SSE converges to a constant. As with ordinary least squares, the final estimates are asymptotically normally distributed as in (A.66), where a consistent estimate of the covariance matrix of the coefficient estimates is provided by (A.72) using $\widehat{\boldsymbol{\Sigma}}_{\boldsymbol{\varepsilon}} = \boldsymbol{G}(\widehat{\boldsymbol{\theta}})$.

Often a relationship between the conditional expectations and the variance of the errors arises from the specification of a parametric model in the population based on an underlying distribution conditional on the values of the covariates. In this case the IRLS estimates of the coefficients equal those obtained from maximum likelihood estimation.

A.6 MAXIMUM LIKELIHOOD ESTIMATION AND EFFICIENT SCORES

A.6.1 Estimating Equation

Another statistical approach to the estimation of parameters and the development of statistical tests is to adopt a likelihood based on an assumed underlying population model. Let $(y_1, y_2, \ldots, y_N)$ refer to a sample of independent and identically distributed ($i.i.d.$) observations drawn at random from a specified distribution $f(y; \theta)$, where the density or probability mass distribution $f(\cdot)$ has unknown parameter θ which may be a p-vector of parameters $\boldsymbol{\theta} = (\theta_1 \ldots \theta_p)^T$. Throughout we assume that θ is the true parameter value, although at times we use the notation $\theta = \theta_0$ to denote that the true value is assumed to be the specified values θ_0. When not ambiguous, results are presented for the scalar case using the parameter θ.

The *likelihood function* then is the total probability of the sample under the assumed model

$$L(y_1, \ldots, y_N; \theta) = \prod_{i=1}^{N} f(y_i; \theta), \tag{A.77}$$

which, for simplicity, is designated as simply $L(\theta)$. Alternately, when a known sufficient statistic, say T, is known to exist for θ, then apart from constants, the likelihood may be expressed in terms of the distribution of T.

The *maximum likelihood estimate* of θ, designated as $\widehat{\theta}$, is that value for which the likelihood is maximized. This value is most easily determined using the log-likelihood, which in the case of (A.77), is represented as a sum rather than a product of terms

$$\ell(\theta) = \log L(\theta) = \sum_{i=1}^{N} \log f(y_i; \theta). \tag{A.78}$$

The *MLE* $\widehat{\theta}$ is then obtained from the calculus of the local extrema (max/min) of a function. Thus in the single parameter case, the maximum likelihood estimator *(MLE)* is that value such that

$$\left.\frac{d\ell(\theta)}{d\theta}\right|_{\theta=\widehat{\theta}} = 0 \quad \text{given that} \quad \left.\frac{d^2\ell(\theta)}{d\theta^2}\right|_{\theta=\widehat{\theta}} < 0. \tag{A.79}$$

The first derivative is the slope of the function $\ell(\theta)$ with respect to θ. The maximum occurs at the point where the tangent to $\ell(\theta)$ is the horizontal line or the point along the likelihood surface where the slope is zero. The second derivative represents the degree of curvature of the function $\ell(\theta)$ and its direction: facing up or down. Thus the condition on the second derivative is that the likelihood function be convex or "concave down."

The *maximum likelihood estimating equations*, therefore, in the scalar or vector parameter cases are

$$\frac{\partial\ell(\theta)}{\partial\theta} = 0, \qquad \text{or} \qquad \begin{pmatrix} \partial\ell(\boldsymbol{\theta})/\partial\theta_1 \\ \vdots \\ \partial\ell(\boldsymbol{\theta})/\partial\theta_p \end{pmatrix} = \begin{pmatrix} 0 \\ \vdots \\ 0 \end{pmatrix}. \tag{A.80}$$

The *MLE* $\widehat{\theta}$ then is the solution for θ in the scalar parameter case, and $\widehat{\boldsymbol{\theta}}$ is the solution for $\boldsymbol{\theta}$ in the multiparameter case, the vector estimating equation applying simultaneously to the p elements of $\boldsymbol{\theta}$. In many cases the solution of these estimating equations requires an iterative procedure such as the Newton-Raphson method.

A.6.2 Efficient Score

The estimating equation for a scalar θ is a function of the *Fisher Efficient Score*

$$U(\theta) = \frac{d\ell(\theta)}{d\theta} = \sum\nolimits_{i=1}^{N} \frac{d\log f(y_i;\theta)}{d\theta} = \sum\nolimits_{i=1}^{N} U_i(\theta)\,, \tag{A.81}$$

where $U_i(\theta)$ is the Fisher efficient score for the ith observation. When $\boldsymbol{\theta}$ is a p-vector then the total score vector is

$$\boldsymbol{U}(\boldsymbol{\theta}) = \left[U(\boldsymbol{\theta})_{\theta_1} \;\ldots\; U(\boldsymbol{\theta})_{\theta_p}\right]^T = \sum_{i=1}^{N} \boldsymbol{U}_i(\theta), \tag{A.82}$$

where for the ith subject,

$$\boldsymbol{U}_i(\boldsymbol{\theta}) = \left[U_i(\boldsymbol{\theta})_{\theta_1} \; \ldots \; U_i(\boldsymbol{\theta})_{\theta_p})\right]^T \qquad \text{(A.83)}$$
$$= \left[\frac{\partial \log f(y_i;\boldsymbol{\theta})}{\partial \theta_1} \cdots \frac{\partial \log f(y_i;\boldsymbol{\theta})}{\partial \theta p}\right]^T.$$

The notation $U(\boldsymbol{\theta})_{\theta j}$ designates that the score for the jth element of the parameter vector is some function of the p-vector $\boldsymbol{\theta}$. The *MLE* is defined as that value of the p-vector $\widehat{\boldsymbol{\theta}}$ for which the total score is the zero vector.

An important property of the efficient score is that $E\left[\boldsymbol{U}(\boldsymbol{\theta})\right] = \mathbf{0}$ when the score is evaluated at the true value of the parameter in the population $\boldsymbol{\theta}$. To derive this result, consider the case where θ is a scalar. Since

$$U_i(\theta) = \frac{d \log f(y_i;\theta)}{d\theta} = \frac{1}{f(y_i;\theta)} \frac{df(y_i;\theta)}{d\theta}, \qquad \text{(A.84)}$$

then

$$E\left[U_i(\theta)\right] = \int \frac{1}{f(y_i;\theta)} \frac{df(y_i;\theta)}{d\theta} f(y_i;\theta) dy = \int \frac{df(y_i;\theta)}{d\theta} dy \qquad \text{(A.85)}$$
$$= \frac{d}{d\theta} \int f(y_i;\theta) dy = \frac{d(1)}{d\theta} = 0, \qquad \forall i.$$

Thus

$$E\left[U(\theta)\right] = E\left[\sum\nolimits_{i=1}^{N} U_i(\theta)\right] = 0. \qquad \text{(A.86)}$$

This result also applies to the multiparameter case where $E\left[\boldsymbol{U}(\boldsymbol{\theta})\right] = \mathbf{0}$.

This property plays a central role in the development of the efficient score test of an hypothesis of the form H_0: $\theta = \theta_0$.

A.6.3 Fisher's Information Function

Again consider the case where θ is a scalar. Since the likelihood in (A.77) is the probability of any observed sample of N $i.i.d.$ observations, then

$$\int \cdots \int L(y_1, \ldots, y_N; \theta) \; dy_1 \cdots dy_N = 1. \qquad \text{(A.87)}$$

Taking the derivative with respect to θ yields

$$\int \cdots \int \frac{dL(\theta)}{d\theta} \; dy_1 \cdots dy_N = 0. \qquad \text{(A.88)}$$

As in (A.84), however,

$$\frac{d\ell(\theta)}{d\theta} = \frac{d \log L(\theta)}{d\theta} = \frac{1}{L(\theta)} \frac{dL(\theta)}{d\theta} \qquad \text{(A.89)}$$

so that

$$\frac{dL(\theta)}{d\theta} = \frac{d\ell(\theta)}{d\theta} L(\theta) \; . \tag{A.90}$$

Then, (A.88) can be expressed as

$$\int \cdots \int \frac{d\ell(\theta)}{d\theta} L(\theta) \; dy_1 \cdots dy_N = E\left[U(\theta)\right] = 0 \; , \tag{A.91}$$

which is a generalization of (A.85).

Differentiating $E\left[U(\theta)\right]$ in (A.91) a second time with respect to θ yields

$$\begin{aligned} \frac{dE\left[U(\theta)\right]}{d\theta} &= \int \cdots \int \left[\frac{d\ell(\theta)}{d\theta}\left(\frac{dL(\theta)}{d\theta}\right) + L(\theta)\left(\frac{d^2\ell}{d\theta^2}\right)\right] dy_1 \cdots dy_N = 0 \\ &= \int \cdots \int \left[\left(\frac{d\ell}{d\theta}\right)^2 L(\theta) + \left(\frac{d^2\ell}{d\theta^2}\right) L(\theta)\right] dy_1 \cdots dy_N = 0 \end{aligned} \tag{A.92}$$

so that

$$\frac{dE\left[U(\theta)\right]}{d\theta} = E\left[\left(\frac{d\ell}{d\theta}\right)^2\right] + E\left[\frac{d^2\ell}{d\theta^2}\right] = 0. \tag{A.93}$$

Since the two terms sum to zero and the first term must be positive, then this yields *Fisher's Information Equality*

$$\begin{aligned} I(\theta) &= E\left[\left(\frac{d\ell}{d\theta}\right)^2\right] = E\left[U(\theta)^2\right] \\ &= E\left[-\frac{d^2\ell}{d\theta^2}\right] = E\left[U'(\theta)\right] . \end{aligned} \tag{A.94}$$

The *Information function* $I(\theta)$ quantifies the expected amount of information in a sample of n observations concerning the true value of θ. The second derivative of the likelihood function, or of the log likelihood, with respect to θ describes the curvature of the likelihood in the neighborhood of θ. Thus the greater the negative derivative, the sharper is the peak of the likelihood function, and the less dispersed the likelihood over the parameter space of θ. Thus the greater is the information about the true value of θ.

In the general multiparameter case, *Fisher's Information function* for a p-vector $\boldsymbol{\theta}$ may be defined in terms of the matrix of mixed partial second derivatives, which is a $p \times p$ matrix defined as

$$\boldsymbol{I}(\boldsymbol{\theta}) = E\left[-\frac{\partial^2\ell}{\partial\boldsymbol{\theta}^2}\right] \quad \text{with elements} \quad \boldsymbol{I}(\boldsymbol{\theta})_{jk} = E\left[-\frac{\partial^2\ell}{\partial\theta_j\partial\theta_k}\right] \tag{A.95}$$

for $1 \leq j \leq k \leq p$. Alternately, $\boldsymbol{I}(\boldsymbol{\theta})$ may be defined from the outer product of the score vector as

$$\boldsymbol{I}(\boldsymbol{\theta}) = E\left[\left(\frac{\partial \ell}{\partial \boldsymbol{\theta}}\right)\left(\frac{\partial \ell}{\partial \boldsymbol{\theta}}\right)^T\right] \quad \text{with elements} \quad \boldsymbol{I}(\boldsymbol{\theta})_{jk} = E\left[\left(\frac{\partial \ell}{\partial \theta_j}\right)\left(\frac{\partial \ell}{\partial \theta_k}\right)\right]. \tag{A.96}$$

These expressions describe the *Expected Information function*.

The matrix of mixed second derivatives is commonly known as the *Hessian* matrix,

$$\boldsymbol{H}(\boldsymbol{\theta}) = \frac{\partial^2 \ell}{\partial \boldsymbol{\theta}^2} = \left\{\frac{\partial^2 \ell}{\partial \theta_j \partial \theta_k}\right\} = \frac{\partial \boldsymbol{U}(\boldsymbol{\theta})}{\partial \boldsymbol{\theta}} = \boldsymbol{U}'(\boldsymbol{\theta}) = \sum_{i=1}^{N} \boldsymbol{U}_i'(\boldsymbol{\theta}). \tag{A.97}$$

Thus from (A.96) $\boldsymbol{I}(\boldsymbol{\theta}) = E[-\boldsymbol{H}(\boldsymbol{\theta})]$. The *observed information*, therefore, is

$$\boldsymbol{i}(\boldsymbol{\theta}) = -\boldsymbol{H}(\boldsymbol{\theta}) = -\boldsymbol{U}'(\boldsymbol{\theta}). \tag{A.98}$$

For a sample of $i.i.d.$ observations

$$\boldsymbol{I}(\boldsymbol{\theta}) = E[\boldsymbol{i}(\boldsymbol{\theta})] = -E\left[\sum_{i=1}^{N} \boldsymbol{U}_i'(\boldsymbol{\theta})\right] = -NE[\boldsymbol{U}_i'(\boldsymbol{\theta})] \tag{A.99}$$

for any randomly selected observation (the ith). Thus if $E[\boldsymbol{U}_i'(\boldsymbol{\theta})]$ exists in closed form, in terms of $\boldsymbol{\theta}$, then one can derive the expression for the expected information $\boldsymbol{I}(\boldsymbol{\theta})$ for any true value $\boldsymbol{\theta}$.

The observed and expected information can also be expressed in terms of the sums of squares and cross-products obtained from the outer product of the vectors of efficient scores. From (A.96) the $p \times p$ matrix of outer products is

$$\begin{aligned}\left(\frac{\partial \ell}{\partial \boldsymbol{\theta}}\right)\left(\frac{\partial \ell}{\partial \boldsymbol{\theta}}\right)^T &= \boldsymbol{U}(\boldsymbol{\theta})\boldsymbol{U}(\boldsymbol{\theta})^T = \left[\sum_{i=1}^{N} \boldsymbol{U}_i(\boldsymbol{\theta})\right]\left[\sum_{i=1}^{N} \boldsymbol{U}_i(\boldsymbol{\theta})^T\right] \\ &= \sum_{i=1}^{N} \boldsymbol{U}_i(\boldsymbol{\theta})\boldsymbol{U}_i(\boldsymbol{\theta})^T + \sum_{i \neq j} \boldsymbol{U}_i(\boldsymbol{\theta})\boldsymbol{U}_j(\boldsymbol{\theta})^T.\end{aligned} \tag{A.100}$$

Therefore,

$$\begin{aligned}E\left[\left(\frac{\partial \ell}{\partial \boldsymbol{\theta}}\right)\left(\frac{\partial \ell}{\partial \boldsymbol{\theta}}\right)^T\right] &= E\left[\sum_{i=1}^{N} \boldsymbol{U}_i(\boldsymbol{\theta})\boldsymbol{U}_i(\boldsymbol{\theta})^T\right] \\ &\quad + E\left[\sum_{i \neq j} \boldsymbol{U}_i(\boldsymbol{\theta})\boldsymbol{U}_j(\boldsymbol{\theta})^T\right].\end{aligned} \tag{A.101}$$

Since the observations are $i.i.d.$, then

$$E\left[\boldsymbol{U}_i(\boldsymbol{\theta})\boldsymbol{U}_j(\boldsymbol{\theta})^T\right] = E[\boldsymbol{U}_i(\boldsymbol{\theta})]\, E\left[\boldsymbol{U}_j(\boldsymbol{\theta})^T\right] = \boldsymbol{0} \tag{A.102}$$

for all $1 \leq i < j \leq N$. Therefore,

$$\boldsymbol{I}(\boldsymbol{\theta}) = E[\boldsymbol{i}(\boldsymbol{\theta})] = E\left[\sum_{i=1}^{N} \boldsymbol{U}_i(\boldsymbol{\theta})\boldsymbol{U}_i(\boldsymbol{\theta})^T\right] = NE\left[\boldsymbol{U}_i(\boldsymbol{\theta})\boldsymbol{U}_i(\boldsymbol{\theta})^T\right] \tag{A.103}$$

for any random observation, arbitrarily designated as the ith.

A.6.4 Cramér-Rao Inequality: Efficient Estimators

Developments similar to the above lead to an important result which establishes the lower bound for the variance of an estimator, the Cramér-Rao lower bound. Consider a statistic T that provides an unbiased estimate of some function $\mu(\theta)$ of the scalar parameter θ such that $E(T \mid \theta) = \mu_T(\theta)$. This statistic may, however, not provide an unbiased estimate of θ itself, as when $\mu_T(\theta) \neq \theta$. Then

$$\mu_T(\theta) = \int \cdots \int T(y_1, \ldots, y_N)\; L(y_1, \ldots, y_N; \theta)\; dy_1 \cdots dy_N \,, \qquad (A.104)$$

where both T and $L(\theta)$ are functions of the observations. Differentiating with respect to θ and substituting (A.90) yields

$$\frac{d\mu_T(\theta)}{d\theta} = \mu_T'(\theta) = \int \cdots \int T\, \frac{d\ell(\theta)}{d\theta} L(\theta)\; dy_1 \cdots dy_N = E\left[TU(\theta)\right]. \quad (A.105)$$

Since $E[U(\theta)] = 0$, then

$$Cov\left[TU(\theta)\right] = E\left\{\left[T - \mu_T(\theta)\right] U(\theta)\right\} = E\left[TU(\theta)\right] = \mu_T'(\theta). \qquad (A.106)$$

We now apply the *Cauchy-Schwartz inequality*

$$\left[E(AB)\right]^2 \leq E(A^2)E(B^2), \qquad (A.107)$$

so that

$$\begin{aligned} \left[\mu_T'(\theta)\right]^2 &= E\left\{\left[T - \mu_T(\theta)\right] U(\theta)\right\} \\ &\leq E\left(\left[T - \mu_T(\theta)\right]^2\right) E\left(\left[U(\theta)\right]^2\right) = V(T)I(\theta). \end{aligned} \qquad (A.108)$$

Therefore,

$$V(T) \geq \frac{\left[\mu_T'(\theta)\right]^2}{I(\theta)} = \frac{\left[dE(T)/d\theta\right]^2}{E\left(\left[d\ell(\theta)/d\theta\right]^2\right)} \,. \qquad (A.109)$$

If $\mu_T'(\theta) = 1$, such as when T is unbiased for θ, then the variance of T is bounded by

$$V(T) \geq I(\theta)^{-1}. \qquad (A.110)$$

Similar developments apply to the multiparameter case where $\boldsymbol{T}$, $\boldsymbol{\mu}_T'(\boldsymbol{\theta})$ and $\boldsymbol{\theta}$ are each a p-vector. In this case, a lower bound can be determined for each element of the covariance matrix for $\boldsymbol{T}$, (*cf.* Rao, 1973, p. 327). For the case where $\boldsymbol{T}$ is unbiased for $\boldsymbol{\theta}$, then the lower bound is provided by the inverse of the information as in (A.110).

From these results we can define an *efficient estimator* $\boldsymbol{T}$ of the parameter $\boldsymbol{\theta}$ as a *minimum variance unbiased estimator (MVUE)* such that $E(\boldsymbol{T}) = \boldsymbol{\theta}$ and $V(\boldsymbol{T}) = \boldsymbol{I}(\boldsymbol{\theta})^{-1}$.

A.6.5 Asymptotic Distribution of the Efficient Score and the MLE

To derive the asymptotic distribution of the maximum likelihood estimate, we first obtain that of the efficient score on which the estimates are based. To simplify matters, we consider only the single parameter case where θ is a scalar.

Since the $\{y_i\}$ are $i.i.d.$, then likewise the scores $U_i(\theta) = \partial \log f(y_i;\theta)/\partial\theta$ are $i.i.d.$ with expectation zero from (A.85). Thus

$$V\left[U(\theta)\right] = \sum_{i=1}^{N} V\left[U_i(\theta)\right] = \sum_{i=1}^{N} E\left[U_i(\theta)^2\right] = I(\theta) = \sum_{i=1}^{N} E\left[-U_i'(\theta)\right] \quad \text{(A.111)}$$

from the Information equality. Further, since the total score $U(\theta)$ is the sum of $i.i.d.$ random variables, and thus can be characterized as a sequence of partial sums as $n \to \infty$, then from the central limit theorem (A.14) it follows that

$$\frac{U(\theta)}{\sqrt{n}} \xrightarrow{d} \mathcal{N}\left(0,\ E\left[-U_i'(\theta)\right]\right) \quad \text{(A.112)}$$

and asymptotically for large N

$$U(\theta) \overset{d}{\approx} \mathcal{N}\left[0,\ I(\theta)\right], \quad \text{(A.113)}$$

where $I(\theta) = NE\left[-U_i'(\theta)\right]$ for any random observation (the ith).

Now a Taylor's expansion of $U(\widehat{\theta})$ about the value $U(\theta)$ yields asymptotically

$$U(\widehat{\theta}) \cong U(\theta) + (\widehat{\theta} - \theta)U'(\theta). \quad \text{(A.114)}$$

Since $U(\widehat{\theta}) = 0$ by definition, then asymptotically

$$\sqrt{n}(\widehat{\theta} - \theta) \cong -\frac{\sqrt{n}U(\theta)}{U'(\theta)} = -\frac{U(\theta)/\sqrt{n}}{U'(\theta)/n}. \quad \text{(A.115)}$$

Since the denominator contains $-U'(\theta) = -\sum_{i=1}^{n} U_i'(\theta)$, then from the law of large numbers, it follows that

$$\frac{-U'(\theta)}{n} \xrightarrow{p} E[-U_i'(\theta)]. \quad \text{(A.116)}$$

Thus the numerator in (A.115) converges in distribution to a normal variate whereas the denominator converges in probability to a constant. Applying Slutsky's Theorem (A.44) we then obtain

$$\begin{aligned} \sqrt{n}(\widehat{\theta} - \theta) &\xrightarrow{d} \mathcal{N}\left[0,\ \frac{E\left[-U_i'(\theta)\right]}{\left(E\left[-U_i'(\theta)\right]\right)^2}\right] \quad \text{(A.117)} \\ &\xrightarrow{d} \mathcal{N}\left[0,\ \left(E\left[-U_i'(\theta)\right]\right)^{-1}\right] \\ &\xrightarrow{d} \mathcal{N}\left[0,\ nI(\theta)^{-1}\right]. \end{aligned}$$

An equivalent result also applies to the multiparameter case.

Thus in the general p-vector parameter case, the large sample distribution of the *MLE* asymptotically is

$$(\widehat{\boldsymbol{\theta}} - \boldsymbol{\theta}) \overset{d}{\approx} \mathcal{N}\left[\mathbf{0},\ \boldsymbol{I}(\boldsymbol{\theta})^{-1}\right]. \tag{A.118}$$

The large sample variance of the jth element of the parameter vector estimate, $V(\widehat{\theta}_j)$ is obtained as the jth diagonal element of the inverse information matrix,

$$V\left(\widehat{\theta}_j\right) = \Sigma_{\widehat{\theta}_j} = \left[\boldsymbol{I}(\boldsymbol{\theta})^{-1}\right]_{jj} = \boldsymbol{I}(\boldsymbol{\theta})^{jj}. \tag{A.119}$$

A.6.6 Consistency and Asymptotic Efficiency of the MLE

From (A.117), the asymptotic variance of the *MLE* is

$$\lim_{n\to\infty} V(\widehat{\theta}) = \lim_{n\to\infty} \frac{1}{nE\left[-U_i'(\theta)\right]} = \lim_{n\to\infty} I(\theta)^{-1} = 0 \tag{A.120}$$

so that the distribution of the *MLE* as $n \to \infty$ is

$$(\widehat{\theta} - \theta) \overset{d}{\to} \mathcal{N}(0,\ 0), \tag{A.121}$$

which is a degenerate normal distribution with variance zero. Thus $\widehat{\theta}$ converges in distribution to the constant θ, which, in turn, implies that $\widehat{\theta} \overset{p}{\to} \theta$ and that the *MLE* is a consistent estimator for θ. A similar result applies in the multiparameter case. The *MLE*, however, is not unbiased with finite samples. In fact, the bias can be substantial when a set of parameters are to be estimated simultaneously; see Cox and Hinkley (1974).

Since the asymptotic variance of the *MLE* is $V(\widehat{\boldsymbol{\theta}}) = \boldsymbol{I}(\boldsymbol{\theta})^{-1}$, which is the Cramer-Rao lower bound for the asymptotic variance of a consistent estimator, then this also establishes that the *MLE* is asymptotically a minimum variance estimator, or is fully efficient.

From these results, an *asymptotically efficient estimator* $\boldsymbol{T}$ of the parameter $\boldsymbol{\theta}$ is one which is consistent, $\boldsymbol{T} \overset{p}{\to} \boldsymbol{\theta}$, and for which the asymptotic variance equals $\boldsymbol{I}(\boldsymbol{\theta})^{-1}$.

A.6.7 Estimated Information

The expressions for the expected and observed information assume that the true value of $\boldsymbol{\theta}$ is known. In practice, these quantities are estimated based on the maximum likelihood estimate $\widehat{\boldsymbol{\theta}}$ of $\boldsymbol{\theta}$. This leads to expressions for the estimated expected information and the estimated observed information.

The estimated Hessian is obtained from (A.97) evaluated at the value of the *MLE* is

$$\boldsymbol{H}(\widehat{\boldsymbol{\theta}}) = \boldsymbol{U}'(\widehat{\boldsymbol{\theta}}) = \boldsymbol{U}'(\boldsymbol{\theta})_{|\boldsymbol{\theta}=\widehat{\boldsymbol{\theta}}}, \tag{A.122}$$

from which the estimated observed information is obtained as $\boldsymbol{i}(\widehat{\boldsymbol{\theta}}) = -\boldsymbol{H}(\widehat{\boldsymbol{\theta}})$.

When the elements of the expected information exist in closed form in either (A.95), (A.96), (A.99), or (A.103), then the estimated expected information, denoted as $\boldsymbol{I}(\widehat{\boldsymbol{\theta}})$, may be obtained by evaluating the resulting expressions at the values of the estimates, that is, $\boldsymbol{I}(\widehat{\boldsymbol{\theta}}) = \boldsymbol{I}(\boldsymbol{\theta})_{|\boldsymbol{\theta}=\widehat{\boldsymbol{\theta}}}$.

Also, since the *MLE* $\widehat{\boldsymbol{\theta}}$ is consistent for $\boldsymbol{\theta}$, it follows from Slutsky's Theorem that $\boldsymbol{i}(\widehat{\boldsymbol{\theta}}) \xrightarrow{p} \boldsymbol{i}(\boldsymbol{\theta})$ and that $\boldsymbol{I}(\widehat{\boldsymbol{\theta}}) \xrightarrow{p} \boldsymbol{I}(\boldsymbol{\theta})$ so that the estimated observed and expected information are also consistent estimates. This provides the basis for large sample confidence interval estimates of the parameters and tests of significance.

A.6.8 Invariance Under Transformations

Finally, an important property of the *MLE* is that it is invariant under one-to-one transformations of θ such as $g(\theta) = \log(\theta)$ for θ non-negative, or $g(\theta) = \sqrt{\theta}$ for θ non-negative, or $g(\theta) = e^{\theta}$ for $\theta \in \mathcal{R}$. In such cases, if $\widehat{\theta}$ is the *MLE* of θ, then $g(\widehat{\theta})$ is the *MLE* of $g(\theta)$. Note that this does not apply to functions such as θ^2, which are not one-to-one.

Example A.9 *Poisson-Distributed Counts*

Consider the estimation and testing of the parameter of a Poisson distribution from a sample of N counts y_i, $i = 1, \ldots, N$. Under the assumed model that the N counts are independently and identically distributed as Poisson with rate parameter θ and probability distribution

$$f(y;\theta) = \frac{e^{-\theta}\theta^{y}}{y!} \qquad y \geq 0, E(y) = \theta \tag{A.123}$$

then the likelihood of the sample of N observations is

$$L(y_1, \ldots, y_N; \theta) = \prod_{i=1}^{N} \frac{e^{-\theta}\theta^{y_i}}{y_i!} \tag{A.124}$$

with

$$\begin{aligned} \ell(\theta) = \log L(\theta) &= \sum_{i=1}^{N} \left[-\theta + y_i \log(\theta) - \log(y_i!)\right] \qquad \text{(A.125)} \\ &= -N\theta + \log(\theta) \sum\nolimits_i y_i - \sum\nolimits_i \log(y_i!) \,. \end{aligned}$$

Therefore,

$$U(\theta) = \frac{d\ell(\theta)}{d\theta} = \frac{\sum_i y_i}{\theta} - N \,, \tag{A.126}$$

which, when set to zero, yields the *MLE*

$$\widehat{\theta} = \frac{\sum_i y_i}{N} \,. \tag{A.127}$$

The observed information function is

$$i(\theta) = -H(\theta) = -\frac{d^2\ell(\theta)}{d\theta^2} = \frac{\sum_i y_i}{\theta^2}. \tag{A.128}$$

Since $E\left[\sum_i y_i\right] = N\theta$ then the expected information function is

$$I(\theta) = E[-H(\theta)] = -E\left[\frac{\sum_i y_i}{\theta^2}\right] = \frac{N}{\theta}. \tag{A.129}$$

Therefore, the large sample variance of the estimate is

$$V\left(\widehat{\theta}\right) = I(\theta)^{-1} = \theta/N \tag{A.130}$$

and asymptotically

$$\widehat{\theta} \overset{d}{\approx} \mathcal{N}[\theta,\ \theta/N]. \tag{A.131}$$

To construct confidence limits one can use either the estimated observed or the estimated expected information, which, in this case, are the same

$$i\left(\widehat{\theta}\right) = \frac{N\widehat{\theta}}{\widehat{\theta}^2} = \frac{N}{\widehat{\theta}} = I\left(\widehat{\theta}\right) = \frac{N}{\widehat{\theta}} \tag{A.132}$$

so that the estimated large sample variance is

$$\widehat{V}\left(\widehat{\theta}\right) = \frac{\widehat{\theta}}{N}. \tag{A.133}$$

Example A.10 *Hospital Mortality*

Consider the following hypothetical data from a sample of ten hospitals. For each hospital, the following are the numbers of deaths among the first 1000 patient admissions during the past year: 8, 3, 10, 15, 4, 11, 9, 17, 6, 8. For this sample of ten hospitals, $\sum_i y_i = 91$ and $\widehat{\theta} = \widehat{\mu} = 9.1$ per 1000 admissions. The estimated information is $I\left(\widehat{\theta}\right) = 10/9.1 = 1.09890$, and the estimated standard error of the estimate is $\sqrt{9.1/10} = .95394$.

A.6.9 Independent But Not Identically Distributed Observations

Similar developments apply to the estimates of the parameters in a likelihood based on a sample of N observations that are statistically independent but are not identically distributed. In particular, suppose that the form of the distribution is the same for all observations, but the moments of the distribution of y_i are a function of covariates $\boldsymbol{x}_i$ through a linear function of a parameter vector $\boldsymbol{\theta}$. Then the conditional distribution of $y|x \sim f(y; \boldsymbol{x}'\boldsymbol{\theta})$, where f is of the same form or family for all observations. Then the likelihood is

$$L(\boldsymbol{\theta}) = \prod_{i=1}^{N} f(y_i; \boldsymbol{x}_i'\boldsymbol{\theta}). \tag{A.134}$$

Although the $\{y_i\}$ are no longer identically distributed, all of the above results still apply to the maximum likelihood estimates of the parameter vector $\boldsymbol{\theta}$ and the associated score statistics. The demonstration of these properties, however, is more tedious than that presented for *i.i.d.* observations. These properties are illustrated by the following example.

Example A.11 *Homoscedastic Normal Errors Regression*
Again, consider the linear model $\boldsymbol{Y} = \boldsymbol{X\theta}+\varepsilon$ of Section A.5.1. There, the ordinary least squares estimates of the parameters were obtained based only on the specification of the first and second moments of the distribution of the errors. Now consider the model where it is also assumed that the errors are independently and identically normally distributed with mean zero and constant variance, $\varepsilon \sim N\left(0, \sigma_\varepsilon^2\right)$. This is called the homoscedastic *i.i.d.* normal errors assumption. Therefore, conditioning on the covariate vector $\boldsymbol{x}_i$, then

$$y_i|\boldsymbol{x}_i \sim \mathcal{N}\left(\boldsymbol{x}_i'\boldsymbol{\theta},\ \sigma_\varepsilon^2\right). \tag{A.135}$$

Thus conditionally, the $y_i|\boldsymbol{x}_i$ are independently but not identically distributed.

The likelihood of a sample of N observations, each with response y_i and a covariate vector $\boldsymbol{x}_{i,}$ is

$$L(y_1,\dots,y_N\,;\,\boldsymbol{\theta},\sigma_\varepsilon^2) = \prod_{i=1}^{N} \frac{1}{\sqrt{2\pi}\sigma_\varepsilon} \exp\left\{-\frac{1}{2}\left(\frac{y_i - \boldsymbol{x}_i'\boldsymbol{\theta}}{\sigma_\epsilon}\right)^2\right\}. \tag{A.136}$$

Thus the log likelihood, up to a constant term that does not depend on $\boldsymbol{\theta}$, is

$$\ell(\boldsymbol{\theta}) = \sum\nolimits_i \frac{-\left(y_i - \boldsymbol{x}_i'\boldsymbol{\theta}\right)^2}{2\sigma_\varepsilon^2} = \sum\nolimits_i \frac{-\left(y_i^2 - 2y_i\boldsymbol{x}_i'\boldsymbol{\theta} + (\boldsymbol{x}_i'\boldsymbol{\theta})^2\right)}{2\sigma_\varepsilon^2}. \tag{A.137}$$

Adopting matrix notation as in Section A.5.1, then

$$\ell(\boldsymbol{\theta}) = \frac{-\mathbf{Y}'\mathbf{Y} + 2\boldsymbol{\theta}'\mathbf{X}'\mathbf{Y} - \boldsymbol{\theta}'\mathbf{X}'\mathbf{X}\boldsymbol{\theta}}{2\sigma_\varepsilon^2}. \tag{A.138}$$

The total score vector then is

$$\boldsymbol{U}(\boldsymbol{\theta}) = \frac{\partial\ell(\boldsymbol{\theta})}{\partial\boldsymbol{\theta}} = \frac{2\mathbf{X}'\mathbf{Y} - 2(\mathbf{X}'\mathbf{X})\boldsymbol{\theta}}{2\sigma_\varepsilon^2}, \tag{A.139}$$

which yields the following estimating equation for $\boldsymbol{\theta}$ when set to zero

$$(\mathbf{X}'\mathbf{X})\boldsymbol{\theta} = \mathbf{X}'\mathbf{Y}. \tag{A.140}$$

Therefore, the *MLE* of the coefficient vector $\boldsymbol{\theta}$ equals the least squares estimate presented in (A.60). Since we condition on the covariate vectors $\{\boldsymbol{x}_i\}$, then the information function is

$$\boldsymbol{I}(\boldsymbol{\theta}) = -E\left[\frac{\partial^2\ell(\boldsymbol{\theta})}{\partial\boldsymbol{\theta}^2}\right] = -E\left[\frac{-\mathbf{X}'\mathbf{X}}{\sigma_\varepsilon^2}\right] = \frac{\mathbf{X}'\mathbf{X}}{\sigma_\varepsilon^2} \tag{A.141}$$

and the large sample variance of the estimates is

$$V\left(\widehat{\boldsymbol{\theta}}\right) = \boldsymbol{I}\left(\boldsymbol{\theta}\right)^{-1} = \left(\mathbf{X}'\mathbf{X}\right)^{-1}\sigma_{\varepsilon}^{2}, \tag{A.142}$$

which equals the variance of the least squares estimates presented in (A.63). Finally, the vector of the *MLEs* of the coefficients are asymptotically normally distributed as

$$\widehat{\boldsymbol{\theta}} \overset{d}{\approx} \mathcal{N}\left[\boldsymbol{\theta},\ \left(\mathbf{X}'\mathbf{X}\right)^{-1}\sigma_{\varepsilon}^{2}\right]. \tag{A.143}$$

A.7 LIKELIHOOD BASED TESTS OF SIGNIFICANCE

The above developments lead to three different approaches to conducting a large sample test for the values of the parameter vector of the assumed model or for elements of the parameter vector. These are the Wald test, the likelihood ratio test and the efficient score test. These tests are described in terms of a p-vector of parameters $\boldsymbol{\theta}$.

A.7.1 Wald Tests

Of the various types of tests, the Wald test requires the least computational effort, and thus is the most widely used.

A.7.1.1 Element-wise Tests Consider that we wish to test H_{0j}: $\theta_j = \theta_0$ for the jth element of the vector $\boldsymbol{\theta}$ versus the alternative hypothesis H_{1j}: $\theta_j \neq \theta_0$. A test can then be based on the fact that the *MLE* of the vector $\boldsymbol{\theta}$ is asymptotically normally distributed as in (A.118) with a large sample variance that can be estimated consistently, $\widehat{\boldsymbol{\Sigma}}_{\widehat{\theta}} = \boldsymbol{I}\left(\widehat{\boldsymbol{\theta}}\right)^{-1} \overset{p}{\to} \boldsymbol{\Sigma}_{\widehat{\theta}}$. From the important work of Wald (1943) and others, a large sample test of H_{0j} for the jth element of $\boldsymbol{\theta}$ is simply

$$X_{W_j}^2 = \frac{\left(\widehat{\theta}_j - \theta_0\right)^2}{\widehat{V}\left(\widehat{\theta}_j\right)}, \tag{A.144}$$

where $\widehat{V}\left(\widehat{\theta}_j\right) = \left[\boldsymbol{I}\left(\widehat{\boldsymbol{\theta}}\right)^{-1}\right]_{jj} = \boldsymbol{I}\left(\widehat{\boldsymbol{\theta}}\right)^{jj}$ and where the test statistic is asymptotically distributed as chi-square on 1 *df*.

A.7.1.2 Composite Test Now assume that we wish to test H_0: $\boldsymbol{\theta} = \boldsymbol{\theta}_0$ for the complete p-vector versus the alternative H_1: $\boldsymbol{\theta} \neq \boldsymbol{\theta}_0$. The Wald large sample test is a T^2-like test statistic

$$X_W^2 = \left(\widehat{\boldsymbol{\theta}} - \boldsymbol{\theta}_0\right)' \widehat{\boldsymbol{\Sigma}}_{\widehat{\theta}}^{-1} \left(\widehat{\boldsymbol{\theta}} - \boldsymbol{\theta}_0\right), \tag{A.145}$$

where $\widehat{\Sigma}_{\widehat{\theta}}^{-1} = \boldsymbol{I}\left(\widehat{\boldsymbol{\theta}}\right)$ and where X_p^2 is asymptotically distributed as chi-square on p *df*.

A.7.1.3 Test of a Linear Hypothesis Sometimes we wish to test a hypothesis about the values of a linear combination or contrast of the elements of $\boldsymbol{\theta}$. In this case we wish to test a null hypothesis of the form

$$H_{0C}\text{: } \boldsymbol{C}'\boldsymbol{\theta} = \boldsymbol{K}\ , \tag{A.146}$$

where $\boldsymbol{C}'$ is an $s \times p$ matrix of rank s $(\leq p)$ and $\boldsymbol{K}$ is the $s \times 1$ solution vector. The ith row of $\boldsymbol{C}'$ is a linear combination of the elements of $\boldsymbol{\theta}$ that yields the solution K_i specified by the ith element of $\boldsymbol{K}$. The Wald test of the hypothesis H_{0C} is then provided by the large sample T^2-like test statistic

$$X_W^2 = \left(\boldsymbol{C}'\widehat{\boldsymbol{\theta}} - \boldsymbol{K}\right)' \left[\boldsymbol{C}'\widehat{\Sigma}_{\widehat{\theta}}\boldsymbol{C}\right]^{-1} \left(\boldsymbol{C}'\widehat{\boldsymbol{\theta}} - \boldsymbol{K}\right), \tag{A.147}$$

which is asymptotically distributed as chi-square on s *df*.

A common special case is where we wish test a hypothesis regarding the values of a subset of the parameter vector $\boldsymbol{\theta}$. In the latter case, let the parameter vector be partitioned as $\boldsymbol{\theta} = (\boldsymbol{\theta}_1 \ // \ \boldsymbol{\theta}_2)$, the two sub-vectors consisting of r and s elements, respectively, $r + s = p$. Then we wish to test H_0: $\boldsymbol{\theta}_1 = \boldsymbol{\theta}_{0(r)}$ irrespective of the values of $\boldsymbol{\theta}_2$, where $\boldsymbol{\theta}_{0(r)}$ is the r-element sub-vector of elements $\boldsymbol{\theta}_0$ specified under the null hypothesis. This hypothesis can be expressed as a simple linear contrast on the elements of $\boldsymbol{\theta}$ of the form H_{0C}: $\boldsymbol{C}'\boldsymbol{\theta} = \boldsymbol{\theta}_{0(r)}$, where $\boldsymbol{C}'$ is a $r \times p$ matrix with elements

$$\boldsymbol{C}' = \left[\boldsymbol{I}_r \ \| \ \boldsymbol{0}_{r\times s}\right]_{r\times p}\ , \tag{A.148}$$

meaning a $r \times r$ identity matrix augmented by a $r \times s$ matrix of zeros. The Wald test of the hypothesis H_{0C} is then provided by

$$X_W^2 = \left(\boldsymbol{C}'\widehat{\boldsymbol{\theta}} - \boldsymbol{\theta}_{0(r)}\right)' \left[\boldsymbol{C}'\widehat{\Sigma}_{\widehat{\theta}}\boldsymbol{C}\right]^{-1} \left(\boldsymbol{C}'\widehat{\boldsymbol{\theta}} - \boldsymbol{\theta}_{0(r)}\right), \tag{A.149}$$

which is asymptotically distributed as chi-square on r *df*.

The most common application of the test of a linear hypothesis is the test of significance of the set of regression coefficients in a regression model. In a model with r-covariates, the parameter vector $\boldsymbol{\theta} = (\alpha\ \beta_1\ \ldots\ \beta_r)$ and we wish to test the significance of the model for which the null hypothesis is H_0: $\beta_1 = \beta_2 = \cdots = \beta_r = 0$. The contrast matrix of this *model chi-square test* is $\boldsymbol{C}' = \left[\boldsymbol{0}_{r\times 1} \ \| \ \boldsymbol{I}_r\right]$.

In most cases, a Wald test uses the variance of the estimates $\widehat{\Sigma}_{\widehat{\theta}}$ evaluated under the alternative hypothesis, not under the null hypothesis of interest. Thus these tests will not be as efficient as would comparable tests for which the covariance matrix of the estimates is evaluated under the null hypothesis. However, all such tests are asymptotically equivalent because the covariance matrix estimated without restriction under the alternative hypothesis is still consistent for the true covariance matrix when the null hypothesis is true. In some cases, however, it is possible to compute a Wald test using the variance estimated under the null hypothesis, $\widehat{\boldsymbol{\Sigma}}_{\boldsymbol{\theta}_0} = \boldsymbol{I}(\boldsymbol{\theta}_0)$, in which case the test statistic comparable to (A.145) is $X^2_{W_0} = \left(\widehat{\boldsymbol{\theta}} - \boldsymbol{\theta}_0\right)' \boldsymbol{I}(\boldsymbol{\theta}_0) \left(\widehat{\boldsymbol{\theta}} - \boldsymbol{\theta}_0\right)$.

$$X^2_{W_0} = \left(\widehat{\boldsymbol{\theta}} - \boldsymbol{\theta}_0\right)' \boldsymbol{I}(\boldsymbol{\theta}_0) \left(\widehat{\boldsymbol{\theta}} - \boldsymbol{\theta}_0\right). \tag{A.150}$$

Likewise the contrast test in (A.147) would employ $\boldsymbol{\Sigma}_{\boldsymbol{\theta}_0} = \boldsymbol{I}(\boldsymbol{\theta}_0)^{-1}$ and the test of an individual parameter in (A.144) would employ $V_0(\widehat{\theta}_j) = \boldsymbol{I}(\boldsymbol{\theta}_0)^{jj}$. This approach is preferable because the size of the test with large samples will more closely approximate the desired Type I error probability level.

A.7.2 Likelihood Ratio Tests

A.7.2.1 Composite Test Another type of test is the likelihood ratio test, which is the uniformly most powerful test of H_0: $\boldsymbol{\theta} = \boldsymbol{\theta}_0$ versus H_1: $\boldsymbol{\theta} = \boldsymbol{\theta}_1$ when both the null and alternative hypothesis values $(\boldsymbol{\theta}_0, \boldsymbol{\theta}_1)$ are completely specified. For a test against the omnibus alternative hypothesis H_1: $\boldsymbol{\theta} \neq \boldsymbol{\theta}_0$, it is necessary that the value of the parameter under the alternative be estimated from the data. When $\boldsymbol{\theta}$ is a p-vector, the null likelihood is

$$L(\boldsymbol{\theta}_0) = \prod_{i=1}^{N} f(y_i; \boldsymbol{\theta})_{|\boldsymbol{\theta}=\boldsymbol{\theta}_0}. \tag{A.151}$$

Under the alternative hypothesis, the likelihood function is estimated using the vector of *MLE*'s $\widehat{\boldsymbol{\theta}}$,

$$L\left(\widehat{\boldsymbol{\theta}}\right) = \prod_{i=1}^{N} f(y_i; \boldsymbol{\theta})_{|\boldsymbol{\theta}=\widehat{\boldsymbol{\theta}}}. \tag{A.152}$$

Then the likelihood ratio test is

$$X^2_{L(p)} = -2\log\left[\frac{L(\boldsymbol{\theta}_0)}{L\left(\widehat{\boldsymbol{\theta}}\right)}\right] = 2\log L\left(\widehat{\boldsymbol{\theta}}\right) - 2\log L(\boldsymbol{\theta}_0), \tag{A.153}$$

which is asymptotically distributed as χ^2_p on p *df* under the null hypothesis.

In general, the quantity $-2\log L(\boldsymbol{\theta})$ is analogous to the SSE in a simple linear regression model, so that $X^2_{L(p)}$ is analogous to the reduction in SSE associated with the addition of the parameter vector $\boldsymbol{\theta}$ to the model.

A.7.2.2 Test of a Sub-Hypothesis The most common application of a likelihood ratio test is to test nested sub-hypotheses which include the model test in a regression model and tests of individual elements. As for the Wald test, assume that the p-vector $\boldsymbol{\theta}$ is partitioned as $\boldsymbol{\theta} = (\boldsymbol{\theta}_1 \;//\; \boldsymbol{\theta}_2)$ of r and s elements, respectively, where we wish to test H_0: $\boldsymbol{\theta}_1 = \boldsymbol{\theta}_{0(r)}$ versus the alternative hypothesis H_1: $\boldsymbol{\theta}_1 \neq \boldsymbol{\theta}_{0(r)}$. This requires that we compare the likelihoods from two fitted models, that using the complete p-vector $\boldsymbol{\theta}$ versus that using only the complement of the sub-vector to be tested, in this case the s-vector $\boldsymbol{\theta}_2$. The likelihood ratio test then is

$$X^2_{L(r)} = -2\log\left[\frac{L\left(\widehat{\boldsymbol{\theta}}_2\right)}{L\left(\widehat{\boldsymbol{\theta}}\right)}\right] = 2\log L\left(\widehat{\boldsymbol{\theta}}\right) - 2\log L\left(\widehat{\boldsymbol{\theta}}_2\right), \tag{A.154}$$

which is asymptotically distributed as χ^2_r on r *df* under the null sub-hypothesis. This can also be viewed as the difference between two independent chi-square statistics relative to the null likelihood $L(\boldsymbol{\theta}_0)$ such as

$$X^2_{L(r)} = -2\log\left[\frac{L(\boldsymbol{\theta}_0)}{L\left(\widehat{\boldsymbol{\theta}}\right)}\right] - \left(-2\log\left[\frac{L(\boldsymbol{\theta}_0)}{L\left(\widehat{\boldsymbol{\theta}}_2\right)}\right]\right) = X^2_{L(p)} - X^2_{L(s)} \tag{A.155}$$

with degrees of freedom equal to $r = p - s$.

To test H_{0j}: $\theta_j = \theta_0$ versus H_{1j}: $\theta_j \neq \theta_0$ for the jth element of $\boldsymbol{\theta}$ requires that one evaluate the difference between the log likelihoods for the complete p-vector $\boldsymbol{\theta}$ and for the subset with the jth element excluded. Thus the computation of the likelihood ratio tests for the elements of a model can be more tedious than the Wald test. However, the likelihood ratio tests, in general, are preferred because they have greater efficiency or power.

A.7.3 Efficient Scores Test

Rao (1963), among others, proposed that the efficient score vector be used as the basis for a statistical test for the assumed parameters. From (A.86), under a hypothesis regarding the true values of the parameter vector $\boldsymbol{\theta}$ such as H_0: $\boldsymbol{\theta} = \boldsymbol{\theta}_0$, then it follows that $E[\boldsymbol{U}(\boldsymbol{\theta}_0)] = 0$. If the data agree with the tested hypothesis, then the score statistic evaluated at $\boldsymbol{\theta}_0$ should be close to zero. If the data do not agree with H_0, then we expect $\boldsymbol{U}(\boldsymbol{\theta}_0)$ to differ from zero.

A.7.3.1 Composite Test To test a composite hypothesis H_0: $\boldsymbol{\theta} = \boldsymbol{\theta}_0$ regarding the elements of the p-vector $\boldsymbol{\theta}$, from (A.113) under H_0 asymptotically

$$\boldsymbol{U}(\boldsymbol{\theta}_0) \overset{d}{\approx} N[0,\ \boldsymbol{I}(\boldsymbol{\theta}_0)]. \tag{A.156}$$

Thus a large sample test of H_0 versus the alternative H_1: $\boldsymbol{\theta} \neq \boldsymbol{\theta}_0$ is provided by

$$X^2_S = \boldsymbol{U}(\boldsymbol{\theta}_0)'\,\boldsymbol{I}(\boldsymbol{\theta}_0)^{-1}\,\boldsymbol{U}(\boldsymbol{\theta}_0), \tag{A.157}$$

which is asymptotically distributed as χ^2 on p *df*. Note that in order to conduct a score test regarding the complete parameter vector $\boldsymbol{\theta}$ does not require that the *MLE* $\widehat{\boldsymbol{\theta}}$ nor the estimated information $\boldsymbol{I}(\widehat{\boldsymbol{\theta}})$ be computed because the score equation and the expected information are evaluated under the null hypothesis parameter values.

A.7.3.2 Test of a Sub-Hypothesis: $C(\alpha)$ Tests Score tests may also be constructed for sub-hypotheses regarding elements of the vector $\boldsymbol{\theta}$. Such tests were originally described by Neyman (1959), who referred to such tests as $C(\alpha)$ tests, α designating the nuisance parameters. Most of the score tests of sub-hypotheses considered herein involve a test for the value of one of two parameters. Thus we first consider the case of a two-parameter vector $\boldsymbol{\theta} = [\alpha\ \beta]^T$, where we wish to test H_{β_0}: $\beta = \beta_0$. This H_{β_0} implies the joint null hypothesis $H_{\boldsymbol{\theta}_0}$: $\boldsymbol{\theta} = \boldsymbol{\theta}_0 = (\alpha, \beta_0)$ where the value of α is unrestricted. Under this hypothesis the bivariate score vector is

$$\boldsymbol{U}(\boldsymbol{\theta}_0) = \left[U(\boldsymbol{\theta})_\alpha\ U(\boldsymbol{\theta})_\beta\right]^T_{|\beta=\beta_0}. \tag{A.158}$$

However, because the hypothesis to be tested makes no restrictions on the value of the nuisance parameter α, then it is necessary to estimate α under the restriction that $\beta = \beta_0$. The *MLE* of α, designated as $\widehat{\alpha}_0$, is obtained as the solution to the estimating equation $U(\boldsymbol{\theta})_\alpha = 0$ evaluated under the null hypothesis H_{β_0}: $\beta = \beta_0$. Thus the estimated parameter vector under $H_{0\beta}$ is $\widehat{\boldsymbol{\theta}}_0 = [\widehat{\alpha}_0\ \beta_0]$.

The resulting score vector can be expressed as

$$\boldsymbol{U}\left(\widehat{\boldsymbol{\theta}}_0\right) = \left[U(\boldsymbol{\theta})_\alpha\ U(\boldsymbol{\theta})_\beta\right]^T_{|\widehat{\alpha}_0, \beta_0}. \tag{A.159}$$

By definition, the first element of the score vector is

$$U\left(\widehat{\boldsymbol{\theta}}_0\right)_\alpha = \left[U(\boldsymbol{\theta})_\alpha\right]_{|\widehat{\alpha}_0, \beta_0} = 0, \tag{A.160}$$

since $\widehat{\alpha}_0$ is the value that satisfies this equality. However, the second element,

$$U\left(\widehat{\boldsymbol{\theta}}_0\right)_\beta = \left[U(\boldsymbol{\theta})_\beta\right]_{|\widehat{\alpha}_0, \beta_0} \tag{A.161}$$

may not equal 0 and, in fact, will only equal 0 when the solution to the score equation for β is $\widehat{\beta} = \beta_0$. Note that we must actually solve for the nuisance parameter α under the restriction that $\beta = \beta_0$, in order to evaluate the score statistic for β under the null hypothesis.

Therefore, the bivariate score vector is

$$\boldsymbol{U}\left(\widehat{\boldsymbol{\theta}}_0\right) = \left[0\ \ U\left(\widehat{\boldsymbol{\theta}}_0\right)_\beta\right]^T, \tag{A.162}$$

which is a random variable augmented by a constant 0. The corresponding estimated information function is

$$\boldsymbol{I}\left(\widehat{\boldsymbol{\theta}}_0\right) = \boldsymbol{I}(\boldsymbol{\theta})_{|\theta=\widehat{\theta}_0}, \tag{A.163}$$

meaning that the elements are evaluated at the values $\widehat{\alpha}_0$ and β_0. Then the score test is the quadratic form

$$X^2_{S(1)} = U\left(\widehat{\theta}_0\right)' I\left(\widehat{\theta}_0\right)^{-1} U\left(\widehat{\theta}_0\right) \tag{A.164}$$
$$= \left[\begin{array}{cc} 0 & U\left(\widehat{\theta}_0\right)_\beta \end{array}\right] \left[\begin{array}{cc} I\left(\widehat{\theta}_0\right)^{\alpha} & I\left(\widehat{\theta}_0\right)^{\alpha\beta} \\ I\left(\widehat{\theta}_0\right)^{\beta\alpha} & I\left(\widehat{\theta}_0\right)^{\beta} \end{array}\right] \left[\begin{array}{c} 0 \\ U\left(\widehat{\theta}_0\right)_\beta \end{array}\right]$$
$$= U\left(\widehat{\theta}_0\right)'_\beta I\left(\widehat{\theta}_0\right)^{\beta} U\left(\widehat{\theta}_0\right)_\beta ,$$

which is asymptotically distributed as χ^2_1. A Wald test for such hypotheses is easy to compute but score tests have the advantage that the variance of the test is evaluated under the null hypothesis for the parameters of interest.

As a special case, this approach also includes a test of an individual element of the parameter vector, such as a test of H_0: $\boldsymbol{\theta}_j = \boldsymbol{\theta}_0$ versus H_1: $\boldsymbol{\theta}_j \neq \boldsymbol{\theta}_0$ for the jth element of $\boldsymbol{\theta}$. This score test is a bit more tedious than a Wald test because the terms which are not restricted by the hypothesis (the $\widehat{\alpha}$) must be estimated and included in the evaluation of the score statistics for the parameters which are restricted, and included in the computation of the estimated information function and its inverse. If we wish to conduct score tests for multiple elements of the parameter vector separately, then a separate model must be fit under each hypothesis.

For the two-parameter example, to also test the H_{α_0}: $\alpha = \alpha_0$ with no restrictions on β requires that we refit the model to obtain estimates of $\widehat{\boldsymbol{\theta}}_0 = \left[\alpha_0 \; \widehat{\beta}_0\right]^T$ and to then compute the score test as $X^2_S = U\left(\widehat{\theta}_0\right)'_\alpha I\left(\widehat{\theta}_0\right)^{\alpha} U\left(\widehat{\theta}_0\right)_\alpha$.

In the more general case, the p-vector $\boldsymbol{\theta}$ may be partitioned as $\boldsymbol{\theta} = (\boldsymbol{\theta}_1 \; // \; \boldsymbol{\theta}_2)$ of r and s elements, respectively, as for the Wald test, and we wish to test H_0: $\boldsymbol{\theta}_1 = \boldsymbol{\theta}_{0(r)}$. Then for $\boldsymbol{\theta}_0 = \left(\boldsymbol{\theta}_{0(r)} \; // \; \boldsymbol{\theta}_2\right)$ the *MLE* under the tested hypothesis is $\widehat{\boldsymbol{\theta}}_0 = \left(\boldsymbol{\theta}_{0(r)} \; // \; \widehat{\boldsymbol{\theta}}_2\right)$ with corresponding score vector $\boldsymbol{U}\left(\widehat{\boldsymbol{\theta}}_0\right) = \left[\boldsymbol{U}(\widehat{\boldsymbol{\theta}}_0)_{\boldsymbol{\theta}_1} \; \boldsymbol{U}(\widehat{\boldsymbol{\theta}}_0)_{\boldsymbol{\theta}_2}\right]^T$. The score test is

$$X^2_{S(r)} = \boldsymbol{U}\left(\widehat{\boldsymbol{\theta}}_0\right)' \boldsymbol{I}\left(\widehat{\boldsymbol{\theta}}_0\right)^{-1} \boldsymbol{U}\left(\widehat{\boldsymbol{\theta}}_0\right) \tag{A.165}$$
$$= \left[\boldsymbol{U}(\widehat{\boldsymbol{\theta}}_0)_{\boldsymbol{\theta}_1} \quad \boldsymbol{U}(\widehat{\boldsymbol{\theta}}_0)_{\boldsymbol{\theta}_2}\right] \boldsymbol{I}\left(\widehat{\boldsymbol{\theta}}_0\right)^{-1} \left[\boldsymbol{U}(\widehat{\boldsymbol{\theta}}_0)_{\boldsymbol{\theta}_1} \quad \boldsymbol{U}(\widehat{\boldsymbol{\theta}}_0)_{\boldsymbol{\theta}_2}\right]^T.$$

However, by definition, $\boldsymbol{U}(\widehat{\boldsymbol{\theta}}_0)_{\boldsymbol{\theta}_2} = [\boldsymbol{U}(\boldsymbol{\theta})_{\boldsymbol{\theta}_2}]_{|\widehat{\boldsymbol{\theta}}_2, \boldsymbol{\theta}_{0(r)}} = \mathbf{0}$, so that

$$X^2_{S(r)} = \left[\boldsymbol{U}(\widehat{\boldsymbol{\theta}}_0)_{\boldsymbol{\theta}_1}\right]^T \boldsymbol{I}\left(\widehat{\boldsymbol{\theta}}_0\right)^{\boldsymbol{\theta}_1} \boldsymbol{U}(\widehat{\boldsymbol{\theta}}_0)_{\boldsymbol{\theta}_1}. \tag{A.166}$$

In this expression, $\boldsymbol{I}\left(\widehat{\boldsymbol{\theta}}_0\right)^{\boldsymbol{\theta}_1}$ is the upper left $r \times r$ submatrix of $\boldsymbol{I}\left(\widehat{\boldsymbol{\theta}}_0\right)^{-1}$ which can be obtained from the expression for the inverse of a patterned matrix in (A.3).

A.7.3.3 Relative Efficiency Versus the Likelihood Ratio Test Score tests are called *efficient score tests* because they can be shown to be asymptotically fully efficient with power approaching that of the UMP likelihood ratio test. For illustration, consider a test of H_0: $\theta = \theta_0$, where θ_0 is a scalar parameter, versus a local alternative hypothesis H_1: $\theta_n = \theta_0 + \delta/\sqrt{n}$ with some fixed quantity δ such that $\theta_n \to \theta_0$ as $n \to \infty$. Then the likelihood ratio statistic is

$$X_L^2 = -2\log\left[\frac{L(\theta_0)}{L(\theta_0+\delta/\sqrt{n})}\right] = 2\log L\left(\theta_0+\frac{\delta}{\sqrt{n}}\right) - 2\log L(\theta_0), \quad \text{(A.167)}$$

which is asymptotically distributed as χ_1^2 under H_0. Now consider a Taylor's expansion of $\log L(\theta_0 + \delta/\sqrt{n})$ about the value θ_0. Then

$$2\log L\left(\theta_0+\frac{\delta}{\sqrt{n}}\right) = 2\log L(\theta_0) + \left(\frac{\delta}{\sqrt{n}}\right)\left[\frac{d[2\log L(\theta_0)]}{d\theta_0}\right] + R_2\,, \quad \text{(A.168)}$$

where R_2 is the remainder involving the term $(\delta/\sqrt{n})^2$ that vanishes in the limit. Therefore, asymptotically

$$X_L^2 \cong (\delta/\sqrt{n})\, 2U(\theta_0) \propto U(\theta_0) \quad \text{(A.169)}$$

and the efficient score test based on the score function $U(\theta_0)$ is a locally optimum test statistic.

In general, therefore, a likelihood ratio test or score test is preferred to a Wald test unless the latter is computed using the variance estimated under the tested hypothesis as in (A.150). Asymptotically, however, it has also be shown that the Wald test is approximately equal to the likelihood ratio test, since under the null hypothesis, the variance estimated under no restrictions converges to the true null variance (*cf.* Cox and Hinkley, 1974).

Example A.12 *Poisson Counts*
For the above example of hospital mortality, suppose we wish to test the hypothesis that the mortality rate in these ten hospitals was 12 deaths per 1000 admissions, or H_0: $\theta = \theta_0 = 12$ deaths/1000 admissions. The Wald test using the estimated information is $X_W^2 = (\widehat{\theta} - \theta_0)^2 I(\widehat{\theta}) = (9.1 - 12)^2 (1.0989) = 9.24$, with $p < 0.0024$.

To compute the likelihood ratio test of H_0: $\theta_0 = 12$ deaths/1000 admissions, the null log likelihood up to an additive constant is $\ell(\theta_0) = [-N\theta_0 + \log(\theta_0)\sum_i y_i] = ([\log(12)](91) - (10)(12)) = 106.127$. The *MLE* is $\widehat{\theta} = 9.1$ and the corresponding log likelihood, up to a constant, is $\ell(\widehat{\theta}) = [-N\widehat{\theta} + \log\left(\widehat{\theta}\right)\sum_i y_i] = ([\log(9.1)](91) - (10)(9.1)) = 109.953$. Thus the likelihood ratio test is $X_L^2 = 2(109.953 - 106.127) = 7.653$, with $p < 0.0057$.

The score test is computed using the score $U(\theta_0) = \sum_i y_i/\theta_0 - N = (91/12) - 10 = 2.4167$ and the information function evaluated under the hypothesis $I(\theta_0) = N/\theta_0 = 10/12 = 0.8333$. Therefore, the score test is $X_S^2 = (2.4167)^2/0.8333 = 7.008$ with $p < 0.0081$.

For this example, the Wald test statistic is greater than both the likelihood ratio test statistic and the score test statistic. Asymptotically all three tests are equivalent, however, with finite samples the three tests will differ, sometimes the Wald test being greater than the likelihood ratio test, sometimes less. However, since the Wald test employs the variance estimated under the alternative, the size or Type I error probability of the Wald test may be affected, and any apparent increase in power may be associated with an inflation in the test size. In general, therefore, either a likelihood ratio or score test is preferred to a Wald test that uses the estimated alternative variance.

However, the Wald test can also be computed using the variance evaluated under the null hypothesis that is obtained as the inverse of the information evaluated under the null hypothesis that forms the basis of the score test, rather than under the information evaluated under the alternative. In this example, the null-variance Wald test is $X^2_{W_0} = (\widehat{\theta} - \theta_0)^2 I(\theta_0) = (9.1 - 12)^2 (0.8333) = 7.008$, which, in this case, equals the score test. However, this will not be the case in general, especially in the multiparameter case.

A.8 EXPLAINED VARIATION

One of the objectives of any model is to describe factors that account for variation among observations. It is also useful, therefore, to quantify the amount or proportion of the total variation in the data that is explained by the model and its components. In ordinary multiple regression, as in Section A.5.1, this is expressed as $R^2 = SS(model)/SS_y$, where $SS(model)$ is the sum of squares for variation in Y explained by the model and SS_y is the total sum of squares of Y. The SS(model) is also obtained as $SS_y - SSE$, where SSE is the residual sum of squares of errors not explained by the model. Analogous measures can be derived for other models with different error structures other than the homoscedastic normal errors assumed in the multiple regression model. Korn and Simon (1991) present an excellent review of this area.

In general, let $\mathcal{L}[y, a(\boldsymbol{x})]$ represent the loss incurred by predicting y using a *prediction function* $a(\boldsymbol{x})$ that is a function of a covariate vector $\boldsymbol{x}$. Then the expected loss associated with $a(\boldsymbol{x})$ is

$$E\left(\mathcal{L}[y, a(\boldsymbol{x})]\right) = \int \mathcal{L}[y, a(\boldsymbol{x})] dF(\boldsymbol{x}, y), \tag{A.170}$$

where $F(\boldsymbol{x}, y)$ is the joint cdf of $\boldsymbol{X}$ and Y. Usually we select $a(\boldsymbol{x}) = \widetilde{y}(\boldsymbol{x})$, which is defined as the prediction function such that $E[\mathcal{L}]$ is minimized. Then the expected loss using the prediction function $\widetilde{y}(\boldsymbol{x})$ is denoted as

$$D_{\mathcal{L}}(\boldsymbol{x}) = E\left(\mathcal{L}[y, \widetilde{y}(\boldsymbol{x})]\right). \tag{A.171}$$

This is then contrasted with the expected loss D_0 using an unconditional prediction function $\widetilde{y}_0$ that does not depend on any covariates and which is constant for all observations, where

$$D_{\mathcal{L}(0)} = E\left(\mathcal{L}[y, \widetilde{y}_0]\right). \tag{A.172}$$

The resulting fraction of expected loss explained by the prediction function then is

$$\rho^2_{\mathcal{L}} = \frac{D_{\mathcal{L}(0)} - D_{\mathcal{L}}(\boldsymbol{x})}{D_{\mathcal{L}(0)}}. \tag{A.173}$$

A.8.1 Squared Error Loss

The most common loss function used to assess the adequacy of predictions, and to derive measures of explained variation, is squared error loss

$$\mathcal{L}\left(y, a(\boldsymbol{x})\right) = \left[y - a(\boldsymbol{x})\right]^2 = \varepsilon(\boldsymbol{x})^2 \tag{A.174}$$

for any prediction function $a(\boldsymbol{x})$. The expected squared error loss then is

$$E\left(\mathcal{L}\right) = E\left[y - a(\boldsymbol{x})\right]^2 = \int \left[y - a(\boldsymbol{x})\right]^2 dF\left(\boldsymbol{x}, y\right). \tag{A.175}$$

From ordinary least squares, $E\left(\mathcal{L}\right)$ is minimized using the prediction function

$$\widetilde{y}\left(\boldsymbol{x}\right) = E\left(y|\boldsymbol{x}\right) = \mu_{y|\boldsymbol{x}} \tag{A.176}$$

with expected squared error loss (ε^2)

$$D_{\varepsilon^2}\left(\boldsymbol{x}\right) = E\left[y - \mu_{y|\boldsymbol{x}}\right]^2 = E\left[V\left(y|\boldsymbol{x}\right)\right]. \tag{A.177}$$

Unconditionally, or not using any covariate information, then

$$\widetilde{y}_0 = E\left(y\right) = \mu_y \tag{A.178}$$

and the expected loss is

$$D_{\varepsilon^2(0)} = E\left(y - \mu_y\right)^2 = V\left(y\right) = \sigma^2_y. \tag{A.179}$$

Thus the fraction of squared error loss (ε^2) explained by the covariates $\boldsymbol{x}$ in the model is

$$\rho^2_{\varepsilon^2} = \frac{D_{\varepsilon^2(0)} - D_{\varepsilon^2}\left(\boldsymbol{x}\right)}{D_{\varepsilon^2(0)}} = \frac{V(y) - E\left[V\left(y|\boldsymbol{x}\right)\right]}{V(y)}. \tag{A.180}$$

Then, using (A.6), we can partition $V\left(y\right) = D_{\varepsilon^2(0)}$ as

$$\begin{aligned} V\left(y\right) &= E\left[V\left(y|\boldsymbol{x}\right)\right] + V\left[E\left(y|\boldsymbol{x}\right)\right] \\ &= E\left[D_{\varepsilon^2}\left(\boldsymbol{x}\right)\right] + V\left[E\left(y|\boldsymbol{x}\right)\right] \end{aligned} \tag{A.181}$$

so that

$$\rho^2_{\varepsilon^2} = \frac{V\left[E\left(y|\boldsymbol{x}\right)\right]}{V\left(y\right)} = \frac{E\left[E(y|\boldsymbol{x}) - E(y)\right]^2}{E\left[y - E(y)\right]^2} . \tag{A.182}$$

This can be estimated as

$$R^2_{\varepsilon^2} = \frac{\widehat{V}\left[E\left(y|\boldsymbol{x}\right)\right]}{\widehat{\sigma}^2_y} = \frac{\widehat{D_{\varepsilon^2(0)} - D_{\varepsilon^2}\left(\boldsymbol{x}\right)}}{\widehat{D}_{\varepsilon^2(0)}} , \tag{A.183}$$

where

$$\widehat{V}\left[E\left(y|\boldsymbol{x}\right)\right] = \frac{1}{N}\sum\nolimits_i \left[\widehat{E}\left(y|\boldsymbol{x}\right) - \overline{\widehat{E}}\left(y|\boldsymbol{x}\right)\right]^2 = \frac{1}{N}\sum\nolimits_i \left[\widehat{E}\left(y|\boldsymbol{x}\right) - \overline{y}\right]^2 , \tag{A.184}$$

$\overline{y}$ being the sample mean of Y.

Example A.13 *Multiple Linear Regression Model*
In the ordinary multiple regression model we assume that $y = \boldsymbol{x}'\boldsymbol{\theta} + \varepsilon$, where $\boldsymbol{x}'\boldsymbol{\theta} = \alpha + x_1\beta_1 + \ldots + x_p\beta_p$, and where the errors are independently and identically distributed with $E\left(\varepsilon\right) = 0$ and $V\left(\varepsilon\right) = \sigma^2_\varepsilon$. Then $E\left(y|\boldsymbol{x}\right) = \boldsymbol{x}'\boldsymbol{\theta} = \widetilde{y}\left(\boldsymbol{x}\right)$ and the expected squared error loss is

$$D_{\varepsilon^2}\left(\boldsymbol{x}\right) = E\left[V\left(y|\boldsymbol{x}\right)\right] = \sigma^2_\varepsilon . \tag{A.185}$$

Since we assume that $\boldsymbol{\theta}$ is known, the term $V\left[E\left(y|\boldsymbol{x}\right)\right]$ equals $V\left[\boldsymbol{x}'\boldsymbol{\theta}\right]$ with respect to the distribution of the covariate vector $\boldsymbol{X}$ with covariance matrix $\Sigma_{\boldsymbol{x}}$. Thus

$$V\left[E\left(y|\boldsymbol{x}\right)\right] = \boldsymbol{\theta}'\Sigma_{\boldsymbol{x}}\boldsymbol{\theta} \tag{A.186}$$

and

$$\rho^2_{\varepsilon^2} = \frac{\boldsymbol{\theta}'\Sigma_{\boldsymbol{x}}\boldsymbol{\theta}}{\boldsymbol{\theta}'\Sigma_{\boldsymbol{x}}\boldsymbol{\theta} + \sigma^2_\varepsilon} \tag{A.187}$$

(Helland, 1987). Note that the first element of $\boldsymbol{x}$ is a constant, so that the first row and column of $\Sigma_{\boldsymbol{x}}$ are 0 and the intercept makes no contribution to this expression.

Now, given the least squares estimate of the parameter vector, $\widehat{\boldsymbol{\theta}}$, then the numerator of (A.187) can be estimated as

$$\widehat{V}\left[E\left(y|\boldsymbol{x}\right)\right] = \frac{1}{N}\sum\nolimits_i \left(\boldsymbol{x}_i'\widehat{\boldsymbol{\theta}} - \overline{\boldsymbol{x}}'\widehat{\boldsymbol{\theta}}\right)^2 = \frac{1}{N}\sum\nolimits_i \left(\widehat{y}_i - \overline{y}\right)^2 , \tag{A.188}$$

since $\widehat{y}_i = \boldsymbol{x}_i'\widehat{\boldsymbol{\theta}}$ and $\overline{y} = \overline{\boldsymbol{x}}'\widehat{\boldsymbol{\theta}}$. Also, from the Gauss-Markov Theorem, an unbiased estimate of σ^2_ε is provided by

$$\widehat{\sigma}^2_\varepsilon = MSE. \tag{A.189}$$

Therefore, a consistent estimate of $\rho^2_{\varepsilon^2}$ is provided by

$$\widehat{\rho}^2_{\varepsilon^2} = R^2_{\varepsilon^2} = \frac{\frac{1}{N}\sum_i \left(\boldsymbol{x}_i'\widehat{\boldsymbol{\theta}} - \overline{y}\right)^2}{\frac{1}{N}\sum_i \left(\boldsymbol{x}_i'\widehat{\boldsymbol{\theta}} - \overline{y}\right)^2 + MSE} . \tag{A.190}$$

This is approximately equal to what is termed the adjusted R^2 in a multiple regression model computed as

$$R^2_{adj} = \frac{S^2_y - MSE}{S^2_y} , \tag{A.191}$$

where S^2_y is the sample variance of Y.

When the estimated model does not explain any of the variation in Y, then $(\widehat{\beta}_1 \dots \widehat{\beta}_p) = \mathbf{0}$ and $\boldsymbol{x}_i'\widehat{\boldsymbol{\theta}} = \alpha = \overline{y}$ for all $i = 1, \dots, N$ and thus $\widehat{\rho}^2_{\varepsilon^2} = 0$. Conversely, when the estimated model explains all the variation in Y, then $\boldsymbol{x}_i'\widehat{\boldsymbol{\theta}} = y_i$ for observations and thus $\widehat{\rho}^2_{\varepsilon^2} = 1$.

A.8.2 Residual Variation

Another estimator of the explained loss using any function $\mathcal{L}(y_i, \widetilde{y}_i)$ is the explained residual variation

$$R^2_{\mathcal{L},resid} = \frac{\sum_i \mathcal{L}\left(y_i, \widehat{\widetilde{y}}_0\right) - \sum_i \mathcal{L}\left(y_i, \widehat{\widetilde{y}}_i\right)}{\sum_i \mathcal{L}\left(y_i, \widehat{\widetilde{y}}_0\right)} , \tag{A.192}$$

where $\widehat{\widetilde{y}}_0$ is the estimated unconditional prediction function free of any covariates, and $\widehat{\widetilde{y}}_i$ the prediction function conditional on covariate vector $\boldsymbol{x}_i$, for which the loss function is minimized. For squared error loss with $\widehat{\widetilde{y}}_i = \widehat{y}_i = \widehat{E}(y|\boldsymbol{x}_i)$ and $\widehat{\widetilde{y}}_0 = \overline{y}$, then

$$R^2_{\varepsilon^2,resid} = \frac{\sum_i (y_i - \overline{y})^2 - \sum_i (y_i - \widehat{y}_i)^2}{\sum_i (y_i - \overline{y})^2} = \frac{SS_y - SSE}{SS_y} . \tag{A.193}$$

This is the traditional measure R^2 employed in multiple linear regression models.

From the partitioning of sums of squares (A.4), it follows that

$$\sum_i (y_i - \overline{y})^2 = \sum_i (y_i - \widehat{y}_i)^2 + \sum_i (\widehat{y}_i - \overline{y})^2 , \tag{A.194}$$

which yields

$$R^2_{\varepsilon^2,resid} = \frac{\sum_i (\widehat{y}_i - \overline{y})^2}{\sum_i (y_i - \overline{y})^2} \tag{A.195}$$

and

$$E\left[R^2_{\varepsilon^2,resid}\right] = \rho^2_{\varepsilon^2}. \tag{A.196}$$

In a multiple regression model $R^2_{\varepsilon^2,resid}$ is asymptotically equal to the $\widehat{\rho}^2_{\varepsilon^2}$ presented in (A.190) above.

A.8.3 Negative Log-Likelihood Loss

Similar methods can be derived for special cases where a loss function other than squared error loss may have special meaning. One such case is the use of the entropy loss function in conjunction with the logistic regression model as described in Chapter 7, Section 7.5.2. In this case, the expression for the fraction of explained residual entropy loss in a logistic regression model reduces to

$$R^2_\ell = \frac{-2\log L(\widehat{\alpha}) - \left[-2\log L(\widehat{\alpha},\widehat{\boldsymbol{\beta}})\right]}{-2\log L(\widehat{\alpha})} = \frac{Model\ X^2_{LR}}{-2\log L(\widehat{\alpha})}. \tag{A.197}$$

Although this measure was originally justified using entropy loss in logistic regression (Efron, 1978), this is a general measure of explained negative log likelihood that can be used with any likelihood-based regression model.

Consider a model with the conditional distribution $f(y;\boldsymbol{x},\boldsymbol{\theta})$ as a function of a parameter vector $\boldsymbol{\theta}$. In logistic regression, for example, Y is a binary variable and $f(y;x,\boldsymbol{\theta})$ is the Bernoulli distribution with probability π that is a function of the parameter vector $\boldsymbol{\theta} = (\alpha\,//\,\boldsymbol{\beta})$. The model negative log likelihood can then be expressed as

$$-\log L(\boldsymbol{\theta}) = \sum\nolimits_i -\log f(y_i;\boldsymbol{x},\boldsymbol{\theta}) = \sum\nolimits_i \mathcal{L}(y_i,\widetilde{y}_i) \tag{A.198}$$

or as a sum of loss with the loss function $\mathcal{L}(y,\widetilde{y}) = -\log f(y;\boldsymbol{\theta})$ where $\widetilde{y}$ is a function of $(\boldsymbol{x},\boldsymbol{\theta})$.

Thus the explained negative log likelihood can be used with any regression model where the corresponding loss function is $-\log f(y;\boldsymbol{\theta})$.

A.8.4 Madalla's R^2_{LR}

Another measure of explained variation initially proposed by Madalla (1983) and reviewed by Magee (1990) can be derived as a function of the likelihood ratio chi-square statistic. In the usual homoscedastic normal errors multiple linear regression model, as shown in the preceding example, the standard definition of R^2 is actually the $R^2_{\varepsilon^2,resid}$ presented in (A.193). In this model, the null and full model log likelihoods are readily shown to be

$$\log\left[L\left(\widehat{\alpha}\right)\right] = C - (N/2)\log\left(SS_y\right) \tag{A.199}$$

$$\log\left[L\left(\widehat{\alpha},\widehat{\boldsymbol{\beta}}\right)\right] = C - (N/2)\log\left(SSE\right),$$

where C is a constant that does not involve α or β. Then the p df model likelihood ratio test for the regression coefficients in this model is

$$X^2_{LR} = -2\log\left[\frac{L(\widehat{\alpha})}{L\left(\widehat{\alpha},\widehat{\boldsymbol{\beta}}\right)}\right] = 2\log L\left(\widehat{\alpha},\widehat{\boldsymbol{\beta}}\right) - 2\log L(\widehat{\alpha}) \tag{A.200}$$

$$= N\log\left(\frac{SS_y}{SSE}\right) = N\log\left(\frac{1}{1-R^2}\right) = -N\log\left(1-R^2\right).$$

Therefore,

$$\exp\left[-X^2_{LR}/N\right] = 1 - R^2 \tag{A.201}$$

and the standard measure of R^2 in normal errors models can also be derived as a *likelihood ratio* R^2

$$R^2_{LR} = 1 - \exp\left[-X^2_{LR}/N\right]. \tag{A.202}$$

This definition of R^2_{LR} can also be applied to other models such as logistic regression, where the likelihood ratio test is employed to test H_0: $\theta = \theta_0$ versus H_1: $\theta \neq \theta_0$ so that

$$R^2_{LR} = 1 - \exp\left(\frac{-X^2_{LR}}{N}\right) = 1 - \exp\left[\frac{2}{N}\log\frac{L(\theta_0)}{L\left(\widehat{\theta}\right)}\right] \tag{A.203}$$

$$= 1 - \left(\frac{L(\theta_0)}{L\left(\widehat{\theta}\right)}\right)^{2/N} = R^2_M.$$

This latter expression is also known as Madalla's R^2_M. This is a generally applicable measure of R^2 that can be applied to any model for a sample of N independent observations.

A.9 ROBUST INFERENCE

A.9.1 Information Sandwich

It is well established that least squares and maximum likelihood estimators are in general robust to specification of the variance structure and remain consistent even when the variance structure is misspecified (Huber, 1967). For example, in the ordinary homoscedastic errors least squares model we assume that the errors are *i.i.d.* with mean zero and common variance σ^2_ϵ. If instead the errors have variance structure with $Cov(\boldsymbol{\varepsilon}) = Cov(y|\boldsymbol{x}) = \boldsymbol{\Sigma}^2_\epsilon \neq \sigma^2_\epsilon \boldsymbol{I}_N$, then the ordinary least squares estimates of the coefficients in the model are still consistent and are still asymptotically normally distributed. The problem is that the correct expression for the variance of the coefficient estimates is the weighted least squares variance, not the OLS variance. That is, $V(\widehat{\boldsymbol{\theta}}) = (\mathbf{X}'\boldsymbol{\Sigma}_\epsilon^{-1}\mathbf{X})^{-1} \neq (\mathbf{X}'\mathbf{X})^{-1}\sigma_\varepsilon^2$. Thus if the error variance is misspecified, then confidence limits and Wald tests that rely on a model-based estimate of the variance of the coefficients are distorted. In such instances it is preferable that the variances be estimated by a procedure that is robust to misspecification of the error variance structure.

Huber (1967), Kent (1982) and White (1982), among others, considered various aspects of robust maximum likelihood inference in which they explored properties of the maximum likelihood estimators when the likelihood is misspecified and suggested robust approximations to the likelihood ratio test. Based on these developments, Royall (1986) described the application of a simple robust estimate of the variance of the *MLEs* that can be used to protect against model misspecification. This estimate has since come to be known as the *Information Sandwich* and is widely used in conjunction with such techniques as quasi-likelihood and generalized estimating equations (GEE) that are described subsequently. The information sandwich can also be used in conjunction with ordinary maximum likelihood estimation.

A.9.1.1 Correct Model Specification Consider a model in a scalar parameter θ where the model is correctly specified, meaning that the correct likelihood is specified. Then as shown in (A.113) of Section A.6.5, given the true value θ, the score $U(\theta) \overset{d}{\approx} \mathcal{N}[0,\ I(\theta)]$. From the Taylor's approximation of $U(\widehat{\theta})$ about the true value θ in (A.114), it follows from (A.115) that asymptotically

$$\sqrt{n}(\widehat{\theta} - \theta) \cong \frac{U(\theta)/\sqrt{n}}{-U'(\theta)/n} = \frac{\sum_i U_i(\theta)/\sqrt{n}}{-\sum_i U_i'(\theta)/n}\,. \tag{A.204}$$

Rather than simplify as in (A.117), consider the limiting distribution of this ratio.

The total score $U(\theta)$ in the numerator is a sum of mean zero *i.i.d.* random variables with variance $V[U(\theta)] = \sum_i V[U_i(\theta)]$, where

$$V[U_i(\theta)] = E[U_i(\theta)^2] - \{E[U_i(\theta)]\}^2 = E[U_i(\theta)^2]. \tag{A.205}$$

Then

$$V[U(\theta)] = nE[U_i(\theta)^2] = \sum\nolimits_i E[U_i(\theta)^2] = E\left[i(\theta)\right] \tag{A.206}$$

from (A.103). Thus the numerator of (A.204) is asymptotically normally distributed with mean zero and variance

$$\lim_{n\to\infty} V\left[\sum\nolimits_i \frac{U_i(\theta)}{\sqrt{n}}\right] = \lim_{n\to\infty} \frac{V[U(\theta)]}{n} = \lim_{n\to\infty} \frac{E\left[i(\theta)\right]}{n}. \tag{A.207}$$

The denominator of (A.204) is the mean of the observed information for each observation in the sample where $i(\theta) = -\sum_i U_i'(\theta)$ is a sum of *i.i.d.* random variables. Thus from the law of large numbers it follows that

$$\frac{i(\theta)}{n} \overset{p}{\to} E[-U_i'(\theta)] = \frac{I(\theta)}{n}\,. \tag{A.208}$$

Using these results with Slutsky's Convergence Theorem (A.44), it follows that the *MLE* is asymptotically distributed as

$$\sqrt{n}(\widehat{\theta} - \theta) \xrightarrow{d} \mathcal{N}[0, \ n\sigma_R^2(\theta)] \tag{A.209}$$

with large sample variance

$$\sigma_R^2(\theta) = \frac{V[U(\theta)]}{I(\theta)^2} = \frac{\sum_i E[U_i(\theta)^2]}{I(\theta)^2} = \frac{E[i(\theta)]}{I(\theta)^2}, \tag{A.210}$$

which is consistently estimated as

$$\widehat{\sigma}_R^2(\widehat{\theta}) = \frac{\sum_i [U_i(\widehat{\theta})^2]}{I(\widehat{\theta})^2}. \tag{A.211}$$

From (A.206) the numerator is the empirical estimate of the observed information whereas the denominator is the square of the model-based estimate of the expected information.

The phrase *Information Sandwich* arises from the corresponding expressions in the vector parameter case. Following similar developments for $\boldsymbol{\theta}$ a p-vector with true value $\boldsymbol{\theta}$, then

$$(\widehat{\boldsymbol{\theta}} - \boldsymbol{\theta}) \overset{d}{\approx} \mathcal{N}[0, \boldsymbol{\Sigma}_R(\widehat{\boldsymbol{\theta}})], \tag{A.212}$$

where the large sample variance–covariance matrix of the estimates is

$$\boldsymbol{\Sigma}_R(\widehat{\boldsymbol{\theta}}) = \boldsymbol{I}(\boldsymbol{\theta})^{-1} \left(\sum\nolimits_i E[\boldsymbol{U}_i(\boldsymbol{\theta}) \boldsymbol{U}_i(\boldsymbol{\theta})^T]\right) \boldsymbol{I}(\boldsymbol{\theta})^{-1}, \tag{A.213}$$

which is consistently estimated as

$$\widehat{\boldsymbol{\Sigma}}_R(\widehat{\boldsymbol{\theta}}) = \boldsymbol{I}(\widehat{\boldsymbol{\theta}})^{-1} \left(\sum\nolimits_i [\boldsymbol{U}_i(\widehat{\boldsymbol{\theta}}) \boldsymbol{U}_i(\widehat{\boldsymbol{\theta}})^T]\right) \boldsymbol{I}(\widehat{\boldsymbol{\theta}})^{-1}. \tag{A.214}$$

The estimator is a "sandwich" where the bread is the model-based variance of the estimates and the meat is the empirical estimate of the observed information.

Thus when the model is correctly specified, the score $\boldsymbol{U}(\boldsymbol{\theta}) \xrightarrow{d} \mathcal{N}[\mathbf{0}, \ \boldsymbol{I}(\boldsymbol{\theta})]$ where the variance can be consistently estimated either as $\boldsymbol{I}(\widehat{\boldsymbol{\theta}})$ or by using the information sandwich $\widehat{\boldsymbol{\Sigma}}_R(\widehat{\boldsymbol{\theta}})$. Thus the covariance matrix of the estimates can also be consistently estimated as either $\boldsymbol{i}(\widehat{\boldsymbol{\theta}})^{-1}$, $\boldsymbol{I}(\widehat{\boldsymbol{\theta}})^{-1}$ or $\widehat{\boldsymbol{\Sigma}}_R(\widehat{\boldsymbol{\theta}})$.

A.9.1.2 Incorrect Model Specification Now consider a "working model" in a parameter θ (possibly a vector) where the likelihood (or quasi-likelihood) used as the basis for developing or fitting the model is not correctly specified so that it differs from the true likelihood in some respects. For example, the true likelihood may involve additional parameters, such as an over-dispersion parameter, that is not incorporated into the working model likelihood. We assume, however, that the parameter of interest θ has the same meaning in both the working and true models. Kent (1982) then shows the following developments.

Given the true value θ we can again apply a Taylor's expansion to yield (A.204). Even though the model is misspecified, $U(\theta) = \sum_i U_i(\theta)$ is still a sum of mean zero $i.i.d.$ random variables, and thus is asymptotically normally distributed as

$$U(\theta) \overset{d}{\approx} \mathcal{N}[0,\ J(\theta)], \tag{A.215}$$

where $J(\theta) = nE[U_i(\theta)^2]$. Here the score is derived under the working model but the expectation is with respect to the correct model. When the model is correctly specified then $J(\theta) = I(\theta)$. However, when the working model is incorrect, then

$$J(\theta) = \sum\nolimits_i V[U_i(\theta)] = \sum\nolimits_i E[U_i(\theta)^2], \tag{A.216}$$

since $E[U_i(\theta)] = 0$ under both the working and true models. Since the correct model is unknown, then the actual expression for $J(\theta)$ is also unknown.

Likewise, the denominator in (A.204) involves the observed information computed using the first derivative of the scores under the working model. Nevertheless, this is the sum of $i.i.d.$ random variables, and from the law of large numbers converges to

$$\lim_{n\to\infty} \sum\nolimits_i -U_i'(\theta) \overset{p}{\to} K(\theta) = nE[-U_i'(\theta)], \tag{A.217}$$

where the expectation is again taken with respect to the correct model. The expression for $K(\theta)$ is also unknown because the correct model is unknown. Then, from (A.204) it follows that asymptotically $\widehat{\theta}$ is distributed as in (A.209) with large sample variance

$$\sigma_R^2(\theta) = \frac{J(\theta)}{K(\theta)^2}$$

For example, if the working model assumes that the distribution of the observations is $f(y;\theta)$, but the correct distribution is $g(y;\theta)$, then the likelihood function and score equations are derived from $f(\theta)$, but $J(\theta)$ and $K(\theta)$ are defined as the expectations of $U_i(\theta)^2$ and $U_i'(\theta)$ with respect to the correct distribution $g(\theta)$. Thus, for example, the term in $J(\theta)$ is of the form

$$E\left[U_i(\theta)^2\right] = \int_y \left(\frac{d\log f(y;\theta)}{d\theta}\right)^2 g(y;\theta)dy\ . \tag{A.218}$$

In this case $J(\theta)$ and $K(\theta)$ are different from the expected information $I(\theta)$ under the working model. Also, note that $K(\theta)$ is the observed information with respect to $g(\theta)$.

Equivalent results are also obtained in the multiparameter case yielding the matrices $\boldsymbol{J}(\boldsymbol{\theta})$ and $\boldsymbol{K}(\boldsymbol{\theta})$. Again, using the Central Limit Theorem and Slutsky's Convergence Theorem, it follows that

$$\sqrt{n}(\widehat{\boldsymbol{\theta}} - \boldsymbol{\theta}) \overset{d}{\to} \mathcal{N}[0,\ n\boldsymbol{K}(\boldsymbol{\theta})^{-1}\boldsymbol{J}(\boldsymbol{\theta})\boldsymbol{K}(\boldsymbol{\theta})^{-1}]. \tag{A.219}$$

Even though the expressions for $\boldsymbol{J}(\boldsymbol{\theta})$ and for $\boldsymbol{K}(\boldsymbol{\theta})$ are unknown, regardless of which model is correct, the variance can be consistently estimated using the empirical estimates

$$\widehat{\boldsymbol{J}}(\boldsymbol{\theta}) = \sum\nolimits_i [\boldsymbol{U}_i(\boldsymbol{\theta})\boldsymbol{U}_i(\boldsymbol{\theta})^T] \tag{A.220}$$
$$\widehat{\boldsymbol{K}}(\boldsymbol{\theta}) = \sum\nolimits_i -\boldsymbol{U}_i'(\boldsymbol{\theta}).$$

In most cases, however, such as an over-dispersed regression model or a quasi-likelihood regression model, it is assumed that the first moment specification is the same for the working and the correct models and that both models belong to the exponential family. In this case Kent (1982) shows that $K(\boldsymbol{\theta}) = \boldsymbol{I}(\boldsymbol{\theta})$ under the assumed model and

$$V(\widehat{\boldsymbol{\theta}}) = \boldsymbol{\Sigma}_R(\widehat{\boldsymbol{\theta}}) = \boldsymbol{I}(\boldsymbol{\theta})^{-1}\boldsymbol{J}(\boldsymbol{\theta})\boldsymbol{I}(\boldsymbol{\theta})^{-1} \tag{A.221}$$

in (A.213), which is consistently estimated using the information sandwich

$$\widehat{\boldsymbol{\Sigma}}_R(\widehat{\boldsymbol{\theta}}) = \boldsymbol{I}(\widehat{\boldsymbol{\theta}})^{-1}\widehat{\boldsymbol{J}}(\widehat{\boldsymbol{\theta}})\boldsymbol{I}(\widehat{\boldsymbol{\theta}})^{-1} \tag{A.222}$$

as in (A.214).

Thus if the first moment model specification is correct, but the second moment specification may be incorrect, the parameter estimates $\widehat{\boldsymbol{\theta}}$ are still consistent estimates even if the working model error structure is incorrect. Further, the information sandwich provides a consistent estimate of the correct variance of the parameter estimates. Thus tests of significance and confidence limits computed using the information sandwich variance estimates are asymptotically correct.

Example A.14 *Poisson-Distributed Counts*
For a sample of count data that is assumed to be distributed as Poisson, then from Example A.9 the score for each observation is

$$U_i(\theta) = \frac{x_i - \theta}{\theta} \tag{A.223}$$

and

$$I(\theta)^{-1} = V\left(\widehat{\theta}\right) = \theta/N. \tag{A.224}$$

Thus the robust information sandwich estimator is

$$\widehat{\sigma}_R^2(\widehat{\theta}) = \frac{\sum_i \left(x_i - \widehat{\theta}\right)^2}{N^2}. \tag{A.225}$$

This estimator is consistent when the data are not distributed as Poisson, but rather have a variance that differs from the Poisson mean.

For the hospital mortality data in Example A.10, this estimator yields $\widehat{\sigma}_R^2(\widehat{\theta}) = 1.77$, which is about double the model based estimate of 0.91, but not large enough to indicate a substantial degree of overdispersion in these data.

Example A.15 *Homoscedastic Normal Errors Regression*
Likewise, in ordinary multiple regression that assumes homoscedastic normally distributed errors, from Example A.11 the score for the ith observation with covariate vector $\boldsymbol{x}_i$, that includes the constant for the intercept, is

$$\boldsymbol{U}_i(\boldsymbol{\theta}) = \frac{\boldsymbol{x}_i(y_i - \boldsymbol{x}_i'\boldsymbol{\theta})}{\sigma_\varepsilon^2} \tag{A.226}$$

and

$$\sum_i [\boldsymbol{U}_i(\widehat{\boldsymbol{\theta}})\boldsymbol{U}_i(\widehat{\boldsymbol{\theta}})^T] = \frac{\mathbf{X}'\{diag[(y_i - \boldsymbol{x}_i'\widehat{\boldsymbol{\theta}})^2]\}\mathbf{X}}{(\sigma_\varepsilon^2)^2} = \frac{\mathbf{X}'\widehat{\boldsymbol{\Sigma}}_{\boldsymbol{\varepsilon}}\mathbf{X}}{(\sigma_\varepsilon^2)^2}\,, \tag{A.227}$$

where

$$\widehat{\boldsymbol{\Sigma}}_{\boldsymbol{\varepsilon}} = diag\left[(y_i - \boldsymbol{x}_i'\widehat{\boldsymbol{\theta}})^2\right]. \tag{A.228}$$

Given the model-based variance

$$V\left(\widehat{\boldsymbol{\theta}}\right) = \boldsymbol{I}\left(\boldsymbol{\theta}\right)^{-1} = \left(\mathbf{X}'\mathbf{X}\right)^{-1}\sigma_\varepsilon^2 \tag{A.229}$$

then the robust information sandwich variance estimate is

$$\widehat{\boldsymbol{\Sigma}}_R(\widehat{\theta}) = \left(\mathbf{X}'\mathbf{X}\right)^{-1}\mathbf{X}'\widehat{\boldsymbol{\Sigma}}_{\boldsymbol{\varepsilon}}\mathbf{X}\left(\mathbf{X}'\mathbf{X}\right)^{-1}\,, \tag{A.230}$$

where the meat of the sandwich is the empirical estimate of the diagonal matrix of the variances of the errors.

This estimate then is consistent when the errors are not homoscedastic. This estimator was derived by White (1980) and is provided by SAS PROC REG using the ACOV specification. White (1980) also described a test of homoscedasticity that is also computed by SAS PROC REG using the SPEC option.

A.9.2 Robust Confidence Limits and Tests

The robust sandwich estimate of the variance of the *MLEs* can be used to provide large sample confidence limits that are robust to departures from the variance structure implied by the model that was used to obtain the *MLEs* of the model parameters. This variance can also be used as the basis for Wald tests and robust efficient score tests. However, because the likelihood ratio test depends explicitly on the complete model specification, including the variance structure, a robust likelihood ratio test cannot be directly computed.

From (A.215), it follows that a robust score test of H_0: $\boldsymbol{\theta} = \boldsymbol{\theta}_0$ for the parameter vector $\boldsymbol{\theta}$, or a $C(\alpha)$-test subset of elements of $\boldsymbol{\theta}$, may be obtained as described in Section A.7.3 using the empirical estimate of the covariance matrix $\widehat{\boldsymbol{J}}(\widehat{\boldsymbol{\theta}}_0)$ evaluated with respect to the parameter vector estimate $\widehat{\boldsymbol{\theta}}_0$ obtained under the tested hypothesis. The properties of the robust score test have not been studied for logistic regression and Poisson regression. However, the robust score test and robust confidence limits in the Cox proportional hazards model were derived by Lin and Wei (1989) and Lin (1994); see Section 9.4.5. These computations are available as part of the SAS PROC PHREG; see Section 9.4.9.

A.10 GENERALIZED LINEAR MODELS AND QUASI-LIKELIHOOD

Generalized Linear Models (*GLMs*) refers to a family of regression models described by Nelder and Wedderburn (1972), which are based on distributions from the exponential family that may be fit using maximum likelihood estimation. This family of *GLMs* includes the normal errors, logistic and Poisson regression models as special cases. For any distribution from the exponential family, the score equation and the observed and expected information are a function of the first and second moments only. This observation led to development of the method of quasi-likelihood by Wedderburn (1974) for fitting models that are not based on an explicit likelihood. Any model based on specification of the first two moments, including models that do not arise from the exponential family, can be fit by the method of quasi-likelihood. The family of *GLMs* and quasi-likelihood estimation are elaborated in the text by McCullagh and Nelder (1989). An excellent general reference is Dobson (1990). Such models can be fit by the SAS PROC GENMOD and other programs.

A.10.1 Generalized Linear Models

In the simple normal errors linear regression model, the structural component is specified to be $y = \boldsymbol{x}'\boldsymbol{\theta}+\varepsilon$ or $y = \mu+\varepsilon$ where $\mu = \boldsymbol{x}'\boldsymbol{\theta}$. The random component of the model that describes the error distribution is then specified to be $\varepsilon \sim \mathcal{N}\left(0,\ \sigma_\epsilon^2\right)$. This, in turn, specifies that the conditional distribution is $y|\boldsymbol{x} \sim \mathcal{N}\left(\mu,\ \sigma_\epsilon^2\right)$. A measure of the goodness of fit of the model is provided by the $Deviance$, which is defined as the difference between the log likelihood of the present model and that of a model with as many parameters as necessary to fit the data perfectly. Thus if there are N independent observations, the deviance is the difference between the log likelihood of a model with N *df* (thus fitting perfectly) and that of the present model with p *df*, the deviance having $N - p$ *df*. For a normal errors linear model the deviance equals the SSE/σ_ϵ^2 for that model.

The *GLM* family of models generalizes this basic model in two ways. First, the conditional distribution of $y|\boldsymbol{x}$ can be any member of the exponential family. This includes the normal, binomial, Poisson, gamma, and so on. Second, the link between the conditional expectation μ and the linear predictor $(\boldsymbol{x}'\boldsymbol{\theta})$ can be any differentiable monotone function $g\left(\mu\right) = \boldsymbol{x}'\boldsymbol{\theta}$, called the link function, that maps the domain of μ onto the real line. Thus $\mu(\boldsymbol{x}) = g^{-1}(\boldsymbol{x}'\boldsymbol{\theta})$ which we designate as $\mu_{\boldsymbol{x}}$. To allow for the variance to be an explicit function of the mean, we denote the error variance as $\sigma_\varepsilon^2(\mu_{\boldsymbol{x}})$.

Therefore, a *GLM* has three components:

1. The systematic component: e.g. $\eta = \boldsymbol{x}'\boldsymbol{\theta} =$ the linear predictor;

2. The link function $g\left(\mu\right) = \eta$ that specifies the form of the relationship between $E(y|\boldsymbol{x}) = \mu_{\boldsymbol{x}}$ and the covariates $\boldsymbol{x}$; and

3. The random component or the conditional distribution specification $y|\boldsymbol{x} \sim f(y;\varphi)$ where $f(\cdot)$ is a member of the exponential family.

The random component specification implies a specific relationship between the conditional mean $\mu_{\boldsymbol{x}}$ and the conditional variance $V(y|\boldsymbol{x})$ that equals the variance of the errors, σ_{ε}^2. The common choices are:

1. *Normal*: With a normal error distribution then it is assumed that σ_{ε}^2 is constant for all $\boldsymbol{x}$ and thus is statistically independent of $\mu_{\boldsymbol{x}}$ for all $\boldsymbol{x}$.

2. *Binomial*: With a binomial distribution it is assumed that error variance for the ith observation with "number of trials" n_i is of the form $\sigma_{\varepsilon}^2(\mu_{\boldsymbol{x}}) = n_i\mu_{\boldsymbol{x}}(1-\mu_{\boldsymbol{x}})$ which is a function of $\mu_{\boldsymbol{x}}$. When $n_i = 1$ the distribution is Bernoulli.

3. *Poisson*: With a Poisson distribution it is assumed that $\sigma_{\varepsilon}^2(\mu_{\boldsymbol{x}}) = \mu_{\boldsymbol{x}}$.

In some cases it is also necessary to incorporate a scale or dispersion factor into the model, designated as ϕ. In addition, the *GLM* allows for observations to be weighted differentially as a function of a weight w_i for the ith observation, as where the ith observation is the mean of n_i measurements in which case $w_i = n_i$.

A.10.2 Exponential Family of Models

The above specifications all fall within the framework of a regression model for the conditional expectation of a member of the exponential family of distributions that includes the normal, binomial and Poisson as special cases. Thus the estimating equations and estimated information for this family of models can be derived from the exponential family of distributions.

The probability function for the canonical form of the exponential family for the ith observation is

$$f(y_i\,;\varphi_i,\phi,w_i) = \exp\left[\frac{y_i\varphi_i - b(\varphi_i)}{a(\phi,w_i)} + c(y_i,\phi,w_i)\right], \tag{A.231}$$

where the parameter of interest for the ith observation is φ_i, ϕ is a scale or dispersion parameter and w_i is a weighting constant. The functions $a(\cdot)$, $b(\cdot)$, and $c(\cdot)$ corresponding to the different members of the exponential family are readily derived and are presented in many basic texts on mathematical statistics. As we show below, the specific expressions are not needed for a family of regression models, the above *GLM* specifications being sufficient.

For a distribution of this form, the score for the ith subject is

$$U_i(\varphi_i) = \frac{d\ell_i}{d\varphi_i} = \frac{y_i - b'(\varphi_i)}{a(\phi,w_i)}, \tag{A.232}$$

where $b'(\varphi) = db(\varphi)/d\varphi$. Likewise, the second derivative of the score for the ith observation is

$$U_i'(\varphi_i) = \frac{-b''(\varphi_i)}{a(\phi, w_i)}, \tag{A.233}$$

where b'' refers to the second derivative. Since $E[U_i(\varphi_i)] = 0$ then (A.232) implies that

$$E(y_i) = \mu_i = b'(\varphi_i). \tag{A.234}$$

To derive the expression for $V(y_i)$ note that Fisher's Information Equality (A.93) states that

$$E[U_i'(\varphi_i)] + E\left[U_i(\varphi_i)^2\right] = 0. \tag{A.235}$$

Since $E[U_i(\varphi_i)] = 0$ for all observations, then $V(U_i(\varphi_i)) = E\left[U_i(\varphi_i)^2\right]$. Also, from (A.232),

$$V[U_i(\varphi_i)] = \frac{V(y_i)}{a(\phi, w_i)^2}. \tag{A.236}$$

Therefore,

$$\frac{-b''(\varphi_i)}{a(\phi, w_i)} + \frac{V(y_i)}{a(\phi, w_i)^2} = 0 \tag{A.237}$$

and thus

$$V(y_i) = b''(\varphi_i)\, a(\phi, w_i). \tag{A.238}$$

Now consider that a *GLM* has been specified such that $E(y|\boldsymbol{x}) = \mu_{\boldsymbol{x}}$ with link function

$$g(\mu_{\boldsymbol{x}}) = \eta_{\boldsymbol{x}} = \boldsymbol{x}'\boldsymbol{\theta} \tag{A.239}$$

in terms of a covariate vector $\boldsymbol{x}$ and coefficient vector $\boldsymbol{\theta}$. From the specified distribution of $y|\boldsymbol{x}$, the conditional variance of $y|\boldsymbol{x}$ equals the variance of the errors under that distribution and may be a function of the conditional expectation, or $V(y|\boldsymbol{x}) = \sigma_\varepsilon^2(\mu_{\boldsymbol{x}})$. For the ith individual with covariate vector $\boldsymbol{x}_i$ we designate the conditional expectation as μ_i and variance as $V(y_i) = \sigma_\varepsilon^2(\mu_i)$. For a distribution from the exponential family, the score vector for the ith observation then is

$$\boldsymbol{U}_i(\boldsymbol{\theta}) = \frac{d\ell_i}{d\varphi_i}\frac{d\varphi_i}{d\mu_i}\frac{\partial\mu_i}{\partial\boldsymbol{\theta}}, \tag{A.240}$$

where $\mu_i \equiv \mu(\boldsymbol{x}_i)$. From (A.232) and (A.234), the first term is

$$\frac{d\ell_i}{d\varphi_i} = \frac{y_i - E(y_i)}{a(\phi, w_i)} = \frac{y_i - \mu_i}{a(\phi, w_i)}. \tag{A.241}$$

The second term is

$$\frac{d\varphi_i}{d\mu_i} = \left(\frac{d\mu_i}{d\varphi_i}\right)^{-1} = \left(\frac{dE(y_i)}{d\varphi_i}\right)^{-1} = \left(\frac{db'(\varphi_i)}{d\varphi_i}\right)^{-1} \tag{A.242}$$
$$= \frac{1}{b''(\varphi_i)} = \frac{a(\phi, w_i)}{\sigma_\varepsilon^2(\mu_i)} .$$

The final term is the vector

$$\frac{\partial \mu_i}{\partial \boldsymbol{\theta}} = \frac{d\mu_i}{d\eta_i}\frac{\partial \eta_i}{\partial \boldsymbol{\theta}} = \boldsymbol{x}_i \left(\frac{d\eta_i}{d\mu_i}\right)^{-1} = \frac{\boldsymbol{x}_i}{g'(\mu_i)} . \tag{A.243}$$

Thus the total score vector is

$$\boldsymbol{U}(\boldsymbol{\theta}) = \frac{\partial \ell}{\partial \boldsymbol{\theta}} = \sum\nolimits_i \left[\frac{(y_i - \mu_i)}{\sigma_\varepsilon^2(\mu_i)}\right] \frac{\boldsymbol{x}_i}{g'(\mu_i)} , \tag{A.244}$$

which provides the *MLE* estimating equation for $\boldsymbol{\theta}$.

The observed information is then obtained as

$$\boldsymbol{i}(\boldsymbol{\theta}) = -\boldsymbol{U}'(\boldsymbol{\theta}) = -\sum\nolimits_i \frac{\partial}{\partial \beta}\left[\frac{d\ell_i}{d\varphi_i}\frac{d\varphi_i}{d\mu_i}\frac{\partial \mu_i}{\partial \boldsymbol{\theta}}\right] . \tag{A.245}$$

After some algebra we obtain

$$\boldsymbol{i}(\boldsymbol{\theta}) = \sum\nolimits_i \left[\frac{\boldsymbol{x}_i \boldsymbol{x}_i'}{\sigma_\varepsilon^2(\mu_i)\left[g'(\mu_i)\right]^2}\right] \tag{A.246}$$
$$+ \sum\nolimits_i \left[(y_i - \mu_i)\, \boldsymbol{x}_i \boldsymbol{x}_i' \left(\frac{\sigma_\varepsilon^2(\mu_i) g''(\mu_i) + \left[\frac{d}{d\mu_i}\sigma_\varepsilon^2(\mu_i)\right] g'(\mu_i)}{\left[\sigma_\varepsilon^2(\mu_i)\right]^2 \left[g'(\mu_i)\right]^3}\right)\right] .$$

Since $E(y_i - \mu_i) = 0$, then the expected information is

$$\boldsymbol{I}(\boldsymbol{\theta}) = E\left[\boldsymbol{i}(\boldsymbol{\theta})\right] = \sum\nolimits_i \left[\frac{\boldsymbol{x}_i \boldsymbol{x}_i'}{\sigma_\varepsilon^2(\mu_i) g'(\mu_i)^2}\right] . \tag{A.247}$$

This expression is also readily obtained as $E[\boldsymbol{U}(\boldsymbol{\theta})\boldsymbol{U}(\boldsymbol{\theta})^T]$. Newton Raphson iteration or Fisher scoring can then be used to solve for $\widehat{\boldsymbol{\theta}}$, to obtain the estimated observed and the estimated expected information. In some programs, such as SAS PROC GENMOD, the estimated covariance matrix of the coefficient estimates is computed using the inverse estimated observed information.

The above models include the ordinary normal errors multiple regression model, the logistic regression model and the homogeneous Poisson regression model as special cases. In a two-parameter exponential distribution model, such as the homoscedastic normal errors model, φ is the mean and the scale parameter ϕ is the variance. In the binomial and Poisson models, the scale parameter is fixed at $\phi = 1$, although in these and other such models it is also possible to specify a scale or dispersion parameter $\phi \neq 1$ to allow for under- or over-dispersion. In such cases, however, we no longer have a "likelihood" in the usual sense. Nevertheless, such a model can be justified as a quasi-likelihood (see the next section).

In the Binomial model, the weight may vary as a function of the number of "trials" (n_i) for the ith observation where $w_i = n_i$. Otherwise, the weight is usually a constant ($w_i = 1$). In these and other cases, $a(\phi, w_i) = \phi / w_i$.

In the exponential family, the expression for $\mu = b'(\varphi_i)$ also implies the *natural or canonical link* for that distribution such that $\varphi_i = g(\mu) = \eta_i = \boldsymbol{x}_i'\boldsymbol{\theta}$, thus leading to major simplifications in the above expressions. For the most common applications, the canonical link is as follows:

Distribution	$\mu = b'(\varphi)$	*Canonical link*, $g(\mu)$
Normal	φ	*identity*
Binomial	$\dfrac{e^{\varphi}}{1 - e^{\varphi}}$	*logit*
Poisson	$\exp(\varphi)$	*log*

Although one could use any differentiable link function with any error distribution, problems may arise in the fitting of the model by Newton-Raphson iteration. For example, one could use the log link with a binomial error distribution in lieu of the usual logistic regression model with a logit link. In this case, the elements of $\boldsymbol{\theta}$ describe the covariate effects on the log risk or have an interpretation as log relative risks. However, this model does not ensure that the estimated probabilities $\pi(\boldsymbol{x}) = \mu(\boldsymbol{x})$ are bounded by (0,1), and the iterative solution of the coefficients and the estimated information may fail unless a method for constrained optimization is used to fit the model.

A.10.3 Deviance and the Chi-Square Goodness of Fit

The model likelihood ratio test is constructed by comparing the $-2\log[L(\alpha)]$ for the null intercept-only model to the $-2\log[L(\alpha, \boldsymbol{\beta})]$ for the $(p+1)$-variate model with parameter vector $\boldsymbol{\theta}$, where the difference is the likelihood ratio test statistic on p *df* under the model null hypothesis H_0: $\boldsymbol{\beta} = \boldsymbol{0}$. In *GLMs* the *deviance* is used to compare the $-2\log[L(\alpha, \boldsymbol{\beta})]$ for the fit of the $(p+1)$-variate model with that of a saturated model that fits the data perfectly on N *df*. The perfect model, therefore, is one where $y_i = \mu_i$ for all $i = 1, \ldots, N$. The deviance is the difference between these values on $(N-p-1)$ *df*. Thus the deviance provides a test of the hypothesis that the additional hypothetical $N - p - 1$ parameters are jointly zero, or that the model with parameters $(\alpha, \boldsymbol{\beta})$ is correct. In most cases, but not all, the deviance is also asymptotically distributed as chi-square under the hypothesis that the model fits the data.

For example, a normal errors linear model has log likelihood

$$\ell = \sum_{i=1}^{N} \left[\frac{-(y_i - \mu_i)^2}{2\sigma_\epsilon^2} - \log(\sqrt{2\pi}\sigma_\varepsilon) \right] \tag{A.248}$$

so that the perfect model has log likelihood $\ell(\mu_1, \ldots, \mu_N) = -N \log(\sqrt{2\pi}\sigma_\varepsilon)$. Then, the deviance for a model where the conditional expectation $\mu(\boldsymbol{x}_i)$ is a function of a covariate vector $\boldsymbol{x}_i$ reduces to

$$\begin{aligned} D(\alpha, \boldsymbol{\beta}) &= -2\ell(\alpha, \boldsymbol{\beta}) - [-2\ell(\mu_1, \ldots, \mu_N)] \\ &= \sum_{i=1}^{N} \frac{[y_i - \mu(\boldsymbol{x}_i)]^2}{\sigma_\epsilon^2} = \frac{SS(errors)}{\sigma_\epsilon^2}. \end{aligned} \tag{A.249}$$

The deviance, therefore, is equivalent to the $-2 \log L(\alpha, \boldsymbol{\beta})$ less any constants. The greater the deviance the greater the unexplained variation or lack of fit of the model.

In a logistic regression model where y_i is a (0,1) binary variable, the perfect model log likelihood is

$$\ell(\mu_1, \ldots, \mu_N) = \sum_{i=1}^{N} y_i \log \mu_i + \sum_{i=1}^{N} (1 - y_i) \log (1 - \mu_i) = 0 \tag{A.250}$$

so that $D(\alpha, \boldsymbol{\beta}) = -2\ell(\alpha, \boldsymbol{\beta})$. However, in a binomial regression model where y_i is also a count, the N *df* $\log L$ is not equal to zero. The same also applies to the Poisson regression model where y_i is a count:

$$\begin{aligned} \ell(\mu_1, \ldots, \mu_N) &= \sum_{i=1}^{N} [-\mu_i + y_i \log[\mu_i] - \log(y_i!)] \\ &= \sum_{i=1}^{N} [-y_i + y_i \log[y_i] - \log(y_i!)] \neq 0. \end{aligned} \tag{A.251}$$

Likelihood ratio tests can also be computed as the difference between the deviances for nested models, the $-2\ell(\mu_1, \ldots, \mu_N)$ cancelling from both model deviances. Thus even though the deviance itself may not be distributed as chi-square, the difference between deviances equals the likelihood ratio test that is asymptotically distributed as chi-square.

When the model is correctly specified and the deviance is asymptotically distributed as chi-square, then $E(\textit{Deviance}) = df = n - p - 1$. This provides a simple assessment of the adequacy of the second moment or mean:variance model specification. Thus when the model variance assumptions apply, $E(\textit{Deviance/df}) = 1$. This also provides for a simple model adjustment to allow for over-dispersion using a quasi-likelihood. However, the adequacy of this approach depends on the adequacy of the chi-square approximation to the large sample distribution of the deviance. For logistic regression models with a binary dependent variable, for example, this may not apply; see McCullagh and Nelder (1989, pp. 118–119). Thus in some such cases where the *Deviance* is not approximately distributed as chi-square, the *Deviance/df* should not be used as an indication of extra-variation.

Rather, it is preferred that the *Pearson chi-square of goodness of fit* for the model be used to assess goodness of fit and the presence of extra-variation. For a *GLM* this statistic is the model test

$$X_P^2 = \sum\nolimits_{i=1}^{N} \frac{\left[y_i - \widehat{\mu}_i\right]^2}{\sigma_\varepsilon^2(\widehat{\mu}_i)}, \tag{A.252}$$

which is asymptotically distributed as chi-square on $N - p - 1$ *df* under the assumption that the model is correctly specified. Thus for any model, $E(X_P^2) = df$ and X_P^2/df provides an estimate of the degree of over- or under-dispersion.

In general, it should be expected that the ratio *Deviance/df* or X_P^2/df will vary by chance about 1 when the model is correctly specified. The variance of the chi-square statistic is $V(X^2) = 2df$, so that $V[X^2/df] = 2/df$. Thus the range of variation expected when the model is correct with 95% confidence is on the order of $\left[1 \pm 1.96\sqrt{2/df}\right] = \left[1 \pm 2.77/\sqrt{df}\right]$, the approximate 95% tolerance limits. Thus with $N = 100$ one should expect the ratio to fall within 1 ± 0.277. One should then only consider adopting an over-dispersed model when the ratio *Deviance/df* or X_P^2/df departs substantially from 1.

A.10.4 Quasi-Likelihood

In an important development, Wedderburn (1974) showed that the score equations in (A.244) could be used as what are termed *quasi-likelihood estimating equations,* even when the precise form of the error distribution is unknown. All that is required for the asymptotic properties to apply is that the mean–variance relationship, or the first two moments of the conditional error distribution, be correctly specified. In this case it can be shown that the parameter estimates are asymptotically normally distributed about the true values with a covariance matrix equal to the inverse expected information, exactly as for maximum likelihood estimates (McCullagh, 1983, among others).

For example, assume that a set of quantitative observations is related to a vector of covariates $\boldsymbol{X}$ with an identity link and the conditional error variance is constant for all values of $\boldsymbol{X}$, but the error distribution and the conditional distribution of $y|\boldsymbol{x}$ is not the normal distribution. Even though the error distribution is not the normal distribution, the quasi-likelihood estimates obtained from a normal errors assumption are asymptotically normally distributed as in (A.118). This is not surprising since the assumptions (excluding normality) are the same as those required for fitting the linear model by ordinary least squares as described in Section A.5.1, which, in turn, are sufficient to provide for the asymptotically normal distribution of the estimates using the Central Limit Theorem.

As a special case of a quasi-likelihood model, consider a *GLM* where the first moment for the ith observation with covariate vector $\boldsymbol{x}_i$ is specified to be of the form $E(y_i|\boldsymbol{x}_i) = \mu_i$, where $g(\mu_i) = \eta_i = \boldsymbol{x}_i'\boldsymbol{\theta}$; and where the second moment is specified to be of the form $V(y_i|\boldsymbol{x}_i) = \phi\sigma^2(\mu_i) = \phi\sigma_i^2$ that can be some function of the conditional expectation with scale or dispersion parameter ϕ. We also allow the ith observation to have weight w_i. Then the quasi-likelihood estimate can also be derived as a *minimum chi-square estimate* of the parameters.

The Pearson chi-square statistic for the goodness of fit of the model, assuming known parameters, is

$$X^2 = \sum_{i=1}^{N} \frac{w_i(y_i - \mu_i)^2}{\phi\sigma_i^2} . \tag{A.253}$$

If we ignore the possible dependence of the conditional variance on the conditional expectations, that is, we treat the $\{\sigma_i^2\}$ as fixed constants, then the minimum chi-square estimates are obtained as the solution to the estimating equation

$$\frac{\partial X^2}{\partial \boldsymbol{\theta}} = \sum_{i=1}^{N} \frac{2w_i(y_i - \mu_i)}{\phi\sigma_i^2} \left(\frac{\partial \mu_i}{\partial \boldsymbol{\theta}}\right) = \mathbf{0}. \tag{A.254}$$

This estimating equation is of the form $\widetilde{\boldsymbol{U}}(\boldsymbol{\theta}) = \mathbf{0}$ in terms of a *quasi-score function* $\widetilde{\boldsymbol{U}}(\boldsymbol{\theta})$, which can be expressed in matrix terms as

$$\widetilde{\boldsymbol{U}}(\boldsymbol{\theta}) = \mathbf{D}'\mathbf{V}^{-1}(\mathbf{Y} - \boldsymbol{\mu})/\phi , \tag{A.255}$$

where $\boldsymbol{D} = (\partial\mu_1/\partial\boldsymbol{\theta} \ \ldots \ \partial\mu_N/\partial\boldsymbol{\theta})^T$, $\mathbf{V} = diag[\sigma_1^2 \ \ldots \ \sigma_N^2]$, $\boldsymbol{Y} = (y_1 \ \ldots \ y_N)$, and $\boldsymbol{\mu} = (\mu_1 \ \ldots \ \mu_N)$.

The family of models considered here is much broader than that in a *GLM* based on an error distribution from the exponential family. Further, in the event that the quasi-likelihood equals an exponential family likelihood, then the quasi-likelihood score equation $\widetilde{\boldsymbol{U}}(\boldsymbol{\theta})$ equals the total score from an exponential family likelihood presented earlier in (A.244) with the simplification that $a(\phi, w_i) = \phi/w_i$.

It also follows that an estimate that minimizes the Pearson chi-square is also a weighted least squares estimate because the chi-square objective function X^2 is equivalent to the weighted sum of squares of errors that is minimized using weighted least squares. Thus quasi-likelihood estimates are usually obtained using iteratively reweighted least squares (IRLS). Algebraically it can also be shown that the systems of equations solved using IRLS are equivalent to those solved using Fisher Scoring iteration. For example, see Hillis and Davis (1994).

Assuming that the model specifications are correct it is then readily shown that

$$\begin{aligned} E[\widetilde{\boldsymbol{U}}(\boldsymbol{\theta})] &= \mathbf{0} \\ E[-\widetilde{\boldsymbol{U}}'(\boldsymbol{\theta})] &= \mathbf{D}'\mathbf{V}^{-1}\mathbf{D}/\phi \\ Cov[\widetilde{\boldsymbol{U}}(\boldsymbol{\theta})] &= \widetilde{\boldsymbol{I}}(\boldsymbol{\theta}) = \mathbf{D}'\mathbf{V}^{-1}\mathbf{D}/\phi . \end{aligned} \tag{A.256}$$

Since the quasi-score function is a sum of $i.i.d.$ random variables, then using the same developments as in Section 6.5, it follows that the quasi score converges in distribution to the normal

$$\sqrt{n}\widetilde{\boldsymbol{U}}(\boldsymbol{\theta}) \xrightarrow{d} \mathcal{N}[\mathbf{0},\ n\mathbf{D}'\mathbf{V}^{-1}\mathbf{D}/\phi] \tag{A.257}$$

and the maximum quasi-likelihood estimates are asymptotically normally distributed as

$$(\widehat{\boldsymbol{\theta}} - \boldsymbol{\theta}) \overset{d}{\approx} \mathcal{N}[\mathbf{0},\ \phi\left(\boldsymbol{D}'\boldsymbol{V}^{-1}\boldsymbol{D}\right)^{-1}]. \tag{A.258}$$

From this asymptotic normal distribution of the estimates, it follows that Wald tests and confidence limits of the parameter estimates are readily obtained. In addition, quasi-score statistics are also readily constructed. Even though we do not start from a full likelihood, a quasi-likelihood ratio test can be constructed using the change in -2 log quasi-likelihood as for a full likelihood.

One advantage of quasi-likelihood estimation is that extra-variation or over-dispersion can readily be incorporated into the model by assuming that a common variance inflation factor ϕ applies to all observations such that $V(y_i|\boldsymbol{x}_i) = \phi\sigma^2(\mu_i) = \phi\sigma_i^2$. As described in Section A.10.3 above, if we first fit a homogeneous (not over-dispersed) model with the scale factor fixed at $\phi = 1$, then $E(X_P^2/df) = \phi$, where X_P^2 is the Pearson chi-square from the homogeneous model. Thus a moment estimator of the variance inflation or over-dispersion factor is $\widehat{\phi} = X_P^2/df$. If this estimate is approximately equal to 1 in a logistic or Poisson regression model, then this indicates that the original model specifications apply. Otherwise, if $\widehat{\phi}$ departs substantially from 1, such as outside the 95% tolerance limits $\left[1 \pm 2.77/\sqrt{df}\right]$, this is an indication that an under- or over-dispersed model may be appropriate. In this case, a model can be refit based on the quasi-likelihood where ϕ is not fixed. An indication of extra-variation, however, may also be due to fundamental model misspecification or the omission of important covariates.

Although it is possible to solve a set of quasi-likelihood estimating equations jointly for $(\phi, \boldsymbol{\theta})$, this is not the approach generally employed. Rather, computer programs such as the SAS PROC GENMOD use the moment estimating equation for ϕ to compute an iterative estimate of the over-dispersion factor.

Finally, we note that the robust variance estimate may also be employed in conjunction with a quasi-likelihood model to obtain estimates of the covariance matrix of the parameter estimates that are robust to misspecification of the first and second moments of the quasi-likelihood. In this case,

$$\widehat{V}(\widehat{\boldsymbol{\theta}}) = \widetilde{\boldsymbol{I}}(\widehat{\boldsymbol{\theta}})^{-1}\widehat{\widetilde{\boldsymbol{J}}}(\widehat{\boldsymbol{\theta}})\widetilde{\boldsymbol{I}}(\widehat{\boldsymbol{\theta}})^{-1} = \widehat{\boldsymbol{\Sigma}}_R(\widehat{\boldsymbol{\theta}}), \tag{A.259}$$

where

$$\widehat{\widetilde{\boldsymbol{J}}}(\widehat{\boldsymbol{\theta}}) = \sum\nolimits_i [\widetilde{\boldsymbol{U}}(\widehat{\boldsymbol{\theta}})\,\widetilde{\boldsymbol{U}}(\widehat{\boldsymbol{\theta}})^T]. \tag{A.260}$$

This provides for robust confidence intervals and Wald tests, and also for a robust quasi-score test.

A.10.5 Conditional GLMs

Chapter 7 describes the conditional logistic regression model for matched sets with binary responses, and Chapter 8 likewise describes the conditional Poisson regression model for count responses. Both of these models are members of the family of *conditional* generalized linear models based on members of the exponential family. These models are also discussed in the general text by McCullagh and Nelder (1989). Another special case is that of a quantitative response that is normally distributed within matched sets, where each matched set has a unique intercept or set effect. Since the sufficient statistic for the matched set intercept is the sum of the observations within the set, the conditional normal likelihood is readily derived, from which estimating equations and inferences for the covariate parameters are readily obtained. Unfortunately, software is not yet available to fit this family of conditional *GLMs* for matched sets.

A.10.6 Generalized Estimating Equations (GEE)

One of the most important developments in recent years is the development of generalized estimating equations (*GEEs*) for the analysis of correlated observations. Liang and Zeger (1986) and Zeger and Liang (1986) proposed fitting *GLM*-like models for correlated observations using a generalization of quasi-likelihood. The use of *GEEs* for the analysis of longitudinal data is reviewed in the text by Diggle, Liang and Zeger (1994), among others. This approach generalizes many of the methods presented in this text to the analysis of longitudinal data with repeated measurements, such as a longitudinal logistic regression model.

In the simplest case, assume that there is a set of repeated measures for each subject. Assume that the structural relationship is correctly specified that relates the expectation to the linear function of the covariates through a link function where the covariates may vary over time. Then, consistent estimates of the coefficients may be obtained using the ordinary quasi-likelihood equations where we act as though the observations are all independent. Since the repeated measures are correlated, then the quasi-likelihood-based estimates of the variance of the coefficients will be invalid. However, a consistent estimate of the covariance matrix of the coefficients can be obtained using the information sandwich. This then provides confidence limits and Wald tests for the coefficients and robust score tests as well.

Further gains in efficiency can be achieved, however, by assuming some "working correlation" structure among the observations. In this case, the information sandwich can also be used to provide a consistent estimate of the covariance matrix of the coefficient estimates and asymptotically valid inference.

References

Aalen, O. (1978). Nonparametric inference for a family of counting processes. *Ann. Statist.,* 6, 701-726.

Agresti, A. (1990). *Categorical Data Analysis.* New York: John Wiley and Sons.

Akaike, H. (1973). Information theory and an extension of the maximum likelihood principle. In, *Second International Symposium on Information Theory*. Petrov, B. N. and Csaki, F. (eds.), 267-281. Budapest: Akademiai Kiado.

Altschuler, B. (1970). Theory for the measurement of competing risks in animal experiments. *Math. Biosci.*, 6, 1-11.

Andersen, P. K. and Rasmussen, N. K. (1986). Psychiatric admissions and choice of abortion. *Stat. Med.*, 5, 243-253.

Andersen, P. K. and Gill, R. D. (1982). Cox's regression model for counting processes: a large sample study. *Ann. Statist.*, 10, 1100-1120.

Andersen, P. K., Borgan, O., Gill, R. D. and Keiding, N. (1982). Linear nonparametric tests for comparison of counting processes, with applications to censored survival data. *Int. Statist. Rev.*, 50, 219-258.

Andersen, P. K., Borgan O., Gill, R. D. and Keiding, N. (1993). *Statistical Models Based on Counting Processes*. New York: Springer-Verlag.

Anderson , J. A. (1972). Separate sample logistic discrimination. *Biometrika* 59, 19-35.

Anderson, T. W. (1984). *An Introduction to Multivariate Analysis*, 2nd edition. New York: John Wiley & Sons.

Anderson, J. R. and Bernstein, L. (1985). Asymptotically efficient two-step estimators of the hazards ratio for follow-up studies and survival data. *Biometrics*, 41, 733-739.

Anscombe, F. J. (1956). On estimating binomial response relations. *Biometrika*, 43, 461-464.

Bailar, J. C., Louis, T. A., Lavori, P. W. and Polansky, M. (1984). A classification for biomedical research reports. *N. Engl. J. Med.*, 311, 1482-1487.

Bancroft, T. A. (1972). Some recent advances in inference procedures using preliminary tests of significance. In, *Statistical Papers in Honor of George W. Snedecor*, Bancroft, T. A. (ed.), 19-30. Ames, Iowa: The Iowa State University Press.

Barnard, G. A. (1945). A new test for 2 x 2 tables. *Nature*, 156, 177.

Barnard, G. A. (1949). Statistical inference. *J. Roy. Statist. Soc., B*, 11, 115-139.

Beach, M. L. and Meier, P. (1989). Choosing covariates in the analysis of clinical trials. *Control. Clin. Trials*, 10, 161S-175S.

Bean, S. J. and Tsokos, C. P. (1980). Developments in non-parametric density estimation. *Int. Statist. Rev.*, 48, 267-287.

Bennett, S. (1983). Log-logistic regression models for survival data. *Appl. Statist.*, 32, 165-171.

Bickel, P. and Doksum, K. (1977). *Mathematical Statistics*. Englewood Cliffs, NJ: Prentice-Hall.

Birch, M. W. (1964). The detection of partial association. I: The 2 X 2 case. *J. Roy. Statist. Soc., B*, 26, 313- 324.

Bloch, D. A. and Moses, L. E. (1988). Nonoptimally weighted least squares. *Am. Statistician*, 42, 50-53.

Blum, A. L. (1982). Principles for selection and exclusion. In, *The Randomized Clinical Trial and Therapeutic Decisions*, Tygstrup, N., Lachin, J. M., and Juhl, E. (eds.), 43-58. New York: Marcel-Dekker.

Breiman, L. (1992). The little bootstrap and other methods for dimensionality selection in regression: X-fixed prediction error. *J. Amer. Statist. Assoc.*, 87, 738-754.

Breiman, L., Friedman, J. H., Olshen, R. A. and Stone, C. J. (1993). *Classification and Regression Trees*. New York: Chapman & Hall.

Breslow, N. E. (1974). Covariance analysis of censored survival data. *Biometrics*, 30, 89-99.

Breslow, N. E. (1975). Analysis of survival data under the proportional hazards model. *Int. Statist. Rev.*, 43, 45-58.

Breslow, N. E. (1981). Odds ratio estimators when the data are sparse. *Biometrika*, 68, 73-84.

Breslow, N. E. (1982). Covariance adjustment of relative-risk estimates in matched studies. *Biometrics*, 38, 661-672.

Breslow, N. E. (1984). Extra-Poisson variation in log-linear models. *Appl. Statist.*, 33, 38-44.

Breslow, N. E. (1996). Statistics in epidemiology: The case-control study. *J. Amer. Statist. Assoc.*, 91, 14-28.

Breslow, N. E. and Day, N. E. (1980). *Statistical Methods in Cancer Research, Volume 1. The Analysis of Case-Control Studies.* Oxford, U.K.: Oxford University Press.

Breslow, N. E. and Day, N. E. (1987). *Statistical Methods in Cancer Research, Volume 2. The Design and Analysis of Cohort Studies.* Oxford, U.K.: Oxford University Press.

Breslow, N. E., Day, N. E., Halvorsen, K. T., Prentice, R. L. and Sabai, C. (1978). Estimation of multiple relative risk functions in matched case-control studies. *Am. J. Epidemiol.*, 108, 299-307.

Bronowski, J. (1973). *The Ascent of Man.* Boston, Mass.: Little Brown and Co.

Byar, D. P. (1985). Prognostic variables for survival in a randomized comparison of treatments for prostatic cancer. In, *Data: A Collection of Problems From Many Fields for the Student and Research Worker.* Herzberg, A. M. and Andrews, D. F. (eds.), 261-274. New York: Springer-Verlag.

Cain, K. C. and Lange, N. T. (1984). Approximate case influence for the proportional hazards regression model with censored data. *Biometrics*, 40, 493-499.

Canner, P. L. (1991). Covariate adjustment of treatment effects in clinical trials. *Control. Clin. Trials*, 12, 359-366.

Chavers, B. M., Bilous, R. W., Ellis, E. N., Steffes, M. W. and Mauer, S. M. (1989). Glomerular lesions and urinary albumin excretion in type I diabetes without overt proteinuria. *N. Engl. J. Med.*, 15, 966-970.

Clayton, D. and Cuzick, J. (1985). Multivariate generalizations of the proportional hazards model. *J. Roy. Statist. Soc., B*, 148, 82-117.

Clopper, C. J. and Pearson, E. S. (1934). The use of confidence or fiducial limits illustrated in the case of the binomial. *Biometrika*, 26, 404-413.

Cochran, W. G. (1954a). Some methods for strengthening the χ^2 tests. *Biometrics*, 10, 417-451.

Cochran, W. G. (1954b). The combination of estimates from different experiments. *Biometrics*, 10, 101-129.

Cochran, W. G. (1983). *Planning and Analysis of Observational Studies.* New York: John Wiley & Sons.

Cohen, J. (1988). *Statistical Power Analysis for the Behavioral Sciences*, 2nd edn. Hillsdale, NJ: Laurence Erlbaum Associates, Publishers.

Collett, D. (1991). *Modelling Binary Data.* London: Chapman and Hall.

Collett, D. (1994). *Modelling Survival Data in Medical Research.* London: Chapman and Hall.

Collins, R., Yusuf, S. and Peto, R. (1985). Overview of randomised trials of diuretics in pregnancy. *Br. Med. J.*, 290, 17-23.

Connett, J. E., Smith, J. A. and McHugh, R. B. (1987). Sample size and power for pair-matched case-control studies. *Stat. Med.*, 6, 53-59.

Connor, R. J. (1987). Sample size for testing differences in proportions for the paired-sample design. *Biometrics*, 43, 207-211.

Conover, W. J. (1974). Some reasons for not using the Yates continuity correction on 2 x 2 contingency tables. (With comments). *J. Amer. Statist. Assoc.*, 69, 374-382.

Cook, R. J. and Sackett, D. L. (1995). The number needed to treat: A clinically useful measure of a treatment effect. *Br. Med. J.*, 310, 452-454.

Cornfield, J. (1951). A method of estimating comparative rates from clinical data. Applications to cancer of the lung, breast, and cervix. *J. Natl. Cancer Inst.*, 11, 1269-1275.

Cornfield, J. (1954). The estimation of the probability of developing a disease in the presence of competing risks. *Am. J. Public Health*, 47, 601-607.

Cornfield, J. (1956). A statistical problem arising from retrospective studies. In, *Proceedings of the Third Berkeley Symposium on Mathematical Statistics and Probability, Vol. 4*, Neyman, J. (ed.), 135-148. Berkeley: University of California Press.

Cornfield, J. (1962). Joint dependence of risk of coronary heart disease on serum cholesterol and systolic blood pressure: A discriminant function analysis. *Federation Proc.*, 21, 58-61.

Coronary Drug Project Research Group (CDP) (1974). Factors influencing long-term prognosis after recovery from myocardial infarction - Three-year findings of the Coronary Drug Project. *J. Chronic Dis.*, 27, 267-285.

Cox, D. R. (1958a). The regression analysis of binary sequences. *J. Roy. Statist. Soc., B*, 20, 215-242.

Cox, D. R. (1958b). Two further applications of a model for binary regression. *Biometrika*, 45, 562-565.

Cox, D. R.. (1970). *The Analysis of Binary Data.* (2nd edn., 1989. Cox, D. R. and Snell, E. J.). London: Chapman and Hall.

Cox, D. R. (1972). Regression models and life-tables (with discussion). *J. Roy. Statist. Soc., B*, 34, 187-220.

Cox, D. R. (1975). Partial likelihood. *Biometrika*, 62, 269-276.

Cox, D. R. and Hinkley, D. V. (1974). *Theoretical Statistics.* London: Chapman and Hall.

Cox, D. R. and Oakes, D. (1984). *Analysis of Survival Data.* London: Chapman and Hall.

Cox, D. R. and Miller, H. D. (1965). *The Theory of Stochastic Processes.* London: Chapman and Hall.

Cramer, H. (1946). *Mathematical Methods of Statistics.* Princeton, NJ: Princeton University Press.

Cutler, S. J. and Ederer, F. (1958). Maximum utilization of the life table method in analyzing survival. *J. Chronic Dis.*, 8, 699-712.

Day, N. E. and Byar, D. P. (1979). Testing hypotheses in case-control studies - Equivalence of Mantel-Haenszel statistics and logit score tests. *Biometrics*, 35, 623-630.

Dean, C. and Lawless, J. F. (1989). Test for detecting over-dispersion in Poisson regression models. *J. Amer. Statist. Assoc.*, 84, 467-472.

Deckert, T., Poulsen, J. E. and Larsen, M. (1978). Prognosis of diabetics with diabetes onset before the age of thirty-one. I: Survival, causes of death, and complications. *Diabetologia*, 14, 363-370.

DerSimonian, R. and Laird, N. (1986). Meta-analysis in clinical trials. *Control. Clin. Trials*, 7, 177-188.

Desu, M. M. and Raghavarao, D. (1990). *Sample Size Methodology.* New York: Academic Press.

Diabetes Control and Complications Trial Research Group (DCCT) (1990). The Diabetes Control and Complications Trial (DCCT): Update. *Diabetes Care*, 13, 427-433.

Diabetes Control and Complications Trial Research Group (DCCT) (1993). The effect of intensive treatment of diabetes on the development and progression of long-term complications in insulin-dependent diabetes mellitus. *N. Engl. J. Med.*, 329, 977-986.

Diabetes Control and Complications Trial Research Group (DCCT) (1995a). Effect of intensive therapy on the development and progression of diabetic nephropathy in the Diabetes Control and Complications Trial. *Kidney Int.*, 47, 1703-1720.

Diabetes Control and Complications Trial Research Group (DCCT) (1995b). Adverse events and their association with treatment regimens in the Diabetes Control and Complications Trial. *Diabetes Care*, 18, 1415-1427.

Diabetes Control and Complications Trial Research Group (DCCT) (1995c). The relationship of glycemic exposure (HbA1c) to the risk of development and progression of retinopathy in the Diabetes Control and Complications Trial. *Diabetes*, 44, 968-983.

Diabetes Control and Complications Trial Research Group (DCCT) (1996). The absence of a glycemic threshold for the development of long-term complications: The perspective of the Diabetes Control and Complications Trial. *Diabetes*, 45, 1289-1298.

Diabetes Control and Complications Trial Research Group (DCCT) (1997). Hypoglycemia in the Diabetes Control and Complications Trial. *Diabetes*, 46, 271-286.

Diabetes Control and Complications Trial Research Group (DCCT) (2000). The effect of pregnancy on microvascular complications in the diabetes control and complications trial. *Diabetes Care*. (To appear.)

Dick, T. D. S. and Stone, M. C. (1973). Prevalence of three cardinal risk factors in a random sample of men and in patients with ischaemic heart disease. *Br. Heart J.*, 35, 381-385.

Diggle, P. J., Liang, K. Y. and Zeger, S. L. (1994). *Analysis of Longitudinal Data*. New York: Oxford University Press.

Dobson, A. (1990). *An Introduction to Generalized Linear Models*. London: Chapman and Hall.

Donner, A. (1984). Approaches to sample size estimation in the design of clinical trials – A review. *Stat. Med.*, 3, 199-214.

Dorn, H. F. (1944). Illness from cancer in the United States. *Public Health Rep.*, 59, Nos. 2, 3, and 4.

Dyke, G. V. and Patterson, H. D. (1952). Analysis of factorial arrangements when the data are proportions. *Biometrics*, 8, 1-12.

Early Breast Cancer Trialists' Collaborative Group (EBCTCG) (1998). Tamoxifen for early breast cancer: an overview of the randomised trials. *Lancet*, 351, 1451-1467.

Efron, B. (1977). The efficiency of Cox's likelihood function for censored data. *J. Amer. Statist. Assoc.*, 72, 557-565.

Efron, B. (1978). Regression and ANOVA with zero-one data: measures of residual variation. *J. Amer. Statist. Assoc.*, 73, 113-121.

Efron, B. and Hinkley, D. V. (1978). Assessing the accuracy of the maximum likelihood estimates: Observed versus expected Fisher information. *Biometrika*, 65, 457-487.

Efroymson, M. A. (1960). Multiple regression analysis. In, *Mathematical Methods for Digital Computers*, Ralston, A. and Wilf, H. S.(eds.), 191-203. New York: Wiley.

Ejigou, A. and McHugh, R. B. (1977). Estimation of relative risk from matched pairs in epidemiological research. *Biometrics*, 33, 552-556.

Elandt-Johnson, R. C. and Johnson, N. L. (1980). *Survival Models and Data Analysis*. New York: John Wiley & Sons.

Epanechnikov, V. A. (1969). Nonparametric estimation of a multivariate probability density. *Theory of Probability and Its Applications*, 14, 153 -158.

Feigl, P. and Zelen, M. (1965). Estimation of exponential survival probabilities with concomitant information. *Biometrics*, 21, 826-838.

Finklestein, D. M. (1986). A proportional hazards model for interval-censored failure time data. *Biometrics*, 42, 845-854.

Fisher, R. A. (1922). On the mathematical foundations of theoretical statistics. *Philos. Trans. R. Soc. London*, A 222, 309-368.

Fisher, R. A. (1925). *Statistical Methods for Research Workers. (*14th edn., 1970*).* Edinburgh: Oliver and Boyd.

Fisher, R. A. (1935). *The Design of Experiments.* (8th edn., 1966). Edinburgh: Oliver and Boyd.

Fisher, R. A. (1956). *Statistical Methods for Scientific Inference.* Edinburgh: Oliver and Boyd.

Fleiss, J. L. (1979). Confidence intervals for the odds ratio in case-control studies: The state of the art. *J. Chronic Dis.*, 32, 69-77.

Fleiss, J. L. (1981). *Statistical Methods for Rates and Proportions.* New York: John Wiley and Sons..

Fleiss, J. L. (1986). *The Design and Analysis of Clinical Experiments.* New York: John Wiley & Sons.

Fleiss, J. L., Bigger, J. T., McDermott, M., Miller, J. P., Moon, T., Moss, A. J., Oakes, D., Rolnitzky, L. M. and Therneau, T. M. (1990). Nonfatal myocardial infarction is, by itself, an inappropriate end point in clinical trials in cardiology. *Circulation*, 81, 684-685.

Fleming, T. R. and Harrington, D. P. (1991). *Counting Processes and Survival Analysis*. New York: John Wiley & Sons, Inc.

Freedman, D. A. (1983). A note on screening regression equations. *Am. Statistician*, 37, 152-155.

Freedman, L. S. (1982). Tables of the number of patients required in clinical trials using the logrank test. *Stat. Med.*, 1, 121-129

Freedman, L. S. and Pee, D. (1989). Return to a note on screening regression equations. *Am. Statistician*, 43, 279-282.

Freireich , E. J., Gehan, E., Frei III, E. Schroeder, L. R., Wolman, L. J., Anbari, R., Burgert, E. O., Mills, S. D., Pinkel, D., Selawry, O. S., Moon, J. H., Gendel, B. R., Spurr, C. L., Storrs, R., Haurani, F., Hoogstraten, B. and Lee, S. (1963). The effect of 6-mercaptopurine on the duration of steroid-induced remissions in acute leukemia: a model for evaluation of other potentially useful therapy. *Blood*, 21, 699-716.

Frick, H. (1995). Comparing trials with multiple outcomes: The multivariate one-sided hypothesis with unknown covariances. *Biom. J.*, 8, 909-917.

Frome, E. L. (1983). The analysis of rates using Poisson regression models. *Biometrics*, 39, 665-674.

Frome, E. L. and Checkoway, H. (1985). Epidemiologic programs for computers and calculators: Use of Poisson regression models in estimating incidence rates and ratios. *Am. J. Epidemiol.*, 121, 309-323.

Gail, M. H. (1973). The determination of sample sizes for trials involving several independent 2 x 2 tables. *J. Chronic Dis.*, 26, 669-673.

Gail, M. H. (1978). The anlaysis of heterogeneity for indirect standardized mortality ratios. *J. Roy. Statist. Soc., A,* 141, 224-234.

Gail, M. H., Santner, T. J. and Brown, C. C. (1980). An analysis of comparative carcinogenesis experiments based on multiple times to tumor. *Biometrics*, 36, 255-266.

Gail, M. H., Wieand., S. and Piantadosi, S. (1984). Biased estimates of treatment effect in randomized experiments with nonlinear regressions and omitted covariates. *Biometrics*, 71, 431-444.

Gart, J. J. (1971). The comparison of proportions: A review of significance tests, confidence intervals and adjustments for stratification. *Rev. Int. Statist. Inst.*, 39, 16-37.

Gart, J. J. (1985). Approximate tests and interval estimation of the common relative risk in the combination of 2x2 tables. *Biometrika*, 72, 673-677.

Gart, J. J. and Tarone, R. E. (1983). The relation between score tests and approximate UMPU tests in exponential models common in biometry. *Biometrics*, 39, 781-786.

Gart, J. J. and Zweifel, J. R. (1967). On the bias of various estimators of the logit and its variance with application to quantal bioassay. *Biometrika*, 54, 181-187.

Gastwirth, J. L. (1966). On robust procedures. *J. Amer. Statist. Assoc.*, 61, 929-948.

Gastwirth, J. L. (1985). The use of maximum efficiency robust tests in combining contingency tables and survival analysis. *J. Amer. Statist. Assoc.*, 80, 380-84.

Gastwirth, J. L. and Greenhouse, S. W. (1995). Biostatistical concepts and methods in the legal setting. *Stat. Med.*, 14, 1641-1653.

Gaynor, J. J., Fener, E. J., Tan, C. C., Wu, D. H., Little, C. R., Straus, D. J., Clarkson, B. D. and Brennan, M. F. (1993). On the use of cause-specific failure and conditional failure probabilities: examples from clinical oncology data. *J. Amer. Statist. Assoc.*, 88, 400-409.

Gehan, E. A. (1965). A generalized Wilcoxon test for comparing arbitrarily singly censored samples. *Biometrika*, 52, 203-223.

George, S. L. and Desu, M. M. (1974). Planning the size and duration of a clinical trial studying the time to some critical event. *J. Chronic Dis.*, 27, 15-29.

Gill, R. D. (1980). *Censoring and Stochastic Integrals*, Mathematical Centre Tracts, 124. Amsterdam: Mathematisch Centrum.

Gill, R. D. (1984). Understanding cox's regression model: A martingale approach. *J. Amer. Statist. Assoc.*, 79, 441-447.

Goodman, L. A. and Kruskal, W. H. (1972). Measures of association for cross-classifications, IV: simplification of asymptotic variances. *J. Amer. Statist. Assoc.*, 67, 415-421.

Gray, R. J. (1988). A class of K-sample tests for comparing the cumulative incidence of a competing risk. *Ann. Statist.*, 16, 1141-1154.

Greenland, S. (1984). A counterexample to the test-based principle of setting confidence limits. *Am. J. Epidemiol.*, 120, 4-7.

Greenwood, M. A. (1926). Report on the natural duration of cancer, *Reports on Public Health and Medical Subjects*, 33, 1-26. London: H. M. Stationery Office.

Grizzle, J. E. (1967). Continuity correction in the χ^2 test for 2 x 2 tables. *Am. Statistician*, 21, 28-32.

Grizzle, J. E., Starmer, C. F. and Koch, G. G. (1969). Analysis of categorical data by linear models. *Biometrics*, 25, 489-503.

Guenther, W. C. (1977). Power and sample size for approximate chi-square tests. *Am. Statistician*, 31, 83-85.

Guilbaud, O. (1983). On the large-sample distribution of the Mantel-Haenszel odds-ratio estimator. *Biometrics*, 39, 523-525.

Hájek, J. and Sidák, Z. (1967). *Theory of Rank Tests.* New York: Academic Press.

Haldane, J. B. S. (1956). The estimation and significance of the logarithm of a ratio of frequencies. *Ann. Human Genet.*, 20, 309- 311.

Halperin, M. (1977). Re: Estimability and estimation in case-referent studies. (Letter). *Am. J. Epidemiol.*, 105, 496-498.

Halperin, M., Ware, J. H., Byar, D. P., Mantel, N., Brown, C. C., Koziol, J., Gail, M. and Green, S. B. (1977). Testing for interaction in an IxJxK contingency table. *Biometrika*, 64, 271-275.

Hardison, C. D., Quade, D. and Langston, R. D. (1986). Nine functions for probability distributions. In, *SUGI Supplemental Library User's Guide*, Version 5 Edition, Hastings, R. P. (ed.), 385-393. Cary, NC: SAS Institute, Inc.

Harrell, F. E. (1986). The PHGLM procedure. In, *SAS Supplemental Library User's Guide*, Version 5. Cary, NC: SAS Institute, Inc.

Harrington, D. P. and Fleming, T. R. (1982). A class of rank test procedures for censored survival data. *Biometrika*, 69, 553-566.

Harris, M. I., Hadden, W. C., Knowler, W. C. and Bennett, P. H. (1987). Prevalence of diabetes and impaired glucose tolerance and plasma glucose levels in US population aged 20-74 yr. *Diabetes*, 36, 523-534.

Harville, D. A. (1977). Maximum likelihood approaches to variance component estimation and to related problems. *J. Amer. Statist. Assoc.*, 72, 320-338.

Hastie, T. J. and Tibshirani, R. J. (1990). *Generalized Additive Models*. New York: Chapman & Hall.

Hauck, W. W. (1979). The large sample variance of the Mantel-Haenszel estimator of a common odds ratio. *Biometrics*, 35, 817-819

Hauck, W. W. (1989). Odds ratio inference from stratified samples. *Comm. Statist., A*, 18, 767-800.

Hauck, W. W. and Donner, A. (1977). Wald's test as applied to hypotheses in logit analysis. *J. Amer. Statist. Assoc.*, 72, 851-853.

Helland, I. S. (1987). On the interpretation and use of R^2 in regression analysis. *Biometrics*, 43, 61-69.

Higgins, J. E. and Koch, G. G. (1977). Variable selection and generalized chi-square analysis of categorical data applied to a large cross-sectional occupational health survey. *Int. Statist. Rev.*, 45, 51-62.

Hillis, S. L. and Davis, C. S. (1994). A simple justification of the iterative fitting procedure for generalized linear models. *Am. Statistician*, 48, 288-289.

Hochberg, Y. (1988). A sharper Bonferroni procedure for multiple tests of significance. *Biometrika*, 75, 800-802.

Holford, T. R. (1980). The analysis of rates and survivorship using log-linear models, *Biometrics*, 36, 299-306.

Holm, S. (1979). A simple sequentially rejective multiple test procedure. *Scand. J. Statist.,* 6, 65-70

Hosmer, D. W. and Lemeshow, S. (1989). *Applied Logistic Regression*. New York: John Wiley.

Huber, P. J. (1967). The behavior of maximum likelihood estimators under nonstandard conditions. *Proceedings of the Fifth Berkeley Symposium on Mathematical Statistics and Probability, Vol. 1*, Neyman, J. (ed.), 221-233. Berkeley: University of California Press.

Irwin, J. O. (1935). Tests of significance for differences between percentages based on small numbers. *Metron*, 12, 83-94.

Johansen, S. (1983). An extension of Cox's regression model. *Int. Statist. Rev.*, 51, 165-174.

Kalbfleisch, J. D. and Prentice, R. L. (1973). Marginal likelihoods based on Cox's regression and life model. *Biometrika*, 60, 267-78.

Kalbfleisch, J. D. and Prentice, R. L. (1980). *The Statistical Analysis of Failure Time Data*. New York: John Wiley & Sons.

Kaplan, E. L. and Meier, P. (1958). Nonparametric estimation from incomplete observations. *J. Amer. Statist. Assoc.*, 53, 457-481.

Karlin, S. and Taylor, H. M. (1975). *A First Course in Stochastic Processes*, 2nd Edition. New York: Academic Press.

Karon, J. M. and Kupper, L. L. (1982). In defense of matching. *Am. J. Epidemiol.*, 116, 852-866.

Katz, D., Baptista, J., Azen, S. P. and Pike, M. C. (1978). Obtaining confidence intervals for the risk ratio in cohort studies. *Biometrics*, 34, 469-474.

Kay, R. (1982). The analysis of transition times in multistate stochastic processes using proportional hazard regression models. *Comm. Statist.*, *A*, 11, 1743-1756.

Kelsey, J. L., Whittemore, A. S., Evans, A. S. and Thompson, W. D. (1996). *Methods in Observational Epidemiology*, 2nd edition. New York: Oxford University Press.

Kendall, Sir M. and Stuart, A. (1979). *The Advanced Theory of Statistics*, Volume 2, 4th edition. New York: Macmillan.

Kenny, S. D., Aubert, R. E. and Geiss, L. S. (1995). Prevalence and incidence of non-insulin-dependent diabetes. In, *Diabetes in America,* 2nd edition, National Diabetes Data Group, 47-67, NIH Publication No. 95-1468. The National Institutes of Health.

Kent, J. T. (1982). Robust properties of likelihood ratio tests. *Biometrika*, 69, 19-27.

Kent, J. T. and O'Quigley, J. (1988). Measure of dependence for censored survival data. *Biometrika*, 75, 525-534.

Kleinbaum, D. G., Kupper, L. L. and Morgenstern, H. (1982). *Epidemiologic Research: Principles and Quantitative Methods.* New York: Van Nostrand Reinhold.

Koch, G. G., McCanless, I., and Ward, J. F. (1984). Interpretation of statistical methodology associated with maintenance trials. *Am. J. Med.*, 77 (supplement 5B), 43-50.

Korn, E. L. (1984). Estimating the utility of matching in case-control studies. *J. Chronic Dis.*, 37, 765-772.

Korn, E. L. and Simon R. (1990). Measures of explained variation for survival data. *Stat. Med.*, 9, 487-503

Korn, E. L. and Simon, R. (1991). Explained residual variation, explained risk, and goodness of fit. *Am. Statistician*, 45, 201-206.

Kudo, A. (1963). A multivariate analogue of the one-sided test. *Biometrika*, 50, 403-418.

Kupper, L. L. and Hafner, K. B. (1989). How appropriate are popular sample size formulas? *Am. Statistician*, 43, 101-105.

Kupper, L. L., Karon, J. M., Kleinbaum, D. G., Morgenstern, H. and Lewis, D. K. (1981). Matching in epidemiologic studies: validity and efficiency considerations. *Biometrics,* 37, 271-292.

Lachin, J. M. (1977). Sample size determinations for r x c comparative trials. *Biometrics*, 33, 315-324.

Lachin, J. M. (1981). Introduction to sample size determination and power analysis for clinical trials. *Control. Clin. Trials*, 2, 93-113.

Lachin, J. M. (1992a). Some large sample distribution-free estimators and tests for multivariate partially incomplete observations from two populations. *Stat. Med.*, 11, 1151-1170.

Lachin, J. M. (1992b). Power and sample size evaluation for the McNemar test with application to matched case-control studies. *Statistics in Medicine*, 11, 1239-1251.

Lachin, J. M. (1996). Distribution-free marginal analysis of repeated measures. *Drug Inform. J.*, 30, 1017-1028.

Lachin, J. M. (1998). Sample size determination. In, *Encyclopedia of Biostatistics*, Armitage P. and Colton T. (eds.), 3892-3903. New York: Wiley.

Lachin, J. M and Bautista, O. M. (1995). Stratified-adjusted versus unstratified assessment of sample size and power for analyses of proportions. In, *Recent Advances in Clinical Trial Design and Analysis*, Thall, P. F. (ed.), 203-223. Boston: Kluwer Academic Publishers.

Lachin, J. M. and Foulkes, M. A. (1986). Evaluation of sample size and power for analyses of survival with allowance for non-uniform patient entry, losses to follow-up, non-compliance and stratification. *Biometrics*, 42, 507-519.

Lachin, J. M., Lan, S. L. and the Lupus Nephritis Collaborative Study Group (1992). Statistical considerations in the termination of a clinical trial with no treatment group difference: The Lupus Nephritis Collaborative Study. *Control. Clin. Trials*, 13, 62-79.

Lachin, J. M. and Wei, L. J. (1988). Estimators and tests in the analysis of multiple nonindependent 2 x 2 tables with partially missing observations. *Biometrics*, 44, 513-528.

Lagakos, S. W. (1978). A covariate model for partially censored data subject to competing causes of failure. *Appl. Statist.*, 27, 235-241.

Lagakos, S. W. (1988). The loss in efficiency from misspecifying covariates in proportional hazards regression models. *Biometrika*, 75, 156-160.

Lagakos, S. W. and Schoenfeld, D. (1984). Properties of proportional-hazards score tests under misspecified regression models. *Biometrics*, 40, 1037-1048.

Lagakos, S. W., Limm L. L-Y. and Robins, J. M. (1990). Adjusting for early treatment termination in comparative clinical trials. *Stat. Med.*, 9, 1417-1424.

Lakatos E. (1988). Sample sizes based on the log-rank statistic in complex clinical trials. *Biometrics*, 44, 229-241.

Laird, N. and Oliver, D. (1981). Covariance analysis of censored survival data using log-linear analysis techniques. *J. Amer. Statist. Assoc.*, 76, 231-240.

Lan, K. K. G. and Wittes, J. T. (1985). Rank tests for survival analysis: A comparison by analogy with games. *Biometrics*, 41, 1063-1069.

Lan, K. K. G. and Lachin, J. M. (1995). Martingales without tears. *Lifetime Data Analysis*, 1, 361-375.

Laupacis, A., Sackett, D. L. and Roberts, R. S. (1988). An assessment of clinically useful measures of the consequences of treatment. *N. Engl. J. Med.*, 318, 1728-1733.

Lawless, J. F. (1982). *Statistical Models and Methods for Lifetime Data.* New York: John Wiley & Sons.

Lee, E. T. (1992). *Statistical Methods for Survival Data Analysis*, 2nd edition. New York: John Wiley & Sons.

Lee, E. W., Wei, L. J. and Amato, D. A. (1992). Cox-type regression analysis for large numbers of small groups of correlated failure time observations. In, *Survival Analysis*, Klein, J. P. and Goel, P. K. (eds.). Netherlands: Kluwer Academic Publishers.

Lehmann, E. L. (1983). *Theory of Point Estimation.* London: Chapman and Hall.

Lehmann, E. L. (1986). *Testing Statistical Hypotheses*, 2nd edition. London: Chapman and Hall.

Lehmann, E. L. (1998). *Elements of Large-Sample Theory*, New York: Springer-Verlag.

Leung, H. K. and Kupper, L. L. (1981). Comparison of confidence intervals for attributable risk. *Biometrics,* 37, 293-302.

Levin, M. L. (1953). The occurrence of lung cancer in man. Acta Unio Internationalis Contra Cancrum, 19, 531-541.

Lewis, E. J., Hunsicker, L. G., Lan, S., Rohde, R. D., Lachin, J. M. and the Lupus Nephritis Collaborative Study Group. (1992). A controlled trial of plasmapheresis therapy in severe lupus nephritis. *N. Engl. J. Med.*, 326, 1373-1379.

Liang, K. Y. and Zeger, S. L. (1986). Longitudinal data analysis using generalized linear models. *Biometrika*, 73, 13-22.

Lin, D. Y. (1991). Goodness-of fit analysis for the Cox regression model based on a class of parameter estimators. *J. Amer. Statist. Assoc.*, 86, 725-728.

Lin, D. Y. (1994). Cox regression analysis of multivariate failure time data: The marginal approach. *Stat. Med.*, 13, 2233-2247.

Lin, D. Y. and Wei, L. J. (1989). The robust inference for the Cox proportional hazards model. *J. Amer. Statist. Assoc.*, 84, 1074-1078.

Lin, D. Y. and Wei, L. J. (1991). Goodness-of-fit tests for the general Cox regression model. *Statistica Sinica*,1, 1-17.

Lin, D. Y., Wei, L. J. and Zing, Z. (1993). Checking the Cox model with cumulative sums of martingale-based residuals. *Biometrika*, 80, 3, 557-72.

Lipsitz, S. H., Fitzmaurice, G. M., Orav, E. J. and Laird, N. M. (1994). Performance of generalized estimating equations in practical situations. *Biometrics*, 50, 270 -278.

Louis, T. A. (1981). Confidence intervals for a binomial parameter after observing no successes. *Am. Statistician*, 35, 154.

Lunn, M. and McNeil, D. (1995). Applying Cox regression to competing risks. *Biometrics*, 51, 524-532.

Machin, D. and Campbell, M. J. (1987). *Statistical Tables for the Design of Clinical Trials.* Oxford: Blackwell Scientific Publications.

Mack, T. M., Pike, M. C., Henderson, B. E., Pfeffer, R. I., Gerkins, V. R., Arthus, B. S., and Brown, S. E. (1976). Estrogens and endometrial cancer in a retirement community. *N. Engl. J. Med.,* 294, 1262-1267.

Madalla, G. S. (1983). *Limited-Dependent and Qualitative Variables in Econometrics*. Cambridge, U.K: Cambridge University Press.

Magee, L. (1990). R^2 measures based on Wald and likelihood ratio joint significance tests. *Am. Statistician*, 44, 250-253.

Mallows, C. L. (1973). Some comments on Cp. *Technometrics*, 15, 661-675.

Mantel, N. (1963). Chi-square tests with one degree of freedom: Extensions of the Mantel-Haenszel procedure. *J. Amer. Statist. Assoc.*, 58, 690-700.

Mantel, N. (1966). Evaluation of survival data and two new rank order statistics arising in its consideration. *Cancer Chemother. Rep.*, 50, 163-170.

Mantel, N. (1967). Ranking procedures for arbitrarily restricted observations. *Biometrics*, 23, 65-78.

Mantel, N. (1970). Why stepdown procedures in variable selection. *Technometrics*, 12, 591-612.

Mantel, N. (1974). Comment and a suggestion. *J. Amer. Statist. Assoc.*, 69, 378-380.

Mantel, N. (1987). Understanding Wald's test for exponential families. *Am. Statistician*, 41, 147-148.

Mantel, N., Brown, C. and Byar, D. P. (1977). Tests for homogeneity of effect in an epidemiologic investigation. *Am. J. Epidemiol.*, 106, 125-129.

Mantel, N. and Greenhouse, S. W. (1968). What is the continuity correction? *Am. Statistician*, 22, 27-30.

Mantel, N. and Haenszel, W. (1959). Statistical aspects of the analysis of data from retrospective studies of disease. *J. Natl. Cancer Inst.*, 22, 719-748.

Marcus, R., Peritz, E. and Gabriel, K. R. (1976). On closed testing procedures with special references to ordered analysis of variance. *Biometrika*, 63, 655-600.

Marubini, E. and Valsecchi, M. G. (1995). *Analysing Survival Data from Clinical Trials and Observational Studies.* New York: John Wiley & Sons.

Maxwell, A. E. (1961). *Analysing Qualitative Data.* London: Methuen & Co., Ltd.

McCullagh, P. (1983). Quasi-likelihood functions. *Ann. Statist.*, 11, 59-67.

McCullagh, P. and Nelder, J. A. (1989). *Generalized Linear Models*, 2nd edition. London: Chapman and Hall.

McHugh, R. B. and Le, C. T. (1984). Confidence estimation and the size of a clinical trial. *Control. Clin. Trials*, 5, 157-163.

McKinlay, S. M. (1978). The effect of non-zero second order interaction on combined estimators of the odds-ratio. *Biometrika*, 65, 191-202.

McNemar, Q. (1947). Note on the sampling error of the difference between correlated proportions or percentages. *Psychometrika*, 12, 153-157.

Mehrotra, K. G., Michalek, J. E. and Mihalko, D. (1982). A relationship between two forms of linear rank procedures for censored data. *Biometrika*, 6, 674-676.

Mehta, C. and Patel, N. (1999). *StatXact 4 for Windows.* Cambridge, Mass.: Cytel Software Corporation.

Meier, P. (1953). Variance of a weighted mean. *Biometrics*, 9, 59-73.

Meng, R. C. and Chapman, D. G. (1966). The power of the chi-square tests for continency tables. *J. Amer. Statist. Assoc.*, 61, 965-975.

Miettinen, O. S. (1968). The matched pairs design in the case of all-or-none responses. *Biometrics*, 24, 339-352.

Miettinen, O. S. (1970). Matching and design efficiency in retrospective studies. *Am. J. Epidemiol.*, 91, 111-118.

Miettinen, O. S. (1974a). Comment. *J. Amer. Statist. Assoc.*, 69, 380-382.

Miettinen, O. S. (1974b). Proportion of disease caused or prevented by a given exposure, trait or intervention. *Am. J. Epidemiol.*, 99, 325.

Miettinen, O. S. (1976). Estimability and estimation in case-referent studies. *Am. J. Epidemiol.*, 103, 226-235.

Miller, A. J. (1984). Selection of subsets of regression variables. *J. Roy. Statist. Soc., A*, 147, 389-425.

Mogensen, C. E. (1984). Microalbuminuria predicts clinical proteinuria and early mortality in maturity-onset diabetes. *N. Engl. J. Med.*, 310, 356-360.

Moore, D. F. (1986). Asymptotic properties of moment estimators for over-dispersed counts and proportions. *Biometrika*, 73, 583-588.

Morris, C. N. (1983). Parametric empirical Bayes inference: Theory and applications. *J. Amer. Statist. Assoc.*, 78, 47-55.

Nagelkerke, N. J. D. (1991). A note on a general definition of the coefficient of determination. *Biometrika*, 78, 691-692.

Nelder, J. A. and Wedderburn, R. W. M. (1972). Generalized linear models. *J. Roy. Statist. Soc.*, 135, 370-384.

Nelson, W. (1972). Theory and applications of hazard plotting for censored failure data. *Technometrics*, 14, 945-965

Neyman, J. (1959). Optimal asymptotic tests of composite statistical hypotheses. In, *Probability and Statistics*, Grenander, U. (ed.), 213-234. Stockholm: Almqvist and Wiksell.

Noether, G. E. (1955). On a theorem of Pitman. *Ann. Math. Statist.*, 26, 64-68.

O'Quigley, J. and Flandre, P. (1994). Predictive capability of proportional hazards regression. *Proc. Natl. Acad. Sci. U. S. A.*, 91, 2310-2314.

O'Quigley, J., Flandre, P. and Reiner, E. (1999). Large sample theory for Schemper's measures of explained variation in the Cox regression model. *Statistician*, 48, 53-62.

Odeh, R. E. and Fox, M. (1991). *Sample Size Choice: Charts for Experiments with Linear Models*, 2nd edn. New York: Marcel Dekker.

Palta, M. and Amini, S. B. (1985). Consideration of covariates and stratification in sample size determination for survival time studies. *J. Chronic Dis.*, 38, 801-809.

Pepe, M. S. (1991). Inference for events with dependent risks in multiple endpoint studies. *J. Amer. Statist. Assoc.*, 86, 770-778.

Pepe, M. S. and Mori, M. (1993). Kaplan-Meier, marginal or conditional probability curves in summarizing competing risks failure time data? *Stat. Med.*, 12, 737-751.

Perlman, M. D. (1969). One-sided testing problems in multivariate analysis. *Ann. Math. Statist.*, 40, 549-567.

Peterson, A. V. (1977). Expressing the Kaplan-Meier estimator as a function of empirical subsurvival functions. *J. Amer. Statist. Assoc.*, 72, 854-858.

Peto, J. (1984). The calculation and interpretation of survival curves. In: *Cancer Clinical Trials, Methods and Practice*, Buyse, M. E., Staquett, M. J. and Sylvester, R. J. (eds.). Oxford: Oxford University Press.

Peto, R. (1972). Contribution to the discussion of paper by D. R. Cox. *J. Roy. Statist. Soc., B*, 34, 205-207.

Peto, R. (1987). Why do we need systematic overviews of randomized trials? *Stat. Med.*, 6, 233-240.

Peto, R. and Lee, P. (1983). Weibull distributions for continuous carcinogenesis experiments. *Biometrics*, 29, 457-470.

Peto, R. and Peto, J. (1972). Asymptotically efficient rank invariant test procedures (with discussion). *J. Roy. Statist. Soc., A*, 135, 185-206.

Peto, R., Pike, M. C., Armitage, P., Breslow, N. E., Cox, D. R., Howard, V., Mantel, N., McPherson, K., Peto, J. and Smith, P. G. (1976). Design and analysis of randomised clinical trials requiring prolonged observation of each patient. Introduction and design. *Br. J. Cancer*, 34, 585-612.

Pettigrew, H. M., Gart, J. J. and Thomas, D. G. (1986). The bias and higher cumulants of the logarithm of a binomial variate. *Biometrika*, 73, 425-435.

Pike, M. C. 1966. A method of analysis of a certain class of experiments in carcinogenesis. *Biometrics*, 22, 142-61.

Pirart, J. (1978a). Diabetes mellitus and its degenerative complications: A prospective study of 4,400 patients observed between 1947 and 1973. *Diabetes Care*, 1, 168-188.

Pirart, J. (1978b). Diabetes mellitus and its degenerative complications: A prospective study of 4,400 patients observed between 1947 and 1973. *Diabetes Care*, 1, 252-263.

Pitman, E. J. G. (1948). *Lecture Notes on Nonparametric Statistics.* New York: Columbia University.

Pregibon, D. (1981). Logistic regression diagnostics. *Ann. Statist.*, 9, 705-724.

Prentice, R. L. (1973). Exponential survivals with censoring and explanatory variables. *Biometrika*, 60, 279-288.

Prentice, R. L. (1978). Linear rank tests with right-censored data. *Biometrika*, 65, 167-179.

Prentice, R. L. and Gloeckler, L. A. (1978). Regression analysis of grouped survival data with application to breast cancer data. *Biometrics*, 34, 57-67.

Prentice, R. L., Kalbfleisch, J. D., Peterson, Jr., A. V., Flournoy, N., Farewell, V. T. and Breslow, N. E. (1978). The analysis of failure times in the presence of competing risks. *Biometrics*, 34, 541-554.

Prentice, R. L. and Marek, P. (1979). A qualitative discrepancy between censored data rank tests. *Biometrics*, 35, 861-7.

Prentice, R. L., Williams, B. J. and Peterson, A. V. (1981). On the regression analysis of multivariate failure time. *Biometrika*, 68, 373-379.

Radhakrishna, S. (1965). Combination of results from several 2 x 2 contingency tables. *Biometrics*, 21, 86-98.

Ramlau-Hansen, H. (1983a). Smoothing counting process intensities by means of kernel functions. *Ann. Statist.*, 11, 453-466.

Ramlau-Hansen, H. (1983b). The choice of a kernel function in the graduation of counting process intensities. *Scand. Actuar. J.*, 165-182.

Rao, C. R. (1963). Criteria of estimation in large samples. *Sankhya, A*, 25, 189-206.

Rao, C. R. (1973). *Linear Statistical Inference and its Application*, 2nd Edition. New York: Wiley.

Robbins, H. (1964). The empirical Bayes approach to statistical decision problems. *Ann. Math. Statist.*, 35, 1-20.

Robins, J. N., Breslow, N. E., and Greenland S. (1986). Estimators of the Mantel-Haenszel variance consistent in both sparse data and large-strata limiting models. *Biometrics*, 42, 311-323.

Robins, J. N., Greenland, S. and Breslow, N. E. (1986). A general estimator for the variance of the Mantel-Haenszel odds ratio. *Am. J. Epidemiol.*, 124, 719-723.

Rochon, J. N. (1989). The application of the GSK method to the determination of minimum sample sizes. *Biometrics*, 45, 193-205.

Ross, S. M. (1983). *Stochastic Processes.* New York: John Wiley & Sons.

Rothman, K. J. (1986). *Modern Epidemiology.* Boston: Little, Brown and Company.

Royall, R. M. (1986). Model robust inference using maximum likelihood estimators. *Int. Statist. Rev.*, 54, 221-226.

Rubenstein L. V., Gail, M. H. and Santner, T. J. (1981). Planning the duration of a comparative clinical trial with losses to follow-up and a period of continued observation. *J. Chronic Dis.*, 34, 469-479.

Sahai, H. and Khurshid, A. (1995). *Statistics in Epidemiology.* Boca Raton: CRC Press.

SAS Institute Inc. (1995). *Logistic Regression Examples Using the SAS system, Version 6.* Cary, NC: SAS Institute, Inc.

SAS Institute Inc. (1997). *SAS/STAT Software: Changes and Enhancements through Release 6.12.* Cary, NC: SAS Institute, Inc.

Schemper M. (1990). The explained variation in proportional hazards regression. *Biometrika*, 77, 216-218. (Correction: 81, 631).

Schemper, M. (1992). Further results on the explained variation in proportional hazards regression. *Biometrika*, 79, 202-204.

Schemper, M. and Stare, J. (1996). Explained variation in survival analysis. *Stat. Med.*, 15, 1999-2012.

Schlesselman, J. J. (1982). *Case-Control Studies: Design, Conduct, Analysis.* New York: Oxford University Press.

Schoenfeld D. (1981). The asymptotic properties of nonparametric tests for comparing survival distributions. *Biometrika*, 68, 316-319.

Schoenfeld, D. (1982). Partial residuals for the proportional hazards regression model. *Biometrika*, 69, 239-241.

Schoenfeld, D. (1983). Sample-size formula for the proportional-hazards regression model. *Biometrics*, 39. 499-503.

Schrek, R., Baker, L. A., Ballard, G. P. and Dolgoff, S. (1950). Tobacco smoking as an etiologic factor in disease. I. Cancer. *Cancer Res.*, 10, 49-58.

Schuster, J. J. (1990). *CRC Handbook of Sample Size Guidelines for Clinical Trials.* Boca Raton, FL: CRC Press.

Seigel, D. G. and Greenhouse, S. W. (1973). Multiple relative risk functions in case-control studies. *Am. J. Epidemiol.*, 97, 324-331.

Selvin, S. (1996). *Statistical Analysis of Epidemiologic Data*, 2nd edition. New York: Oxford University Press.

Serfling, R. J. (1980). *Approximation Theorems of Statistics.* New York: John Wiley and Sons.

Snedecor, G. W. and Cochran, W. G. (1967). *Statistical Methods*, 6th Edition. Ames, IA: The Iowa State University Press.

Starmer, C. F., Grizzle, J. E. and Sen, P. K. (1974). Comment. *J. Amer. Statist. Assoc.*, 69, 376-378.

Steffes, M. W., Chavers, B. M., Bilous, R. W. and Mauer, S. M. (1989). The predictive value of microalbuminuria. *Am. J. Kidney Dis.*, 13, 25-28.

Stokes, M. E., Davis, C. S. and Koch, G. G. (1995). *Categorical Data Analysis Using the SAS System*. Cary, NC: SAS Institute, Inc.

Tang, D. I., Gnecco, C. and Geller, N. L. (1989). An approximate likelihood ratio test for a normal mean vector with nonnegative components with application to clinical trials. *Biometrika*, 76, 577-583.

Tarone, R. E. (1985). On heterogeneity tests based on efficient scores. *Biometrika*, 72, 91-95.

Tarone, R. E. and Ware J. (1977). On distribution free tests for equality of survival distributions. *Biometrika*, 64, 156-160.

Thall, P. F. and Lachin, J. M. (1986). Assessment of stratum-covariate interactions in Cox's proportional hazards regression model. *Stat. Med.*, 5, 73-83.

Thall, P. F., Russell, K. E. and Simon, R. M. (1997). Variable selection in regression via repeated data splitting. *J. Comput. Graph. Statist.*, 6, 416-434.

Thall, P. F., Simon, R. and Grier, D. A. (1992). Test-based variable selection via cross-validation. *J. Comput. Graph. Statist.*, 1, 41-61.

Therneau, T. M., Grambsch, P. M. and Fleming, T. R. (1990). Martingale hazards regression models and the analysis of censored survival data. *Biometrika*, 77, 147-160.

Thisted, R. A. (1988). *Elements of Statistical Computing*. New York: Chapman and Hall.

Thomas, D. G. and Gart, J. J. (1977). A table of exact confidence limits for differences and ratios of two proportions and their odds ratios. *J. Amer. Statist. Assoc.*, 72, 73-76.

Thomas, D. C. and Greenland, S. (1983). The relative efficiencies of matched and independent sample designs for case-control studies. *J. Chronic Dis.*, 36, 685-697.

Tocher, K. D. (1950). Extension of the Neyman-Pearson theory of tests to discontinuous variates. *Biometrika*, 37, 130-144.

Truett, J., Cornfield, J. and Kannel, W. (1967). A multivariate analysis of the risk of coronary heart disease in Framingham. *J. Chronic Dis.*, 20, 511-524.

Tsiatis, A. A. (1981). A large sample study of Cox's regression model. *Ann. Statist.*, 9, 93-108.

Turnbull, B. W. (1976). The empirical distribution function with arbitrarily censored and truncated data. *J. Roy. Statist. Soc., B*, 38, 290-295.

University Group Diabetes Program (UGDP) (1970). A study of the effects of hypoglycemic agents on vascular complications in patients with adult-onset diabetes. *Diabetes*, 19 (Suppl. 2), Appendix A, 816-830.

US Surgeon General. (1964). *Smoking and Health.* Publication No. (PHS) 1103. US Department of Health, Education, & Welfare.

US Surgeon General. (1982). *The Health Consequences of Smoking: Cancer.* Publication No. (PHS) 82-50179. Rockville, MD: US Department of Health and Human Services.

Vaeth, M. (1985). On the use of Wald's test in exponential families. *Int. Statist. Rev.*, 53, 199-214.

Wacholder, S. and Weinberg, C. R. (1982). Paired versus two-sample design for a clinical trial of treatments with dichotomous outcome: Power considerations. *Biometrics*, 38, 801-812.

Wald, A. (1943). Tests of statistical hypotheses concerning several parameters when the number of observations is large. *Trans. Am. Math. Soc.*, 54, 426-482.

Wallenstein, S. and Wittes, J. (1993). The power of the Mantel-Haenszel test for grouped failure time data. *Biometrics*, 49, 1077-1087.

Walter, S. D. (1975). The distribution of Levin's measure of attributable risk. *Biometrika*, 62, 371-375.

Walter, S. D. (1976). The estimation and interpretation of attributable risk in health research. *Biometrics*, 32, 829-849.

Walter, S. D. (1978). Calculation of attributable risks from epidemiological data. *Int. J. Epidemiol.*, 7, 175-182.

Wedderburn, R. W. M. (1974). Quasilikelihood functions, generalized linear models and the Guass-Newton method. *Biometrika*, 63, 27-32.

Wei, L. J. and Lachin, J. M. (1984). Two-sample asymptotically distribution-free tests for incomplete multivariate observations. *J. Amer. Statist. Assoc.*, 79, 653-661.

Wei, L. J., Lin, D. Y. and Weissfeld, L. (1989). Regression analysis of multivariate incomplete failure time data by modeling marginal distributions. *J. Amer. Statist. Assoc.*, 84, 1065 -1073.

White, H. (1980). A heteroskedasticity-consistent covariance matrix estimator and a direct test for heteroskedasticity. *Econometrica*, 48, 817-838.

White, H. (1982). Maximum likelihood estimation of misspecified models. *Econometrica*, 50, 1-25.

Whitehead, A. and Whitehead, J. (1991). A general parametric approach to the meta-analysis of randomized clinical trials. *Stat. Med.*, 10, 1665-1677.

Whitehead, J. (1989). The analysis of relapse clinical trials, with application to a comparison of two ulcer treatments. *Stat. Med.*, 8, 1439-1454.

Whitehead, J. (1992). *The Design and Analysis of Sequential Clinical Trials*, 2nd Edition. New York: Ellis Horwood.

Whittemore, A. (1981). Sample size for logistic regression with small response probability. *J. Amer. Statist. Assoc.*, 76, 27-32.

Wilks, S. S. (1962). *Mathematical Statistics.* New York: John Wiley and Sons, Inc.

Wittes, J. and Wallenstein, S. (1987). The power of the Mantel-Haenszel test. *J. Amer. Statist. Assoc.*, 82, 400, 1104-1109.

Woolf, B. (1955). On estimating the relation between blood group and disease. *Ann. Human Genet.*, 19, 251-253.

Younes, N. and Lachin, J. M. (1997). Link-based models for survival data with interval and continuous time censoring. *Biometrics*, 53, 1199-1211.

Yusuf, S., Peto, R., Lewis, J., Collins, R. and Sleight, T. (1985). Beta blockade during and after myocardial infarction: An overview of the randomized trials. *Progress in Cardiovascular Disease*, 27, 335-371.

Zeger, S. L. and Liang, K. Y. (1986). Longitudinal data analysis for discrete and continuous outcomes. *Biometrics*, 42, 121-130.

Zelen, M. (1971). The analysis of several 2 x 2 contingency tables. *Biometrika*, 58, 129-137.

Zuckerman, D. M., Kasl, S. V. and Ostfeld, A. M. (1984). Psychosocial predictors of mortality among the elderly poor: the role of religion, well-being, and social contacts. *Am. J. Epidemiol.*, 119, 419-423.

Author Index

Index

WILEY SERIES IN PROBABILITY AND STATISTICS

ESTABLISHED BY WALTER A. SHEWHART AND SAMUEL S. WILKS

Editors
Vic Barnett, Noel A. C. Cressie, Nicholas I. Fisher,
Iain M. Johnstone, J. B. Kadane, David G. Kendall, David W. Scott,
Bernard W. Silverman, Adrian F. M. Smith, Jozef L. Teugels;
Ralph A. Bradley, Emeritus, J. Stuart Hunter, Emeritus

Probability and Statistics Section

*ANDERSON · The Statistical Analysis of Time Series
ARNOLD, BALAKRISHNAN, and NAGARAJA · A First Course in Order Statistics
ARNOLD, BALAKRISHNAN, and NAGARAJA · Records
BACCELLI, COHEN, OLSDER, and QUADRAT · Synchronization and Linearity: An Algebra for Discrete Event Systems
BASILEVSKY · Statistical Factor Analysis and Related Methods: Theory and Applications
BERNARDO and SMITH · Bayesian Statistical Concepts and Theory
BILLINGSLEY · Convergence of Probability Measures, *Second Edition*
BOROVKOV · Asymptotic Methods in Queuing Theory
BOROVKOV · Ergodicity and Stability of Stochastic Processes
BRANDT, FRANKEN, and LISEK · Stationary Stochastic Models
CAINES · Linear Stochastic Systems
CAIROLI and DALANG · Sequential Stochastic Optimization
CONSTANTINE · Combinatorial Theory and Statistical Design
COOK · Regression Graphics
COVER and THOMAS · Elements of Information Theory
CSÖRGŐ and HORVÁTH · Weighted Approximations in Probability Statistics
CSÖRGŐ and HORVÁTH · Limit Theorems in Change Point Analysis
*DANIEL · Fitting Equations to Data: Computer Analysis of Multifactor Data, *Second Edition*
DETTE and STUDDEN · The Theory of Canonical Moments with Applications in Statistics, Probability, and Analysis
DEY and MUKERJEE · Fractional Factorial Plans
*DOOB · Stochastic Processes
DRYDEN and MARDIA · Statistical Shape Analysis
DUPUIS and ELLIS · A Weak Convergence Approach to the Theory of Large Deviations
ETHIER and KURTZ · Markov Processes: Characterization and Convergence
FELLER · An Introduction to Probability Theory and Its Applications, Volume 1, *Third Edition,* Revised; Volume II, *Second Edition*
FULLER · Introduction to Statistical Time Series, *Second Edition*
FULLER · Measurement Error Models
GHOSH, MUKHOPADHYAY, and SEN · Sequential Estimation
GIFI · Nonlinear Multivariate Analysis
GUTTORP · Statistical Inference for Branching Processes
HALL · Introduction to the Theory of Coverage Processes
HAMPEL · Robust Statistics: The Approach Based on Influence Functions
HANNAN and DEISTLER · The Statistical Theory of Linear Systems
HUBER · Robust Statistics
HUSKOVA, BERAN, and DUPAC · Collected Works of Jaroslav Hajek—with Commentary

*Now available in a lower priced paperback edition in the Wiley Classics Library.

Probability and Statistics (Continued)

IMAN and CONOVER · A Modern Approach to Statistics
JUREK and MASON · Operator-Limit Distributions in Probability Theory
KASS and VOS · Geometrical Foundations of Asymptotic Inference
KAUFMAN and ROUSSEEUW · Finding Groups in Data: An Introduction to Cluster Analysis
KELLY · Probability, Statistics, and Optimization
KENDALL, BARDEN, CARNE, and LE · Shape and Shape Theory
LINDVALL · Lectures on the Coupling Method
McFADDEN · Management of Data in Clinical Trials
MANTON, WOODBURY, and TOLLEY · Statistical Applications Using Fuzzy Sets
MORGENTHALER and TUKEY · Configural Polysampling: A Route to Practical Robustness
MUIRHEAD · Aspects of Multivariate Statistical Theory
OLIVER and SMITH · Influence Diagrams, Belief Nets and Decision Analysis
*PARZEN · Modern Probability Theory and Its Applications
PRESS · Bayesian Statistics: Principles, Models, and Applications
PUKELSHEIM · Optimal Experimental Design
RAO · Asymptotic Theory of Statistical Inference
RAO · Linear Statistical Inference and Its Applications, *Second Edition*
RAO and SHANBHAG · Choquet-Deny Type Functional Equations with Applications to Stochastic Models
ROBERTSON, WRIGHT, and DYKSTRA · Order Restricted Statistical Inference
ROGERS and WILLIAMS · Diffusions, Markov Processes, and Martingales, Volume I: Foundations, *Second Edition;* Volume II: Îto Calculus
RUBINSTEIN and SHAPIRO · Discrete Event Systems: Sensitivity Analysis and Stochastic Optimization by the Score Function Method
RUZSA and SZEKELY · Algebraic Probability Theory
SCHEFFE · The Analysis of Variance
SEBER · Linear Regression Analysis
SEBER · Multivariate Observations
SEBER and WILD · Nonlinear Regression
SERFLING · Approximation Theorems of Mathematical Statistics
SHORACK and WELLNER · Empirical Processes with Applications to Statistics
SMALL and McLEISH · Hilbert Space Methods in Probability and Statistical Inference
STAPLETON · Linear Statistical Models
STAUDTE and SHEATHER · Robust Estimation and Testing
STOYANOV · Counterexamples in Probability
TANAKA · Time Series Analysis: Nonstationary and Noninvertible Distribution Theory
THOMPSON and SEBER · Adaptive Sampling
WELSH · Aspects of Statistical Inference
WHITTAKER · Graphical Models in Applied Multivariate Statistics
YANG · The Construction Theory of Denumerable Markov Processes

Applied Probability and Statistics Section

ABRAHAM and LEDOLTER · Statistical Methods for Forecasting
AGRESTI · Analysis of Ordinal Categorical Data
AGRESTI · Categorical Data Analysis
ANDERSON, AUQUIER, HAUCK, OAKES, VANDAELE, and WEISBERG · Statistical Methods for Comparative Studies
ARMITAGE and DAVID (editors) · Advances in Biometry

*Now available in a lower priced paperback edition in the Wiley Classics Library.

*ARTHANARI and DODGE · Mathematical Programming in Statistics
ASMUSSEN · Applied Probability and Queues
*BAILEY · The Elements of Stochastic Processes with Applications to the Natural Sciences
BARNETT · Comparative Statistical Inference, *Third Edition*
BARNETT and LEWIS · Outliers in Statistical Data, *Third Edition*
BARTHOLOMEW, FORBES, and McLEAN · Statistical Techniques for Manpower Planning, *Second Edition*
BATES and WATTS · Nonlinear Regression Analysis and Its Applications
BECHHOFER, SANTNER, and GOLDSMAN · Design and Analysis of Experiments for Statistical Selection, Screening, and Multiple Comparisons
BELSLEY · Conditioning Diagnostics: Collinearity and Weak Data in Regression
BELSLEY, KUH, and WELSCH · Regression Diagnostics: Identifying Influential Data and Sources of Collinearity
BHAT · Elements of Applied Stochastic Processes, *Second Edition*
BHATTACHARYA and WAYMIRE · Stochastic Processes with Applications
BIRKES and DODGE · Alternative Methods of Regression
BLISCHKE AND MURTHY · Reliability: Modeling, Prediction, and Optimization
BLOOMFIELD · Fourier Analysis of Time Series: An Introduction, *Second Edition*
BOLLEN · Structural Equations with Latent Variables
BOULEAU · Numerical Methods for Stochastic Processes
BOX · Bayesian Inference in Statistical Analysis
BOX and DRAPER · Empirical Model-Building and Response Surfaces
*BOX and DRAPER · Evolutionary Operation: A Statistical Method for Process Improvement
BUCKLEW · Large Deviation Techniques in Decision, Simulation, and Estimation
BUNKE and BUNKE · Nonlinear Regression, Functional Relations and Robust Methods: Statistical Methods of Model Building
CHATTERJEE and HADI · Sensitivity Analysis in Linear Regression
CHERNICK · Bootstrap Methods: A Practitioner's Guide
CHILÈS and DELFINER · Geostatistics: Modeling Spatial Uncertainty
CHOW and LIU · Design and Analysis of Clinical Trials: Concepts and Methodologies
CLARKE and DISNEY · Probability and Random Processes: A First Course with Applications, *Second Edition*
*COCHRAN and COX · Experimental Designs, *Second Edition*
CONOVER · Practical Nonparametric Statistics, *Second Edition*
CORNELL · Experiments with Mixtures, Designs, Models, and the Analysis of Mixture Data, *Second Edition*
*COX · Planning of Experiments
CRESSIE · Statistics for Spatial Data, *Revised Edition*
DANIEL · Applications of Statistics to Industrial Experimentation
DANIEL · Biostatistics: A Foundation for Analysis in the Health Sciences, *Sixth Edition*
DAVID · Order Statistics, *Second Edition*
*DEGROOT, FIENBERG, and KADANE · Statistics and the Law
DODGE · Alternative Methods of Regression
DOWDY and WEARDEN · Statistics for Research, *Second Edition*
DUNN and CLARK · Applied Statistics: Analysis of Variance and Regression, *Second Edition*
*ELANDT-JOHNSON and JOHNSON · Survival Models and Data Analysis
EVANS, PEACOCK, and HASTINGS · Statistical Distributions, *Second Edition*
*FLEISS · The Design and Analysis of Clinical Experiments
FLEISS · Statistical Methods for Rates and Proportions, *Second Edition*
FLEMING and HARRINGTON · Counting Processes and Survival Analysis

*Now available in a lower priced paperback edition in the Wiley Classics Library.

Applied Probability and Statistics (Continued)

GALLANT · Nonlinear Statistical Models
GLASSERMAN and YAO · Monotone Structure in Discrete-Event Systems
GNANADESIKAN · Methods for Statistical Data Analysis of Multivariate Observations, *Second Edition*
GOLDSTEIN and LEWIS · Assessment: Problems, Development, and Statistical Issues
GREENWOOD and NIKULIN · A Guide to Chi-Squared Testing
*HAHN · Statistical Models in Engineering
HAHN and MEEKER · Statistical Intervals: A Guide for Practitioners
HAND · Construction and Assessment of Classification Rules
HAND · Discrimination and Classification
HEIBERGER · Computation for the Analysis of Designed Experiments
HEDAYAT and SINHA · Design and Inference in Finite Population Sampling
HINKELMAN and KEMPTHORNE: · Design and Analysis of Experiments, Volume 1: Introduction to Experimental Design
HOAGLIN, MOSTELLER, and TUKEY · Exploratory Approach to Analysis of Variance
HOAGLIN, MOSTELLER, and TUKEY · Exploring Data Tables, Trends and Shapes
HOAGLIN, MOSTELLER, and TUKEY · Understanding Robust and Exploratory Data Analysis
HOCHBERG and TAMHANE · Multiple Comparison Procedures
HOCKING · Methods and Applications of Linear Models: Regression and the Analysis of Variables
HOGG and KLUGMAN · Loss Distributions
HOSMER and LEMESHOW · Applied Logistic Regression
HØYLAND and RAUSAND · System Reliability Theory: Models and Statistical Methods
HUBERTY · Applied Discriminant Analysis
JACKSON · A User's Guide to Principle Components
JOHN · Statistical Methods in Engineering and Quality Assurance
JOHNSON · Multivariate Statistical Simulation
JOHNSON and KOTZ · Distributions in Statistics
Continuous Multivariate Distributions
JOHNSON, KOTZ, and BALAKRISHNAN · Continuous Univariate Distributions, Volume 1, *Second Edition*
JOHNSON, KOTZ, and BALAKRISHNAN · Continuous Univariate Distributions, Volume 2, *Second Edition*
JOHNSON, KOTZ, and BALAKRISHNAN · Discrete Multivariate Distributions
JOHNSON, KOTZ, and KEMP · Univariate Discrete Distributions, *Second Edition*
JUREČKOVÁ and SEN · Robust Statistical Procedures: Aymptotics and Interrelations
KADANE · Bayesian Methods and Ethics in a Clinical Trial Design
KADANE AND SCHUM · A Probabilistic Analysis of the Sacco and Vanzetti Evidence
KALBFLEISCH and PRENTICE · The Statistical Analysis of Failure Time Data
KELLY · Reversability and Stochastic Networks
KHURI, MATHEW, and SINHA · Statistical Tests for Mixed Linear Models
KLUGMAN, PANJER, and WILLMOT · Loss Models: From Data to Decisions
KLUGMAN, PANJER, and WILLMOT · Solutions Manual to Accompany Loss Models: From Data to Decisions
KOVALENKO, KUZNETZOV, and PEGG · Mathematical Theory of Reliability of Time-Dependent Systems with Practical Applications
LACHIN · Biostatistical Methods: The Assessment of Relative Risks
LAD · Operational Subjective Statistical Methods: A Mathematical, Philosophical, and Historical Introduction
LANGE, RYAN, BILLARD, BRILLINGER, CONQUEST, and GREENHOUSE · Case Studies in Biometry
LAWLESS · Statistical Models and Methods for Lifetime Data

*Now available in a lower priced paperback edition in the Wiley Classics Library.

Applied Probability and Statistics (Continued)

LEE · Statistical Methods for Survival Data Analysis, *Second Edition*
LEPAGE and BILLARD · Exploring the Limits of Bootstrap
LINHART and ZUCCHINI · Model Selection
LITTLE and RUBIN · Statistical Analysis with Missing Data
LLOYD · The Statistical Analysis of Categorical Data
MAGNUS and NEUDECKER · Matrix Differential Calculus with Applications in Statistics and Econometrics, *Revised Edition*
MALLER and ZHOU · Survival Analysis with Long Term Survivors
MANN, SCHAFER, and SINGPURWALLA · Methods for Statistical Analysis of Reliability and Life Data
McLACHLAN and KRISHNAN · The EM Algorithm and Extensions
McLACHLAN · Discriminant Analysis and Statistical Pattern Recognition
McNEIL · Epidemiological Research Methods
MEEKER and ESCOBAR · Statistical Methods for Reliability Data
*MILLER · Survival Analysis, *Second Edition*
MONTGOMERY and PECK · Introduction to Linear Regression Analysis, *Second Edition*
MYERS and MONTGOMERY · Response Surface Methodology: Process and Product in Optimization Using Designed Experiments
NELSON · Accelerated Testing, Statistical Models, Test Plans, and Data Analyses
NELSON · Applied Life Data Analysis
OCHI · Applied Probability and Stochastic Processes in Engineering and Physical Sciences
OKABE, BOOTS, and SUGIHARA · Spatial Tesselations: Concepts and Applications of Voronoi Diagrams
PANKRATZ · Forecasting with Dynamic Regression Models
PANKRATZ · Forecasting with Univariate Box-Jenkins Models: Concepts and Cases
PIANTADOSI · Clinical Trials: A Methodologic Perspective
PORT · Theoretical Probability for Applications
PUTERMAN · Markov Decision Processes: Discrete Stochastic Dynamic Programming
RACHEV · Probability Metrics and the Stability of Stochastic Models
RÉNYI · A Diary on Information Theory
RIPLEY · Spatial Statistics
RIPLEY · Stochastic Simulation
ROLSKI, SCHMIDLI, SCHMIDT, and TEUGELS · Stochastic Processes for Insurance and Finance
ROUSSEEUW and LEROY · Robust Regression and Outlier Detection
RUBIN · Multiple Imputation for Nonresponse in Surveys
RUBINSTEIN · Simulation and the Monte Carlo Method
RUBINSTEIN and MELAMED · Modern Simulation and Modeling
RYAN · Statistical Methods for Quality Improvement, *Second Edition*
SCHUSS · Theory and Applications of Stochastic Differential Equations
SCOTT · Multivariate Density Estimation: Theory, Practice, and Visualization
*SEARLE · Linear Models
SEARLE · Linear Models for Unbalanced Data
SEARLE, CASELLA, and McCULLOCH · Variance Components
SENNOTT · Stochastic Dynamic Programming and the Control of Queueing Systems
STOYAN, KENDALL, and MECKE · Stochastic Geometry and Its Applications, *Second Edition*
STOYAN and STOYAN · Fractals, Random Shapes and Point Fields: Methods of Geometrical Statistics
THOMPSON · Empirical Model Building
THOMPSON · Sampling
THOMPSON · Simulation: A Modeler's Approach

*Now available in a lower priced paperback edition in the Wiley Classics Library.

Applied Probability and Statistics (Continued)

TIJMS · Stochastic Modeling and Analysis: A Computational Approach
TIJMS · Stochastic Models: An Algorithmic Approach
TITTERINGTON, SMITH, and MAKOV · Statistical Analysis of Finite Mixture Distributions
UPTON and FINGLETON · Spatial Data Analysis by Example, Volume 1: Point Pattern and Quantitative Data
UPTON and FINGLETON · Spatial Data Analysis by Example, Volume II: Categorical and Directional Data
VAN RIJCKEVORSEL and DE LEEUW · Component and Correspondence Analysis
VIDAKOVIC · Statistical Modeling by Wavelets
WEISBERG · Applied Linear Regression, *Second Edition*
WESTFALL and YOUNG · Resampling-Based Multiple Testing: Examples and Methods for p-Value Adjustment
WHITTLE · Systems in Stochastic Equilibrium
WOODING · Planning Pharmaceutical Clinical Trials: Basic Statistical Principles
WOOLSON · Statistical Methods for the Analysis of Biomedical Data
*ZELLNER · An Introduction to Bayesian Inference in Econometrics

Texts and References Section

AGRESTI · An Introduction to Categorical Data Analysis
ANDERSON · An Introduction to Multivariate Statistical Analysis, *Second Edition*
ANDERSON and LOYNES · The Teaching of Practical Statistics
ARMITAGE and COLTON · Encyclopedia of Biostatistics: Volumes 1 to 6 with Index
BARTOSZYNSKI and NIEWIADOMSKA-BUGAJ · Probability and Statistical Inference
BENDAT and PIERSOL · Random Data: Analysis and Measurement Procedures, *Third Edition*
BERRY, CHALONER, and GEWEKE · Bayesian Analysis in Statistics and Econometrics: Essays in Honor of Arnold Zellner
BHATTACHARYA and JOHNSON · Statistical Concepts and Methods
BILLINGSLEY · Probability and Measure, *Second Edition*
BOX · R. A. Fisher, the Life of a Scientist
BOX, HUNTER, and HUNTER · Statistics for Experimenters: An Introduction to Design, Data Analysis, and Model Building
BOX and LUCEÑO · Statistical Control by Monitoring and Feedback Adjustment
BROWN and HOLLANDER · Statistics: A Biomedical Introduction
CHATTERJEE and PRICE · Regression Analysis by Example, *Third Edition*
COOK and WEISBERG · Applied Regression Including Computing and Graphics
COOK and WEISBERG · An Introduction to Regression Graphics
COX · A Handbook of Introductory Statistical Methods
DILLON and GOLDSTEIN · Multivariate Analysis: Methods and Applications
*DODGE and ROMIG · Sampling Inspection Tables, *Second Edition*
DRAPER and SMITH · Applied Regression Analysis, *Third Edition*
DUDEWICZ and MISHRA · Modern Mathematical Statistics
DUNN · Basic Statistics: A Primer for the Biomedical Sciences, *Second Edition*
FISHER and VAN BELLE · Biostatistics: A Methodology for the Health Sciences
FREEMAN and SMITH · Aspects of Uncertainty: A Tribute to D. V. Lindley
GROSS and HARRIS · Fundamentals of Queueing Theory, *Third Edition*
HALD · A History of Probability and Statistics and their Applications Before 1750
HALD · A History of Mathematical Statistics from 1750 to 1930
HELLER · MACSYMA for Statisticians

*Now available in a lower priced paperback edition in the Wiley Classics Library.

Texts and References (Continued)

HOEL · Introduction to Mathematical Statistics, *Fifth Edition*
HOLLANDER and WOLFE · Nonparametric Statistical Methods, *Second Edition*
HOSMER and LEMESHOW · Applied Survival Analysis: Regression Modeling of Time to Event Data
JOHNSON and BALAKRISHNAN · Advances in the Theory and Practice of Statistics: A Volume in Honor of Samuel Kotz
JOHNSON and KOTZ (editors) · Leading Personalities in Statistical Sciences: From the Seventeenth Century to the Present
JUDGE, GRIFFITHS, HILL, LÜTKEPOHL, and LEE · The Theory and Practice of Econometrics, *Second Edition*
KHURI · Advanced Calculus with Applications in Statistics
KOTZ and JOHNSON (editors) · Encyclopedia of Statistical Sciences: Volumes 1 to 9 wtih Index
KOTZ and JOHNSON (editors) · Encyclopedia of Statistical Sciences: Supplement Volume
KOTZ, REED, and BANKS (editors) · Encyclopedia of Statistical Sciences: Update Volume 1
KOTZ, REED, and BANKS (editors) · Encyclopedia of Statistical Sciences: Update Volume 2
LAMPERTI · Probability: A Survey of the Mathematical Theory, *Second Edition*
LARSON · Introduction to Probability Theory and Statistical Inference, *Third Edition*
LE · Applied Categorical Data Analysis
LE · Applied Survival Analysis
MALLOWS · Design, Data, and Analysis by Some Friends of Cuthbert Daniel
MARDIA · The Art of Statistical Science: A Tribute to G. S. Watson
MASON, GUNST, and HESS · Statistical Design and Analysis of Experiments with Applications to Engineering and Science
MURRAY · X-STAT 2.0 Statistical Experimentation, Design Data Analysis, and Nonlinear Optimization
PURI, VILAPLANA, and WERTZ · New Perspectives in Theoretical and Applied Statistics
RENCHER · Linear Models in Statistics
RENCHER · Methods of Multivariate Analysis
RENCHER · Multivariate Statistical Inference with Applications
ROSS · Introduction to Probability and Statistics for Engineers and Scientists
ROHATGI · An Introduction to Probability Theory and Mathematical Statistics
RYAN · Modern Regression Methods
SCHOTT · Matrix Analysis for Statistics
SEARLE · Matrix Algebra Useful for Statistics
STYAN · The Collected Papers of T. W. Anderson: 1943–1985
TIERNEY · LISP-STAT: An Object-Oriented Environment for Statistical Computing and Dynamic Graphics
WONNACOTT and WONNACOTT · Econometrics, *Second Edition*

*Now available in a lower priced paperback edition in the Wiley Classics Library.

WILEY SERIES IN PROBABILITY AND STATISTICS
ESTABLISHED BY WALTER A. SHEWHART AND SAMUEL S. WILKS

Survey Methodology Section

BIEMER, GROVES, LYBERG, MATHIOWETZ, and SUDMAN · Measurement Errors in Surveys
COCHRAN · Sampling Techniques, *Third Edition*
COUPER, BAKER, BETHLEHEM, CLARK, MARTIN, NICHOLLS, and O'REILLY (editors) · Computer Assisted Survey Information Collection
COX, BINDER, CHINNAPPA, CHRISTIANSON, COLLEDGE, and KOTT (editors) · Business Survey Methods
*DEMING · Sample Design in Business Research
DILLMAN · Mail and Telephone Surveys: The Total Design Method
GROVES and COUPER · Nonresponse in Household Interview Surveys
GROVES · Survey Errors and Survey Costs
GROVES, BIEMER, LYBERG, MASSEY, NICHOLLS, and WAKSBERG · Telephone Survey Methodology
*HANSEN, HURWITZ, and MADOW · Sample Survey Methods and Theory, Volume 1: Methods and Applications
*HANSEN, HURWITZ, and MADOW · Sample Survey Methods and Theory, Volume II: Theory
KISH · Statistical Design for Research
*KISH · Survey Sampling
KORN and GRAUBARD · Analysis of Health Surveys
LESSLER and KALSBEEK · Nonsampling Error in Surveys
LEVY and LEMESHOW · Sampling of Populations: Methods and Applications, *Third Edition*
LYBERG, BIEMER, COLLINS, de LEEUW, DIPPO, SCHWARZ, TREWIN (editors) · Survey Measurement and Process Quality
SIRKEN, HERRMANN, SCHECHTER, SCHWARZ, TANUR, and TOURANGEAU (editors) · Cognition and Survey Research

*Now available in a lower priced paperback edition in the Wiley Classics Library.